Birkhäuser

ISNM
International Series of Numerical Mathematics

Volume 165

More information about this series at
http://www.springer.com/series/4819

Günter Leugering • Peter Benner •
Sebastian Engell • Andreas Griewank •
Helmut Harbrecht • Michael Hinze •
Rolf Rannacher • Stefan Ulbrich
Editors

Trends in PDE Constrained Optimization

Birkhäuser

Editors
Günter Leugering
Department Mathematik
Universität Erlangen-Nürnberg
Erlangen
Germany

Peter Benner
Institut für Dynamik komplexer technischer Systeme
Max-Planck-Institut
Magdeburg
Germany

Sebastian Engell
Department of Biochemical and Chemical Engineering
Technische Universität Dortmund
Dortmund
Germany

Andreas Griewank
Department of Mathematics
Humboldt-Universität zu Berlin
Berlin
Germany

Helmut Harbrecht
Mathematisches Institut
Universität Basel
Basel
Switzerland

Michael Hinze
Fachbereich Mathematik Optimierung und Approximation
Universität Hamburg
Hamburg
Germany

Rolf Rannacher
Institut für Angewandte Mathematik
Ruprecht-Karls-Universität Heidelberg
Heidelberg
Germany

Stefan Ulbrich
Fachbereich Mathematik
Technische Universität Darmstadt
Darmstadt
Germany

ISSN 0373-3149 ISSN 2296-6072 (electronic)
ISBN 978-3-319-05082-9 ISBN 978-3-319-05083-6 (eBook)
DOI 10.1007/978-3-319-05083-6
Springer Cham Heidelberg New York Dordrecht London

Library of Congress Control Number: 2014958298

Mathematics Subject Classification (2010): 35K55, 35Q93, 49J20, 49K20, 49M25, 65K10, 65M60, 65N15, 65N30, 76D55, 90C30

Printed on acid-free paper

Springer is part of Springer Science+Business Media (www.springer.com)

Contents

Contributors

Thomas Apel Institute for Mathematics and Civil Engineering Informatics, Universität der Bundeswehr München, Germany

Eberhard Bänsch Lehrstuhl für Angewandte Mathematik 3, Friedrich-Alexander-Universität Erlangen-Nürnberg, Erlangen, Germany

Richard C. Barnard Institute for Mathematics and Scientific Computing, University of Graz, Graz, Austria

Malte Behrens Department of Biochemical and Chemical Engineering, TU Dortmund, Dortmund, Germany

Peter Benner Research Group Computational Methods in Systems and Control Theory (CSC), Max Planck Institute for Dynamics of Complex Technical Systems Magdeburg, Magdeburg, Germany

Luise Blank Fakultät für Mathematik, Universität Regensburg, Regensburg, Germany

Hans Georg Bock Interdisciplinary Center for Scientific Computing, Heidelberg University, Heidelberg, Germany

Torsten Bosse Department of Mathematics, Humboldt-Universität zu Berlin, Berlin, Germany

Nikolai D. Botkin Zentrum Mathematik, Technische Universität München, München, Germany

Stefanie Bott Department of Mathematics and Graduate School of Computational Engineering, Technische Universität Darmstadt, Darmstadt, Germany

Malte Braack Mathematical Seminar, Christian-Albrechts-University of Kiel, Kiel, Germany

Martin Burger Institut für Numerische und Angewandte Mathematik, Westfälische Wilhelms-Universität Münster, Münster, Germany

Debora Clever Department of Mathematics, Technische Universität Darmstadt, Darmstadt, Germany

Sergio Conti Institut für Angewandte Mathematik, Universität Bonn, Bonn, Germany

Klaus Deckelnick Institut für Analysis und Numerik, Otto–von–Guericke–Universität Magdeburg, Magdeburg, Germany

Markus Dick Department of Mathematics, University of Erlangen-Nuremberg, Erlangen, Germany

Sebastian Engell Department of Biochemical and Chemical Engineering, TU Dortmund, Dortmund, Germany

M. Hassan Farshbaf-Shaker Weierstraß-Institut, Berlin, Germany

Thomas G. Flaig Institut für Mathematik und Bauinformatik, Universität der Bundeswehr München, Neubiberg, Germany

Marcisse Fouego Institut für Numerische und Angewandte Mathematik, Westfälische Wilhelms-Universität Münster, Münster, Germany

Martin Frank MATHCCES, Department of Mathematics, RWTH Aachen University, Aachen, Germany

Thomas Franke Institute of Physics, Universität Augsburg, Augsburg, Germany

Harald Garcke Fakultät für Mathematik, Universität Regensburg, Regensburg, Germany

Nicolas R. Gauger Department of Mathematics and Center for Computational Engineering Science, RWTH Aachen University, Aachen, Germany

Benedict Geihe Institut für Numerische Simulation, Universität Bonn, Bonn, Germany

Andreas Griewank Department of Mathematics, Humboldt-Universität zu Berlin, Berlin, Germany

Michael Gröschel Institute of Applied Mathematics 2, Friedrich-Alexander University Erlangen-Nürnberg, Erlangen, Germany

Martin Gugat Department of Mathematics, University of Erlangen-Nuremberg, Erlangen, Germany

Stefanie Günther Department of Mathematics and Center for Computational Engineering Science, RWTH Aachen University, Aachen, Germany

Helmut Harbrecht Mathematisches Institut, Universität Basel, Basel, Schweiz

Claudia Hecht Fakultät für Mathematik, Universität Regensburg, Regensburg, Germany

Michael Herty IGPM, Department of Mathematics, RWTH Aachen University, Aachen, Germany

Roland Herzog Faculty of Mathematics, Technische Universität Chemnitz, Chemnitz, Germany

Michael Hintermüller Institut für Mathematik, Humboldt-Universität zu Berlin, Berlin, Germany

Michael Hinze Schwerpunkt Optimierung und Approximation, Universität Hamburg, Hamburg, Germany

Karl-Heinz Hoffmann Zentrum Mathematik, Technische Universität München, München, Germany

Ronald H. W. Hoppe Institute of Mathematics, Universität Augsburg, Augsburg, Germany

Department of Mathematics, University of Houston, Houston, TX, USA

Lena Kaland Department of Mathematics and Center for Computational Engineering Science, RWTH Aachen University, Aachen, Germany

Phawitphorn Khobkhun Department of Biochemical and Chemical Engineering, TU Dortmund, Dortmund, Germany

Markus Klein Mathematical Institute, Tübingen University, Tübingen, Germany

Kristina Kohl Fachbereich Mathematik, Institut für Angewandte Analysis und Numerische Simulation, Universität Stuttgart, Stuttgart, Germany

Michael Köster Institute of Applied Mathematics, TU Dortmund, Dortmund, Germany

Claudia Kratzenstein Christian-Albrechts-Universität zu Kiel, Kiel, Germany

Jens Lang Department of Mathematics and Graduate School of Computational Engineering, Technische Universität Darmstadt, Darmstadt, Germany

Antoine Laurain Institut für Mathematik, Technische Universität Berlin, Berlin, Germany

Lutz Lehmann Department of Mathematics, Humboldt-Universität zu Berlin, Berlin, Germany

Günter Leugering Department of Mathematics, Universität Erlangen-Nürnberg, Erlangen, Germany

Christopher Linsenmann Institute of Mathematics, Universität Augsburg, Augsburg, Germany

Caroline Löbhard Humboldt-Universität zu Berlin, Berlin, Germany

Dominik Meidner Fakultät für Mathematik, Lehrstuhl für Optimale Steuerung, Technische Universität München, Garching b. München, Germany

Christian Meyer Faculty of Mathematics, Technical Universität Dortmund, Dortmund, Germany

Josef Michl Institut für Theoretische Physik, Universität Regensburg, Regensburg, Germany

Anil Nemili Department of Mathematics and Center for Computational Engineering Science, RWTH Aachen University, Aachen, Germany

Emre Özkaya Department of Mathematics and Center for Computational Engineering Science, RWTH Aachen University, Aachen, Germany

Wolfgang Peukert Institute of Particle Technology, Friedrich-Alexander University Erlangen-Nürnberg, Erlangen, Germany

Sebastian Pfaff Department of Mathematics, Technische Universität Darmstadt, Darmstadt, Germany

Johannes Pfefferer Universität der Bundeswehr München, München, Germany

René Pinnau Fachbereich Mathematik, Technische Universität Kaiserslautern, Kaiserslautern, Germany

Andreas Potschka Interdisciplinary Center for Scientific Computing, Heidelberg University, Heidelberg, Germany

Andreas Prohl Mathematical Institute, Tübingen University, Tübingen, Germany

Rolf Rannacher Institut für angewandte Mathematik, Universität Heidelberg, Heidelberg, Germany

Sebastian Rau Fachbereich Mathematik, Technische Universität Kaiserslautern, Kaiserslautern, Germany

Carlos N. Rautenberg Department of Mathematics and Scientific Computing, Karl-Franzens-University of Graz, Graz, Austria

Arnd Rösch Fakultät für Mathematik, Universität Duisburg–Essen, Essen, Germany

Martin Rumpf Institut für Numerische Simulation, Universität Bonn, Bonn, Germany

Christoph Rupprecht Fakultät für Mathematik, Universität Regensburg, Regensburg, Germany

Jens Saak Research Group Computational Methods in Systems and Control Theory (CSC), Max Planck Institute for Dynamics of Complex Technical Systems Magdeburg, Magdeburg, Germany

Ekkehard Sachs FB 4–Mathematik, University of Trier, Trier, Germany

Lothar Schmid Institute of Physics, Universität Augsburg, Augsburg, Germany

Marina Schneider FB IV – Department of Mathematics, University of Trier, Trier, Germany

Dirk Schröder Department of Mathematics, Technische Universität Darmstadt, Darmstadt, Germany

Matthias Schu FB IV – Department of Mathematics, University of Trier, Trier, Germany

Rüdiger Schultz Fakultät für Mathematik, Universität Duisburg-Essen, Duisburg, Germany

Volker Schulz Universität Trier, Trier, Germany

Kunibert G. Siebert Fachbereich Mathematik, Institut für Angewandte Analysis und Numerische Simulation, Universität Stuttgart, Stuttgart, Germany

Thomas Slawig Christian-Albrechts-Universität zu Kiel, Kiel, Germany

Sonja Steffensen IGPM, RWTH Aachen University, Aachen, Germany

Vanessa Styles Department of Mathematics, University of Sussex, Brighton, UK

Thomas M. Surowiec Institut für Mathematik, Humboldt-Universität zu Berlin, Berlin, Germany

Johannes Tausch Department of Mathematics, Southern Methodist University, Dallas, TX, USA

Benjamin Tews Mathematical Seminar, Christian-Albrechts-University of Kiel, Kiel, Germany

Stefan Turek Institute of Applied Mathematics, TU Dortmund, Dortmund, Germany

Varvara D. Turova Zentrum Mathematik, Technische Universität München, München, Germany

Stefan Ulbrich Fachbereich Mathematik, Technische Universität Darmstadt, Darmstadt, Germany

Boris Vexler Fakultät für Mathematik, Lehrstuhl für Optimale Steuerung, Technische Universität München, Garching b. München, Germany

Stefan Volkwein Department of Mathematics and Statistics, University of Konstanz, Konstanz, Germany

Gerd Wachsmuth Faculty of Mathematics, Technische Universität Chemnitz, Chemnitz, Germany

Ke Wang School of Mathematical Sciences, Fudan University, Shanghai, China

Heiko K. Weichelt Research Group Mathematics in Industry and Technology (MiIT), Technische Universität Chemnitz, Chemnitz, Germany

Achim Wixforth Institute of Physics, Universität Augsburg, Augsburg, Germany

Winnifried Wollner Department of Mathematics, Universität Hamburg, Hamburg, Germany

Jan Carsten Ziems Department of Mathematics, Technische Universität Darmstadt, Darmstadt, Germany

Introduction

Günter Leugering, Andreas Griewank, Stefan Ulbrich, Helmut Harbrecht, Peter Benner, Rolf Rannacher, Michael Hinze, and Sebastian Engell

G. Leugering (✉)
Department Mathematik, Universität Erlangen-Nürnberg, Erlangen, Germany
e-mail: leugering@math.fau.de

A. Griewank
Department of Mathematics, Humboldt-Universität zu Berlin, Unter den Linden 6, 10099 Berlin, Germany
e-mail: griewank@math.hu-berlin.de

S. Ulbrich
Fachbereich Mathematik, Technische Universität Darmstadt, Darmstadt, Germany
e-mail: ulbrich@mathematik.tu-darmstadt.de

H. Harbrecht
Mathematisches Institut, Universität Basel, Basel, Switzerland
e-mail: helmut.harbrecht@unibas.ch

P. Benner
Institut für Dynamik komplexer technisch, Max-Planck-Institut, Magdeburg, Germany
e-mail: benner@mpi-magdeburg.mpg.de

R. Rannacher
Institut für Angewandte Mathematik, Ruprecht-Karls-Universität Heidelberg, Heidelberg, Germany
e-mail: rannacher@iwr.uni-heidelberg.de

M. Hinze
Fachbereich Mathematik Optimierung und Approximation, Universität Hamburg, Hamburg, Germany
e-mail: michael.hinze@uni-hamburg.de

S. Engell
Department of Biochemical and Chemical Engineering, TU Dortmund, Emil-Figge-Str. 70, 44227 Dortmund, Germany
e-mail: sebastian.engell@bci.tu-dortmund.de

G. Leugering et al. (eds.), *Trends in PDE Constrained Optimization*, International Series of Numerical Mathematics 165, DOI 10.1007/978-3-319-05083-6_1

Problems of optimization and optimal control subject to constraints governed by partial differential equations (PDEs) arise in a huge variety of industrial, technological, economic, medical and environmental applications. The questions that appear in this field range from shape and topology optimization problems, such as optimal design of the wing of an aircraft, to optimal control of water flow in irrigation canals and medical separation processes of biological cells. For a fruitful and comprehensive study of such optimization problems, it is crucial to combine innovative optimization techniques with new algorithmic and numerical approaches.

The contributors to this special volume are mostly members of the Deutsche Forschungsgemeinschaft (DFG) priority program 1253 "Optimization with Partial Differential Equations" which was active from 2006 until 2013. This priority program, in which forty research projects were involved, brought together specialists in the field of PDE constrained optimization and optimal control from more than fifteen German universities. The results of their work, which are presented in this special volume, received a matchless benefit from the synergetic collaboration and networking within the priority program. Landmark results with respect to solving the underlying industrial, technological and medical problems have been achieved.

The topics presented in this special volume cover almost the entire field of recent research in PDE constrained optimization and optimal control. The results range from derivation of new mathematical paradigms to design and analysis of new numerical approaches. Innovative algorithmic schemes have been developed, implemented and validated in the context of real-world applications. The book is organized in the following five thematic parts:

- Constrained Optimization, Identification and Control
- Shape and Topology Optimization
- Adaptivity and Model Reduction
- Discretization: Concepts and Analysis
- Applications

Research articles present recent results in almost the entire range of PDE constrained optimization and optimal control. Survey articles give an overview of central topics which set sustainable trends for future research.

The editors and authors would like to thank the DFG for their financial support and all the referees who were involved in this special volume and in the priority program. The work presented in this book has benefitted enormously from their helpful comments and suggestions. The editors and authors also would like to thank Dr. Markus Dick for his continuous work in organizing this special volume. In the following we give a short summary of the five thematic parts of this book. A more detailed description can be found in the introduction that precedes each part.

1 Constrained Optimization, Identification and Control

This first part contains eight progress reports and one survey article. The latter describes and analyses a general one-shot methodology for solving already discretized design optimization problems for which explicitly forming and directly solving the already discretized KKT conditions is impossible. Bott et al. developed a general framework for the solution of PDAE constrained optimal control problems by an adaptive SQP method in Hilbert spaces. Several other reports establish existence, uniqueness and regularity results for particular problem classes, often deriving optimality conditions for regularized problems and then driving the regularization or penalty parameters to zero. Of particular concern is the treatment of state constraints and variational inequalities, sometimes in a bilevel setting. All papers report numerical results, often using second-order methods with trust region stabilization.

2 Shape and Topology Optimization

Shape and topology optimization is indispensable for designing and constructing industrial components. Many problems that arise in application, particularly in structural mechanics and in the optimal control of distributed parameter systems, can be formulated as the minimization of functionals defined over a class of admissible domains. In this part of the book, novel approaches are presented to deal with such optimization problems. First, a two-scale model for elastic shape optimization within a stochastic framework is considered. Then, results on shape optimization with parabolic state equation are given. Finally, multi-material structural topology and shape optimization problems are formulated and solved within a phase field approach.

3 Adaptivity and Model Reduction

In the part *Adaptivity and Model Reduction*, different techniques for reducing the complexity of solving PDE constrained optimization problems numerically are discussed. One possibility is to use tailored discretizations that adapt the meshsize according to the optimization goal. A different approach consists in model order reduction, i.e., application of mathematical methods for automatically reducing the state-space dimension of the control problem while preserving accuracy in the map from the input functions or parameters to the optimized quantity-of-interest. Both approaches to complexity reduction are discussed in corresponding survey chapters.

4 Discretization: Concepts and Analysis

The chapter *Discretization: Concepts and Analysis* summarizes recent trends and addresses future research directions in the field of discrete concepts for PDE constrained optimization with elliptic and parabolic PDEs in the presence of pointwise constraints. It covers the range from tailored discrete concepts over adaptive a posteriori finite element approaches, to the modern algorithmical treatment of challenging optimal control applications with fluid flows.

5 Applications

The work of researchers in the priority program was driven not only by challenging mathematical problems but also by fascinating future applications. This section is dedicated to a presentation of recent application results in the field of PDE constrained optimization and optimal control. The results range from optimal treatment planning in radiotherapy to stabilization of gas transportation networks. Optimization of particle synthesis in chemical industry is studied as well as optimal control of self-consistent classical and quantum particle systems which play an important role in the design of semiconductor devices. Another focus of this part is optimal control of biomedical and -technological separation processes.

Part I
Constrained Optimization, Identification and Control

Introduction to Part I Constrained Optimization, Identification and Control

Stefan Ulbrich and Andreas Griewank

In the article *Optimal Control of Allen-Cahn Systems* the authors Luise Blank, M. Hassan Farshbaf-Shaker, Claudia Hecht, Josef Michl and Christoph Rupprecht report results of their project regarding a control problem for a multi-component Allen-Cahn equation that incorporates elastic effects. This gives rise to a coupled elliptic-parabolic system. Distributed control of the Allen-Cahn equation and Neumann boundary for the stress tensor on a part of the boundary are considered. Existence results and optimality conditions are given for multi-component systems and smooth potential without distributed control. Concerning obstacle potentials, first-order conditions for the limiting systems of approximating problems are given for the boundary control using a penalization approach and for distributed control by a relaxation technique. Finally, there are some numerical results obtained with Trust-Region-Newton-Steihaug-CG method.

In the article *Optimal Control of Elastoplastic Processes: Analysis, Algorithms Numerical Results* the authors Roland Herzog, Christian Meyer and Gerd Wachsmuth report their results on the optimal control of elastoplasticity systems, where they concentrate on static elastoplasticity with small strains in its so-called dual (stress-based) formulation with linear kinematic hardening; in addition, also some results for the quasi-static case are stated. They consider the forward problem as well as a regularized problem and show the existence of optimal solutions. Moreover, an optimality system of C-stationary type is derived and also strong stationarity as well as B-stationarity are discussed. The authors consider the

S. Ulbrich (✉)
Fachbereich Mathematik, Technische Universität Darmstadt, Darmstadt, Germany
e-mail: ulbrich@mathematik.tu-darmstadt.de

A. Griewank
Department of Mathematics, Humboldt-Universität zu Berlin, Unter den Linden 6, 10099 Berlin, Germany
e-mail: griewank@math.hu-berlin.de

G. Leugering et al. (eds.), *Trends in PDE Constrained Optimization*, International Series of Numerical Mathematics 165, DOI 10.1007/978-3-319-05083-6_2

computational approximation of stationary points and present numerical results for the quasi-static case.

In the survey article *One-Shot Approaches to Design Optimization* the authors Torsten Bosse, Nicolas R. Gauger, Andreas Griewank, Stefanie Günther and Volker Schulz describe a general methodology for solving design optimization problems for which explicitly forming and directly solving the KKT conditions is impossible. Given an iterative state equation solver a corresponding adjoint solver can be obtained by algorithmic differentiation or hand coding. Certain Seidel and Jacobi type variants and their asymptotic rate of convergence to second order sufficient constraint optimizers are presented and analyzed. In particular it is shown that the Seidel variant achieves 2-cycle quadratic convergence in the limiting case where the state equation solver is close to Newton's method and the reduced Hessian is available. The methods are verified on the incompressible Navier Stokes equation with a boundary control and a tracking type objective.

In the article *Optimal Design with Bounded Retardation for Problems with Non-separable Adjoints* the authors Torsten Bosse, Nicolas R. Gauger, Andreas Griewank, Stefanie Günther, Lena Kaland, Claudia Kratzenstein, Lutz Lehmann, Anil Nemili, Emre Özkaya and Thomas Slawig report results from their SPP project. Problems ranging from oceanography to aerodynamics, and Burgers equation were attacked as test cases, using a one shot approach specifically geared to problems with a nonlinear interaction between state and control, which leads to nonseparable Jacobians. In special cases asymptotic convergence rates and thus the retardation factor compared to the underlying state space iteration are derived.

In the article *On a Fully Adaptive SQP Method for PDAE-Constrained Optimal Control Problems with Control and State Constraints* the authors Stefanie Bott, Debora Clever, Jens Lang, Stefan Ulbrich, Jan Carsten Ziems and Dirk Schröder present an adaptive multilevel SQP method to solve complex optimal control problems for time-dependent nonlinear partial PD(A)Es with control and state constraints. The multilevel method generates adaptive finite-element approximations during the optimization, where the refinement strategy is based on a posteriori error estimators for the PDE-constraint, the adjoint equation and the criticality measure. The resulting optimization method allows to use existing adaptive PD(A)E-solvers and error estimators in a modular way. State constraints are handled by Moreau-Yosida regularization and convergence results for the resulting algorithm are presented. The multilevel SQP method is combined with the space-time adaptive PD(A)E-solver Kardos. The efficiency of the method is demonstrated for a 3-D radiative heat transfer problem modeling the cooling process in glass manufacturing and a 2-D thermistor problem modeling the heating process in steel hardening.

In the article *Optimal Control of Nonlinear Hyperbolic Conservation Laws with Switching* the authors Sebastian Pfaff, Stefan Ulbrich and Günter Leugering consider optimal control problems governed by nonlinear hyperbolic conservation laws at junctions, where switching between different states occurs in node or boundary controls. The authors analyze in particular the Fréchet-differentiability of the reduced objective functional with respect to switching times of the controls. This is done by showing that the control-to-state mapping of the considered problems satisfies a generalized notion of differentiability. They consider both, the case where the controls are the initial and the boundary data as well as the case where the system is controlled by the switching times of the node condition. Differentiability results are presented for the considered problems in a quite general setting including an adjoint-based gradient representation of the reduced objective function.

In the article *Elliptic Mathematical Programs with Equilibrium Constraints in Function Space: Optimality Conditions and Numerical Realization* the authors Michael Hintermüller, Antoine Laurain, Caroline Löbhard, Carlos Rautenberg and Thomas Surowiec report their results on elliptic mathematical programs with equilibrium constraints (MPECs) in function space. They derive stationarity conditions for control problems with point tracking objectives and subject to the obstacle problem as well as for optimization problems with variational inequality constraints and pointwise constraints on the gradient of the state. A bundle-free solution method as well as adaptive finite element discretizations are introduced and verified. Moreover, shape design problems subject to elliptic variational inequality constraints are treated analytically and numerically. Finally, the authors propose a fixed-point-Moreau-Yosida-based semismooth Newton solver for a class of nonlinear elliptic quasi-variational inequality problems involving gradient constraints.

In the article *Models and Optimal Control in Freezing and Thawing of Living Cells and Tissues* the authors Karl-Heinz Hoffmann, Nikolai D. Botkin and Varvara L. Turova outline the results obtained in their SPP project. They apply the theory of partial differential equations and optimal control techniques to minimize damaging factors in cryopreservation of living cells and tissues in order to increase the survival rate of frozen and subsequently thawed out cells. The authors present mathematical models of the processes of freezing and thawing and describe the application of optimal control theory to the design of optimal cooling and warming protocols which reduce damaging effects and improve the survival rate of cells.

In the article *Optimal Control-Based Feedback Stabilization of Multi-field Flow Problems* the authors Eberhard Bänsch, Peter Benner, Jens Saak and Heiko K. Weichelt consider the numerical solution of the feedback stabilization problem for multi-field flow problems. The approach is based on an analytical Riccati feedback concept derived by Raymond for the incompressible Navier-Stokes equations. The approach uses a linear-quadratic regulator (LQR) approach for the linearized Navier-Stokes equations. The authors extend the approach to the Navier-Stokes equations coupled with a diffusion-convection equation describing the transport of a reactive species in a fluid. The feedback LQR-control is obtained via solving an operator Riccati equation. The authors describe a numerical procedure to solve this Riccati equation and illustrate the performance of the proposed method by a numerical example.

Optimal Control of Allen-Cahn Systems

Luise Blank, M. Hassan Farshbaf-Shaker, Claudia Hecht, Josef Michl, and Christoph Rupprecht

Abstract Optimization problems governed by Allen-Cahn systems including elastic effects are formulated and first-order necessary optimality conditions are presented. Smooth as well as obstacle potentials are considered, where the latter leads to an MPEC. Numerically, for smooth potential the problem is solved efficiently by the Trust-Region-Newton-Steihaug-cg method. In case of an obstacle potential first numerical results are presented.

Keywords Allen-Cahn system • Parabolic obstacle problems • Linear elasticity • Mathematical programs with complementarity constraints • Optimality conditions • Trust-Region-Newton method

Mathematics Subject Classification (2010). Primary 49J40; Secondary 49K20, 49J20, 49M15, 74P99.

1 Introduction and Problem Formulation

Optimization problems with interfaces and free boundaries frequently appear in materials science, fluid dynamics and biology (see i.e. [6] and the references therein). In this paper we concentrate on a phase field approach, more precisely on a multi-component Allen-Cahn model, to describe the dynamics of the interface.

L. Blank (✉) • C. Hecht • C. Rupprecht
Fakultät für Mathematik, Universität Regensburg, 93040 Regensburg, Germany
e-mail: Luise.Blank@mathematik.uni-regensburg.de; claudia.hecht@mathematik.uni-regensburg.de; christoph.rupprecht@mathematik.uni-regensburg.de

M.H. Farshbaf-Shaker
Weierstraß-Institut, Mohrenstr. 39, 10117 Berlin, Germany
e-mail: MohammadHassanFarshbaf.Shaker@wias-berlin.de

J. Michl
Institut für Theoretische Physik, Universität Regensburg, 93040 Regensburg, Germany
e-mail: josef.michl@physik.uni-regensburg.de

G. Leugering et al. (eds.), *Trends in PDE Constrained Optimization*, International Series of Numerical Mathematics 165, DOI 10.1007/978-3-319-05083-6_3

This allows complex topological changes. The possibly sharp interface between the phases is replaced by a thin transitional layer of width $\mathcal{O}(\varepsilon)$ where $\varepsilon > 0$ is a small parameter, and the N different phases are described by a phase field variable $\boldsymbol{c} = (c_1, \dots, c_N)^T$, where c_i denotes the fraction of the i-th material. The underlying non-convex interfacial energy is based on the generalized Ginzburg-Landau energy, see [13],

$$E(\boldsymbol{c}, \boldsymbol{u}) := \int_\Omega \left\{ \frac{\varepsilon}{2} |\nabla \mathbf{c}|^2 + \frac{1}{\varepsilon} \Psi(\mathbf{c}) + W(\mathbf{c}, \mathcal{E}(\boldsymbol{u})) \right\} \, \mathrm{d}x, \tag{1.1}$$

where $\Omega \subset \mathbb{R}^d$, $1 \le d \le 3$, is a bounded domain with either convex or $C^{1,1}$-boundary. Moreover, $\boldsymbol{u}$ is the displacement field mapping into $\mathbb{R}^d$ and Ψ is the bulk potential. In general the potential Ψ is assumed to have global minima at the pure phases and in physical situations there are many choices possible, see [5]. Here we consider two different cases: a smooth double-well potential in Sects. 2.1 and 3.1, and a nonsmooth obstacle potential in Sects. 2.2 and 3.2. The latter ensures in particular that the pure phases correspond exactly to $c_i = 1$, whereas in the smooth case those are given by $c_i \approx 1$. The term $W(\mathbf{c}, \mathcal{E}(\boldsymbol{u}))$ in (1.1) is the elastic free energy density. Since in phase separation processes of alloys the deformations are typically small we choose a theory based on the linearized strain tensor (see [7]) given by $\mathcal{E} := \mathcal{E}(\boldsymbol{u}) = \frac{1}{2}(\nabla \boldsymbol{u} + \nabla \boldsymbol{u}^T)$ and

$$W(\boldsymbol{c}, \mathcal{E}) = \frac{1}{2}(\mathcal{E} - \mathcal{E}^*(\boldsymbol{c})) : \mathcal{C}(\mathcal{E} - \mathcal{E}^*(\boldsymbol{c})). \tag{1.2}$$

Here $\mathcal{C}$ is the symmetric, positive definite, possibly anisotropic elasticity tensor mapping from symmetric tensors in $\mathbb{R}^{d \times d}$ into itself. The quantity $\mathcal{E}^*(\boldsymbol{c})$ is the eigenstrain at concentration $\boldsymbol{c}$ and following Vegard's law we choose $\mathcal{E}^*(\boldsymbol{c}) = \sum_{i=1}^N c_i \mathcal{E}^*(\boldsymbol{e}_i)$, where $\mathcal{E}^*(\boldsymbol{e}_i)$ is the value of the strain tensor when the material consists only of component i and is unstressed. Here $(\boldsymbol{e}_i)_{i=1}^N$ denote the standard coordinate vectors in $\mathbb{R}^N$. The dynamics of the interface motion can be modelled by the steepest descent of (1.1) with respect to the L^2-norm, see [4,12]. The mechanical equilibrium is obtained on a much faster time scale and therefore we assume quasi-static equilibrium for the mechanical variable $\boldsymbol{u}$. For a smooth potential Ψ this results after suitable rescaling of time in the following elastic Allen-Cahn equation

$$\begin{pmatrix} \varepsilon \partial_t \boldsymbol{c} \\ \mathbf{0} \end{pmatrix} = \begin{pmatrix} \varepsilon \Delta \boldsymbol{c} - \frac{1}{\varepsilon} D\Psi(\boldsymbol{c}) - D_c W(\boldsymbol{c}, \mathcal{E}(\boldsymbol{u})) \\ -\nabla \cdot D_{\mathcal{E}} W(\boldsymbol{c}, \mathcal{E}(\boldsymbol{u})) \end{pmatrix}. \tag{1.3}$$

We denote by D_c and $D_{\mathcal{E}}$ the differentials with respect to $\boldsymbol{c}$ and $\mathcal{E}$, respectively. In the case of a nonsmooth obstacle potential, Ψ is given as the sum of a differentiable and a non-differentiable convex function and the derivative $D\Psi(\boldsymbol{c})$ has to be understood as the sum of the differentiable part plus the subdifferential of the non-differentiable convex summand, and so the first component of (1.3) will result in a variational inequality, see Sect. 2.2. We have $D_c W(\boldsymbol{c},\mathcal{E}) = -\mathcal{E}^* : \mathcal{C}(\mathcal{E}-\mathcal{E}^*(\boldsymbol{c}))$ and $D_{\mathcal{E}} W(\boldsymbol{c},\mathcal{E}) = \mathcal{C}(\mathcal{E}-\mathcal{E}^*(\boldsymbol{c}))$.

We assume now that a volume force $\boldsymbol{f}$ acts on $\Omega_T := \Omega \times (0,T)$ and a surface load $\boldsymbol{g} \in L^2(0,T;L^2(\Gamma_g,\mathbb{R}^d))$ acts on $\Gamma_g \subset \Gamma := \partial\Omega$ until a given time $T > 0$. Then with $\Gamma_D := \Gamma \setminus \Gamma_g$, $\Gamma_T := \Gamma \times (0,T)$ and the outer unit normal $\boldsymbol{n}$ the mechanical system is given by

$$\begin{cases} -\nabla \cdot D_{\mathcal{E}} W(\boldsymbol{c},\mathcal{E}(\boldsymbol{u})) = \boldsymbol{0} & \text{in } \Omega, \\ \boldsymbol{u} = \boldsymbol{0} & \text{on } \Gamma_D, \\ D_{\mathcal{E}} W(\boldsymbol{c},\mathcal{E}(\boldsymbol{u}))\cdot \boldsymbol{n} = \boldsymbol{g} & \text{on } \Gamma_g \end{cases} \tag{1.4}$$

which has to hold for a.e. $t \in (0,T)$, and the Allen-Cahn system is given by

$$\begin{cases} \varepsilon\partial_t \boldsymbol{c} - \varepsilon\Delta\boldsymbol{c} + \frac{1}{\varepsilon}D\Psi(\boldsymbol{c}) + D_c W(\boldsymbol{c},\mathcal{E}(\boldsymbol{u})) = \boldsymbol{f} & \text{in } \Omega_T, \\ \nabla\boldsymbol{c}\cdot\boldsymbol{n} = \boldsymbol{0} & \text{on } \Gamma_T, \\ \boldsymbol{c}(0) = \boldsymbol{c}_0 & \text{in } \Omega \end{cases} \tag{1.5}$$

in case of a smooth potential Ψ. Our aim in this paper is to transform an initial phase distribution $\boldsymbol{c}_0 : \Omega \to \mathbb{R}^N$ with minimal cost of the controls to some desired phase pattern $\boldsymbol{c}_T \in \boldsymbol{L}^2(\Omega) := L^2(\Omega,\mathbb{R}^N)$ at a given final time $T > 0$ while tracking a desired evolution $\boldsymbol{c}_d \in \boldsymbol{L}^2(\Omega_T) := L^2(0,T;\boldsymbol{L}^2(\Omega))$. Hence we consider the following objective functional:

$$\begin{aligned} J(\boldsymbol{c},\boldsymbol{f},\boldsymbol{g}) :=& \frac{\nu_T}{2}\|\boldsymbol{c}(T,\cdot)-\boldsymbol{c}_T\|^2_{\boldsymbol{L}^2(\Omega)} + \frac{\nu_d}{2}\|\boldsymbol{c}-\boldsymbol{c}_d\|^2_{\boldsymbol{L}^2(\Omega_T)} + \\ &+ \frac{\nu_f}{2\varepsilon}\|\boldsymbol{f}\|^2_{\boldsymbol{L}^2(\Omega_T)} + \frac{\nu_g}{2}\|\boldsymbol{g}\|^2_{L^2(0,T;L^2(\Gamma_g,\mathbb{R}^d))}. \end{aligned} \tag{1.6}$$

This leads to the following optimal control problem:

$$(\mathcal{P}) \begin{cases} \min & J(\boldsymbol{c},\boldsymbol{f},\boldsymbol{g}) \\ \text{over} & (\boldsymbol{c},\boldsymbol{f},\boldsymbol{g}) \in \boldsymbol{\mathcal{V}} \times \boldsymbol{L}^2(\Omega_T) \times L^2(0,T;L^2(\Gamma_g,\mathbb{R}^d)) \\ \text{s.t.} & (1.4) \text{ and } (1.5) \text{ hold} \end{cases} \tag{1.7}$$

with $\boldsymbol{\mathcal{V}} := L^\infty(0,T;\boldsymbol{H}^1(\Omega)) \cap H^1(0,T;\boldsymbol{L}^2(\Omega)) \cap L^2(0,T;\boldsymbol{H}^2(\Omega))$. We assume, that the Dirichlet part Γ_D has positive $(d-1)$-dimensional Hausdorff measure and introduce the notation $H^1_D(\Omega,\mathbb{R}^d) := \{\boldsymbol{u} \in H^1(\Omega,\mathbb{R}^d) \mid \boldsymbol{u}|_{\Gamma_D} = \boldsymbol{0}\}$. Later on we will use also the space $\boldsymbol{\mathcal{W}}(0,T) := L^2(0,T;\boldsymbol{H}^1(\Omega)) \cap H^1(0,T;\boldsymbol{H}^1(\Omega)^*)$.

2 Existence Theory and First-Order Optimality Conditions

In this section we discuss the existence of a minimum and the derivation of first-order necessary optimality systems. First we present the smooth potential case. Here, the standard optimization theory in function spaces is applicable and delivers a first-order necessary optimality system. Afterwards, we focus on the control problem with an obstacle potential leading to an optimal control problem with variational inequalities. Hence this belongs to the class of MPECs, where the standard control theory is in general not applicable. Here we employ a penalty approach for the problem without distributed control and a relaxation approach for the model without elasticity.

2.1 *Smooth* Ψ

We start by considering the setting without volume force, i.e. $\boldsymbol{f} \equiv \mathbf{0}$. In a system with two phases, i.e. $N = 2$, the problem can be reduced to a single unknown by defining $c := c_1 - c_2$, which results in a scalar problem. One typical choice of a smooth potential is then the double-well potential $\Psi(c) = \frac{1}{4}(c^2 - 1)^2$. The scalar case with this Ψ is studied extensively in [15] without tracking c_d, i.e. $\nu_d = 0$. For a regularized obstacle potential Ψ_σ (see Sect. 2.2.1) the vector-valued case with possibly $\nu_d \neq 0$ is discussed in [11]. However, Ψ_σ is not a physical potential. The following theorem summarizes the results of [11, 15].

Theorem 2.1. *Let* $(\mathcal{P})$ *be given as a scalar problem for* $N = 2$ *with potential* $\Psi = \frac{1}{4}\left(c^2 - 1\right)^2$ *and* $\nu_d = 0$ *or for* $N \geq 2$ *and* $\nu_d \geq 0$ *arbitrary with a regularized obstacle potential* Ψ_σ *as mentioned above. For fixed initial distribution* $\boldsymbol{c}_0 \in \boldsymbol{H}^1(\Omega)$ *and given surface load* $\boldsymbol{g} \in L^2(0,T;L^2(\Gamma_g,\mathbb{R}^d))$ *there exists a unique solution* $(\boldsymbol{c},\boldsymbol{u}) \in \boldsymbol{\mathcal{V}} \times L^2(0,T;H^1_D(\Omega,\mathbb{R}^d))$ *of* (1.4)–(1.5) *and hence the solution operator* $\boldsymbol{S} : L^2(0,T;L^2(\Gamma_g,\mathbb{R}^d)) \to \boldsymbol{\mathcal{V}} \times L^2(0,T;H^1_D(\Omega,\mathbb{R}^d))$ *with its components* $\boldsymbol{S}(\boldsymbol{g}) := (\boldsymbol{S}_1(\boldsymbol{g}), \boldsymbol{S}_2(\boldsymbol{g})) = (\boldsymbol{c},\boldsymbol{u})$ *is well-defined.*

Then the control problem $(\mathcal{P})$ is equivalent to minimizing the reduced cost functional $j(\boldsymbol{g}) := J(\boldsymbol{S}_1(\boldsymbol{g}), \boldsymbol{g})$ over $L^2(0,T;L^2(\Gamma_g,\mathbb{R}^d))$. This result is established by applying energy methods to a time-discretized version of (1.4)–(1.5) and showing a series of uniform a priori estimates for the time discretized solutions, where one has to consider the particular functions Ψ and Ψ_σ, respectively, and the coupling of the systems. By the direct method in the calculus of variations one can then show existence of a minimizer for $(\mathcal{P})$. The differentiability of the solution operator can be shown by an implicit function argument and thus we can differentiate the reduced cost functional to obtain the following necessary optimality condition:

Theorem 2.2. *Every minimizer* $\boldsymbol{g} \in L^2(0,T;L^2(\Gamma_g,\mathbb{R}^d))$ *of* j *fulfills the following optimality system:* (1.4), (1.5) *and*

$$q + \nu_g g = 0 \quad \textit{a.e. on } (0,T)\times\Gamma_g, \tag{2.1}$$

$$\begin{cases} -\varepsilon\partial_t p - \varepsilon\Delta p + \frac{1}{\varepsilon}D^2\Psi(c)p + D_p W(p,\mathcal{E}(q)) = \nu_d(c - c_d) & \textit{in } \Omega_T, \\ \nabla p\cdot n = 0 & \textit{on } \Gamma_T, \\ \varepsilon p(T) = \nu_T(c(T) - c_T) & \textit{in } \Omega, \end{cases} \tag{2.2}$$

$$\begin{cases} -\nabla\cdot D_{\mathcal{E}}W(p,\mathcal{E}(q)) = 0 & \textit{in } \Omega, \\ q = 0 & \textit{on } \Gamma_D, \\ D_{\mathcal{E}}W(p,\mathcal{E}(q))\cdot n = 0 & \textit{on } \Gamma_g. \end{cases} \tag{2.3}$$

For a setting without elasticity but with distributed control, i.e. $f \not\equiv 0$ and arbitrary $\nu_d, \nu_T \geq 0$, we refer for instance to [10]. There, the scalar case, i.e. $N = 2$ as above, is considered with a penalized double obstacle potential Ψ_σ. Moreover, the optimality system is investigated rigorously and is given by (1.5), (2.2) without elastic energy together with the gradient equation

$$p + \frac{\nu_f}{\varepsilon} f = 0 \quad \text{a.e. in } \Omega_T. \tag{2.4}$$

2.2 Obstacle Potential

In the case of an obstacle potential each component of c stands, in contrast to the smooth potential, exactly for the fraction of one phase. Hence the phase space is the Gibbs simplex $G := \{v \in \mathbb{R}^N \mid v_i \geq 0, \sum_{i=1}^N v_i = 1\}$ and the bulk potential $\Psi : \mathbb{R}^N \to \mathbb{R}\cup\{\infty\}$ is the multi-obstacle potential $\Psi(v) := \Psi_0(v) + I_G(v)$, where e.g. $\Psi_0(v) := -\frac{1}{2}\|v\|^2$, which we consider, and I_G is the indicator function of the Gibbs simplex. The differential of the indicator function has to be understood in the sense of subdifferentials, and thus the Allen-Cahn system (1.5) results in a variational inequality, which can also be written in the following form (see [3]):

$$\begin{cases} \varepsilon\partial_t c - \varepsilon\Delta c - P_\Sigma\left(\frac{1}{\varepsilon}(c + \xi) - D_c W(c,\mathcal{E}(u))\right) = f & \text{in } \Omega_T, \\ \nabla c\cdot n = 0 & \text{on } \Gamma_T, \\ c(0) = c_0 & \text{in } \Omega, \end{cases} \tag{2.5}$$

together with the complementarity conditions

$$c \geq 0 \text{ a.e. in } \Omega_T,\ \xi \geq 0 \text{ a.e. in } \Omega_T,\ (\xi, c)_{L^2(\Omega_T)} = 0, \tag{2.6}$$

the additional constraint $\boldsymbol{c} \in \boldsymbol{\Sigma} := \{\boldsymbol{v} \in \mathbb{R}^N \mid \sum_{i=1}^N v_i = 1\}$ a.e. in Ω_T and the requirement $\boldsymbol{f} \in \boldsymbol{T\Sigma} := \{\boldsymbol{v} \in \mathbb{R}^N \mid \sum_{i=1}^N v_i = 0\}$ a.e. in Ω_T. Here $\boldsymbol{P}_\Sigma : \mathbb{R}^N \to \boldsymbol{T\Sigma}$ is the projection operator defined by $\boldsymbol{P}_\Sigma \boldsymbol{v} := \boldsymbol{v} - \boldsymbol{1}\frac{1}{N}\sum_{i=1}^{N} v_i$. The variable $\boldsymbol{\xi}$ can be interpreted as a Lagrange multiplier corresponding to the constraint $\boldsymbol{c} \geq \boldsymbol{0}$, and as a slack variable used for reformulating the variational inequality into a standard MPEC problem. Denoting $\boldsymbol{L}^2_{\boldsymbol{T\Sigma}}(\Omega_T) := \left\{\boldsymbol{v} \in \boldsymbol{L}^2(\Omega_T) \mid \boldsymbol{v} \in \boldsymbol{T\Sigma} \text{ a.e. in } \Omega_T\right\}$ and $\boldsymbol{\mathcal{V}}_{\boldsymbol{T\Sigma}}$, $\boldsymbol{\mathcal{V}}_{\boldsymbol{\Sigma}}$ respectively, the optimal control problem in the case of the obstacle potential is given by

$$(\mathcal{P}_0) \begin{cases} \min & J(\boldsymbol{c}, \boldsymbol{f}, \boldsymbol{g}) \\ \text{over} & (\boldsymbol{c}, \boldsymbol{f}, \boldsymbol{g}) \in \boldsymbol{\mathcal{V}}_{\boldsymbol{\Sigma}} \times \boldsymbol{L}^2_{\boldsymbol{T\Sigma}}(\Omega_T) \times L^2(0,T; L^2(\Gamma_g, \mathbb{R}^d)) \\ \text{s.t.} & (1.4), (2.5) \text{ and } (2.6) \text{ hold.} \end{cases} \tag{2.7}$$

The optimization problem $(\mathcal{P}_0)$ belongs to the problem class of so-called MPECs (Mathematical Programs with Equilibrium Constraints) which violate classical NLP constraint qualifications. In the next two subsections we present results concerning first-order necessary optimality systems obtained by the penalization approach, see [11], or the relaxation approach, see [9]. These techniques have been discussed also in [2, 16, 17].

2.2.1 Penalization Approach Without Distributed Control

In this section we discuss the penalization approach for the case $\boldsymbol{f} \equiv \boldsymbol{0}$. For the scalar Allen-Cahn case with $\boldsymbol{f} \not\equiv \boldsymbol{0}$ but without elasticity we refer the reader to [10]. Following [11] we replace the indicator function for the Gibbs simplex by a convex function $\tilde{\psi}_\sigma \in C^2(\mathbb{R})$, $\sigma > 0$, given by $\tilde{\psi}_\sigma(r) := 0$ for $r \geq 0$, $\tilde{\psi}_\sigma(r) := -\frac{1}{6\sigma^2}r^3$ for $-\sigma < r < 0$ and $\tilde{\psi}_\sigma(r) := \frac{1}{2\sigma}\left(r + \frac{\sigma}{2}\right)^2 + \frac{\sigma}{24}$ for $r \leq -\sigma$, and define the regularized potential function by $\Psi_\sigma(\boldsymbol{c}) = \Psi_0(\boldsymbol{c}) + \hat{\Psi}_\sigma(\boldsymbol{c})$ with $\hat{\Psi}_\sigma(\boldsymbol{c}) := \sum_{i=1}^{N} \tilde{\psi}_\sigma(c_i)$. For the resulting penalized optimal control problem denoted by $(\mathcal{P}_\sigma)$, exploiting techniques as in Sect. 2.1, we derive for $\sigma > 0$ first-order necessary optimality conditions. Proving a priori estimates, uniformly in $\sigma > 0$, employing compactness and monotonicity arguments and using the definition $\boldsymbol{\mathcal{W}}_0(0,T) = \{\boldsymbol{v} \in \boldsymbol{\mathcal{W}}(0,T) : \boldsymbol{v}(0,\cdot) = \boldsymbol{0}\}$ with dual space $\boldsymbol{\mathcal{W}}_0(0,T)^*$, we are able to show the following existence and approximation result:

Theorem 2.3. *Whenever $\{\boldsymbol{g}_\sigma\} \subset L^2(0,T; L^2(\Gamma_g, \mathbb{R}^d))$ is a sequence of optimal controls for $(\mathcal{P}_\sigma)$ with the sequence of corresponding states $(\boldsymbol{c}_\sigma, \boldsymbol{u}_\sigma, \boldsymbol{\xi}_\sigma) \in \boldsymbol{\mathcal{V}}_{\boldsymbol{\Sigma}} \times L^2(0,T; H^1_D(\Omega, \mathbb{R}^d)) \times \boldsymbol{L}^2(\Omega_T)$, where $-\boldsymbol{\xi}_\sigma := D\hat{\Psi}_\sigma(\boldsymbol{c}_\sigma)$, and adjoint variables $(\boldsymbol{p}_\sigma, \boldsymbol{q}_\sigma, \boldsymbol{\zeta}_\sigma) \in \boldsymbol{\mathcal{V}}_{\boldsymbol{T\Sigma}} \times L^2(0,T; H^1_D(\Omega, \mathbb{R}^d)) \times \boldsymbol{L}^2(\Omega_T)$, where $-\boldsymbol{\zeta}_\sigma := D^2\hat{\Psi}_\sigma(\boldsymbol{c}_\sigma)\boldsymbol{p}_\sigma$, there exists a subsequence, which is denoted again by $\{\boldsymbol{g}_\sigma\}$, that converges weakly to $\boldsymbol{g}$ in $L^2(0,T; L^2(\Gamma_g, \mathbb{R}^d))$. Moreover, $\boldsymbol{g}$ is an optimal control* *of* $(\mathcal{P}_0)$ *with corresponding states* $(\boldsymbol{c}, \boldsymbol{u}, \boldsymbol{\xi}) \in \boldsymbol{\mathcal{V}}_{\boldsymbol{\Sigma}} \times \boldsymbol{L}^2(\Omega_T) \times L^2(0, T; H^1_D(\Omega, \mathbb{R}^d))$ *and adjoint variables* $(\boldsymbol{p}, \boldsymbol{q}, \boldsymbol{\zeta}) \in L^2(0, T; \boldsymbol{H}^1(\Omega)) \times L^2(0, T; H^1_D(\Omega, \mathbb{R}^d)) \times \boldsymbol{\mathcal{W}}_0(0, T)^*$ *and we have for* $\sigma \searrow 0$*:*

$$
\begin{array}{llll}
\boldsymbol{c}_\sigma & \longrightarrow \boldsymbol{c} & \textit{weakly} & \textit{in } H^1(0,T;\boldsymbol{L}^2(\Omega)) \cap L^2(0,T;\boldsymbol{H}^2(\Omega)),\\
\boldsymbol{u}_\sigma & \longrightarrow \boldsymbol{u} & \textit{weakly} & \textit{in } L^2(0,T;H^1_D(\Omega,\mathbb{R}^d)),\\
\boldsymbol{\xi}_\sigma & \longrightarrow \boldsymbol{\xi} & \textit{weakly} & \textit{in } \boldsymbol{L}^2(\Omega_T),\\
\boldsymbol{p}_\sigma & \longrightarrow \boldsymbol{p} & \textit{weakly} & \textit{in } L^2(0,T;\boldsymbol{H}^1(\Omega)),\\
\boldsymbol{q}_\sigma & \longrightarrow \boldsymbol{q} & \textit{weakly} & \textit{in } L^2(0,T;H^1_D(\Omega,\mathbb{R}^d)),\\
\boldsymbol{P}_{\boldsymbol{\Sigma}}(\boldsymbol{\zeta}_\sigma) & \longrightarrow \boldsymbol{\zeta} & \textit{weakly-star} & \textit{in } \boldsymbol{\mathcal{W}}_0(0,T)^*.
\end{array}
\tag{2.8}
$$

Furthermore we obtain first order conditions:

Theorem 2.4. *The following optimality system holds for the limit elements* $(\boldsymbol{g}, \boldsymbol{c}, \boldsymbol{u}, \boldsymbol{\xi})$ *with adjoint variables* $(\boldsymbol{p}, \boldsymbol{q}, \boldsymbol{\zeta})$ *of Theorem 2.3:*
(1.4), (2.1), (2.3), (2.5), (2.6), $\boldsymbol{c} \in \boldsymbol{\Sigma}$, $\boldsymbol{f} \in \boldsymbol{T\Sigma}$ *a.e. in* Ω_T *and*

$$
\begin{aligned}
&-\frac{1}{\varepsilon}\boldsymbol{\zeta}(\boldsymbol{v}) + \varepsilon \int_0^T \langle \partial_t \boldsymbol{v}, \boldsymbol{p}\rangle \, \mathrm{d}t + \varepsilon \int_0^T \int_\Omega \nabla \boldsymbol{p} \cdot \nabla \boldsymbol{v} \, \mathrm{d}x \, \mathrm{d}t + \\
&-\frac{1}{\varepsilon}\int_0^T \int_\Omega \boldsymbol{p} \cdot \boldsymbol{v} \, \mathrm{d}x \, \mathrm{d}t + \int_0^T \int_\Omega \boldsymbol{P}_{\boldsymbol{\Sigma}}(D_p W(\boldsymbol{p}, \mathcal{E}(\boldsymbol{q}))) \cdot \boldsymbol{v} \, \mathrm{d}x \, \mathrm{d}t + \\
&-\int_0^T \int_\Omega \nu_d (\boldsymbol{c} - \boldsymbol{c}_d) \cdot \boldsymbol{v} \, \mathrm{d}x \, \mathrm{d}t - \int_\Omega \nu_T (\boldsymbol{c}(T, \cdot) - \boldsymbol{c}_T) \cdot \boldsymbol{v}(T) \, \mathrm{d}x = 0,
\end{aligned}
\tag{2.9}
$$

which has to hold for all $\boldsymbol{v} \in \boldsymbol{\mathcal{W}}_0(0, T)$*. Moreover, the limit elements satisfy some sort of complementarity slackness conditions:*

$$\lim_{\sigma \searrow 0} (\boldsymbol{\zeta}_\sigma, \boldsymbol{p}_\sigma)_{\boldsymbol{L}^2(\Omega_T)} \leq 0, \tag{2.10}$$

$$\lim_{\sigma \searrow 0} (\boldsymbol{\zeta}_\sigma, \max(\boldsymbol{0}, \boldsymbol{c}_\sigma))_{\boldsymbol{L}^2(\Omega_T)} = 0, \tag{2.11}$$

$$\lim_{\sigma \searrow 0} (\boldsymbol{p}_\sigma, \boldsymbol{\xi}_\sigma)_{\boldsymbol{L}^2(\Omega_T)} = 0. \tag{2.12}$$

2.2.2 Relaxation Approach with Distributed Control and Without Elasticity

Studying the control problem with distributed control, i.e. $\boldsymbol{f} \not\equiv \boldsymbol{0}$ in general, and without elasticity we use a relaxation approach. Details for our presented results can be found in [9]. After reformulating as in (2.5) and (2.6) the Allen-Cahn system with the help of a slack variable $\boldsymbol{\xi}$ into an MPEC, we add to the problem $(\mathcal{P}_0)$ an additional constraint $\frac{1}{2}\|\boldsymbol{\xi}\|^2_{\boldsymbol{L}^2(\Omega_T)} \leq R$ and denote this modified optimization problem by $(\mathcal{P}_R)$. The constant R is sufficiently large. This approach is also used in

[2] where the control of an obstacle problem is considered. As a first step we treat the state constraint $\boldsymbol{c} \geq \mathbf{0}$, which usually raises problems concerning regularity, by adding a regularization term to J. I.e. we define $J_\gamma(\boldsymbol{c}, \boldsymbol{f}) = J(\boldsymbol{c}, \boldsymbol{f}) + \frac{1}{2\gamma\varepsilon}\sum_{i=1}^{N} \|\max(0, \overline{\lambda} - \gamma c_i)\|^2_{L^2(\Omega_T)}$ where $\overline{\lambda} \in L^2(\Omega_T)$ is fixed, nonnegative and corresponds to a regular version of the multiplier associated to $\boldsymbol{c} \geq \mathbf{0}$. Next we relax the complementarity condition to $(\boldsymbol{\xi}, \boldsymbol{c})_{\boldsymbol{L}^2(\Omega_T)} \leq \varepsilon\alpha_\gamma$ for some $\alpha_\gamma > 0$. We denote this regularized relaxed version of $(\mathcal{P}_R)$ as $(\mathcal{P}_{R,\gamma})$. Subsequently we are interested in $\gamma \nearrow \infty$ where simultaneously $\alpha_\gamma \searrow 0$. We are able to use techniques from mathematical programming in Banach spaces, see [18], and get an optimality system for $(\mathcal{P}_{R,\gamma})$, where γ is fixed. Considering $\gamma \nearrow \infty$ we then obtain optimality conditions for problem $(\mathcal{P}_R)$. Similar to the process in Sect. 2.2.1 we have: for any $\gamma > 0$ there exists a minimizer $(\boldsymbol{c}_\gamma, \boldsymbol{f}_\gamma, \boldsymbol{\xi}_\gamma) \in \boldsymbol{V}_{\boldsymbol{\Sigma}} \times \boldsymbol{L}^2(\Omega_T) \times \boldsymbol{L}^2(\Omega_T)$ of $(\mathcal{P}_{R,\gamma})$ with corresponding adjoint variables. Using the Lagrange multiplier $r_\gamma \in \mathbb{R}$ of the constraint $(\boldsymbol{\xi}_\gamma, \boldsymbol{c}_\gamma)_{\boldsymbol{L}^2(\Omega_T)} \leq \varepsilon\alpha_\gamma$ one defines $\zeta_{\gamma,i} := r_\gamma \xi_{\gamma,i} - \max(0, \overline{\lambda} - \gamma c_{\gamma,i})$ and $\boldsymbol{\zeta}_\gamma := (\zeta_{\gamma,i})_{i=1}^N$. Then we obtain:

Theorem 2.5. *Whenever* $\{\boldsymbol{f}_\gamma\}$ *is a sequence of optimal controls* $(\mathcal{P}_{R,\gamma})$ *with the sequence of corresponding states* $(\boldsymbol{c}_\gamma, \boldsymbol{\xi}_\gamma)$ *and adjoint variables* $(\boldsymbol{p}_\gamma, \boldsymbol{\zeta}_\gamma)$, *there exists a subsequence, which is denoted the same, with* $\boldsymbol{f}_\gamma \to \boldsymbol{f}$ *weakly in* $\boldsymbol{L}^2(\Omega_T)$ *and* $\boldsymbol{\zeta}_\gamma \to \boldsymbol{\zeta}$ *weakly-star in* $\boldsymbol{\mathcal{W}}_0(0,T)^*$ *as* $\gamma \nearrow \infty$. *The convergence of the variables* $\boldsymbol{c}_\gamma$, $\boldsymbol{\xi}_\gamma$ *and* $\boldsymbol{p}_\gamma$ *is as in* (2.8). *These limits fulfill the corresponding optimality system for* $(\mathcal{P}_R)$ *as in Theorem 2.4 without elasticity system but with distributed control, i.e.* (2.4), (2.5), (2.6), (2.9), $\boldsymbol{c} \in \boldsymbol{\Sigma}$, $\boldsymbol{f} \in \boldsymbol{T\Sigma}$ *a.e. in* Ω_T *and the limits with* $(\boldsymbol{p}_\gamma, \boldsymbol{\zeta}_\gamma)$ *satisfy the complementarity slackness conditions* (2.10)–(2.12) *for* $\gamma \nearrow \infty$ *instead of* $\sigma \searrow 0$. *In addition we have the constraint* $\frac{1}{2}\|\boldsymbol{\xi}\|^2_{\boldsymbol{L}^2(\Omega_T)} \leq R$.

The last inequality is in practice inactive using R large enough.

3 Numerics

In this section we neglect elastic effects, but study smooth as well as nonsmooth obstacle potentials with distributed control numerically.

3.1 *Smooth Potential*

3.1.1 Newton's Method

For smooth Ψ we obtain an unconstrained optimal control problem when eliminating the state equation. Hence, numerical methods for unconstrained problems can be applied to the reduced problem

$$\min j(\boldsymbol{f}) := J(S(\boldsymbol{f}), \boldsymbol{f}), \quad \boldsymbol{f} \in \boldsymbol{L}^2(\Omega_T).$$

We choose the Trust-Region-Newton-Steihaug-cg (TRN) method, see [8], since it is capable of solving large scale optimization problems very efficiently because the underlying cg-solver is matrix-free and it attains the local convergence properties of Newton's method. Iteratively the model $m_k(\boldsymbol{\delta f}) = j(\boldsymbol{f}_k) + (\nabla j(\boldsymbol{f}_k), \boldsymbol{\delta f})_{L^2(\Omega_T)} + \frac{1}{2}(\nabla^2 j(\boldsymbol{f}_k)\boldsymbol{\delta f}, \boldsymbol{\delta f})_{L^2(\Omega_T)}$ is minimized within a trust-region and the method is stopped if $\|\nabla j(\boldsymbol{f}_k)\|_{L^2(\Omega_T)} < tol$.

Based on Sect. 2.1 the L^2-gradient is given by $\nabla j(\boldsymbol{f}) = \frac{\nu_f}{\varepsilon}\boldsymbol{f} + \boldsymbol{p}$. The Hessian we derived formally for $\nu_d = 0$, see [23], and is given by $\nabla^2 j(\boldsymbol{f})\boldsymbol{\delta f} = \frac{\nu_f}{\varepsilon}\boldsymbol{\delta f} + \boldsymbol{\delta p}$, where $\boldsymbol{\delta p}$ can be calculated by first solving the linear forward equation $\varepsilon\partial_t\boldsymbol{\delta c} - \varepsilon\Delta\boldsymbol{\delta c} + \frac{1}{\varepsilon}D^2\Psi(\boldsymbol{c})\boldsymbol{\delta c} = \boldsymbol{\delta f}$ in Ω_T, $\nabla(\boldsymbol{\delta c})\cdot\boldsymbol{n} = \boldsymbol{0}$ on Γ_T and $\boldsymbol{\delta c}(0) = \boldsymbol{0}$ in Ω and then solving the linear backward equation $-\varepsilon\partial_t\boldsymbol{\delta p} - \varepsilon\Delta\boldsymbol{\delta p} + \frac{1}{\varepsilon}D^2\Psi(\boldsymbol{c})\boldsymbol{\delta p} = -\frac{1}{\varepsilon}D^3\Psi(\boldsymbol{c})[\boldsymbol{\delta c}, \boldsymbol{p}, .]$ in Ω_T, $\nabla(\boldsymbol{\delta p})\cdot\boldsymbol{n} = \boldsymbol{0}$ on Γ_T and $\varepsilon\boldsymbol{\delta p}(T) = \nu_T\boldsymbol{\delta c}(T)$ in Ω. The cost of one iteration of the algorithm consists in evaluating j, which means solving the nonlinear state equation, in calculating $\nabla j(\boldsymbol{f})$, which means solving the linear adjoint equation, and in performing the Steihaug-cg method, where in each cg-iteration $\nabla^2 j(\boldsymbol{f})$ has to be evaluated in some direction $\boldsymbol{\delta f}$. For similar control problems gradient type methods have been used, see e.g. [14, 22]. However, they cannot solve our problems in reasonable time.

The following numerical results summarize the investigations in [23].

3.1.2 Discretization and Error Estimation

We consider an implicit and a semi-implicit Euler scheme in time. Although solving the semi-implicit discrete equations is much faster, it has the disadvantage that the two approaches "first discretize then optimize" and "first optimize then discretize" do not commute. This has been shown by looking upon the implicit discretization as a discontinuous Galerkin ansatz [23]. Thus we use semi-implicit discretization only in an initialization phase to compute an approximative optimal control, and use implicit discretization in the main phase.

In space we discretize with standard P1-elements. For equidistant meshes we implemented the TRN method with the toolbox FEniCS [19], exploiting the structure of the arising systems for equidistant meshes. The existing adaptive strategy for the Allen-Cahn equation without control uses a fine mesh on the interface and coarse mesh on the bulk regions, see e.g. [1]. However, with control, nucleation of a phase may appear. This cannot be resolved using the concept in [1]. Moreover, a method of adaptively controlling the time steps for Allen-Cahn equations is not available. Hence, for studying adaptive meshes we use the toolbox RoDoBo, where the TRN method together with a dual weighted residual (DWR) error estimator is implemented, see [20]. In our applications the DWR error estimator establishes both: adequate adaptive spatial meshes and adaptive time steps. For example in a nucleation situation the mesh in [1] is only fine when the new phase was already created, whereas the DWR mesh is also fine at timesteps before the nucleation process starts.

3.1.3 Numerical Results

In all experiments we choose $d = N = 2$, $\nu_T = 1$, $\nu_d = 0$, $\nu_f = 0.01$, $\varepsilon = (14\pi)^{-1}$, $tol = 10^{-13}$ and $\Omega = (-1, 1)^2$. As mentioned above we reduce the problem to a scalar problem and use $\Psi(c) = \frac{1}{4}(c^2 - 1)^2$. Figure 1 depicts the large speed up using the TRN method instead of the gradient method. Here the Newton residual $\|\nabla j(f_k)\|_{L^2(\Omega_T)}$ for the TRN method and the gradient method are listed for an example where $c(T)$ shall be the same circle as c_0. The cpu-time is still large for the TRN method using RoDoBo. However, using an equidistant mesh and therefore being able to exploit the structure of the problem, our implementation in FEniCS is significantly faster. Already the adjoint equation can be solved 25 times faster.

In order to get quadratic convergence of the Newton-cg method for smooth problems the inner tolerance tol_{cg} has to be appropriate. While one can decrease tol_{cg} with the number of iterations, we set $tol_{cg} = 10^{-13}$ in order to solve the inner problem nearly exact and the resulting numerical error does not influence the performance of the Newton-method. In most experiments the Newton method converged just superlinearly which reveals that the problem is not smooth enough. Only in an experiment where a vertical interface is moved from left to right we could

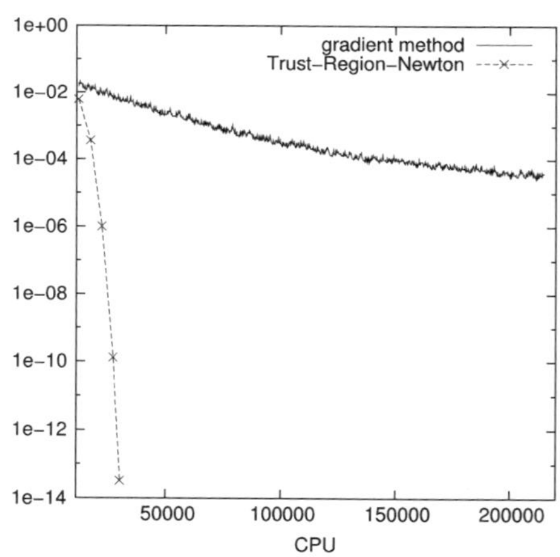

Fig. 1 $\|\nabla j(f_k)\|_{L^2(\Omega_T)}$ depending on cpu-time for the TRN and the gradient method applied to have $c(T) = c_0$

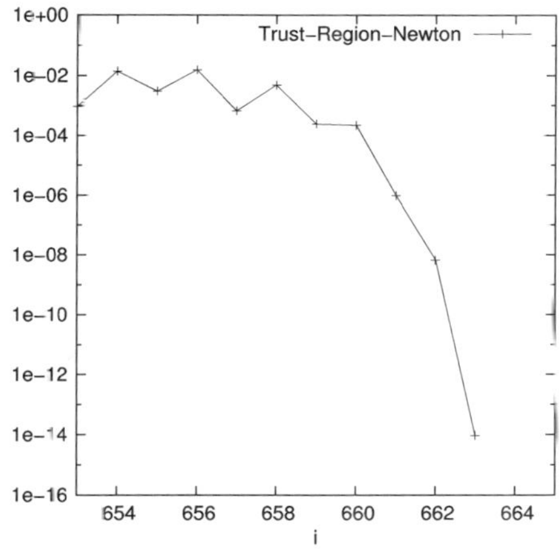

Fig. 2 Newton residual for moving a vertical interface from left to right

observe quadratic convergence in $\nabla j(f_k)$, see Fig. 2. For the first 660 iterations in this example, an approximation of the model problem is computed with less than 40 Steihaug-cg-iterations. They always lie on the boundary of the trust-region. In the last three iterations the trust-region constraint stays inactive and then about 600 cg-iterations are necessary to solve the quadratic subproblem. In these last three outer iterations the convergence rate of Newton's method can be observed. Also in the other experiments in [23] the Steihaug-cg method performs only few inner cg-iterations when the trust-region constraint is active. In the last few steps the calculation of the unconstrained minimizer of m_k is much more expensive.

Next we consider the situation where a circle in the center shall be split into two circles next to each other. Figure 3 shows the optimal state and control. The circle is stretched horizontally until it separates into two circles. In Fig. 4 the plot of $t \mapsto \|f(t)\|_{L^2(\Omega)}$ is depicted. The peak is at the time when the topological change occurs. The large increase of the cost at the end time is due to the fact that c_T has a smaller interface thickness than proposed by the model with ε.

In the following we investigate the temporal mesh. Figure 5 shows the time steps created by the DWR error estimator together with the value $c(t)$ at the location $\boldsymbol{x} = \boldsymbol{0}$. We can see that the time steps are small before the pinching occurs, attain their minimum in the middle of the pinching process and are larger in the second half of the pinching process. To study how the time steps depend on the interface velocity we consider an experiment where a circle shrinks and vanishes at finite

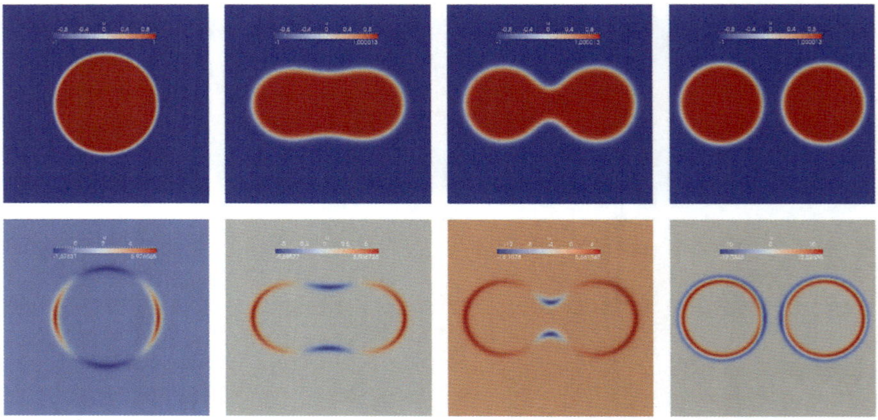

Fig. 3 Optimal state (*top*) and optimal control (*bottom*) at times $t = 0, \frac{1}{2}T, \frac{3}{4}T, T$, for a splitting circle scenario

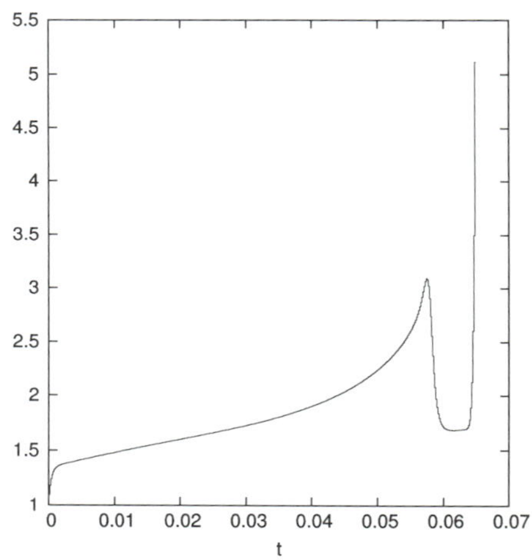

Fig. 4 $L^2(\Omega)$-norm of the control corresponding to Fig. 3

time. The end time is chosen in such a way that $f \equiv 0$ is the optimal control, i.e. the interface evolution is given by the Allen-Cahn equation without outer force. Figure 6 depicts the interface velocity together with the time steps. As expected, the larger the interface velocity becomes, the smaller the time steps have to be chosen.

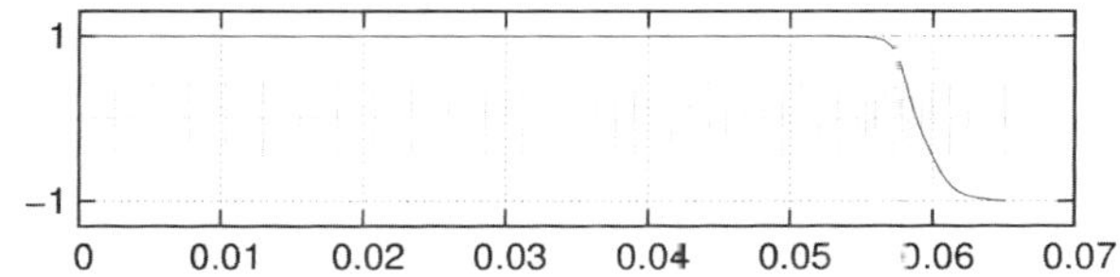

Fig. 5 Temporal mesh for a splitting circle scenario

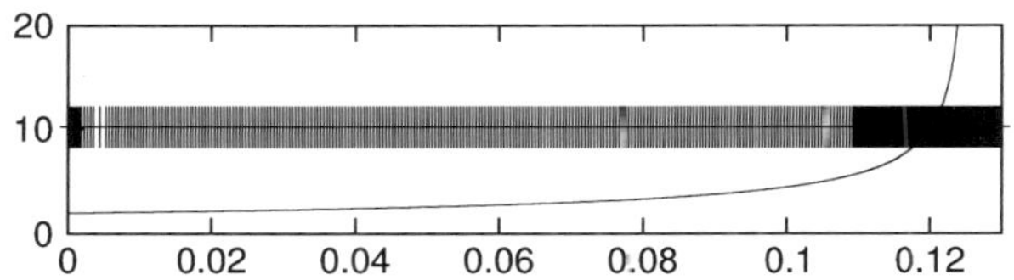

Fig. 6 Temporal mesh for the time evolution without control

3.2 *Obstacle Potential*

In the case of an obstacle potential we studied first the differences in the approaches "first discretize then optimize" and "first optimize then discretize". As in the smooth case the choice of discretization is essential. We chose again an implicit discretization of the Allen-Cahn system, which is understood as a discontinuous Galerkin discretization in time. Hence the time integrals of functions are discretized by an iterated rectangle rule using the right endpoints. This approximation is also used in the cost function. We compared the discretized optimality system of the ansatz presented in Sect. 2.2.2 with the optimality system arising for the discretized optimization problem, where the first order conditions (C-stationarity) are derived by the relaxation approach in [24] for finite dimensional MPECs assuming MPEC-LICQ. In the latter only the complementarity condition is relaxed as in Sect. 2.2.2 to $(\boldsymbol{\xi}_\alpha, \boldsymbol{c}_\alpha)_{L^2(\Omega_T)} \leq \alpha$. The systems are identical apart from the additional constraint on $\boldsymbol{\xi}$, which is inactive in the numerics, and, as expected, the complementarity slackness conditions, which hold pointwise for the ansatz, where the problem is discretized first [21].

Our first numerical experiments are based on the MATLAB solver *fmincon* where the discretized, relaxed optimization problem is solved—due to the memory limitations—using an interior point algorithm with internal cg-solver for decreasing α. The initial $\alpha_0 = 1$ is successively divided by 10 and the solutions for α_i are used as initial data for the problem with relaxation parameter α_{i+1}. In the first example with $N = 3$ the goal is to keep the initial setting unchanged for the time interval $[0, 0.0005]$, where one phase in a circle is surrounded by an annulus with a second phase and a third phase in the remainder of the domain $\Omega = (0, 1)^2$. Without any control the two inner phases would vanish due to the curvature. We set $\nu_T = 1$, $\nu_d = 10^4$, $\nu_f = 0.001$, $\varepsilon = 0.1$ and the time step $\tau = 10^{-4}$ while the equidistant mesh size in space is $h = 1/59$. The phases stay nearly constant as do the controls which we therefore list only for $T = 0.0003$ and $\alpha = 10^{-9}$ in Fig. 7. The control f_1 is positive on the innermost interface to ensure that this circle does not shrink. However, noticeable is that f_1 is negative on the other interface, where it seems that c_1 would otherwise increase, i.e. phase one would develop. In the same way

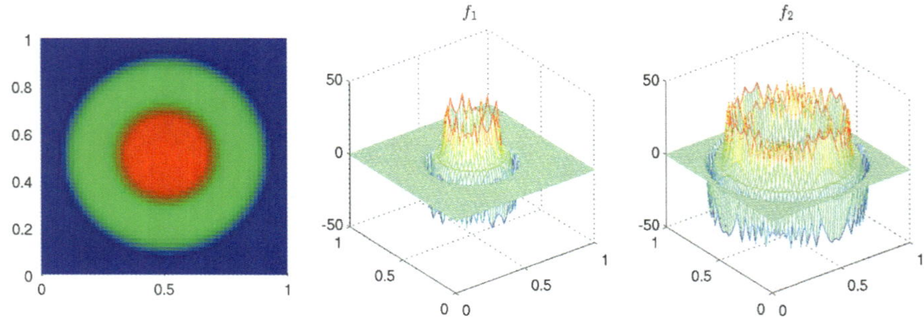

Fig. 7 The state $\mathbf{c}$ and the control functions f_1 and f_2 for three phases, which shall stay constant, at time $T = 0.0003$

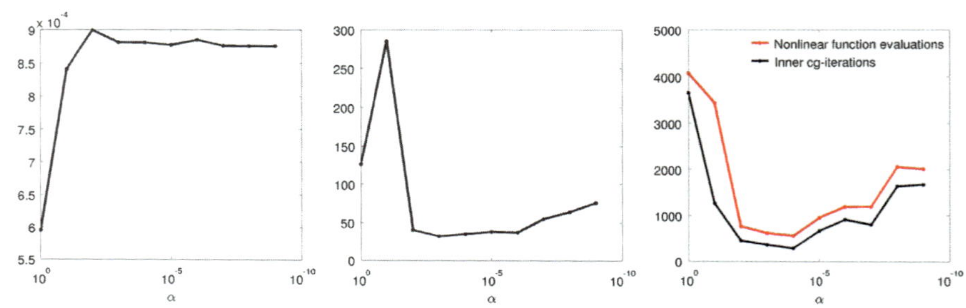

Fig. 8 Results for varying α for the example in Fig. 7

f_2 is negative on the innermost circle while positive to hold the interface constant on the outer circle. Correspondingly f_3 behaves. In Fig. 8 the first plot shows the values of the cost function J neglecting the constant part for decreasing α. For $\alpha \leq 10^{-3}$ it changes only mildly. The main effort of calculating the optimal control is used for large α as the other two plots in Fig. 8 indicate, which list the number of interior point iterations and the number of nonlinear function evaluations together with the cg-iterations. They indicate also the expected cost if a more sophisticated implementation of an optimization solver is employed.

In the next example three phases are vertically aligned. Since the interfaces have no curvature the phases would stay constant without control. However, in this experiment we set the target $\mathbf{c}_d$ such that in the end the enclosed phase occupies a larger rectangle than the others as the numerical result shows in Fig. 9 for $\alpha = 10^{-9}$. Hence the controls are now time dependent. In the first row of Fig. 10 f_1 is depicted and in the second f_2 while $f_3 = -f_1 - f_2$ is neglected. As expected the controls work mainly on the interfaces. The control f_2 is positive at both interfaces while the other two controls support the movement by negative force. Like in the first

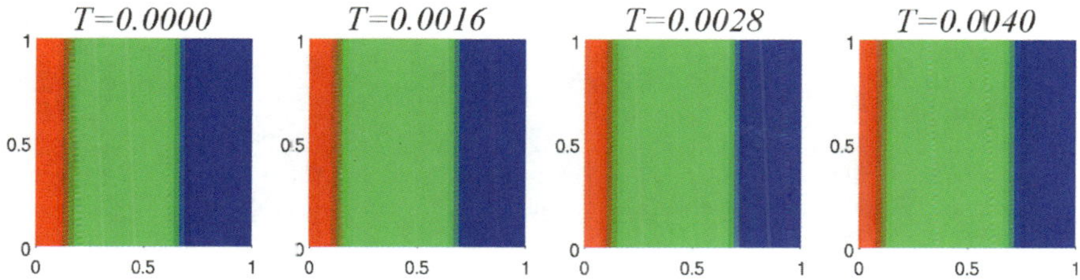

Fig. 9 The state **c** for three phases for moving walls

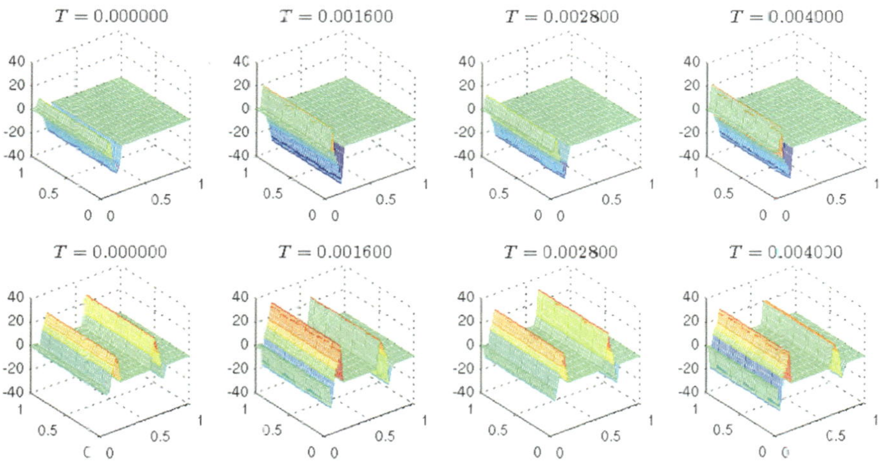

Fig. 10 Control functions f_1 in the *first* and f_2 in the *second row* corresponding to Fig. 9

example the value of the cost function stays nearly constant for $\alpha \leq 10^{-3}$ and the substantial work of determining the optimal control is done for large α. We therefore omit the figures.

References

1. J.W. Barrett, R. Nürnberg, V. Styles, Finite element approximation of a phase field model for void electromigration. SIAM J. Numer. Anal. **42**(2), 738–772 (2004)
2. M. Bergounioux, Optimal control of an obstacle problem. Appl. Math. Optim. **36**(2), 147–172 (1997)
3. L. Blank, H. Garcke, L. Sarbu, V. Styles, Nonlocal Allen-Cahn systems: analysis and a primal-dual active set method. IMA J. Numer. Anal. **33**(4), 1126–1155 (2013)
4. T. Blesgen, U. Weikard, Multi-component Allen-Cahn equation for elastically stressed solids. Electron. J. Differ. Equ. **2005**(89), 1–17 (2005)
5. J.F. Blowey, C.M. Elliott, The Cahn-Hilliard gradient theory for phase separation with non-smooth free energy Part I: mathematical analysis. Eur. J. Appl. Math. **2**, 233–280 (1991)

6. A. Bossavit, A. Damlamian, *Free Boundary Problems: Applications and Theory*, vol. 1 (Pitman Advanced Publishing Program, London, 1985)
7. P.G. Ciarlet, *Three-Dimensional Elasticity*. Number 1 in Studies in Mathematics and Its Applications (Elsevier, Amsterdam, 1988)
8. A.R. Conn, N.I.M. Gould, P.L. Toint, *Trust Region Methods* (SIAM, Philadelphia, 2000)
9. M.H. Farshbaf-Shaker, A relaxation approach to vector-valued Allen-Cahn MPEC problems, Universität Regensburg, Mathematik, (Preprint-SPP1253-127-2, 2011)
10. M.H. Farshbaf-Shaker, A penalty approach to optimal control of Allen-Cahn variational inequalities: MPEC-view. Numer. Funct. Anal. Optim. **33**(11), 1321–1349 (2012)
11. M.H. Farshbaf-Shaker, C. Hecht, Optimal control of elastic vector-valued Allen-Cahn variational inequalities, (Preprintreihe der Fakultät Mathematik 16/2013, Universität Regensburg, 2013)
12. H. Garcke, *On Mathematical Models for Phase Separation in Elastically Stressed Solids*, Habilitation thesis, University Bonn, 2000
13. H. Garcke, B. Nestler, A mathematical model for grain growth in thin metallic films. Math. Models Methods Appl. Sci. **10**(06), 895–921 (2000)
14. F. Haußer, S. Rasche, A. Voigt, The influence of electric fields on nanostructures – simulation and control. Math. Comput. Simul. **80**(7), 1449–1457 (2010)
15. C. Hecht, Existence theory and necessary optimality conditions for the control of the elastic Allen-Cahn system, Diplomarbeit, Universität Regensburg, 2011
16. M. Hintermüller, I. Kopacka, Mathematical programs with complementarity constraints in function space: C- and strong stationarity and a path-following algorithm. SIAM J. Optim. **20**(2), 868–902 (2009)
17. M. Hintermüller, D. Wegner, Distributed optimal control of the Cahn-Hilliard system including the case of a double-obstacle homogeneous free energy density. SIAM J. Control Optim. **50**(1), 388–418 (2012)
18. S. Kurcyusz, J. Zowe, Regularity and stability for the mathematical programming problem in Banach spaces. Appl. Math. Optim. **5**(1), 49–62 (1979)
19. A. Logg, K. Mardal, G. Wells et al., *Automated Solution of Differential Equations by the Finite Element Method* (Springer-Verlag Berlin Heidelberg, 2012). http://www.fenicsproject.org/
20. D. Meidner, B. Vexler, Adaptive space-time finite element methods for parabolic optimization problems. SIAM J. Control Optim. **46**(1), 116–142 (2007)
21. J. Michl, Optimale Steuerung von Phasengrenzen mit Optimierungsmethoden für MPECs. Diplomarbeit, Universität Regensburg, 2013
22. T. Ohtsuka, K. Shirakawa, N. Yamazaki, Optimal control problem for Allen-Cahn type equation associated with total variation energy. Discret. Contin. Dyn. Syst. Ser. S **5**(1), 159–181 (2012)
23. C. Rupprecht, Numerische Lösung optimaler Steuerung der Allen-Cahn Gleichung inklusive Adaptivität, Diplomarbeit, Universität Regensburg, 2011
24. S. Scholtes, Convergence properties of a regularization scheme for mathematical programs with complementarity constraints. SIAM J. Optim. **11**(4), 918–936 (2001)

Optimal Control of Elastoplastic Processes: Analysis, Algorithms, Numerical Analysis and Applications

Roland Herzog, Christian Meyer, and Gerd Wachsmuth

Abstract An optimal control problem is considered for the variational inequality representing the stress-based (dual) formulation of static elastoplasticity. The linear kinematic hardening model and the von Mises yield condition are used. The forward system is reformulated such that it involves the plastic multiplier and a complementarity condition. In order to derive necessary optimality conditions, a family of regularized optimal control problems is analyzed. C-stationarity type conditions are obtained by passing to the limit with the regularization. Numerical results are presented.

Keywords Mathematical programs with complementarity constraints in function space • Variational inequalities • Elastoplasticity • Regularization • Optimality conditions

Mathematics Subject Classification (2010). Primary 49K20, 70Q05, 74C05; Secondary 90C33, 35R45.

1 Introduction

Solid bodies depart from their rest shape under the influence of applied loads. In case the applied loads or stresses are sufficiently small, many solids exhibit a linearly elastic and reversible behavior. If, however, the stress induced by the applied loads exceeds a certain threshold (the yield stress), the material behavior switches from

R. Herzog (✉) • G. Wachsmuth
Faculty of Mathematics, Technische Universität Chemnitz, 09126 Chemnitz, Germany
e-mail: roland.herzog@mathematik.tu-chemnitz.de;
gerd.wachsmuth@mathematik.tu-chemnitz.de

C. Meyer
Faculty of Mathematics, Technical Universität Dortmund, Vogelpothsweg 87, 44227 Dortmund, Germany
e-mail: christian.meyer@math.tu-dortmund.de

G. Leugering et al. (eds.), *Trends in PDE Constrained Optimization*, International Series of Numerical Mathematics 165, DOI 10.1007/978-3-319-05083-6_4

the elastic to the so-called plastic regime. In this state, the overall loading process is no longer reversible and permanent deformations remain even after the loads are withdrawn. Mathematically, this leads to a description involving *variational inequalities* (VIs), or equivalently, *complementarity conditions*.

Plastic deformation is desired for instance as an industrial shaping technique of metal workpieces, as e.g. by deep-drawing of body sheets in the automotive industry. The task of finding appropriate time-dependent loads which effect a desired final deformation leads to *optimal control problems for elastoplasticity systems*. These problems are also motivated by the desire to reduce the amount of *springback*, i.e., the partial reversal of the final material deformation due to a release of the stored elastic energy once the loads are removed.

In this review, we concentrate for the sake of brevity on the model of *static* elastoplasticity with small strains in its so-called *dual (stress-based) formulation*, and with linear kinematic hardening. Within the project, similar results were achieved also for the more challenging *quasi-static* model, see [24–26] and the dissertation [23]. The system describing the quasi-static forward problem is given in Sect. 2.3. We also refer to [3] for the analysis of an optimal control problem involving the static *primal (strain-based)* counterpart.

According to the standard approach for infinite strains we do not distinguish between reference and actual configuration and identify Ω with the workpiece under consideration. In its strong form, the static problem of elastoplasticity in its *dual* formulation with linear kinematic hardening reads

$$\left.\begin{aligned}
\mathbb{C}^{-1}\boldsymbol{\sigma} + \boldsymbol{\varepsilon}(\boldsymbol{u}) + \lambda\,(\boldsymbol{\sigma}^D + \boldsymbol{\chi}^D) &= \mathbf{0} && \text{in } \Omega,\\
\mathbb{H}^{-1}\boldsymbol{\chi} \qquad\quad + \lambda\,(\boldsymbol{\sigma}^D + \boldsymbol{\chi}^D) &= \mathbf{0} && \text{in } \Omega,\\
\operatorname{div}\boldsymbol{\sigma} &= -\boldsymbol{f} && \text{in } \Omega,\\
\text{with complement. conditions} \quad 0 \le \lambda \perp \phi(\boldsymbol{\Sigma}) &\le 0 && \text{in } \Omega,\\
\text{and boundary conditions} \quad \boldsymbol{u} &= \mathbf{0} && \text{on } \Gamma_D,\\
\boldsymbol{\sigma}\cdot\boldsymbol{n} &= \boldsymbol{g} && \text{on } \Gamma_N.
\end{aligned}\right\} \tag{1.1}$$

The state variables consist of the stress $\boldsymbol{\sigma}$ and back stress $\boldsymbol{\chi}$, combined into the generalized stresses $\boldsymbol{\Sigma} = (\boldsymbol{\sigma}, \boldsymbol{\chi})$, plus the displacement $\boldsymbol{u}$ and the plastic multiplier λ associated with the yield condition $\phi(\boldsymbol{\Sigma}) \le 0$ which we assume to be of von Mises type, see (2.1). The first two equations in (1.1), together with the complementarity conditions, represent the material law of static elastoplasticity. The tensors $\mathbb{C}^{-1}$ and $\mathbb{H}^{-1}$ are the inverses of the elasticity tensor (the compliance tensor) and of the hardening modulus, respectively, $\boldsymbol{\sigma}^D$ denotes the deviatoric part of $\boldsymbol{\sigma}$, while $\boldsymbol{\varepsilon}(\boldsymbol{u})$ is the linearized strain. The third equation in (1.1) represents the equilibrium of forces. The boundary conditions correspond to clamping on Γ_D and the prescription of boundary loads $\boldsymbol{g}$ on the remainder $\Gamma_N = \Gamma \setminus \Gamma_D$.

Due to the complementarity between the plastic multiplier λ and the yield condition $\phi(\boldsymbol{\Sigma})$, the optimal control of (1.1) leads to a *mathematical program with complementarity constraints* (MPCC) in function space. As is known already for finite dimensional MPCCs, classical constraint qualifications such as MFCQ fail to hold. To overcome these difficulties, several competing stationarity concepts tailored for MPCCs have been developed, see for instance [15, 21] for an overview in the finite dimensional case. For the infinite dimensional case, we refer to the classical works [1, 19, 20] and the recent contributions [14, 16, 22].

2 Optimal Control Problems in Small-Strain Static Elastoplasticity

In this section we present the optimal control problem under consideration. We set up some notation in Sect. 2.1. Afterwards, we discuss the static forward problem in Sect. 2.2 and state the quasi-static forward problem in Sect. 2.3. The optimal control problem of static plasticity is considered in Sect. 2.4. As mentioned in the introduction, we concentrate on the model of *static* elastoplasticity with small strains in its so-called *dual formulation*, and with linear kinematic hardening.

2.1 Notation and Standing Assumptions

2.1.1 Variables

Our notation follows [8] for the forward problem. Since the presentation of optimality conditions relies on adjoint variables and Lagrange multipliers associated with inequality constraints, additional variables are needed.

2.1.2 Function Spaces

Let $\Omega \subset \mathbb{R}^d$ be a bounded domain with Lipschitz boundary $\Gamma = \partial\Omega$ in dimension $d \in \{2, 3\}$. We point out that the presented analysis is not restricted to the case $d \le 3$, but for reasons of physical interpretation we focus on the two and three dimensional cases. The boundary consists of two disjoint parts Γ_N and Γ_D. We denote by $\mathbb{S} := \mathbb{R}^{d\times d}_{\text{sym}}$ the space of symmetric d-by-d matrices, endowed with the inner product $\boldsymbol{A} : \boldsymbol{B} = \sum_{i,j=1}^{d} A_{ij} B_{ij}$, and we define

$$V = H^1_D(\Omega; \mathbb{R}^d) = \{\boldsymbol{u} \in H^1(\Omega; \mathbb{R}^d) : \boldsymbol{u} = 0 \text{ on } \Gamma_D\},$$

$$S = L^2(\Omega; \mathbb{S})$$

as the spaces for the displacement $\boldsymbol{u}$, stress $\boldsymbol{\sigma}$, and back stress $\boldsymbol{\chi}$, respectively. The control $(\boldsymbol{f}, \boldsymbol{g})$ belongs to the space

$$U = L^2(\Omega; \mathbb{R}^d) \times L^2(\Gamma_N; \mathbb{R}^d).$$

2.1.3 Yield Function and Admissible Stresses

We restrict our discussion to the von Mises yield function. In the context of linear kinematic hardening, it reads

$$\phi(\boldsymbol{\Sigma}) = \left(|\boldsymbol{\sigma}^D + \boldsymbol{\chi}^D|^2 - \tilde{\sigma}_0^2\right)/2 \tag{2.1}$$

for $\boldsymbol{\Sigma} = (\boldsymbol{\sigma}, \boldsymbol{\chi}) \in S^2$, where $|\cdot|$ denotes the pointwise Frobenius norm of matrices,

$$\boldsymbol{\sigma}^D = \boldsymbol{\sigma} - \frac{1}{d}\,(\operatorname{trace} \boldsymbol{\sigma})\, \boldsymbol{I}$$

is the deviatoric part of $\boldsymbol{\sigma}$, and $\tilde{\sigma}_0$ is the yield stress. The yield function gives rise to the set of admissible generalized stresses

$$\mathcal{K} = \{\boldsymbol{\Sigma} \in S^2 : \phi(\boldsymbol{\Sigma}) \leq 0 \quad \text{a.e. in } \Omega\}. \tag{2.2}$$

Due to the structure of the yield function, $\boldsymbol{\sigma}^D + \boldsymbol{\chi}^D$ appears frequently and we abbreviate it and its adjoint by

$$\mathcal{D}\boldsymbol{\Sigma} = \boldsymbol{\sigma}^D + \boldsymbol{\chi}^D \quad \text{and} \quad \mathcal{D}^\star \boldsymbol{\sigma} = \begin{pmatrix} \boldsymbol{\sigma}^D \\ \boldsymbol{\sigma}^D \end{pmatrix}$$

for matrices $\boldsymbol{\Sigma} \in \mathbb{S}^2$ as well as for functions $\boldsymbol{\Sigma} \in S^2$. When considered as an operator in function space, $\mathcal{D}$ maps $S^2 \to S$. For later reference, we also remark that

$$\mathcal{D}^\star \mathcal{D}\boldsymbol{\Sigma} = \begin{pmatrix} \boldsymbol{\sigma}^D + \boldsymbol{\chi}^D \\ \boldsymbol{\sigma}^D + \boldsymbol{\chi}^D \end{pmatrix}$$

holds.

2.1.4 Operators and Forms

We begin by defining the bilinear forms associated with (1.1). For $\boldsymbol{\Sigma} = (\boldsymbol{\sigma}, \boldsymbol{\chi}) \in S^2$ and $\boldsymbol{T} = (\boldsymbol{\tau}, \boldsymbol{\mu}) \in S^2$, let

$$a(\boldsymbol{\Sigma}, \boldsymbol{T}) = \int_\Omega \boldsymbol{\sigma} : \mathbb{C}^{-1} \boldsymbol{\tau} \,\mathrm{d}x + \int_\Omega \boldsymbol{\chi} : \mathbb{H}^{-1} \boldsymbol{\mu} \,\mathrm{d}x. \tag{2.3}$$

Here $\mathbb{C}^{-1}(x)$ and $\mathbb{H}^{-1}(x)$ are maps from $\mathbb{S}$ to $\mathbb{S}$ which may depend on the spatial variable x. For $\boldsymbol{\Sigma} = (\boldsymbol{\sigma}, \boldsymbol{\chi}) \in S^2$ and $\boldsymbol{v} \in V$, let

$$b(\boldsymbol{\Sigma}, \boldsymbol{v}) = -\int_\Omega \boldsymbol{\sigma} : \boldsymbol{\varepsilon}(\boldsymbol{v})\,\mathrm{d}x. \tag{2.4}$$

We recall that $\boldsymbol{\varepsilon}(\boldsymbol{v}) = \frac{1}{2}\big(\nabla \boldsymbol{v} + (\nabla \boldsymbol{v})^\top\big)$ denotes the (linearized) strain tensor.

The bilinear forms induce operators

$$A : S^2 \to S^2, \quad \langle A\boldsymbol{\Sigma},\, \boldsymbol{T}\rangle = a(\boldsymbol{\Sigma}, \boldsymbol{T}),$$
$$B : S^2 \to V', \quad \langle B\boldsymbol{\Sigma},\, \boldsymbol{v}\rangle = b(\boldsymbol{\Sigma}, \boldsymbol{v}).$$

Here and throughout, $\langle\cdot,\ \cdot\rangle$ denotes the dual pairing between V and its dual V', or the scalar products in S or S^2, respectively.

Assumptions.

1. The domain $\Omega \subset \mathbb{R}^d$, $d \geq 2$ is a bounded domain with Lipschitz boundary in the sense of [4, Chapter 1.2]. The boundary of Ω, denoted by Γ, consists of two disjoint measurable parts Γ_N and Γ_D such that $\Gamma = \Gamma_N \cup \Gamma_D$. While Γ_N is a relatively open subset, Γ_D is a relatively closed subset of Γ. Furthermore Γ_D is assumed to have positive measure. In addition, the set $\Omega \cup \Gamma_N$ is regular in the sense of Gröger, cf. [6]. A characterization of regular domains for the case $d \in \{2, 3\}$ can be found in [7, Section 5]. This class of domains covers a wide range of geometries.

   We make these assumptions in order to apply the regularity results in [10] pertaining to systems of nonlinear elasticity. The latter appear in the forward problem and its regularizations. Additional regularity leads to a norm gap, which is needed to prove the differentiability of the control-to-state map.
2. The yield stress $\tilde{\sigma}_0$ is assumed to be a positive constant. It equals $\sqrt{2/3}\,\sigma_0$, where σ_0 is the uni-axial yield stress.
3. $\mathbb{C}^{-1}$ and $\mathbb{H}^{-1}$ are elements of $L^\infty(\Omega; \mathcal{L}(\mathbb{S}, \mathbb{S}))$, where $\mathcal{L}(\mathbb{S}, \mathbb{S})$ denotes the space of linear operators $\mathbb{S} \to \mathbb{S}$. Both $\mathbb{C}^{-1}(x)$ and $\mathbb{H}^{-1}(x)$ are assumed to be uniformly coercive. Standard examples are isotropic and homogeneous materials, where

$$\mathbb{C}^{-1}\boldsymbol{\sigma} = \frac{1}{2\,\mu}\boldsymbol{\sigma} - \frac{\lambda}{2\,\mu\,(2\,\mu + d\,\lambda)}\,\mathrm{trace}(\boldsymbol{\sigma})\,\boldsymbol{I}$$

   with Lamé constants μ and λ. (These constants appear only here and there is no risk of confusion with the plastic multiplier λ or the Lagrange multiplier μ.) In this case $\mathbb{C}^{-1}$ is coercive, provided that $\mu > 0$ and $d\,\lambda + 2\,\mu > 0$ hold. A common example for the hardening modulus is given by $\mathbb{H}^{-1}\boldsymbol{\chi} = \boldsymbol{\chi}/k_1$ with hardening constant $k_1 > 0$, see [8, Section 3.4].

Assumption (3) shows that $a(\boldsymbol{\Sigma}, \boldsymbol{\Sigma}) \geq \underline{\alpha}\,\|\boldsymbol{\Sigma}\|_{S^2}^2$ for some $\underline{\alpha} > 0$.

2.2 The Forward Problem and Its Regularization

In this section, we address the lower-level problem of static plasticity. The weak formulation of (1.1) is given by

$$a(\boldsymbol{\Sigma}, \boldsymbol{T}) + b(\boldsymbol{T}, \boldsymbol{u}) + \int_\Omega \lambda\, \mathcal{D}\boldsymbol{\Sigma} : \mathcal{D}\boldsymbol{T} \,\mathrm{d}x = 0 \qquad \text{for all } \boldsymbol{T} \in S^2, \tag{2.5a}$$

$$b(\boldsymbol{\Sigma}, \boldsymbol{v}) = \langle \ell, \boldsymbol{v} \rangle \qquad \text{for all } \boldsymbol{v} \in V, \tag{2.5b}$$

$$0 \le \lambda \quad \perp \quad \phi(\boldsymbol{\Sigma}) \le 0 \qquad \text{a.e. in } \Omega, \tag{2.5c}$$

and it represents an energy minimization problem subject to a feasibility constraint for the generalized stresses. Here, $\lambda \perp \phi(\boldsymbol{\Sigma})$ represents the pointwise a.e. complementarity condition $\lambda\, \phi(\boldsymbol{\Sigma}) = 0$. It is well known that given $\ell \in V'$, (2.5) has a unique solution $(\boldsymbol{\Sigma}, \boldsymbol{u}, \lambda)$, see, e.g., [9, Proposition 3.1] and [12, Theorem 2.2]. The components $(\boldsymbol{\Sigma}, \boldsymbol{u}) \in S^2 \times V$ of the solution depend Lipschitz continuously on $\ell \in V'$. For the equivalence of (2.5) with a mixed VI of first kind, we refer to [11, Theorem 1.4] and [12, Theorem 2.2].

A standard way to derive qualified optimality conditions for the *upper-level* problem is based on the differentiability of the load-to-state map $\ell \mapsto (\boldsymbol{\Sigma}, \boldsymbol{u})$. However, the load-to-state operator associated with problem (2.5) is not Gâteaux-differentiable, since the directional derivative in turn involves a complementarity system and is thus not linear w.r.t. the direction, see [13]. What one can show is that the load-to-state map is Bouligand-differentiable under additional smoothness assumptions, see [2], but the nonlinearity of the directional derivative precludes the application of the standard adjoint approach.

To remedy the lack of Fréchet differentiability, we regularize the complementarity condition of the lower-level problem. This regularization is two-fold. First, the constraint $\phi(\boldsymbol{\Sigma}) \le 0$ is replaced by a quadratic penalty term in the lower-level objective. Second, the occurring $\max\{0, \cdot\}$-term is locally smoothed. We require that the smooth replacement $\max_\varepsilon$ of $\max\{0, \cdot\}$ satisfies the following conditions: for all $\varepsilon > 0$, the function $\max_\varepsilon : \mathbb{R} \to \mathbb{R}$ is of class $C^{1,1}$ and satisfies

1. $\max_\varepsilon(x) \ge \max\{0, x\}$,
2. $\max_\varepsilon$ is monotone increasing and convex,
3. $\max_\varepsilon(x) = \max\{0, x\}$ for $|x| \ge \varepsilon$.

It is easy to see that there exists a class of functions satisfying these requirements, and we refrain from fixing a certain choice of $\max_\varepsilon$ here. This leaves a choice for numerical implementations.

It is convenient to define

$$J_{\gamma,\varepsilon}(\boldsymbol{\Sigma}) = p_{\gamma,\varepsilon}(|\mathcal{D}\boldsymbol{\Sigma}|)\, \mathcal{D}^\star \mathcal{D}\boldsymbol{\Sigma} \quad \text{where} \quad p_{\gamma,\varepsilon}(x) = \max_\varepsilon\big(\gamma\,(1 - \tilde{\sigma}_0/x)\big), \tag{2.6}$$

which acts pointwise on functions in S^2. Here, $\gamma > 0$ is the penalty parameter.

In [12, Section 2.2] we obtained the following smoothed version of the optimality condition (2.5):

$$a(\boldsymbol{\Sigma}_{\gamma,\varepsilon}, \boldsymbol{T}) + b(\boldsymbol{T}, \boldsymbol{u}_{\gamma,\varepsilon}) + \langle J_{\gamma,\varepsilon}(\boldsymbol{\Sigma}_{\gamma,\varepsilon}), \boldsymbol{T}\rangle = 0 \quad \text{for all } \boldsymbol{T} \in S^2, \tag{2.7a}$$

$$b(\boldsymbol{\Sigma}_{\gamma,\varepsilon}, \boldsymbol{v}) = \langle \ell, \boldsymbol{v}\rangle \quad \text{for all } \boldsymbol{v} \in V. \tag{2.7b}$$

Note that the expression $\langle J_{\gamma,\varepsilon}(\boldsymbol{\Sigma}_{\gamma,\varepsilon}), \boldsymbol{T}\rangle$ is well defined for $\boldsymbol{T} \in S^2$, since $J_{\gamma,\varepsilon}(\boldsymbol{\Sigma}_{\gamma,\varepsilon}) \in S^2$ due to $p_{\gamma,\varepsilon}(|\mathcal{D}\boldsymbol{\Sigma}_{\gamma,\varepsilon}|) \in L^\infty(\Omega)$. The existence and uniqueness of a solution can be shown by the theory of monotone operators. We obtain that for any $\ell \in V'$, (2.7) has a unique solution

$$G_{\gamma,\varepsilon}(\ell) = (G^{\boldsymbol{\Sigma}}_{\gamma,\varepsilon}(\ell), G^{\boldsymbol{u}}_{\gamma,\varepsilon}(\ell)) = (\boldsymbol{\Sigma}_{\gamma,\varepsilon}, \boldsymbol{u}_{\gamma,\varepsilon}) \in S^2 \times V. \tag{2.8}$$

Moreover, $\boldsymbol{\Sigma}_{\gamma,\varepsilon}$ and $\boldsymbol{u}_{\gamma,\varepsilon}$ depend Lipschitz continuously on ℓ, with a Lipschitz constant L independent of γ and ε.

By using the L^p-regularity result (with $p > 2$) of [10], we obtain the Fréchet differentiability of $G_{\gamma,\varepsilon}$. The derivative at $(\boldsymbol{\Sigma}_{\gamma,\varepsilon}, \boldsymbol{u}_{\gamma,\varepsilon}) = G_{\gamma,\varepsilon}(\ell)$ in the direction $\delta\ell \in U$ is given by the unique solution $(\delta\boldsymbol{\Sigma}, \delta\boldsymbol{u})$ of

$$(A + J'_{\gamma,\varepsilon}(\boldsymbol{\Sigma}_{\gamma,\varepsilon}))\,\delta\boldsymbol{\Sigma} + B^\star\delta\boldsymbol{u} = \boldsymbol{0}, \tag{2.9a}$$

$$B\,\delta\boldsymbol{\Sigma} = \delta\ell. \tag{2.9b}$$

Here, $J'_{\gamma,\varepsilon}$ is the derivative of $J_{\gamma,\varepsilon}(\boldsymbol{\Sigma})$ given by

$$J'_{\gamma,\varepsilon}(\boldsymbol{\Sigma})\,\boldsymbol{T} = p'_{\gamma,\varepsilon}(|\mathcal{D}\boldsymbol{\Sigma}|)\,\frac{\mathcal{D}\boldsymbol{\Sigma} : \mathcal{D}\boldsymbol{T}}{|\mathcal{D}\boldsymbol{\Sigma}|}\,\mathcal{D}^\star\mathcal{D}\boldsymbol{\Sigma} + p_{\gamma,\varepsilon}(|\mathcal{D}\boldsymbol{\Sigma}|)\,\mathcal{D}^\star\mathcal{D}\boldsymbol{T} \tag{2.10}$$

with

$$p'_{\gamma,\varepsilon}(x) = \max{}'_\varepsilon\left(\gamma\,(1 - \tilde{\sigma}_0\,x^{-1})\right)\gamma\,\tilde{\sigma}_0\,x^{-2}.$$

Let us remark that the differentiability of the solution operator of (2.7) is a non-trivial result. This can be appreciated when we reformulate (2.7) as the following quasi-linear system in $\boldsymbol{u}$, where the principal part depends nonlinearly on the gradient of $\boldsymbol{u}$:

$$B\,(A + J_{\gamma,\varepsilon})^{-1}(-B^\star\boldsymbol{u}_{\gamma,\varepsilon}) = \ell.$$

General differentiability results for such systems can be found in [27].

Finally, we obtain the convergence of the regularization. Let us denote by $(\boldsymbol{\Sigma}, \boldsymbol{u}, \boldsymbol{\lambda})$ the solution of (2.5) with right hand side $\ell \in V'$ and by $(\boldsymbol{\Sigma}_{\gamma,\varepsilon}, \boldsymbol{u}_{\gamma,\varepsilon})$ the solutions of the regularized problems (2.7) with right hand side $\ell_{\gamma,\varepsilon}$ for $\gamma, \varepsilon > 0$. Then we obtain

$$\|\boldsymbol{\Sigma} - \boldsymbol{\Sigma}_{\gamma,\varepsilon}\|_{S^2}^2 \le C\left(\|\ell - \ell_{\gamma,\varepsilon}\|_{V'}\|\boldsymbol{u} - \boldsymbol{u}_{\gamma,\varepsilon}\|_V + \gamma^{-1}\|\ell\|_{V'}\|\ell_{\gamma,\varepsilon}\|_{V'} + \varepsilon\right),$$
$$\|\boldsymbol{u} - \boldsymbol{u}_{\gamma,\varepsilon}\|_V \le C\left(\|\ell - \ell_{\gamma,\varepsilon}\|_{V'}\|\boldsymbol{u} - \boldsymbol{u}_{\gamma,\varepsilon}\|_V + \gamma^{-1}\|\ell\|_{V'}\|\ell_{\gamma,\varepsilon}\|_{V'} + \varepsilon + \|\boldsymbol{\Sigma} - \boldsymbol{\Sigma}_{\gamma,\varepsilon}\|_{S^2}\right),$$

where C is independent of ℓ, $\ell_{\gamma,\varepsilon}$, γ and ε. In particular, we find $(\boldsymbol{\Sigma}_{\gamma,\varepsilon}, \boldsymbol{u}_{\gamma,\varepsilon}) \to (\boldsymbol{\Sigma}, \boldsymbol{u})$ if $\gamma \to \infty$, $\varepsilon \to 0$ and $\ell_{\gamma,\varepsilon} \to \ell$ in V'.

The comparison of (2.5a) and (2.7a) gives rise to the definition

$$\lambda_{\gamma,\varepsilon} := p_{\gamma,\varepsilon}(|\mathcal{D}\boldsymbol{\Sigma}_{\gamma,\varepsilon}|). \tag{2.11}$$

From the definition of $p_{\gamma,\varepsilon}$, we see that $0 \le \lambda_{\gamma,\varepsilon} \le \max\{\gamma, \varepsilon\}$ holds. Finally, we obtain the convergence $\lambda_{\gamma,\varepsilon} \to \lambda$ in $L^2(\Omega)$ under the same assumptions as for the convergence of $(\boldsymbol{\Sigma}, \boldsymbol{u})$.

Similar results are obtained in the case of *quasi-static* plasticity in [23, 25].

2.3 The Quasi-static Forward Problem

For convenience of the reader, we state the forward problem of quasi-static plasticity. This problem is time-dependent but rate-independent. We denote by $H^1(0, T; X)$ the standard Bochner-Sobolev space of functions which map the interval $[0, T]$ into the Banach space X and which possess a square-integrable weak derivative in time.

The time-dependent load $\ell \in H^1(0, T; V')$ satisfies $\ell(0) = 0$. The associated states $(\boldsymbol{\Sigma}, \boldsymbol{u}) \in H^1(0, T; S^2 \times V)$ also satisfy homogeneous initial conditions $(\boldsymbol{\Sigma}(0), \boldsymbol{u}(0)) = \boldsymbol{0}$. In the case of a pre-loaded workpiece, non-zero initial conditions apply. Together with the plastic multiplier $\lambda \in L^2(0, T; L^2(\Omega))$, the system

$$A\dot{\boldsymbol{\Sigma}} + B^\star\dot{\boldsymbol{u}} + \lambda\,\mathcal{D}^\star\mathcal{D}\boldsymbol{\Sigma} = \boldsymbol{0} \quad \text{in } L^2(0, T; S^2), \tag{2.12a}$$
$$B\boldsymbol{\Sigma} = \ell \quad \text{in } L^2(0, T; V'), \tag{2.12b}$$
$$0 \le \lambda \quad \perp \quad \phi(\boldsymbol{\Sigma}) \le 0 \quad \text{a.e. in } (0, T) \times \Omega. \tag{2.12c}$$

constitutes the forward problem. The existence and uniqueness of solutions can be found in [8, Sec. 8], regularity of the plastic multiplier was proved in [11], and continuity results are given in [5, 24].

2.4 An Optimal Control Problem

As was mentioned before, the volume and boundary forces $\boldsymbol{f}$ and $\boldsymbol{g}$ act as control variables. They induce in the forward system (2.5) the load $\ell = R(\boldsymbol{f}, \boldsymbol{g})$ defined by

$$\langle \ell, \boldsymbol{v}\rangle = \langle R(\boldsymbol{f}, \boldsymbol{g}), \boldsymbol{v}\rangle := -\int_\Omega \boldsymbol{f}\cdot\boldsymbol{v}\,\mathrm{d}x - \int_{\Gamma_N} \boldsymbol{g}\cdot\boldsymbol{v}\,\mathrm{d}s, \quad \boldsymbol{v}\in V \tag{2.13}$$

for $(\boldsymbol{f}, \boldsymbol{g}) \in U$. The optimal control, or upper-level problem under consideration reads

$$\left.\begin{aligned} \text{Minimize}\quad & \frac{1}{2}\|\boldsymbol{u}-\boldsymbol{u}_d\|^2_{L^2(\Omega;\mathbb{R}^d)} + \frac{\nu_1}{2}\|\boldsymbol{f}\|^2_{L^2(\Omega;\mathbb{R}^d)} + \frac{\nu_2}{2}\|\boldsymbol{g}\|^2_{L^2(\Gamma_N;\mathbb{R}^d)} \\ \text{where}\quad & (\boldsymbol{\Sigma}, \boldsymbol{u}, \lambda) \text{ solves the static plasticity problem (2.5)} \\ & \text{with right-hand side } \ell = R(\boldsymbol{f}, \boldsymbol{g}). \end{aligned}\right\} \qquad \textbf{(P)}$$

The desired displacement $\boldsymbol{u}_d$ is an element of $L^2(\Omega;\mathbb{R}^d)$. Moreover, ν_1 and ν_2 are positive constants. The objective expresses the goal of reaching as closely as possible a desired deformation $\boldsymbol{u}_d$. In the interest of not further complicating the presentation, control constraints are not considered but they could be easily included with obvious modifications.

The optimal control problem in the quasi-static case reads

$$\left.\begin{aligned} \text{Minimize}\quad & \frac{1}{2}\|\boldsymbol{u}(T)-\boldsymbol{u}_d\|^2_{L^2(\Omega;\mathbb{R}^d)} + \frac{\nu}{2}\|\boldsymbol{g}\|^2_{H^1(0,T;L^2(\Gamma_N;\mathbb{R}^d))} \\ \text{where}\quad & (\boldsymbol{\Sigma}, \boldsymbol{u}, \lambda) \text{ solves the quasi-static problem (2.12)} \\ & \text{with right-hand side } \ell(t) = R(\boldsymbol{0}, \boldsymbol{g}(t)) \\ \text{and}\quad & \boldsymbol{g}(0) = \boldsymbol{g}(T) = \boldsymbol{0}. \end{aligned}\right\} \qquad (\mathbf{P_q})$$

Note that volume forces are not present. The control constraints on $\boldsymbol{g}$ refer to an unloaded initial and terminal state. We mention that optimal control problems with more general objectives and additional control constraints are considered in [24].

Existence of solutions for problem (**P**) was proved in [9, Proposition 3.6] by using the compactness of $R : U \to V'$. The existence result in the quasi-static variant is a little bit more involved, since the pointwise application of R considered as a mapping $H^1(0,T;U) \to H^1(0,T;V')$ is not compact. However, one can show that the solution mapping is weakly continuous, which yields the existence of solutions, see [24, Theorem 2.9].

Additionally, one can show that local solutions of (**P**) can be approximated by solutions of the regularized versions of (**P**), where (2.5) is replaced by (2.7). For the precise formulation of this approximation result, consult [12, Section 3.2]. Similar results in the case of quasi-static plasticity are obtained in [25, Section 4], see also [23].

3 Optimality Conditions

As was mentioned in the introduction, minimizers of MPCCs often do not fulfill the KKT conditions, and thus alternative stationarity concepts must be devised, along with tailored constraint qualifications. Briefly speaking, one disposes of the Lagrange multiplier pertaining to the complementarity conditions. One also redefines those multipliers belonging to the inequalities involved in the complementarity relation. In our setting, the latter comprise the multiplier μ (associated with the non-negativity of the plastic multiplier $\lambda \geq 0$) and θ (associated with the yield condition $\phi(\boldsymbol{\Sigma}) \leq 0$). Existing stationarity concepts differ in what conditions are imposed for μ and θ.

Our first result provides an optimality system of *C-stationary* type. It is characteristic for this class that a sign is known only for the product $\theta\,\mu$, in the sense that $\theta\,\mu \geq 0$ holds a.e. in Ω.

$$A\boldsymbol{\Sigma} + \lambda\,\mathcal{D}^\star\mathcal{D}\boldsymbol{\Sigma} + B^\star\boldsymbol{u} = \boldsymbol{0}, \tag{3.1a}$$

$$B\boldsymbol{\Sigma} = R(\boldsymbol{f}, \boldsymbol{g}), \tag{3.1b}$$

$$0 \leq \lambda \quad \perp \quad \phi(\boldsymbol{\Sigma}) \leq 0, \tag{3.1c}$$

$$A\boldsymbol{\Upsilon} + \lambda\,\mathcal{D}^\star\mathcal{D}\boldsymbol{\Upsilon} + \theta\,\mathcal{D}^\star\mathcal{D}\boldsymbol{\Sigma} + B^\star\boldsymbol{w} = \boldsymbol{0}, \tag{3.2a}$$

$$B\boldsymbol{\Upsilon} = -(\boldsymbol{u} - \boldsymbol{u}_d), \tag{3.2b}$$

$$(\nu_1\,\boldsymbol{f}, \nu_2\,\boldsymbol{g}) - R^\star\boldsymbol{w} = \boldsymbol{0}, \tag{3.3}$$

$$\mathcal{D}\boldsymbol{\Sigma} : \mathcal{D}\boldsymbol{\Upsilon} - \mu = 0, \tag{3.4a}$$

$$\mu\,\lambda = 0, \tag{3.4b}$$

$$\theta\,\phi(\boldsymbol{\Sigma}) = 0, \tag{3.4c}$$

$$\theta\,\mu \geq 0. \tag{3.4d}$$

The following result was proved in [12, Theorem 3.16] by means of a family of regularized optimal control problems, wherein the lower-level static plasticity problems are replaced by their approximations (2.7), and passage to the limit.

Theorem 3.1. *Let* $(\boldsymbol{f}, \boldsymbol{g})$ *be a local optimal solution of* (**P**). *Let* $(\boldsymbol{\Sigma}, \boldsymbol{u})$ *and* λ *denote the associated stresses, displacements, and plastic multiplier. Then there exist adjoint stresses and displacements* $(\boldsymbol{\Upsilon}, \boldsymbol{w}) \in S^2 \times V$ *and Lagrange multipliers* $\theta, \mu \in L^2(\Omega)$ *such that the C-stationarity system* (3.1)–(3.4) *is satisfied.*

C-stationarity was also obtained in the quasi-static setting for a semi-discretized in time problem, see [26, Section 2]. In passing to the limit in the time discretization parameter, the sign condition corresponding to $\theta\,\mu \geq 0$ is lost. What remains is weak stationarity, see [26, Section 3].

A stronger stationarity concept than C-stationarity is *strong stationarity*, which asks for $\theta \geq 0$ and $\mu \geq 0$ on the so-called biactive set, defined by $\mathcal{B} := \{x \in \Omega : \phi(\boldsymbol{\Sigma}(x)) = \lambda(x) = 0\}$. Results for various MPCC control problems in the literature which imply the strong stationarity of local minimizers have in common that the control functions must be sufficiently rich. A long-standing open question whether or not control constraints impede strong stationarity was recently resolved in [28]. Nevertheless, it still stands as a conjecture that the controls need to be distributed controls in the range space of the differential operators defining the forward problem, see (2.5a)–(2.5b). In accordance with this, we proved in [13, Theorem 4.5] a strong stationarity result for local minimizers of a modified problem with richer controls.

Moreover, we also obtained optimality conditions from the class of *B-stationarity* conditions. Rather than working with dual quantities, these conditions state that at a local minimizer, directional derivatives of the objective are non-negative in all directions from certain cones. By showing the weak directional differentiability of the control-to-state map, we obtained in [13, Corollary 3.12] the non-negativity of all directional derivatives of the reduced objective in tangential directions. It is noteworthy that the cone of tangential directions is taken to be the closure of the cone of feasible directions w.r.t. the *weak topology*. For the precise formulation of these results, we refer to [13].

Finally, sufficient second-order optimality conditions for the static problem were derived in [2]. For this purpose, the weak differentiability results from [13] had to be sharpened. To be more precise, it was shown that, under mild additional assumptions on the integrability of the hardening variable $\boldsymbol{\chi}$, the control-to-state mapping is Bouligand differentiable from $W_D^{1,p}(\Omega)'$ to $S^2 \times V$, where $p > 2$. The associated remainder term property allows to deduce sufficient conditions by means of a second-order Taylor expansion of a particularly chosen Lagrange functional. The obtained sufficient conditions are comparable to the ones known from finite dimensional MPCCs, see e.g. [21]. However, one observes a substantial gap to the necessary optimality conditions, since the sufficient conditions involve a system which is even more rigorous compared to strong stationarity.

4 Numerical Results

Within this project, we also developed some algorithms to solve the optimal control problems. For the *quasi-static variant* of the optimal control problem ($\mathbf{P_q}$), we built a solver using the finite element library FEniCS, see [17]. The results shown below are based on a discretization by continuous, piecewise quadratic functions for the displacement, whereas the stresses are discretized only at the quadrature points. The temporal derivatives were replaced by an implicit Euler scheme. We used a globalized Newton-CG approach to compute stationary points of the discretized and regularized problem.

In Fig. 1 we present the computed (optimal) state for a problem with 96 time steps and 50,115 DoFs (per time step) for the displacement and 460,800 DoFs (per time step) for the stresses. The control boundary is located in the middle of the upper boundary, as can be seen from the red (pressure) and green (tension) arrows in Fig. 1. The observation boundary coincides with the control boundary. The desired final deformation is a deflection of the observation boundary by -0.1 in z-direction. The final deformation approaches this desired deformation very well, see Fig. 2.

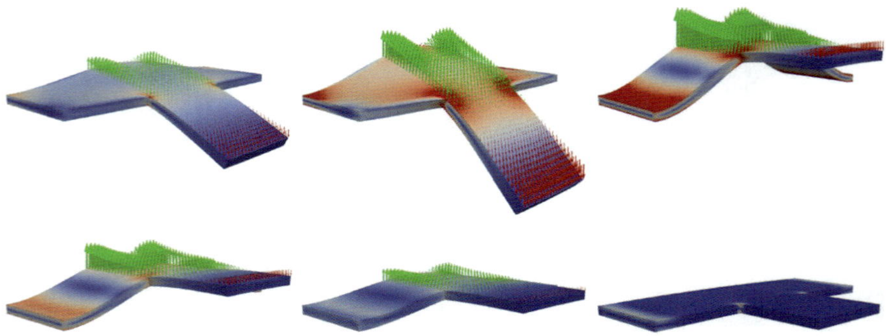

Fig. 1 The computed optimal state for different time steps $t = i\,T/6$, $i = 1, \ldots, 6$, where T is the final time. The control is shown by *red* (pressure) and *green* (tension) *arrows*. The workpiece is colored by the von Mises equivalent stress. The deformation is 20 times enlarged

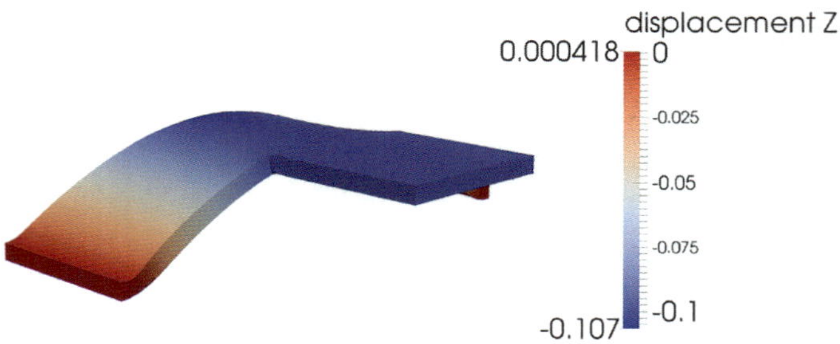

Fig. 2 Final deformation, 500 times enlarged

5 Further Results of the Project and Ongoing Work

We finally mention some further results of the project and related ongoing work in this section. In a recent manuscript [3] we considered an optimal control problem similar to (**P**), but with the static forward problem given in its *primal (strain-based)* formulation. The latter involves a variational inequality of the second kind in place of a complementarity system. Instead of the generalized stresses $\boldsymbol{\Sigma}$ and plastic multiplier λ, the plastic strain $\boldsymbol{p}$ appears as a state variable. By means of regularization, we obtained in [3, Theorem 1.1] a certain system of first-order necessary optimality conditions. Since a classification paralleling the notions of B-, C- or strong stationarity for optimal control problems involving variational inequalities of the second kind is not available in the literature, it is a priori not clear how strong this result is. Interestingly, we were able to show that the optimality system obtained is precisely equivalent to the C-stationarity conditions for the optimal control problem (**P**), i.e., when the formulation is replaced by the corresponding dual system (2.5).

Concerning the finite element error analysis for MPCCs in function space, optimal control problems governed by the obstacle problem were investigated in [18]. Quasi-optimal a priori error estimates for state and control were derived and confirmed by numerical examples. At the moment, a posteriori error representations based on the dual weighted residual approach are being developed in cooperation with A. Rademacher (TU Dortmund) and W. Wollner (University of Hamburg). The transfer of the a priori and a posteriori results to optimal control of elastoplastic deformation processes will be the subject of future research.

In the paper [28], we considered the distributed control of the obstacle problem subject to control constraints. As already mentioned, it was an long-standing open problem whether local minimizers (together with suitable multipliers) satisfy the strong stationarity conditions. We were able to prove that the answer is affirmative if certain mild conditions on the control constraints are satisfied. Moreover, it is possible to construct counter-examples when these conditions are violated.

Preprints and technical reports can be found at our publication page http://www.tu-chemnitz.de/mathematik/part_dgl/publications.php and at the preprint page of the DFG priority program SPP 1253 http://www.am.uni-erlangen.de/home/spp1253/wiki/index.php/Preprints.

References

1. V. Barbu, *Optimal Control of Variational Inequalities*. Research Notes in Mathematics, vol. 100. (Pitman, Boston, 1984)
2. T. Betz, C. Meyer, Second-order sufficient optimality conditions for optimal control of static elastoplasticity with hardening. To appear in ESAIM J. Control Optim. Calc. Var.
3. J.C. de los Reyes, R. Herzog, C. Meyer, Optimal control of static elastoplasticity in primal formulation. Technical report SPP1253-151, Priority Program 1253, German Research Foundation, 2013

4. P. Grisvard, *Elliptic Problems in Nonsmooth Domains* (Pitman, Boston, 1985)
5. K. Gröger, Initial value problems for elastoplastic and elastoviscoplastic systems, in *Nonlinear Analysis, Function Spaces and Applications (Proceedings of Spring School, Horni Bradlo, 1978)* (Teubner, Leipzig, 1979), pp. 95–127
6. K. Gröger, A $W^{1,p}$-estimate for solutions to mixed boundary value problems for second order elliptic differential equations. Mathematische Annalen **283**, 679–687 (1989). doi:10.1007/BF01442860
7. R. Haller-Dintelmann, C. Meyer, J. Rehberg, A. Schiela, Hölder continuity and optimal control for nonsmooth elliptic problems. Appl. Math. Optim. **60**(3), 397–428 (2009). doi:10.1007/s00245-009-9077-x
8. W. Han, B.D. Reddy, *Plasticity* (Springer, New York, 1999)
9. R. Herzog, C. Meyer, Optimal control of static plasticity with linear kinematic hardening. J. Appl. Math. Mech. **91**(10), 777–794 (2011). doi:10.1002/zamm.200900378
10. R. Herzog, C. Meyer, G. Wachsmuth, Integrability of displacement and stresses in linear and nonlinear elasticity with mixed boundary conditions. J. Math. Anal. Appl. **382**(2), 802–813 (2011). doi:10.1016/j.jmaa.2011.04.074
11. R. Herzog, C. Meyer, G. Wachsmuth, Existence and regularity of the plastic multiplier in static and quasistatic plasticity. GAMM Rep. **34**(1), 39–44 (2011). doi:10.1002/gamm.201110006
12. R. Herzog, C. Meyer, G. Wachsmuth, C-stationarity for optimal control of static plasticity with linear kinematic hardening. SIAM J. Control Optim. **50**(5), 3052–3082 (2012). doi:10.1137/100809325
13. R. Herzog, C. Meyer, G. Wachsmuth, B- and strong stationarity for optimal control of static plasticity with hardening. SIAM J. Optim. **23**(1), 321–352 (2013). doi:10.1137/110821147
14. M. Hintermüller, I. Kopacka, Mathematical programs with complementarity constraints in function space: C- and strong stationarity and a path-following algorithm. SIAM J. Optim. **20**(2), 868–902 (2009). ISSN 1052-6234. doi:10.1137/080720681
15. T. Hoheisel, C. Kanzow, A. Schwartz, Theoretical and numerical comparison of relaxation methods for mathematical programs with complementarity constraints. Math. Program. **137** (1–2), 257–288 (2013). doi:10.1007/s10107-011-0488-5
16. K. Ito, K. Kunisch, Optimal control of elliptic variational inequalities. Appl. Math. Optim. **41**, 343–364 (2000)
17. A. Logg, K.-A. Mardal, G. N. Wells et al., *Automated Solution of Differential Equations by the Finite Element Method* (Springer, Berlin/New York, 2012). ISBN 978-3-642-23098-1. doi:10.1007/978-3-642-23099-8
18. C. Meyer, O. Thoma, A priori finite element error analysis for optimal control of the obstacle problem. SIAM J. Numer. Anal. **51**(1), 605–628 (2013)
19. F. Mignot, Contrôle dans les inéquations variationelles elliptiques. J. Funct. Anal. **22**(2), 130–185 (1976)
20. F. Mignot, J.-P. Puel, Optimal control in some variational inequalities. SIAM J. Control Optim. **22**(3), 466–476 (1984)
21. H. Scheel, S. Scholtes, Mathematical programs with complementarity constraints: Stationarity, optimality, and sensitivity. Math. Oper. Res. **25**(1), 1–22 (2000). doi:10.1287/moor.25.1.1.15213
22. A. Schiela, D. Wachsmuth, Convergence analysis of smoothing methods for optimal control of stationary variational inequalities with control constraints. ESAIM Math. Model. Numer. Anal. **47**(3), 771–787 (2013). doi:10.1051/m2an/2012049
23. G. Wachsmuth, Optimal control of quasistatic plasticity – An MPCC in function space. PhD Thesis, Chemnitz University of Technology, 2011
24. G. Wachsmuth, Optimal control of quasistatic plasticity with linear kinematic hardening, Part I: existence and discretization in time. SIAM J. Control Optim. **50**(5), 2836–2861 (2012). doi:10.1137/110839187
25. G. Wachsmuth, Optimal control of quasistatic plasticity with linear kinematic hardening, Part II: regularization and differentiability. Technical report, TU Chemnitz, 2012

26. G. Wachsmuth, Optimal control of quasistatic plasticity with linear kinematic hardening, Part III: optimality conditions. Technical report, TU Chemnitz, 2012
27. G. Wachsmuth, Differentiability of implicit functions. J. Math. Anal. Appl. **414**(1), 259–272 (2014), doi: 10.1016/j.jmaa.2014.01.007
28. G. Wachsmuth, Strong stationarity for optimal control of the obstacle problem with control constraints. To appear SIAM J. Optim.

One-Shot Approaches to Design Optimzation

Torsten Bosse, Nicolas R. Gauger, Andreas Griewank, Stefanie Günther, and Volker Schulz

Abstract The paper describes general methodologies for the solution of design optimization problems. In particular we outline the close relations between a fixed point solver based piggy back approach and a Reduced SQP method in Jacobi and Seidel variants. The convergence rate and general efficacy is shown to be strongly dependent on the characteristics of the state equation and the objective function. In the QP scenario where the state equation is linear and the objective quadratic, finite termination in two steps is obtained by the Seidel variant with Newton state solver and perfect design space preconditioning. More generally, it is shown that the retardation factor between simulation and optimization is bounded below by 2 with the difference depending on a cross-term representing the total sensitivity of the adjoint equation with respect to the design.

Keywords Simulation • Optimization • PDE • Automatic differentiation • Fixed-point solver • Retardation factor • Convergence • Numerics

Mathematics Subject Classification (2010). Primary: 90C30; 68U20
Secondary: 35Q68; 35Q90; 35Q93.

N.R. Gauger • S. Günther
Department of Mathematics and Center for Computational Engineering Science, RWTH Aachen University, Schinkelstraße 2, 52062 Aachen, Germany
e-mail: gauger@mathcces.rwth-aachen.de; guenther@mathcces.rwth-aachen.de

T. Bosse • A. Griewank (✉)
Department of Mathematics, Humboldt-Universität zu Berlin, Unter den Linden 6, 10099 Berlin, Germany
e-mail: bosse@math.hu-berlin.de; griewank@math.hu-berlin.de

V. Schulz
Universität Trier, Universitätsring 15, 54286 Trier, Germany
e-mail: Volker.Schulz@uni-trier.de

G. Leugering et al. (eds.), *Trends in PDE Constrained Optimization*, International Series of Numerical Mathematics 165, DOI 10.1007/978-3-319-05083-6_5

1 From Simulation to Optimization

In many applications of scientific computing one wishes to convert a system simulation code into one that optimizes certain performance indices with respect to design and control variables. We consider primarily the case where system simulation means solving a coupled systems of partial differential equations (PDEs) to obtain the system state for a given design. Within the SPP 1253 many such applications were considered and described in the first volume of [42]. In virtually all such cases an adjoint state equation was set up explicitly or implicitly to yield sensitivities of objectives and global constraint functions w.r.t. to design and control. If the state equations can be solved rather rapidly by Newton-like methods, the same is usually true for the adjoint equation, though consistency of the discretization and error control may be issues. In that Newton-like scenario recovering primal and dual feasibility after each design update more or less exactly is a valid and widely applied algorithmic strategy, which one may refer to as hierarchical design optimization.

In some important applications like shape optimization in aerodynamics even the concerted effort of a large expert community over several decades has only yielded primal solver of fixed-point type with rather sluggish linear convergence. Rather than expanding a large effort to recover primal and adjoint feasibility at far from optimal design point it then is rather promising to pursue a *One-shot* approach, where feasibility and optimality is pursued simultaneously.

Several authors have developed simultaneous optimization strategies where state and adjoint feasibility is never fully achieved during the optimization process until the optimal design point is reached. In the literature this strategy is variously known as *all-at-once approach* [12, 16, 33, 56], *Simultaneous Optimization Approach* [54] or *simultaneous analysis and design (SAND)* [30] and also *optimization boundary value approach* [6,7]. The *One-shot* method for simultaneous optimization was first proposed by Ta'asan et al. [53] in a multi-grid framework. In this approach, only few iterations of the primal and adjoint solvers are performed in each optimization cycle. Since then, the One-shot method has been further developed with special application in aerodynamic shape optimization [23,43]. In practical terms, this strategy requires the coordination of suitable primal and adjoint correction steps with the design changes. The latter typically require grid modifications and other additional efforts, so their appropriate spacing is rather important.

The structure of this paper is as follows: In Sect. 2 we lay out our framework of design optimization problems including a user provided state equation solver. We will quantify certain key characteristics of given problems, which will determine the efficacy of optimization schemes and their appropriate tuning. In the central Sect. 3 we develop and analyze one-shot schemes of Jacobi and Seidel type with the aim of guaranteeing local convergence and estimating the asymptotic convergence rate. These methods differ in that the Jacobi variant utilizes everywhere previous iteration values whereas the Seidel variant always utilizes the latest information.

Throughout we will illustrate the close relation between a reduced SQP approach based on block preconditioning the KKT optimality system (see e.g. [36]) and a piggy back approach based on augmenting the primal solver with a dual iteration and a design optimization step (see e.g. [14]). In Sect. 4 we quote some comparative numerical results from [29] and very briefly discuss other one-shot approaches like those proposed in [6, 7] and [56]. The survey is concluded with a Summary and an Outlook.

2 Problem Formulation and Optimality Conditions

The focus of this paper is on numerical methods for simulation-driven design optimization. In contrast to general non-linear optimization tasks we can assume that the variables are a-priori partitioned into a state vector $y \in Y$ and a design vector $u \in U$, where Y and U are closed convex subsets of Hilbert spaces. The problem can be then stated as

$$\min_{(u,y)\in U\times Y} f(u,y) \quad \text{subject to} \quad c(u,y)=0. \tag{2.1}$$

where $f : U \times Y \to \mathbb{R}$ is the objective function and $c : U \times Y \to Y^*$ is the state equation with Y^* being the topological dual of Y. Then the set of feasible points is given by

$$\mathcal{F} \equiv \{(u,y) \in U \times Y : c(u,y)=0\}. \tag{2.2}$$

The simulations and applications can be found in various fields such as in marine-science (cf. [40, 48]), geo-science (cf. [15, 18, 22, 41]) and aerodynamics. For example in aerodynamics, one wants to optimize a variable airfoil shape in order to reduce the drag (cf. [37, 43, 45, 46]). In this case the constraint could be some variant of the Navier-Stokes equation that simulates the airflow around the airfoil with an appropriate turbulence model.

We assume that all functions are at least twice continuously Fréchet differentiable and denote their partial derivatives with respect to the variable x by f_x, c_x etc.

We restrict our attention to cases with a minimizer (u_*, y_*) in the interior of $U \times Y$, where second order sufficiency conditions hold. Furthermore, we require that for any $u \in U$ there is a non-singular solution $y \in Y$ of the state equations. Thus, we assume throughout the paper, that $c_y(u,y)$ has a bounded inverse for all points of interest. The implicit function theorem then ensures the existence of a locally unique state $y = y(u) \in Y$ with $c(u, y(u)) = 0$. The optimization parameters of (2.1) can therefore be reduced to the design space U by introducing the reduced objective function $\hat{f} : U \to \mathbb{R}$ with $\hat{f}(u) \equiv f(u, y(u))$ and considering

$$\min_{u\in U} \hat{f}(u). \tag{2.3}$$

Any design change causes a change in the state variable through the state condition $c(u, y) = 0$. The sensitivity of the state variable with respect to design changes can be computed from the identity

$$c_y(u, y)y_u(u) + c_u(u, y) = 0. \tag{2.4}$$

Introducing the adjoint variable $\lambda \in Y^*$ via the adjoint equation

$$-f_y(u, y) = \langle \lambda, c_y(u, y) \rangle \tag{2.5}$$

the sensitivity of $\hat{f}$ with respect to design changes, the so called reduced gradient, can be derived using the chain rule

$$\begin{aligned} \frac{\mathrm{d}\hat{f}(u)}{\mathrm{d}u} &= f_u(u, y) + \langle f_y(u, y), y_u(u) \rangle \\ &= f_u(u, y) - \langle f_y(u, y), c_y^{-1}(u, y) c_u(u, y) \rangle \\ &= f_u(u, y) + \langle \lambda, c_u(u, y) \rangle. \end{aligned} \tag{2.6}$$

Any local minimizer $u_* \in U$ of (2.3) is a root of the reduced gradient resulting in the first order necessary optimality condition, i.e. there exists $\lambda_* \in Y^*$ and $y_* \in Y$ such that the following equalities hold

$$\begin{aligned} 0 &= f_u(u_*, y_*) + \langle \lambda_*, c_u(u_*, y_* \rangle \\ 0 &= c(u_*, y_*), \\ 0 &= f_y(u_*, y_*) + \langle \lambda_*, c_y(u_*, y_* \rangle. \end{aligned} \tag{2.7}$$

The optimality condition can equivalently be formulated in terms of the Lagrangian function (cf. [44])

$$\mathcal{L}(u, y, \lambda) \equiv f(u, y) + \langle \lambda, c(u, y) \rangle. \tag{2.8}$$

whose gradient must vanish so that

$$\mathcal{L}_u(u_*, y_*, \lambda_*) = 0, \quad \mathcal{L}_\lambda(u_*, y_*, \lambda_*) = 0, \quad \mathcal{L}_y(u_*, y_*, \lambda_*) = 0. \tag{2.9}$$

Note, that the above KKT-conditions in terms of the Lagrangian formulation are equivalent to the first order necessary optimality conditions (2.7) of the reduced problem. Rather than solving (2.7) or (2.9) by a pure first order method one usually needs to collect or approximate some second order information at least on the tangent space of (2.2).

In PDE constrained optimal control, as discussed in [11, 13, 35, 55], often the reduced Hessian is a compact perturbation of the identity. Then, this information has to be reflected also in the update technique as presented in [28]. In PDE constrained shape optimization, the reduced Hessian has been studied in detail in [20, 50] in terms of the so-called shape Hessian. Because a low-cost direct implementation of the shape Hessian is not possible, Fourier analysis has been used, in order to obtain information on the symbol of the Hessian. This operator symbol guides the way to preconditioners with mesh independent contraction properties. It has been found in [50] for aerodynamic shape optimization problems involving the Stokes and the incompressible Navier-Stokes equations that the symbol is a first order operator on the shape to be optimized involving the Dirichlet-to-Neumann-map which is a non-local operator. However, this operator can be efficiently approximated by a linear combination of the identity with the Laplace-Beltrami-operator on the boundary, thus providing very fast convergence results in combination with a one-shot strategy within a realistic framework in [51].

The adjoint variable as introduced in (2.5) provides us with an efficient way of computing the reduced gradient of the optimization problem. In contrast to direct sensitivity computation, only one primal state solution y and one solution of the adjoint equation for λ is needed. The cost of solving the adjoint equation is roughly the same as that of solving the primal so that the reduced gradient is obtained at about twice the cost of just simulating the system.

The adjoint approach was first introduced by Pironneau in control theory [49] and has since been widely used for sensitivity computations in optimization with state constraints.

Two approaches where developed, namely the discrete adjoint approach, where the adjoint equation is derived from the discretized optimization problem and the continuous adjoint approach where an adjoint operator is derived from the continuous optimization problem in an appropriate function space. The discrete approach can be automated by applying Automatic Differentiation (AD) to the state equation solver and the objective function.

For notational simplicity, we will assume from now on as in [14, 36] that the problem has already been discretized and denote by m the number of design variables $u \in U \subseteq \mathbb{R}^m$ and n the dimension of the primal state $y \in Y \subseteq \mathbb{R}^n$. Provided $\lambda \in Y^*$ or $\bar{y} \in Y^*$ are initialized to 0 the algorithms discussed in this article are invariant w.r.t. to linear transformations on $Y^* \subseteq \mathbb{R}^n$. Hence, we may interpret λ and $\bar{y}$ as row vectors in $\mathbb{R}^n$ and write the duality pairing in matrix vector notation as $\langle \bar{y}, y \rangle = \bar{y}^\top y$.

The second derivative of the reduced objective function, the so called reduced Hessian, can analogously be computed using the adjoint variable resulting in

$$\mathcal{H} \equiv \frac{\mathrm{d}^2 \hat{f}(u)}{\mathrm{d}u^2} = [\mathcal{Z}^\top, I] \begin{bmatrix} f_{yy} + \lambda^\top c_{yy} & f_{yu} + \lambda^\top c_{yu} \\ f_{uy} + \lambda^\top c_{uy} & f_{uu} + \lambda^\top c_{uu} \end{bmatrix} \begin{bmatrix} \mathcal{Z} \\ I \end{bmatrix} \tag{2.10}$$

where $\mathcal{Z} = -c_y^{-1} c_u$. The columns of $[\mathcal{Z}^\top, I]^\top$ span the tangent space of $\mathcal{F}$.

The second order sufficiency optimality condition ensures that any stationary point (u_*, y_*, λ_*) satisfying the first order optimality conditions is a strict local minimizer of (2.3) if the reduced Hessian is positive definite at u_*.

In terms of the Lagrangian function $\mathcal{L}$ the above reduced Hessian (2.10) is the projection of the full Hessian $\nabla^2_{y,u}\mathcal{L}$ onto the range of $[\mathcal{Z}^\top, I]^\top \in \mathbb{R}^{(n+m)\times m}$ spanning the tangent space of $\mathcal{F}$.

Many gradient-based optimization strategies perform preconditioned design updates while Newton's method indicates that the preconditioner should approximate the reduced Hessian. Often, secant updates of the reduced Hessian from the Broyden-class are used such that positive definiteness and symmetry of the approximation is preserved as long as a proper initialization is implemented (cf. [44]).

2.1 The Fixed-Point Solver Paradigm

In view of practical scenarios, where simulation codes have been developed and refined over long periods of time, we also cast the state equation in terms of a fixed-point equation $y = G(u, y)$ with a contractive function $G : U \times Y \rightarrow Y$. In particular, we assume that any solution of the state equation is a solution of the fixed-point equation and vice versa, i.e.

$$c(u, y) = 0 \Longleftrightarrow y = G(u, y). \tag{2.11}$$

The function G represents one step of an iterative solver with a contraction rate that is smaller than 1 close to a solution for any fixed design u. Then we can reformulate the original non-linear problem (2.1) as

$$\min_{(u,y)\in U\times Y} f(u, y), \text{ s.t. } G(u, y) - y = 0. \tag{2.12}$$

with the Lagrangian function

$$L(u, y, \lambda) \equiv f(u, y) + \bar{y}^\top (G(u, y) - y). \tag{2.13}$$

In terms of the new Lagrangian the first order optimality conditions for a stationary point (u_*, y_*) are given by

$$L_u(u_*, y_*, \bar{y}_*) = 0, \quad L_{\bar{y}}(u_*, y_*, \bar{y}_*) = 0, \quad L_y(u_*, y_*, \bar{y}_*) = 0 \tag{2.14}$$

with some Lagrange multiplier $\bar{y}_* \in Y^*$. The reduced Hessian for the second order optimality conditions is

$$H \equiv [Z^\top, I] \begin{bmatrix} L_{yy} & L_{yu} \\ L_{uy} & L_{uu} \end{bmatrix} \begin{bmatrix} Z \\ I \end{bmatrix} \tag{2.15}$$

where we now have $Z = (I - G_y)^{-1} G_u$. The columns of $[Z^\top, I]^\top \in \mathbb{R}^{(n+m)\times m}$ span again the tangent space of the feasible set $\mathcal{F}$, which is the same for (2.1) and (2.12). In particular, it holds that $Z = \mathcal{Z}$ and $H = \mathcal{H}$ at (u_*, y_*) since both have to coincide with the Hessian of the reduced problem (2.3).

Often the state equation iteration will take the Newton-like form

$$y_+ = G(u, y) = y - A^{-1}c(u, y) \iff A(y_+ - y) = -c(u, y)$$

where $A \in \mathbb{R}^{n\times n}$ is an approximation to the Jacobian $c_y(u, y)$ serving as a preconditioner. In case of Newton's iteration $G(u, y) = y - c_y(u, y)^{-1}c(u, y)$ we have $G_y(y_*, u_*) = 0$, which is also the limiting scenario for multi-grid and other fast solvers.

For simplicity we will later assume that the corresponding adjoint solver is based on the transposed $A^\top$ even though in some applications their may be some discrepancy, which is for example accounted for in the theoretical development of [36]. Naturally, A like G will depend on u, y but we will mostly not mark that dependency explicitly and also neglect its derivatives with respect to y and u, since they are multiplied by the hopefully small residual $c(u, y)$.

Then we effectively solve the preconditioned problem

$$\min f(u, y) \quad \text{such that} \quad A^{-1}c(u, y) = 0$$

which is mathematically, but not algorithmically equivalent to (2.1). As we will see later, there is a close relation between the quantities associated with the two Lagrangians $\mathcal{L}$ and L, in particular the adjoint vectors λ and $\bar{y}$.

2.2 *Problem and Solver Characteristics*

Obviously, some problems are easier to solve than others, e.g. a quadratic problem with a quadratic goal functional and linear constraints is not as challenging as some problem with highly non-linear functions. This fact should be reflected in the required effort needed to solve the underlying problem. Moreover, the choice of the optimization scheme and its parameters is important, i.e. a bad choice for the solver might result in excruciatingly slow convergence, whereas a suitable choice can reduce the computational effort drastically.

We will quantify the efficiency of an optimization method for a specific fixed-point problem in terms of the retardation factor

$$r \equiv \frac{1 - \rho(G_y(x_*))}{1 - \rho(J_*)} \approx \frac{\ln(\rho(G_y))}{\ln(\rho(J_*))}. \tag{2.16}$$

This ratio reflects the increase in the number of steps needed for a comparable reduction in the residuals when going from simulation to optimization. Here, J_* denotes the Jacobian of the outer optimization scheme iteration

$$J_* = \partial(u_+, y_+, \bar{y}_+)/\partial(u, y, \bar{y}) \quad \text{at} \quad (u_*, y_*, \bar{y}_*) \tag{2.17}$$

and $\rho(M)$ an estimate of the spectral radius of a square matrix M. Then $\rho(G_y) = \rho(I - A^{-1}c_y) < 1$ implies for a suitable inner product norm $\|\cdot\|$ on the state space that

$$\left\|G_y(u, y)\right\| \leq \rho_0 < 1, \qquad \text{for all } (u, y) \in U \times Y. \tag{2.18}$$

In general, we cannot expect that such contractivity will be obtained w.r.t. more or less natural norms on the state space. Pseudo-time-stepping with specialized Runge-Kutta schemes can achieve spectral radii below 1 in the case that the Courant-Friedrichs-Lewy condition is ensured. However, the typical plots of successive residual norms show non-monotonic decline with periodic structure (cf. Fig. 5, Sect. 5). Therefore, estimates of the contraction factors should be based on geometric averages over a large number of iterations, which are asymptotically identical for all equivalent norms. The results in Sect. 3 are formulated in terms of a norm that ensures contractivity of the given state equation solver for fixed design. This norm will generally be unknown so that the estimates of related quantities must be based on topologically equivalent norms, which is likely to degrade the choice of method parameters.

Apart from ρ_0 the following characteristics of the optimization problem are important.

- The partial derivatives f_y, f_u, c_y and c_u represent the sensitivity of the objective function and the primal state equation w.r.t. state and design changes. We assume that c_y is invertible.
- The columns of the matrix

$$Z = (I - G_y)^{-1} G_u = -c_y^{-1} c_u$$

  span the tangent space of $\{y \in Y : c(u, y) = 0, \ u \in U\}$. With $G_u \approx A^{-1}c_u$ the sensitivity of the fixed-point solver and preconditioned state equation w.r.t. changes in the design variable u we find the bound $\|Z\| \leq \|(I - G_y)^{-1}\| \, \|G_u\| \leq \|G_u\|/(1 - \rho_0)$.
- The partial Hessian $\mathcal{L}_{yy}$ and L_{yy} represent the sensitivity of the adjoint w.r.t. to the primal variable. The norm of these derivatives may be viewed as a measure of the coupling between the primal and adjoint variables. When the state space

equation is linear we have $\mathcal{L}_{yy} = f_{yy} = L_{yy}$ independently of the adjoints λ and $\bar{y}$. Throughout we will abbreviate $p \equiv \|L_{yy}\|$ and $d = p\|Z\|^2$.

- The mixed derivative $\mathcal{L}_{yu}$ and L_{yu} represent the sensitivity of adjoint equation with respect to design variables. Often one has a separable adjoint in that $\mathcal{L}_{yu} = L_{yu} = 0$, which holds for many standard test problems. As a measure of seperability we use the ratio $q \equiv \max_v \|L_{yu}v\|/\|G_u v\|$ and we define $c \equiv \|L_{yu} + L_{yy}Z\| \, \|Z\|$, analoguous to d.
- The partial Hessian $\mathcal{L}_{uu}$ and L_{uu} need not be positive definite for second order sufficiency condition $H \succ 0$. However, this property is often guaranteed by a regularization of the design vector u, e.g. when $\mathcal{L}_{uu}$ and L_{uu} is gradually scaled up to infinity we have likely $u_* \to 0$ and $r \to 0$. Conversely one might have $\|u_*\| \to \infty$ and $r \to \infty$ as $\mathcal{L}_{uu}$ and L_{uu} becomes small and H nearly singular.
- The reduced Hessian $\mathcal{H} = H$ and the positive definiteness condition $H \succ 0$ for $(u, y) \approx (u_*, y_*)$ representing second order sufficiency is completely independent of the chosen state equation formulation and iteration function G.
- The quality of the preconditioner B occurring later in the stepping schemes (3.1) and (3.6) is measured by the spectral norm

$$\gamma \equiv \|I - \tau H^{\frac{1}{2}} B^{-1} H^{\frac{1}{2}}\|,$$

so that $\gamma = 0$ in the ideal case where B is equal to the reduced Hessian H and the step-size τ equals 1.

When the *adjoint is separable* in that $L_{yu} = 0$ it follows from the consistency of the operator norms that $c \leq d$. In contrast to Z and H, which are independent of the state fixed-point solver G, the constants defined in items 2 and 3 are slightly dependent on G. Their values are crucial for the estimation of the retardation factor and the selection of the number of primal and dual steps between design updates. The minimal values of c, d and thus the tightest bound on the retardation factor are attained when the projected Hessian at the minimizer (u_*, y_*, λ_*) is the identity, i.e.

$$I \equiv [H_*^{-\frac{1}{2}} Z^\top, I] \begin{bmatrix} L_{yy} & L_{y\tilde{u}} \\ L_{\tilde{u}y} & L_{\tilde{u}\tilde{u}} \end{bmatrix} \begin{bmatrix} Z H_*^{-\frac{1}{2}} \\ I \end{bmatrix}.$$

This ideal situation can be theoretically always achieved by changing variables to $\tilde{u} = H_*^{\frac{1}{2}} u$ on the design space. It then follows by elementary arguments that

$$|2c - d| \lesssim \|I - L_{\tilde{u}\tilde{u}}\| \leq 2c + d\ .$$

holds for the parameters c and d. Assuming that $L_{\tilde{u}\tilde{u}}$ is positive definite but not too large, which indicates some degree of regularization, we find that approximately $2c - d \leq 1$. In the separable case we have $c \leq d$ so that we may assume both of them to be roughly of order $O(1)$. In [14] it is described how c and d can be estimated from various numerical quantities obtained during the coupled iteration such that the ideal scaling is implicitly realized.

3 Jacobi and Seidel One-Shot

3.1 Jacobi One-Shot Iteration

In [25, 26] the authors considered a Jacobi type One-shot method to solve the problem (2.12). The basic idea is to augment the given fixed-point solver for the underlying PDE by an adjoint and design step based on the Lagrangian (2.13). Using the necessary condition for a stationary $0 = L_y = f_y + \bar{y}^\top\big(G_y(u, y) - I\big)$ one can formulate a fixed-point iteration for the adjoint vector $\bar{y} \in Y^*$. Moreover, the contractivity of G_y implies that this adjoint equation is also non-singular yielding a unique solution $\bar{y} = \bar{y}(u, y)$. For the iterative solution of the design equation $0 = L_u$ one can use any pre-conditioned gradient descent method of the form $u_+ = u - \tau B_{Jac}^{-1} L_u$. Here B_{Jac} is a symmetric positive definite matrix and τ a suitable step multiplier. We will refer to B_{Jac} as the design space preconditioner. The partial derivatives L_u and L_y of the Lagrangian can be efficiently evaluated by applying common techniques from Automatic Differentiation (AD) [24] on the functions f and G.

All three update steps can be combined into one simultaneous update yielding the iteration of the single step One-shot method as explored in [24]:

$$\begin{aligned} u_+ &= u - \tau B_{Jac}^{-1} L_u(u, y, \bar{y}), \\ y_+ &= G(u, y), \\ \bar{y}_+ &= \bar{y} + L_y(u, y, \bar{y}) . \end{aligned} \tag{3.1}$$

starting from some initial value $(u_0, y_0, \bar{y}_0)$. Since the coupled iteration updates all variables simultaneously it is referred to as Jacobi One-shot method and abbreviated (see [14]) by

$$\cdots \rightarrow (\text{ DESIGN, STATE, ADJOINT}) \rightarrow \ldots \text{ or } \cdots \rightarrow (\text{DSA}) \rightarrow \ldots .$$

Following the notation of [36] the stepping scheme (3.1) can be written as

$$\begin{bmatrix} B_{Jac} & 0 & 0 \\ 0 & A & 0 \\ 0 & 0 & A^\top \end{bmatrix} \begin{bmatrix} \Delta u \\ \Delta y \\ \Delta \lambda \end{bmatrix} = - \begin{bmatrix} \mathcal{L}_u(u, y, \lambda) \\ \mathcal{L}_\lambda(u, y, \lambda) \\ \mathcal{L}_y(u, y, \lambda) \end{bmatrix}, \tag{3.2}$$

where $\Delta u = u_+ - u$, $\Delta y = y_+ - y$, and $\Delta\lambda = -A^{-\top}(\bar{y}_+ - \bar{y})$, if we assume that $\tau = 1$, $G(u, y) = y - A^{-1}c(u, y)$, and $\lambda = -A^{-\top}\bar{y}$. It can easily be seen by some inductive argument that the later equivalence holds true for all subsequent iterations.

3.2 Augmented Lagrangian Preconditioning

First differentiating the new iterate $(u_+, y_+, \bar{y}_+)$ with respect to the old $(u, y, \bar{y})$, and then evaluating at $(u_*, y_*, \bar{y}_*)$ we obtain the coupled Jacobian

$$J_* = \begin{bmatrix} (I - B_{Jac}^{-1} L_{uu}) & -B_{Jac}^{-1} L_{uy} & -B_{Jac}^{-1} G_u^\top \\ G_u & G_y & 0 \\ L_{yu} & L_{yy} & G_y^\top \end{bmatrix} \in \mathbb{R}^{(2n+m)\times(2n+m)} \tag{3.3}$$

Block elimination yields for $\sigma \in \text{spect}(J_*) \setminus \text{spect}(G_y)$ the characterization

$$\det[P(\sigma)] = 0 \quad \text{with} \quad P(\sigma) = (\sigma - 1) B_{Jac} + H(\sigma) \tag{3.4}$$

where

$$H(\sigma) = \begin{bmatrix} Z(\sigma)^\top & I \end{bmatrix} \begin{bmatrix} L_{yy} & L_{yu} \\ L_{uy} & L_{uu} \end{bmatrix} \begin{bmatrix} Z(\sigma) \\ I \end{bmatrix} \quad \text{for } Z(\sigma) \equiv (\sigma I - G_y)^{-1} G_u \tag{3.5}$$

Note that $H(1) = H$ is the reduced Hessian of the optimization problem. It was shown in [27] that

Proposition 3.1. *The preconditioner*

$$B_{Jac} \equiv L_{uu} + \alpha\, G_u^\top G_u + \beta\, L_{uy} L_{yu} \succ H(1)$$

where

$$\alpha = \frac{p}{(1-\rho_0)^2} + \frac{q}{(1-\rho_0)} \quad \textit{and} \quad \beta = \frac{1}{q(1-\rho_0)}$$

ensures that

$$\det P(\lambda) \neq 0 \quad \textit{for all} \quad \lambda \in \mathbb{R} \quad \textit{with} \quad |\lambda| \geq 1.$$

The preconditioner B_{Jac} given above can be viewed as an approximation to the partial Hessian of the augmented Lagrangian w.r.t. the design variable u

$$L^a(u, y, \bar{y}) \;=\; L(u, y, \bar{y}) + \tfrac{\alpha}{2}\|G(u, y) - y\|^2 + \tfrac{\beta}{2}\|L_y(u, y, \bar{y})\|^2 .$$

where weighted residuals of the primal and adjoint variables are added to the standard Lagrangian function. It has properties close to $H(-1) \prec B_{Jac}$, which has some theoretical advantages but is rather costly to evaluate. Rather than evaluating the derivative matrices $L_{uu}, G_u = -A^{-1}c_u$ and L_{yu} explicitly one may use secant updating to compute B_{Jac} approximately. That has worked quite well, though even approximating L_{yu} may be costly unless the problem is separable. While L_{uu} suggest a Newton like step the additional terms $G_u^\top G_u$ and $L_{uy} L_{yu}$ caution effectively against design changes that strongly perturb the primal and adjoint state equation.

The choice of the penalty parameters α and β is quite critical. In [31] it was shown that the values of α and β given above imply that the fixed points of the coupled Jacobi one-shot iterations are exactly the stationary points of L^a. They also ensure that the Hessian of the augmented Langrangian is positive definite at all local minimizers (u_*, y_*) where second order sufficiency is satisfied. These properties are also sought in standard strategies for the selection of penalty parameters, as described for example in Chapter 17 of [44]. The updating rules usually require an increase in the parameters whenever primal or dual feasibility is slipping and a reduction when the primal and dual residuals are sufficiently small. The estimates given in the proposition are definitely on the conservative side with a small estimate for $(1 - \rho_0)$ leading to rather large penalty factors.

The Jacobi One-shot approach combined with an approximate augmented Hessian preconditioning has been implemented for various numerical applications especially in the field of aerodynamic shape optimization. In [45, 47] the shape of an airfoil under transonic flight conditions governed by the 2D Euler equations was optimized. A drag reduction up to 40 % was achieved while a retardation factor of 4 was measured compared to a primal flow computation. Shape optimization with Reynolds-averaged Navier-Stokes equations (RANS) using $k - \omega$ turbulence model was applied in [46] for an airfoil in subsonic flow. The measured retardation factor $r \approx 3$ clearly demonstrates the efficiency gain of the One-shot method over hierarchical optimization methods.

3.3 Seidel One-Shot Approach

As observed in [27] a rather disappointing aspect of the Jacobi One-shot approach is that even when G represents Newton's method on the state equation and $B_{Jac} = H(1)$ is the reduced Hessian only rather slow linear convergence is obtained. Also, while our choice precludes the existence of real eigenvalues outside the unit circle, it was observed in [10] that complex eigenvalues with modulus greater than 1 may indeed occur.

Alternatively, one may follow the philosophy of the Seidel method (cf. [38, 39]) by always using the most recent variable values. Then we obtain the Seidel variant:

$$\begin{aligned} u_+ &= u - \tau B_{Seid}^{-1} L_u(u, y, \bar{y}) \\ y_+ &= G(u_+, y) \\ \bar{y}_+ &= \bar{y} + L_y(u_+, y_+, \bar{y}), \end{aligned} \tag{3.6}$$

or in short

$$\cdots \to \text{DESIGN} \to \text{STATE} \to \text{ADJOINT} \to \ldots \text{ or } \cdots \to \text{D} \to \text{S} \to \text{A} \to \ldots .$$

Approximating by Taylor

$$G(u_+, y) = y - A^{-1}c(u_+, y) \approx y - A^{-1}[c(u, y) + c_u(u, y)\Delta u]$$

and

$$\mathcal{L}_y(u_+, y_+, \lambda) \approx \mathcal{L}_y(u, y, \lambda) + \mathcal{L}_{yu}(u, y, \lambda)\Delta u + \mathcal{L}_{yy}(u, y, \lambda)\Delta y$$

we obtain for the full step method where $\tau = 1$ the matrix form

$$\begin{bmatrix} B_{Seid} & 0 & 0 \\ c_u & A & 0 \\ \mathcal{L}_{yu} & \mathcal{L}_{yy} & A^\top \end{bmatrix} \begin{bmatrix} \Delta u \\ \Delta y \\ \Delta \lambda \end{bmatrix} = - \begin{bmatrix} \mathcal{L}_u(u, y, \lambda) \\ \mathcal{L}_\lambda(u, y, \lambda) \\ \mathcal{L}_y(u, y, \lambda) \end{bmatrix}. \tag{3.7}$$

Also, the update of the preconditioner B_{Seid}, to be discussed later will be based on the new information y_+ and $\bar{y}_+$. The resulting method is similar to the one proposed in [32] for the specific aerodynamic context. The matrix in Eq. (3.7) has also been used as preconditioner for Lagrange-Newton-Krylov methods as in [1, 4]. More general preconditioning studies for KKT systems can be found, e.g., in [3, 34, 52].

The Seidel One-shot methods can be visualized as in Fig. 1. It shows the sequence of primal and adjoint states $(y, \bar{y})$ coupled with the design vector u converging towards a stationary point $(u_*, y_*, \bar{y}_*)$ of (2.1). It satisfies the optimality condition $L_u(u_*, y_*, \bar{y}_*) = 0$ and lies on the manifold S of feasible primal and dual points, i.e. the set where $y_* = G(\cdot, y_*)$ and $L_y(\cdot, y_*, \bar{y}_*) = 0$ holds.

In terms of function and derivative evaluations the computational effort of the Seidel variant can be almost twice that of the Jacobi variant. The iteration function G must now be evaluated at two different combinations (u, y) and (u_+, y), though only one of them involves a design change compared to the previous cycle. Only at (u, y) the trajectory of all intermediate values must be stored for the subsequent accumulation of the gradient components $L_u(u, y, \bar{y})$ and $L_y(u, y, \bar{y}_+)$. Notice that the adjoint arguments $\bar{y}$ and $\bar{y}_+$ differ so that two reverse sweeps are required. In summary the Seidel variant requires one additional forward and one additional reverse sweep, but both are a little cheaper than the original ones, because there is no design change and no need for a new trajectory, respectively.

One advantage of the Seidel type method demonstrated in [24] (relying on [25]) is that the Lagrangian converges twice as fast to the optimal value of the objective than in the Jacobi variant. As observed in [27] another important distinction is that in the Newton case $G_y(u, y_*) = 0$ two step quadratic convergence can be achieved

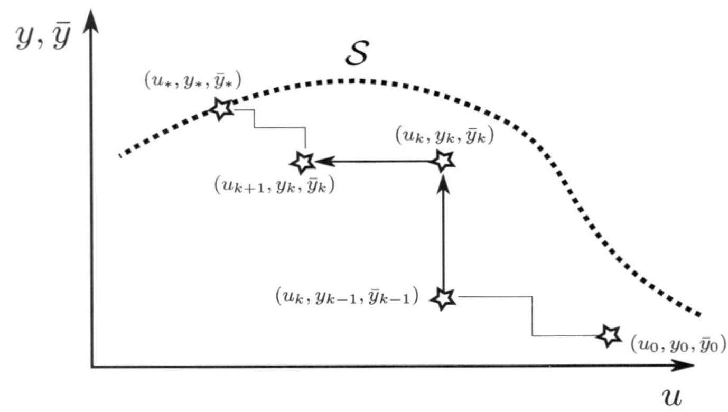

Fig. 1 Geometry of Seidel one-shot methods

by the Seidel variant, but no choice of B_{Jac} yields super-linear convergence of the Jacobi variant except when $G_{yu}(u, y_*) = 0$.

3.4 Finite Termination on QPs

Of particular theoretical interest is the QP case when the objective $f(u, y)$ is quadratic and the constraint $c(u, y)$ is linear. Here we would expect finite convergence from an optimization scheme that is given all the required information.

In [36] three-step quadratic convergence was shown for the update sequence

$$\cdots \to \text{ADJOINT} \to \text{DESIGN} \to \text{STATE} \to \ldots \text{ or } \cdots \to \text{A} \to \text{D} \to \text{S} \to \ldots .$$

More specifically it was shown that on a linear quadratic problem (QP) the Jacobian corresponding to a (Newton) adjoint step, followed by a (reduced Hessian) design step and a subsequent (Newton) state step has the nil-potency degree 3. Earlier in [1], it has been observed that exact preconditioners derived from this iterations lead to convergence of appropriate Krylov methods in three steps. Here, we will see that this sequence has a Jacobian even with nil-potency 2 and that the Jacobian of the sequence S $\to$ A $\to$ D $\to$ S $\to$ A vanishes already. This means that with regards to the number of design steps we can achieve one-step quadratic convergence if we interleave two S $\to$ A pairs in between them.

To obtain this result we may break (3.3) into three pieces using $G_y = 0$ for Newtons step

$$J_D = \begin{bmatrix} (I - B_{Seid}^{-1} L_{uu}) & -B_{Seid}^{-1} L_{uy} & -B_{Seid}^{-1} G_u^\top \\ 0 & I & 0 \\ 0 & 0 & I \end{bmatrix}, \tag{3.8}$$

$$J_S = \begin{bmatrix} I & 0 & 0 \\ G_u & 0 & 0 \\ 0 & 0 & I \end{bmatrix} \quad \text{and} \quad J_A = \begin{bmatrix} I & 0 & 0 \\ 0 & I & 0 \\ L_{yu} & L_{yy} & 0 \end{bmatrix}. \tag{3.9}$$

Multiplication yields immediately that

$$J_A\, J_S\, J_D\, J_A\, J_S \;=\; \begin{bmatrix} R & 0 & 0 \\ (L_{yy} G_u + L_y u) R & 0 & 0 \\ G_u R & 0 & 0 \end{bmatrix}, \tag{3.10}$$

where $R = I - B_{Seid}^{-1} H$, which vanishes if $B_{Seid} = H$. In other words in order to obtain fast convergence when the state equation can be solved in a Newton like fashion the reduced Hessian H is the only choice for the preconditioner.

When the optimization problem is sufficiently smooth and the state equation is efficiently solvable quite accurately, e.g. by multi-grid method, one would still expect fast convergence using the Seidel approach with an approximation $B_{Seid} \approx H$. It seems natural to base this approximation on secant updating with respect to differences in the reduced gradient $L_u(u, y, \bar{y})$ (from (2.6)), but care must be taken that inaccuracies in its evaluation do not lead to instabilities in the updating process. Also we cannot expect that the steps $(\Delta u, \Delta y)$ really become tangential to the feasible set $\mathcal{F}$. Consequently we cannot expect that B_{Seid} converges to H but rather another Schur complement as observed in [36].

3.5 Asymptotic Convergence Rate

In [36] an exact perturbation argument relative to the three-step termination result is used to quantify a linear convergence rate in the QP case with approximate primal and adjoint solver. The iteration discussed there is of the type (3.7) or equivalently (3.6). Convergence of this iteration is shown, if the forward and adjoint iterative method is sufficiently contractive and the matrix B_{Seid} is sufficiently close to a so-called consistent reduced Hessian, where $\mathcal{Z}$ in Eq. (2.10) is based on the approximate solver rather than the matrix c_y. The bounds on the perturbations are rather involved and difficult to verify in practice. The convergence analysis cannot be compared to the discussion in this paper because of this different approach.

Now we present a single iteration analysis which also yields a bound on the spectral radius of the coupled iteration at a second order sufficient minimum. So strictly speaking we are again talking about the QP scenario, but with inaccurate solvers and reduced Hessian approximation.

While the contraction rate ρ_0 is considered a constant we can reduce it to $\eta = \rho_0^s$ by performing $s > 0$ primal followed by s adjoint steps before each design change, i.e. we consider the Multistep One-shot method

$$\cdots \rightarrow \text{DESIGN} \rightarrow \text{STATE}^s \rightarrow \text{ADJOINT}^s \rightarrow \ldots, \quad s \in \mathbb{N}.$$

The s adjoint steps can all be based on the last primal step, whose trajectory of intermediate values can be reused s times. This multi-step strategy has been advocated already by S. Ta'asan and is being used by many practitioners. In the paper [14] there has been an attempt to optimize the s in view of the efforts for the various sub-steps. For example, if design changes entail expensive regriding operations s should be selected rather large. We also expect the optimal s to grow when the problem becomes more QP like as we approach an optimal point.

In [14] the following bound on the eigenvalues of the coupled iteration was obtained based on the problem quantities that were defined in Sect. 2.2:

Proposition 3.2. *Under the stated assumptions all eigenvalues $\sigma \in \mathbb{C}$ of J_* for the Multi-step one-shot iteration with the preconditioner matrix B_{Seid} satisfy*

$$|\sigma| \leq \eta \quad or \quad |\sigma| \leq \gamma + (1+\gamma)\left[d\left(\eta\frac{|\sigma|+1}{|\sigma|-\eta}\right)^2 + 2c\left(\eta\frac{|\sigma|+1}{|\sigma|-\eta}\right)\right] \tag{3.11}$$

For $|\sigma| > \eta$, the rational function $\varphi(|\sigma|, \eta)$ on the right hand side has continuous positive and negative derivatives w.r.t. η and $|\sigma|$, respectively.

In the Newton case $\eta = 0$ the bound given in Proposition 3.11 is tight in that we simply have $|\sigma| \leq \gamma = \|I - \tau H^{\frac{1}{2}} B_{Seid}^{-1} H^{\frac{1}{2}}\|$. This is the convergence rate of a quasi-Newton or variable metric method on the reduced system, where primal and adjoint feasibility is recuperated exactly after each design step. Otherwise the inequality is likely to hold strictly since the bound on the right hand side can only be reached if, for example, the eigenmodes of $(I - G_y^s)^{-1}$ associated with its largest eigenvalues in modulus are also the dominant eigenvectors of L_{yy}. Hence we expect that the ratio between the right and left hand side at the largest eigenvalue can be sizable for small η close to 1 but tends to 1 as η tends to 0. These ranges of $\eta = \rho_0^s$ can be achieved by selecting s sufficiently large or rather small, respectively.

Due to the monotonicity properties, the implicit function theorem ensures the existence of a unique solution $|\sigma^*| = |\sigma^*(\eta)|$ at the intersection of the diagonal with the right hand side $\varphi(|\sigma|, \eta)$ of (3.11) which is differentiable and strictly increasing on the interval (γ, ∞). This situation is depicted in the Fig. 2. When η is sufficiently small, e.g. $\eta_1 = 0.12$ one obtains an upper bound σ_1^* on the spectral radius of the coupled iteration below 1 so that we have contraction. When $\eta = \eta_2 = 0.35$ the resulting bound σ_2^* is greater than 1 and thus useless.

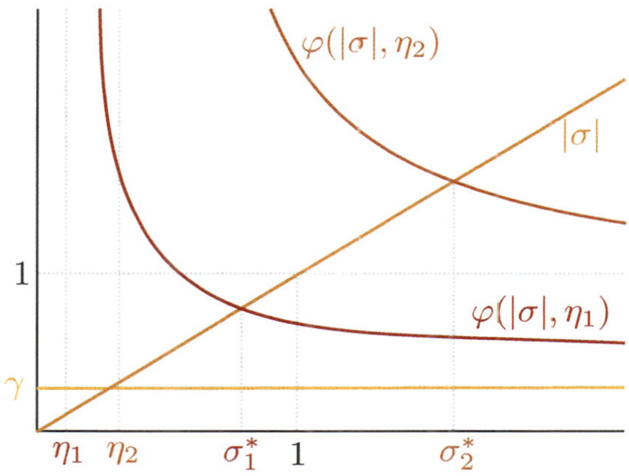

Fig. 2 Left side $|\sigma|$ and right side $\varphi(|\sigma|,\eta)$ of (3.11) for two different values $\eta_1 = \rho_0^{s_1}$ and $\eta_2 = \rho_0^{s_2}$ with $\eta_1 < \eta_2$ and two arbitrarily chosen positive constants c and d

Unfortunately, the above bound cannot be solved explicitly for $|\sigma|$. However given a desired overall contraction $\rho \geq |\sigma^*|$ we obtain a quadratic equation for η whose solution is given by

$$\eta(\rho) = \frac{\rho(\rho-\gamma)}{(\rho-\gamma)+(1-\gamma)(1+\rho)\left(\sqrt{d(\rho-\gamma)/(1-\gamma)+c^2}+c\right)} \tag{3.12}$$

$$\leq \frac{\rho^2}{\rho+(1+\rho)\left(\sqrt{d\,\rho+c^2}+c\right)} \approx \frac{\rho^2}{2\,c} \quad \text{if} \quad \rho \approx 0 \tag{3.13}$$

where the second bound is obtained by setting γ to 0 and its approximation by considering rather small ρ. Taking the logarithm of that inequality we obtain a lower bound for the retardation factor estimate namely

$$\kappa \equiv \frac{\log(\eta(\rho))}{\log(\rho)} \gtrsim 2 + \frac{2\,c}{|\log(\rho)|} \to 2 \quad \text{for} \quad \rho \to 0$$

Hence we see that the retardation factor is always bounded below by 2 and with good preconditioning that lower bound is almost achieved as both the primal rate $\eta = \rho_0^s$ and the combined rate ρ tend to 0, i.e. we recover feasibility quite well before after each design step. The predicted retardation is plotted in Fig. 3.

It should be noted that adaptivity issues for optimization methods of (reduced) SQP type have been studied, e.g. in [33, 56] and in a more general framework adaptivity aspects of Newton methods for nonlinear equations are discussed, e.g., in [17, 19].

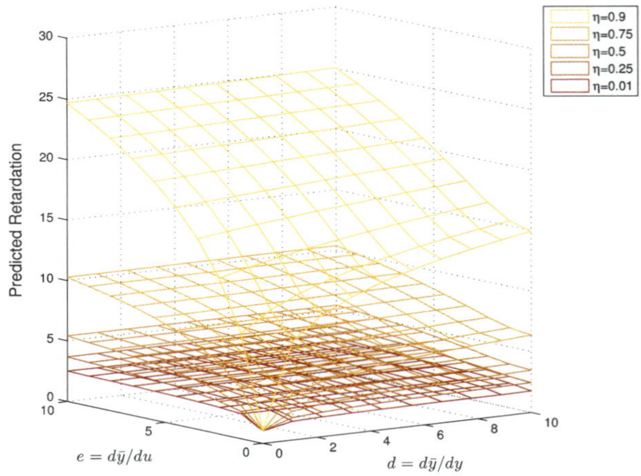

Fig. 3 Retardation factor surface as a function of c and d

4 Other Approaches

A particular variant of one-shot methods is the *optimization boundary value approach* originating from [6, 7]. It is representative also for other approaches like the simultaneous optimization approach [54] or SAND [30]. They have in common that they form a nonlinear set of equations related to the KKT conditions, which is then solved by Newton's method. A separation of steps in design, adjoint and forward step in the fashion above based on fixed point iterations is not explicitly made, although this distinction is blurred to some extent depending on the particular choice of the approximation of the KKT matrix used. In particular, in [8], the explicit treatment of inexact linear solves in the forward and adjoint problem is taken into account and generalized to inequality constraints. In combination with a multiple-shooting parametrization of the time history of dynamical systems, the boundary value problem approach allows for very flexible adaptivity with the respect to the discretization accuracy. This accuracy and the Newton step size selection is controlled by the restrictive monotonicity test [9]. Similarly, in [4, 5] a large scale KKT system for a PDE constrained optimization problem is formed and solved by Newton-Krylov methods. Furthermore, multi-grid optimization techniques as surveyed in [12] fall in this class of setting up a large KKT system and solving it as a whole.

There are several contributions on inexact SQP methods in the literature (see e.g. [33, 56]). There not only discretization errors but also residuals of inexact equation solvers are coordinated to achieve global convergence to the continuous solution. The grid adaption is based on suitable error estimators, especially the dual weighted residual approach of Becker and Rannacher [2]. In the papers of Ziems and Ulbrich the effects of non-linearity are controlled using a trust region.

5 Numerical Results

In this section we present numerical results that demonstrate the efficiency of the Jacobi as well as the Seidel variant of the One-shot method. As a test case we choose an inverse design optimization problem subject to the incompressible Navier-Stokes equations. By controlling the boundary conditions a prescribed flow field has to be reproduced.

As PDE-constraint of the optimization problem, we investigate buoyancy driven flow in a cavity $\Omega = [0,1]^2$ governed by the incompressible Navier-Stokes equations. The left wall $\Gamma_1 = \{0\} \times [0,1]$ is isothermal, while the temperature distribution at the right wall $\Gamma_2 = \{1\} \times [0,1]$ can be altered choosing a variable temperature to control the fluid flow, these temperature values serve as design variables in the optimization problem. While cool fluid is falling along the left wall, the heated and thus hotter fluid is rising along the right wall producing a counterclockwise flow in Ω. Figure 4 shows the temperature distribution of a steady-state reference flow field that we want to reproduce as well as the corresponding reference temperature at the right wall.

As an objective function of the optimization problem we choose the following tracking type functional

$$f(u,y) \equiv \|y - y_{\text{ref}}\|_2^2 + \mu\|u\|_2^2. \tag{5.1}$$

where $y = \left[\vec{v}, p, T\right]$ denotes the state vector consisting of the velocities $\vec{v}$, pressure p and temperature T of the fluid while y_{ref} denotes the reference flow field. A regularization term consisting of the weighted norm of the design u is added with a factor $\mu = 0.001$.

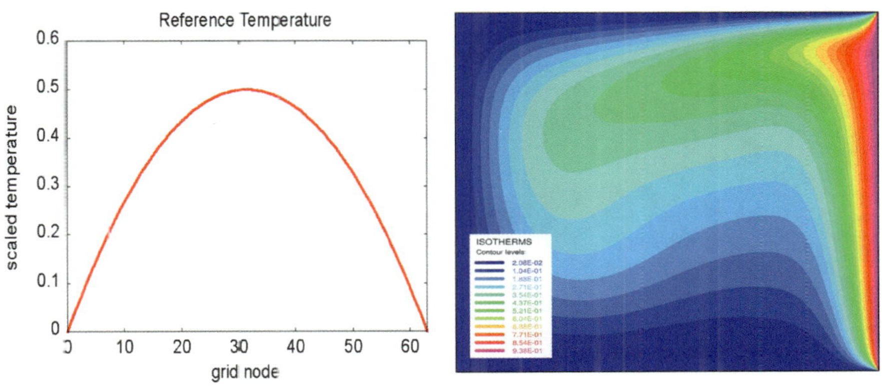

Fig. 4 Reference temperature distribution (*left*) and resulting flow field for the temperature of reference at the right wall Γ_2(*right*)

The optimization problem is defined by

$$\min_{u,y} f(u, y) \quad \text{subject to} \tag{5.2}$$

$$\begin{aligned} \frac{\partial \vec{v}}{\partial t} + (\vec{v} \cdot \nabla)\vec{v} - \Delta \vec{v} + \nabla p &= \vec{b} \text{ in } \Omega \\ \operatorname{div} \vec{v} &= 0 \text{ in } \Omega \\ \frac{\partial T}{\partial t} + \vec{v} \cdot \nabla T - \Delta T &= 0 \text{ in } \Omega \\ \vec{v} &= 0 \text{ on } \partial\Omega \\ \frac{\partial T}{\partial n} &= 0 \text{ on } \partial\Omega \setminus (\Gamma_1 \cup \Gamma_2) \\ T &= 0 \text{ on } \Gamma_1 \\ T &= u \text{ on } \Gamma_2 \end{aligned} \tag{5.3}$$

where the buoyancy force is given as $\vec{b} = [0, \mathrm{RaPr}\, T]$ with the dimensionless Rayleigh and Prandtl number $\mathrm{Ra} = 10^5$, $\mathrm{Pr} = 0.1$ (cf. [21]).

The governing incompressible Navier-Stokes equations are solved using an implicit finite volume method that performs pressure-correction iterations until a steady-state of the system is reached. In other words the iteration function G represents one step of the SIMPLE-method (c.f. [21]). In Fig. 5, the 2-norm of the corresponding residuum of the primal equations during the iterative flow computation is plotted. As mentioned in Sect. 2.2, we observe a monotonic decline only for geometric averages.

Automatic Differentiation in reverse mode is applied to the iteration function G and evaluating the objective function f in order to generate an iterative adjoint solver and compute the reduced gradient. The design space preconditioners B_{Jac}, B_{Seid} are approximated using secant updates of the gradient of the augmented Lagrangian and the gradient of the standard Lagrangian, respectively. As stopping criterion we used the condition $\|L_u\|_2 < 10^{-5}$ on the reduced gradient.

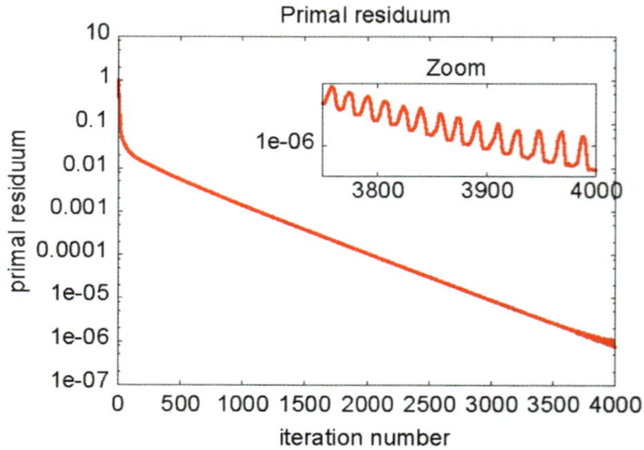

Fig. 5 Residuum of the incompressible Navier-Stokes equations during flow computation

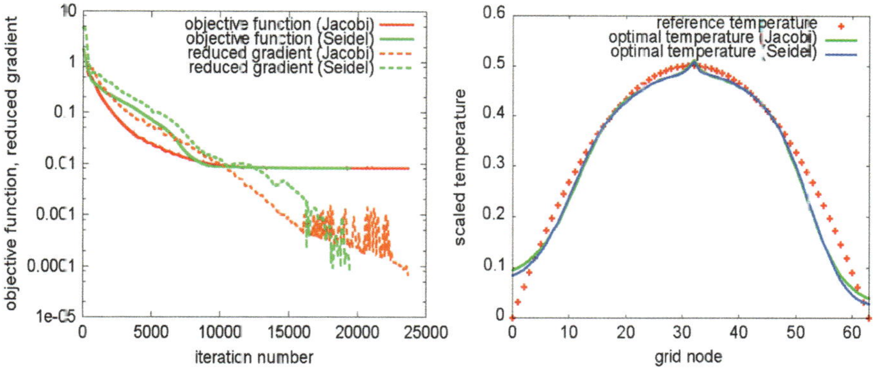

Fig. 6 Convergence history of Jacobi- and Seidel-type one-shot optimization (*left*) and optimized temperature distribution with reference profile (*right*)

Figure 6 shows the objective function $f(u_k, y_k)$ as well as the norm of the reduced gradient $\|L_u(u_k, y_k, \bar{y}_k)\|_2$ during the optimization for both the Jacobi and the Seidel type One-shot method. From the reduction of the reduced gradient it can be seen that both approaches converge with a similar convergence rate. Since the Seidel variant hits the stopping criterion slightly earlier a retardation factor of approximately $r = 7.2$ is achieved whereas a retardation factor of $r = 7.9$ is measured for the Jacobi type One-shot method. Both optimization approaches recover the reference temperature distribution quite well as can be seen from Fig. 6. Furthermore, the residuum of the Navier-Stokes equations and its adjoint is reduced to the order of 10^{-6} at the optimal point, thus optimality and feasibility is achieved simultaneously during the One-shot iterations.

The numerical results demonstrate, that both variants of the One-shot method yield an efficient optimization of the inverse design problem while the cost for an optimization is only a small multiple of the cost of a primal simulation. The considered test case indicates a slight advantage of the Seidel type One-shot approach, which is in good agreement with numerical results for solving linear equations with the Seidel and the Jacobi method.

6 Summary and Outlook

In this paper we mainly surveyed two closely related approaches for one-shot optimization. As it turns out the main theoretical difference is a different representation of the adjoint states, which are however closely related, though a comparative investigation of their regularity property in function space remains to be done. On a numerical test for the incompressible Navier Stokes equation both approaches were found to be similarly effective. An asymptotic analysis of the retardation factor between simulation and optimization confirmed the usual understanding that the latter is at least twice as expensive as the former.

The ideal ratio is reached if either the state space solver is Newton-like, or the cross term c representing the total sensitivity of the adjoint equation with respect to design vector is quite small. Generally speaking, the success of the one-shot approach depends on a good synchronization between primal, dual and optimization steps. As in other reported implementations of inexact SQP methods the synchronization is based on several tolerances, norm estimates and method parameters. Their appropriate setting remains to be a serious challenge and for the time being require a tuning process by an expert user for a particular class of design optimization problems.

References

1. A. Battermann, E. Sachs, Block preconditioners for KKT systems in PDE-governed optimal control problems, in *Fast Solution of Discretized Optimization Problems*, ed. by K.-H. Hoffmann, R.H.W. Hoppe, V. Schulz. Number 138 in ISNM (Birkhaeuser, Basel, 2001), pp. 1–18
2. R. Becker, R. Rannacher, An optimal control approach to a posteriori error estimation in finite element methods. Acta Numer. **10**(1), 1–102 (2001)
3. M. Benzi, A.J. Wathen, Some preconditioning techniques for saddle point problems, in *Model Order Reduction: Theory, Research Aspects and Applications*, ed. by W.H.A. Schilders, H.A. Vorst, J. Rommes. Mathematics in Industry, vol. 13 (Springer, Berlin/Heidelberg, 2008), pp. 195–211
4. G. Biros, O. Ghattas, Parallel lagrange-Newton-Krylov-Schur methods for PDE-constrained optimization. Part I: the Krylov-Schur solver. SIAM J. Sci. Comput. **27**(2), 687–713 (2005)
5. G. Biros, O. Ghattas, Parallel Lagrange-Newton-Krylov-Schur methods for pde-constrained optimization. Part II: the Lagrange-Newton solver, and its application to optimal control of steady viscous flows. SIAM J. Sci. Comput. **27**(2), 714–739 (2005)
6. H.G. Bock, Numerical treatment of inverse problems in chemical reaction kinetics, in *Modelling of Chemical Reaction Systems*, ed. by K.H. Ebert, P. Deuflhard, W. Jäger. Number 18 in Springer Series in Chemical Physics (Springer-Verlag Berlin Heidelberg, 1981)
7. H.G. Bock, *Randwertproblemmethoden zur Parameteridentifizierung in Systemen nichtlinearer Differentialgleichungen*, Bonner Mathematische Schriften, Vol. 183, (Rheinische Friedrich-Wilhelms-Universität, Bonn, 1987)
8. H.G. Bock, V.H. Schulz, Mathematical aspects of CFD-model based optimization, in *Optimization and Computational Fluid Dynamics*, ed. by D. Thevenin, G. Janiga (Springer-Verlag Berlin Heidelberg, 2008), pp. 61–78
9. H.G. Bock, E. Kostina, J.P. Schöder, On the role of natural level functions to achieve global convergence for damped Newton methods. Syst. Model. Optim. IFIP Int. Fed. Inf. Process. **46**, 51–74 (2000)
10. H.G. Bock, A. Potschka, S. Sager, J.P. Schlöder, On the connection between forward and optimization problem in one-shot one-step methods, in *Constrained Optimization and Optimal Control for Partial Differential Equations*, ed. by G. Leugering, S. Engell, A. Griewank, M. Hinze, R. Rannacher, V. Schulz, M. Ulbrich, S. Ulbrich. International Series of Numerical Mathematics, vol. 160 (Springer, Basel, 2012), pp. 37–49
11. J.F. Bonnans, A. Shapiro, *Perturbation Analysis of Optimization Problems*. Springer Series in Operations Research (Springer, New York, 2000)

12. A. Borzì, V. Schulz, Multigrid methods for PDE optimization. SIAM Rev. **51**(2), 361–39 (2009)
13. A. Borzì, V. Schulz, *Computational Optimization of Systems Governed by Partial Differential Equations*. Number 08 in SIAM Book Series on Computational Science and Engineering (SIAM, Philadelphia, 2012)
14. T. Bosse, L. Lehmann, A. Griewank, Adaptive sequencing of primal, dual, and design steps in simulation based optimization. Comput. Optim. Appl. **57**(3), 731–760 (2014)
15. G. Chavent, J. Jaffré, *Mathematical Models and Finite Elements for Reservoir Simulation: Single Phase, Multiphase, and Multicomponent Flows Through Porous Media*. Studies in Mathematics and Its Applications (Elsevier, Amsterdam, 1986)
16. E.J. Cramer, P.D. Frank, G.R. Shubin, J.E. Dennis, R Lewis, On alternative problem formulations for multidisciplinary design optimization, in *Proceedings of the Fourth AIAA/USAF/NASA/OAI Symposium on Multidisciplinary Analysis and Optimization*, Cleveland, 1992, AIAA-Paper 92-4572, 518–530 (1992)
17. P. Deuflhard, *Newton Methods for Nonlinear Problems, Affine Invariance and Adaptive Algorithms*. Number 35 in Springer Series in Computational Mathematics (Springer, Berlin/Heidelberg, 2004)
18. M.A. Diaz Viera, P. Sahay, M. Coronado, A.O. Tapia, *Mathematical and Numerical Modeling in Porous Media: Applications in Geosciences*, 1st edn. (CRC, Boca Raton, 2012)
19. S.E. Eisenstat, H.F. Walker, Globally convergent inexact Newton methods. SIAM J. Optim. **4**(2), 393–422 (1994)
20. K. Eppler, S. Schmidt, V. Schulz, C. Ilic, Preconditioning the pressure tracking in fluid dynamics by shape hessian information. J. Optim. Theory Appl. **141**(3), 513–531 (2009)
21. J.H Ferziger, M. Perić, *Computational Methods for Fluid Dynamics*, vol. 3 (Springer, Berlin, 1996)
22. A. Fowler, *Mathematical Geoscience*. Interdisciplinary Applied Mathematics (Springer, London, 2011)
23. N. Gauger, A. Griewank, A. Hamdi, C. Kratzenstein, E. Oezkaya, T. Slawig, Automated extension of fixed point PDE solvers for optimal design with bounded retardation. Int. Ser. Numer. Math. **106**, 99–122 (2012)
24. A. Griewank, Projected Hessians for preconditioning in one-step one-shot design optimization, in *Large-Scale Nonlinear Optimization*. Nonconvex Optimization and Its Applications, vol. 83 (Springer, New York, 2006), pp. 151–171
25. A. Griewank, C. Faure, Reduced functions, gradients and Hessians from fixed-point iterations for state equations. Numer. Algorithms **30**(2), 113–139 (2002)
26. A. Griewank, C. Faure, Piggyback differentiation and optimization, in *Large-Scale PDE-Constrained Optimization (Santa Fe, NM, 2001)*. Lecture Notes in Computational Science and Engineering, vol. 30 (Springer, Berlin, 2003), pp. 148–164
27. A. Griewank, E. Özkaya, Quantifying retardation in simulation based optimization, in *Optimization, Simulation, and Control*. Springer Optimization and Its Application, vol. 76 (Springer, New York, 2013), pp. 79–96
28. A. Griewank, A. Walther, How up-to-date are low-rank updates? Rev. Investig. Oper. **25**(2), 137–147 (2004)
29. S. Günther, Simultane Optimierung unter PDE-Nebenbedingungen. Ein Vergleich zweier One-Shot-Methoden (Simultaneous optimization under PDE-constraints. A comparison of two One-shot methods). Diploma thesis, University of Trier, 2012
30. R.T. Haftka, Z. Gurdal, *Elements of Structural Optimization*. Contributions to Phenomenology (Kluwer Academic, Dordrecht, 1992)
31. Ad Hamdi, A. Griewank, Properties of an augmented lagrangian for design optimization. Optim. Methods Softw. **25**(4), 645–664 (2010)
32. S.B. Hazra, V. Schulz, J. Brezillon, N.R. Gauger, Aerodynamic shape optimization using simultaneous pseudo-timestepping. J. Comput. Phys. **204**(1), 46–64 (2005)
33. M. Heinkenschloss, L.N. Vicente, Analysis of inexact trust-region sqp algorithms. SIAM J. Optim. **12**(2), 283–302 (2002)

34. R. Herzog, E. Sachs, Preconditioned conjugate gradient method for optimal control problems with control and state constraints. SIAM J. Matrix Anal. Appl. **31**(5), 2291–2317 (2010)
35. M. Hinze, R. Pinnau, M. Ulbrich, S. Ulbrich, *Optimization with PDE Constraints*. Number 23 in Mathematical Modelling: Theory and Applications (Springer, Cham, 2009)
36. K. Ito, K. Kunisch, V. Schulz, I. Gherman, Approximate nullspace iterations for KKT systems. SIAM J. Matrix Anal. Appl. **31**(4), 1835–1847 (2010)
37. A. Jameson, Aerodynamic design via control theory, in *Recent Advances in Computational Fluid Dynamics (Princeton, NJ, 1988)*. Lecture Notes in Engineering, vol. 43 (Springer, Berlin, 1989), pp. 377–401
38. C. Kanzow, *Numerik Linearer Gleichungssysteme – Direkte und Iterative Verfahren*, 1. aufl. edn. (Gabler Wissenschaftsverlage, Wiesbaden, 2004)
39. C.T. Kelley, *Iterative Methods for Optimization*. Frontiers in Applied Mathematics, vol. 18 (SIAM, Philadelphia, 1999)
40. C. Kratzenstein, T. Slawig, Simultaneous model spin-up and parameter identification with the one-shot method in a climate model example. Int. J. Optim. Control **3**(2), 99–110 (2013)
41. A. Lashanizadegan, Sh. Ayatollahi, M. Homayoni, Simultaneous heat and fluid flow in porous media: case study: steam injection for tertiary oil recovery. Chem. Eng. Commun. **195**(5), 521–535 (2008)
42. G. Leugering, S. Engell, A. Griewank, M. Hinze, R. Rannacher, V. Schulz, M. Ulbrich, S. Ulbrich, *Constrained Optimization and Optimal Control for Partial Differential Equations*. International Series of Numerical Mathematics (Springer, Basel, 2012)
43. A. Nemili, E. Özkaya, N.R. Gauger, A. Carnarius, F. Thiele, Optimal control of unsteady flows using discrete adjoints, AIAA-Paper 3720, 1–14 (2011)
44. J. Nocedal, S.J. Wright, *Numerical Optimization*. Springer Series in Operations Research and Financial Engineering, 2nd edn. (Springer, New York, 2006)
45. E. Özkaya, N.R. Gauger, Single-step one-shot aerodynamic shape optimization, in *Optimal Control of Coupled Systems of Partial Differential Equations*. International Series of Numerical Mathematics, vol. 158 (Birkhäuser Verlag, Basel, 2009), pp. 191–204
46. E. Özkaya, N.R. Gauger, Automatic transition from simulation to one-shot shape optimization with Navier-Stokes equations. GAMM-Mitt. **33**(2), 133–147 (Springer-Verlag Berlin Heidelberg, 2010)
47. E. Özkaya, N. R. Gauger, An efficient one-shot algorithm for aerodynamic shape design, in *New Results in Numerical and Experimental Fluid Mechanics VII* (2010), pp. 35–42
48. P. Parekh, M.J. Follows, E.A. Boyle, Decoupling of iron and phosphate in the global ocean. Glob. Biogeochem. Cycles **19**(2), GB2020 (2005)
49. O. Pironneau, On optimum profiles in Stokes flow. J. Fluid Mech. **59**(01), 117–128 (1973)
50. S. Schmidt, V. Schulz, Impulse response approximations of discrete shape Hessians with application in CFD. SIAM J. Control Optim. **48**(4), 2562–2580 (2009)
51. S. Schmidt, C. Ilic, V. Schulz, N. Gauger, Three-dimensional large-scale aerodynamic shape optimization based on shape calculus. AIAA J. **51**(11), 2615–2627 (2013)
52. J. Schöberl, W. Zulehner, Symmetric indefinite preconditioners for saddle point problems with applications to PDE-constrained optimization problems. SIAM J. Matrix Anal. Appl. **29**(3), 752–773 (2007)
53. S. Ta'asan, 'One Shot'methods for optimal control of distributed parameter systems i: finite dimensional control. Technical report, DTIC Document, 1991
54. I.B. Tjoa, L.T. Biegler, Simultaneous solution and optimization strategies for parameter estimation of differential-algebraic equation systems. Ind. Eng. Chem. Res. **30**(2), 376–385 (1991)
55. F. Tröltzsch, *Optimale Steuerung partieller Differentialgleichungen* (Vieweg, Wiesbaden, 2005)
56. J.C. Ziems, S. Ulbrich, Adaptive multilevel inexact sqp methods for pde-constrained optimization. SIAM J. Optim. **21**(1), 1–40 (2011)

Optimal Design with Bounded Retardation for Problems with Non-separable Adjoints

Torsten Bosse, Nicolas R. Gauger, Andreas Griewank, Stefanie Günther, Lena Kaland, Claudia Kratzenstein, Lutz Lehmann, Anil Nemili, Emre Özkaya, and Thomas Slawig

Abstract In the natural and enginiering sciences numerous sophisticated simulation models involving PDEs have been developed. In our research we focus on the transition from such simulation codes to optimization, where the design parameters are chosen in such a way that the underlying model is optimal with respect to some performance measure. In contrast to general non-linear programming we assume that the models are too large for the direct evaluation and factorization of the constraint Jacobian but that only a slowly convergent fixed-point iteration is available to compute a solution of the model for fixed parameters.

Therefore, we pursue the so-called One-shot approach, where the forward simulation is complemented with an adjoint iteration, which can be obtained by handcoding, the use of Automatic Differentiation techniques, or a combination thereof. The resulting adjoint solver is then coupled with the primal fixed-point iteration and an optimization step for the design parameters to obtain an optimal solution of the problem. To guarantee the convergence of the method an appropriate sequencing of these three steps, which can be applied either in a parallel (Jacobi) or in a sequential (Seidel) way, and a suitable choice of the preconditioner for the design step are necessary. We present theoretical and experimental results for two

---

The work was funded by the DFG (Deutsche Forschungsgesellschaft) as part of SPP1253.

T. Bosse (✉) • A. Griewank • L. Lehmann
Department of Mathematics, Humboldt-Universität zu Berlin, Unter den Linden 6, 10099 Berlin, Germany
e-mail: mail@torsten-bosse.de; griewank@math.hu-berlin.de; llehmann@math.hu-berlin.de

N.R. Gauger • L. Kaland • S. Günther • E. Özkaya • A. Nemili
Department of Mathematics and Center for Computational Engineering Science, RWTH Aachen University, Schinkelstr. 2, 52062 Aachen, Germany
e-mail: gauger@mathcces.rwth-aachen.de; kaland@mathcces.rwth-aachen.de; guenther@mathcces.rwth-aachen.de; ozkaya@mathcces.rwth-aachen.de; nemili@mathcces.rwth-aachen.de;

T. Slawig • C. Kratzenstein
Christian-Albrechts-Universität zu Kiel, Christian-Albrechts-Platz 4, 24118 Kiel, Germany
e-mail: ts@informatik.uni-kiel.de; ctu@informatik.uni-kiel.de

G. Leugering et al. (eds.), *Trends in PDE Constrained Optimization*, International Series of Numerical Mathematics 165, DOI 10.1007/978-3-319-05083-6_6

choices, one based on the reduced Hessian and one on the Hessian of an augmented Lagrangian. Furthermore, we consider the extension of the One-shot approach to the infinite dimensional case and problems with unsteady PDE constraints.

Keywords Simulation • Optimization • PDE • Automatic differentiation • Fixed-point solver • Retardation factor • One-shot • Piggyback • Numerics

1 Introduction

In the research project *Automated Extension of Fixed Point PDE Solvers for Optimal Design with Bounded Retardation* we focus on design optimization problems of the form

$$\min_{(u,y)\in U\times Y} f(u,y) \quad \text{s. t.} \quad c(u,y)=0 \tag{DOP}$$

where $f : U \times Y \to \mathbb{R}$ denotes an objective function and $c : U \times Y \to H$ with $\dim H = \dim Y = n$ represents some state equation. This scenario has been approached by many computational scientist with inexact variants of large-scale SQP methods. For a partial survey we recommend [1–3, 12, 21].

As a key assumption we require that for any control $u \in U$ there is a non-singular solution $y(u) \in Y$ of the state equation $c(u, y) = 0$. Moreover, we assume that the state constraint can be equivalently written as a fixed-point equation with some contractive function $G : U \times Y \to Y$, i.e.

$$\|G_y(u,y)\| \le \rho_0 < 1 \qquad \text{for all } (u,y) \in U \times Y,$$

such that the fixed-point iteration $y_{k+1} = G(u, y_k)$ provides a solution $y(u) = \lim_{k\to\infty} y_k$ of the original state equation for any fixed control $u \in U$ and initial state $y_0 \in Y$. We also assume in the statement and for the execution of numerical algorithms that the functions are at least once continuously differentiable to guarantee well posedness of the problem and twice continuously differentiable for the convergence theory.

Thus, by standard results from nonlinear optimization [16] we see in the finite dimensional case ($n < \infty$) that for any local minimum (u_*, y_*) of (DOP) in the interior of $U \times Y$ there exists a Lagrange multiplier $\bar{y}_* \in \mathbb{R}^n$ such that the first order necessary optimality conditions

$$0 = L_u(u_*, y_*, \bar{y}_*), \;\; y_* = G(u_*, y_*), \;\text{and}\; 0 = L_y(u_*, y_*, \bar{y}_*)$$

hold, where $L : U \times Y \times \mathbb{R}^n \to \mathbb{R}$ denotes the Lagrangian function

$$L(u,y,\bar{y}) = f(u,y) + \bar{y}^\top (G(u,y) - y)\,.$$

Assuming that second order sufficient optimality conditions are satisfied we find that the projected Hessian of the Lagrangian

$$H_* := \left[I,\, Z^\top\right] \begin{bmatrix} L_{uu} & L_{uy} \\ L_{yu} & L_{yy} \end{bmatrix} \begin{bmatrix} I \\ Z \end{bmatrix} \quad \text{with} \quad Z := (I - G_y)^{-1}\, G_u \tag{1.1}$$

evaluated at a strict local minimum $(u_*, y_*, \bar{y}_*)$ is positive definite and the same holds true in an neighborhood of the minimizer.

In the first part of our project we pursued a so-called (Jacobi) One-shot strategy [4, 8, 9, 11]

$$\begin{aligned} u_+ &= u - \alpha_{step} B_{Jac}^{-1} L_u(u, y, \bar{y}) \\ y_+ &= G(u, y) \\ \bar{y}_+ &= \bar{y} + L_y(u, y, \bar{y}) \end{aligned} \tag{1.2}$$

or in short

$$\ldots \rightarrow (\text{ DESIGN, STATE, ADJOINT}) \rightarrow \ldots$$

to find first order optimal points. Here $\alpha_{step} \in \mathbb{R}$ denotes some step-multiplier and B is a suitable symmetric positive definite preconditioner, which may depend on the variables $(u, y, \bar{y})$, the given functions f, G and their derivatives. As a special choice we investigate the augmented Lagrangian preconditioner

$$B_{Jac} = L_{uu} + \alpha\, G_u G_u^\top + \beta\, L_{uy} L_{yu}$$

and BFGS approximations of it with some suitable coefficients $\alpha, \beta \in \mathbb{R}$.

Beside the original (Jacobi) one-step One-shot method [8], several other stepping schemes can be found. Therefore, we also propose the Multistep-Seidel-version

$$\ldots \rightarrow (\text{DESIGN}) \rightarrow (\text{STATE})^s \rightarrow (\text{ADJOINT})^s \rightarrow \ldots,$$

where after one design update several repeated state updates are followed by the same number of repeated adjoint updates, or in detail,

$$\begin{aligned} u_+ &= u - \alpha B_{Seid}^{-1} L_u(u, y, \bar{y}) && \text{single design update,} \\ y_+ &= G^s(u_+, y) && s \text{ state updates,} \\ \bar{y}_+ &= \bar{G}^s(u_+, y_+, \bar{y}) && s \text{ adjoint updates}. \end{aligned} \tag{1.3}$$

where

$$\begin{aligned} G^{k+1}(u, y) &:= G\left(u, G^k(u, y)\right) \quad \text{and} \\ \bar{G}^{k+1}(u, y, \bar{y}) &:= \bar{G}\left(u, y, \bar{G}^k\,(u, y, \bar{y})\right) \end{aligned}$$

for $k = 1, \ldots, s - 1$ with $\bar{G}(u, y, \bar{y}) := L_y(u, y, \bar{y}) + \bar{y}$.

In contrast to before, the preconditioner $B_{Seid} \approx H_*$ may also depend on the number of state/adjoint updates s. We present the basic ideas (cf. [5]) needed to prove that the Multistep One-shot method is locally convergent for a sufficient choice of α, B_{Seid} and the step number s which is mainly depending on the contraction rate ρ_0 of G and problem specific derivative information.

In the sequel, we will give a short summary of our project for the last research period of the DFG SPP-1253 project. The structure is as follows:

In Sects. 2–4 we present some of our results for the Jacobi method containing the findings for the exact quantification of the retardation factor, an application in marine science and the extension of the approach to function space. For the Multistep One-shot method we will state sufficient conditions for the convergence of the method in terms of problem dependent quantities and present some numerical examples for an application in aerodynamic shape optimization, which is done in Sects. 5 and 6, respectively. Furthermore, we will consider in Sect. 7 the case where the constraint mapping c represents a PDE only allowing for unsteady solutions.

2 Exact Quantification of Retardation

In One-shot methods, retardation refers to the increase of steps needed for a comparable reduction in the residuals when going from simulation to optimization in the coupled iteration. *Bounded* retardation, i.e., a limited increase of these steps, has been achieved by many groups in the priority program. However, a general theoretical statement to quantify the factor of retardation for the Jacobi method has not been achieved yet. In the second period, we obtained theoretical results for separable problems [9], where $L_{yu} = L_{uy} = 0$. We investigated:

1. A Newton scenario for separable problems,
2. Jacobi and multigrid scenarios for a standard elliptic problem.

In the Newton scenario for the separable case, we have $G_y = 0$ and thus $G_u = dy/du$. We expect the observed results to remain valid also in the case when G represents an inner iteration. We tested the example of several multigrid cycles that resolve the state equation with higher accuracy before a change of the design variables. In this case, the retardation factor was found to be $\gamma/3$, where $\gamma = \|\Gamma\|$ is the weighted Euclidean norm of $G_u^\top L_{yy} G_u$ w.r.t. to the projected Hessian H.

In the Jacobi and multigrid scenarios, we consider an elliptic boundary value problem with a tracking type objective function and Tikhonov regularization on the L_2 norm of the control with the weighting parameter μ. This standard test problem was solved by the rather slow Jacobi method and the rather fast multigrid method. Here, we find that the preconditioner should be a multiple of the identity and its optimal scaling can be found by solving a system of three cubic polynomials, which can be reduced to a single polynomial in the convergence factor ρ_0.

In Fig. 1, the retardation factors as a function of the reciprocal $1/\mu$ for three different grid sizes N are shown. As one can see, the retardation factor for the

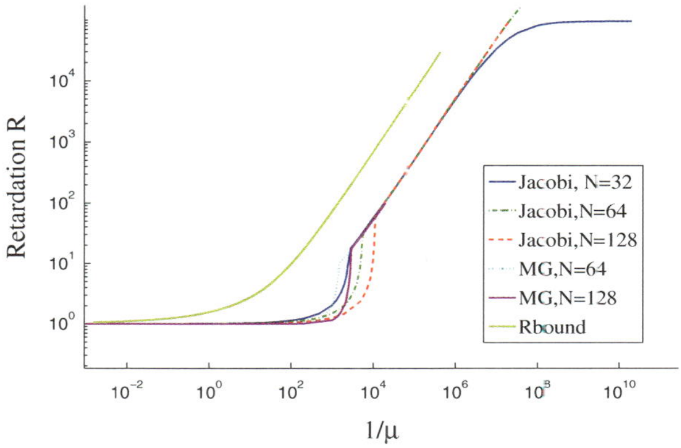

Fig. 1 Retardation factor for Jacobi and multigrid methods

Jacobi scenario is very small until $1/\mu$ is about 10^2, then grows quite rapidly until it becomes a linear function of $1/\mu$, and finally for very large $1/\mu$ it becomes constant. The same behavior is also observed for the V-cycle multigrid case with Jacobi smoother. In all cases, we observed a much better retardation factor than the theoretical upper bound without optimized step multiplier (yellow line).

3 Application in Marine Science

Parameter optimization is an important task in all models that simulate parts of the climate system, as for example ocean or atmosphere models. In these models, many processes are not well-understood or cannot be resolved. These processes are *parametrized* using simplified model functions with parameters that have to be optimized for calibration according to measurements or other models' data. The parameters appear as factors of the state variables, thus leading to nonseparability in the state equations. Often, calibration is performed for a steady stationary or periodic solution, the latter representing a stable annual cycle. Computation of a steady state is usually the result of a *spin-up*, i.e., a time integration until no significant changes are observed. For ocean models, the spin-up needs thousands of years of model time, which reflects the long time scales of the global ocean circulation. In three space dimensions, the pure simulation of the ocean circulation is a challenging computational task which requires considerable time. As a consequence, the One-shot method is a promising approach for parameter optimization in ocean models. However, the additional computational effort of simultaneous update of the state and parameter corrections must not be ignored and we propose simplifications of the strategy. We considered two examples.

3.1 Calibration of a Box Model of the North Atlantic Circulation

At first, we calibrated a conceptual box model of the North-Atlantic thermohaline circulation by Rahmstorf [20]. It has eight nonlinear ODEs and a global warming parameter that varies in a given range and is not to be optimized. For each value of this parameter f_1, the amount of water overturning $m(u, y(f_1))$ is obtained as an aggregated quantity from the state variables and the parameters. The model is numerically integrated into a steady state where $c(u, y(f_1)) = 0$ by an explicit Euler scheme. Since it is computationally cheap and has been calibrated using other methods (see [18]), we used it to investigate the applicability of the One-shot method and to compare results and performance in a real world problem. Data m_d from a more complex model (see [18]) are used as desired state in a tracking type functional with regularization term incorporating a prior guess u_{guess} for the six parameters to be optimized:

$$\min_{u,y} f(u, y) := \frac{1}{2}\|m(u, y(f_1)) - m_d\|_2^2 + \frac{\epsilon}{2}\|u - u_{guess}\|_2^2,$$
$$\text{s.t.} \quad 0 = c(u, y(f_{1,i})), \quad i = 1, \ldots, l.$$

The parameters are subject to box constraints, which were not treated explicitly in the One-shot method. Without regularization, typically several local minima occur.

We compared the One-shot results both with full computation of the preconditioner B_{Jac} and using its BFGS approximation on the one hand with results obtained by direct optimization using a full spin-up in every function evaluation on the other hand. For the direct optimization we applied our own BFGS implementation as well as the L-BFGS and L-BFGS-B codes from [19].

As summarized in [14], the One-shot method was successful, even though no contractivity, but only quasi-contractivity (see [7]) is given. Simplifications of the algorithm as fixing the parameter ρ representing the contraction factor to 0.9 and limiting the exact computation of B_{Jac} to every 1,000th iteration was adequate. The latter reduced computational time to about half of the time needed in optimization runs with computation of B_{Jac} in each iteration. The final states obtained by the two One-shot variants are close to the data and to the ones obtained by the direct methods, also with small regularization parameter ϵ. The parameters computed by One-shot were to some extent similar to those of the direct optimization with L-BFGS-B. They stayed in acceptable ranges without any explicit constraint treatment, but differ among the chosen methods when $\epsilon < 1$, which is due to the ill-posedness of the problem.

As can be seen in Fig. 2, the One-shot strategy showed good performance: The number of iterations was about 10–40 times larger than those for a spin-up. Direct optimization strategies needed at least 30 optimization steps, each requiring several complete spin-ups. Using full computation of B_{Jac} performs well for most regularization parameters ϵ, whereas the One-shot-BFGS strategy does not show

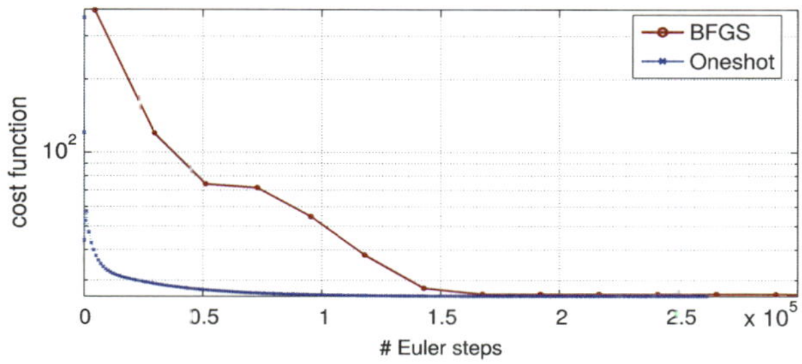

Fig. 2 Typical optimization run for parameter optimization of the box model: comparison of total necessary Euler steps by direct BFGS optimization and One-shot method with full computation of the preconditioner

good performance. This behavior also varies with respect to the global warming parameter, likely because the model itself has difficulties finding the steady state for high values of this parameter.

3.2 *Calibration of a 3-D Marine Ecosystem Model*

Marine ecosystem models describe the physical and bio-geochemical processes that determine the oceanic part of the global carbon cycle. They are non-linearly coupled transport or advection-diffusion-reaction equations, with ocean circulation data as forcing. In three dimensions, the computation of a steady annual cycle of such models takes several days on a parallel machine.

We performed parameter optimization for a characteristic model (see [17]) consisting of two spatially distributed state variables (tracers), namely phosphate and dissolved organic phosphorus. The parameter optimization problem is of tracking type including a regularization term with an initial parameter guess:

$$\min_{u,y} f(u,y) := \frac{1}{2}\|y - y_{data}\|^2 + \frac{\epsilon}{2}\|u - u_{guess}\|^2 \text{ s.t. } 0 = c(u,y)$$

At first synthetic data created by the model were used as desired state, tests with real data taken from the World Ocean Atlas are work in progress. Direct optimization runs that are still possible in coarse resolutions suggest that for this configuration several local minima exist. Nevertheless, the One-shot optimization method without regularization found the correct parameters u_* for synthetic data. Figure 3 shows an example with regularization parameter $\epsilon = 0.01$, but where the initial guess u_{guess} did not equal the value u_* used to create the synthetic desired state. The convergence of the parameters differs. The cost function is significantly reduced, as can be seen in Fig. 4. Comparing performance, the One-shot method leads to results comparable with a direct optimization after about 15,000 steps (equal model years). A usual

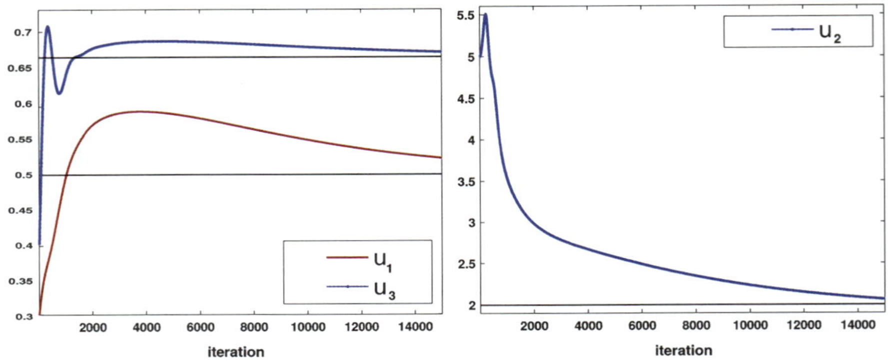

Fig. 3 Some parameters during optimization, $u_{guess} \neq u_*$, $\epsilon = 0.01$. *Straight lines* represent optimal values u_*

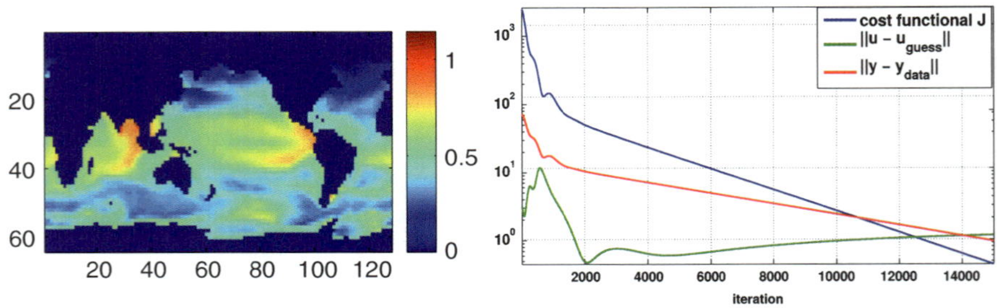

Fig. 4 Typical tracer distribution at the ocean surface (*left*) and cost function f during One-shot optimization ($u_{guess} \neq u_*$, $\epsilon = 0.01$)

Table 1 Computation of derivatives using different approaches

Derivative	Mode of Computation
f_y, f_u, f_{yu}	Analytically
G_u	Forward mode of AD + analytically for linear parts
$\bar{y}^\top G_y, \bar{y}^\top G_u$	One reverse sweep of AD + analytically for linear parts
$\bar{y}^\top G_{yu}$	Finite differences applied to $\bar{y}^\top G_y$

spin-up takes about 5,000 years, but it has to be noticed that a One-shot iteration requires additional effort due to adjoint and parameter updates. In this example, the One-shot iteration step requires about 23 times the computational time needed for one step of the spin-up. The costs can be reduced to a factor of only eight, if the update of B_{Jac} is performed only every fifth iteration step.

Table 1 summarizes the use of Automatic Differentiation (AD) in the realization of the One-shot method.

4 One-Shot in Function Spaces

For the treatment of the One-shot method in function spaces we again consider problem (DOP) for general Hilbert spaces U and Y. Here, $c(u, y) = 0$ with $c : U \times Y \to Y^*$ denotes the governing equations in form of a PDE. In order to define the Lagrange function with respect to the fixed point operator $G(u, y)$ in a Hilbert space setting correctly we need to consider the transition from the PDE to the fixed-point formulation. According to [13], this is given in terms of a linear, bounded and bijective operator $F(y) : Y \to Y^*$ so that

$$c(u, y) = F(y)[y - G(u, y)].$$

For the sake of simplicity, we assume $F(y)$ to be independent of u. The Lagrangian is now defined incorporating the fixed-point formulation as follows

$$\begin{aligned} L(u, y, \bar{y}) &= f(u, y) - \langle \bar{y}, c(u, y) \rangle_{Y,Y^*} \\ &= f(u, y) - \langle F(y)^* \bar{y}, y - G(u, y) \rangle_{Y^*,Y}. \end{aligned} \tag{4.1}$$

Computing the KKT system based on (4.1) yields a fixed-point formulation of the optimality system and a simultaneous update of state, adjoint and design equation

$$y_+ = G(u, y) \tag{4.2}$$

$$\bar{y}_+ = \Phi(u, y, \bar{y}) \tag{4.3}$$

$$u_+ = u - B^{-1} L_u(u, y, \bar{y}). \tag{4.4}$$

with an appropriate preconditioner B. Here, the operator $\Phi(u, y, \bar{y})$ in (4.3) is the fixed-point operator of the adjoint equation and defined by (see [13])

$$\begin{aligned} \langle F(y)^* \Phi(u, y, \bar{y}), w \rangle_{Y^*,Y} := f_y(u, y)w - \langle \bar{y}, F_y(y)w[y - G(u, y)] \rangle_{Y,Y^*} \\ + \langle F(y)^* \bar{y}, G_y(u, y)w \rangle_{Y^*,Y} \end{aligned}$$

for all $w \in Y$. Note that it holds

$$L_y(u, y, \bar{y})w = \langle F(y)^* \Phi(u, y, \bar{y}), w \rangle_{Y^*,Y} - \langle F(y)^* \bar{y}, w \rangle_{Y^*,Y}.$$

In [13] a convergence proof is given for the general case and specified for model problems including the solid fuel ignition model and the viscous Burgers equations. In the following, we only note the leading steps of the general convergence proof. Therefore, consider the augmented Lagrangian defined as

$$L^a(u, y, \bar{y}) = L(u, y, \bar{y}) + \frac{\alpha}{2} \|G(u, y) - y\|_Y^2 + \frac{\beta}{2} \|\Phi(u, y, \bar{y}) - \bar{y}\|_Y^2 \tag{4.5}$$

with the penalty parameters $\alpha, \beta > 0$. The convergence proof follows the idea of the finite dimensional setting to show that the augmented Lagrangian acts as a penalty function, i.e. that every local minimum of the original optimization problem (DOP) is also a local minimum of L^a. Further, we show that the One-shot method yields descent on L^a and therefore reaches the minimum correctly. The next theorem (cf. [13]) is the main result in this procedure and ensures the equivalence of the stationary points as well as the descent condition.

Theorem 4.1. *If there exist constants $\alpha > 0$ and $\beta > 0$ such that the following conditions are fulfilled*

$$\alpha(1-\rho_0) - \tfrac{\alpha^2}{2\tilde{\gamma}}\|G_u\|^2 > \|F(y)\| + \tfrac{\beta}{2}\|\Phi_y\|,$$
$$\beta(1-\rho_0) > \|F(y)\| + \tfrac{\beta}{2}\|\Phi_y\|, \quad \textit{and} \quad \gamma > \tfrac{\tilde{\gamma}}{2},$$

for a positive preconditioner B with $(Bh,h)_U \geq \gamma\|h\|_U^2$, $\|\Phi_{\bar{y}}\| \leq \rho_0 < 1$ and a constant $\tilde{\gamma} > 0$, then a point is a stationary point of L^a if and only if it is a solution of the KKT-system to (DOP)*. Additionally, the increment vector of the One-shot method is a descent direction for L^a.*

These general conditions are difficult to verify. Nevertheless, for specific model problem they can be simplified and tested [13]. Numerical investigations of the method, with a preconditioner chosen as a scaled identity operator, show a mesh-independent behavior for several model problems:

Example (see [13]). Consider the minimization of the tracking type functional

$$\min f(y,u) := \frac{1}{2}\int_\Omega |y - z_d|^2\,dx + \frac{\epsilon}{2}\int_\Omega |u|^2\,dx$$

subject to $(y,u) \in H_0^1(\Omega) \times L^2(\Omega)$ fulfilling the Viscous Burgers equation

$$-\nu\Delta y + (y\cdot\nabla)y = u \quad \text{in } \Omega, \qquad \text{and} \qquad y = 0 \quad \text{on } \Gamma.$$

The corresponding first order optimality system

$$\begin{aligned}
-\nu\Delta y + (y\cdot\nabla)y - u &= 0, & y|_\Gamma &= 0\\
-\nu\Delta\bar{y} - (y\cdot\nabla)\bar{y} - \operatorname{div}(y)\bar{y} + (\nabla y)^T\bar{y} - y + z_d &= 0, & \bar{y}|_\Gamma &= 0\\
\epsilon u + \bar{y} &= 0 & \text{a.e. in } \Omega&.
\end{aligned}$$

was solved by the One-shot iteration:

$$\begin{aligned}
y_+ &= G(y,u) = (-\nu\Delta + y\cdot\nabla)^{-1}(u)\\
\bar{y}_+ &= (-\nu\Delta - y\cdot\nabla - \operatorname{div}(y))^{-1}(y - z_d - (\nabla y)^T\bar{y})\\
u_+ &= u - \frac{1}{\gamma}(\epsilon u + \bar{y})
\end{aligned}$$

The resulting number of iterations for the 2D case are given in Table 2.

Table 2 2D Burgers equation with $\epsilon = 0.01$, $\nu = 0.1$ and $z_d \equiv 1$

# Degree of freedom	$\gamma = 0.5$	$\gamma = 0.55$	$\gamma = 0.6$	$\gamma = 0.65$	$\gamma = 0.7$
882	389	424	460	497	534
1922	∞	429	459	494	530
3362	∞	∞	460	493	529
5202	∞	∞	463	493	528

It is important to note that in this example the focus does not lie on the efficiency of the method, it rather demonstrates the mesh-independency. Therefore, the total number of iterations necessary for the optimization does not increase significantly. The formulation and analysis of the method in function spaces as well as the numerical mesh-independent behavior motivates an extension of the method in terms of an additional adaptive step (cf. [13]).

5 Adaptive Sequencing of Primal, Adjoint and Control Updates

As a part of our research we also considered various stepping schemes, one of them being the Multistep One-shot (1.3). Assuming for the analysis that the design variables were transformed in such a way that the projected Hessian is the identity, i.e., we set $u = T\tilde{u}$ and $\tilde{u} = T^{-1}u$ if $T^{-\top}T^{-1} := H_* = H(1)$, we were able (see [5]) to bound all complex eigenvalues of the Jacobian

$$J_* = \frac{\partial(u_+, y_+, \bar{y}_+)}{\partial(u, y, \bar{y})}$$

for the coupled iteration in terms of the problem dependent quantities

$$d \equiv \|L_{yy}\| \, \|\tilde{Z}\|^2, \quad e \equiv \left\|L_{y\tilde{u}} + L_{yy}\tilde{Z}\right\| \, \|\tilde{Z}\|, \quad \text{and} \quad \gamma \equiv \|I - \alpha_{step}\tilde{B}_{Seid}^{-1}\|.$$

Here $\tilde{Z} = Z\,T$ and $\tilde{B}_{Seid} = T^\top B_{Seid} T$ represent the transformed quantities.

Proposition 5.1. *Under the stated assumptions all eigenvalues $\lambda \in \mathbb{C}$ of J_* for the Multistep One-shot iteration with the preconditioner matrix B_{Seid} satisfy*

$$|\lambda| \leq \eta \quad or \quad |\lambda| \leq \gamma + \nu\left[d\,\mu^2(\eta, |\lambda|) + 2\,e\,\mu(n, |\lambda|)\right]. \tag{5.1}$$

where $\eta = \rho_0^s$, $\nu = \alpha_{step}\|\tilde{B}_{Seid}^{-1}\|$ and $\mu(\eta, |\lambda|) = \eta(|\lambda| + 1)/(|\lambda| - \eta)$.

Note that $L_{yy} = \partial L_y/\partial y$ is the partial derivative of the adjoint equation w.r.t. y and $L_{yy}Z + L_{yu} = dL_y/du$ is the total derivative of the adjoint equation w.r.t.

u. thus, e and d can be understood as a measure for the sensitivity of the adjoint equation with respect to state and design, respectively. Moreover, we have:

Proposition 5.2. *If $\gamma < 1$, then by adjusting s and thus $\eta = \rho_0^s$, any rate $\rho \in (\gamma, 1)$ can be attained as upper bound of the spectrum of J_*. The following relation between s, η and ρ for given e, d, γ and ν is sufficient:*

$$\rho_0^s = \eta \leq \eta_*(\rho) = \frac{\rho(\rho - \gamma)}{(\rho - \gamma) + \nu(1 + \rho)\left(\sqrt{d(\rho - \gamma)/\nu + e^2} + e\right)} \tag{5.2}$$

In other words, we found a sufficient condition on the number s of primal and adjoint iterations that ensures the local convergence of the approach in terms of the above mentioned quantities.

Corollary 5.3. *The spectral radius ρ of J_* is less than 1 if the number of inner iterations $s \in \mathbb{N}$ satisfies*

$$s > \underline{s} = \log_{(1/\rho_0)}\left[1 + 2\left(\sqrt{d\,(1-\gamma)/\nu + e^2} + e\right)\nu/(1-\gamma)\right]$$

This theoretical lower bound on the number s of primal and adjoint updates was used to implement an self-adapting algorithm ABOSO. Within the algorithm all required quantities, such as e, δ, γ, and ρ_0, are approximated by difference quotients and other already computed information instead of the exact calculation which is in general too expensive. Also, the measurements are averaged over the last iterations to have more reliable estimates.

Example. The self-adapting algorithm was verified on various examples, e.g. on the non-linear problem Bratu problem (see [15])

$$\min_{(u,y)} \frac{1}{2}\|\partial_2 y(\cdot, 1) - \phi_1\|^2_{\mathcal{L}^2(\Omega)} + \frac{\mu}{2}\|u\|^2_{\mathcal{H}^1(\Omega)}, \quad (u, y) \in \mathcal{H}^1(\Omega) \times \mathcal{H}^1_0(\Omega) \text{ s.t.}$$

$$-\Delta y = \lambda \exp(y) \text{ in } \Omega, \quad y(s, 1) = u(s), \quad y(s, 0) = \phi_2(s), \quad y(1, t) = y(0, t)$$

that describes the combustion of solids over the unit square $\Omega = [0, 1]^2 \subset \mathbb{R}^2$ for given functions ϕ_1 and ϕ_2. The fixed-point function $y_+ = z = G(u, y)$ was computed on purpose in a Seidel type iteration by solving the implicit univariate equations

$$z_{ij} - \frac{h^2}{4} exp(z_{ij}) = y_{mean} = \frac{1}{4}(y_{i,j-1} + y_{i,j+1} + y_{i-1,j} + y_{i+1,j})$$

using the equidistant grid points $(i/m, j/m)$ with $m = 12$ so that $y_{m,j} = y_{0,j}$ and copying the values u_i into $z_{i,m}$ after each inner iteration. Naturally, there are faster solvers for this elliptic problem, but we deliberately wished to mimic slow fixed point solvers in more complicated application areas. The behavior of the algorithm

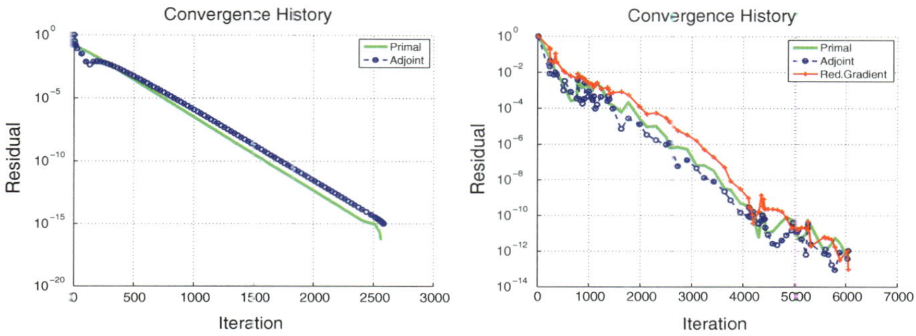

Fig. 5 Residuals for the simulation (*right*)and optimization(*left*)

is displayed in Fig. 5 for the parameters $\lambda = 1$ and $\mu = 10^{-4}$. In particular, it can be seen by comparing the residual of the simulation without design changes on the left side and the residual of the optimization on the right that the retardation factor is approximately 2.5, i.e. for achieving the same residual in the optimization only an small number of additional simulation steps by a factor of 2.5 is required.

6 Application in Aerodynamic Shape Optimization

We have applied the Multistep One-shot method for the shape optimization of a NACA0012 airfoil at transonic flow conditions using Euler equations. As the shape parameterization, the free-node parametrization is chosen, in which all the mesh points on the airfoil surface are taken as shape parameters. This type of parameterization enables that maximum degree of freedom can be given to the optimization algorithm. The shape sensitivities, which are required for the One-shot method, are computed using the consistent discrete adjoint approach based on Automatic Differentiation [6]. Although this approach is slower than the continuous and hand-discrete adjoint approaches, it has been chosen because of its robustness and its ability to compute exact derivative information without utilizing any approximations. The test case is chosen as the inviscid drag minimization scenario at constant lift. The Mach number and the angle of attack for this case are chosen as 0.85 and 2 respectively. The grid used for the study is the 325×65 C-type grid with 196 grid points on the airfoil surface. As it can be seen in the Fig. 6, the initial NACA0012 airfoil creates a strong inviscid shock on the suction side of the airfoil, which leads to a high amount of drag in the transonic flow regime (left figure). In the right figure, the pressure distribution for the optimized shape is illustrated. It can be observed that inviscid shock disappears in the optimized airfoil, which leads to a substantial drag reduction of 60 % while maintaining the lift.

In order to assess the performance of Multistep One-shot method, we have made a comparison between a nested optimization approach using BFGS method with

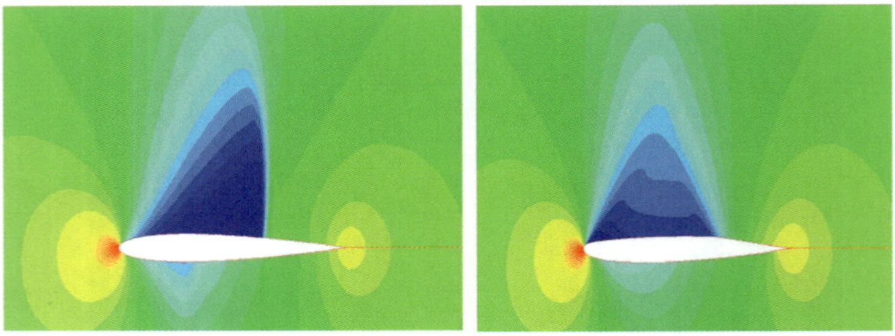

Fig. 6 Initial NACA0012 and optimized airfoils with pressure contours for the transonic case

Table 3 Iteration count and run-time measurements for the primal simulation, nested optimization and One-shot optimization

Case	Iteration counts	Ret. factor	Run-time (s)	Ret. factor
Primal simulation	2613	1	107	1
Nested opt.	192788	73.8	102016	953
One-shot opt.	10140	3.9	3867	36.1

line searches and One-shot method with $s = 10$. The performance results of both methods compared to a single primal simulation are presented in Table 3. The nested approach takes totally 11 adjoint and 65 primal solver evaluations. Note that in the nested approach, the number of iterations taken by the primal and adjoint solvers for each run vary since five decade residue fall is set as the stopping criteria. The nested approach takes totally 192,788 primal/dual steps and the optimization takes ca. 28 h on a 2.4 GHz Intel machine. The retardation factor of the nested approach is measured as 73.8 in iteration counts and 107 in run-time. As it can be observed from the results, the One-shot method is significantly faster than the nested approach and has a factor of retardation 3.9 in iteration counts and 36.1 in run-time.

7 One-Shot Optimization with Unsteady PDE Constraints

For time-dependent PDEs, the state variable varies with time and thus is a function $y: [0, T] \to Y$. The objective function to be minimized is typically given by some time averaged quantity. The general formulation of the optimization problem with unsteady PDEs reads

$$\min_{u,y} \frac{1}{T} \int_0^T f(u, y(t))\, dt \quad \text{s.t.} \quad \begin{cases} \frac{\partial y(t)}{\partial t} + c(u, y(t)) = 0 & \forall t \in [0, T] \\ y(0) = y_0 \, . \end{cases} \tag{7.1}$$

Unsteady PDEs are typically discretised by an implicit time marching scheme. The resulting implicit equations are solved iteratively by applying a fixed point solver at each physical time step until a steady-state solution at that time step is achieved:

$$\text{for } i = 1, \ldots, N: \quad y^i_{k+1} = G(u, y^i_k, y^{i-1}_*, y^{i-2}_*, \ldots) \xrightarrow{k\to\infty} y^i_* \tag{7.2}$$

Here, y^i_* denotes the converged steady-state at the discrete time step $t_i = i\Delta t$. N is the total number of time steps, given by $T = N\Delta t$. The contractive fixed-point iterator G not only depends on the design variable but also on the converged state solutions at previous time steps.

In order to extend to One-shot, where one incorporates design updates already during the primal flow computation, the time marching scheme (7.2) is modified as

$$\text{for } k = 1, 2, \ldots: \quad y^i_{k+1} = G(u, y^i_k, y^{i-1}_{k+1}, y^{i-2}_{k+1}, \ldots) \quad \forall i \in \{1, \ldots, N\}. \tag{7.3}$$

In contrast to (7.2), where fixed point iterations are performed at each time step for a state y^i, in the One-shot framework (7.3) the complete trajectory of the unsteady solution is updated within one iteration. Interpreting the state as a discrete vector from the product space $y \in Y^N := Y \times \ldots \times Y$ with state components y^i, we can write (7.3) in terms of an update function

$$y_{k+1} = \mathcal{H}(u, y_k) \tag{7.4}$$

where $\mathcal{H}: U \times Y^N \to Y^N$ performs the update formulas (7.3) for all time steps. Using the contractivity of the fixed point iterator G it can be shown, that $\mathcal{H}$ is contractive with respect to $y \in Y^N$ and, thus, y_k converges to the unsteady solution of the PDE (cf. [10]).

Replacing the unsteady PDE constraint by the fixed point equation $y = \mathcal{H}(u, y)$, the Lagrangian function corresponding to the unsteady optimization problem is defined as

$$L(u, y, \bar{y}) := I(u, y) + \bar{y}^T\left(\mathcal{H}(u, y) - y\right), \tag{7.5}$$

where $I(u, y) = \frac{1}{N}\sum_{i=1}^{N} f(u, y^i)$ approximates the objective function. This formulation has the same structure as the definition of the Lagrangian in Sect. 1. Thus, the concept of One-shot optimization can be applied in the same way by replacing the fixed point iterator with the mapping $\mathcal{H}$ and the objective function with the approximation I.

For a fixed design $u \in U$, iterating only in the state and the adjoint variable simultaneously in the so called piggy-back iteration is implemented for the problem of optimal active flow control around a 2D cylinder. Eight actuation slits are installed on the surface of the cylinder where sinusoidal blowing and suction is applied in order to reduce vorticity downstream the cylinder. Amplitude and phase

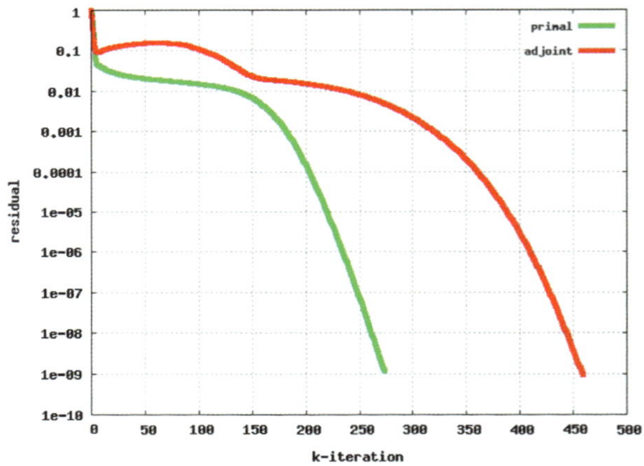

Fig. 7 Convergence history of primal and adjoint states for incompressible URANS in One-shot framework

shift of the actuation are used as design variables. The governing incompressible URANS (unsteady Reynolds-averaged Navier-Stokes) equations are solved by applying the new approach to a second order implicit finite volume code. To study the convergence behavior, the L_2-norm of the state and the adjoint residuals $\|y - \mathcal{H}(u, y)\|_2$, $\|L_y(u, y, \bar{y})\|_2$ are computed. From Fig. 7 it can be observed, that both variables converge with the same asymptotic convergence rate. In future, a preconditioned control update will be incorporated in the piggy-back iteration for the implementation of One-shot in unsteady framework.

Conclusion

In the second phase of the project, the theoretical results and the applications from the first one have been extended in different ways. First of all, it was possible to quantify the retardation factor of some test problems and Newton, Jacobi and multigrid iterations. Moreover, the application of the One-shot method in its Jacobi variant was shown to be feasible and successful for parameter optimization in complex, spatially three-dimensional climate models using a fixed-point type iteration to compute steady seasonal cycles. These results show a high potential for application on various real-world problems in climate research, thus emphasizing the interdisciplinary benefit of the project.

Whereas these theoretical results and applications are based on the finite-dimensional setting of the method, we additionally extended the theory for the One-shot Jacobi variant on two prominent infinite-dimensional problems, namely the viscous Burger and the solid fuel ignition model. For both cases also numerical studies were performed.

(continued)

Furthermore, we developed the Multistep One-shot method that uses an adaptive sequencing or adjustment of the number of primal, adjoint and control updates used during the algorithm. For this method, we provide a theory relating the number of necessary primal and adjoint steps per control update to the spectral radius of the Jacobian and thus the convergence speed of the coupled iteration. This modified method was applied successfully in shape optimization in Computational Fluid Dynamics. In this context, we also extended the method to non-linear (inner) iterations in non-stationary flow solvers.

Acknowledgements The work was funded by the DFG (Deutsche Forschungsgesellschaft) as part of the *DFG Schwerpunktprogramm 1253 – Optimization with partial differential equations*. Our sincere thanks are due to several other groups of the SPP 1253 (Bock et al., Schulz et al.) for their stimulating comments and discussions. We are especially grateful for the many helpful suggestions and for the encouraging interest shown by other research groups outside of the SPP 1253: Alonso et al., Farrell et al., Koziel et al., Oschlies et al., De los Reyes et al., Thiele et al., and Wang et al.

References

1. G. Biros, O. Ghattas, Parallel Lagrange-Newton-Krylov-Schur methods for PDE-constrained optimization. Part I: the Krylov-Schur solver. SIAM J. Sci. Comput. **27**(2), 687–713 (2005)
2. G. Biros, O. Ghattas, Parallel lagrange-Newton-Krylov-Schur methods for pde-constrained optimization. part i: the krylov-schur solver. SIAM J. Sci. Comput. **27**(2), 687–713 (2005)
3. A. Borzì, V. Schulz, Multigrid methods for PDE optimization. SIAM Rev. **51**(2), 361–39 (2009)
4. T. Bosse, A. Griewank, N.R. Gauger, S. Günther, V. Schulz, One-shot approaches to design optimzation, in *Trends in PDE Constrained Optimization*, ed. by P. Benner, G. Leugering, S. Engell, A. Griewank, H. Harbrecht, M. Hinze, R. Rannacher, S. Ulbrich. International Series of Numerical Mathematics (Springer, Basel, 2014). To appear
5. T. Bosse, L. Lehmann, A. Griewank, Adaptive sequencing of primal, dual, and design steps in simulation based optimization. Comput. Optim. Appl. (2013). doi:10.1007/s10589-013-9606-z
6. A. Carnarius, F. Thiele, E. Özkaya, A. Nemili, N.R. Gauger, Optimal control of unsteady flows using a discrete and a continuous adjoint approach, in *System Modelling and Optimization*. IFIP Advances in Information and Communication Technology, vol. 391, ed. by D. Hömberg, F. Tröltzsch (Springer, Berlin/Heidelberg, 2011), pp. 318–327
7. L.B. Ciric, A generalization of Banach's contraction principle. Proc. Am. Math. Soc. **45**(2), 267–273 (1974)
8. N. Gauger, A. Griewank, A. Hamdi, C. Kratzenstein, E. Özkaya, T. Slawig, Automated extension of fixed point pde solvers for optimal design with bounded retardation, in *Constrained Optimization and Optimal Control for Partial Differential Equations*, ed. by G. Leugering, S. Engell, A. Griewank, M. Hinze, R. Rannacher, V. Schulz, M. Ulbrich, S. Ulbrich. International Series of Numerical Mathematics, vol. 160 (Springer, Basel, 2012). pp. 99–122
9. A. Griewank, E. Özkaya, Quantifying retardation in simulation based optimization, in *Optimization, simulation, and control*. Springer Optimization and its Application, vol. 76 (Springer, New York, 2013), pp. 79–96

10. S. Günther, N.R. Gauger, Q. Wang, Extension of the One-shot method for optimal control with unsteady PDEs, in *Proceedings of the International Conference on Evolutionary and Deterministic Methods for Design, Optimization and Control with Applications to Industrial and Societal Problems (EUROGEN)*, Spain, 2013
11. A. Hamdi, A. Griewank, Reduced quasi-Newton method for simultaneous design and optimization. Comput. Optim. Appl. **49**(3), 521–548 (2011)
12. M. Heinkenschloss, L.N. Vicente, Analysis of inexact trust-region sqp algorithms. SIAM J. Optim. **12**(2), 283–302 (2002)
13. L. Kaland, J.C. De Los Reyes, N.R. Gauger, One-shot methods in function space for PDE-constrained optimal control problems. Optim. Methods Softw. 1–30 (2013). doi:10.1080/10556788.2013.774397
14. C. Kratzenstein, T. Slawig, Simultaneous model spin-up and parameter identification with the One-shot method in a climate model example. Int. J. Optim. Control **3**(2), 99–110 (2013)
15. U. Naumann, *The art of differentiating computer programs: an introduction to algorithmic differentiation*. Software, Environments and Tools (Society for Industrial and Applied Mathematics, Philadelphia, 2011)
16. J. Nocedal, S.J. Wright, *Numerical Optimization*. Springer Series in Operations Research and Financial Engineering, 2nd edn. (Springer, New York, 2006)
17. P. Parekh, M.J. Follows, E.A. Boyle, Decoupling of iron and phosphate in the global ocean. Glob. Biogeochem. Cycles **19**(2), GB2020 (2005)
18. T. Slawig, K. Zickfeld, Parameter optimization using algorithmic differentiation in a reduced-forms model of the atlantic thermohaline circulation. Nonlinear Anal. Real World Appl. **5/3**, 501–518 (2004)
19. C. Zhu, R.H. Byrd, J. Nocedal, L-bfgs-b: algorithm 778: L-bfgs-b, fortran routines for large scale bound constrained optimization. ACM Trans. Math. Softw. **23**(4), 550–560 (1997)
20. K. Zickfeld, T. Slawig, S. Rahmstorf, A low-order model for the response of the atlantic thermohaline circulation to climate change. Ocean. Dyn. **54**, 8–26 (2004)
21. J.C. Ziems, S. Ulbrich, Adaptive multilevel inexact sqp methods for pde-constrained optimization. SIAM J. Optim. **21**(1), 1–40 (2011)

On a Fully Adaptive SQP Method for PDAE-Constrained Optimal Control Problems with Control and State Constraints

Stefanie Bott, Debora Clever, Jens Lang, Stefan Ulbrich, Jan Carsten Ziems, and Dirk Schröder

Abstract We present an adaptive multilevel optimization approach which is suitable to solve complex real-world optimal control problems for time-dependent nonlinear partial differential algebraic equations with point-wise constraints on control and state. Relying on Moreau-Yosida regularization, the multilevel SQP method presented in Clever et al. (Generalized multilevel SQP-methods for PDAE-constrained optimization based on space-time adaptive PDAE solvers. In: Constrained optimization and optimal control for partial differential equations. Volume 160 of International series of numerical mathematics. Springer, Basel, pp 37–60, 2012) is extended to the state-constrained case. First-order convergence results are shown. The new multilevel SQP method is combined with the state-of-the-art software package KARDOS to allow the efficient resolution of different space and time scales in an adaptive manner. The numerical performance of the method is demonstrated and analyzed for a real-life three-dimensional radiative heat transfer problem modeling the cooling process in glass manufacturing and a two-dimensional thermistor problem modeling the heating process in steel hardening.

Keywords PDAE-constrained optimization • Multilevel optimization • Generalized SQP method • Trust region methods • Control constraints • State constraints • Moreau-Yosida regularization • Adaptive finite elements • Rosenbrock methods • Glass cooling • Radiation • Steel hardening

---

S. Bott • J. Lang • S. Ulbrich (✉)
Department of Mathematics, Graduate School of Computational Engineering, Technische Universität Darmstadt, Dolivostr. 15, 64293 Darmstadt, Germany
e-mail: bott@gsc.tu-darmstadt.de; lang@mathematik.tu-darmstadt.de; ulbrich@mathematik.tu-darmstadt.de

D. Clever • J.C. Ziems • D. Schröder
Department of Mathematics, Technische Universität Darmstadt, Dolivostr. 15, 64293 Darmstadt, Germany
e-mail: clever@mathematik.tu-darmstadt.de; ziems@mathematik.tu-darmstadt.de; schroeder@mathematik.tu-darmstadt.de

G. Leugering et al. (eds.), *Trends in PDE Constrained Optimization*, International Series of Numerical Mathematics 165, DOI 10.1007/978-3-319-05083-6_7

Mathematics Subject Classification (2010). 49J20, 49M25, 65C20, 65K10, 65M60, 90C46, 90C55

1 Introduction

To explore the fundamental scientific issues of high dimensional complex engineering applications such as optimal control problems with time-dependent partial differential algebraic equations (PDAEs) scalable numerical algorithms are requested. This means that the work necessary to solve increasingly larger problems should grow all but linearly – the optimal rate. Therefore, we combine modern numerical solution strategies to solve time-dependent systems of PDAEs such as adaptive multilevel finite element methods and error-controlled linearly implicit time integrators of higher order with novel generalized adaptive multilevel SQP methods. The environment is used to solve two showcase engineering applications arising during the manufacturing process of glass and steel.

In this paper, we present a multilevel algorithm for optimal control problems governed by a system of partial differential equations (PDEs) or partial differential algebraic equations (PDAEs), which do not only contain control constraints but also state constraints. More specifically, we consider

$$\min_{y\in Y, u\in U} J(y,u) \quad \text{s.t.} \quad C(y,u)=0, \quad u\in U_{ad}, \quad y\in Y_{ad}, \tag{P}$$

where $J : Y \times U \to \mathbb{R}$ is the objective function and $C : Y \times U \to V^*$ is the state equation. The control space is a Hilbert space U, for concreteness $U = L^2$, the state space Y and the space of test functions V are Banach spaces. The state equation $C(y,u) = 0$ is assumed to be a system of PDEs or PDAEs in weak form and the control and state constraints are defined by

$$\begin{aligned} Y_{ad} &:= \{y = (y_C, y_F) \in Y \mid y_a \le y_C \le y_b\}, \\ U_{ad} &:= \{u \in U \mid u_a \le u \le u_b\}, \end{aligned}$$

with $y_a, y_b, u_a, u_b \in L^\infty$. Here, y_C are the constrained components of y and we assume that $y \in Y$ implies that $y_C \in C^0$.

For simplicity of presentation, we assume in the following, that all state variables are constrained, i.e. $Y \subset C^0$ and $Y_{ad} := \{y \in Y \mid y_a \le y \le y_b\}$. Our considerations are based on the generalized multilevel adaptive SQP method presented in [5, 31]. It was developed in the first funding period of this program for control-constrained problems, see also the dissertation of Ziems [29] and the papers [30, 32]. The main idea of this "*first-optimize-then-discretize*" – approach is to combine a trust-region SQP method with a successive adaptive grid refinement based on appropriate error estimators. In the generalized multilevel SQP method, the discretized state and adjoint PDE are solved exactly in contrast to the inexact SQP method of Ziems and Ulbrich [29, 30, 32].

In [5,31], the multilevel SQP algorithm has been designed for control constrained optimal control problems. Here, we will involve state constraints by combining this method with the Moreau-Yosida regularization. More precisely, we solve the Moreau-Yosida regularized subproblems with the multilevel SQP method. In order to fulfill the assumptions of the multilevel SQP method, we choose for simplicity a cubic regularization of the state constraints. However, also a quadratic regularization can be used, which leads to a nonquadratic SQP subproblem

Moreover, we extend this approach to a new multilevel SQP method for state constraints. To this end the error control criteria invoke now the penalty parameter in addition to the criticality measure and we couple the penalty parameter update with the criticality measure. This allows us to show a first-order convergence result.

We realize the algorithm by coupling it with the software package KARDOS based on a semi-discretization of Rothe's type. For each PDAE the space-time mesh is refined adaptively to meet a predefined accuracy depending on the optimization progress. Due to the modular structure of the algorithm, we can exploit the structure of each PDE separately. This reduces the computational costs in such a way that we obtain a highly efficient optimization method.

In [5], we have treated control constraints only in a theoretical way. In order to numerically cope with them we use an inexact variant of the projected Newton method [1, 13] that uses an ϵ-active set strategy and a projected BiCGstab method [5, 29]. Our numerical examples, the glass cooling problem and the thermistor problem, contain in parts control and state constraints, PDAEs as state equation and additionally difficult computational domains. Despite these obstacles of real-world applications we see that our algorithm works very well in practice.

The paper is organized as follows. In Sect. 2 we introduce the basic concepts and develop the adaptive multilevel SQP method for state constraints based on Moreau-Yosida regularization. Moreover, we summarize first-order convergence results. In Sect. 3 we describe the realization of the algorithm by using the PD(A)E-solver KARDOS. Finally we present numerical results for the glass cooling problem and the thermistor problem.

2 Adaptive Multilevel Generalized SQP Method

Our goal is to extend the multilevel SQP method of [5, 31] for control-constrained problems to state constrained problems of the form (P). It is well known that the efficient numerical treatment of state constraints is involved, since the Lagrange multipliers for the state constraints are Radon measures. Hence, the complementarity conditions cannot be written in a pointwise fashion. Therefore, we apply Moreau-Yosida regularization to approximate the state constraints by a penalty term. The regularized problem can be solved by the adaptive SQP method of Ziems and Ulbrich [29, 31]. By coupling the adaptive mesh refinement with the adjustment of the penalty parameter we will obtain a new multilevel SQP method for state constraints.

Assumptions We make the assumptions of [31].

- The functionals J and C are twice continuously Fréchet differentiable
- The derivative $C_y(u)$ has a bounded inverse for all $u \in U$
- The functionals J, J_x, J_{xx}, C, C_u are bounded

see [31] for more details. Moreover, we assume that every local solution $\bar{u}$ of (P) satisfies the Slater-type condition

$$\exists u_0 \in U_{ad} : G(\bar{u}) + G'(\bar{u})(u_0 - \bar{u}) \in \operatorname{int}(Y_{ad}).$$

These assumptions ensure in particular that first order necessary conditions hold at local solutions of (P). We assume that they hold throughout this section.

2.1 Moreau-Yosida Regularization

An efficient approach to treat state constraints is a quadratic penalty approach, which is often called Moreau-Yosida regularization in this context. The basic idea is to replace the state constraints by a penalty term [9, 12]. For shorter notation, we consider only state constraints from above, the bilateral case can be treated analogously. We use the penalty function

$$R : Y \to \mathbb{R}, \quad R(y) = \frac{1}{p} \int_{\Omega} (y(\omega) - y_b(\omega))_+^p \, d\omega, \tag{2.1}$$

where $p \geq 2$ and $(\cdot)_+ = \max(0, \cdot)$. While the usual choice $p = 2$ would also be possible in our multilevel method, we use for simplicity the choice $p = 3$, since R is then twice continuously differentiable which allows to work with a quadratic subproblem in the SQP method. We now approximate (P) by the following regularized subproblems for $\gamma > 0$, $p = 3$.

$$\min_{y \in Y, u \in U} J^\gamma(y, u) = J(y, u) + \gamma R(y) \quad \text{s.t.} \quad C(y, u) = 0, \ u \in U_{ad}. \tag{2.2}$$

The basic procedure is now to solve this subproblem for a fixed $\gamma > 0$. Then γ is increased and the new subproblem is solved where the starting iterate is the solution of the previous subproblem. We will embed this approach in a multilevel scheme with error control.

Neitzel and Tröltzsch [24] showed for semilinear parabolic optimal control problems that there exists a sequence of global solutions $(x_\gamma)_{\gamma > 0}$ of (2.2) that converges strongly in L^2 to the original solution. Meyer and Yousept proved in [23] for an optimal control problem governed by the stationary heat equation with radiation that a similar result holds also for a sequence of local solutions.

It can be shown by standard arguments that for a local solution (y_γ, u_γ) of (2.2) the first order necessary optimality conditions are satisfied:

$$\begin{aligned} C(y_\gamma, u_\gamma) &= 0, \\ L_y^\gamma(y_\gamma, u_\gamma, \lambda_\gamma) &= 0, \\ P_{U_{ad}-u_\gamma}(-\nabla_u L^\gamma(y_\gamma, u_\gamma, \lambda_\gamma)) &= 0, \end{aligned} \tag{2.3}$$

where $L^\gamma(y, u, \lambda) = J^\gamma(y, u) + \langle \lambda, C(y, u)\rangle_{V,V^*}$, $\nabla_u L^\gamma(y_\gamma, u_\gamma, \lambda_\gamma) \in U$ denotes the Riesz representation of $L_u^\gamma(y_\gamma, u_\gamma, \lambda_\gamma)$ and $P_{U_{ad}-u_\gamma}$ is the orthogonal projection onto $U_{ad} - u_\gamma$.

2.2 Multilevel Generalized Adaptive SQP Algorithm

Since the penalized subproblems are no longer state constrained we can apply the adaptive SQP method. In this subsection we will give a short summary of this method for the penalized subproblem (2.2) with fixed $\gamma > 0$.

Discretization For simplicity of presentation, we choose a conformal finite element discretization, i.e. the discretized spaces Y_h, U_h, and V_h are subspaces of the infinite dimensional spaces Y, U and V and $U_{ad}^h \subset U_{ad}$, see [29, 31]. We introduce the discretized optimization problem of (2.2) by

$$\min_{y^h \in Y_h, u^h \in U_h} J^\gamma(y^h, u^h) \quad \text{s.t. } C^h(y^h, u^h) = 0, \; u^h \in U_{ad}^h, \tag{2.4}$$

where h indicates the current grid. Denote by $y = G(u)$ and $y^h = G^h(u^h)$ the solution operators of the state equation $C(y, u) = 0$ and the discretized state equation $C^h(y^h, u^h) = 0$, respectively, and by $\lambda = G_{ad,\gamma}(y, u)$ and $\lambda^h = G_{ad,\gamma}^h(y^h, u^h)$ the solution operators of the adjoint equation and the discretized adjoint equation. We assume that the discretization schemes are convergent, i.e., for any $u \in U$ we can generate a sequence of grids h_k such that $\|G^{h_k}(u) - G(u)\|_Y \to 0$ and $\|G_{ad,\gamma}^{h_k}(G^{h_k}(u), u) - G_{ad,\gamma}(G(u), u)\|_V \to 0$ as $k \to \infty$. For simplicity we assume that the control spaces are nested, i.e., $U_{h_k} \subset U_{h_{k+1}} \subset U$.

Reduced trust-region SQP subproblem Let h_k denote the current grid, $u_k \in U_{h_k}$ the current control and $y_k = G^{h_k}(u_k)$, $\lambda_k = G_{ad,\gamma}^{h_k}(y_k, u_k)$ the corresponding discrete state and adjoint state. We obtain the reduced SQP subproblem of (2.4) by approximating the objective function quadratically and the constraints linearly. Thus, the next iterate is computed by $x_{k+1} = (y_{k+1}, u_{k+1})$ with $u_{k+1} = u_k + s_{u,k}$, $y_{k+1} = G^{h_k}(u_{k+1})$, where $s_{u,k}$ is the solution of the reduced SQP problem at (x_k, λ_k) with trust region radius $\Delta_k > 0$

$$\min_{s_{u,k}\in U_{h_k}} q_k^\gamma(s_{u,k}) := J_k^\gamma + \left(\hat{g}_\gamma^{h_k}, s_{u,k}\right)_U + \frac{1}{2}\left\langle s_{u,k}, \hat{H}_k^\gamma s_{u,k}\right\rangle_{U,U^*}$$

$$\text{s.t. } u_k + s_{u,k} \in U_{ad}^{h_k}, \ \|s_{u,k}\|_U \le \Delta_k, \tag{2.5}$$

where $J_k^\gamma = J^\gamma(x_k), \hat{g}_\gamma^{h_k} = \nabla_u L^\gamma(x_k, \lambda_k)$ denotes the Riesz representation of $L_u^\gamma(x_k, \lambda_k)$ and $\hat{H}_k^\gamma$ is an approximation of the reduced Hessian. The discretized state and adjoint equation lead to inexact solutions of the state and adjoint equation, i.e., $\|C(y_k, u_k)\|_{V^*} \le \varepsilon_k^y$, $\|L_y^\gamma(x_k, \lambda_k)\|_{Y^*} \le \varepsilon_k^\lambda$, and we will reduce these errors appropriately within the multilevel method.

2.3 *Multilevel SQP Method for State Constraints*

The aim is to solve the penalized subproblems (2.2) with an adaptive SQP method. Recall that the original adaptive multilevel SQP method of [29, 31] solves problems of the form (2.2) by using subproblems similar to (2.5) but without penalty term in the objective function. It can be shown that the convergence theory of [29, 31] remains valid for the regularized problem (2.2) with fixed γ.

Additionally we will integrate a penalty parameter update and modify the refinement conditions of the grid. For that, let us introduce the function $a : \mathbb{R}^+ \to \mathbb{R}^+$ which is monotone decreasing and which satisfies

$$\lim_{\gamma\to\infty} a(\gamma) = 0. \tag{2.6}$$

2.3.1 Refinement Conditions

The refinement conditions rely on the adaptive SQP method of Ziems and Ulbrich, see [29, 31]. Thus we assume that we have reliable error estimators $\eta_y^{h_k}, \eta_\lambda^{h_k}, \eta_u^{h_k}$ available such that with constants $C_1, C_2, C_3 > 0$ holds

$$\|C(y_k, u_k)\|_{V^*} \le C_1 \eta_y^{h_k}(y_k, u_k), \tag{2.7}$$

$$\|L_y^\gamma(y_k, u_k, \lambda_k)\|_{Y^*} \le C_2 \eta_\lambda^{h_k}(y_k, u_k, \lambda_k), \tag{2.8}$$

$$\|P_{U_{ad}-u_k}(-\hat{g}_\gamma^{h_k}) - P_{U_{ad}^{h_k}-u_k}(-\hat{g}_\gamma^{h_k})\|_U \le C_3 \eta_u^{h_k}(\hat{g}_\gamma^{h_k}, u_k). \tag{2.9}$$

These error estimators can be regarded as a measure for the quality of the discretization. In contrast to the former adaptive SQP method without state constraints, the new refinement conditions are not only dependent on the criticality measure but also on the penalty parameter and the stopping parameter. We now check the following refinement criteria:

$$\eta_y^{h_k}(y_k,u_k) \le \max(\tilde{c}_1 \| P_{U_{ad}^{h_k}-u_k}(-\hat{g}_\gamma^{h_k})\|_U, \tilde{c}_2 a(\gamma), \tilde{c}_3 \varepsilon_{\text{term}}), \tag{2.10}$$

$$\eta_\lambda^{h_k}(y_k,u_k,\lambda_k) \le \max(\tilde{c}_4 \| P_{U_{ad}^{h_k}-u_k}(-\hat{g}_\gamma^{h_k})\|_U, \tilde{c}_5 a(\gamma), \tilde{c}_6 \varepsilon_{\text{term}}), \tag{2.11}$$

$$\eta_u^{h_k}(\hat{g}_\gamma^{h_k},u_k) \le \max(\tilde{c}_7 \| P_{U_{ad}^{h_k}-u_k}(-\hat{g}_\gamma^{h_k})\|_U, \tilde{c}_8 a(\gamma), \tilde{c}_9 \varepsilon_{\text{term}}), \tag{2.12}$$

with fixed $\varepsilon_{\text{term}} > 0$ and fixed constants $\tilde{c}_i > 0, i = 1,\ldots 9$. Thus, the grid will be refined adaptively if the criticality measure is small and the penalty parameter is large compared to the accuracy of the discretization.

2.3.2 Penalty Parameter Update

Each request of a penalty parameter update in one fixed iteration k is called *trial iteration*. We denote the number of trial iterations by l. We denote γ_k as the last updated penalty parameter in a fixed iteration k and $\tilde{\gamma}_{k,l}$ as the trial penalty parameter in the k-th iteration and l-th trial iteration. We set $\tilde{\gamma}_{k,0} = \gamma_{k-1}$.

For the penalty parameter update we check if the criticality measure and the stop tolerance $\varepsilon_{\text{term}}$ are smaller than $a(\tilde{\gamma}_{k,l})$, more precisely

$$a(\tilde{\gamma}_{k,l}) > \max(\| P_{U_{ad}^h - u_k}(-\hat{g}_{\tilde{\gamma}_{k,l}}^{h_k})\|_U, \varepsilon_{\text{term}}). \tag{2.13}$$

If condition (2.13) holds, then we increase the trial penalty parameter, e.g., we set $\tilde{\gamma}_{k,l+1} = \tau \tilde{\gamma}_{k,l}$, for a fixed $\tau > 1$, $l := l+1$ and we go back to the beginning of the algorithm. Otherwise we set $\gamma_k = \tilde{\gamma}_{k,l}$. In order to indicate that the penalty parameter can differ for every SQP subproblem we write (SQP_{γ_k}).

As in the multilevel SQP method of [29, 31] we require that the approximate solution of (2.5) satisfies a generalized Cauchy decrease condition and we evaluate the step depending on the ratio of actual and predicted reduction. If the step is rejected, the inexactness of the reduced gradient $\hat{g}_{\gamma_k}^{h_k}$ is controlled by the gradient accuracy condition

$$\begin{aligned} &| \langle J_x^\gamma(x_k), \hat{s}_k \rangle_{X^*,X} - \left(\hat{g}_\gamma^{h_k}, s_{u,k}\right)_U | \\ &\le \xi_1 \min\{\| P_{U_{ad}^{h_k}-u_k}(-\hat{g}_\gamma^{h_k})\|_U, \Delta_k\} \|s_{u,k}\|_U \end{aligned} \tag{2.14}$$

for some $\xi_1 > 0$ and with the tangential step $\hat{s}_k = ((G^{h_k}(u_k))' s_{u,k}, s_{u,k})$. More details can be found in [29, 31].

2.3.3 Algorithm

By combining these ingredients, we arrive at the following multilevel SQP method, which is also visualized in Fig. 1.

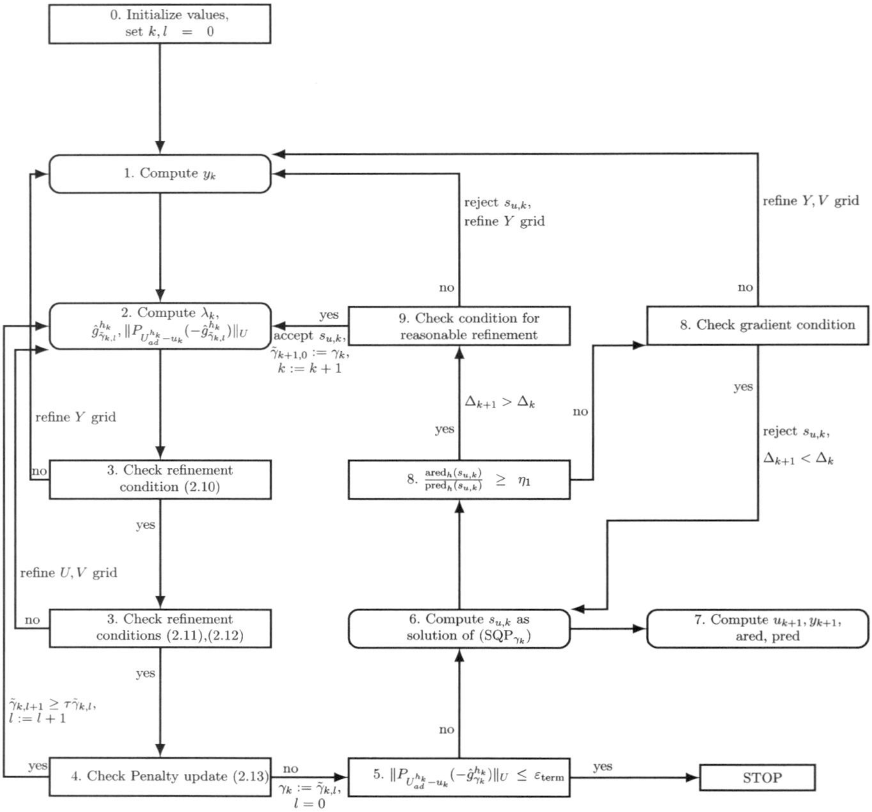

Fig. 1 Flowchart of the multilevel SQP Algorithm 2.1

Algorithm 2.1 (Multilevel SQP algorithm for state constraints).

0. Initialization: Choose $\varepsilon_{\text{term}} > 0$, $\eta_1 \in (0, 1)$, $\tilde{c}_i > 0, i = 1, \ldots 9$, $\tau > 1$, a function a with the property (2.6) and an initial discretization h_0. Set $k, l := 0$. Choose $\gamma_0 > 0$ and a starting control $u_0 \in U^{h_0}_{ad}$. Set $\tilde{\gamma}_{0,0} := \gamma_0$.
   For $k = 0, 1, 2, \ldots$
1. Compute the solution $y_k = G^{h_k}(u_k)$ of the discretized state equation.
2. Compute the solution $\lambda_k = G^{h_k}_{ad,\tilde{\gamma}_{k,l}}(y_k, u_k)$ of the discretized adjoint equation, $\hat{g}^{h_k}_{\tilde{\gamma}_{k,l}} = \nabla_u L^{\tilde{\gamma}_{k,l}}(y_k, u_k, \lambda_k)$ and $\|P_{U^{h_k}_{ad}-u_k}(-\hat{g}^{h_k}_{\tilde{\gamma}_{k,l}})\|_U$.
3. If the refinement conditions (2.10)–(2.12) hold for $\gamma = \tilde{\gamma}_{k,l}$, then go to step 4. Otherwise refine the Y-, V- or U- grid adaptively leading to an updated grid h_k and go to step 1 respectively 2.
4. If (2.13) holds then choose $\tilde{\gamma}_{k,l+1} \ge \tau\tilde{\gamma}_{k,l}$, set $l := l + 1$ and go to step 2. Otherwise set $\gamma_k := \tilde{\gamma}_{k,l}$, $l = 0$ and go to step 5.

5. If $\| P_{U_{ad}^{h_k}-u_k}(-\hat{g}_{\gamma_k}^{h_k})\|_U \leq \varepsilon_{\text{term}}$ then stop and return x_k as approximate solution for (P).
6. Compute the step $s_{u,k}$ as an inexact solution of (SQP$_{\gamma_k}$) satisfying a generalized Cauchy decrease condition.
7. Compute $u_{k+1} := u_k + s_{u,k}$, $y_{k+1} := G^{h_k}(u_{k+1})$, $\text{ared}_h(s_{u,k})$, $\text{pred}_h(s_{u,k})$.
8. If $\frac{\text{ared}_h(s_{u,k})}{\text{pred}_h(s_{u,k})} \geq \eta_1$ then update the trust-region radius (as in [31]), accept the step and go to step 9. Otherwise check the gradient condition (2.14). If it is fulfilled then reject the step $s_{u,k}$ and reduce the trust-region radius (as in [31]) and go back to step 6. Otherwise refine the Y- and V-grid and go to step 1.
9. If a condition for reasonable refinement (see [31]) is satisfied then accept the step $s_{u,k}$ and go to step 2 with $\tilde{\gamma}_{k+1,0} := \gamma_k$ and $k := k+1$. Otherwise reject $s_{u,k}$, refine the Y- grid properly and go to step 1.

Remark 2.2. In contrast to [29], the refinement conditions (2.10)–(2.12) in the multilevel SQP method for state constraints involve additionally $\tilde{c}_i \cdot a(\tilde{\gamma}_{k,l})$ for $i = 2, 5, 8$. Under the assumptions on the discretization (see Assumption 3.2.1 of [31]) step 3 of Algorithm 2.1 is satisfied after finitely many refinements, since $\tilde{\gamma}_{k,l}$ is constant in step 3 and the error estimators become smaller than $\tilde{c}_i \cdot a(\tilde{\gamma}_{k,l})$ for $i = 2, 5, 8$ after finitely many refinements.

Remark 2.3. If we do not update the penalty parameter γ and if we set $\tilde{c}_2, \tilde{c}_3, \tilde{c}_5, \tilde{c}_6, \tilde{c}_8, \tilde{c}_9 = 0$ and $\tilde{c}_1, \tilde{c}_4, \tilde{c}_7 > 0$ arbitrary in Algorithm 2.1 then it coincides with the adaptive SQP method of [29, 31] applied to (2.2). Now the convergence result of [29,31] yields for $\varepsilon_{\text{term}} = 0$ and $x_k = (y_k, u_k)$

$$\liminf_{k\to\infty} \Big(\|C(x_k)\|_{V^*} + \|L_y^\gamma(x_k, \lambda_k)\|_{Y^*} + \|P_{U_{ad}-u_k}(-\nabla_u L^\gamma(x_k, \lambda_k))\|_U \Big) = 0.$$

2.4 Auxiliary Lemmas

We analyze now the interplay of the mesh refinement with the adjustment of the penalty parameter γ_k. The proofs build on the results in [29–32] and will be published in a forthcoming paper.

Lemma 2.4. *For $\varepsilon_{term} > 0$, step 4 of Algorithm 2.1 is satisfied after finitely many trial iterations, i.e. for all $k \in \mathbb{N}$ there exists $l \in \mathbb{N}$ with*

$$a(\tilde{\gamma}_{k,l}) \leq \max(\| P_{U_{ad}^{h_k}-u_k}(-\hat{g}_{\tilde{\gamma}_{k,l}}^{h_k})\|_U, \varepsilon_{term}). \tag{2.15}$$

For $\varepsilon_{term} = 0$, we either obtain the same result as in the first case or we obtain an iterate k with

$$\lim_{l\to\infty} \| P_{U_{ad}^{h_k}-u_k}(-\hat{g}_{\tilde{\gamma}_{k,l}}^{h})\|_U = \lim_{l\to\infty} a(\tilde{\gamma}_{k,l}) = 0. \tag{2.16}$$

The next lemma is needed in order to apply convergence results of Moreau-Yosida-type penalization to the above algorithm.

Lemma 2.5. *Let* $\varepsilon_{term} = 0$. *Then Algorithm 2.1 produces infinitely many penalty parameter updates, i.e. the sequence* $(\tilde{\gamma}_{k,l})_{k,l}$ *is increasing and unbounded.*

Theorem 2.6. *Let* γ_k *be fixed in Algorithm 2.1. If* $\varepsilon_{term} = 0$ *then the algorithm terminates finitely or*

$$\liminf_{k\to\infty} \| P_{U_{ad}^{h_k}-u_k}(-\hat{g}_{\gamma_k}^{h_k})\|_U = 0.$$

If $\varepsilon_{term} > 0$ *then the algorithm stops finitely with* $\| P_{U_{ad}^{h_k}-u_k}(-\hat{g}_{\gamma_k}^{h_k})\|_U \le \varepsilon_{term}$.

2.5 Main Convergence Results

The following convergence results extend our previous results in [29–32] to the case of state constraints. The proofs will be published in a forthcoming paper.

Let us define the residual of the optimality system (2.3) for (2.2)

$$K(y,u,\lambda,\gamma) := \|C(y,u)\|_{V^*} + \|L_y^{\gamma}(y,u,\lambda)\|_{Y^*} + \|P_{U_{ad}-u}(-\nabla_u L^{\gamma}(y,u,\lambda))\|_U.$$

Our first convergence result addresses the case $\varepsilon_{\text{term}} > 0$.

Theorem 2.7. *If* $\varepsilon_{term} > 0$, *Algorithm 2.1 terminates finitely. The last iterate* $(y_k, u_k, \lambda_k) \in Y_{h_k} \times U_{h_k} \times V_{h_k}$ *satisfies with* $x_k = (y_k, u_k)$

$$\begin{aligned} \| P_{U_{ad}^{h_k}-u_k}(-\nabla_u L^{\gamma_k}(y_k,u_k,\lambda_k))\|_U &\le \varepsilon_{term}, \\ \|C(y_k,u_k)\|_{V^*} &\le C_1 \max(\tilde{c}_1,\tilde{c}_2,\tilde{c}_3)\varepsilon_{term}, \\ \|L_y^{\gamma_k}(x_k,\lambda_k)\|_{Y^*} &\le C_2 \max(\tilde{c}_4,\tilde{c}_5,\tilde{c}_6)\varepsilon_{term}, \\ a(\gamma_k) &\le \varepsilon_{term}. \end{aligned}$$

In particular, there holds

$$\begin{aligned} K(y_k,u_k,\lambda_k,\gamma_k) \le &(1 + C_1 \max(\tilde{c}_1,\tilde{c}_2,\tilde{c}_3) \\ &+ C_2 \max(\tilde{c}_4,\tilde{c}_5,\tilde{c}_6) + C_3 \max(\tilde{c}_7,\tilde{c}_8,\tilde{c}_9))\varepsilon_{term}. \end{aligned}$$

Next we state the convergence of the criticality measure along a subsequence of the iterates in the case that $\varepsilon_{\text{term}}$ is zero.

Theorem 2.8. *If* $\varepsilon_{term} = 0$, *the multilevel SQP algorithm for state constraints generates a sequence of iterates that satisfies*

$$\liminf_{k+l\to\infty} \| P_{U_{ad}^{h_k}-u_k}(-\hat{g}^{h_k}_{\tilde{\gamma}_{k,l}})\|_U = 0, \tag{2.17}$$

$$\liminf_{k+l\to\infty} K(y_k, u_k, \lambda_k, \tilde{\gamma}_{k,l}) = 0. \tag{2.18}$$

In order to obtain stronger convergence results we need additional assumptions. Let $f = J \circ G$ denote the reduced objective function. Let $x_{\gamma_k} = (y_{\gamma_k}, u_{\gamma_k})$ denote a sequence of solutions of (2.2) that converges strongly to a solution $\bar{x} = (\bar{y}, \bar{u})$ of the original problem (P), see also Sect. 2.1.

Assumptions

- For $\gamma_k > 0$ sufficiently large, there holds in the solution $(y_{\gamma_k}, u_{\gamma_k})$ of (2.2) a quadratic growth condition:

$$J^{\gamma_k}(y,u) \geq J^{\gamma_k}(y_{\gamma_k}, u_{\gamma_k}) + \beta_P \|u - u_{\gamma_k}\|_U^2,$$

  for all $(y,u) \in Y \times U_{ad}$ with $C(y,u) = 0$, $\|u - u_{\gamma_k}\|_U \leq \delta_P$, where β_P and δ_P are independent of γ_k.
- The sequence u_k of the adaptive SQP method for state constraints satisfies

$$\|u_k - \bar{u}\|_U \leq \frac{\tilde{\delta}_P}{2},$$

  with $\tilde{\delta}_P \leq \delta_P$.
- Furthermore the second-order sufficient condition for u_k holds

$$\left\langle f^{\gamma_k}_{uu}(u_k)(u-u_k), u-u_k \right\rangle_{U^*,U} \geq \epsilon \|u-u_k\|_U^2 \text{ for all } u \in U_{ad},$$

  for $\epsilon > 0$.
- The second derivative of the reduced objective function $f^{\gamma_k}_{uu}(u)$ is bounded for all $u \in U_{ad}$.
- The functions $f^{\gamma_k}_{uu}(u)$, $J^{\gamma_k}_x(y,u)$ and $C_y^{-1}(y,u)C(y,u)$ are Lipschitz continuous.

The first, statement is for example shown for the elliptic case in [16] and [15]. The third statement holds in the elliptic case for u_{γ_k}, see e.g. [16] and [15]. Thanks to this, we obtain the third statement with the aid of the second one.

Now we state a convergence result in the case that there are no control constraints.

Theorem 2.9. *If $U_{ad} = U$ and the above assumptions hold then there exists a $C > 0$ such that it holds*

$$\|u_k - u_{\gamma_k}\|_U \leq C\gamma_k \|\hat{g}^{h_k}_{\gamma_k}\|_{U^*}. \tag{2.19}$$

Thus if $\varepsilon_{term} = 0$ and if we set $a(\gamma_k) = \frac{b(k)}{C\gamma_k}$ with $b : \mathbb{N} \to \mathbb{R}^+$ bounded, then for a subsequence u_{k_n} of u_k it holds

$$\|u_{k_n} - u_{\gamma_{k_n}}\|_U \leq b(k_n).$$

If b is a null sequence then for this subsequence it holds

$$\|u_{k_n} - \bar{u}\|_U \to 0 \, for \, \gamma_{k_n} \to \infty.$$

We consider now the case of control constraints. For that purpose, we additionally need the following requirement.

Assumptions There exists $C > 0$ such that

$$U_{ad} \subset B_C(0) = \{u \in U | \; \|u\|_U \leq C\}.$$

Furthermore, $C_u^*(y, u)$ is bounded for all $(y, u) \in Y \times U_{ad}$ and C_u^*, $J_u^{\gamma_k}$, G_{ad,γ_k} are locally Lipschitz continuous w.r.t. y.

Theorem 2.10. *Let all assumptions above hold then we have*

$$\|u_{\gamma_k} - u_k\|_U^2 \leq C\gamma_k \|P_{U_{ad}^{h_k} - u_k}(-\hat{g}_{\gamma_k}^{h_k})\|_U.$$

Additionally, let a be defined as in Theorem 2.9 then it holds for a subsequence u_{k_n} of u_k

$$\|u_{k_n} - u_{\gamma_{k_n}}\|_U^2 \leq b(k_n). \tag{2.20}$$

If b is a null sequence then for this subsequence it holds

$$\|u_{k_n} - \bar{u}\|_U \to 0 \, for \, \gamma_{k_n} \to \infty.$$

3 Numerical Experiments

To realize the presented multilevel SQP algorithm we couple it with the fully space-time adaptive PDAE-solving environment KARDOS, which is based on a semi-discretization of Rothe's type. Here, we rely on linearly implicit one-step methods of Rosenbrock type to integrate in time and multilevel finite elements to discretize in space. In Sect. 3.1 we give a short overview about the coupling. For more details on the KARDOS-based optimization environment we refer to [2,5]. For a detailed description of KARDOS we refer to [7, 17].

To study the performance of the developed optimization environment, we focus on two real-world applications. In Sect. 3.2 we present the optimization of the cooling process in glass manufacturing [8, 22] and in Sect. 3.3 we discuss the optimization of the heating process in steel hardening [11]. Using an appropriate reduced model, the first application is a suitable test case for an optimal boundary control problem on a three-dimensional computational domain with point-wise constraints on the control. The second application describes a control and state-constrained optimal control problem on a two-dimensional non-convex domain, where the control only acts on a small sub-interval of the boundary.

In both cases we consider a quadratic objective functional, where state and control occur in separated terms, which are twice continuously differentiable on the state and control space, respectively. The glass cooling problem is analyzed in [25–27] with the state space $Y = (W \times V) \cap L^\infty((0,t_e) \times \Omega)^2$, where $V = L^2(0,t_e;H^1(\Omega))$, $W = \{v \in V : \partial_t v \in L^2(0,t_e;V^*)\}$, and the control space $U = H^1(0,t_e)$ of spatially constant boundary controls. For more details we refer to [25–27]. For the Thermistor problem it is not sufficient to consider classical L^2-theory. It is necessary to use L^r-spaces in time and $W^{1,q}$-spaces in space with $r > 2q/(q-2)$ and $q > d = 2$ for the two-dimensional case. Furthermore we have to take into account that the control $u \in L^\infty((0,t_e), L^2(\Gamma_N))$ only acts on a part of the spatial boundary and that it vanishes on another part. For more details see [11].

3.1 Optimization Environment

Reporting on the results of the first funding period, we have presented the strategy of coupling the multilevel SQP algorithm with the space-time adaptive PDAE solver KARDOS in [5]. To tailor the grid refinement in accordance to the current level of accuracy, we build on local error estimates and rely on the principle of tolerance proportionality in time and hierarchical basis techniques in space. However, the implementation we have presented in [5] only allows for problems without additional constraints on control and state.

To include point-wise constraints on the control within the KARDOS-based environment we use a projected Newton method for the approximate solution of the SQP subproblem based on the ϵ-active set strategy presented in [5], which has been first introduced by Bertsekas [1] and further developed by Kelley [13] and Ziems [29]. The main idea is to solve the Newton problem only on an underestimated inactive set and combine it with a gradient step on the active part. Since the implemented SQP method is matrix free, the reduced Hessian can only be applied to a direction. Therefore, the reduced Hessian cannot be restricted directly, as it has been suggested in [13]. To this end, we modify the linear solver as suggested in [29] for CG. However, as in [5] we use BICGSTAB as inner iteration, since we use an optimize-then-discretize approach which can lead to a slightly non-symmetric reduced Hessian on the discrete level.

To include point-wise constraints on the state, we take advantage of a cubic Moreau-Yosida regularization as described in Algorithm 2.1. The regularized objective and all resulting modifications in adjoint system, adjoint-for-Hessian system, and sensitivities are handled by the PDAE-solver KARDOS directly. Hence, except for the refinement strategy (2.10)–(2.12), there is no difference of the actual optimization code for the case with or without point-wise constraints on the state. In particular there is no modification of the inner iteration. Note, that in the KARDOS-implementation we divide the global estimates $\eta_y^{h_k}, \eta_\lambda^{h_k}$ for state and adjoint error in a temporal and a spatial part and that the control discretization is inherited by the state discretization. Therefore we can neglect (2.12) and implement (2.10) and (2.11) by the four checkable conditions

$$\eta_y^{\{t/x\}} \leq \max\{c_y^{\{t/x\}} \| P_{U_{ad}^{h_k} - u_k}(-\hat{g}_\gamma^{h_k}) \|_U, \tilde{c}_2 a(\gamma), \tilde{c}_3 \varepsilon_{\text{term}}\} \tag{3.1a}$$

$$\eta_\lambda^{\{t/x\}} \leq \max\{c_\lambda^{\{t/x\}} \| P_{U_{ad}^{h_k} - u_k}(-\hat{g}_\gamma^{h_k}) \|_U, \tilde{c}_5 a(\gamma), \tilde{c}_6 \varepsilon_{\text{term}}\}. \tag{3.1b}$$

3.2 Glass Cooling Problem

The aim of the cooling process in glass manufacturing is to track the glass temperature distribution as close as possible to a desired profile, for which good performance of the involved chemical processes is known. The cooling process itself takes place in a preheated furnace, where the spatially constant furnace temperature acts as control on the glass surface. Due to the physically given operation interval of the furnace the control has to be restricted to the feasible set $U_{\text{ad}} := [300, 1{,}200]$. For regularization reasons, we additionally include a tracking of the glass temperature at final time and a tracking of the control to a given guideline. Due to the high temperatures that occur especially at the beginning of the cooling process, the direction- and frequency-dependent thermal radiation field and the spectral radiative properties of semi-transparent glass play a dominant role.

As state system, which describes the behavior of the glass temperature with respect to the control, we consider the well studied Gray-Scale-Model on a space-time cylinder with three dimensional spatial domain. In this model, radiation is described by a mean intensity, averaging the dependency on wavelength and frequency [14]. It leads to a system of PDEs of differential-algebraic type with a highly nonlinear coupling of state and control. For an analysis of the well-posedness of the problem, we refer to [25]. For more details on the model, the objective function and the parameters, we refer to [2].

In the following, we examine the performance of the 3d-optimization environment on the computational domain $\Omega_{3\text{d}}$, which is given by the convex hull of the eight points

$$p_1 = (-1, -1, -1), \quad p_2 = (1, -1, -1), \quad p_3 = (1, 1, -1), \quad p_4 = (-1, 1, -1),$$
$$p_5 = (0.5, 0.5, 1), \quad p_6 = (1, 0.5, 1), \quad p_7 = (1, 1, 1), \quad p_8 = (0.5, 1, 1),$$

see Fig. 2a. We use linear finite elements and the third-order Rosenbrock method ROS3PL [18,19] to solve the underlying PDAEs. On each level of accuracy the grids are adaptively refined to meet the accuracy requirements of the current level, using

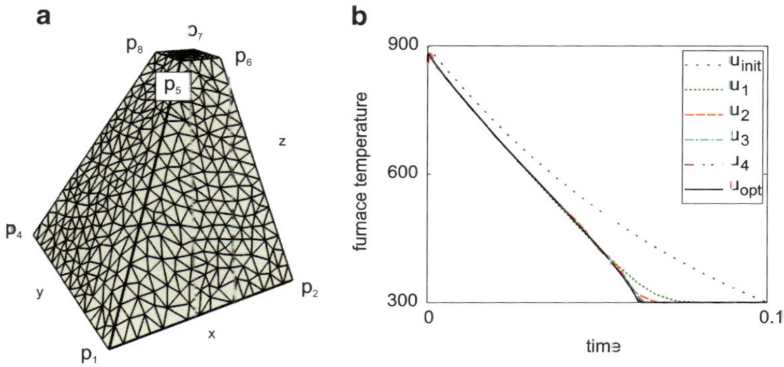

Fig. 2 Spatial domain and optimal control. (**a**) Three dimensional computational domain Ω_{3d} with initial grid. (**b**) Control iterates and optimal control

the refinement constants $c_y^t = 5.0\text{e}{-3}$, $c_y^x = 2.5\text{e}{-2}$, $c_\lambda^t = 5.0\text{e}{-2}$, $c_\lambda^x = 5.0\text{e}{-1}$, and $\tilde{c}_2 = \tilde{c}_3 = \tilde{c}_5 = \tilde{c}_6 = 1.0\text{e}{-5}$. The optimization starts on a common time grid of 17 nodes and an initial spatial grid with 1,399 spatial nodes for the state system. To meet the required accuracy in the adjoint system, the number of spatial nodes varies between 1,399 and 3,672. The optimization algorithm terminates after five optimization iterations, on a space-time mesh with 73 time steps and up to 72,987 spatial nodes. Then, the objective value is reduced by more than 50 % and the criticality measure by approximately three orders of magnitude.

The optimal control $u_{\text{opt}} := u_5$ and all previous control iterates are shown in Fig. 2b. The level lines of the glass temperature at final time resulting from the optimal control u_{opt} through three representative cuts through the geometry are shown in Fig. 3. On the left (Fig. 3a) we study the glass temperature evolution at the bottom of the pyramid. In the middle we see the cut through $z = \frac{1}{3}$ (Fig. 3b) and on the right we cut the geometry at $y = 0.5$ which is a vertical cross section through the points p_5 and p_6 (Fig. 3c). In all figures, the level lines have a distance of ten. Hence, the closer the lines are in a certain region, the greater are the temperature changes within this region. It can be seen that the optimal control enforces a final state that is close to the desired value of 300 in terms of an integral mean and is distributed quite homogeneously.

In Table 1 we present the reduction of objective values and criticality measures during the optimization. In column 4 and 5, we show the number of grid points of the state and adjoint space-time mesh, autonomously controlled by the multilevel strategy. Comparing the coarsest level of accuracy to the finest, we see an increase of the number of grid points by a factor of 60. In the fifth column we present the number of inner iterations, which we restrict to a maximum of 5. Whereas in the first two outer iterations the inner iteration terminates due to a small residuum in the last three outer ones it terminates due to the number of maximum iterations. In the last column, we show the numerical effort in each optimization iteration. It is

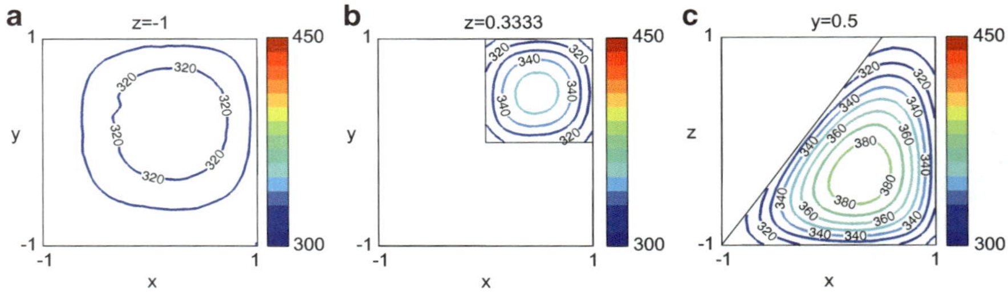

Fig. 3 Glass temperature distribution resulting from optimal control u_{opt}. (**a**) Bottom. (**b**) Cross section. (**c**) Vertical cut

Table 1 Optimization protocol, Glass Cooling Problem

Opt. it.	Objective	Crit. meas.	dof_y	dof_λ	#it. BiCGStab	cpt (%)
0	1.4602e+03	4.0245e+01	23,783	26,056	–	0.001
1	5.4880e+02	2.8733e+00	23,783	25,683	2	0.04
1	5.5664e+02	2.9758e+00	49,869	53,775	–	0.04
2	5.4908e+02	3.4451e−01	49,869	51,704	5	0.35
2	5.4507e+02	9.5386e−01	139,208	144,550	–	0.07
3	5.4478e+02	3.0787e−01	148,263	151,285	5	1.05
3	5.5319e+02	5.4405e−01	426,208	426,192	–	0.50
4	5.5302e+02	1.0118e−01	428,554	428,538	5	9.96
4	5.6995e+02	8.5377e−01	1,716,852	1,713,806	–	2.84
5	5.6997e+02	6.8337e−02	1,756,763	1,753,717	5	85.10

remarkable that the last optimization iteration on the finest level requires more than 85 % of the entire computing time, whereas the other four iterations need less than 15 %.

For more details on the model and the involved parameters and on further numerical experiments including a wider class of objective functionals and different approximative models for the radiative heat transfer equations we refer to [2–4].

3.3 Thermistor Problem

The task is to heat the teeth of a steel rack up to a desired temperature profile by induction of a direct current on a part of the boundary. In a second step the steel rack is cooled immediately such that it gets hardened.

The non-convex computational space-domain together with the initial grid is shown in Fig. 4a, top. The current is induced on Γ_N. The temperature is tracked

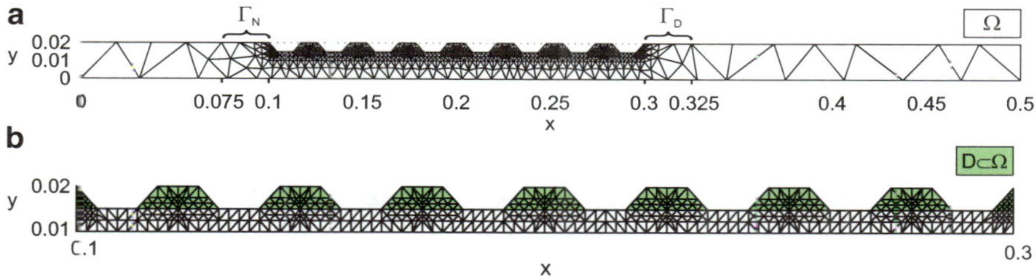

Fig. 4 Steel rack with initial grid. (**a**) Domain Ω with boundary Γ_N and Γ_D. (**b**) Zoom including sub-domain $D := \Omega \cap ((0.1, 0.3) \times (0.015, 0.02))$ (*highlighted*)

within $D := \Omega \cap [(0.1, 0.3) \times (0.015, 0.02)]$, which is the domain covered by the teeth, see Fig. 4b. The rack is quite similar to the one considered in [11]. However, we cut the tips of the teeth to obtain a more realistic shape and substitute the 19 teeth by just 7 teeth, which is only motivated by reasons of visualization. Note, that the initial grid is significantly refined in the region of the teeth.

In the following we determine an optimal control for the first part of the process, where heating takes place. The state $y := (\theta, \varphi)$ consists of the steel temperature θ and the electrical potential φ. The control is the current u. The state system, which describes the temperature evolution in accordance to the induced current, is given by the following system of partial differential algebraic equations:

$$C\rho\partial_t\theta - \nabla\cdot(\kappa\nabla\theta) = (\sigma(\theta)\nabla\varphi)\cdot\nabla\varphi \quad \text{in } Q \tag{3.2a}$$

$$-\nabla\cdot(\sigma(\theta)\nabla\varphi) = 0 \quad \text{in } Q \tag{3.2b}$$

$$n\cdot(\kappa\nabla\theta) = \alpha(\theta_l - \theta) \quad \text{in } \Sigma \tag{3.2c}$$

$$n\cdot(\sigma(\theta)\nabla\varphi) = C_u u \quad \text{in } \Sigma_N \tag{3.2d}$$

$$n\cdot(\sigma(\theta)\nabla\varphi) = 0 \quad \text{in } \Sigma_R \tag{3.2e}$$

$$\varphi = 0 \quad \text{in } \Sigma_D \tag{3.2f}$$

$$\theta(x, 0) = \theta_0 \quad \text{in } \Omega \tag{3.2g}$$

It couples the instationary heat equation (3.2a) with the quasi-static potential equation (3.2b), both defined on the space-time cylinder $Q := \Omega \times (0, t_e)$. In contrast to the boundary conditions for the heat equation (3.2c), which are defined on $\Sigma = \partial\Omega \times (0, t_e)$, the boundary conditions for the potential equation are defined piecewise. To model the induction of a direct current on the boundary part Γ_N, we use a Neumann condition with right hand side $C_u u$ on $\Sigma_N := \Gamma_N \times (0, t_e)$. The anode on the boundary part Γ_D is modeled by a Dirichlet condition on $\Sigma_D := \Gamma_D \times (0, t_e)$. The remaining boundary $\Gamma_R := \partial\Omega \backslash (\bar{\Gamma}_N \cup \Gamma_D)$ is assumed to be isolated on $\Sigma_R := \Gamma_R \times (0, t_e)$.

Here, $C = 470\,\text{J/kg K}$ is the heat capacity, $\rho = 7{,}900\,\text{kg/m}^3$ the density, $\kappa = 50\,\text{W/m K}$ the heat conduction coefficient, and $\alpha = 20\,\text{W/m}^2\,\text{K}$ the heat transfer coefficient. The electric conductivity σ depends non-linearly on θ and is modeled by $\sigma(\theta) = (a + b\theta + c\theta^2 + d\theta^3)^{-1}$, with $a = 4.9659\text{e}{-7}$, $b = 8.4121\text{e}{+10}$, $c = -3.7246\text{e}{-13}$, $d = 6.1960\text{e}{-17}$.

As a modification to the model presented in [11], we substitute u_{orig} by $C_u u$, with $C_u = 1.0e + 3$. The constant C_u is introduced to balance the orders of magnitude of the involved quantities, especially the control u and the reduced gradient $\nabla \hat{J}(u)$.

The optimal control problem is then given by

$$\min_{(y,u)} J(y,u) = \frac{\delta_y}{2} \int_0^{t_e} \int_{D\subset\Omega} (\theta(x,t) - \theta_d(t))^2 \mathrm{d}x\mathrm{d}t + \frac{\delta_e}{2} \int_{D\subset\Omega} (\theta(x,t_e) - \theta_d(t_e))^2 \mathrm{d}x + \frac{\delta_u}{2} \int_0^{t_e} \int_{\Gamma_N} u(t)^2 \mathrm{d}x\mathrm{d}t, \tag{3.3}$$

subject to the state system (3.2), and

$$\theta(x,t) \le \theta_{\max}(x,t), \quad \text{a.e. in } Q \tag{3.4}$$

$$0 \le u(t) \le u_{\max}(t), \quad \text{a.e. in } [0, t_e]. \tag{3.5}$$

The steel temperature θ is tracked to a desired profile θ_d on the open part $D \subset \Omega$ with special weighting at final time t_e. Furthermore, we include a Tikhonov regularization for the control u, where u describes a scaled current induced to the Neumann boundary Γ_N. The upper control bound $u_{\max}$ reflects the technical limits for the control, the upper state bound $\theta_{\max}$ prevents the material from melting. The desired state θ_d is set spatially constant to enforce a homogeneous temperature distribution, which is especially important at final time t_e, as the terminal state of the heating phase serves as initial state for the cooling phase.

The non-negative parameters $\delta_y, \delta_e, \delta_u$, allow for a different weighting within the objective. Whereas we assume $\delta_e > 0$ and $\delta_u > 0$, the parameter δ_y may be set to zero. For an analysis of the well-posedness of the problem we refer to [11].

As described in Sects. 1 and 2 we rely on the Moreau-Yosida regularization to define a regularized objective $J_\gamma(y,u)$, with upper bound $\theta_{\max}$. Furthermore, we consider the initial value $\gamma_{0,0} = 1.0$, the factor $\tau = 10.0$ and the function $a(\gamma) = 500\gamma^{-\frac{1}{3}}$ to control the autonomous increase of the penalty parameter $\gamma_{k,l}$.

In [11] it has been observed, that if an upper state constraint of 1,800 K is considered, a process time of 2 s is too short to force the final steel temperature to the desired profile. Since we have made the same observation for our setting, we consider a final time of $t_e = 4\,\text{s}$.

Using the parameters given in Table 2, the optimization starts on a space-time mesh with 35 nodes in time and 677 nodes in space. During the optimization the number of nodes is increased by approximately one order of magnitude. However, due to the fact that the initial grid is significantly refined in the region of the teeth already, there is no additional mesh refinement necessary. As in the previous example we set $\tilde{c}_2 = \tilde{c}_3 = \tilde{c}_5 = \tilde{c}_6 = 1.0e{-}5$.

The penalty term $R(\theta) = \frac{1}{3}\int_0^{t_e}\int_\Omega(\max(0, \theta(x,t) - \theta_{max}(x,t)))^3 dx dt$ has an initial value of 3.2761e+5 and is reduced by more than nine orders of magnitude to a value of 2.5014e−4, see Table 3.

We terminate the optimization if the criticality measure cm_k descends below $\varepsilon_{term} = 10^{-2}cm_0 + 10^{-2}$ and if the maximal penalization of the state constraints is less then 1 % of the maximal value of the upper bound θ_{max}. In this setting this means less than 18 K.

The resulting optimal control is visualized in Fig. 5a. Comparing the control profile obtained with state constraints to that without state constraints, it can be seen that the maximal induced current is reduced. To achieve a proper final state, the current is hold almost constant between 2.5 and 3.5 s, whereas in the unconstrained case it decreases faster.

Heating the steel rack with the optimal control presented in Fig. 5a the maximal violation of the upper state bound occurs at time $t = 2.10$ s in the point $(2.93e-01, 1.50e-02)$, which is located on the right of the teeth, see Fig. 6a. There, the steel takes a temperature of 1,817.3 K, which violates the upper bound by 0.96 %.

Whereas in the state constrained case the optimal control results in a penalty term of $R(\theta) = 2.50e{-}4$ and a maximal violation of 1.73e+1 K, in the case without state

Table 2 Problem and model specific qualities

U_{ad}	Feasible control set	$[0, 5.0e+4]$ K	$\gamma_{0,0}$	Initial penalty para.	1.0e+0
θ_{max}	Upper state bound	1.8e+3 K	$tol^t_{y,0}$	Initial time tol., state	5.0e−5
t_e	Final time	4.0e+0 s	$tol^x_{y,0}$	Initial space tol., state	5.0e−3
$\theta_0(x)$	Initial steel temp.	2.9e+2 K	$tol^t_{\lambda,0}$	Initial time tol., adj.	5.0e−5
$\theta_d(t_e)$	Desired final steel temp.	1.5e+3 K	$tol^x_{\lambda,0}$	Initial space tol., adj.	1.0e−2
			c^t_y	Time ref. const., state	1.0e−6
$u_0(t)$	Initial control	$5.0e+4(t_e - t)/t_e$ K	c^x_y	Space ref. const.,state	1.0e−4
δ_y	State reg. weight	0.0	c^t_λ	Time ref. const., adj.	1.0e−5
δ_e	Final value weight	1.0e+0	c^x_λ	Space ref. const., adj.	5.0e−4
δ_u	Control reg. weight	1.0e−10	$\varepsilon_{term,a}$	Abs. term. parameter	1.0e−2
Δ_k	Trust region	5.0e+4	$\varepsilon_{term,r}$	Rel.term. parameter	1.0e−2

Table 3 Optimization protocol, Thermistor Problem

γ	Oit	Iit	Objective	Pen. term	Max. pen.	Crit. meas.	dof_y	cpt (%)
1e+0	0	0	3.2769e+5	3.2761e+5	9.2458e+2	1.4979e+2	2.37e+4	0.17
1e+1	2	10	1.2535e+5	1.2534e+4	4.9127e+2	1.2846e+2	2.37e+4	2.66
1e+2	3	15	1.3885e+4	1.3884e+2	2.4801e+2	3.4204e+1	2.64e+4	4.33
1e+3	2	10	1.3390e+4	1.3389e+1	1.5877e+2	3.8208e+1	7.24e+4	8.85
1e+4	3	15	2.5161e+3	2.5155e−1	6.3363e+1	1.5264e+1	6.97e+4	11.45
1e+5	3	15	2.6217e+2	2.6115e−3	2.4467e+1	3.9273e+0	2.34e+5	44.57
1e+6	2	10	2.5128e+2	2.5014e−4	1.7350e+1	3.3702e+0	2.32e+5	27.97

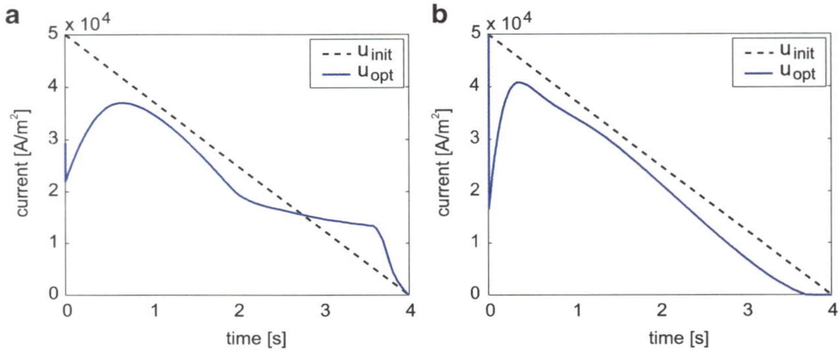

Fig. 5 Initial and optimal control with and without state constraints. (**a**) State constrained. (**b**) Without state constraints

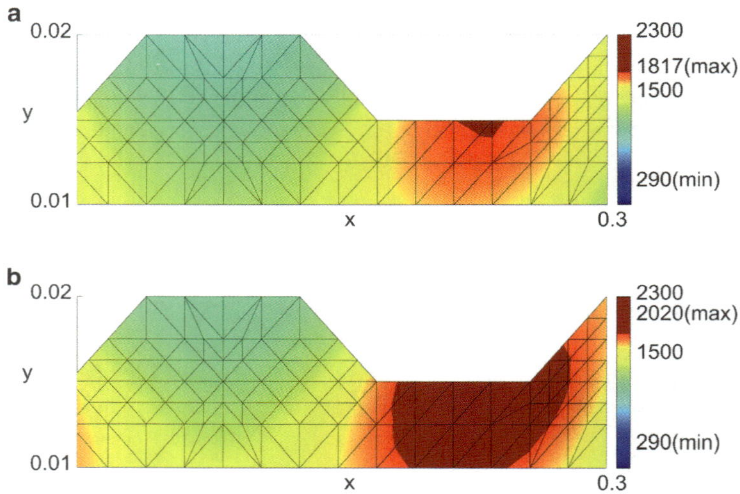

Fig. 6 Steel temperature (in (K)) at the point of maximal violation $x = (2.93e{-}01, 1.50e{-}02)$ at time $t = 2.1$ s with state constraints and $t = 2.16$s without state constraints. (**a**) State constrained. (**b**) Without state constraints

constraints we have $R(\theta) = 3.61\text{e}{-}1$ and a maximal violation of 2.20e+2 K. This maximal violation of more than 10 % occurs at time $t = 2.16$ s, also in the point $(2.93e-01, 1.50e-02)$, see Fig. 6b. It can also be seen, that the area of violation is much greater in the unconstrained case than in the constrained case, compare dark regions around $(2.93e-01, 1.50e-02)$ in Fig. 6a, b.

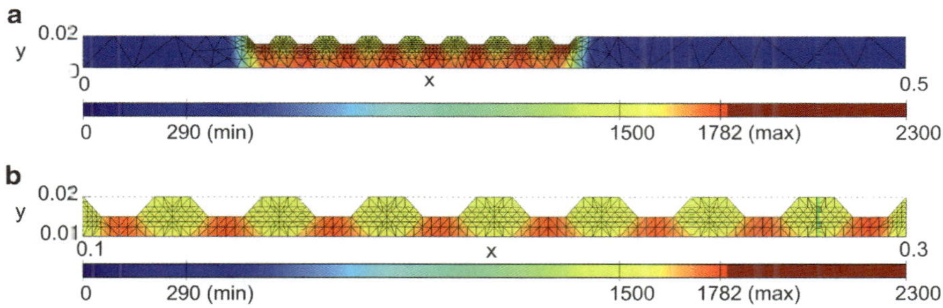

Fig. 7 Optimal steel temperature (in (K)) at final time in the state-constrained case. Note, that D covers only the region of the teeth. (**a**) Computational domain Ω. (**b**) Zoom including subdomain D

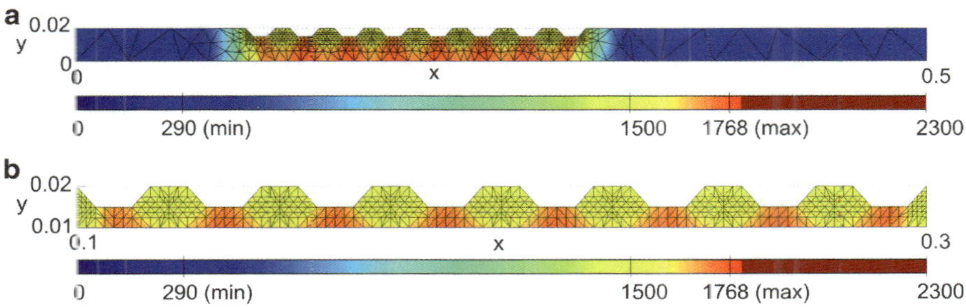

Fig. 8 Optimal steel temperature (in (K)) at final time in the unconstrained case. Note, that D covers only the region of the teeth. (**a**) Computational domain Ω. (**b**) Zoom including subdomain D

In both cases the desired temperature of 1,500 K on the subdomain D is reached quit well, see Figs. 7b and 8b. Note, that D covers only the region of the teeth, which is the area between $y = 0.015$ and $y = 0.02$. The value of the tracking term $\frac{\delta_e}{2} \int_{D \subset \Omega} (\theta(x, t_e) - \theta_d(t_e))^2 \mathrm{d}x$ in the state-constrained case is only less than one order of magnitude higher than in the unconstrained case. This is achieved by the optimized control profile in the constrained case (Fig. 5a) and the sufficient long process time.

A detailed protocol of the optimization including the number of outer and inner iterations (oit and iit), objective values, penalty terms, criticality measure and relative computing time (cpt) is given in Table 3. Since in the presented computations the dimension of the space-time state-grid coincides with the space-time adjoint-grid, we only include the value for dof_y within the table. As in the first example, we again restrict the number of inner iterations to five BiCGStab-iterations, which is exploited in all iterations. It can be observed that for each new γ_k the sub-optimization only needs two to three iterations to achieve a proper decrease.

Conclusions

We have presented an adaptive multilevel generalized SQP-method to solve PDAE-constrained optimal control problems with point-wise constraints on control and state. To this end we combine the Moreau-Yosida regularization [23, 24] with the adaptive multilevel trust-region SQP-method of Ziems and Ulbrich, see [5, 29, 31]. In order to apply the adaptive SQP-method to the regularized subproblems (P_γ) we ensure that (P_γ) satisfies the assumptions of the adaptive SQP-method. We introduce an augmented refinement criteria and a penalty parameter update, such that global convergence of the finite dimensional iterates to an infinite dimensional stationary point can be shown.

Relying on continuous adjoint calculus, we realize the multilevel SQP-method for state constraints by coupling it with the space-time adaptive PDAE-solving environment KARDOS [7]. To meet the accuracy requirements given by the multilevel strategy, we use global a posteriori error estimators based on local strategies and the principle of tolerance proportionality in time and hierarchical basis techniques in space [6, 20, 28].

Two real-world applications serve as test cases to study the performance of the developed optimization environment: the cooling process in glass manufacturing [8, 22] and the heating process in steel hardening [11]. The numerical experiments show, that the combination of space-time adaptivity and the multilevel strategy has a high potential to solve real-world applications even in three spatial dimensions and on complicated spatial domains. The inclusion of the Moreau-Yosida regularization results in a robust and reliable implementation of point-wise state constraints. The efficiency is improved by approaching feasibility mainly on coarse meshes. In the case of state constraints of order 0, the cubic penalization has turned out to be a convenient tool to ensure twice continuous Fréchet differentiability. However, the situation gets more complicated in the case of state constraints of order one, as they naturally occur for the glass cooling problem. In this case, it seems reasonable to combine a quadratic regularization with a semi-smooth Newton method, see e.g. [10].

Further improvements with respect to the numerical efficiency can be obtained by using reduced order models (ROMs) in the multilevel method. While the FEM-solutions on the sequence of suitably refined grids serve as an approximation of the PD(A)E-solutions, the ROM-solutions serve as an approximation of their FEM-counterparts. A sophisticated coupling of full space-time adaptivity and ROMs is work in progress.

Even though the presented multilevel SQP-method for state constraints controls the inexact reduction by a multilevel strategy, it is of numerical benefit with respect to efficiency and robustness to apply discrete adjoint discretization techniques, such that discrete-adjoint and adjoint-discretization

(continued)

commute. Promising progress has, for example, been made in the context of discrete-adjoint W-methods, see [21].

Acknowledgements The authors gratefully acknowledge the support of the German Research Foundation (DFG) within the Priority Program 1253 "Optimization with Partial Differential Equations" under grants LA1372/6-2 and UL158/7-2 and the support by the 'Excellence Initiative' of the German Federal and State Governments within the Graduate School of Computational Engineering at Technische Universität Darmstadt.

References

1. D.P. Bertsekas, Projected Newton methods for optimization problems with simple constraints. SIAM J. Control Optim. **20**, 221–246 (1982)
2. D. Clever, Adaptive multilevel methods for PDAE-constrained optimal control problems. PhD thesis, Technische Universität Darmstadt, 2013. Verlag Dr. Hut
3. D. Clever, J. Lang, Optimal control of radiative heat transfer in glass cooling with restrictions on the temperature gradient. Optim. Control Appl. Methods **33**(2), 157–175 (2012)
4. D. Clever, J. Lang, D. Schröder, Model hierarchy based optimal control of radiative heat transfer. Int. J. Comput. Sci. Eng. **9**(5/6), 509–525 (2014)
5. D. Clever, J. Lang, S. Ulbrich, J.C. Ziems, Generalized multilevel SQP-methods for PDAE-constrained optimization based on space-time adaptive PDAE solvers, in *Constrained Optimization and Optimal Control for Partial Differential Equations*. Volume 160 of International Series of Numerical Mathematics (Springer, Basel, 2012), pp. 37–60
6. K. Debrabant, J. Lang, On global error estimation and control of finite difference solution for parabolic equations, in *Adaptive Modeling and Simulation* (International Center for Numerical Methods in Engineering (CIMNE), Barcelona, 2013), pp. 187–198
7. B. Erdmann, J. Lang, R. Roitzsch, KARDOS-User's Guide. Manual, Konrad-Zuse-Zentrum Berlin, 2002
8. M. Frank, A. Klar, Radiative heat transfer and applications for glass production processes, in *Mathematical Models in the Manufacturing of Glass*. Lecture Notes in Mathematics (Springer, Berlin/Heidelberg, 2011), pp. 57–134
9. M. Hintermüller, K. Kunisch, Path-following methods for a class of constrained minimization problems in function space. SIAM J. Optim. **17**(1), 159–187 (2006)
10. M. Hintermüller, K. Kunisch, PDE-constrained optimization subject to pointwise constraints on the control, the state, and its derivative. SIAM J. Optim. **20**(3), 1133–1156 (2009)
11. D. Hömberg, C. Meyer, J. Rehberg, W. Ring, Optimal control for the thermistor problem. SIAM J. Control Optim. **48**(5), 3449–3481 (2010)
12. K. Ito, K. Kunisch, Semi-smooth Newton methods for state-constrained optimal control problems. Syst. Control Lett. **50**(3), 221–228 (2003)
13. C.T. Kelley, *Iterative Methods for Optimization* (SIAM, Philadelphia, 1999)
14. A. Klar, E.W. Larsen, G. Thömmes, New frequency-averaged approximations to the equations of radiative heat transfer. SIAM J. Appl. Math. **64**(2), 565–582 (2003)
15. K. Krumbiegel, I. Neitzel, A. Rösch, Sufficient optimality conditions for the Moureau-Yosida-type regularization concept applied to semilinear elliptic optimal control problems with pointwise state constraints. Math. Appl. Ann. AOSR **2**(2), 222–246 (2010)
16. K. Krumbiegel, I. Neitzel, A. Rösch, Regularization for semilinear elliptic optimal control problems with pointwise state and control constraints. Comput. Optim. Appl. **52**(1), 181–207 (2012)

17. J. Lang, *Adaptive Multilevel Solution of Nonlinear Parabolic PDE Systems* (Springer, Berlin/New York, 2001)
18. J. Lang, D. Teleaga, Towards a fully space-time adaptive FEM for magnetoquasistatics. IEEE Trans. Magn. **44**, 1238–1241 (2008)
19. J. Lang, J. Verwer, ROS3P – an accurate third-order Rosenbrock solver designed for parabolic problems. BIT Numer. Math. **41**(4), 731–738 (2001)
20. J. Lang, J. Verwer, On global error estimation and control for initial value problems. SIAM J. Sci. Comput. **29**, 1460–1475 (2007)
21. J. Lang, J. Verwer, W-methods in optimal control. Numer. Math. **124**, 337–360 (2013)
22. E.W. Larsen, G. Thömmes, A. Klar, M. Seaïd, T. Götz, Simplified P_N approximations to the equations of radiative heat transfer and applications. J. Comput. Phys. **183**, 652–675 (2002)
23. C. Meyer, I. Yousept, State-constrained optimal control of semilinear elliptic equations with nonlocal radiation interface conditions. SIAM J. Control Optim. **48**(2), 734–755 (2009)
24. I. Neitzel, F. Tröltzsch, On convergence of regularization methods for nonlinear parabolic optimal control problems with control and state constraints. Control Cybern. **37**(4), 1013–1043 (2008)
25. R. Pinnau, Analysis of optimal boundary control for radiative heat transfer modeled by the SP_n-system. Commun. Math. Sci. **5**(4), 951–969 (2007)
26. R. Pinnau, A. Schulze, Newton's method for optimal temperature-tracking of glass cooling processes. IPSE **15**(4), 303–323 (2007)
27. R. Pinnau, G. Thömmes, Optimal boundary control of glass cooling processes. Math. Methods Appl. Sci. **27**(11), 1261–1281 (2004)
28. L.F. Shampine, Tolerance proportionality in ODE codes, in *Numerical Methods for Ordinary Differential Equations*. Lecture Notes in Mathematics, vol. 1386 (Springer, Berlin/Heidelberg, 1989), pp. 118–136
29. J.C. Ziems, Adaptive multilevel SQP-methods for PDE-constrained optimization. PhD thesis, Technische Universität Darmstadt, 2010. Verlag Dr. Hut
30. J.C. Ziems, Adaptive multilevel inexact SQP-methods for PDE-constrained optimization with control constraints. SIAM J. Optim. **23**(2), 1257–1283 (2013)
31. J.C. Ziems, S. Ulbrich, Adaptive multilevel generalized SQP-methods for PDE-constrained optimization. Technical report, Department of Mathematics, TU Darmstadt, 2011, submitted
32. J.C. Ziems, S. Ulbrich, Adaptive multilevel inexact SQP-methods for PDE-constrained optimization. SIAM J. Optim. **21**(1), 1–40 (2011)

Optimal Control of Nonlinear Hyperbolic Conservation Laws with Switching

Sebastian Pfaff, Stefan Ulbrich, and Günter Leugering

Abstract We consider optimal control problems governed by nonlinear hyperbolic conservation laws at junctions and analyze in particular the Fréchet-differentiability of the reduced objective functional. This is done by showing that the control-to-state mapping of the considered problems satisfies a generalized notion of differentiability. We consider both, the case where the controls are the initial and the boundary data as well as the case where the system is controlled by the switching times of the node condition. We present differentiability results for the considered problems in a quite general setting including an adjoint-based gradient representation of the reduced objective function.

Keywords Optimal control • Scalar conservation law • Network

1 Introduction

This paper serves as a final report of the project *Optimal Control of Switched Networks for Nonlinear Hyperbolic Conservation Laws*. In this work we consider optimal control problems for entropy solutions of hyperbolic conservation laws involving objective functionals of the form

$$J(y(u)) := \int_a^b \psi(y(\bar{t}, x; u), y_d(x))\, \mathrm{d}x, \tag{1.1}$$

S. Pfaff (✉) • S. Ulbrich
Department of Mathematics, Technische Universität Darmstadt, Dolivostr. 15, 64293 Darmstadt, Germany
e-mail: pfaff@mathematik.tu-darmstadt.de; ulbrich@mathematik.tu-darmstadt.de

G. Leugering
Department of Mathematics, Universität Erlangen-Nürnberg, Cauerstr. 11, 91058 Erlangen, Germany
e-mail leugering@math.fau.de

G. Leugering et al. (eds.), *Trends in PDE Constrained Optimization*, International Series of Numerical Mathematics 165, DOI 10.1007/978-3-319-05083-6_8

where $\psi \in C^{1,1}_{\mathrm{loc}}(\mathbb{R}^2)$ and $y_d \in BV([a,b])$ is a desired state. The state y is the entropy solution of either an initial-boundary value problem for a scalar conservation law

$$y_t + f(y)_x = g(\cdot, y, u_1)$$

or of a traffic light problem, as we will call it throughout this paper, where two conservation laws are coupled through a node at which the switching times between red and green phases is controlled. Our motivation is to develop a variational calculus for initial, boundary and switching time control that has the potential to be extended to the optimal control of networks with switching control at nodes.

It is well-known that weak solutions to Cauchy problems of nonlinear hyperbolic conservation laws are in general not unique and that one has to consider entropy solutions, that can be obtained as the vanishing viscosity limit of a parabolic regularization [4, 26]. Even for smooth initial and boundary data entropy solutions can develop discontinuities (shocks) after finite time [7]. This leads to fundamental difficulties for the sensitivity analysis and optimal control theory, since the shock locations depend on the control. Hence, the control-to-state mapping $u \mapsto y(\bar{t}, \cdot; u)$ is at best differentiable with respect to the weak topology of measures and sensitivities are necessarily measures with singular part along the shock curves. For networks, where the solutions on two or more intervals are connected by a (possibly controlled) node, the situation gets even more involved.

Motivated by its practical relevance, despite these difficulties the analysis and numerical solution of optimal control problems for hyperbolic conservation laws has become an active research field in recent years.

The existence of optimal controls for the Cauchy problem and the initial-boundary value problem was discussed for example in [1, 2, 35, 36].

The issue of non-differentiability of the solution operator was treated by different authors by introducing generalized notions of differentiability, e.g. [5, 8, 9, 11, 15, 36, 37]. The present work is based on the notion of *shift-differentiability*, that was introduced in [36], where it was also shown to hold for the Cauchy problem. Here the theory of generalized characteristics by Dafermos [17] is a crucial instrument. This approach also includes an adjoint calculus for the reduced objective function, see also [19, 20, 38].

Networks for hyperbolic conservation laws have been considered in various contexts in recent years. Several node conditions have been discussed, most of them are tailored to specific applications, such as traffic modeling [10, 13, 22, 25], gas pipelines [3, 16, 23] or supply chains [21]. The conditions are mostly formulated for Riemann problems and then generalized by wave front tracking. Besides the question of well-posedness also aspects from the optimal control viewpoint have been considered. But these approaches often either consider the linear case or assume the existence of a strong solution. Conservation laws with modal switching have been discussed for the first time in [24], where switching is considered in the fluxes, the boundary condition and the coupling condition at the nodes of a network.

This paper is organized as follows. In Sect. 2 we introduce the two considered problems, the initial-boundary value problem and the traffic light problem. In Sect. 3 we collect results on the well-posedness for these problems and structural properties of the corresponding solutions. The main results will be presented in Sect. 4, where we show the generalized differentiability of the solution operator and the resulting Fréchet-differentiability of the reduced objective function.

2 The Models

In this paper we focus on two types of problems for a scalar conservation law. The first one is the initial-boundary value problem (IBVP), the second is the traffic light problem (TLP). We will also consider the pure initial value problem (IVP), since on one hand the IVP is helpful to understand the more involved IBVP and on the other hand the traffic light problem is a combination of both. While the IBVP is an important step towards node conditions on networks, the traffic light problem can be seen as a relevant node condition with switching in a traffic network.

2.1 *Initial-Boundary Value Problem*

The first model problem under consideration is an initial-boundary value problem (IBVP) on an interval $\Omega = (a, b)$, where we explicitly allow for a, b to be $\pm\infty$, respectively. The IBVP is then given by

$$y_t + f(y)_x = g(\cdot, y, u_1), \quad \text{on } \Omega_T, \tag{2.1a}$$

$$y(0, \cdot) = u_0, \quad \text{on } \Omega, \tag{2.1b}$$

$$y(\cdot, a+) = u_{B,a}, \quad \text{in the sense of (2.4a)} \quad (\text{if } a > -\infty), \tag{2.1c}$$

$$y(\cdot, b-) = u_{B,b}, \quad \text{in the sense of (2.4b)} \quad (\text{if } b < \infty), \tag{2.1d}$$

where $\Omega_T := [0, T] \times \Omega$. In order to show existence of a unique solution, following [4, 26], the conservation law (2.1a) has to be understood in sense of an entropic solution, which can be characterized by requiring that for every (Kružkov-) entropy $\eta_c(\lambda) := |\lambda - c|$, $c \in \mathbb{R}$, and associated entropy flux $q_c(\lambda) := \operatorname{sgn}(\lambda - c)(f(\lambda) - f(c))$ the following entropy inequality holds in the sense of distributions

$$(\eta_c(y))_t + (q_c(y))_x \leq \eta_c'(y) g(\cdot, y, u_1) \qquad \text{in } \mathcal{D}'(\Omega_T). \tag{2.2}$$

The initial data in (2.1b) have to be understood in the weak L^1_{loc}-sense, which means that for every $R > 0$

$$\operatorname*{esslim}_{t \to 0+} \|y(t, \cdot) - u_0\|_{1, \Omega \cap (-R, R)} = 0 \tag{2.3}$$

is fulfilled. The Dirichlet-like boundary conditions (2.1c), (2.1d) must not be understood literally. Rather, the solution y of (2.1a)–(2.1d) has to be interpreted as the limit of its parabolic regularization, that is (2.1a)–(2.1d) with the term εy_{xx} added on the right hand side of (2.1a). The boundary condition of the limit solution can then be characterized, as shown in [4], by

$$\min_{k\in I(y(\cdot,a+),u_{B,a})} \operatorname{sgn}(u_{B,a}-y(\cdot,a+))(f(y(\cdot,a+))-f(k))=0, \quad \text{a.e. on } [0,T], \tag{2.4a}$$

$$\min_{k\in I(y(\cdot,b-),u_{B,b})} \operatorname{sgn}(y(\cdot,b-)-u_{B,b})(f(y(\cdot,b-))-f(k))=0, \quad \text{a.e. on } [0,T], \tag{2.4b}$$

with $I(\alpha,\beta) := [\min(\alpha,\beta),\max(\alpha,\beta)]$, see also [18, 27, 30, 31]. In the literature the above formulation of the boundary condition from [4] is sometimes called the *BLN-condition*.

2.2 *Traffic Lights*

The second topic of this work is a problem that is motivated by a traffic flow problem. This special type of problem is also of great interest because it is a simple example for a network of conservation laws with modular switchings in the node condition. Before we formulate the mathematical problem, we give a short overview on traffic flow modeling by hyperbolic conservation laws.

2.2.1 Macroscopic Model for Traffic Flow

In the mid 1950s, Lighthall and Whitham [29] and Richards [34] proposed a continuum model for heavy traffic. The traffic is described by means of a traffic density ρ and the conservation of cars is ensured by

$$\rho_t + f(\rho)_x = 0, \qquad f(\rho) := \rho v(\rho),$$

where the velocity v of the traffic depends only on the density. This model is widely used and is known as the *LWR*-model. For a detailed overview on traffic flow modeling by partial differential equations we refer to [25]. Usually one assumes that f is a concave function, but since most theoretical results on conservation laws work with convex fluxes, we will make a change of signs and work with a convex flux function. In the following the state y can be interpreted as the negative traffic density $-\rho$. We further assume that the road reaches its maximum density when $y = -1$ and is empty for $y = 0$. The flux $f(y)$ is equal to 0 for these two values and strictly convex in between. In particular, f is negative on $(-1, 0)$.

2.2.2 A Traffic Light on an Open Road

We consider a long unidirectional road $I = \mathbb{R}$ that has to be closed for some reason (e.g. because of pedestrian or railway crossings) for some time periods at a specific point $x = 0$, most likely by a traffic light. So the considered time interval $[0, T]$ is split into two different types of phases, namely green $[\sigma_g^{i-1}, \sigma_r^i)$, $i = 1, \ldots, n_\sigma + 1$ and red phases $[\sigma_r^i, \sigma_g^i)$, $i = 1, \ldots, n_\sigma$ where the incoming traffic at $x = 0$ is or is not allowed to cross respectively. A similar problem was already briefly introduced in [29].

For the sequel we assume $\sigma = (\sigma_g^0, \sigma_r^1, \sigma_g^1, \ldots, \sigma_g^{n_\sigma}, \sigma_r^{n_\sigma+1}) \in \Sigma$, where

$$\Sigma := \left\{ \nu \in \mathbb{R}^{2(n_\sigma+1)} \ : \ 0 = \nu_1 < \nu_2 < \ldots < \nu_{2n_\sigma+1} < \nu_{2n_\sigma+2} = T \right\}. \tag{2.5}$$

for the sake of simplicity. The presented analysis can also be carried over to the case where the first and/or the final phase is a red phase.

During the i-th green phase a solution y of such a traffic light problem (TLP) is determined by solving a Cauchy problem on $\Omega_{g,i} := [\sigma_g^{i-1}, \sigma_r^i] \times \mathbb{R}$ with initial data

$$u_0 = y(\sigma_g^{i-1}-, \cdot), \quad i = 2, \ldots, n_\sigma + 1.$$

Here, $y(\sigma_g^{i-1}-, \cdot)$ is the final state of the previous red phase.

For the i-th red phase the solution y consists of two parts, namely y_1 and y_2, its restriction to the incoming and outgoing part $I_1 := (-\infty, 0)$ and $I_2 := (0, \infty)$ of the road. The restriction y_1 is the solution of an initial-boundary value problem on $\Omega_{r,i}^1 := [\sigma_r^i, \sigma_g^i] \times I_1$ with initial value $y(\sigma_r^i-, \cdot)$ and boundary data $u_{B,0} \equiv -1$. Similarly, y_2 solves an IBVP on $\Omega_{r,i}^2 := [\sigma_r^i, \sigma_g^i] \times I_2$ with $u_{B,0} \equiv 0$. For the first green phase, i.e. the first IVP, the initial data are given by some function u_I. The traffic light problem can then formulated in the following way:

$$y_t + f(y)_x = g(\cdot, y, u_1), \qquad \text{on } \Omega_{g,i+1}, \quad i = 0, \ldots, n_\sigma, \quad j = 1, 2, \tag{2.6a}$$

$$y_t + f(y)_x = g(\cdot, y, u_1), \qquad \text{on } \Omega_{r,i}^j, \quad i = 1, \ldots, n_\sigma, \quad j = 1, 2, \tag{2.6b}$$

$$y(0, \cdot) = u_I, \qquad \text{on } I, \tag{2.6c}$$

$$y(\sigma_g^i, \cdot)\Big|_{I_j} = y_j(\sigma_g^i-, \cdot), \qquad \text{on } I_j, \quad i = 1, \ldots, n_\sigma, \quad j = 1, 2, \tag{2.6d}$$

$$y_j(\sigma_r^i, \cdot) = y(\sigma_r^i-, \cdot)\Big|_{I_j}, \qquad \text{on } I_j, \quad i = 1, \ldots, n_\sigma, \quad j = 1, 2, \tag{2.6e}$$

$$y_1(\cdot, 0-) = -1, \qquad \text{on } [\sigma_r^i, \sigma_g^i], \quad i = 1, \ldots, n_\sigma, \tag{2.6f}$$

$$y_2(\cdot, 0+) = 0, \qquad \text{on } [\sigma_r^i, \sigma_g^i], \quad i = 1, \ldots, n_\sigma. \tag{2.6g}$$

The conservation laws (2.6a), (2.6b) model the conservation of cars. The source term g can be seen as additional traffic that enter or leave the road from minor roads or parking lots, that are not modeled in detail. The boundary conditions that model the red lights (red light conditions) (2.6f), (2.6g) guarantee, that during these periods no cars enter or leave the two roads over the artificial boundary, since, as stated in Sect. 2.2.1, the flux $f(y)$ is equal to zero for $y \in \{-1, 0\}$. Moreover, even if formally one has to interpret these boundary conditions in the BLN-sense, we will see that under mild assumptions they may be considered literally. We will discuss these conditions more detailed in Sect. 3.2. The continuity conditions between the phases (2.6d), (2.6e) describe the transition from one phase into another.

3 Properties of Entropy Solutions

In this section we collect important properties of the solutions to (2.1) and (2.6).

3.1 General and Structural Properties of Solutions to IBVPs

First we consider the initial value problem (2.1a)–(2.1b) for $\Omega = \mathbb{R}$. We make the following assumptions:

(A1) The flux function satisfies $f \in C^2(\mathbb{R})$ and there exists $m_{f''} > 0$ such that $f'' \geq m_{f''}$. The source term satisfies $g \in L^\infty\left(\Omega_T; C^{0,1}_{\text{loc}}(\mathbb{R} \times \mathbb{R}^m)\right) \cap L^\infty\left(0, T; C^1_{\text{loc}}(\mathbb{R} \times \mathbb{R} \times \mathbb{R}^m)\right)$ and for all $M_u > 0$ there exist constants $C_1, C_2 > 0$ such that for all $(t, x, y, u_1) \in \Omega_T \times \mathbb{R} \times [-M_u, M_u]^m$ it holds that

$$g(t, x, y, u_1)\text{sgn}(y) \leq C_1 + C_2|y|.$$

(A2) The set of admissible controls U_{ad} is bounded in $U_\infty := L^\infty(\mathbb{R}) \times L^\infty(\Omega_T)^m$ by some constant M_u and closed in $U_1 := L^1_{\text{loc}}(\mathbb{R}) \times L^1_{\text{loc}}(\Omega_T)^m$.

We recall Proposition 1 from [38], that covers some of the most important properties of the solution to the IVP.

Proposition 3.1 (Existence and Uniqueness for Cauchy problems). *Let (A1) and (A2) hold. Then for every* $u = (u_0, u_1) \in U_\infty$ *there exists a unique entropy solution* $y = y(u) \in L^\infty(\Omega_T)$ *of* (2.1a)–(2.1b) *on* $\Omega = \mathbb{R}$. *After a possible modification on a set of measure zero it even holds* $y \in C([0, T]; L^1_{\text{loc}}(\mathbb{R}))$. *There are constants* $M_y, L_y > 0$ *such that for every* $u, \hat{u} \in U_{\text{ad}}$ *and all* $t \in [0, T]$ *the following estimates hold:*

$$\|y(t, \cdot; u)\|_\infty \leq M_y,$$
$$\|y(t, \cdot; u) - y(t, \cdot; \hat{u})\|_{1,[a,b]} \leq L_y(\|u_0 - \hat{u}_0\|_{1,I_t} + \|u_1 - \hat{u}_1\|_{1,[0,t]\times I_t}),$$

where $a < b$ *and* $I_t := [a - tM_{f'}, b + tM_{f'}]$, $M_{f'} := \max_{|y| \le M_y} |f'(y)|$.

Set $\hat{U}_{\mathrm{ad}} := \{u \in U_{\mathrm{ad}} : \|u_1\|_{L^\infty(0,T;C^1(\Omega_T)^m)} \le M_u\}$. *Then there is a constant* $M > 0$ *such that for all* $u \in \hat{U}_{\mathrm{ad}}$ *and all* $t \in (0, T]$ *Oleinik's entropy condition*

$$y_x(t, \cdot; u) \le \left((1 - e^{-m_{f''}Mt})M^{-1} + e^{-m_{f''}Mt}(C_{u,M})^{-1}\right)^{-1}$$

holds with $C_{u,M} := \max\left\{M, \operatorname{esssup}_{x \neq z} \frac{u_0(x)-u_0(z)}{x-z}\right\}$. *In particular* $y(t, \cdot) \in BV_{\mathrm{loc}}(\mathbb{R})$ *for all* $t \in (0, T]$ *and* $y \in BV([s, T] \times [-R, R])$ *for all* $s, R > 0$.

For the case of an initial-boundary value problem we have a similar result. We restrict ourselves to the case of $\Omega = (0, \infty)$. The first thing to mention here is the fact that the BLN-condition (2.4a) involves the boundary trace $y(\cdot, 0+)$. When Bardos, le Roux and Nédélec stated this formulation they only considered the case where the solution has bounded total variation, see Remark 3.3. In order to also allow for L^∞-data in [30, 31] Otto proposed another characterization of the boundary condition that is equivalent to the one in [4] if the boundary trace exists. But Vasseur showed [39] that under mild assumptions even for L^∞-entropy solutions there always exist boundary traces. Therefore the formulation in (2.4a) (and (2.4b)) is valid even in the L^∞-setting, see also [14].

We make the following assumptions:

(A1′) The flux function satisfies $f \in C^2(\mathbb{R})$ and there exists $m_{f''} > 0$ such that $f'' \ge m_{f''}$. The source term is non-negative and satisfies $g \in C\left(\Omega_T; C^{0,1}_{\mathrm{loc}}(\mathbb{R} \times \mathbb{R}^m)\right) \cap C^1\left([0, T]; C^1_{\mathrm{loc}}(\Omega \times \mathbb{R} \times \mathbb{R}^m)\right)$ and for all $M_u > 0$ there exist constants $C_1, C_2 > 0$ such that for all $(t, x, y, u_1) \in \Omega_T \times \mathbb{R} \times [-M_u, M_u]^m$ holds:

$$g(t, x, y, u_1)\operatorname{sgn}(y) \le C_1 + C_2|y|.$$

(A2′) The set of admissible controls U_{ad} is bounded in $U_\infty := L^\infty(\mathbb{R}) \times L^\infty(0, T) \times L^\infty([0, T] \times \mathbb{R})^m$ by some constant M_u and closed in $U_1 := L^1_{\mathrm{loc}}(\mathbb{R}) \times L^1(0, T) \times L^1_{\mathrm{loc}}([0, T] \times \mathbb{R})^m$.

For technical reasons we consider the source term and the corresponding control u_1 not only for the considered spatial domain. Of course the solution depends only on its restriction to Ω_T.

Under the above assumptions we get the following properties of a solution to (2.1), cf. [4, 14, 31].

Proposition 3.2 (Existence and Uniqueness for IBVPs). *Let (A1′) and (A2′) hold. Then for every* $u = (u_0, u_B, u_1) \in U_\infty$ *there exists a unique entropy solution* $y = y(u) \in L^\infty(\Omega_T)$ *of (2.1) on* $\Omega = (0, \infty)$. *After a possible modification on a set of measure zero it even holds that* $y \in C([0, T]; L^1_{\mathrm{loc}}(\Omega))$. *Moreover, there are constants* $M_y, L_y > 0$ *such that for every* $u, \hat{u} \in U_{\mathrm{ad}}$ *and all* $t \in [0, T]$ *the following estimates hold:*

$$\|y(t,\cdot;u)\|_\infty \le M_y,$$
$$\|y(t,\cdot;u) - y(t,\cdot;\hat{u})\|_{1,[a,b]} \le L_y(\|u_0 - \hat{u}_0\|_{1,I_t} + \|u_B - \hat{u}_B\|_{1,[0,t]} + \|u_1 - \hat{u}_1\|_{1,[0,t]\times I_t}),$$

where $a < b$ *and* $I_t := [a - tM_{f'}, b + tM_{f'}] \cap \Omega$, $M_{f'} := \max_{|y|\le M_y} |f'(y)|$.

Remark 3.3. Under the stronger assumptions $u_0 \in BV_{\text{loc}}(\Omega)$ and $u_B \in BV([0,T])$, (2.1) admits a solution satisfying $y \in BV([0,T] \times [0,R])$ for all $R > 0$ (cf. [4, 28]).

The basic idea behind the proof of the main result of this work is the theory of generalized characteristics from [17], which will be considered in the remaining part of this section. We will assume that in addition to (A1)–(A2), (A1′)–(A2′) respectively, the following assumption holds.

(A3) g is globally Lipschitz w.r.t. x and y.

Furthermore we will only consider $(u_0, u_1) \in \hat{U}_{\text{ad}}$ (see Proposition 3.1), $u_0 \in BV_{\text{loc}}(\Omega)$ and boundary data $u_B \in PC^1([0,T]; t_1, \ldots, t_{n_t})$, that is a piecewise continuously differentiable function with possible kinks or discontinuities at $0 < t_1 < \ldots < t_{n_t}$ for some $n_t \in \mathbb{N}$.

Using the properties collected in Propositions 3.1 and 3.2, we conclude that $y \in L^\infty(\Omega_T) \cap C([0,T]; L^1_{\text{loc}}(\Omega))$ has the following properties: For all $(t,x) \in (0,T] \times \Omega$ the one-sided limits $y(t,x-)$ and $y(t,x+)$ exist and satisfy $y(t,x-) \ge y(t,x+)$. It will be convenient to work with a pointwise defined representative of $y \in C([0,T]; L^1_{\text{loc}}(\Omega))$ where $y(t,x)$ is identified with one of the limits $y(t,x-)$ or $y(t,x+)$.

We now recall the definition of a generalized characteristic in the sense of Dafermos from [17].

Definition 3.4 (Generalized characteristics). A Lipschitz curve

$$[\alpha,\beta] \subset [0,T] \to \Omega_T,\ t \mapsto (t, \xi(t))$$

is called a *generalized characteristic* on $[a,b]$ if

$$\dot{\xi}(t) \in [f'(y(t,\xi(t)+)), f'(y(t,\xi(t)-))], \quad \text{a.e. on } [\alpha,\beta]. \tag{3.1}$$

The generalized characteristic is called *genuine* if the lower and upper bound in (3.1) coincide for almost all $t \in [\alpha,\beta]$.

In the following we will also call ξ a (generalized) characteristic instead of $t \mapsto (t,\xi(t))$. It will also be useful to introduce notions of *extreme* or *maximal/minimal characteristics* $\xi_\pm$, that satisfy

$$\dot{\xi}_\pm(t) = f'(y(t,\xi(t)\pm)).$$

Since by Propositions 3.1 or 3.2 y is (essentially) bounded on Ω_T, an a priori bound on the speed of all generalized characteristic is known. Therefore, characteristics do not escape and they either exist for the whole time period $[0, T]$ or (in the bounded case) leave the spatial domain at some point $(\theta, \xi(\theta)) \in [0, T] \times \partial\Omega$. Moreover it can be shown [17] that (3.1) can be restricted to

$$\dot{\xi}(t) = \begin{cases} f'(y(t, \xi(t))) & \text{if } f'(y(t, \xi(t)+)) = f'(y(t, \xi(t)-)) \\ \frac{[f(y(t,\xi(t)))]}{[y(t,\xi(t))]} & \text{if } f'(y(t, \xi(t)+)) \neq f'(y(t, \xi(t)-)) \end{cases}, \quad \text{a.e. on } [\alpha, \beta],$$

where for $\varphi \in BV(\mathbb{R})$ the expression

$$[\varphi(x)] := \varphi(x-) - \varphi(x+)$$

denotes the height of the jump of φ across x.

Based on the notion of generalized characteristics in [17] Dafermos exploits structural properties of BV-solutions that are essential for the analysis in the present paper.

Proposition 3.5 (Structure of BV-Solutions). *Let (A1)–(A3) hold. Consider an entropy solution $y = y(u)$ of the Cauchy problem* (2.1a)–(2.1b) *on $\Omega = \mathbb{R}$ for controls $u = (u_0, u_1) \in \hat{U}_{\text{ad}}$, $u_0 \in BV_{\text{loc}}(\mathbb{R})$.*

For $(\bar{t}, \bar{x}) \in \Omega_T$ fixed denote by ξ a backward characteristic on $[0, \bar{t}]$ through $(\bar{t}, \bar{x})$. Then ξ has the following properties:

1. *If ξ is an extreme backward characteristic, i.e. $\xi = \xi_{\pm}$, then ξ is genuine, i.e. $y(t, \xi_{\pm}(t)-) = y(t, \xi_{\pm}(t)+)$ for all $t \in (0, \bar{t})$.*
2. *If ξ is genuine, i.e. $y(t, \xi(t)-) = y(t, \xi(t)+)$, $t \in (0, \bar{t})$, then it satisfies*

$$\xi(t) = \zeta(t), \quad t \in [0, \bar{t}], \qquad y(t, \xi(t)) = v(t), \quad t \in (0, \bar{t}), \tag{3.2a}$$

$$u_0(\xi(0)-) \le v(0) \le u_0(\xi(0)+), \quad y(\bar{t}, \xi(\bar{t})-) \ge v(\bar{t}) \ge y(\bar{t}, \xi(\bar{t})+), \tag{3.2b}$$

where (ζ, v) is a solution of the characteristic equation

$$\dot{\zeta}(t) = f'(v(t)), \tag{3.3a}$$

$$\dot{v}(t) = g(t, \zeta(t), v(t), u_1(t, \zeta(t))). \tag{3.3b}$$

For extreme characteristics $\xi_{\pm}$ the initial values are given by

$$(\zeta, v)(\bar{t}) = (\bar{x}, y(\bar{t}, \bar{x}\pm)). \tag{3.3c}$$

Although this classical result by Dafermos is widely-known and gives important information about the inner structure of entropy solutions, the earliest extension to be found in the literature to a bounded spatial domain is in a work of Perrollaz in [32] published in 2013. Here the situation for characteristics ξ that stay inside the spatial domain for the whole considered time interval is exactly the same as in Proposition 3.5. But there are two cases that require special consideration. On the one hand there are backward characteristics through some $(\bar{t}, \bar{x}) \in \Omega_T$ that leave the spatial domain at some time, say $\theta \in (0, \bar{t})$, on the other hand one has to consider characteristics that enter the spatial domain at some time θ.

It turns out that for $\Omega = (0, \infty)$ the non-negativity condition on g is crucial for the second and third part of Proposition 3.6, since it avoids some degeneracy of the characteristics near the boundary. For a spatial domain $(-\infty, 0)$ the condition on the source term becomes a non-positivity condition and consequently this leads to the requirement that g has to vanish if one considers general intervals (a, b) of finite length. But as mentioned, this property is only important near the boundary and can therefore be weakened to a local condition.

The following proposition collects the results of section 3 in [32].

Proposition 3.6. *Let (A1′), (A2′) and (A3) hold. Consider an entropy solution* $y = y(u)$ *of the mixed initial-boundary value problem* (2.1) *on* $\Omega = (0, \infty)$ *for controls* $u = (u_0, u_B, u_1) \in U_{\text{ad}}$ *with* $(u_0, u_1) \in \hat{U}_{\text{ad}}$, $u_0 \in BV_{\text{loc}}(\mathbb{R})$ *and* $u_B \in PC^1([0, T]; t_1, \dots, t_{n_t})$. *Then the following holds:*

1. *Consider* $\theta \in (0, T)$ *with* $f'(y(\theta, 0+)) < 0$, *then there exists a genuine backward characteristic* ξ *through* $(\theta, 0)$ *with* $\dot{\xi}(\theta) = f'(y(\theta, 0+))$.
2. *Let* ξ *be a genuine characteristic through* $(\bar{t}, \bar{x}) \in \Omega_T$ *satisfying* $\xi(t) \in \Omega$ *for* $t \in (\theta, \bar{t}] \subset [0, T]$ *and* $\lim_{t \searrow \theta} \xi(t) = 0$. *Denote by* (ζ, v) *the solution of the characteristic equation* (3.3a)–(3.3b) *associated to* ξ *by Proposition 3.5 on every* $[\tilde{t}, \bar{t}] \subset (\theta, \bar{t}]$. *Then with* $v(\theta) := \lim_{t \searrow \theta} v(t)$ *it holds*
$$u_B(\theta+) \leq v(\theta) \leq u_B(\theta-). \tag{3.4}$$
3. *Let* ξ *be a forward characteristic in* $[0, \tilde{t}] \times \Omega$ *for every* $\tilde{t} \in (0, \theta)$ *and* (ζ, v) *be the associated solution of the characteristic equation. If now* $\lim_{t \nearrow \theta} \xi(t) = 0$ *then*
$$f'(\bar{v}) \leq 0 \quad \text{and} \quad f(\bar{v}) \geq f(u_B(\theta-)), \tag{3.5}$$
where $\bar{v} := \lim_{t \nearrow \theta} v(t)$.

This connection between the genuine characteristics and the characteristic equation is very useful, since by the following lemma, that is a consequence of a result on ordinary differential equations (cf. Proposition 3.4.5 and Lemma 3.4.6 in [36] or chapter 5.6 in [33]) this yields some important information on the local differentiability properties of a solution y of the I(B)VP.

Lemma 3.7. *Let (A1′) and (A3) hold and denote for* $(\theta, z, w, u_1) \in [0, T] \times \mathbb{R}^2 \times C^1([0, T] \times \mathbb{R}^m)$ *by* $(\zeta, v)(\,\cdot\,, \theta, z, w, u_1)$ *the solution of* (3.3a)–(3.3b) *for initial data*
$$(\zeta, v)(\theta) = (z, w).$$

Let $M_w, M_u > 0$ be given and set

$$\mathcal{B}_i := [0,T] \times \mathbb{R}^2 \times L^2(0,T;C^i(\mathbb{R})^m), \quad i = 0,1,$$
$$\bar{\mathcal{B}} := \left\{(\theta,z,w,u_1) \in \mathcal{B}_1 \ : \ |w| < M_w, \quad \|u_1\|_{C^1([0,T]\times\mathbb{R}^m)} < M_u\right\}.$$

Then the mapping

$$(\theta,z,w,u_1) \in (\bar{\mathcal{B}}, \|\cdot\|_{\mathcal{B}_i}) \longmapsto (\zeta,v)(\,\cdot\,,\theta,z,w,u_1) \in C([0,T])^2$$

is Lipschitz continuous for $i = 0$ and continuously Fréchet-differentiable for $i = 1$ and on $\bar{\mathcal{B}}$ the right hand side is uniformly Lipschitz w.r.t. t.

Lemma 3.7 is a direct generalization of the first assertion of Lemma 3.4.6 in [36] to the case where the dependence on the time θ where the initial datum is specified, is considered, too. The remaining statements of Lemma 3.4.6 can also be carried over to this generalized case.

3.2 General and Structural Properties of Solutions to Traffic Light Problems

In this section we analyze the structure of solutions to traffic light problems. We consider $u_I \in BV_{\text{loc}}(\mathbb{R})$ and u_1 bounded in $C^1([0,T]\times\mathbb{R})^m$. Since a solution of a TLP is a concatenation of solutions to IVPs and IBVPs on a finite number of time slabs, the existence, uniqueness and stability properties can easily be transferred to such solutions.

We add the following requirements to our setting.

(A4) g is non-positive on $(-\infty,0)$, non-negative on $(0,\infty)$ and vanishes on $(-\varepsilon,\varepsilon)$ for some $\varepsilon > 0$. In addition g is chosen such that $-1 \le y \le 0$ is guaranteed. Furthermore, let $U_{\text{ad}} \subset \{(u_0,u_1) \in U_\infty \ : \ -1 \le u_0 \le 0\}$ and let $\Sigma_{\text{ad}} \subset \Sigma$ be a closed set in $[0,T]$, with Σ defined in (2.5).

Remark 3.8. The condition on g in (A4) holds clearly for the choice $g \equiv 0$.

Corollary 3.9 (Existence and Uniqueness for traffic light problems). *Let (A1′), (A2′) and (A4) hold. Then for every $u = (u_0,u_1) \in U_\infty$ and $\sigma \in \Sigma_{\text{ad}}$ there exists a unique entropy solution $y = y(u,\sigma) \in L^\infty(\Omega_T)$ of (2.1) on $\Omega = (0,\infty)$. After a possible modification on a set of measure zero it even holds $y \in C([0,T];L^1_{\text{loc}}(\Omega))$.*

Moreover for every $t \in [0,T]$ and $a < b$ we have the following stability estimates:

1. *For fixed $u \in U_{\text{ad}}$ there is $L_\Sigma > 0$ such that for all $\tilde{\sigma},\hat{\sigma} \in \Sigma_{\text{ad}}$ holds*

$$\|y(t,\,\cdot\,;u,\tilde{\sigma}) - y(t,\,\cdot\,;u,\hat{\sigma})\|_{1,(a,b)} \le L_\Sigma \|\tilde{\sigma} - \hat{\sigma}\|.$$

2. *For fixed* $\sigma \in \Sigma_{\text{ad}}$ *there is* $L_U > 0$ *such that for all* $\tilde{u}, \hat{u} \in U_{\text{ad}}$ *holds*

$$\|y(t,\cdot;\tilde{u},\sigma) - y(t,\cdot;\hat{u},\sigma)\|_{1,(a,b)} \le L_U\left(\|\tilde{u}_0 - \hat{u}_0\|_{1,I_t} + \|\tilde{u}_1 - \hat{u}_1\|_{1,[0,t]\times I_t}\right),$$

where $I_t := [a - tM_{f'}, b + tM_{f'}] \cap \Omega$, $M_{f'} := \max_{|y| \le M_y} |f'(y)|$.

One can easily verify that the structural features for solutions to IBVPs provided by Propositions 3.5 and 3.6 also hold for genuine backward characteristics ξ that correspond to the solution $y = (y_1, y_2)$ of a TLP as long as they do not touch the switching points $(\sigma, 0)$.

We now discuss what happens to the solution $y = (y_1, y_2)$ during a red phase and at the beginning of the green phase. The following considerations are illustrated in Fig. 1.

First we consider the initial-boundary value problem for y_1 during a red phase $[\sigma_{\mathrm{r}}^i, \sigma_{\mathrm{g}}^i]$. Here especially the situation on the boundary is of interest. We recall, that the boundary data $u_{B,0}$ are chosen to be equal to -1 and that by assumption (A4) $-1 \le y = (y_1, y_2) \le 0$ holds. Therefore the BLN-boundary condition (2.4b) becomes

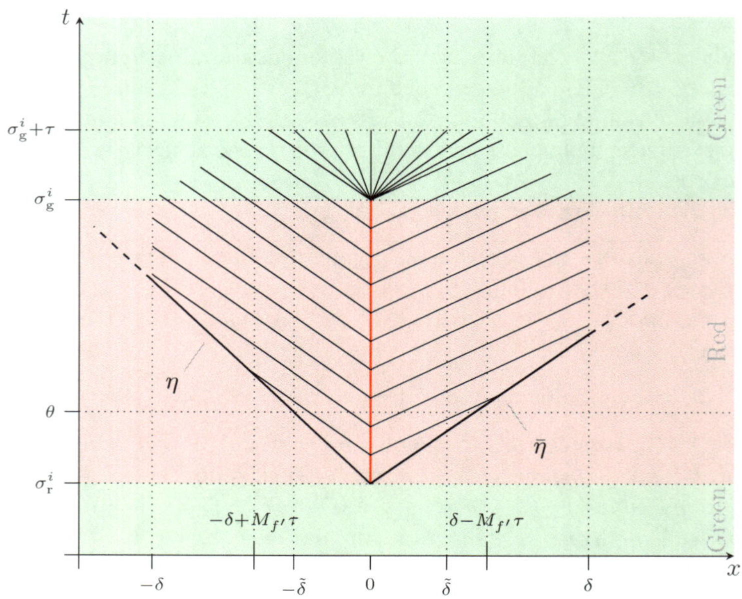

Fig. 1 Characteristics in a neighborhood of a red phase

$$\min_{k\in[-1,y(\cdot,0-)]} \operatorname{sgn}(y(\cdot,0-)+1)(f(y(\cdot,0-))-f(k))=0.$$

Since k may be chosen equal to $y(\cdot,0-)$, the condition is equivalent to

$$\operatorname{sgn}(y(t,0-)+1)(f(y(t,0-))-f(k))\geq 0, \quad \forall k\in[-1,y(t,0-)].$$

Here the first factor is strictly positive whenever $y(\cdot,0-)\neq -1$ and for $k=-1$, the second factor is negative if $f(y(\cdot,0-))\neq 0$. Hence we can deduce that there are only two possibilities for the boundary trace, namely $y(t,0-)\in\{0,-1\}$ for almost all $t\in[\sigma_{\mathrm{r}}^i,\sigma_{\mathrm{g}}^i]$. If $y(\theta,0-)=0$ for some $\theta\in(\sigma_{\mathrm{r}}^i,\sigma_{\mathrm{g}}^i)$, then the existence of a backward characteristic ξ satisfying $\dot{\xi}(\theta)=f'(0)<0$ can be deduced from Proposition 3.6. Since by the sign condition on the source term, all genuine characteristics on $\Omega_{\mathrm{r},i}^1$ are concave and since two genuine characteristics may not intersect each other, this implies that $y(t,0-)=0$ must hold for all $t\in(\sigma_{\mathrm{r}}^i,\theta]$. Conversely speaking, this means, that if $y(\tilde{\theta},0-)=-1$ for some $\tilde{\theta}\in(\sigma_{\mathrm{r}}^i,\sigma_{\mathrm{g}}^i)$, then $y(t,0-)=-1$ holds for all $t\in[\tilde{\theta},\sigma_{\mathrm{g}}^i]$. Consequently, if the initial data of the IBVP on $\Omega_{\sigma_{\mathrm{g}}^i}^1$ are bounded away from 0 in a small neighborhood of the right boundary at $x=0$, $y(\cdot,0-)=-1$ holds during the whole time slab. We will assume this property for the sequel.In this case a generalized characteristic η emanates from $(\sigma_{\mathrm{r}}^i,0)$ having strictly negative speed at least for a small time period $(\sigma_{\mathrm{r}}^i,\tilde{\tau})$, see Fig. 1. (More precisely, η is either a shock or a characteristic traveling with speed $f'(-1)$.) After that period, it keeps traveling with non-positive speed at least up to $t=\sigma_{\mathrm{g}}^i$. The solution y_1 is constantly equal to -1 on the nonempty set $\{(t,x)\in[\sigma_{\mathrm{r}}^i,\sigma_{\mathrm{g}}^i]\times[-\varepsilon,0) \;:\; \eta(t)<x\}$ with ε from assumption (A4). The situation for y_2 is completely analogous. If the initial data of the IBVP on $\Omega_{\mathrm{r},i}^2$ are bounded away from -1 in a small neighborhood of the boundary at $x=0$, $y(\cdot,0+)=0$ holds during the whole time slab and we conclude that $y_2=0$ on a set $\{(t,x)\in[\sigma_{\mathrm{r}}^i,\sigma_{\mathrm{g}}^i]\times(0,\varepsilon] \;:\; \bar{\eta}(t)>x\}$. Therefore, for every $t\in(\sigma_{\mathrm{r}}^i,\sigma_{\mathrm{r}}^i)$ the solution $y(t,\cdot)$ is known at least in a small neighborhood of $x=0$, compare the area filled with characteristics in Fig. 1. We now examine the solution y on the subsequent green phase $[\sigma_{\mathrm{g}}^i,\sigma_{\mathrm{r}}^{i+1}]$. Here the situation at $x=0$ is again of special interest. By the previous considerations we know that there is a $\delta>0$ such that with u_0 being the initial data of the considered Cauchy problem on $\Omega_{\mathrm{g},i+1}$, $u_0(x)=\frac{1}{2}(\operatorname{sgn}(x)-1)$ for all $x\in(-\delta,\delta)$. Together with the finite propagation speed this implies that locally y is the solution of a Riemann problem producing a rarefaction wave.

We subsume the previous considerations in the following lemma. The assertions and occurring quantities are also illustrated in Fig. 1.

Lemma 3.10. *Let (A1′), (A2′) and (A4) hold and let* $u_0\in BV_{\mathrm{loc}}(\mathbb{R})$, $u_1\in C^1([0,T]\times\mathbb{R})^m$. *Consider for* $i=1,\ldots,n_\sigma$ *the i-th red phase of the traffic light problem* (2.6). *Assume that the final state of the i-th green phase* $y(\sigma_{\mathrm{r}}^i,\cdot)$ *is bounded away from 0 on* $(-\tilde{\varepsilon},0]$ *and bounded away from −1 on* $[0,\tilde{\varepsilon})$ *for some* $\tilde{\varepsilon}>0$. *Then the solution of* (2.6) *satisfies the following equations.*

1. *For every* $\theta \in (\sigma_{\mathrm{r}}^i, \sigma_{\mathrm{g}}^i)$ *there exists* $\tilde{\delta} > 0$ *such that there holds*

$$y_1(t,x) = -1, \qquad (t,x) \in (\theta, \sigma_{\mathrm{g}}^i) \times (-\tilde{\delta}, 0),$$

$$y_2(t,x) = 0, \qquad (t,x) \in (\theta, \sigma_{\mathrm{g}}^i) \times (0, \tilde{\delta}).$$

2. *There exists* $\delta > 0$ *such that for all* $0 < \tau < \frac{\delta}{2M_{f'}}$ *there holds*

$$y(\sigma_{\mathrm{g}}^i + \tau, x) = \begin{cases} f'^{-1}\left(\frac{x}{\tau}\right), & \text{if } x \in [f'(-1)\tau, f'(0)\tau], \\ 0, & \text{if } x \in (f'(0)\tau, \delta - M_{f'}\tau), \\ -1, & \text{if } x \in (-\delta + M_{f'}\tau, f'(-1)\tau), \end{cases}$$

with $M_{f'}$ *from Corollary* 3.9.

4 Shift-Differentiability

In this section we present the main results of this paper, namely the shift-differentiable dependence of the control-to-state mapping for the considered problems (2.1) and (2.6).

4.1 Motivation and Preliminary Work

One of the main difficulties that arise when one considers optimal control problems concerning entropy solutions of hyperbolic conservation laws is, that the control-to-state mapping $u \mapsto y(u)$ is generally not differentiable in a sense, that is strong enough in order to simply deduce Fréchet-differentiability of the reduced objective functional. This issue of non-differentiability is caused by the presence of shocks in the entropic solution, even for smooth (e.g. C^∞) data. However we illustrate the situation by means of an example where the data are discontinuous, namely a Riemann problem.

Example. Consider the parametrized Cauchy problem

$$y_t^\varepsilon + \left(\tfrac{1}{2}(y^\varepsilon)^2\right)_x = 0 \qquad \text{on } [0,T] \times \mathbb{R}$$

$$y^\varepsilon(0,\cdot) = \varepsilon - \mathrm{sgn} \qquad \text{on } \mathbb{R}.$$

Then the entropy solution is almost everywhere given by

$$y^{\varepsilon}(t,x) = \begin{cases} \varepsilon + 1, & \text{if } x \le \varepsilon t, \\ \varepsilon - 1, & \text{if } x > \varepsilon t. \end{cases}$$

Furthermore, consider the mapping $S : \mathbb{R} \to L^1([a,b]),\ \varepsilon \mapsto y^{\varepsilon}(\bar{t}, \cdot)$. Clearly S is not differentiable in 0, since the obvious candidate for the derivative, $1 + 2\bar{t}\delta_0$, where δ_0 denotes the Dirac measure at $x = 0$, does not belong to $\mathcal{L}(\mathbb{R}, L^1([a,b]))$. In fact, differentiability does only hold in the weak topology of the measure space $\mathcal{M}([a\ b])$.

In order to still achieve a differentiability result for the reduced objective, a non-standard variational calculus was introduced in [9] and [36, 37]. The so called shift-variations mimic the observed behavior of the solution in the neighborhood of discontinuities. Shift-variations consist of an additive part (in L^1) and a second part that allows for horizontal shifts of discontinuities. We recall the definitions of the notions of shift-variations and shift-differentiability.

Definition 4.1 (Shift-variations, shift-differentiability).

1. Let $a < b$ and $v \in BV([a,b])$. For $a < x_1 < x_2 < \ldots < x_N < b$ we associate with $(\delta v, \delta x)$ the *shift-variation* $S_v^{(x_i)}(\delta v, \delta x) \in L^1([a,b])$ of v by

$$S_v^{(x_i)}(\delta v, \delta x)(x) := \delta v(x) \cdot \sum_{i=1}^{n} [v(x_i)] \operatorname{sgn}(\delta x_i) \mathbb{1}_{I(x_i, x_i + \delta x_i)}(x),$$

   where $[v(x_i)] := v(x_i-) - v(x_i+)$ and $I(\alpha, \beta) := [\min(\alpha,\beta), \max(\alpha,\beta)]$.
2. Let U be a real Banach space and $D \subset U$ open. Consider a locally bounded mapping $D \to L^{\infty}(\mathbb{R}),\ u \mapsto v(u)$. For $\bar{u} \in U$ with $v(\bar{u}) \in BV([a,b])$, we call v *shift-differentiable at* $\bar{u}$ if there exist $a < x_1 < x_2 < \ldots < x_N < b$ and $D_s v(\bar{u}) \in \mathcal{L}(U, L^r([a,b]) \times \mathbb{R}^N)$ for some $r \in (1, \infty]$, such that for $\delta u \in U$, $(\delta v, \delta x) := D_s v(\bar{u}) \cdot \delta u$

$$\left\| v(u + \delta u) - v(u) - S_v^{(x_i)}(\delta v, \delta x) \right\|_{1,[a,b]} = o(\|\delta u\|_U).$$

The utility of this variational concept lies in the feature that it implies the Fréchet-differentiability of tracking type functionals as in (1.1) (see Lemma 3.2.3 in [36]) as long as y_d and $y(\bar{t}, \cdot)$ do not share discontinuities on $[a,b]$. The derivative is given by

$$d_F J(y(u)) \cdot \delta u = \left(\psi_y(y(\bar{t}, \cdot; u), y_d), \delta y\right)_{2,[a,b]} + \sum_{i=1}^{N} \bar{\psi}_y(x_i)[y(\bar{t}, \cdot; u)] \delta x_i,$$

with

$$\bar{\psi}_y(x) := \int_0^1 \psi_y\left(y(\bar{t}, x+; u)) + \tau[y(\bar{t}, x; u)], y_d(x+) + \tau[y_d(x)]\right) \, d\tau.$$

4.2 Shift-Differentiability of Solutions to IBVPs and Traffic Light Problems

We now state the main results. First we consider the differentiability of the solution operator for the initial-boundary value problem. We restrict ourselves to the case $\Omega = (0,\infty)$, where the result for general intervals is similar. A reinspection of the formulation of the boundary condition (2.4a) motivates to only consider boundary data with $u_B \geq f'^{-1}(0)$, since both the choices u_B and $\max(u_B, f'^{-1}(0))$ as boundary data will yield the same solution. Therefore it is useful to define the space

$$U_B^\alpha := \{\varphi \in PC^1([0,T]; t_1 \ldots, t_K) \ : \ f'(\varphi) \geq \alpha\} \tag{4.1}$$

for given $0 < t_1 < t_2 < \ldots < t_K$. Consider $u = (u_0, u_B, u_1)$ where $u_B \in U_B^\alpha$ for some small $\alpha > 0$, $u_0 \in PC^1(\Omega; x_1, \ldots, x_N)$ for some $0 < x_1 < x_2 < \ldots < x_N$ and $u_1 \in C([0,T]; C^1(\mathbb{R})^m)$. We want to investigate the shift-differentiable dependence of $\delta u \mapsto y(\bar{t}, \cdot; u + \delta u)$ on δu. In addition to usual variations in the controls, we additionally consider some shift-variations of the initial and the boundary data. This means that we consider explicit shifts of discontinuities that create shocks, but no rarefactions. For this purpose we define

$$\begin{aligned}
\mathbf{S}_{(x_i)} &:= \{s \in \mathbb{R}^N \ : \ u_0(x_i-) < u_0(x_i+) \ \Rightarrow s_i = 0, \ i = 1, \ldots, N\},\\
\mathbf{S}_{(t_j)} &:= \{s \in \mathbb{R}^K \ : \ u_B(t_j-) > u_B(t_j+) \ \Rightarrow s_j = 0, \ j = 1, \ldots, K\}
\end{aligned}$$

and consider variations in

$$\begin{aligned}
W := {} & PC^1(\Omega; x_1, \ldots, x_N) \times \mathbf{S}_{(x_i)}\\
& \times PC^1([0,T]; t_1, \ldots, t_K) \times \mathbf{S}_{(t_j)} \times C([0,T]; C^1(\mathbb{R})^m).
\end{aligned} \tag{4.2}$$

Under a nondegeneracy condition on the shocks (see Definition 3.6.1 in [36]) we get the following result.

Theorem 4.2 (Shift-Differentiability for IBVPs). *Let (A1′) and (A3) hold and let in addition g be affine linear w.r.t. y. Let* $\Omega = (0,\infty)$ *and* $0 < x_1 < x_2 < \ldots < x_N$, $0 < t_1 < t_2 < \ldots < t_K$ $u_0 \in PC^1(\Omega; x_1, \ldots, x_N)$, $u_B \in U_B^\alpha$ *for some* $\alpha > 0$ *and* $u_1 \in C([0,T]; C^1(\mathbb{R})^m)$. *For* $u = (u_0, u_B, u_1)$ *denote by* $y = y(u) \in L^\infty(\Omega_T) \cap C([0,T]; L^1_{\text{loc}}(\Omega))$ *the entropy solution of the initial-boundary value problem* (2.1) *on* Ω_T. *Let* $0 < a < b$ *and* $\bar{t} \in (0,T)$ *such that on* $[a,b]$ $y(\bar{t}, \cdot; u)$ *has no shock generation points and only a finite number of shocks at* $a < \bar{x}_1 < \ldots < \bar{x}_{\bar{N}} < b$, *that all are neither degenerated nor shock interaction points. Further assume that for almost all* $t \in [0,T]$ *the boundary trace* $y(\cdot, 0+; u) \in L^\infty(0,T)$ *satisfies* $u_B(t) \neq y(t, 0+; u) \ \Rightarrow \ f(u_B(t)) \neq f(y(t, 0+; u))$.

For W from (4.2) *we consider the mapping*

$$(\delta u_0, \delta x, \delta u_B, \delta t, \delta u_1) \in W \longmapsto$$

$$y\left(\bar{t}, \cdot; u_0 + S_{u_0}^{(x_i)}(\delta u_0, \delta x), u_B + S_{u_B}^{(t_j)}(\delta u_B, \delta t), u_1 + \delta u_1\right) \in L^1(a,b). \tag{4.3}$$

If $(x_i), (t_j)$ *are real discontinuities of* u_0, u_B, *i.e.* $u_0(x_i-) \neq u_0(x_i+)$ *and* $u_B(t_j-) \neq u_B(t_j+)$, *respectively, then the mapping* (4.3) *is continuously shift-differentiable on a sufficiently small neighborhood* $B_\rho^W(0) := \{\delta_u \in W : \|\delta u\|_W \leq \rho\}$. *The shift-derivative satisfies* $T_s(0) = D_s y(\bar{t}, \cdot; u) \in \mathcal{L}(W, PC([a,b]; \bar{x}_1, \ldots, \bar{x}_{\bar{N}}) \times \mathbb{R}^{\bar{N}})$.

Remark 4.3. If u_0 or u_B are continuous at some x_i or t_j, respectively, similarly to the second assertion of Theorem 3.3.2 in [36], the shift-differentiability of (4.3) in 0 is preserved. The shift-derivative satisfies $T_s(0) \in \mathcal{L}(W, PC([a,b]; \bar{x}_1, \ldots, \bar{x}_{\bar{N}}, \tilde{x}_1, \ldots, \tilde{x}_{\tilde{N}}) \times \mathbb{R}^{\bar{N}})$, where the set of discontinuities of $y(u)$ is augmented by continuity points $\tilde{x}_k$ that are starting points of genuine backward characteristics that end in a (pseudo-) discontinuity x_i or t_j.

The proof can be obtained by a very careful extension of the proof of Theorem 3.3.2 in [36]. This requires a proper analysis of the solution y in small neighborhoods of different types of generalized backward characteristics. A detailed proof will be presented in a forthcoming paper.

The following corollary is a simple consequence of the above theorem and Lemma 3.2.3 in [36].

Corollary 4.4. *Let the assumptions of Theorem 4.2 hold and consider J defined as in* (1.1). *If* y_d *is continuous in a small neighborhood of* $\{\bar{x}_1, \ldots, \bar{x}_{\bar{N}}\}$, *then the reduced objective functional* $\delta u \in W \mapsto J(y(u + \delta u))$ *is continuously Fréchet-differentiable on* $B_\rho^W(0)$ *for* $\rho > 0$ *small enough.*

An adjoint-based formula for the gradient of the considered mapping will be presented in Theorem 4.8.

For the traffic light problem we have a very similar result.

Theorem 4.5 (Shift-Differentiability for traffic light problems). *Let (A1′) and (A4) hold and let in addition g be affine linear w.r.t. y. Let* $x_1 < x_2 < \ldots < x_N$, $\sigma = (\sigma_g^0, \sigma_r^1, \sigma_g^1, \ldots, \sigma_g^{n_\sigma}, \sigma_r^{n_\sigma+1}) \in \Sigma_{ad}$, $PC^1(\mathbb{R}; x_1, \ldots, x_N)$ *and* $u_1 \in C([0,T]; C^1(\mathbb{R})^m)$. *For* $\sigma \in \Sigma_{ad}$ *denote by* $y = y(\sigma) \in L^\infty(\Omega_T) \cap C([0,T]; L^1_{loc}(\Omega))$ *the solution of the traffic light problem* (2.6). *Let* $a < b$ *and* $\bar{t} \in (\sigma_g^{n_\sigma}, \sigma_r^{n_\sigma-1})$ *such that on* $[a,b]$ $y(\bar{t}, \cdot; \sigma)$ *has no shock generation points and only a finite number of shocks at* $a < \bar{x}_1 < \ldots < \bar{x}_{\bar{N}} < b$, *that all are neither degenerated nor shock interaction points. Furthermore assume that for almost all* $t \in [\sigma_g^i, \sigma_r^i]$, $i = 1, \ldots, n_\sigma$ *the boundary traces* $(y(\cdot, 0-; \sigma), y(\cdot, 0+; \sigma)) \in L^\infty(0,T)^2$ *are equal to* $(-1, 0)$.

Finally let $\Sigma_0 := \{\nu \in \mathbb{R}^{2(n_\sigma+1)} \,:\, \nu_1 = \nu_{2(n_\sigma+1)} = 0\}$ *then the mapping*

$$\delta\sigma \in \Sigma_0 \longmapsto y\,(\bar{t}, \cdot; \sigma + \delta\sigma) \in L^1(a,b)$$

is continuously shift-differentiable on a sufficiently small neighborhood $B_\rho^\Sigma(0) := \{\delta\sigma \in \Sigma_0 \,:\, \|\delta\sigma\| \le \rho\}$. *The shift-derivative satisfies* $T_s(0) = D_s y(\bar{t}, \cdot; \sigma) \in \mathcal{L}(\Sigma_0, PC([a,b]; \bar{x}_1, \cdot, \bar{x}_{\bar{N}}) \times \mathbb{R}^{\bar{N}})$.

It is important to emphasize that in comparison to the result for the initial (-boundary) value problem, also green switching times, i.e. rarefaction centers, may explicitly be shifted. This is because the solution in a neighborhood of such points is thoroughly known for TLPs, see Lemma 3.10, whereas the structure for general rarefaction waves may be more delicate.

As for the IBVP, one can deduce from Lemma 3.2.3 in [36] the total differentiability for reduced objective functionals.

Corollary 4.6. *Let the assumptions of Theorem 4.5 hold and consider* J *defined as in* (1.1). *If* y_d *is continuous in a small neighborhood of* $\{\bar{x}_1, \ldots, \bar{x}_{\bar{N}}\}$, *then the reduced functional* $\delta\sigma \in \Sigma_0 \mapsto J(y(\sigma + \delta\sigma))$ *is continuously differentiable on* $B_\rho^\Sigma(0)$ *for* $\rho > 0$ *small enough.*

One also may consider the optimal control of the traffic light problem for fixed switching times where the source term and the initial data is controlled. Here one can obtain similar results as for the initial (-boundary) value problem without any traffic lights.

4.3 Adjoint Equation

The sensitivity of the shock position, that is needed in order to obtain the shift-differentiability result of Theorem 4.2, is based on an adjoint-argument. As already discussed in [36] for the Cauchy problem, the classical adjoint calculus is not applicable for problems concerning discontinuous solutions of hyperbolic equations. Nevertheless one can define an adjoint state as a solution of the following equation

$$p_t + f'(y)p_x = -g_y(\cdot, y, u_1)p, \qquad \text{on } \Omega_{\bar{t}}, \tag{4.4a}$$

$$p(\bar{t}, \cdot) = p^{\bar{t}}, \qquad \text{on } \Omega. \tag{4.4b}$$

The adjoint equation (4.4) is a linear transport equation with discontinuous coefficients, since y may contain shocks. In [6] Bouchut and James showed that for $\Omega = \mathbb{R}$, $g \equiv 0$ and Lipschitz continuous end data $p^{\bar{t}}$ Eq. (4.4) does not admit a unique solution within the space of Lipschitz continuous functions. Nevertheless they give a definition and a characterization of a *reversible solution* for (4.4), which satisfies a crucial duality relation. In [36, 38] this notion was extended to more general source terms g and discontinuous end data. In this case the reversible solution p can be characterized as the solution along generalized characteristics of the state y. For the IBVP on $\Omega = (0, \infty)$ we have to deal with the fact that this definition might lead to an underdetermined problem, since not all characteristics on $\Omega_{\bar{t}}$ intersect the line $\{\bar{t}\} \times \Omega$, where the initial (or terminal) condition acts. One can show by the theory of generalized characteristics that the set D of points that lie on a genuine characteristic that does not reach the line $\{\bar{t}\} \times \Omega$ is a connected set that lies in the lower left corner of the space-time cylinder $\Omega_{\bar{t}}$.

Definition 4.7. Let $p^{\bar{t}}$ be a bounded function that is the pointwise everywhere limit of a sequence (w_n) in $C^{0,1}(0, \infty)$, with (w_n) bounded in $C(0, \infty) \cap W^{1,1}_{\text{loc}}(0, \infty)$. The adjoint state p associated to (4.4) for $\Omega = (0, \infty)$ is characterized by the following requirements:

1. For every generalized characteristic ξ of y through $(\bar{t}, \bar{x}) \in \Omega_T$

$$t \mapsto p^{\xi}(t) = p(t, \xi(t))$$

is the solution of the ordinary differential equation

$$\dot{p}^{\xi}(t) = -g_y(t, \xi(t), y(t, \xi(t)), u_1(t, \xi(t)))p^{\xi}(t), \quad t \in (0, \bar{t}] \ : \ \xi(t) > 0,$$
$$p^{\xi}(\bar{t}) = p^{\bar{t}}(\bar{x}).$$

2. For every $(t, x) \in D$ there holds $p(t, x) = 0$, where

$$D := \{(t, x) \in \Omega_{\bar{t}} \ : \ t \in [0, \tau], \ x \leq \tilde{\xi}(t)\}.$$

Here $\tilde{\xi}$ denotes the maximal backward characteristic through $(\tau, 0)$, where $\tau := \operatorname{esssup}\{t \in [0, \bar{t}] \ : \ f'(y(t, 0+)) < 0\}$.

Using the above definition of an adjoint state, we are now able to formulate a representation of the gradient of the reduced objective function.

Theorem 4.8. *Let the assumptions of Corollary* 4.4 *hold and let the terminal data in* (4.4) *be given by*

$$p^{\bar{t}}(t, x) := \gamma(x) \int_0^1 \psi_y(y(\bar{t}, x+) + \tau[y(\bar{t}, x)], x) \, d\tau.$$

Then there exists an adjoint state p according to Definition 4.7, *satisfying*

$$p \in B((0, \bar{t}) \times (0, \infty)) \cap BV_{\text{loc}}([0, \bar{t}] \times [0, \infty)),$$

where $B((0, \bar{t}) \times (0, \infty))$ *denotes the space of measurable bounded functions (defined pointwise everywhere).*

The derivative of the reduced functional $\delta u \in W \mapsto \hat{J}(\delta u) = J(y(u + \delta u))$ *for* $\rho > 0$ *small enough is given by*

$$\begin{aligned}\hat{J}'(0) \cdot \delta u = (p, g_{u_1}(\cdot, y, u_1)\delta u_1)_{2,(0,\bar{t})\times\mathbb{R}+} \\ + (p(0, \cdot), \delta u_0)_{2,\mathbb{R}+} + \big(p(\cdot, 0), f'(u_B)\delta u_B\big)_{2,(0,\bar{t})} \\ + \sum_{i=1}^{N} p(0, x_i)[u_0(x_i)]\delta x_i + \sum_{j=1}^{K} p(t_j, 0)[f(u_B(t_j))]\delta t_j .\end{aligned}$$

Conclusion and Outlook

We have presented a generalized differentiability result for an initial-boundary value problem for a nonlinear hyperbolic conservation law on an interval by using the theory of generalized characteristics. This property implies the Fréchet-differentiability of the reduced objective functional, for which we also presented an adjoint-based gradient representation. The result is an important step to make such problems accessible to gradient based optimization algorithms. Furthermore we have discussed the dependence of the state on the switching times of a traffic light on a single road. Also in this case we were able to show shift-differentiability by similar arguments. The considered problem for the traffic light can also be seen as a network problem involving one node and two edges and can be in a straight-forward manor extended to the case of multiple incoming and outgoing roads that are connected by a similar modular node condition that time dependently connects some pairs of incoming and outgoing roads and closes others. If one chooses the sequence of modes in such a way, that no road is open for two or more consecutive time phases, the same arguments as for the traffic light problem can be used. Questions for future research will be whether one may drop the latest assumption. Moreover we will have to investigate the case when the boundary data of the red light condition (2.6f), (2.6g) is not assumed by the boundary trace, which means that the traffic light turns red, when either the incoming road is empty near the traffic light or the outgoing road has already reached its maximum capacity. This becomes more important, if one considers multiple traffic lights in a row. Another interesting modification of the traffic problem is the case where the flux functions on the two sides of the junction are not necessarily the same. Moreover, it will be of interest how the shift-differentiability concept applies to networks of three edges, that are connected by more common node conditions, as those from [10] and [13].

Finally, our results form the basis for the convergence analysis of numerical approximations of the considered optimal control problems. So far, there exist several results in the context of initial value problems with initial control and

(continued)

sometimes also with control in the source term. The convergence of optimal solutions of discretized optimal control problems was considered e.g. in [11, 35]. The convergence of sensitivities, adjoints and reduced gradients was analyzed in [19, 20, 36–38], see also [11] for an alternating descent method. We are currently investigating the extension of these results to the case of the initial-boundary value problem with boundary control and to the traffic light problem. Here, we follow the approach in [12] for the discrete approximation of the boundary condition, where the convergence to the unique entropy solution of the initial-boundary value problem according to [4] is shown. A particular issue will be the appropriate discrete approximation of shift variations for boundary controls. We plan to consider the variation of the times step sizes between switching times as well as discretization techniques with fixed time steps.

Acknowledgements The authors gratefully acknowledge the support of the German Research Foundation (DFG) within the Priority Program 1253 "Optimization with Partial Differential Equations" under grant UL158/8-1. Moreover, we gratefully acknowledge discussions with T. I. Seidman.

References

1. F. Ancona, G.M. Coclite, On the attainable set for Temple class systems with boundary controls. SIAM J. Control Optim. **43**(6), 2166–2190 (2005). (electronic)
2. F. Ancona, A. Marson, On the attainable set for scalar nonlinear conservation laws with boundary control. SIAM J. Control Optim. **36**(1), 290–312 (1998). (electronic)
3. M.K. Banda, M. Herty, A. Klar, Coupling conditions for gas networks governed by the isothermal Euler equations. Netw. Heterog. Media **1**(2), 295–314 (2006)
4. C. Bardos, A.Y. le Roux, J.-C. Nédélec, First order quasilinear equations with boundary conditions. Commun. Partial Differ. Equ. **4**(9), 1017–1034 (1979)
5. S. Bianchini, On the shift differentiability of the flow generated by a hyperbolic system of conservation laws. Discret. Contin. Dyn. Syst. **6**(2), 329–350 (2000)
6. F. Bouchut, F. James, One-dimensional transport equations with discontinuous coefficients. Nonlinear Anal. **32**(7), 891–933 (1998)
7. A. Bressan, *Hyperbolic Systems of Conservation Laws*. Volume 20 of Oxford Lecture Series in Mathematics and Its Applications (Oxford University Press, Oxford, 2000). The one-dimensional Cauchy problem.
8. A. Bressan, G. Guerra, Shift-differentiability of the flow generated by a conservation law. Discret. Contin. Dynam. Syst. **3**(1), 35–58 (1997)
9. A. Bressan, A. Marson, A variational calculus for discontinuous solutions of systems of conservation laws. Commun. Partial Differ. Equ. **20**(9–10), 1491–1552 (1995)
10. G. Bretti, R. Natalini, B. Piccoli, Numerical approximations of a traffic flow model on networks. Netw. Heterog. Media **1**(1), 57–84 (2006)
11. C. Castro, F. Palacios, E. Zuazua, An alternating descent method for the optimal control of the inviscid Burgers equation in the presence of shocks. Math. Models Methods Appl. Sci. **18**(3), 369–416 (2008)

12. B. Cockburn, F. Coquel, P.G. LeFloch, Convergence of the finite volume method for multidimensional conservation laws. SIAM J. Numer. Anal. **32**(3), 687–705 (1995)
13. G.M. Coclite, M. Garavello, B. Piccoli, Traffic flow on a road network. SIAM J. Math. Anal. **36**(6), 1862–1886 (2005)
14. G.M. Coclite, K.H. Karlsen, Y.-S. Kwon, Initial-boundary value problems for conservation laws with source terms and the Degasperis-Procesi equation. J. Funct. Anal. **257**(12), 3823–3857 (2009)
15. R.M. Colombo, A. Groli, On the optimization of the initial boundary value problem for a conservation law. J. Math. Anal. Appl. **291**(1), 82–99 (2004)
16. R.M. Colombo, G. Guerra, M. Herty, V. Schleper, Optimal control in networks of pipes and canals. SIAM J. Control Optim. **48**(3), 2032–2050 (2009)
17. C.M. Dafermos, Generalized characteristics and the structure of solutions of hyperbolic conservation laws. Indiana Univ. Math. J. **26**(6), 1097–1119 (1977)
18. F. Dubois, P. LeFloch, Boundary conditions for nonlinear hyperbolic systems of conservation laws. J. Differ. Equ. **71**(1), 93–122 (1988)
19. M. Giles, S. Ulbrich, Convergence of linearized and adjoint approximations for discontinuous solutions of conservation laws. Part 1: linearized approximations and linearized output functionals. SIAM J. Numer. Anal. **48**(3), 882–904 (2010)
20. M. Giles, S. Ulbrich, Convergence of linearized and adjoint approximations for discontinuous solutions of conservation laws. Part 2: adjoint approximations and extensions. SIAM J. Numer. Anal. **48**(3), 905–921 (2010)
21. S. Göttlich, M. Herty, A. Klar, Modelling and optimization of supply chains on complex networks. Commun. Math. Sci. **4**(2), 315–330 (2006)
22. M. Gugat, M. Herty, A. Klar, G. Leugering, Optimal control for traffic flow networks. J. Optim. Theory Appl. **126**(3), 589–616 (2005)
23. M. Gugat, M. Herty, V. Schleper, Flow control in gas networks: exact controllability to a given demand. Math. Methods Appl. Sci. **34**(7), 745–757 (2011)
24. F.M. Hante, G. Leugering, T.I. Seidman, Modeling and analysis of modal switching in networked transport systems. Appl. Math. Optim. **59**(2), 275–292 (2009)
25. H. Holden, N.H. Risebro, A mathematical model of traffic flow on a network of unidirectional roads. SIAM J. Math. Anal. **26**(4), 999–1017 (1995)
26. S.N. Kružkov, Quasilinear parabolic equations and systems with two independent variables. Trudy Sem. Petrovsk. **5**, 217–272 (1979)
27. P. LeFloch, Explicit formula for scalar nonlinear conservation laws with boundary condition. Math. Methods Appl. Sci. **10**(3), 265–287 (1988)
28. A.Y. le Roux, Étude du problème mixte pour une équation quasi-linéaire du premier ordre. C. R. Acad. Sci. Paris Sér. A-B **285**(5), A351–A354 (1977)
29. M.J. Lighthill, G.B. Whitham, On kinematic waves. II. A theory of traffic flow on long crowded roads. Proc. R. Soc. Lond. Ser. A. **229**, 317–345 (1955)
30. J. Málek, J. Nečas, M. Rokyta, M. Ruužička, *Weak and Measure-Valued Solutions to Evolutionary PDEs*. Volume 13 of Applied Mathematics and Mathematical Computation (Chapman & Hall, London, 1996)
31. F. Otto, Initial-boundary value problem for a scalar conservation law. C. R. Acad. Sci. Paris Sér. I Math. **322**(8), 729–734 (1996)
32. V. Perrollaz, Asymptotic stabilization of entropy solutions to scalar conservation laws through a stationary feedback law. Ann. Inst. H. Poincaré Anal. Non Linéaire **30**(5), 879–915 (2013)
33. E. Polak, *Optimization*. Volume 124 of Applied Mathematical Sciences (Springer, New York, 1997). Algorithms and consistent approximations
34. P.I. Richards, Shock waves on the highway. Oper. Res. **4**, 42–51 (1956)
35. S. Ulbrich, On the existence and approximation of solutions for the optimal control of nonlinear hyperbolic conservation laws, in *Optimal Control of Partial Differential Equations* (Chemnitz, 1998). Volume 133 of International Series of Numerical Mathematics (Birkhäuser, Basel, 1999), pp. 287–299

36. S. Ulbrich, Optimal control of nonlinear hyperbolic conservation laws with source terms. Habilitation, Zentrum Mathematik, Technische Universität München, 2001
37. S. Ulbrich, A sensitivity and adjoint calculus for discontinuous solutions of hyperbolic conservation laws with source terms. SIAM J. Control Optim. **41**(3), 740–797 (2002). (electronic)
38. S. Ulbrich, Adjoint-based derivative computations for the optimal control of discontinuous solutions of hyperbolic conservation laws. Syst. Control Lett. **48**(3–4), 313–328 (2003). Optimization and control of distributed systems
39. A. Vasseur, Strong traces for solutions of multidimensional scalar conservation laws. Arch. Ration. Mech. Anal. **160**(3), 181–193 (2001)

Elliptic Mathematical Programs with Equilibrium Constraints in Function Space: Optimality Conditions and Numerical Realization

Michael Hintermüller, Antoine Laurain, Caroline Löbhard, Carlos N. Rautenberg, and Thomas M. Surowiec

Abstract Recent advances in the analytical as well as numerical treatment of classes of elliptic mathematical programs with equilibrium constraints (MPECs) in function space are discussed. In particular, stationarity conditions for control problems with point tracking objectives and subject to the obstacle problem as well as for optimization problems with variational inequality constraints and pointwise constraints on the gradient of the state are derived. For the former problem class including the case of L^2-tracking-type objectives (rather than pointwise ones) a bundle-free solution method as well as adaptive finite element discretizations are introduced. Moreover, the analytical and numerical treatment of shape design problems subject to elliptic variational inequality constraints is highlighted. With respect

This work was completed with the support of DFG-Project "Elliptic Mathematical Programs with Equilibrium Constraints (MPECs) in Function Space: Optimality Conditions and Numerical Realization" within the DFG Priority Program SPP 1253 on "Optimization with Partial Differential Equations".

M. Hintermüller (✉) • C. Löbhard • T.M. Surowiec
Institut für Mathematik, Humboldt-Universität zu Berlin, Unter den Linden 6, D-10099 Berlin, Germany
e-mail: hint@math.hu-berlin.de; loebhard@math.hu-berlin.de; surowiec@mathematik.hu-berlin.de

A. Laurain
Institut für Mathematik, Technische Universität Berlin, Straße des 17. Juni 136, D-10623 Berlin, Germany
e-mail: laurain@math.tu-berlin.de

C.N. Rautenberg
Department of Mathematics and Scientific Computing, Karl-Franzens-University of Graz, Heinrichstrasse 36, A-8010 Graz, Austria
e-mail: carlos.rautenberg@uni-graz.at

G. Leugering et al. (eds.), *Trends in PDE Constrained Optimization*, International Series of Numerical Mathematics 165, DOI 10.1007/978-3-319-05083-6_9

to problems involving gradient constraints, the paper ends with a fixed-point-Moreau-Yosida-based semismooth Newton solver for a class of nonlinear elliptic quasi-variational inequality problems.

Keywords MPECs and MPCCs in function space • Gradient constraints • Point-tracking • Adaptive finite element methods • Nonlinear elliptic quasivariational inequality problem • Optimal shape design subject to elliptic variational inequalities

Mathematics Subject Classification (2010). 49K20, 49M25, 65K10, 90C33.

1 Introduction

A variety of phenomena in engineering, life sciences, mathematical finance, economics, and physics can be modeled by variational and quasi-variational inequalities. Among the many applications within these areas one finds contact problems in elasticity, torsion problems in plasticity, option pricing in finance, the magnetization of superconductors, ionization problems in electrostatics as well as economic phenomena such as the behavior of oligopolies and coalitions. In addition to the challenging aspects of analysis and numerical treatment of these "equilibrium problems", one is often interested in influencing the system under consideration in order to optimize a certain output quantity. The resulting optimization/optimal control problems fall under the category of mathematical programs with equilibrium constraints (MPECs).

In the following, we use as a prototype the following class of problems to represent a general MPEC:

$$\text{Minimize } J(u, y) \qquad \text{over } (u, y) \in U \times Y, \tag{1.1a}$$

$$\text{subject to (s.t.) } y \in S(u), \text{ and } u \in \mathcal{U}_{ad} \subset U \tag{1.1b}$$

Here, J is an objective functional dependent on both the control variable $u \in U$ and the state variable y in a reflexive Banach space Y with Y^* the associated dual space. Duality pairs are typically denoted by $\langle \cdot, \cdot \rangle$. We denote the set of admissible controls by the nonempty subset $\mathcal{U}_{ad}$ and we let $S : U \to 2^Y$ be the possibly set-valued solution mapping for a given equilibrium problem

$$\text{Find } y \in K : \langle Ay - Bu - f, z - y \rangle \geq 0, \ \forall z \in K. \tag{1.2}$$

Above $K \subset Y$ is nonempty, closed, and convex; $A \in \mathcal{L}(Y, Y^*)$ is coercive or $A : Y \to Y^*$ is a strongly monotone, hemicontinuous operator; and $f \in Y^*$. Here, the control u enters via Bu, where $B \in \mathcal{L}(U, Y^*)$. However, one may also encounter situations in which a domain Ω acts as a control (see Sect. 4). Problems of this class are called elliptic variational inequalities (VIs) for a fixed feasible set K

as given above, and (1.2) becomes a so-called elliptic quasi-variational inequality (QVI) when K depends on y. The study of such problems has a long history going back to the influential work of Fichera, Lions, Stampacchia and Brezis in the 1960s, cf. [7, 3, 10, 29] and the references in [7].

In most cases, the solution mapping, even when it is single-valued, fails to have the necessary properties needed for the derivation of a classical multiplier-based first-order optimality system, e.g. Gâteaux differentiability. Even in settings where one can introduce a slack variable λ and reformulate the problem (1.2) e.g. as

$$Ay - Bu - f = \lambda, \qquad y \geq 0, \qquad \lambda \geq 0, \qquad \langle \lambda, y \rangle = 0, \tag{1.3}$$

the resulting problem has an inherently degenerate feasible set, for which classical KKT theory cannot be applied. These problems are often referred to as mathematical programs with complementarity constraints (MPCC) in the literature, cf. [32,37,38] in a finite dimensional setting, or [14, 15, 17] for problems posed in function space.

Using penalty and regularization techniques, much work was done in the 1970s and 1980s concerning the optimal control of elliptic VIs as can be seen in the monograph by Barbu [2], in which the 'adapted penalty' approach of Lions [30, 31] and Yvon [43] is generalized to a larger setting. Without using this penalty approach, Mignot and Puel [34] derive stronger optimality conditions than those in [2] for the optimal control of the obstacle problem by relying on earlier work by Mignot on the generalized differentiability of S, cf. [33].

Nevertheless, the results mentioned in the works above, with the exception of those of Yvon in [43, 44], stop short of developing numerical methods and were mainly concerned with the derivation of first-order optimality conditions. In [44], the solution operator S was smoothed and a sequence of related optimal control problems was solved, thus foreshadowing some of the more popular methods currently in use, cf. [17, 18, 39]. Later in [25], conditions similar to those in [2] are rederived and a numerical method via a Gauss-Seidel-type iteration, as suggested by Barbu [2, p. 89], is implemented. However, no convergence results were provided.

This paper highlights some important recent results on the development of a suitable optimality theory as well as the design and implementation of efficient solution algorithms for classes of MPECs in function space, as well as for Quasi Variational Inequalities (QVIs) with gradient constraints.

2 Analytical Methods for the Derivation of Stationarity Conditions for MPECs in Function Space

In this section, we discuss different techniques to provide new stationarity conditions for two classes of MPECs which are modified variants of those derived in [34] and similar to those in [17, 18].

For an open bounded domain $\Omega \subset \mathbb{R}^n$, $n \in \mathbb{N}$, we consider $a_{ij} \in L^\infty(\Omega)$ $(i, j \in \{1, \ldots, n\})$ such that the linear operator $A : Y = H_0^1(\Omega) \to H^{-1}(\Omega)$, defined by $A = -\mathrm{div}((a_{ij})\nabla \cdot)$ is uniformly coercive and bounded. Further, we assume $K \subset H_0^1(\Omega)$, $u, f \in L^2(\Omega)$, and $B = i_{L^2 \hookrightarrow H^{-1}}$ the embedding of $L^2(\Omega)$ into $H^{-1}(\Omega)$ in the variational inequality problem (1.2) with the (thus well-defined) solution operator $\mathrm{S} : L^2(\Omega) \to H_0^1(\Omega)$. In Sect. 2.1, the objective is defined on $L^2(\Omega) \times \mathcal{C}(\overline{\Omega})$ and the solution operator S of an obstacle type constraint has to be considered as a mapping into $\mathcal{C}(\overline{\Omega})$. In Sect. 2.2 we consider the optimal control of a variational inequality with gradient constraints.

2.1 Optimal Control of an Obstacle Problem with Pointwise Tracking Term

By adapting the proof of [27, IV, Thm.2.3], it was shown in [6] that if $\Omega \subset \mathbb{R}^2$ is a Lipschitz domain, then the solution operator S of (1.2) with $K = \{y \in H_0^1(\Omega) \mid y \geq 0 \text{ a.e. on } \Omega\}$ maps $W^{-1,q}(\Omega)$ into $W_0^{1,q}(\Omega)$ for some $q > 2$. Furthermore, $S(\cdot)$ is a singleton. Since $W_0^{1,q}(\Omega) \subset C(\overline{\Omega})$ for $q > 2$, we may thus consider the following optimal control problem:

$$\text{Minimize } J(u, y) = \frac{1}{2}\sum_{w \in I}(y(w) - y_w)^2 + \frac{\alpha}{2}\|u\|_{L^2(\Omega)}^2 \tag{2.1a}$$

$$\text{over } (y, u) \in W_0^{1,q}(\Omega) \times L^2(\Omega), \tag{2.1b}$$

$$\text{s.t. } y = S(u) \text{ solves (1.2)}, \tag{2.1c}$$

$$\text{and } u \in \mathcal{U}_{ad} = \{u \in L^2(\Omega) \mid \underline{u} \leq u \leq \overline{U} \text{ a.e. in } \Omega\}. \tag{2.1d}$$

Here, $w \in \mathbb{R}^2$ denotes an evaluation point in a finite set $I \subset \Omega$, $y_w \in \mathbb{R}$ is a desired (or measured) value of the state y at w, $\alpha > 0$ represents the cost of the control and the bounds satisfy $\underline{u}, \overline{U} \in L^2(\Omega)$ and $\underline{u} < \overline{U}$ a.e. in Ω or $\underline{u} = -\infty$, $\overline{U} = \infty$. The point evaluation of $y \in W_0^{1,q}(\Omega)$ in w is also denoted by $\langle \delta_w, y\rangle_{-1,q'}$, where $\delta_w \in W^{-1,q'}(\Omega) = (W_0^{1,q}(\Omega))^*$, the dual of $W_0^{1,q}(\Omega)$, with $q' = \frac{q}{q-1}$. The weak continuity of $S : L^2(\Omega) \to W_0^{1,q}(\Omega)$ with respect to subsequences, the weak closedness of $\mathcal{U}_{ad}$, and the weak lower semi-continuity of the objective functional $J : L^2(\Omega) \times W_0^{1,q}(\Omega) \to \mathbb{R}$ yield the existence of a solution of (2.1). In a physical interpretation of this model problem, one would aim to control the deflection of a membrane, which is clamped at the boundary of Ω, such that it is as close as possible to certain values of interest (e.g. measurements in an inverse problem context), and $\alpha > 0$ is the cost of the control.

Auxiliary Problem We approximate the variational inequality in the constraints by a sequence of semi-linear partial differential equations and replace the point evaluation in the objective by an averaging integral. In this way, we obtain the following auxiliary optimal control problem:

$$\text{Minimize } J_r(u,y) = \sum_{w\in I} \frac{1}{2|B_r(w)|} \|y - y_w\|^2_{L^2(B_r(w))} + \frac{\alpha}{2} \|u\|^2_{L^2(\Omega)} \tag{2.2a}$$

$$\text{over } (y,u) \in H^1_0(\Omega) \times \mathcal{U}_{ad}, \quad \text{s.t. } Ay - \gamma \max{}^l_\epsilon(0,-y) = u + f. \tag{2.2b}$$

Here, for $r > 0$, $B_r(w) = \{x \in \Omega \,|\, |x - w| < r\}$, $\max^l_\epsilon$ is for example the locally smoothed max-operator from [18, Eq. (2.4)], $\gamma > 0$ is the penalization and $\epsilon > 0$ the smoothing parameter. Moreover, for the ease of notation, we do ncte write B. One can prove the following convergence result, cf. [6]:

Theorem 2.1. *Let $\gamma_k \to \infty$, $r_k \to 0$ be positive sequences and let $(\epsilon_k)_{k\in\mathbb{N}} = (\epsilon(\gamma_k))_{k\in\mathbb{N}} \subset \mathbb{R}_+\setminus\{0\}$ satisfy $\lim_{k\to\infty} \gamma_k\epsilon_k = 0$. For all $k \in \mathbb{N}$ let (u_k, y_k) denote a solution of the smoothed penalized problem (2.2) with parameters $(\gamma,\epsilon,r) = (\gamma_k,\epsilon_k,r_k)$. Then there exists a subsequence of $(u_k, y_k)_{k\in\mathbb{N}}$ (denoted the same) and a solution (u,y) of (2.1) such that $y_k \to y$ in $W^{1,q}_0(\Omega)$, $u_k \rightharpoonup u$ in $L^2(\Omega)$, and $\gamma_k \max^l_{\epsilon_k}(0,-y_k) \rightharpoonup \lambda := Ay - u - f$ in $L^2(\Omega)$.*

Resulting stationarity system By applying Theorem 3.1 in [45], one can derive a stationarity system for solutions of (2.2). We utilize Theorem 2.1 and a technique from [1] (which requires the averaging of the point evaluations in the objective functional) to show that stationary points of the auxiliary problem converge to solutions of the system stated in Theorem 2.2 below, see [6] for details. In its formulation we use the inactive set $\mathcal{I}(y) := \{x \in \Omega \,|\, y(x) > 0\}$.

Theorem 2.2. *An optimal solution solution $(y,u,\lambda) \in W^{1,q}_0(\Omega)\times\mathcal{U}_{ad}\times W^{-1,q}(\Omega)$ of (2.1) satisfies the so-called* limiting ε-almost C-stationary conditions, *i.e., there exist multipliers $p \in W^{1,q'}_0(\Omega)$ and $\mu \in (L^\infty(\Omega))^*$ and sequences $(p_k)_{k\in\mathbb{N}} \subset H^1_0(\Omega)$, $(\mu_k)_{k\in\mathbb{N}} \subset H^{-1}(\Omega)$ such that $p_k \rightharpoonup p$ in $W^{1,q'}_0(\Omega)$, and $\mu_k \rightharpoonup^* \mu$ in $(L^\infty(\Omega))^*$, and the following conditions are satisfied:*

$$u = \mathrm{Proj}_{\mathcal{U}_{ad}}(\frac{1}{\alpha}p), \qquad A^*p - \mu - \sum_{w\in I}(y(w) - y_w)\delta_w = 0, \tag{2.3a}$$

$$\langle \mu, y\rangle_{-1,q'} = 0, \qquad \langle \lambda, p\rangle_{-1,q} = 0, \tag{2.3b}$$

$$\forall \tau > 0\, \exists E_\tau \subset \mathcal{I}(y) \text{ s.t. } |\mathcal{I}(y)\setminus E_\tau| < \tau \text{ and}$$

$$\forall \varphi \in L^\infty(\Omega), \varphi|_{\Omega\setminus E_\tau} = 0, \quad \langle \mu, \varphi\rangle_{(L^\infty(\Omega))^*} = 0, \tag{2.3c}$$

$$\limsup\{\langle \mu_k, p_k\rangle_{-1,2} \,|\, k \in \mathbb{N}\} \le 0. \tag{2.3d}$$

Here, $\mathrm{Proj}_{\mathcal{U}_{ad}}$ is the $L^2(\Omega)$-projection onto $\mathcal{U}_{ad}$. Note that in finite dimensional subspaces, condition (2.3c) yields $\mu = 0$ on the inactive set, whereas the limiting sign property in line (2.3d) reduces to $\langle \mu, p\rangle \le 0$. In particular, when using a conforming discretization, solutions of (2.1) satisfy the C-stationarity conditions.

2.2 Optimal Control of a Variational Inequality with Gradient Constraints

Another possible variational inequality of practical interest arises when one considers the state constraint $K := \{y \in H_0^1(\Omega) \,||\nabla y| \leq \psi, \text{ a.e. on } \Omega\}$. When $\psi \equiv 1$ and $f = c \in \mathbb{R}$, the variational inequality models the elastoplastic torsion of a cylinder. However, there are a number of more complex phenomena, which can be modeled with such a framework; see Sect. 5 below. This highly nonlinear constraint adds additional difficulties when considering the MPEC and certain assumptions on regularity of the data are currently needed for the derivation of an optimality system. We briefly sketch two methods here.

Method 1 Recall that the tangent and normal cones to a closed convex set $C \subset X$ in a normed linear space at a point $z \in X$ are given by

$$T_C(z) := \left\{ d \in X \,\middle|\, \exists t_k \downarrow 0,\ d_k \overset{X}{\to} d : z + t_k d_k \in C \right\},$$

$$N_C(z) := \{ v \in X^* \,|\langle v, z' - z\rangle \leq 0,\ \forall z' \in C \}.$$

In addition, we define the active and the inactive set belonging to $y \in K$ by $\mathcal{A} := \{x \in \Omega \,|\, |\nabla y(x)| = \psi\}$ and $\mathcal{I} := \Omega \backslash \mathcal{A}$, respectively. It was shown in [21], that if

1. $y \in H_0^1(\Omega)$ solves the variational inequality,
2. $T_K(y) = \{d \in H_0^1(\Omega) \,|\nabla y \cdot \nabla d \leq 0, \text{ a.e. on } \mathcal{A}\}$,
3. $N_K(y) = \{v = -\text{div}(\lambda \nabla y) \in H^{-1}(\Omega) \,|\lambda \in L^2(\Omega) : 0 \leq \lambda \perp \psi - |\nabla y| \geq 0\}$,
4. For $v \in N_K(y)$ such that $Ay + v = u + f$ and $\lambda \in L^\infty(\Omega)$,

then the solution operator S for the variational inequality is (Hadamard) directionally differentiable from the control space into $H_0^1(\Omega)$. The directional derivative of S at u in direction h is given by the unique solution d of the following variational inequality:

$$\text{Find } d \in T_K(y) \cap \{v\}^\perp : \langle Ad - 2\,\text{div}(\lambda \nabla d) - h, d' - d\rangle \geq 0,\ \forall d' \in T_K(y) \cap \{v\}^\perp.$$

This in turn leads to the following first order optimality conditions, which in the finite dimensional literature would be known as strong stationarity conditions:

Theorem 2.3. *Let (u, y) be a (locally) optimal solution to the corresponding MPEC. Under the assumptions above, there exist multipliers $p \in H_0^1(\Omega)$ and $\mu \in L^2(\Omega)$, such that*

$$0 = \nabla_u J(u, y) - p, \tag{2.4}$$

$$0 = \nabla_y J(u, y) + A^* p - 2\text{div}(\lambda \nabla p) + 2\text{div}(\mu \nabla y), \tag{2.5}$$

$$0 = Ay - u - 2\text{div}(\lambda \nabla y). \tag{2.6}$$

Defining the strongly active and the biactive set belonging to y and λ by $\mathcal{A}^+ := \{x \in \mathcal{A} \mid \lambda(x) > 0\}$ and $\mathcal{B} := \{x \in \mathcal{A} \mid \lambda(x) = 0\}$, the multipliers satisfy the following sign conditions,

$$\begin{aligned}
&\nabla y \cdot \nabla p \geq 0 \ \text{a.e. on } \mathcal{B}, \quad \mu \leq 0 \ \text{a.e. on } \mathcal{B},\ \lambda \geq 0 \ \text{a.e. on } \mathcal{A}, \\
&\nabla y \cdot \nabla p = 0 \ \text{a.e. on } \mathcal{A}^+,\ \mu = 0 \ \text{a.e. on } \mathcal{I},\ \lambda = 0 \ \text{a.e. on } \mathcal{I}.
\end{aligned}$$

This assumption on the form of the tangent cone is known in the optimization literature as *Abadie's constraint qualification*. Though it is one of the weakest possible conditions to require in regard to the tangent cone, it can sometimes be very difficult to verify. The additional requirement on the regularity of the multiplier λ is somewhat strong, though examples can be found where λ enjoys this increased regularity.

Method 2 In order to avoid the assumptions made above, which guarantee strong stationarity of an optimal solution, one can approach the problem as in [2]. Here, an 'adapted penalty' approach as mentioned in the introduction is applied in which the variational inequality is replaced by a quasi-linear PDE in divergence form. However, this method also requires further assumptions, in particular, the boundary of Ω must be $\mathcal{C}^2$ and the regularity of the solutions of the quasi-linear PDE are required to have a higher regularity ($W^{2,q}(\Omega) \cap H_0^1(\Omega)$ for $q > n$). Upon passing to the limit, a much weaker form of stationarity is obtained than in (2.4)–(2.6); see the diploma thesis [26].

3 Numerical Treatment of MPECs in Function Space

Suitable stationarity conditions, like those in Sect. 2, can be used to design efficient mesh independent solvers for the MPEC. This section provides a globally convergent function-space-based descent method for the solution of B- and C-stationarity systems (Sect. 3.1) as well as a goal-oriented mesh refinement technique in the spirit of the dual weighted residual approach (cf. [4]) for the adaptive discretization of MPECs in function space (Sect. 3.2).

3.1 An Algorithm for the Solution of C- or B-Stationarity Systems

The differentiability results on the solution operator S corresponding to (1.2) of [33] were extended to a much wider class of problems in [21] using techniques from non-smooth analysis. Moreover, in the absence of biactivity, S is typically Gâteaux differentiable. These facts lead one to naturally consider the following subclass of MPECs:

$$\text{Minimize } \mathcal{J}(u) := J(u, S(u)) \text{ over } u \in U, \tag{3.1a}$$

where, in addition to the usual assumptions needed to prove the existence of an optimal control, we assume that $S : U \to Y$ is Lipschitz continuous and directionally differentiable and J is Fréchet differentiable in both arguments.

We demonstrate in [22] that it is theoretically possible to obtain a descent direction h for $\mathcal{J}$ at u by solving the following regularized auxiliary problem:

$$\text{Minimize } \frac{1}{2}q(h, h) + \mathcal{J}'(u; h) \text{ over } h \in U, \tag{3.2}$$

where q is a coercive quadratic form. Of course, when S is smooth, (3.2) has a unique solution and a descent direction can be obtained by solving the first-order optimality conditions associated with (3.2). Otherwise, we proposed in [22] a new method for obtaining a descent direction in nonsmooth settings when $S'(u; h)$ has an explicit form. With these ideas, we develop a first order method for solving (3.1) and discuss its convergence properties.

Optimal Control of an Obstacle Problem Throughout the rest of this section, let $\Omega \subset \mathbb{R}^n$, $n \in \{1, 2, 3\}$, be either convex polyhedral or have a $\mathcal{C}^{1,1}$-boundary. In addition, we set $\mathcal{U}_{ad} = L^2(\Omega)$, $Y = H_0^1(\Omega)$ and $K = \{y \in H_0^1(\Omega) \mid y \geq 0 \text{ a.e. on } \Omega\}$ as in Sect. 2.1. The solution operator of (1.2) is denoted by S, $B \in \mathcal{L}(L^2(\Omega), H^{-1}(\Omega))$ and $f \in L^2(\Omega)$. A is a symmetric second-order linear elliptic operator associated with the bilinear form $a : H_0^1(\Omega) \times H_0^1(\Omega) \to \mathbb{R}$ defined by

$$\langle Av, w\rangle := a(v, w) = \sum_{i,j=1}^{n} \int_\Omega a_{ij} \frac{\partial v}{\partial x_j} \frac{\partial w}{\partial x_i} dx + \int_\Omega cvwdx, \ \ \forall v, w \in H_0^1(\Omega),$$

with $c \in L^\infty(\Omega)$, $c \geq 0$, and $a_{ij} \in \mathcal{C}^{0,1}(\overline{\Omega})$, i.e., Lipschitz continuous on the closure of Ω, with $\sum_{i,j} w_i a_{ij} w_j \geq \xi |w|^2_{\mathbb{R}^n}$ for all $w \in \mathbb{R}^n$ and some real $\xi > 0$. A solution y of the variational inequality satisfies $y \in H^2(\Omega) \cap H_0^1(\Omega)$.

We define the active set $\mathcal{A} := \{x \in \Omega \mid y(x) = 0\}$ and the inactive set $\mathcal{I} := \Omega \setminus \mathcal{A}$. Moreover, using λ from the complementarity formulation (1.3) we define the strongly active set $\mathcal{A}^+ := \{x \in \mathcal{A} \mid \lambda(x) > 0\}$ and the biactive set $\mathcal{B} := \{x \in \mathcal{A} \mid \lambda(x) = 0\}$.

It is possible to show that $d = S'(u; h)$ if and only if d is the unique solution to the optimization problem

$$\begin{aligned} &\min \tfrac{1}{2}\langle Ad, d\rangle_{H^{-1}, H_0^1} - (Bh, d)_{L^2} \text{ over } d \in H_0^1(\Omega) \\ &s.t. \quad d = 0, \text{ a.e. on } \mathcal{A}^+, \\ &\qquad\ \ d \geq 0, \text{ q.e. on } \mathcal{A}. \end{aligned} \tag{3.3}$$

Here, 'q.e' stands for 'quasi-everywhere' and refers to a relation which is satisfied up to a set of zero capacity; see [13] for details in this concept. If strict complementarity holds, $\mathcal{A} = \overline{\text{int}(\mathcal{A})}$, and the free boundary is sufficiently regular, then it can be argued that we may write (3.3) in the more compact form:

$$d = 0, \text{ q.e. on } \mathcal{A}^+, \quad Ad = Bh, \text{ in } \mathcal{I}. \tag{3.4}$$

Otherwise, i.e., if biactivity is present, d solves a VI. By letting $\max_\varepsilon^g(0, \cdot)$ be a global C^2-smoothing of the pointwise $\max(0, \cdot)$-operator, we can approximate $S'(u; \cdot)$ by the solution operator of the semilinear equation:

$$Ad + \gamma\chi_{\mathcal{A}^+} d - \gamma\chi_{\mathcal{A}} \max_\varepsilon^g(0, -d)' = Bh. \tag{3.5}$$

Here, χ represents the standard characteristic function whereas $\gamma > 0$ and $\varepsilon > 0$ are penalty and smoothing parameters, respectively. We let $S'_{\gamma,\varepsilon}(u; \cdot)$ represent the solution operator associated with (3.5).

For a tracking type objective J defined by

$$J(u, y) := \frac{1}{2}\|y - y_d\|^2_{L^2(\Omega)} + \frac{\alpha}{2}\|u\|^2_{L^2(\Omega)}, \; \alpha > 0, y_d \in L^2(\Omega), \tag{3.6}$$

define

$$F_{\gamma,\varepsilon}(h) := \frac{1}{2}q(h, h) + \alpha(u, h)_{L^2} + (y - y_d, S'_{\gamma,\varepsilon}(u; h))_{L^2},$$

where $(\cdot, \cdot)_{L^2}$ or sometimes just $(\cdot, \cdot)$ is the usual L^2-inner product. One can show that if $\gamma > 0$ is large enough then $h = -F'_{\gamma,\varepsilon}(0)$ is a proper descent direction for the original reduced objective functional $\mathcal{J}$ at u. Note that the calculation of $F'_{\gamma,\varepsilon}(0)$ requires the solution of the adjoint equation

$$Ap + \gamma\chi_{\mathcal{A}^+} p = -\nabla_y J(u, y) \tag{3.7}$$

where $y = S(u)$ and $\nabla_y J(u, y)$ is the partial derivative of J with respect to y. These results and observations lead to Algorithm 1, which, if the step sizes τ_k are generated by an Armijo line search and $\tau_k \not\to 0$ as $k \to \infty$, can be shown to provide the following:

1. If strict complementarity holds for all sufficiently large k, then a type of C-stationarity is obtained. Moreover, a B-stationary point is obtained if the function $\mathcal{J}(u)$ is Clarke-regular at the solution $u^\star$.
2. Otherwise, one has ϵ-almost-C-stationarity of weak accumulation points.

Moreover, by replacing S with the solution operator for

$$Ay + \delta \max_\eta^g(0, -y) = Bu + f, \; \delta > 0, \eta > 0,$$

Algorithm 1 A first-order method for MPECs

Input: $u_0, \gamma_0, \varepsilon_0 > 0, \beta > 1, k := 0$

while Convergence criterion is not fulfilled **do**
  Calculate $y_k = S(u_k)$
  if $m(\mathcal{B}) = 0$ **then**
    Calculate descent direction h_k by solving (3.2) with $u = u_k$.
  else
    while descent criterion violated **do**
      Set $\gamma_k \geq \beta\gamma_k$
    end while
    Set $h_k = -F'_{\gamma_k,\varepsilon_k}(0)$.
  end if
  Determine a step size $\tau_k > 0$ according to Armijo line search.
  Set $u_{k+1} := u_k + \tau_k h_k$, $k := k + 1$.
end while

whenever τ_k appears to be rapidly converging to zero, one can embed Algorithm 1 into an outer loop that uses a standard steepest descent method for the smoothed MPEC in order to generate a new $\tilde{u}_k$. Afterwards, Algorithm 1 can be restarted with $k = 0$, $u_k = \tilde{u}_k$. In such a case, the conditions on τ_k can be dropped and the enhanced algorithm will provide sequences whose weak accumulation points are ε-almost-C-stationary, cf. [17, 18, 22]

Finally, we mention that although the methods in [17, 18], when used in conjunction with a non-linear PATH strategy or a heuristic line search argument exhibit globally convergent behavior experimentally, Algorithm 1 is the only proven globally convergent function-space-based algorithm for this problem class.

Example 1. This example is taken from the literature specifically due to the presence of a large biactive set at the solution, see [18]. Here, we let $\Omega = (0,1) \times (0,1)$, J is a tracking type functional (3.6) with $\alpha = 1$ and given

$$y^\dagger(\mathbf{x}_1, \mathbf{x}_2) = \begin{cases} 1600(\mathbf{x}_1^3 - \mathbf{x}_1^2 + 0.25\mathbf{x}_1)(\mathbf{x}_2^3 - \mathbf{x}_2^2 + 0.25\mathbf{x}_2) & \text{in } (0, 0.5)^2, \\ 0 & \text{else,} \end{cases}$$

$$\lambda^\dagger(\mathbf{x}_1, \mathbf{x}_2) = \max(0, -2|\mathbf{x}_1 - 0.8| - 2|\mathbf{x}_1\mathbf{x}_2 - 0.3| + 0.5)$$

we set $f = -\Delta y^\dagger - y^\dagger - \lambda^\dagger$, and $y_d = y^\dagger + \lambda^\dagger - \alpha\Delta y^\dagger$. Let further $A = -\Delta$, $B = i_{L^2 \hookrightarrow H^{-1}}$, and discretize A using the standard five-point finite difference stencil. For the numerical solution, a nested grid strategy is applied. The algorithm was stopped when either the maximum residual of the C-stationarity system reached 10^{-6} or the direction $\|h_k\|_{L^2} \leq 10^{-6}$. An Armijo line search was used in which we set $\tau = 0.5\tau$ until

$$\mathcal{J}(u_k + \tau h_k) \leq \mathcal{J}(u_k) + 0.01\tau\mathcal{J}'(u_k; h_k).$$

Table 1 Performance of Algorithm 1 for Example 1

dof	lnsrch	C-stat.	$\|h\|$	lnslve	nsstep	sstep
9	3	6.0848e−09	6.0848e−09	16	2	2
49	3	1.7046e−09	1.7046e−09	16	2	2
225	3	1.1502e−09	1.1502e−09	24	2	2
961	3	5.5962e−10	5.5962e−10	21	2	2
3,969	3	4.1594e−10	4.1594e−10	18	2	2
16,129	3	3.7574e−10	3.7574e−10	16	2	2
65,025	1	1.9297e−06	7.2621e−10	10	2	0
261,121	2	8.3587e−10	2.59e−12	14	3	0

The underlying VI was solved using a primal-dual active set strategy, which is known to be locally superlinearly convergent for each mesh, [23]. The semilinear equation in the nonsmooth step is solved using a standard Newton step. The performance of the algorithm can be seen in Table 1. We use the notation: dof = degrees of freedom, lnsrch = number of line searches, C-stat = maximum residual of C-stationarity, $\|h\|$ norm of step, lnslve = total number of linear systems solved, nsstep = number of nonsmooth steps, sstep = number of smooth steps.

3.2 An Adaptive Finite Element Method for a Class of MPECs

Assuming that we can find solutions of suitable stationarity systems with the algorithm from Sect. 3.1 above, this section suggests a mesh adaption technique from [24] that allows us to find a discrete space that fits to special properties of the solution. The dual weighted residual approach to a posteriori error estimation for optimal control problems has been pioneered in [4, 12, 16, 42]. A residual type estimator for the optimal control of an obstacle problem can be found in [11].

Based on the notion of a modified Lagrangian function associated with the MPEC (MPEC-Lagrangian) one employs suitable first order stationarity conditions for the function space setting, see e.g. (2.3) for the point evaluation case, or [17] for the L^2-tracking objective functional, as well as for a discretized version of the MPEC. Then using Taylor's expansion of the MPEC-Lagrangian one can derive the error representation formula as stated in Theorem 3.1 below, see [6, 24] for details.

For the state y_h and the adjoint state p_h, we use a conforming discretization of Y with P^1 finite elements on a regular triangulation $\mathcal{T}_h$ of the domain Ω. The discrete multipliers λ_h and μ_h are defined in the discrete dual space and prolongated to linear continuous mappings on the full space Y, whereas the control u_h as well as the respective multipliers σ_{ah}, σ_{bh} result from an $L^2(\Omega)$-projection of p_h onto the admissible set $\mathcal{U}_{ad}$. In this section, we assume $(u_h, y_h, \lambda_h, p_h, \mu_h, \sigma_{ah}, \sigma_{bh})$ to be C-stationary for the discrete MPEC in the resulting discrete space, i.e., it satisfies conditions (2.1c), (2.3) (or those in [17]) tested in the finite element space.

Theorem 3.1. *For* $(u, y, \lambda, p, \mu, \sigma_a, \sigma_b)$ *satisfying* (2.1c), (2.3) *or the respective C-stationarity system from* [17] *and every* $\delta x_h = (\delta u_h, \delta y_h, \delta\lambda_h, \delta p_h)$ *in the discrete space, it holds that*

$$
\begin{aligned}
&2(J(u, y) - J(u_h, y_h)) = \\
&a(y_h, p - \delta p_h) - (u_h + f, p - \delta p_h) - (\lambda_h, p - \delta p_h) \\
&+ a(y - \delta y_h, p_h) + \langle J_y(u_h, y_h) - \mu_h, y - \delta y_h \rangle \\
&+ (J_u(u_h, y_h) - p_h - \sigma_{ah} + \sigma_{bh}, u - \delta u_h) \\
&- \langle \mu, y_h \rangle + \langle \mu_h, y \rangle + \langle \lambda_h, p \rangle - (\lambda, p_h) \\
&- (u_h - a, \sigma_a) + (u - a, \sigma_{ah}) - (b - u_h, \sigma_b) + (b - u, \sigma_{bh}) .
\end{aligned}
$$

This error representation involves primal residuals weighted by dual variables and vice versa as well as error terms covering the mismatch in complementarity. The latter error indicators are relevant in the location of the coincidence or active sets which arise due to the variational inequality constraint.

We suggest a heuristic way to replace the continuous solutions, e.g. y, which typically arise in goal oriented dual weighted error estimation, locally by a quadratic function that minimizes the least square distance to the values of the respective discrete function (e.g. y_h) in the midpoints of the edges of a triangle T and in the nodes of its neighbors sharing an edge with T which are no nodes of T. The formula can then be written as a sum over terms that reflect the error contribution on each triangle in $\mathcal{T}_h$.

Example 2. We consider $A = -\Delta$ on the L-shaped domain $\Omega = (-1, 0) \times (-1, 1) \cup (0, 1) \times (0, 1)$, and the MPEC (1.1) with the L^2-tracking type objective functional (3.6), $U = \mathcal{U}_{ad} = L^2(\Omega)$, and obstacle problem constraint as in Sect. 3.1. Furthermore, we set $\alpha = 0.01$ and define y_d and f by

$$
y_d(x) = \begin{cases} -1 & \text{if } |x| \geq \frac{1}{10}, \\ 1 - 100x_1^2 - 50x_2^2 & \text{else,} \end{cases} \qquad f(x) = \frac{1}{2} + \frac{1}{2}(x_1 - x_2).
$$

Figure 1 compares the errors $|J(u_h, y_h) - J(u, y)|$ for discrete solutions on uniform meshes (solid blue lines) with the total values of the respective estimators (dashed blue lines), and the errors in the objective for discrete solutions on adaptively refined meshes (solid red lines) with the corresponding total value of the estimator (dashed red lines). Here, the superscript 'A' denotes quantities related to the adaptive refinement, whereas 'U' refers to the uniform mesh refinement. Since the exact solution (y, u) is not known, we estimate it by a discrete solution on a sufficiently fine mesh. The associated objective value is J^*. The convergence plot shows the reliability of the estimator and the faster convergence of the adaptive method. The right part shows that the adaptively generated mesh exhibits a higher

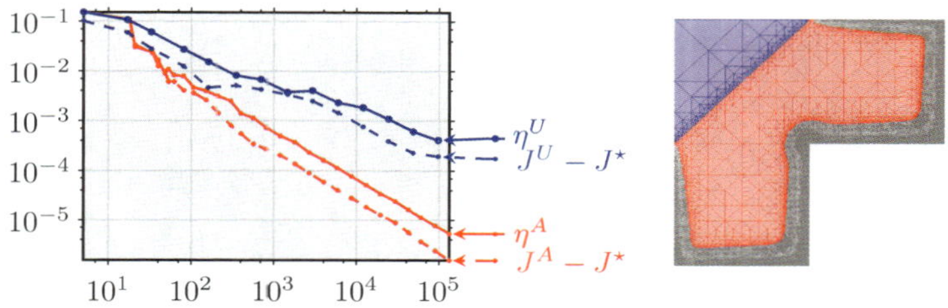

Fig. 1 *Left*: Convergence of the estimators η^A as η^U as well as the errors in the objective values $|J^A - J^*|$ and $|J^U - J^*|$ w.r.t. the number of degrees of freedom for Example 2. *Right*: Adaptively generated mesh with strongly active set (*blue*) and biactive set (*red*)

density of nodes in the region around the non-convexity of Ω as well as at the free boundary between strongly active (blue) and biactive (red) and inactive sets.

4 Shape Design for a Variational Inequality

In contrast to the previous sections, we now consider shape and topology optimization problems subject to VIs. This means that the domain Ω acts as the control. These problems have received considerably less attention than problems governed by elliptic partial differential equations; see [3, 35, 36, 40, 41], and the literature on solution algorithms in function space is even more scarce. In addition to the challenges with VI constraints discussed in the previous sections, in the present context VI constraints pose several additional difficulties in analysis and numerics, since the shape derivative is typically non-linear, which is, as before, related to the presence of biactive sets. We present some of the key aspects of the paper [19].

Let $\Omega \subset \mathbb{R}^2$ be a bounded domain with a $C^{1,1}$-boundary Σ. Assume $\overline{\Omega} \subset D$ where D is a bounded domain and let $\psi \in H^4(D)$ with $0 < \underline{M} < \psi \leq \overline{M}$ and $f \in H^2(D)$. We set $K = \{y \in H_0^1(\Omega) \mid y \leq \psi \text{ a.e. in } \Omega\}$ and $B \equiv 0$ in (1.2). For the unique solution $y = y(\Omega) \in H^2(\Omega) \cap H_0^1(\Omega)$ of (1.2) there exists a Lagrange multiplier $\lambda \in L^2(\Omega)$ satisfying the complementarity formulation

$$-\Delta y + \lambda = f, \qquad y \leq \psi,\ \lambda \geq 0,\ \lambda(y - \psi) = 0 \quad \text{a.e. in } \Omega. \tag{4.1}$$

Define the active, inactive and biactive sets with respect to the solution y and the multiplier λ as

$$\mathcal{A} = \{x \in \Omega \ :\ y(x) = \psi(x)\}, \qquad \mathcal{I} = \Omega \setminus \mathcal{A}, \qquad \mathcal{B} = \{x \in \mathcal{A} \ :\ \lambda(x) = 0\},$$

respectively. The continuity of $y - \psi$ on $\overline{\Omega}$ and of λ on $\mathcal{A}$ (note that λ enjoys extra regularity on $\mathcal{A}$ due to the invoked data regularity and the first equation in (4.1)) yield that $\mathcal{A}$ and $\mathcal{B}$ are closed in Ω and $\mathcal{I}$ is open. Let $V \in C([0, T], C^2(\Omega, \mathbb{R}^2))$ and $T_t(V)(X) = x(t)$ be the solution of

$$\frac{dx}{dt}(t) = V(t, x(t)), \quad 0 < t < \tau, \qquad x(0) = X \in \mathbb{R}^2$$

for $\tau > 0$ sufficiently small, i.e. such that $\overline{\Omega_t(V)} \subset D$, where $\Omega_t(V) := T_t(V)(\Omega)$. The *shape derivative* y' of $y = y(\Omega)$ at Ω in direction V is defined (formally) as

$$y'(\Omega; V) := \lim_{t \downarrow 0} (y(\Omega_t(V)) - y(\Omega))/t. \tag{4.2}$$

Here, we provide a characterization of $y'(\Omega; V)$ for obstacle problems. It is known that if $\mathcal{B}$ is non-empty, then one has to solve an obstacle problem to obtain y'. We introduce the function $v_\nu := V(0) \cdot \nu$ defined on $\partial\Omega$ (with the outer unit normal ν on $\partial\Omega$) and the set $\mathcal{I}^+ := \mathcal{I} \cup \text{int}(\mathcal{B})$ which is assumed to be a Lipschitz domain, and use Corollary 4.17 of [40, p. 183].

Theorem 4.1. *The shape derivative $y' \in H^1(\Omega)$ is the unique solution of*

$$y' \in S_v(\Omega) : \int_\Omega \nabla y' \cdot \nabla(\theta - y') \geq 0 \quad \forall \theta \in S_v(\Omega), \tag{4.3}$$

where the cone $S_v(\Omega) \subset H^1(\Omega)$ is of the form

$$S_v(\Omega) = \left\{ \theta \in H^1(\Omega) : \theta = -\partial_\nu y \, v_\nu \text{ on } \partial\Omega,\ \theta \leq 0 \text{ q.e. in } \mathcal{A}, \int_\Omega \nabla y \cdot \nabla\theta = \int_\Omega f\theta \right\}.$$

To circumvent the non-linearity of y' with respect to θ, we regularize (4.1). This allows to compute the shape derivative for the regularized problem on all of Ω and to use standard numerical algorithms. For this purpose, consider

$$-\Delta y_\gamma + \lambda_\gamma = f \ \text{ in } \Omega, \quad y_\gamma = 0 \ \text{ on } \partial\Omega, \tag{4.4}$$

$$\lambda_\gamma = \max(0, \bar{\lambda} + \gamma(y_\gamma - \psi))^2, \tag{4.5}$$

with $\gamma > 0$. We choose $\bar{\lambda} \in L^4(\Omega)$ such that

$$\bar{\lambda} \geq 0, \qquad \bar{\lambda}^2 - (f + \Delta\psi) \geq 0 \text{ a.e. in } \Omega. \tag{4.6}$$

Corollary 4.2 (cf. Cor.2 in [19]). *If* (4.6) *holds we have $\lambda_\gamma \to \lambda$ in $L^2(\Omega)$ and $y_\gamma \to y$ in $H^2(\Omega)$ as $\gamma \to \infty$, respectively.*

According to [40], the shape derivative $y'_\gamma \in H_0^1(\Omega)$ of y_γ solves:

$$-\Delta y'_\gamma + 2\gamma\sqrt{\lambda_\gamma}\, y'_\gamma = 0 \ \text{ in } \Omega, \quad y'_\gamma = -\partial_\nu y_\gamma \, v_\nu \ \text{ on } \partial\Omega. \tag{4.7}$$

Introduce $\mathcal{A}^+ := \{\lambda > 0\}$ which is assumed to be Lipschitz and define y'_∞ as the solution of

$$-\Delta y'_\infty = 0 \text{ in } \Omega \setminus \overline{\mathcal{A}^+} =: \mathcal{I}^+, \tag{4.8}$$

$$y'_\infty = -\partial_\nu y\, v_\nu \text{ on } \partial\Omega, \tag{4.9}$$

$$y'_\infty = 0 \text{ on } \partial\mathcal{A}^+, \tag{4.10}$$

$$y'_\infty \equiv 0 \text{ in } \overline{\mathcal{A}^+}. \tag{4.11}$$

Theorem 4.3. *If $\mathcal{A}^+$ and $\mathcal{I}^+$ have a Lipschitz boundary, the solution y'_γ of* (4.7) *converges to the solution y'_∞ of* (4.8)–(4.11) *strongly in $H^1(\Omega)$ as $\gamma \to \infty$.*

Example 3. We consider an application to electrochemical machining (ECM) [3, 9, 41]. Let $B \subset E \subset \Omega \subset D$ be smooth domains. The set Ω is the control domain, D is fixed, and E is a target shape for the active set $\mathcal{A}$. The aim is to minimize

$$\mathcal{J}(\Omega) = \int_{E\setminus B} y^2(x)\, dx + \int_{\Omega\setminus E} (y - y_l)^2(x)\, dx, \tag{4.12}$$

where y is the solution of the obstacle problem

$$\begin{cases} -\Delta y + \lambda = f \quad \text{in } \Omega \setminus B, \quad \text{with } y = 1 \text{ on } \partial\Omega \text{ and } y = 0 \text{ on } \partial B, \\ y \geq 0,\ \lambda \leq 0,\ \lambda y = 0 \quad \text{a.e. in } \Omega \setminus B, \end{cases} \tag{4.13}$$

and y_l is the solution of the linear problem

$$-\Delta y_l = f \quad \text{in } \Omega \setminus E, \quad y_l = 1 \quad \text{on } \partial\Omega, \quad y_l = 0 \quad \text{on } \partial E.$$

The penalized version of problem (4.13) corresponds to minimizing

$$J_\gamma(\Omega) = \int_{E\setminus B} y_\gamma^2(x)\, dx + \int_{\Omega\setminus E} (y_\gamma - y_l)^2(x)\, dx, \tag{4.14}$$

where y_γ is the solution of

$$-\Delta y_\gamma - \lambda_\gamma = f \text{ in } \Omega \setminus B, \quad y_\gamma = 0 \text{ on } \partial B, \quad y_\gamma = 1 \text{ on } \partial\Omega \tag{4.15}$$

with $\gamma > 0$ and $\lambda_\gamma := \min(0, \bar{\lambda} + \gamma y_\gamma)^2$. We introduce the adjoint states p_1 and p_2 as the solutions of

$$-\Delta p_1 + 2\gamma\sqrt{\lambda_\gamma}\, p_1 = 2y_\gamma \chi_{E\setminus B} + 2(y_\gamma - y_l)\chi_{\Omega\setminus E} \text{ in } \Omega \setminus B,\ p_1 = 0 \text{ on } \partial\Omega \cup \partial B,$$

$$-\Delta p_2 = -2(y_\gamma - y_l) \quad \text{in } \Omega \setminus E, \quad p_2 = 0 \quad \text{on } \partial\Omega \cup \partial E.$$

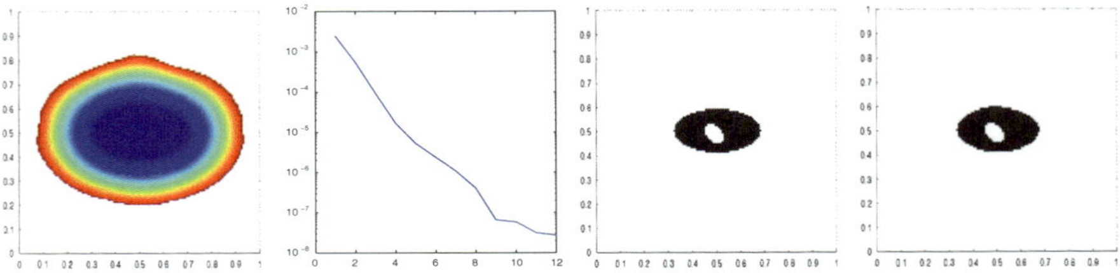

Fig. 2 From *left* to *right*: solution y, history of the cost J_γ, target set $E \setminus B$ and solution active set $\mathcal{A} \setminus B$

where χ stands for the indicator function of a set. We obtain for the shape derivative of (4.14)

$$dJ_\gamma(\Omega, V) = \int_{\partial\Omega} \left(\nabla p_1 \cdot \nabla y_\gamma + \nabla p_2 \cdot \nabla y_l\right) v_\nu.$$

We use the descent direction

$$v_\nu = -\nabla p_1 \cdot \nabla y_\gamma - \nabla p_2 \cdot \nabla y_l$$

in a level-set based steepest descent method for minimizing J_γ. Choosing $\gamma = 10^4$ for the penalization, Fig. 2 depicts the numerical results for the ECM problem with the penalization approach and shows a good match between the target $E \setminus B$ and the solution active set $\mathcal{A} \setminus B$.

5 Stationarity and a Solution Algorithm for QVIs with Gradient Constraints

Quasi-variational inequalities (QVIs), introduced by Bensoussan and Lions in [5] and [30], arise as mathematical models of various phenomena in the applied sciences. These involve, for instance, game theory, solid and continuum mechanics or superconductivity. QVIs generalize VIs in the sense that the constraint set is no longer a constant set but a set-valued mapping. More precisely, the constraint set depends on the solution. In this section, we briefly sketch the results of [20].

Let Ω be a bounded domain of $\mathbb{R}^n$, $n \in \mathbb{N}$ and $p \geq 2$. Suppose that C is a closed and convex subset of $W_0^{1,p}(\Omega)$ such that $0 \in C$ and we are given a nonlinear completely continuous mapping $\Phi : C \subset W_0^{1,p}(\Omega) \to L_\nu^\infty(\Omega) \subset L^\infty(\Omega)$, where $L_\nu^\infty(\Omega) = \{\varphi \in L^\infty(\Omega) : \varphi(x) \geq \nu \text{ a.e. on } x \in \Omega\}$ with $\nu > 0$. The set-valued constraint map $K : C \subset W_0^{1,p}(\Omega) \to 2^{W_0^{1,p}(\Omega)}$ is given by

$$K(z) = \{y \in W_0^{1,p}(\Omega) : |\nabla y(x)| \leq \Phi(z)(x) \text{ a.e. on } \Omega\}.$$

Note that for every $z \in C$, $K(z)$ contains the zero element and is closed and convex.

Suppose that $f \in W^{-1,p'}(\Omega)$ and $A : W_0^{1,p}(\Omega) \to W^{-1,p'}(\Omega)$ is a strongly monotone, hemicontinuous operator and $A(0) = 0$. Then we define the problem (P_{QVI}) as the following QVI:

$$\text{Find } y \in C \text{ s.t. } y \in K(y) : \langle A(y) - f, z - y\rangle \geq 0, \quad \forall z \in K(y). \tag{P_{QVI}}$$

Note that if $C \ni z \mapsto S(z)$ is the solution mapping of problem (1.2) with $K = K(z)$, then solutions to (P_{QVI}) are equivalently solutions to $y = S(y)$. A prototypical example for A is given by the p-Laplacian, given by $\langle -\Delta_p(y), z\rangle = \int_\Omega |\nabla y|^{p-2}\nabla y \cdot \nabla z$, which reduces to the usual Laplacian when $p = 2$. Existence of solutions for (P_{QVI}) can be proven by the application of well-known fixed-point theorems (see [28]), e.g., Schauder and Leray-Schauder, however the derivation of numerical methods requires further results.

In what follows, assume that $A(ty) = t^k A(y)$ for each $y \in W_0^{1,p}(\Omega)$, $t > 0$ and some $k \in \mathbb{N}$, and that the Gâteaux derivative of $\tilde{a}(y) := \frac{1}{k+1}\langle A(y), y\rangle$ is given by $\tilde{a}'(y) = A(y)$. Note that (P_{QVI}) is in general *not* equivalent (see [20]) to

$$\min j(y) := \frac{1}{k+1}\langle A(y), y\rangle - \langle f, y\rangle, \quad \text{s.t. } y \in K(y), \tag{5.1}$$

unless $C = W_0^{1,p}(\Omega)$ and $\Phi(z) = \varphi \in L_\nu^\infty(\Omega)$ for all $z \in W_0^{1,p}(\Omega)$. The latter fact can be utilized to derive that $z \mapsto K(z) = K$ is constant and then the problem reduces to a VI.

Returning to a non-trivial QVI setting, we note that if $C = W_0^{1,p}(\Omega)$, Φ is Lipschitz with constant L_Φ, $f \in L^q(\Omega)$ for large enough q, and A satisfies certain conditions regarding its strong monotonicity (see [20]), then it can be proven that

$$\|S(y) - S(z)\|_{W_0^{1,p}(\Omega)} \leq L_S(L_\Phi, f)\|y - z\|_{W_0^{1,p}(\Omega)}, \tag{5.2}$$

such that $\lim_{L_\Phi \downarrow 0} L_S(L_\Phi, f) = 0$ and $\lim_{\|f\|_{L^q(\Omega)} \downarrow 0} L_S(L_\Phi, f) = 0$. This implies that, under certain conditions, the mapping $z \mapsto S(z)$ is contractive and the iteration $y_n = S(y_{n-1})$ converges strongly to the unique solution to (P_{QVI}). The above contractive property proves that there exist non-trivial cases for which a numerical method with guaranteed convergence can be developed. In fact, given y_{n-1}, an approximation of $y_n = S(y_{n-1})$ can be obtained by solving the penalized problem

$$\min j_\gamma(y) := j(y) + \frac{\gamma}{2}\|(|\nabla y| - \Phi(y_{n-1}))^+\|^2_{L^2(\Omega)} \text{ over } y \in W_0^{1,p}(\Omega). \tag{P^γ}$$

Here, the solution y_n^γ satisfies $\lim_{\gamma\to\infty} y_n^\gamma = y_n = S(y_{n-1})$ as well as the optimality system

$$(\gamma(|\nabla y_n^\gamma| - \Phi(y_{n-1}))^+, q_\gamma \nabla z) = \langle f, z\rangle - \langle A(y_n^\gamma), z\rangle \; \forall z \in W_0^{1,p}(\Omega);$$
$$q_\gamma(x) \in \begin{cases} \frac{\nabla y_n^\gamma}{|\nabla y_n^\gamma|}(x), \text{ if } |\nabla y_n^\gamma(x)| > 0; \\ \bar{B}_1(0)^l, \quad \text{otherwise}, \end{cases} \qquad (\mathrm{OS}^\gamma)$$

where $\bar{B}_1(0)^l$ denotes the usual closed unit ball in $\mathbb{R}^l$. The above can be written as $F(y) = A(y) - f + \gamma\mathfrak{P}(y)$ with $\mathfrak{P}(y) := \nabla^*(|\nabla y| - \Phi(y_{n-1}))^+ q(\nabla y)\nabla$. Then, provided that $y \mapsto A(y)$ is differentiable, F is Newton (or slantly) differentiable, i.e., semismooth, as a mapping from $W_0^{1,p}(\Omega)$ to $(W_0^{1,s'})'$ for $3 \le 3s \le p < \infty$ where $1/s + 1/s' = 1$. This can be utilized to derive a *semismooth Newton solver* for each $\gamma > 0$ and an algorithm arises for the approximation of solutions to $(\mathrm{P_{QVI}})$. The algorithm can be compactly summarized as follows: (i) Choose $y_0 \in W_0^{1,p}(\Omega)$ and set $n = 1$. (ii) Approximate $y_n = S(y_{n-1})$ by solving (OS^γ) for large γ by means of a continuation technique in combination with a semismooth Newton solver. (iii) Unless stopping criteria are satisfied, set $n := n + 1$ and go to (ii). For further details, we refer to [20].

Example 4. Let $p = 3$, $\Omega = (0,1) \times (0,1)$, $A = -\Delta_p$ (the p-Laplacian), $f(x_1, x_2) = \sin(2\pi x_1)\sin(\pi x_2)$ and $\Phi(z)(x) = 10\phi_1(x)|\int_\Omega \phi_2(w)z(w)dw| + 0.05$ with the functions ϕ_1 and ϕ_2 defined by $\phi_1(x_1, x_2) := \exp(-5((x_1 - 1/3)^2 + (x_2 - 1/3)^2))$ and $\phi_2(x_1, x_2) := 10\chi_{\Omega_0}(x_1, x_2) + 0.1$, where χ_{Ω_0} denotes the characteristic function of the set $\Omega_0 = (0.5, 1) \times (0.5, 1)$. The approximation of the solution to the QVI and its active set are depicted in Fig. 3.

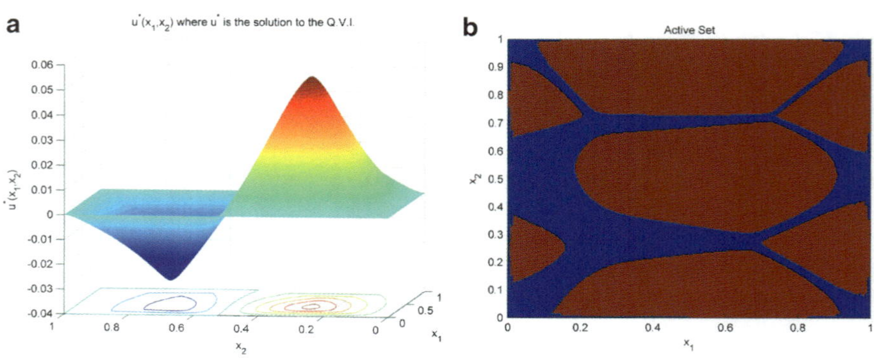

Fig. 3 (**a**): approximate solution to the QVI. (**b**): approximation of the active set (*red*) $\mathcal{A} = \{x \in \Omega : |\nabla y| = \Phi(y)\}$

Acknowledgements The authors acknowledge support by DFG-Project "Elliptic Mathematical Programs with Equilibrium Constraints (MPECs) in Function Space: Optimality Conditions and Numerical Realization" within the DFG Priority Program SPP 1253 on "Optimization with Partial Differential Equations", project C28 of the DFG Research Center "Matheon" as well as the Austrian Science Fund FWF under START-Project Y305 "Interfaces and Free Boundaries".

References

1. Y. Achdou, An inverse problem for a parabolic variational inequality arising in volatility calibration with american options. SIAM J. Control Optim. **43**(5), 1583–1615 (2005)
2. V. Barbu, *Optimal Control of Variational Inequalities*. Volume 100 of Research Notes in Mathematics (Pitman Advanced Publishing, Boston, 1984)
3. V. Barbu, A. Friedman, Optimal design of domains with free-boundary problems. SIAM J. Control Optim. **29**(3), 623–637 (1991)
4. R. Becker, H. Kapp, R. Rannacher, Adaptive finite element methods for optimal control of partial differential equations: basic concept. SIAM J. Control Optim. **39**(1), 113–132 (2000)
5. A. Bensoussan, J.-L. Lions, Controle impulsionnel et inéquations quasi-variationnelles d'évolutions. C. R. Acad. Sci. Paris **276**, 1333–1338 (1974)
6. C. Brett, C. Elliott, M. Hintermüller, C. Löbhard, A dual-weighted residual approach to adaptivity for optimal control of elliptic variational inequalities with pointwise objective functionals. IFB-Report, **67**, 1–28 (2013)
7. H. Brézis, Monotonicity methods in Hilbert spaces and some applications to nonlinear partial differential equations, in *Contributions to Nonlinear Functional Analysis* (Academic, New York, 1971), pp. 101–156
8. H. Brézis, G. Stampacchia, Sur la regularité de la solution d'inéquations elliptiques. Bulletin de la Société Mathématique de France **96**, 153–180 (1968)
9. C.M. Elliott, On a variational inequality formulation of an electrochemical machining moving boundary problem and its approximation by the finite element method. J. Inst. Math. Appl. **25**(2), 121–131 (1980)
10. G. Fichera, Problemi elettrostatici con vincoli unilaterali: il problema di Signorini con ambigue condizioni al contorno. Memorie dell'Accademia Nazionale dei Lincei **8**, 91–140 (1964)
11. A. Gaevskaya, M. Hintermüller, R. Hoppe, Adaptive finite elements for optimally controlled elliptic variational inequalities of obstacle type. IFB-Report No. 68. 09 2013
12. A. Günther, M. Hinze, A posteriori error control of a state constrained elliptic control problem. J. Numer. Math. **16**(4), 307–322 (2008)
13. A. Henrot, M. Pierre, *Variation et optimisation de formes*. Volume 48 of Mathématiques et Applications (Springer, Berlin, 1992). Une analyse géométrique
14. R. Herzog, C. Meyer, G. Wachsmuth, C-stationarity for optimal control of static plasticity with linear kinematic hardening. SIAM J. Control Optim. **50**(5), 3052–3082 (2012)
15. R. Herzog, C. Meyer, G. Wachsmuth, B- and strong stationarity for optimal control of static plasticity with hardening. SIAM J. Optim. **23**(1), 321–352 (2013)
16. M. Hintermüller, R. Hoppe, Goal-oriented adaptivity in control constrained optimal control of partial differential equations. SIAM J. Control Optim. **47**, 1721–1743 (2008)
17. M. Hintermüller, I. Kopacka, Mathematical programs with complementarity constraints in function space: C- and strong stationarity and a path-following algorithm. SIAM J. Optim. **20**, 868–902 (2009)
18. M. Hintermüller, I. Kopacka, A smooth penalty approach and a nonlinear multigrid algorithm for elliptic MPECs. Comput. Optim. Appl. **50**(1), 111–145 (2011)

19. M. Hintermüller, A. Laurain, Optimal shape design subject to elliptic variational inequalities. SIAM J. Control Optim. **49**(3), 1015–1047 (2011)
20. M. Hintermüller, C.N. Rautenberg, A sequential minimization technique for elliptic quasi-variational inequalities with gradient constraints. SIAM J. Optim. **22**(4), 1224–1257 (2012)
21. M. Hintermüller, T. Surowiec, First order optimality conditions for elliptic mathematical programs with equilibrium constraints via variational analysis. SIAM J. Optim. **21**(4), 1561–1593 (2011)
22. M. Hintermüller, T. Surowiec, A bundle-free implicit programming approach for a class of MPECs in function space. IFB-Report, 60, 2012
23. M. Hintermüller, K. Ito, K. Kunisch, The primal-dual active set strategy as a semismooth Newton method. SIAM J. Optim. **13**, 865–888 (2002)
24. M. Hintermüller, R. Hoppe, C. Löbhard, ESAIM: Control, Optimization and Calculus of Variations **20**(2), 524–546 (2014)
25. K. Ito, K. Kunisch, An active set strategy based on the augmented lagrangian formulation for image restoration. ESAIM: Math. Model. Numer. Anal. **33**(1), 1–21 (1999)
26. T. Keil, A smooth penalty approach for MPECs with gradient constrained lower-level problems in Banach spaces. Master's thesis, Humboldt-Universität zu Berlin, 2013
27. D. Kinderlehrer, G. Stampacchia, *An Introduction to Variational Inequalities and Their Applications* (Academic, New York, 1980)
28. M. Kunze, J. Rodrigues, An elliptic quasi-variational inequality with gradient constraints and some of its applications. Math. Methods Appl. Sci. **23**, 897–908 (2000)
29. J. Lions, G. Stampacchia, Variational inequalities. Commun. Pure Appl. Math. **20**, 493–519 (1967)
30. J.-L. Lions, Sur le côntrole optimal des systemes distribuées. Enseigne **19**, 125–166 (1973)
31. J.-L. Lions, Various topics in the theory of optimal control of distributed systems, in optimal control theory and its applications, part I. Lect. Notes Econ. Math. Syst. **105**, 166–309 (1974)
32. Z. Luo, J. Pang, D. Ralph, *Mathematical Programs with Equilibrium Constraints* (Cambridge University Press, Cambridge/New York, 1996)
33. F. Mignot, Controle dans les inéquations variationnelles elliptiques. Funct. Anal. **22**, 130–185 (1976)
34. F. Mignot, J. Puel, Optimal control in some variational inequalities. SIAM J. Control Optim. **22**(3), 466–476 (1984)
35. P. Neittaanmäki, J. Sokołowski, J.-P. Zolésio, Optimization of the domain in elliptic variational inequalities. Appl. Math. Optim. **18**(1), 85–98 (1988)
36. P. Neittaanmaki, J. Sprekels, D. Tiba, *Optimization of Elliptic Systems: Theory and Applications*. Springer Monographs in Mathematics (Springer, New York, 2006)
37. J. Outrata, M. Kočvara, J. Zowe, *Nonsmooth Approach to Optimization Problems with Equilibrium Constraints: Theory, Applications, and Numerical Results*. No. 152 in Nonconvex Optimization and Its Applications (Kluwer Academic, Dordrecht/Boston, 1998)
38. H. Scheel, S. Scholtes, Mathematical programs with complementarity constraints: stationarity, optimality, and sensitivity. Math. Oper. Res. **25**(1), 1–22 (2000)
39. A. Schiela, D. Wachsmuth, Convergence analysis of smoothing methods for optimal control of stationary variational inequalities with control constraints. ESAIM: Math. Model. Numer. Anal. **47**(5), 771–787 (2013)
40. J. Sokołowski, J.-P. Zolésio, *Introduction to Shape Optimization: Shape Sensitivity Analysis*. Volume 16 of Springer Series in Computational Mathematics (Springer, Berlin, 1992)
41. D. Tiba, Propriétés de controlabilité pour les systèmes elliptiques, la méthode des domaines fictifs et problèmes de design optimal, in *Optimization, Optimal Control and Partial Differential Equations* (Iaşi, 1992). Volume 107 of International Series Numerical Mathematics (Birkhäuser, Basel, 1992), pp. 251–261
42. B. Vexler, W. Wollner, Adaptive finite elements for elliptic optimization problems with control constraints. SIAM J. Control Optim. **47**(1), 509–534 (2008)

43. J. Yvon, Contrôle optimal de systèmes gouvernés par des inéquations variationnelles. PhD thesis, Université de Compiegne, Paris, 1973
44. J. Yvon, Optimal control of systems governed by variational inequalities, in *5th Conference on Optimization Techniques*. Lecture Notes in Computer Science (Springer, Berlin/Heidelberg, 1973), Rome, Italy, pp. 265–275
45. J. Zowe, S. Kurcyusz, Regularity and stability for the mathematical programming problem in Banach spaces. Appl. Math. Optim. **5**, 49–62 (1979)

Models and Optimal Control in Freezing and Thawing of Living Cells and Tissues

Karl-Heinz Hoffmann, Nikolai D. Botkin, and Varvara L. Turova

Abstract This paper outlines results obtained by the authors in the framework of the DFG Priority Program "Optimization with partial differential equations" (SPP 1253). The intention of the authors was related to the application of the theory of partial differential equations and optimal control techniques to the minimization of damaging factors in cryopreservation of living cells and tissues in order to increase the survival rate of frozen and subsequently thawed out cells. The paper presents mathematical models of the processes of freezing and thawing and describes the application of optimal control theory to the design of optimal cooling and warming protocols which reduce damaging effects and improve the survival rate of cells.

Keywords Freezing and thawing of biological cells • Damaging factors • Mathematical model • Control system • Hamilton-Jacobi equations • Grid methods • Optimal cooling and warming rates

Mathematics Subject Classification (2010). Primary 92B05, 35Q92, 35F21, 65M06; Secondary 49L20.

1 Introduction

The authors of patent [1] have found that certain tooth follicles contain the so called pluripotent (able to develop into multiple types) stem cells. The patent also outlines the application of such stem cells in tissue engineering, gene therapy, and in identifying, assaying or screening with respect to cell-cell interactions.

These technologies involve freezing and thawing out of small tissue samples in such a manner that the cells preserve their functional properties. Optimization and control are necessary here because of several competitive effects of cooling. Slow

K.-H. Hoffmann • N.D. Botkin (✉) • V.L. Turova
Zentrum Mathematik, Technische Universität München, Boltzmannstr. 3, 85748 München, Germany
e-mail: hoffmann@ma.tum.de; botkin@ma.tum.de; turova@ma.tum.de

G. Leugering et al. (eds.), *Trends in PDE Constrained Optimization*, International Series of Numerical Mathematics 165, DOI 10.1007/978-3-319-05083-6_10

cooling causes slow freezing of the extracellular fluid, which results in an increase in the concentration of salt in the remaining unfrozen part of the extracellular solution. Since the intracellular liquid remains unfrozen relatively long, the osmotic effect leads to the cellular dehydration and shrinkage. Another effect of this process is a large stress which can damage the integrity of cell membranes. If cooling is rapid, the water inside the cells forms small, irregularly-shaped ice crystals (dendrites) that are relatively unstable. If frozen cells are subsequently thawed out too slowly, dendrites will aggregate to form larger, more stable crystals that may cause damage. Maximum viability is obtained by cooling at a rate in a transition zone in which the combined effect of both these mechanisms is minimized. Thus, an optimization problem can be formulated for mathematical models describing the processes of freezing and thawing. Moreover, some general arguments and our freezing experiments show that better results can be obtained if the ambient temperature falls not monotonically in time, especially in the temperature range where the latent heat is released. Thus, time dependent optimal controls (optimal cooling protocols) are reasonable. Our numerical simulations and experiments show that positive effects can be achieved by creating temperature gradients in the freezing area or by forcing ice nucleation through mechanical vibration or some temperature shocks localized in a small area (seeding). Another control tool is related to cryoprotective agents which vary eutectic properties of solutions.

Additional difficulties arise when preserving structured solid tissues. Significant problems are associated with the difficulty of controlling heat transfer in a large object with a complex internal structure. The presence of different cell types, each with its own requirements for optimal cryopreservation, limits cell survival when a single thermal protocol is imposed on all of the cells. Extracellular ice can cause damage of the structural integrity of the tissue. Mechanical stresses caused by delayed freezing of the intracellular water are dangerous for cell membranes. Each of these is an additional source of damage, over and above those that are already known from studies of cells in suspension. Therefore, optimization and optimal control are necessary in this case.

2 Mathematical Models

Usually, tissue samples are being frozen using special plants e.g. of the IceCube family developed by SY-LAB, Geräte GmbH (Austria), see Fig. 1. The main part of such plants is a freezing chamber supplied by a cooling system. The plant is controlled by a computer that allows the user to prescribe a cooling protocol to be tracked. Tissue samples are put into plastic ampoules containing solutions. The ampoules are placed into a rack located in the freezing chamber.

The main objective of this report is to outline mathematical models describing processes running in the ampoules during freezing and to sketch the application of optimal control theory to the design of improved cooling protocols that reduce

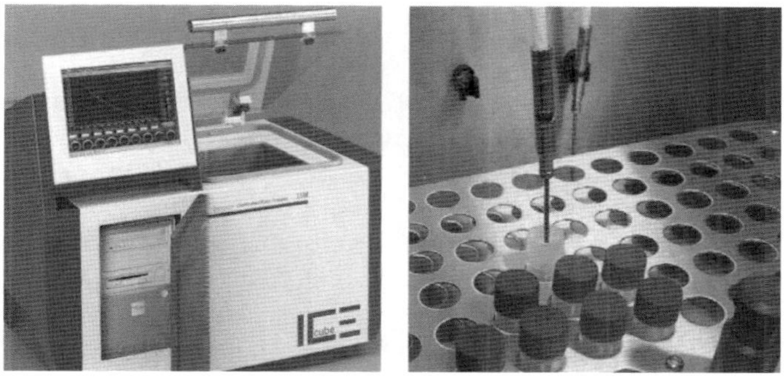

Fig. 1 Outlook of the IceCube plant (*to the left*) and its freezing chamber with two temperature sensors and ampoules containing tissue samples (*to the right*)

damaging effects caused by the release of the latent heat and by stresses arising due to delayed freezing of the intracellular water.

Such models can be classified as follows.

- The first model utilizes mean values of thermodynamical parameters to describe the mean boundary temperature of the ampoule. The control here is the temperature regime in the freezing chamber. A version of such a model including the design of improved cooling protocols is implemented in IceCube plants and tested in experiments on freezing of dental tissues.
- The second model of freezing deals with spatially distributed parameters and describes ice formation in the liquid surrounding the tissue sample. This approach is based on the so-called phase-field models described by partial differential equations. They have been introduced by Caginalp (see [2]) and studied by many scientists. We base the study here on the results of [3] where an optimal control problem for a phase-field model is considered and investigated both from the mathematical and algorithmic points of view. The design of optimal controls utilizes gradient descent methods and techniques of adjoint equations.
- The third model should describe ice formation on the cellular level. This includes modeling of phase changes in the extracellular liquid confined inside of small pores of the extracellular matrix and computing of mechanical stresses exerted on cell boundaries due to delayed freezing of the intracellular water. Another effect is dehydration of cells due to the osmotic outflow caused by the increase of the salt concentration in the extracellular liquid during its freezing. The opposite effect, rehydration, occurs during thawing. Formation of dendrites that can aggregate into lagre sharp ice crystals should also be accounted for in this model.

The basic tools here are the theory of ice formation in porous media and Stefan type models, see [4–6].

3 Mean Value Temperature Response Model

A preliminary approach to the control of global (averaged) thermodynamical parameters was elaborated and verified using a Freezer IceCube 15M (SY-LAB, Geräte GmbH, Austria). IceCube 15M is developed for controlled freezing of small tissue samples put into plastic ampoules (see Fig. 1).

The main parts of the plant are a freezing chamber containing a cooling system based on gas nitrogen, a rack for placing ampoules, and two temperature sensors that measure the chamber and sample temperatures, respectively. The plant is equipped with a computer that allows the user to input a cooling protocol either manually or as a file prepared in off-line regime. The computer controls the cooling system and forces the chamber temperature to track the prescribed cooling protocol.

Freezing experiments show an irregular behavior of the temperature near the freezing point because of the release of the latent heat and the crystallization. A typical response of the object to the cooling with a constant rate is shown in Fig. 2.

The supercooling (S) is not too dangerous itself but in combination with the latent heat release (L). This yields the creation of damaging dendrites: the longer runs the latent heat release (L), the more dendrites appear. A sudden drop of the temperature (D) causes a temperature shock to cells. Therefore, it would be preferable to reduce both the duration of the latent heat release (L) and the temperature drop (D).

The following simple thermodynamical model is based on averaged values of parameters (see [7] and [8]):

$$\frac{d}{dt}H = -\alpha\big(T(H) - T_e\big) \tag{3.1}$$

$$\dot{T}_e = u, \quad |u| \le \mu. \tag{3.2}$$

Here H and T are the averaged enthalpy density and the temperature, respectively; T_e the chamber temperature; u the chamber temperature rate whose maximal absolute value μ is about 20°C/min; $\alpha = |S||V|^{-1}h$, where S, V, and h are the surface area, the volume, and the overall heat transfer coefficient of the ampoule,

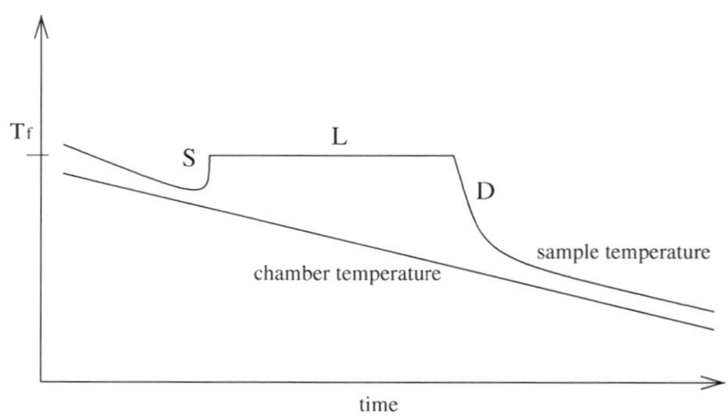

Fig. 2 Idealized typical temperature response of the sample when the chamber temperature falls linearly in time. Three dangerous processes are pointed out

respectively. The main part of the model is the constitutive law $T = T(H)$, which is specific for the sample. Nevertheless such a relation is robust for a series of similar samples. The function $T(H)$ can be obtained approximatively by recovering the mean surface temperature $T(t)$ of the ampoule from measurements for a given temperature profile $T_e(t)$. Substituting the functions $T(t)$ and $T_e(t)$ into (3.1) and computing $H(t)$ yields the pair $\{T(t), H(t)\}$, which defines implicitly the desired function.

The aim of the control is to smooth the temperature response of the sample during the production of the latent heat (see Fig. 2). Formally, this is expressed through the minimization of the following performance index:

$$J = \int_{t_1}^{t_2} \Big(\frac{d}{dt}T(H(t)) - \theta_0\Big)^2 dt \equiv \int_{t_1}^{t_2} \Big(\alpha\frac{\partial T}{\partial H}(H)\cdot(T(H) - T_e) + \theta_0\Big)^2 dt, \quad (3.3)$$

where θ_0 is a desirable slope of the temperature curve. Ordinary differential equations (3.1), (3.2) and functional (3.3) form a controlled system, where the control variable u is the rate of the chamber temperature. This model possesses the following nice property: the functional J does not depend (up to a positive multiplier) on the choice of α, whenever the constitutive law $T(H)$ is restored using (3.1), and $H(t)$ is found from (3.1) with the same value of α. Therefore, the value of α can be simply chosen as $\alpha = 1$, which saves the trouble of measuring the physical parameters $|V|$, $|S|$, and h. Application of optimal control theory allows us to find cooling profiles which essentially improve the temperature response of samples.

A very important question is the robustness of optimal control if the chamber and sample temperatures are measured with an error. This has been investigated using the theory of optimal conflict control. The disturbances were considered as controls of an opposite player maximizing the objective functional. New numerical methods for solving Hamilton-Jacobi equations arising from conflict control problems with state constraints were (see [9–11]) applied, and the robustness of solutions was validated.

The above sketched optimization techniques are implemented in IceCube freezers. With the graphical interface of an IceCube plant, users can choose an optimization option to compute the optimizing cooling impulse. The corresponding optimized temperature response is much better than a non-optimized one (see [7]).

4 Distributed Modeling of Ice Formation in the Liquid Milieu

Nowadays, phase field techniques for modeling of solidification and freezing processes become very popular. They are based on the consideration of the Gibbs free energy which depends on an order parameter that assumes values from -1 (solid) to 1 (liquid) and changes sharply but smoothly over the solidification front so that the sharp liquid/solid interface becomes smoothed. The rate of smoothing is controlled by a small parameter, which enables to reach arbitrary approximation of the sharp interface.

We use a phase field model (see [2, 3, 5], and [12]) to describe phase changes in the milieu containing a tissue. The control parameter is the temperature in the chamber. Note that the heat flux on the boundary of the ampoule is proportional to the temperature jump on the boundary. For simplicity, we do not include the solid part (i.e. the tissue and the walls of the ampoule) into the following description bearing in mind that they are accounted for in numerical simulations presented in [8]. Therefore, Ω being the interior of the ampoule, Γ the boundary of Ω. The equations read as follows:

$$u_t + \frac{\ell}{2}\phi_t - K\Delta u = 0, \qquad x \in \Omega, \tag{4.1}$$

$$\tau\phi_t - \xi^2\Delta\phi - \frac{1}{2}(\phi - \phi^3) - 2u = 0, \qquad x \in \Omega, \tag{4.2}$$

$$-K\frac{\partial u}{\partial n} = h(u - u_e(t) - g), \quad \frac{\partial \phi}{\partial n} = 0, \qquad x \in \Gamma, \tag{4.3}$$

$$u|_{t=0} = u_0 \equiv \text{ const } > 0, \quad \phi|_{t=0} = \phi_0 \equiv -1. \tag{4.4}$$

Here, u is the scaled distribution of the temperature; ϕ the phase function: $\phi = 1$ for the frozen state and $\phi = -1$ for the liquid state; ℓ the scaled latent heat; K the scaled heat conductivity coefficient; h the scaled overall heat conductivity; g the boundary control (add-on to the nominal cooling protocol $u_e(t) = u_0 + \theta_0 t$, where $\theta_0 < 0$ is a given slope). In contrast to the work [3], we do not assume C^2 regularity of Γ. It is supposed that Γ is of the class $C^{0,1}$, i.e. Lipschitz continuous. Such an assumption covers various technical designs of ampoules and permits a direct extension of the result to the case where a solid part (tissue) immersed into the fluid is present.

The regularity of solutions is investigated in [13] and [14]. In particular, the existence and continuity of solutions in time under discontinuous initial dates, $\phi_0 \in L^2(\Omega)$, is proved.

First, the following functional that estimates the mean quadratic deviation from the nominal cooling protocol $u_e(t)$ was considered:

$$J = \frac{1}{2}\int_0^{t_f}\int_\Omega (u - u_e(t))^2\, dxdt. \tag{4.5}$$

The adjoint system is derived as in [3] with some modifications related to the boundary method of control:

$$
\begin{aligned}
&-p_t - K\Delta p - 2q = h(u - u_e(t)), && x \in \Omega,\\
&-\tau q_t - \frac{\ell}{2}p_t - \xi^2\Delta q - \frac{1}{2}(1 - 3\phi^2)q = 0, && x \in \Omega,\\
&-K\frac{\partial p}{\partial n} = hp, \quad \frac{\partial q}{\partial n} = 0, && x \in \Gamma,\\
&p(t_f) = 0, \quad q(t_f) = 0.
\end{aligned}
\tag{4.6}
$$

Here, p is the adjoint variable corresponding to u; q the adjoint variable related to ϕ. The appropriate regularity of solutions of (4.6) was established (see [7]) so that the derivative of the functional J with respect to the control g is defined by the formula:

$$
J'(u,\phi)(\delta g) = \int_0^{t_f}\int_\Gamma \delta g(t,s)p(t,s)dsdt, \quad \text{for all} \quad \delta g \in L^2\big((t_0,t_f)\times\Gamma\big). \tag{4.7}
$$

Therefore, we can identify $J'(u,\phi)$ with $p|_{(0,t_f)\times\Gamma}$. The method of conjugate gradients looks as follows. Consider the n th step. Assume that the control g^n is already known. Compute then the states u^n, ϕ^n, and the adjoint states p^n and q^n. Compute g^{n+1} as $g^{n+1} = g^n + \alpha^n d^n$, where α^n is an approximate solution of the line search problem $\alpha^n \to \min_\alpha J(g^n + \alpha d^n)$, and the conjugate direction d^n is found from the relation (see [15] for finite dimensional case)

$$
d^n = -p^n + \beta^n d^{n-1}, \quad \beta^n = \left[\int_0^{t_f}\int_\Gamma (p^{n-1})^2 dsdt\right]^{-1}\int_0^{t_f}\int_\Gamma (p^n)^2 dsdt.
$$

The numerical results obtained by minimizing the functional (4.5) show oscillations around the nominal protocol $u_e(t)$ (see [8]), which makes this functional practically unusable. To avoid such effects, it is necessary to include the time derivative of the temperature into the functional. Several functionals of such a type have been considered and rejected because of essential technical difficulties carefully discussed in [8].

An appropriate solution is the use of the following functional that estimates the deviation of the slope of the mean temperature from a given slope:

$$
J = \frac{1}{2}\int_C^{t_f}([u]_t - \theta_0)^2 dt, \quad \text{where} \quad [u] = \frac{1}{|\Omega|}\int_\Omega u dx.
$$

The idea is to express $[u]_t$ through values that do not contain time derivatives. Integrating the model equations (4.1) and (4.2) over Ω yields:

$$[u]_t - \theta_0 = \gamma_{u,\phi,g}(t) :=$$
$$-\frac{1}{|\Omega|}\left[h\int_\Gamma [u - u_e(t) - g]ds + \frac{1}{4\tau}\int_\Omega (\phi - \phi^3)dx + \frac{1}{\tau}\int_\Omega u dx\right] - \theta_0. \tag{4.8}$$

It was shown that the corresponding adjoint system is well-posed, and the derivative of the last functional J is correctly defined. Nevertheless, the implementation of such method is not simple. First, the computation of an optimized cooling protocol is time consuming. Second, optimized controls obtained have a complex structure on the boundary, which can hardly be implemented technically. Therefore, it is reasonable to look for a control that is constant on the boundary so that g is a function of t only. Such a control $g(t)$ can be computed from the condition $\gamma_{u,\phi,g}(t) = 0$, i.e. $[u]_t - \theta_0 = 0$ (see relation (4.8)). This yields the following feedback rule:

$$g(t) = \frac{1}{h|\Gamma|}\left[h\int_\Gamma [u - u_e(t)]ds + \frac{1}{4\tau}\int_\Omega (\phi - \phi^3)dx + \frac{1}{\tau}\int_\Omega u dx\right] + \frac{|\Omega|}{h|\Gamma|}\theta_0. \tag{4.9}$$

The result of such a heuristic rule seems to be comparable with that obtained by using adjoint equations, but the run time is sufficiently shorter (see [7]).

4.1 Improvement of the Heat Conductivity of Ampoule Walls

The authors of patent [16] have shown that the temperature gradient inside the ampoule during freezing is favorable for the homogeneous ice nucleation and avoiding the dendritic crystal growth. The idea of the invention [16] is to cover the bottom of ampoules with a metal, say copper, and to place the tissue sample at a certain distance from the bottom. Computations performed with the use of the above-described phase field model (4.1)–(4.4) confirm this idea and indirectly show the improvement of the phase transition process (see [7]). Paper [17] confirms the idea to improve the heat conductivity of the ampoule walls by placing of micro rods on its surfaces.

4.2 Ice Formation in the Extracellular Matrix

The description of ice formation in the extracellular matrix of a tissue is based on the thermodynamics of porous media (see e.g. [4]). The main feature of porous or soil media is that the unfrozen water content is a function of the temperature only. Such a function can be considered as material property that depends on the pore size distribution and the material of the solid matrix. The interaction of the liquid with the solid matrix causes as a rule the homogenous nucleation of the liquid and, therefore, avoids it from supercooling. Thus, the unfrozen water content is not defined from any equation but is a given function of the temperature. In freezing soil science, it is measured directly by NMR (Nuclear Magnetic Resonance). There are a lot of theoretical works on derivation of this function (see e.g. [4] and [18]).

The model looks as follows:

$$\rho C \frac{\partial \theta}{\partial t} + \rho L \frac{\partial \beta_\ell}{\partial t} - \mathcal{K} \Delta \theta = 0, \quad -\mathcal{K} \frac{\partial \theta}{\partial n} = \lambda(\theta - \theta_0 t - g). \tag{4.10}$$

Here, ρ is the density (assume that the densities of the liquid and ice are equal), L is the phase change latent heat, C is the specific heat capacity (assume they are equal for the liquid and ice), β_ℓ is the liquid water volume fraction (unfrozen water content), and g is the boundary control (add-on to the nominal cooling protocol $\theta_0 t$, where $\theta_0 < 0$ being a given slope). The function β_ℓ was recovered from data obtained in experiments with tissue samples on an IceCube plant (see [7, 8] for β_ℓ, simulations with system (4.10), and optimization problems).

It is well known that the increase of the volume during the water to ice phase change is rather large. If the phase change occurs in a pore of a porous material, a very large stress can be exerted on the pore walls. The computation of stresses in porous media is based on the theory of linear elasticity, homogenization theory, and the model of ice formation from the previous section. Simulations show that the stress inside porous media can rise up to 2 Bar. The stress exerted on a sole cell located in a pore with freezing water is about 1 Bar (see [7]).

5 Consideration of Damaging Processes on Cellular Level

Processes of freezing and thawing inside of pores of the extracellular matrix play an imported role in cryopreservation. One of the objectives of our study was the development of mathematical models of such processes and elaboration of control procedures to reduce damaging factors arising during freezing and thawing.

Three injuring factors are considered. One of them is a large stress exerted on cell membranes. Another factor is excessive shrinkage and swelling due to osmotic dehydration and rehydration occurring during the freezing and warming phases, respectively. The third effect is related to the growth of dendrites that appear during freezing and can aggregate into large crystals during thawing. The growth of dendritic seeds occurs at rapid cooling rates.

Large stresses exerted on cell membranes occur at slow cooling because of non-simultaneous freezing of extracellular and intracellular fluids. The use of rapid cooling rates is limited by dendritic growth. Therefore, it is reasonable to apply control theory to provide simultaneous freezing of extracellular and intracellular fluids even for slow cooling rates. To this end, mathematical models containing control variables and optimization criteria were formulated. We have started with spatially distributed models describing the dynamics of phases in each spatial point (see e.g. [2, 4], and [3]), and then an averaging was applied to reduce partial differential equations to a few ordinary differential equations with control parameters and uncertainties (see [19]). These equations contain nonlinear dependencies given by tabular data, which complicates the application of traditional control design methods based on Pontryagin's maximum principle. Nevertheless, dynamic programming methods related to Hamilton-Jacobi-Bellman-Isaacs (HJBI) equations are suitable. Stable grid procedures that enable to design optimized controls (cooling protocols) for ODE systems describing competitive ice formation inside and outside of living cells have been developed (see [9, 10], and [8]). It should be noticed that the formation of dendritic seeds are also included into our models by utilizing of the corresponding thermodynamical relations.

Cell dehydration and rehydration occur because of abnormal water transport across cell membranes. This effect is caused by the osmotic pressure arising because of different salt concentrations in intra- and extracellular fluids. Conventional models of cell dehydration during freezing (see e.g. [20] and [21]) describe the change of the cell volume. The cell shape is supposed to be spherical or cylindrical. However, as it is reported by biologists (see e.g. [22]), controlling cell shape is also important for the survival rate of cells. For this reason, mathematical models concerned with the evolution of cell shape depending on the temperature distribution and the amount of intra- and extracellular ice have been developed in our project. The study was based on the theory of ice formation in porous media (see [4]) and Stefan-type models (see [5]) describing the motion of the cell membrane due to osmotic flow into and out of the cell. The evolution of cell shape was described with Hamilton-Jacobi type equations solved using both finite-difference schemes and reachable set methods (see [23] and [24]).

5.1 Balance of Ice Formation in Intra- and Extracellular Liquid

The cells of a tissue are surrounded by an extracellular liquid confined in small vessels of an extracellular matrix. Biologists suppose that cells may communicate through these vessels. Since the mechanism and the role of cell-cell interactions are still not well understood (see e.g. [25]), we have considered a simplified model which does not take into account possible cell-cell interactions. For this reason, the extracellular liquid is assumed to be confined in small non-communicating cavities or pores of the extracellular matrix (see a sketch in the left part of Fig. 3). Each cell has a membrane that provides a physical separation between the intra- and extracellular environments, which may cause delayed freezing of intracellular liquid. This effect results in a very large stress exerted on the cell membrane. The magnitude of this effect can be approximately estimated as follows:

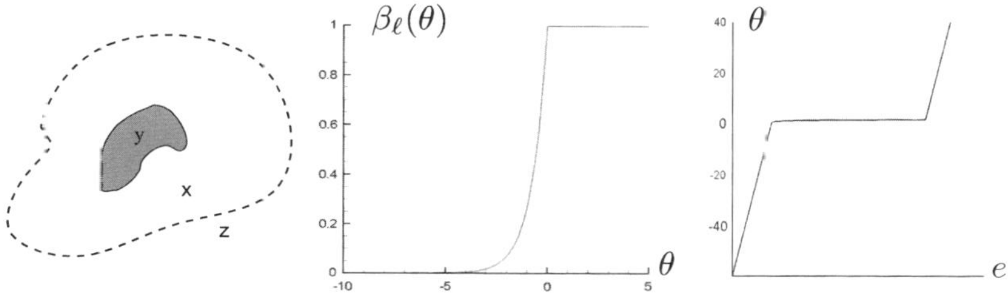

Fig. 3 *To the left*: two-dimensional sketch of a cell located inside a pore of the extracellular matrix. The pore is filled with an extracellular fluid, whereas the cell contains intracellular liquid. *At the center*: a typical form of the function defining the fraction of unfrozen liquid. The graphs of these functions can be shifted to the left or right according to the freezing points of the extra- and intracellular liquids. *To the right*: the inverse to the function $e = \rho C\theta + \rho L\beta_\ell(\theta)$ (expression of the temperature through the internal energy)

$p \approx E^{ice}\,æ \cdot (1 - \beta_\ell)$, where p is the pressure, E^{ice} is the elastic modulus of ice, æ is the ratio of volume expansion due to the water-to-ice phase transition, and β_ℓ is the unfrozen water fraction so that $1 - \beta_\ell$ is the ice content. A rough estimate yields $p \approx 1$Bar, which may be dangerous for cell membranes. To reduce this effect, the fluids inside and outside the cell must freeze simultaneously. This can be achieved by lowering the freezing point of the extracellular fluid using a cryoprotector, say dimethyl sulfoxide, and optimizing cooling protocols.

Some averaging technique described in [8, 19, 24] yields the following ODE model:

$$\begin{aligned} \dot{x} &= -\alpha_1[\Theta_1(x) - \Theta_2(y)] - \lambda[\Theta_1(x) - z] + v_1, \\ \dot{y} &= -\alpha_2[\Theta_2(y) - \Theta_1(x)] + v_2, \\ \dot{z} &= u. \end{aligned} \tag{5.1}$$

Here, x is the averaged density of the internal energy of the extracellular liquid, y is the same for the intracellular fluid, z is the temperature outside the pore (chamber temperature), u is the cooling rate, and v_1, v_2 are disturbances interpreted as data errors. The control variable u (cooling rate) is restricted by $|u| \le \mu$, the disturbances v_1, v_2 are bounded by $|v_1| \le \nu$, $|v_2| \le \nu$. The functions $\Theta_1(x)$ and $\Theta_2(y)$ are the inverse to the functions $x = \rho C\theta_1 + \rho L\beta_\ell^1(\theta_1)$ and $y = \rho C\theta_2 + \rho L\beta_\ell^2(\theta_2)$, respectively. The functions β_ℓ^1 and β_ℓ^2 express the unfrozen liquid fractions for extra- and intracellular fluids, respectively, ρ is density, C the specific heat capacity, and L the specific latent heat. Thus, $\Theta_1(x)$ and $\Theta_2(y)$ express the temperatures in the extra- and intracellular liquids through the internal energies x and y, respectively. See also Fig. 3 for the further explanations.

According to the meaning of the functions $\beta_\ell^i, i = 1, 2$, exact simultaneous freezing of the extra- and intracellular liquids can be expressed as vanishing of the following functional:

$$J = \int_0^{t_f} \left|\beta_\ell^1(\Theta_1(x(t))) - \beta_\ell^2(\Theta_2(y(t)))\right|^2 dt \tag{5.2}$$

that estimates the difference of the ice fractions in the extra- and intracellular regions.

Differential game (5.1) and (5.2) assumes that the objective of the control u is to minimize the functional (5.2), whereas the objective of the disturbance is opposite. Moreover, the trajectories should remain in a state constraint set represented in the form of inequalities defined on trajectories of (5.1).

The value function of differential game (5.1) and (5.2) has been computed as a viscosity solution (see [26]) to the corresponding HJBI equation using upwind grid methods developed in [9–11]. The optimal feedback control was designed by applying the procedure of extremal aiming (see [27]). The computations have been performed on a Linux computer admitting 64 GB memory and 32 threads. The coefficient of parallelization was equal to 0.7 per thread (23 times speedup totally). The grid size was equal 300^3, the number of time steps equaled 30000. The runtime is approximately 60 min.

The simulation presented in Fig. 4 shows the case of different freezing points for the pore, θ_{1s}, and the cell, θ_{2s}, with $\theta_{1s} - \theta_{2s} = -13°\text{C}$. Thus, the freezing point of the extracellular fluid is lowered, e.g. by adding a cryoprotector. This enables us to freeze the intracellular fluid using temperatures laying above the freezing point of the extracellular liquid, which makes possible simultaneous freezing.

The effect of supercooling of the intracellular fluid can be accounted for by introducing a kink into the dependence of the temperature on the internal energy at the freezing point (see [28] and [8] for the exact definition and the corresponding simulations).

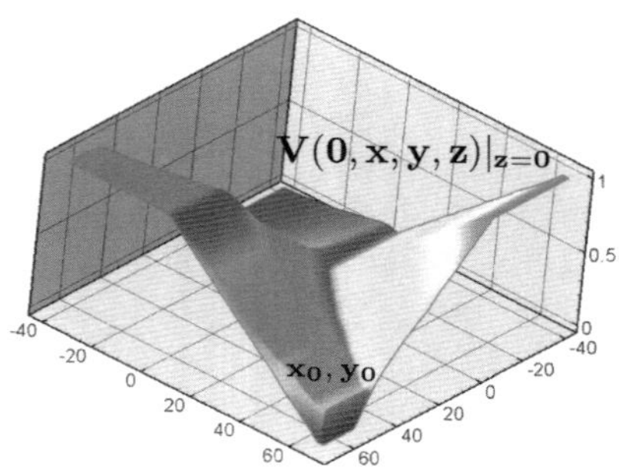

Graph of the value function at $t = 0$, $z = 0$

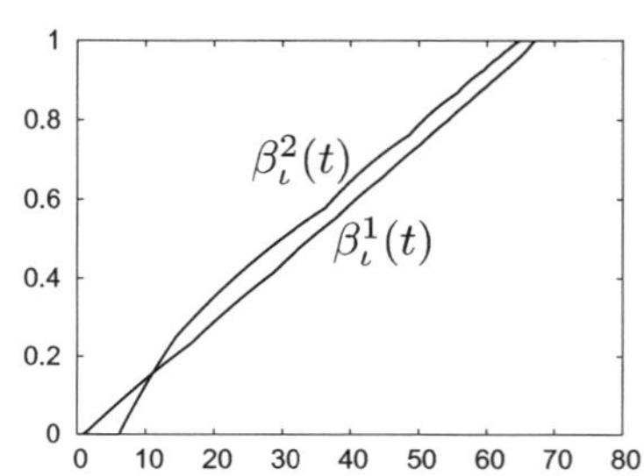

Ice fractions in the extracellular space and in the cell versus time

Fig. 4 Almost simultaneous freezing of extra- and intracellular liquids

5.2 Balance of Ice Formation with Accounting for Dendrite Growth

Accounting for the growth of dendrite seeds is done by the following modification of the internal energy inside the cell: $y = \rho C\theta_2 + \rho L\beta_\ell^2(\theta_2) + \rho D\,\kappa\,\beta_\ell^2(\theta_2)$, where the last term is treated as a dendrite generation energy (D is the specific latent heat of dendrite growth). The value κ is computed as follows: $\kappa = \kappa_0(\dot{y})^- = -\kappa_0\alpha_2(\theta_2 - \theta_1)^+$, where $(a)^- = \min\{a, 0\}$ and $(a)^+ = \max\{a, 0\}$. The objective of the control is to minimize the generation of dendrite seeds along with the balance of ice formation so that the following extended functional is considered:

$$J = \int_0^{t_f} \left|\beta_\ell^1(\Theta_1) - \beta_\ell^2(\Theta_2)\right|^2 d\tau + \kappa_0\rho D\alpha_2 \int_0^{t_f} (\Theta_2 - \Theta_1)^+ \beta_\ell^2(\Theta_2)d\tau. \tag{5.3}$$

Here $\Theta_1 = \Theta_1(x(t))$ and $\Theta_2 = \Theta_2(x(t), y(t))$. Additionally, the state constraint $z \le 2°\text{C}$ (remember that z is the temperature outside the pore) is imposed.

It was observed in simulations (see [8]) that 30% less dendritic seeds are formed if functional (5.3) is used instead of functional (5.2).

5.3 Simulation of Cell Thawing with Optimization of the Osmotic Inflow and Accounting for the Dendrite Growth

Optimization of the Osmotic Inflow First note that, compared to the freezing procedure, the dependence of the unfrozen water fraction on the temperature is modified (the graph of the function is shifted to the right to account for the delay of thawing). The functional to minimize expresses the amount of liquid moving into the cell, which is proportional to the difference between the concentration of the physiological salt solution and salt concentration in the cell

$$J = \alpha \int_0^{t_f} \left|c_0 - g\Big(\frac{m_0}{W_0\,\beta_\ell^2(\Theta_2(y(\tau)))}\Big)\right| d\tau. \tag{5.4}$$

Here, c_0 is the concentration of the physiologic salt solution, m_0 the salt amount in the cell, W_0 the initial cell volume, g a function defining the saturation of salt concentration in the cell. Thus, $W_0\beta_\ell^2$ is the unfrozen water volume inside the cell, and $g(m_0/(W_0\beta_\ell^2))$ the salt concentration in the cell. Additionally, the state constraints

$$-50°\text{C} \leq z \leq 40°\text{C} \quad \text{and} \quad \Theta_2(y) \leq 20°\text{C} \tag{5.5}$$

are imposed to prevent excessive warming. Optimization results and optimized warming protocols are presented in [8].

Accounting for the Dendrite Growth Dendrite seeds formed during freezing can form large ice crystals at the thawing stage, which may be dangerous for cells. To take into account the growth of dendrites, the following functional is considered:

$$\begin{aligned} J = \alpha \int_0^{t_f} \left| c_0 - g\Big(\frac{m_0}{W_0\, \beta_\ell^2(\Theta_2(y(\tau)))}\Big)\right| d\tau + \\ \max_{t\in[0,t_f]} \big[\big(1-\beta_\ell^2(\Theta_2(y(t)))\big)\, p(t)\beta_\ell^2(\Theta_2(y(t)))\big]. \end{aligned} \tag{5.6}$$

The first term in (5.6) coincides with (5.4), and the second one expresses the amount of ice formed due to aggregation of dendrite seeds according to the Poisson law (see [29])

$$p(t) = 1 - e^{-\lambda(\theta,\Delta E, D(\theta))t}.$$

The nucleation rate λ is a function of the temperature, the activation energy, and the diffusive mobility of dendritic seeds, respectively. In the simulation, the nucleation rate λ is supposed to be constant.

Since both terms in the functional are non-antagonistic, it is clear that rapid thawing is preferable. To prevent excessive warming, state constraints (5.5) were applied.

Two simulations of system (5.1) with state constraints (5.5) were performed: the first one with functional (5.4), and the second one with functional (5.6). The comparison of the simulation results shows that 7% less dendrites are formed in the second case, see [8].

5.4 *Mathematical Models of Dehydration and Rehydration of Cells*

Each biological cell is located inside a pore or channel filled with a saline solution called extracellular fluid. The cell interior is separated from the outer liquid by a cell membrane whose structure ensures a very good permeability for water. This leads to abnormal water transport across the cell membrane in the presence of osmotic pressure caused by different salt concentrations in intra- and extracellular fluids.

5.4.1 Dehydration of Cells

In the freezing phase, the mechanism of the osmotic effect is the following. Ice formation occurs initially in the extracellular solution. Since ice is practically free of salt, the water-to-ice phase change results in the increase of the salt concentration (c_{out}) in the remaining extracellular liquid. The osmotic pressure forces the outflow of water from the cell to balance the intracellular (c_{in}) and extracellular (c_{out}) salt concentrations. Modeling of cell shrinkage is based on free boundary problem techniques. The main relation here is the so-called Stefan condition: $\mathcal{V} = \alpha(c_{\text{out}} - c_{\text{in}})$, where $\mathcal{V}$ is the normal velocity of the cell boundary (directed to the cell interior), and the right-hand-side represents the osmotic flux that is proportional to the difference of the salt concentrations. The coefficient α is the product of the Boltzmann constant, the temperature, and the hydraulic conductivity of the membrane (see e.g. [21]). Note that α is practically constant in our case. The extracellular salt concentration c_{out} depends on the unfrozen fraction $\beta_\ell(\theta)$ of the extracellular liquid.

The intracellular and extracellular salt concentrations are estimated using the mass conservation law as follows:

$$c_{\text{in}} = c_{\text{in}}^0 W_c^0 / W_c(t), \quad c_{\text{out}} = c_{\text{out}}^0 W^0 / W(t), \quad W(t) = \int_{W^0} \beta_\ell(\theta(t,x))dx,$$

where W_c^0 and $W_c(t)$ are the initial and current cell volumes, respectively, W^0 and $W(t)$ are the initial and current volumes of the unfrozen part of the pore. The distribution of the temperature $\theta(t,x)$ is found from the phase field model (4.10).

The cell region $\Sigma(t)$ is searched as the level set of a function $\Psi(t,x)$, i.e.

$$\Sigma(t) = \{x \colon \Psi(t,x) \leq 1\}, \qquad x \in R^3 \text{ (or } R^2).$$

Assuming that the cell boundary propagates with the normal velocity $\mathcal{V}$ yields the following Hamilton-Jacobi equation for the function $\Psi(t,x)$:

$$\Psi_t - \alpha(c_{\text{out}} - c_{\text{in}})|\nabla\Psi| = 0, \quad \Psi(0,x) = \inf\{\lambda > 0 \colon x \in \lambda \cdot \Sigma(0)\}. \qquad (5.7)$$

Here $|\nabla\Psi|$ denotes the Euclidean norm of the gradient. Examples of the corresponding computer simulations can be found in [8, 23, 24].

5.4.2 Rehydration of Cells

In the thawing phase, the osmotic effect results in the inflow of water into cells and therefore causes cell swelling. We use the following mass conservation law for the salt content:

$$W_c^0\, c_{\text{in}}^0 = W_s\, c_{\text{in}}^0 + W_\ell\, c_{\text{in}}, \quad W_c = W_s + W_\ell, \quad W_\ell(t) \approx \int\limits_{W_c^0} \beta_\ell\,(\theta(t,x))dx, \qquad (5.8)$$

where W_c^0 is the initial volume of the frozen cell, W_c the current volume of the cell, W_s and W_ℓ are volumes of the frozen and unfrozen parts of the cell, respectively, c_{in}^0 and c_{in} the salt concentrations in the frozen and unfrozen parts of the cell, respectively. Relation (5.8) yields:

$$c_{\text{in}} = c_{\text{in}}^0 \left(1 + (W_c^0 - W_c)/\,W_\ell\right),$$

where W_c is calculated from the current cell shape. The salt concentration c_{out} outside the cell is supposed to be a constant, and finally we arrive at an equation of the form of (5.7). The corresponding computer simulation can be found in [8].

5.4.3 Accounting for the Membrane Tension Using Reachable Set Approach

In reality, the deformation of the cell membrane depends on the membrane tension which is a function of the curvature. Therefore, a more realistic expression for the normal velocity of the cell boundary would be:

$$\mathcal{V}(t,x) = \alpha(c_{\text{out}}(t) - c_{\text{in}}(t)) + \gamma\sigma(x),$$

where $\sigma(x)$ is the angular curvature at the current point x of the cell boundary (see [8]), and γ is a constant. The resulting equation reads

$$\Psi_t - \big(\alpha(c_{\text{out}} - c_{\text{in}}) + \gamma\sigma(x)\big)|\nabla\Psi| = 0, \quad \Psi(0,x) = \inf\{\lambda > 0\colon\, x \in \lambda\cdot\Sigma(0)\}. \qquad (5.9)$$

Note that accounting for the curvature can alter the convexity/concavity structure of the Hamiltonian depending on the state x.

The problem was treated using the method of reachable sets (see [27, 30]). Examples of the application of reachable sets method to (5.9) in R^2 can be found in [8, 23, 24]).

References

1. J. Schierholz, N. Brenner, H.-F. Zeilhofer, K.-H. Hoffmann, C. Morsczeck, *Pluripotent embryonic-like stem cells derived from teeth and uses thereof.* European Patent. Date of publication and mention of the grant of the patent 04.06.2008. Application number: 03704549.9, International application number: PCT/EP2003/001131. International publication number: WO 2003/066840 (140.8.2003 Gazetta 2003/33)

2. G. Caginalp, An analysis of a phase field model of a free boundary. Arch. Rat. Mech. Anal. **92**, 205–245 (1986)
3. K.-H. Hoffmann, J. Lishang, Optimal control of a phase field model for solidification. Numer. Funct. Anal. Optimiz. **13**(1&2), 11–27 (1992)
4. M. Frémond, *Non-Smooth Thermomechanics* (Springer, Berlin, 2002)
5. G. Caginalp, X. Chen, Convergence of the phase field model to its sharp interface limit. Eur. J. Appl. Math. **9**, 417–445 (1998)
6. A.M. Meirmanov, *The Stefan Problem* (Walter de Gruyter, Berlin, 1992)
7. K-H. Hoffmann, N.D. Botkin, Optimal control in cryopreservation of cells and tissues. Adv. Math. Sci. Appl. **29**, 177–200 (2008)
8. N.D. Botkin, K.-H. Hoffmann, V.L. Turova, Mathematical modeling and simulations in cryopreservation of living cells, in *Cryopreservation: Technologies, Applications and Risks/Outcomes*, ed. by A. Colvert, H. Coty (Nova Science Publishers Inc, 2013)
9. N.D. Botkin, K-H. Hoffmann, V.L. Turova, Stable numerical schemes for solving Hamilton-Jacobi-Bellman-Isaacs equations. SIAM J. Sci. Comput. **33**(2), 992–1007 (2011)
10. N.D. Botkin, K.-H. Hoffmann, N. Mayer, V.L. Turova, Approximation schemes for solving disturbed control problems with non-terminal time and state constraints. Analysis **31**, 355–379 (2011)
11. N. Botkin, K.-H. Hoffmann, N. Mayer, V. Turova, Computation of value functions in nonlinear differential games with state constraints, in *System Modeling and Optimization* ed. by D. Hömberg, F. Tröltzsch. Proceedings of the 25th IFIP TC7 Conference (2013), 235–244
12. Y. Xu, J.M. McDonough, K.A. Tagavi, D. Gao, Two-dimensional phase-field model applied to freezing into supercooled melt. Cell Preserv. Technol. **2**(2), 113–124 (2004)
13. K.-H. Hoffmann, T.G. Amler, N.D. Botkin, K. Ruf, Regularity of solutions of a phase field model. Preprint-Nr. SPP1253-141, 2012. http://www.am.uni-erlangen.de/home/spp1253/wiki/index.php/Preprints
14. T.G. Amler, N.D. Botkin, K.-H. Hoffmann, I. Hoteit, Continuity in time of solutions of a phase-field model. Preprint-Nr. SPP1253-142, 2012. http://www.am.uni-erlangen.de/home/spp1253/wiki/index.php/Preprints
15. R. Fletcher, C.M. Reeves, Function minimization by conjugate gradients. Comput. J. **7**, 149–154 (1964)
16. N. Botkin, O. Degistirici, B. Faßbender, J. Siemonsmeier, M. Thie, *Zahn-Einfrier-Behälter.* Deutsches Patent- und Markenamt. Reference number DE 10 2005 047 438 A1. Date of publication and mention of the grant of the patent 05.04.2007
17. N.D. Botkin, K-H. Hoffmann, A. Frackowiak, A.M. Cialkowski, *Study of the Heat Transfer Between Gases and Solid Surfaces Covered With Micro Rods*, Preprint-Nr. SPP1253-10-05, 2008
18. P. Haeupl, Y. Xu, Numerical simulation of freezing melting in porous materials under the consideration of the coupled heat and moisture transport. J. Therm. Envel. Bilding Sci. **25**(1), 4–31 (2001)
19. N.D. Botkin, K.-H. Hoffmann, Optimal control in cryopreservation of living cells, in *Proceedings of the International Conference "Actual Problems of Stability and Control Theory" (APSCT2009),* Russia, Ekaterinburg, 233–240 (2010)
20. R.P. Batycky, R. Hammerstedt, D.A. Edwards, Osmotically driven intracellular transport phenomena. Phil. Trans. R. Soc. Lond. A **355**, 2459–2488 (1997)
21. L. Mao, H.S. Udaykumar, J.O.M. Karlsson, Simulation of micro scale interaction between ice and biological cells. Int. J. Heat Mass Transf. **46**, 5123–5136 (2003)
22. S.C. Chen, M. Mrksich, S. Huang, G.M. Whitesides, D.E. Ingber, Geometric control of cell life and death. Science **276**, 1425–1428 (1997)
23. V.L. Turova, Modeling osmotic de- and rehydration of living cells using Hamilton-Jacobi equations and reachable set techniques, in *Proceedings of the International Conference "Actual Problems of Stability and Control Theory" (APSCT2009),* Russia, Ekaterinburg, 308–315 (2010)

24. K.-H. Hoffmann, N.D. Botkin, V.L. Turova, Freezing of living cells: mathematical models and design of optimal cooling protocols. Int. Ser. Numer. Math. **160**, 521–540 (2012)
25. J. Tchir, J. Acker, Mitochondria and membrane cryoinjury in micropatterned cells: effects of cell-cell interactions. Cryobiology **61**, 100–107 (2010)
26. M.G. Crandall, P.L. Lions, Viscosity solutions of Hamilton – Jacobi equations. Trans. Am. Math. Soc. **277**, 1–47 (1983)
27. N.N. Krasovskii, A.I. Subbotin, *Game-Theoretical Control Problems* (Springer, New York, 1988)
28. N.D. Botkin, K-H. Hoffmann, V.L. Turova, Optimal control of ice formation in living cells during freezing. Appl. Math. Model. **35**, 4044–4057 (2011)
29. J.O.M. Karlsson, A theoretical model of intracellular devitrification. Cryobiology **42**, 154–169 (2001)
30. A.I. Subbotin, *Generalized Solutions of First Order PDEs* (Birkhäuser, Boston, 1995)

Optimal Control-Based Feedback Stabilization of Multi-field Flow Problems

Eberhard Bänsch, Peter Benner, Jens Saak, and Heiko K. Weichelt

Abstract We discuss the numerical solution of the feedback stabilization problem for multi-field flow problems. Our approach is based on an analytical Riccati feedback concept derived by Raymond which allows to steer a perturbed flow back to its desired state, assumed to be a stationary, possibly unstable, flow profile. This concept, originally derived for incompressible flow fields described by the Navier-Stokes equations, uses a linear-quadratic regulator (LQR) approach for the linearized Navier-Stokes equations formulated on the space of divergence-free velocity fields. We extend this approach to a setting where the Navier-Stokes equations are coupled to a diffusion-convection equation describing the transport of a reactive species in a fluid. The stabilizing feedback control resulting from the LQR problem is obtained via solving the associated operator Riccati equation. We describe a numerical procedure to solve this Riccati equation numerically. This involves several technical difficulties on the algebraic level that we address in this report. We illustrate the performance of our method by a numerical example.

Keywords Coupled flow control • Navier-Stokes equations • Diffusion-convection equation • Riccati-based feedback

---

E. Bänsch
Lehrstuhl für Angewandte Mathematik 3, Friedrich-Alexander-Universität Erlangen-Nürnberg, Cauerstr. 11, 91058 Erlangen, Germany
e-mail: baensch@am.uni-erlangen.de

P. Benner • J. Saak
Research Group Computational Methods in Systems and Control Theory (CSC), Max Planck Institute for Dynamics of Complex Technical Systems Magdeburg, Sandtorstr. 1, 39106 Magdeburg, Germany

Faculty of Mathematics, Research Group Mathematics in Industry and Technology (MiIT), Technische Universität Chemnitz, Reichenhainer Str. 39/41, 09126 Chemnitz, Germany
e-mail: benner@mpi-magdeburg.mpg.de; saak@mpi-magdeburg.mpg.de

H. K. Weichelt (✉)
Faculty of Mathematics, Research Group Mathematics in Industry and Technology (MiIT), Technische Universität Chemnitz, Reichenhainer Str. 39/41, 09126 Chemnitz, Germany
e-mail weichelt@mpi-magdeburg.mpg.de

G. Leugering et al. (eds.), *Trends in PDE Constrained Optimization*, International Series of Numerical Mathematics 165, DOI 10.1007/978-3-319-05083-6_11

Mathematics Subject Classification (2010). 76D55,93D15,93C20,15A24.

1 Introduction

In this article we extend the ideas from [4, 5, 8, 18] for a closed-loop (boundary) control of the linearized Navier-Stokes equations to a coupled flow problem consisting of the Navier-Stokes equations and a diffusion-convection problem. The latter models passive transport of a reactive species by the flow field. A homogeneous Dirichlet condition on part of the boundary of the domain is used as a toy model for surface reaction. Although this setting is rather academic, it may serve as a paradigm to solve more involved problems.

In the present paper we focus on the computational realization of the feedback control for the coupled flow problem. More details about the underlying mathematical basis may be found in [5, 18]. The present article builds upon [4] and demonstrates how the numerical solution concept outlined there is realized for a multi-field flow problem.

Let us mention some related work. The numerical realization of a linear-quadratic regularization process applied to the Stokes equations is discussed in [8], where the focus lies on efficient solution strategies for the arising saddle point systems. These ideas are extended to the more general Navier-Stokes equations in [5]. Furthermore, a different approach to stabilize Navier-Stokes flow problems via boundary influence is shown in [1]. Extending these ideas and numerical techniques to a more general coupled multi-field flow problem is the main issue of this paper.

The rest of this article is organized as follows. In Sect. 2 the coupled flow problem is stated. Section 2.1 outlines the necessary concept of linearization. Discretization by finite elements leads to a finite dimensional system of differential-algebraic equations (DAEs), which is described in Sect. 2.2. Before we introduce the problem setting used for numerical testing in the Sect. 3, we discuss some details about the block structured saddle point systems in Sect. 2.4. The paper is concluded in Section "Conclusions and Outlook", where we discuss some further ideas that are part of ongoing research.

2 The Coupled Flow Problem

The basis for the coupled flow problems is described by the incompressible Navier-Stokes equations defined on a spatial inflow/outflow domain $\Omega \subset \mathbb{R}^2$ and the time interval of interest $\mathcal{I} \subset [0, \infty)$. For $x \in \Omega$, $t \in \mathcal{I}$, the velocity $\vec{v}(t, \vec{x}) \in \mathbb{R}^2$ and pressure $p(t, \vec{x}) \in \mathbb{R}$ satisfy

$$\frac{\partial}{\partial t}\vec{v} - \frac{1}{\mathrm{Re}}\Delta\vec{v} + (\vec{v}\cdot\nabla)\vec{v} + \nabla\chi = \vec{f}, \tag{2.1a}$$

$$\operatorname{div}\vec{v} = 0 \tag{2.1b}$$

with the Reynolds number Re, see [5].

Moreover, we consider the passive transport of some concentration field $c(t, \vec{x}) \in \mathbb{R}$ described by a diffusion-convection equation (DCE):

$$\frac{\partial}{\partial t} c - \frac{1}{\mathrm{ReSc}} \Delta c + (\vec{v} \cdot \nabla) c = 0 \tag{2.2}$$

with the Schmidt number Sc.

The joint evolution for both systems takes place within $\Omega \times \mathcal{I}$ with the boundary

$$\partial\Omega =: \Gamma = \Gamma_{in} \cup \Gamma_{wall} \cup \Gamma_{out} \cup \Gamma_r.$$

In this decomposition of the boundary, Γ_r denotes the boundary of an obstacle within the domain. For (2.1), we prescribe a parabolic inflow at Γ_{in}, "no-slip" boundary conditions at Γ_{wall} and Γ_r, and "do-nothing" conditions for the outflow. The initial condition is given by a stationary solution to (2.1). We impose the following boundary and initial conditions for the concentration:

$$c(t, \vec{x}) = h_{in}(\vec{x}) \quad \text{on } \Gamma_{in}, \tag{2.3a}$$

$$\frac{\partial c(t, \vec{x})}{\partial \vec{n}} = 0 \quad \text{on } \Gamma_{wall} \cup \Gamma_{out}, \tag{2.3b}$$

$$c(t, \vec{x}) = 0 \quad \text{on } \Gamma_r, \tag{2.3c}$$

$$c(0, \vec{x}) = 0 \quad \text{in } \Omega \tag{2.3d}$$

with $\vec{n}$ the outward normal to Γ_{out} and Γ_{wall}. The initial and boundary conditions can be interpreted as follows. The concentration enters the domain through Γ_{in} (2.3a) and leaves the domain, only convection driven, via Γ_{out} (2.3b). As soon as the concentration reaches the obstacle, a fast reaction is assumed absorbing the species. In the case considered here, the reaction is much faster than the transport in Ω and can thus be modeled by a homogeneous Dirichlet condition for $c(t, \vec{x})$ (2.3c). We will omit the arguments $t, \vec{x}$ hereafter for better readableness.

2.1 *Linearization*

In this section we show how to linearize equations (2.1) and (2.2). Linearization is necessary for applying the Riccati based feedback stabilization approach to the coupled system. Using the linearization idea [5, Section 2.1], we define

$$\vec{z} := \vec{v} - \vec{w}, \tag{2.4a}$$

$$p := \chi - \chi_s \tag{2.4b}$$

that fulfill the linearized equations

$$\frac{\partial}{\partial t}\vec{z} - \frac{1}{\text{Re}}\Delta\vec{z} + (\vec{w}\cdot\nabla)\vec{z} + (\vec{z}\cdot\nabla)\vec{w} + \nabla p = 0, \tag{2.5a}$$

$$\text{div}\vec{z} = 0 \tag{2.5b}$$

with the same boundary and initial conditions, as in [5]. Furthermore, we define the stationary diffusion-convection equation

$$-\frac{1}{\text{ReSc}}\Delta c_{\vec{w}} + (\vec{w}\cdot\nabla)c_{\vec{w}} = 0 \tag{2.6}$$

with similar boundary conditions like those in (2.3). The goal now is to stabilize $c_{\vec{w}}$, when this field is subject to (small) perturbations. We may for instance imagine the situation when the field arises from an open-loop controller [13] and thus is a desired state to be maintained. Let us define

$$c_{\vec{z}} = c - c_{\vec{w}} \tag{2.7}$$

as the difference between the actual concentration c and the stationary concentration $c_{\vec{w}}$. Using the linearization points (2.4), (2.7) together with (2.6) yields the linearized diffusion-convection equation

$$\frac{\partial}{\partial t}c_{\vec{z}} - \frac{1}{\text{ReSc}}\Delta c_{\vec{z}} + (\vec{w}\cdot\nabla)c_{\vec{z}} + (\vec{z}\cdot\nabla)c_{\vec{w}} = 0, \tag{2.8}$$

defined for $t \in \mathcal{I}$ and $\vec{x} \in \Omega$ with boundary and initial conditions

$$\begin{aligned} c_{\vec{z}} &= 0 \quad \text{on } \Gamma_{in} \cup \Gamma_r, \\ \frac{\partial c_{\vec{z}}}{\partial \vec{n}} &= 0 \quad \text{on } \Gamma_{wall} \cup \Gamma_{out}, \\ c_{\vec{z}}(0,\vec{x}) &= 0 \quad \text{in } \Omega. \end{aligned}$$

The main goal of boundary feedback stabilization is the asymptotic stabilization of $\vec{z}$ and $c_{\vec{z}}$, which implies that the actual velocity field fulfills $\vec{v} \approx \vec{w}$ and the actual concentration $c \approx c_{\vec{w}}$, respectively. In the following we are going to apply a finite dimensional LQR approach, based on a spatial semi-discretization of the linearized Navier-Stokes (2.5) and diffusion-convection (2.8) equations. The discretization by finite elements is described in the following subsection.

2.2 Discretization

We use the same discretization idea for (2.5) as described in [5, Section 2.2] using the $\mathcal{P}_2 - \mathcal{P}_1$ *Taylor-Hood* [14] element and end up with the discretized linearized Navier-Stokes equations

$$M_{\mathbf{z}}\frac{d}{dt}\mathbf{z}(t) = A_{\mathbf{z}}\mathbf{z}(t) + G\mathbf{p}(t) + \mathbf{f}_{\mathbf{z}}(t), \tag{2.9a}$$

$$0 = G^T\mathbf{z}(t) \tag{2.9b}$$

with n_v degrees of freedom for the velocity space and n_p degrees of freedom for the pressure space. Equation (2.8) is discretized in space by linear finite elements yielding

$$M_{\mathbf{c}}\frac{d}{dt}\mathbf{c}(t) = A_{\mathbf{c}}\mathbf{c}(t) - R_{\vec{w}}\mathbf{z} + \mathbf{f}_{\mathbf{c}}(t) \tag{2.10}$$

with the nodal vector of discretized concentrations $\mathbf{c}(t) \in \mathbb{R}^{n_c}$, the concentration mass matrix $M_{\mathbf{c}} = M_{\mathbf{c}}^T \succ 0 \in \mathbb{R}^{n_c \times n_c}$, the concentration system matrix $A_{\mathbf{c}} \in \mathbb{R}^{n_c \times n_c}$, and the reaction term $R_{\vec{w}}$ that depends on the stationary velocity $\vec{w}$ and couples (2.9) and (2.10). Similar to the velocity field, the concentration field may be subject to a control $\mathbf{u}_{\mathbf{c}}$ acting through the source term $\mathbf{f}_{\mathbf{c}}(t) := B_{\mathbf{c}}\mathbf{u}_{\mathbf{c}}(t)$.

After reordering (2.9) and (2.10), we obtain the system of DAEs [15]

$$\begin{bmatrix} M_{\mathbf{z}} & 0 & 0 \\ 0 & M_{\mathbf{c}} & 0 \\ 0 & 0 & 0 \end{bmatrix} \frac{d}{dt} \begin{bmatrix} \mathbf{z} \\ \mathbf{c} \\ \mathbf{p} \end{bmatrix} = \begin{bmatrix} A_{\mathbf{z}} & 0 & G \\ -R_{\vec{w}} & A_{\mathbf{c}} & 0 \\ G^T & 0 & 0 \end{bmatrix} \begin{bmatrix} \mathbf{z} \\ \mathbf{c} \\ \mathbf{p} \end{bmatrix} + \begin{bmatrix} B_{\mathbf{z}} & 0 \\ 0 & B_{\mathbf{c}} \\ 0 & 0 \end{bmatrix} \begin{bmatrix} \mathbf{u}_{\mathbf{z}} \\ \mathbf{u}_{\mathbf{c}} \end{bmatrix}. \tag{2.11a}$$

We assume that only parts of the states $\mathbf{z}$ and $\mathbf{c}$ are observed. Therefore, we add the observation equations

$$\begin{bmatrix} \mathbf{y}_{\mathbf{z}} \\ \mathbf{y}_{\mathbf{c}} \end{bmatrix} = \begin{bmatrix} C_{\mathbf{z}} & 0 \\ 0 & C_{\mathbf{c}} \end{bmatrix} \begin{bmatrix} \mathbf{z} \\ \mathbf{c} \end{bmatrix}. \tag{2.11b}$$

The DAE system (2.11) is of differential index 2 if G has full rank [19]. Since we use the inf-sup stable Taylor–Hood element, the latter condition is fulfilled. Defining the block matrices

$$\begin{aligned} M &= \begin{bmatrix} M_{\mathbf{z}} & 0 \\ 0 & M_{\mathbf{c}} \end{bmatrix}, \quad A = \begin{bmatrix} A_{\mathbf{z}} & 0 \\ -R_{\vec{w}} & A_{\mathbf{c}} \end{bmatrix}, \quad \tilde{G} = \begin{bmatrix} G \\ 0 \end{bmatrix}, \quad \mathbf{x} = \begin{bmatrix} \mathbf{z} \\ \mathbf{c} \end{bmatrix}, \\ B &= \begin{bmatrix} B_{\mathbf{z}} & 0 \\ 0 & B_{\mathbf{c}} \end{bmatrix}, \quad C = \begin{bmatrix} C_{\mathbf{z}} & 0 \\ 0 & C_{\mathbf{c}} \end{bmatrix}, \quad \mathbf{u} = \begin{bmatrix} \mathbf{u}_{\mathbf{z}} \\ \mathbf{u}_{\mathbf{c}} \end{bmatrix}, \quad \mathbf{y} = \begin{bmatrix} \mathbf{y}_{\mathbf{z}} \\ \mathbf{y}_{\mathbf{c}} \end{bmatrix}, \end{aligned} \tag{2.12}$$

(2.9) can be written as

$$\begin{bmatrix} M & 0 \\ 0 & 0 \end{bmatrix} \frac{d}{dt} \begin{bmatrix} \mathbf{x} \\ \mathbf{p} \end{bmatrix} = \begin{bmatrix} A & \tilde{G} \\ \tilde{G}^T & 0 \end{bmatrix} \begin{bmatrix} \mathbf{x} \\ \mathbf{p} \end{bmatrix} + \begin{bmatrix} B \\ 0 \end{bmatrix} \mathbf{u}, \tag{2.13a}$$

$$\mathbf{y} = C\mathbf{x}. \tag{2.13b}$$

The matrix pencil

$$\left(\begin{bmatrix} A & \tilde{G} \\ \tilde{G}^T & 0 \end{bmatrix}, \begin{bmatrix} M & 0 \\ 0 & 0 \end{bmatrix} \right)$$

of the DAE (2.13) has $n_v + n_c - n_p$ finite eigenvalues $\lambda_i \in \mathbb{C}$ and $2n_p$ infinite eigenvalues $\lambda_\infty = \infty$ [10].

The DAE system (2.13) has the same structure as the DAE system arising from the Navier-Stokes equations in [5]; there, the projection approach of [12] is applied to transform the DAE [5, Equation (6)] into a generalized state space system. The main difficulty here is that we only want to project the velocity part $\mathbf{z}$ of the state variable $\mathbf{x}$. In the next subsection we show that this is indeed possible by adapting the projection idea from [12] to our block structured DAE system (2.13).

2.3 Projection Method

In order to adapt the projector definition of [12] to the case of our block matrices (2.12), we define

$$\begin{aligned} \tilde{\Pi} &:= I_{\mathbf{x}} - \tilde{G}(\tilde{G}^T M^{-1} \tilde{G})^{-1} \tilde{G}^T M^{-1} \\ &= \begin{bmatrix} I_{\mathbf{v}} & 0 \\ 0 & I_{\mathbf{c}} \end{bmatrix} - \begin{bmatrix} G \\ 0 \end{bmatrix} \left([G^T \ 0] \begin{bmatrix} M_{\mathbf{z}}^{-1} & 0 \\ 0 & M_{\mathbf{c}}^{-1} \end{bmatrix} \begin{bmatrix} G \\ 0 \end{bmatrix} \right)^{-1} [G^T \ 0] \begin{bmatrix} M_{\mathbf{z}}^{-1} & 0 \\ 0 & M_{\mathbf{c}}^{-1} \end{bmatrix} \\ &= \begin{bmatrix} I_{\mathbf{v}} & 0 \\ 0 & I_{\mathbf{c}} \end{bmatrix} - \begin{bmatrix} G(G^T M_{\mathbf{z}}^{-1} G)^{-1} G^T M_{\mathbf{z}}^{-1} & 0 \\ 0 & 0 \end{bmatrix} = \begin{bmatrix} \Pi & 0 \\ 0 & I_{\mathbf{c}} \end{bmatrix} \end{aligned}$$

with the discrete *Helmholtz* projector $\Pi \in \mathbb{R}^{n_v \times n_v}$ as defined in [5]. Note that we have the following equivalences:

$$\Pi^T \mathbf{z} = \mathbf{z} \wedge \mathbf{c} = \mathbf{c} \quad \Leftrightarrow \quad \begin{bmatrix} \Pi & 0 \\ 0 & I_{\mathbf{c}} \end{bmatrix}^T \begin{bmatrix} \mathbf{z} \\ \mathbf{c} \end{bmatrix} = \begin{bmatrix} \mathbf{z} \\ \mathbf{c} \end{bmatrix} \quad \Leftrightarrow \quad \tilde{\Pi}^T \mathbf{x} = \mathbf{x}.$$

Using (formally) the decomposition $\Pi = \Theta_l \Theta_r^T$ as in [12, Section 3], the projection matrix $\tilde{\Pi}$ can be decomposed into

$$\tilde{\Pi} = \tilde{\Theta}_l \tilde{\Theta}_r^T \quad \Leftrightarrow \quad \begin{bmatrix} \Pi & 0 \\ 0 & I_{\mathbf{c}} \end{bmatrix} = \underbrace{\begin{bmatrix} \Theta_l & 0 \\ 0 & I_{\mathbf{c}} \end{bmatrix}}_{=:\tilde{\Theta}_l} \underbrace{\begin{bmatrix} \Theta_r^T & 0 \\ 0 & I_{\mathbf{c}} \end{bmatrix}}_{=:\tilde{\Theta}_r^T}$$

with $\tilde{\Theta}_l^T \tilde{\Theta}_r = I_{\mathbf{x}}$. This decomposition is used to project the discretized velocity $\mathbf{z}$ onto the $n_v - n_p$ dimensional subspace of discretely divergence-free functions as in [12], without changing the discrete concentration $\mathbf{c}$. Substituting

$$\tilde{\mathbf{x}} = \tilde{\Theta}_l^T \mathbf{x} = \begin{bmatrix} \Theta_l^T & 0 \\ 0 & I_{\mathbf{c}} \end{bmatrix} \begin{bmatrix} \mathbf{z} \\ \mathbf{c} \end{bmatrix} = \begin{bmatrix} \Theta_l^T \mathbf{z} \\ \mathbf{c} \end{bmatrix} = \begin{bmatrix} \tilde{\mathbf{z}} \\ \mathbf{c} \end{bmatrix} \in \mathbb{R}^{(n_v - n_p) + n_c}$$

in (2.13) yields

$$\tilde{\Theta}_r^T M \tilde{\Theta}_r \frac{d}{dt} \tilde{\mathbf{x}} = \tilde{\Theta}_r^T A \tilde{\Theta}_r \tilde{\mathbf{x}} + \tilde{\Theta}_r^T B \mathbf{u},$$
$$\mathbf{y} = C \tilde{\Theta}_r \tilde{\mathbf{x}}.$$

We define the projected block structured matrices

$$\mathcal{M} = \begin{bmatrix} \Theta_r^T M_{\mathbf{z}} \Theta_r & 0 \\ 0 & M_{\mathbf{c}} \end{bmatrix}, \quad \mathcal{A} = \begin{bmatrix} \Theta_r^T A_{\mathbf{z}} \Theta_r & 0 \\ -R_{\vec{w}} \Theta_r & A_{\mathbf{c}} \end{bmatrix},$$
$$\mathcal{B} = \begin{bmatrix} \Theta_r^T B_{\mathbf{z}} & 0 \\ 0 & B_{\mathbf{c}} \end{bmatrix}, \quad \mathcal{C} = \begin{bmatrix} C_{\mathbf{z}} \Theta_r & 0 \\ 0 & C_{\mathbf{c}} \end{bmatrix} \tag{2.14}$$

and end up with the generalized state space system

$$\mathcal{M} \frac{d}{dt} \tilde{\mathbf{x}} = \mathcal{A} \tilde{\mathbf{x}} + \mathcal{B} \mathbf{u}, \tag{2.15a}$$

$$\mathbf{y} = \mathcal{C} \tilde{\mathbf{x}} \tag{2.15b}$$

with $\mathcal{M} = \mathcal{M}^T \succ 0 \in \mathbb{R}^{(n_v - n_p) + n_c}$.

2.4 The Linear-Quadratic Regulator Approach

To test the feedback stabilization approach for a coupled flow problem let us define

$$q := \int_{\Gamma_r} \partial_{\vec{n}} c_{\vec{w}} \, \mathrm{ds} \tag{2.16}$$

as the total flux of the stationary concentration $c_{\vec{w}}$ through the obstacle boundary Γ_r. Analogously to the LQR approach in [5] we define the cost functional

$$\mathcal{J}(c, \mathbf{u}(t)) := \frac{1}{2} \int_0^\infty \lambda \left| \int_{\Gamma_r} \partial_{\vec{n}} c \text{ ds } - q \right|^2 + |\mathbf{u}(t)|^2 \text{ dt} \tag{2.17}$$

measuring the difference of the actual flux of c through Γ_r and q, as well as the control costs $\mathbf{u}$, in the square of the Euclidean norm. Using the definition (2.16) in (2.17) we obtain

$$\int_{\Gamma_r} \partial_{\vec{n}} c \text{ ds } - q = \int_{\Gamma_r} \partial_{\vec{n}} (c - c_{\vec{w}}) \text{ ds} = \int_{\Gamma_r} \partial_{\vec{n}} c_{\vec{z}} \text{ ds.}$$

After discretization, this yields

$$C_{\mathbf{c}} \mathbf{c} = \mathbf{y}_{\mathbf{c}}$$

as the observation equation in (2.11b). We do not consider any observation of the velocity field such that we set $C_{\mathbf{z}} = 0$ and reduce the output Equation (2.11b) to $\mathbf{y} = C_{\mathbf{c}}\mathbf{c}$. Using this output equation, the minimization problem for the LQR approach can be written as:

Minimize

$$\mathcal{J}(\mathbf{c}(t), \mathbf{u}(t)) := \frac{1}{2} \int_0^\infty \lambda \left(\mathbf{c}(t)^T C_{\mathbf{c}}^T C_{\mathbf{c}} \mathbf{c}(t) \right) + \mathbf{u}(t)^T \mathbf{u}(t) \text{ dt,} \tag{2.18}$$

subject to (2.15a).

Minimizing this cost functional subject to (2.15a) forces the discrete velocity field $\mathbf{z}$ and concentration $\mathbf{c}$ asymptotically to zero for $t \to \infty$ so that the actual flow field $\vec{v}$ and concentration c are expected to approach the stationary velocity field $\vec{w}$ and the concentration $c_{\vec{w}}$, respectively. Introducing the regularization parameter λ in the first term of (2.18) provides the possibility to achieve qualitatively different results.

On the one hand, we observe only parts of the concentration $\mathbf{c}$. On the other hand, we want to influence the whole system only via a control influence on the velocity field $\mathbf{z}$; that means we define $B_{\mathbf{c}} = 0$ and reduce the control input to $\left[B_{\mathbf{z}}^T \; 0 \; 0\right]^T \mathbf{u}_z$ in (2.11). We will skip details about the realization of $B_{\mathbf{z}}$.

Starting from the setting to minimize (2.18) subject to (2.15a), the whole process to compute the optimal control $\mathbf{u}_*(t) = -\mathcal{K}\tilde{x}_*(t)$ with the feedback $\mathcal{K}$ via a generalized Newton-ADI iteration analogous to [5] is used. In short, this method consists in applying Newton's method to the algebraic Riccati equation obtained from the LQR problem after (implicitly) projecting onto the space of discretely divergence-free functions. In each step of Newton's method applied to an algebraic Riccati equation, a Lyapunov equation has to be solved. This is a linear system of equations having tensor structure. As suggested in [6], we employ the alternating directions implicit (ADI) method for this purpose. This requires the solution of a linear system of equations involving the projected system matrices (2.14) in each ADI step, see also [9]. How to avoid the explicit formation of the projected matrices following the approach from [12] is discussed in detail in [5] for the Navier-Stokes case. In the following, we will adapt this to the case of the multi-field flow problem discussed here. Therefore, we consider a projected system of the form

$$\left((\mathcal{A}^{(m)})^T + q_i \mathcal{M}\right) \Lambda = \mathcal{Y} \tag{2.19}$$

in the innermost step of the nested Newton-ADI iteration. Equation (2.19) is of the same structure as in the Navier-Stokes case and the approach in [12] to avoid this explicit projection can be applied in a similar way for the coupled flow problem. To this end we observe that the solution Λ is determined by solving the linear system

$$\tilde{\Pi}\left((A - BK^{(m)})^T + q_i M\right) \tilde{\Pi}^T \Lambda = \tilde{\Pi} Y,$$

which is equivalent to solving the saddle point system

$$\begin{bmatrix} (A - BK^{(m)})^T + q_i M & \tilde{G} \\ \tilde{G}^T & 0 \end{bmatrix} \begin{bmatrix} \Lambda \\ * \end{bmatrix} = \begin{bmatrix} Y \\ 0 \end{bmatrix} \tag{2.20}$$

with the feedback matrix $K^{(m)}$ in the m-th Newton step and q_i the ADI shift in the i-th ADI step. (“$*$” denotes an auxiliary quantity not further used.)

The feedback matrix K can then be computed via the generalized low-rank Cholesky factor Newton method as it is shown in Algorithm 1. The whole algorithm uses the original large-scale and sparse matrices from (2.11). We will skip details about shift selection in this paper and refer to [5, 16].

The Newton iteration consists of a number of Newton steps, each of which requires a certain amount of ADI steps to determine the update for the Newton iteration [5] and, in our formulation, to directly update the feedback matrix K. In turn, the saddle point system (2.20) has to be solved for different ADI shifts q_i in every ADI step for a couple of right hand sides during the Newton-ADI iteration. Solving the large-scale saddle point system efficiently is crucial for a suitable computation time.

Following the algebra in [5], where the *Sherman-Morrison-Woodbury* formula is exploited, one eventually ends up with a system to be solved having the form

$$\begin{bmatrix} A^T + q_i M & \tilde{G} \\ \tilde{G}^T & 0 \end{bmatrix} \begin{bmatrix} \Lambda \\ * \end{bmatrix} = \begin{bmatrix} \tilde{Y} \\ 0 \end{bmatrix}.$$

Using the block matrix definitions (2.12) yields

$$\underbrace{\begin{bmatrix} A_{\mathbf{z}}^T + q_i M_{\mathbf{z}} & -R_{\vec{w}}^T & G \\ 0 & A_{\mathbf{c}}^T + q_i M_{\mathbf{c}} & 0 \\ G^T & 0 & 0 \end{bmatrix}}_{=:\mathbf{A}} \begin{bmatrix} \Lambda_{\mathbf{z}} \\ \Lambda_{\mathbf{c}} \\ * \end{bmatrix} = \begin{bmatrix} \tilde{Y}_{\mathbf{z}} \\ \tilde{Y}_{\mathbf{c}} \\ 0 \end{bmatrix}$$

Algorithm 1 Generalized low-rank Cholesky factor Newton method for coupled flow problems

Input: $M_{\mathbf{z}}, M_{\mathbf{c}}, A_{\mathbf{z}}, A_{\mathbf{c}}, G, R_{\vec{w}}, B_{\mathbf{z}}, C_{\mathbf{c}}$, initial feedback $K_{\mathbf{z}}^{(0)}$,
ADI shift parameters $q_i \in \mathbb{C}^- : i = 1, \ldots, n_{\text{ADI}}$,
tol_{ADI}, tol_{Newton}, and regularization parameter λ

Output: feedback operator K

for $m = 1, 2, \ldots, n_{\text{Newton}}$ **do**

$$W^{(m)} = \begin{bmatrix} 0 & \sqrt{\lambda}\, C_{\mathbf{c}} & 0 \\ (K_{\mathbf{z}}^{(m-1)}) & 0 & 0 \end{bmatrix}$$

Get $\left[V_{1,\mathbf{z}}^T \; V_{1,\mathbf{c}}^T\right]^T$ by solving

$$\begin{bmatrix} A_{\mathbf{z}}^T - (K_{\mathbf{z}}^{(m-1)})^T B_{\mathbf{z}}^T + q_1 M_{\mathbf{z}} & -R_{\vec{w}}^T & G \\ 0 & A_{\mathbf{c}}^T + q_1 M_{\mathbf{c}} & 0 \\ G^T & 0 & 0 \end{bmatrix} \begin{bmatrix} V_{1,\mathbf{z}} \\ V_{1,\mathbf{c}} \\ * \end{bmatrix} = \sqrt{-2\mathrm{Re}\,(q_1)}\,(W^{(m)})^T$$

$$K_{1,\mathbf{z}}^{(m)} = \begin{bmatrix} B_{\mathbf{z}}^T V_{1,\mathbf{z}} \overline{V}_{1,\mathbf{z}}^T M_{\mathbf{z}} & B_{\mathbf{z}}^T V_{1,\mathbf{z}} \overline{V}_{1,\mathbf{c}}^T M_{\mathbf{c}} \end{bmatrix}$$

for $i = 2, 3, \ldots, n_{\text{ADI}}$ **do**

Get $\left[\tilde{V}_{\mathbf{z}}^T \; \tilde{V}_{\mathbf{c}}^T\right]^T$ by solving

$$\begin{bmatrix} A_{\mathbf{z}}^T - (K_{\mathbf{z}}^{(m-1)})^T B_{\mathbf{z}}^T + q_i M_{\mathbf{z}} & -R_{\vec{w}}^T & G \\ 0 & A_{\mathbf{c}}^T + q_i M_{\mathbf{c}} & 0 \\ G^T & 0 & 0 \end{bmatrix} \begin{bmatrix} \tilde{V}_{\mathbf{z}} \\ \tilde{V}_{\mathbf{c}} \\ * \end{bmatrix} = \begin{bmatrix} M_{\mathbf{z}}\, V_{i-1,\mathbf{z}} \\ M_{\mathbf{c}}\, V_{i-1,\mathbf{c}} \\ 0 \end{bmatrix}$$

$$\begin{bmatrix} V_{i,\mathbf{z}} \\ V_{i,\mathbf{c}} \end{bmatrix} = \sqrt{\mathrm{Re}\,(q_i)/\mathrm{Re}\,(q_{i-1})} \left(\begin{bmatrix} V_{i-1,\mathbf{z}} \\ V_{i-1,\mathbf{c}} \end{bmatrix} - (q_i + \overline{q_{i-1}}) \begin{bmatrix} \tilde{V}_{\mathbf{z}} \\ \tilde{V}_{\mathbf{c}} \end{bmatrix} \right)$$

$$K_{i,\mathbf{z}}^{(m)} = K_{i-1,\mathbf{z}}^{(m)} + \begin{bmatrix} B_{\mathbf{z}}^T V_{i,\mathbf{z}} \overline{V}_{i,\mathbf{z}}^T M_{\mathbf{z}} & B_{\mathbf{z}}^T V_{i,\mathbf{z}} \overline{V}_{i,\mathbf{c}}^T M_{\mathbf{c}} \end{bmatrix}$$

if $\left(\frac{\|K_{i,\mathbf{z}}^{(m)} - K_{i-1,\mathbf{z}}^{(m)}\|_F}{\|K_{i,\mathbf{z}}^{(m)}\|_F} < tol_{\text{ADI}} \right)$ **then**

break

end if

end for

$$K_{\mathbf{z}}^{(m)} = K_{n_{\text{ADI}},\mathbf{z}}^{(m)}$$

if $\left(\frac{\|K_{\mathbf{z}}^{(m)} - K_{\mathbf{z}}^{(m-1)}\|_F}{\|K_{\mathbf{z}}^{(m)}\|_F} < tol_{\text{Newton}} \right)$ **then**

break

end if

end for

$$K = \begin{bmatrix} K_{\mathbf{z}}^{(n_{\text{Newton}})} & 0 \end{bmatrix}$$

with $M_{\mathbf{z}}, M_{\mathbf{c}}$ symmetric positive definite, $G, R_{\vec{w}}$ full rank, and $q_i \in \mathbb{C}^-$. The full system matrix $\mathbf{A}$ is indefinite $\forall q_i \in \mathbb{C}^-$.

Details about the solution strategy for such block structured indefinite saddle point systems are not part of this article. Note that the use of preconditioned iterative solvers is necessary if the dimension of the system grows. In this case the block preconditioning ideas in [5, 8] can be extended. However, more details regarding this have to be postponed to future publications due to space limitations.

In the next section the configurations for the numerical tests used to illustrate our numerical procedure are introduced.

3 Numerical Examples

The main focus of this section is to verify the usability of the Newton-ADI iteration described in the previous section to compute the optimal control $\mathbf{u}(t)$ for the linearized version of a diffusion-convection equation coupled with the linearized Navier-Stokes equations. All computations are based on the finite element discretization of the reactor model shown in Fig. 1.

The reactor consists of an inflow channel on the left and an outflow channel on the right. Both have a diameter of 1.0 and a length of 3.0. Inside the reactor of dimension 5.0×7.0 there is a quadratic obstacle of dimension 1.5×1.5. The fluid flows around the obstacle and transports the concentration via the convection through the domain. Additionally, the concentration is spread due to a diffusion process. As described above, we assume a fast reaction of the concentration at the surface Γ_r of the obstacle, such that concentration that arrives at the obstacle is absorbed immediately.

The coarse discretization depicted in Fig. 1 is refined using a *Bänsch refinement* [2]. We apply this refinement strategy as a threefold bisection in the whole domain, ninefold bisection in the outflow channel and elevenfold bisection on the boundary of the obstacle, yielding the dimensions in Table 1b. Furthermore, we use heuristic Penzl ADI shifts for all configurations [16].

The FORTRAN90 based finite element software NAVIER [3] was used to assemble the matrices representing the finite element discretization. The computations for the resulting matrix equations were executed in MATLAB® R2012b on a a 64-bit server with Intel® Xeon® X5650 @2.67 GHz, with 2 CPUs, 12 Cores (6

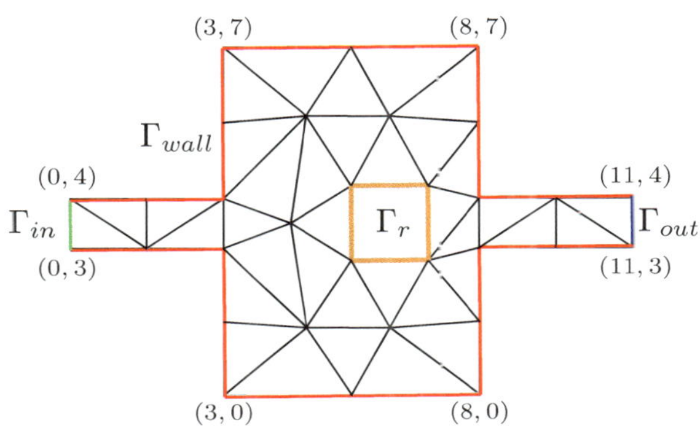

Fig. 1 Initial triangulation of the reactor model with coordinates and boundary conditions

Cores per CPU) and 48 GB main memory available. We refer the reader to [5] for more details regarding the interaction of both software packages.

3.1 Reynolds Number and Regularization Parameter λ

The Newton-ADI method is tested for five different combinations of Reynolds number Re and Schmidt number Sc, as given in Table 1a. Figure 2 depicts the convergence behavior of the Newton-ADI method clustered in subfigures for different regularization parameters. Each subfigure shows the evolution of the relative change

$$\frac{||K^{(m)} - K^{(m-1)}||_F}{||K^{(m)}||_F}$$

of the feedback matrix dependent on the Newton step m for all different sets in Table 1a. It shows that the graphs group for the different products of $\mathrm{Re}\,\mathrm{Sc} \in \{1, 10, 100\}$. If the product becomes larger, the Newton-ADI iteration needs more steps to converge. Note that the convergence for the set with larger Reynolds number is slightly slower within the group.

The regularization parameter λ penalizes the output $\mathbf{y}$ in the cost functional (2.18); that means the computed feedback K should stabilize our system more efficiently. This is reflected in the increasing number of Newton steps for increasing λ. Nevertheless, the Newton-ADI method computes the feedback matrix K for all settings to a suitable accuracy.

The quadratic convergence of the Newton iteration highly depends on the accuracy of the ADI method. In Fig. 2e stagnation appears for Set V during the iteration. In that case we would need to use a higher ADI accuracy. We analyze this phenomenon in more detail in the next subsection.

Table 1 Test parameter settings

Set	Re	Sc
I	1	1
II	1	10
III	10	1
IV	1	100
V	10	10

(a) Different parameter settings.

Variable	Dimension
n_v	9 092
n_c	1 187
n_p	1 276
n_x	11 555

(b) Different dimensions of FE space.

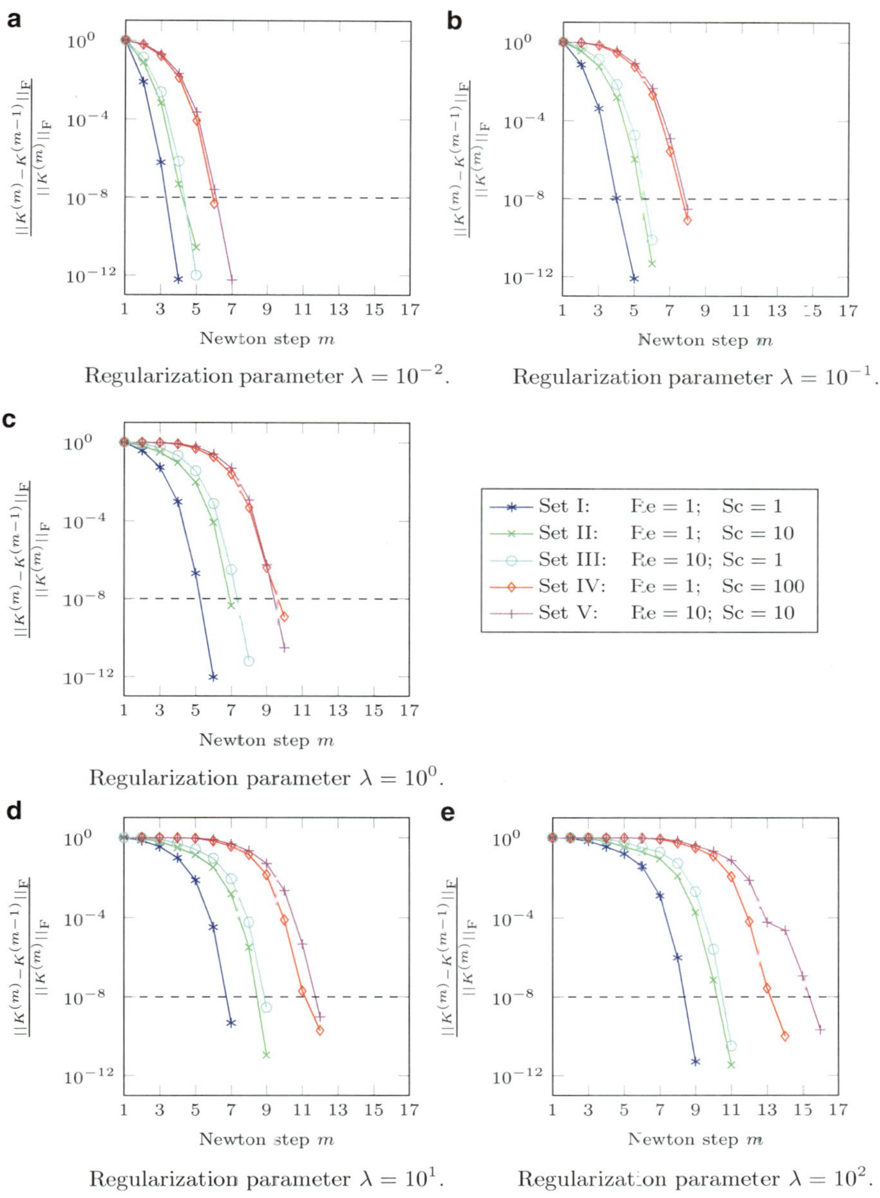

Fig. 2 Influence of Reynolds/Schmidt numbers and regularization parameter λ on the Newton-ADI convergence ($tol_{Newton} = 10^{-8}, tol_{ADI} = 10^{-7}$)

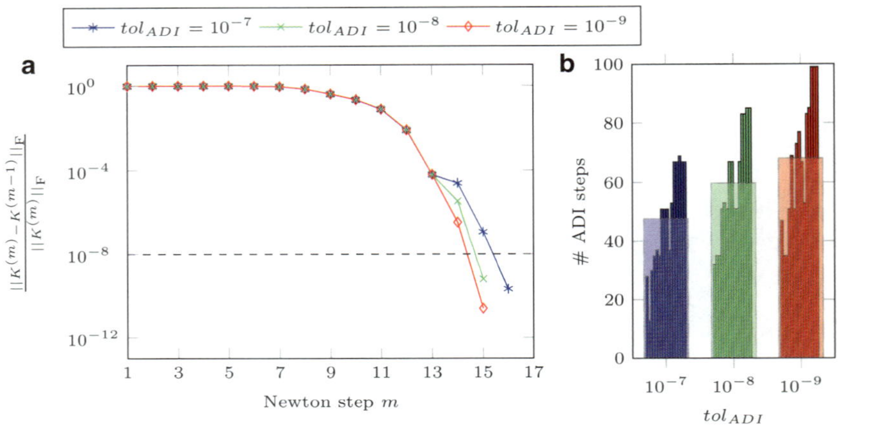

Convergence behavior of Newton iteration for different ADI accuracies.

Number of ADI steps per Newton step (small bars) and average over all Newton steps (wide bars).

Fig. 3 Influence of tol_{ADI} for Newton-ADI convergence ($tol_{Newton} = 10^{-8}$, Set V: Re = 10, Sc = 10)

3.2 ADI Tolerance vs. Newton Convergence

Figure 3 illustrates the influence of the ADI accuracy. We increase the ADI accuracy to avoid the observed stagnation for Set V. Figure 3a shows that for $tol_{ADI} = 10^{-8}$ stagnation still occurs. For $tol_{ADI} = 10^{-9}$ the Newton iteration converges quadratically. The higher ADI accuracy implies more ADI steps, as it is depicted in Fig. 3b. In total, the Newton-ADI with a higher ADI accuracy needs more time, although we can save one Newton step. To avoid these problems we will extend the idea of inexact Newton methods for the standard state space case worked out theoretically in [11] and for practical implementations in [7] to the structured DAE problems in the future. The difficulty here is the necessity for the projected ADI residuals in order to perform the accuracy control but also avoiding the explicit projection. A formulation of the index-2 ADI that has this capability is currently being investigated.

Conclusions and Outlook

In this report, we have extended the Riccati feedback stabilization approach for incompressible Navier-Stokes flows developed by Raymond in [17, 18] to a multi-field flow setting. For this purpose, we have coupled the Navier-

(continued)

Stokes equations with a convection-diffusion equation modeling the passive transport of a reactive species in a fluid.

We have extended the numerical method detailed in [5] for stabilization of the perturbed Navier-Stokes equations to this setting. For a proof-of-concept, we have used a merely academic problem configuration and tested our algorithm on this setting. The numerical results indicate that the Newton-ADI framework from [5] extended to the coupled problem can be used to robustly solve for the Riccati feedback in a regime of modest Reynolds and Schmidt numbers.

Future work will include the extension of our approach in several directions. This includes the development of efficient preconditioners for the saddle point problems to be solved in the innermost step of the Newton-ADI method for the coupled setting, the extension to higher Reynolds/Schmidt numbers (though highly turbulent flow can probably not be tackled by this stabilization method), and the adaptation of our approach to more complicated multi-field flow problems. The latter also requires the extension of Raymond's functional analysis framework to coupled stabilization problems, as well as a convergence analysis for the computed finite-dimensional feedback operators and an investigation of their stabilization properties for the infinite-dimensional system.

Acknowledgements We would like to thank Stephan Weller for his helpful advices regarding the handling of the FEM software NAVIER and René Schneider for many useful discussion throughout the whole project time. In addition, we would like to thank Martin Stoll, Andy Wathen, Matthias Heinkenschloss, and Jan Heiland for various discussions that extended our knowledge about optimal control for fluid mechanics.

References

1. L. Amodei, J.-M. Buchot, A stabilization algorithm of the Navier-Stokes equations based on algebraic Bernoulli equation. Numer. Lin. Alg. Appl. **19**, 700–727 (2012)
2. E. Bänsch, Local mesh refinement in 2 and 3 dimensions. IMPACT Comput. Sci. Eng. **3**, 181–191 (1991)
3. E. Bänsch, Simulation of instationary, incompressible flows. Acta Math. Univ. Comenianae **67**, 101–114 (1998)
4. E. Bänsch, P. Benner, Stabilization of incompressible flow problems by Riccati-based feedback, in *Constrained Optimization and Optimal Control for Partial Differential Equations*, ed. by G. Leugering, S. Engell, A. Griewank, M. Hinze, R. Rannacher, V. Schulz, M. Ulbrich, S. Ulbrich. Vol. 160 of International Series of Numerical Mathematics (Birkhäuser, Basel, 2012), pp. 5–20
5. E. Bänsch, P. Benner, J. Saak, H. K. Weichelt, Riccati-based boundary feedback stabilization of incompressible Navier-Stokes flow, Preprint SPP1253-154, DFG-SPP1253, 2013. http://www.am.uni-erlangen.de/home/spp1253/wiki/images/b/ba/Preprint-SPP1253-154.pdf

6. P. Benner, J.-R. Li, T. Penzl, Numerical solution of large Lyapunov equations, Riccati equations, and linear-quadratic control problems. Numer. Lin. Alg. Appl. **15**, 755–777 (2008)
7. P. Benner, J. Saak, Numerical solution of large and sparse continuous time algebraic matrix Riccati and Lyapunov equations: a state of the art survey. GAMM Mitteilungen **36**, 32–52 (2013)
8. P. Benner, J. Saak, M. Stoll, H. K. Weichelt, Efficient solution of large-scale saddle point systems arising in Riccati-based boundary feedback stabilization of incompressible Stokes flow. SIAM J. Sci. Comput. **35**, S150–S170 (2013)
9. P. Benner, T. Stykel, Numerical solution of projected algebraic Riccati equations. SIAM J. Numer. Anal. **52**(2), 581–600 (2014)
10. K. A. Cliffe, T. J. Garratt, A. Spence, Eigenvalues of block matrices arising from problems in fluid mechanics. SIAM J. Matrix Anal. Appl. **15**, 1310–1318 (1994)
11. F. Feitzinger, T. Hylla, E. W. Sachs, Inexact Kleinman-Newton method for Riccati equations. SIAM J. Matrix Anal. Appl. **31**, 272–288 (2009)
12. M. Heinkenschloss, D. C. Sorensen, K. Sun, Balanced truncation model reduction for a class of descriptor systems with application to the Oseen equations. SIAM J. Sci. Comput. **30**, 1038–1063 (2008)
13. M. Hinze, K. Kunisch, Second order methods for boundary control of the instationary Navier-Stokes system. Z. Angew. Math. Mech. **84**, 171–187 (2004)
14. P. Hood, C. Taylor, Navier-Stokes equations using mixed interpolation, in *Finite Element Methods in Flow Problems*, ed. by J. T. Oden, R. H. Gallagher, C. Taylor, O. C. Zienkiewicz (University of Alabama in Huntsville Press, Huntsville, 1974), pp. 121–132
15. P. Kunkel, V. Mehrmann, *Differential-Algebraic Equations: Analysis and Numerical Solution*. Textbooks in Mathematics (EMS Publishing House, Zürich, 2006)
16. T. Penzl, LYAPACK *users guide*, Tech. Report SFB393/00-33, Sonderforschungsbereich 393 *Numerische Simulation auf massiv parallelen Rechnern*, TU Chemnitz, 09107 Chemnitz, 2000. Available from http://www.tu-chemnitz.de/sfb393/sfb00pr.html
17. J.-P. Raymond, Local boundary feedback stabilization of the Navier-Stokes equations, in Control Systems: Theory, Numerics and Applications, Rome, 30 March–1 April 2005, Proceedings of Science, SISSA, http://pos.sissa.it, 2005
18. J.-P. Raymond, *Feedback boundary stabilization of the two-dimensional Navier-Stokes equations*. SIAM J. Control Optim. **45**, 790–828 (2006)
19. J. Weickert, Navier-Stokes equations as a differential-algebraic system, Preprint SFB393/96-08, Preprint Series of the SFB 393 "Numerische Simulation auf massiv parallelen Rechnern", Chemnitz University of Technology, Aug 1996

Part II
Shape and Topology Optimization

Introduction to Part II
Shape and Topology Optimization

Helmut Harbrecht

This part contains several results of recent research in shape and topology optimization. It consists of the following three independent sections:

Sergio Conti, Benedikt Geihe, Martin Rumpf, and Rüdiger Schultz combine, in *Two-stage stochastic optimization meets two-scale simulation*, a two-scale model in elastic shape optimization with a stochastic framework. The microstructured material to be optimized is composed of an elastic material with geometrically simple perforations located on a regular periodic lattice, whose parameters depend on the macroscopic position.

Helmut Harbrecht and Johannes Tausch review, in *On shape optimization with parabolic state equation*, their results on numerical methods for the efficient solution of shape optimization problems with parabolic state equation. For a specific parabolic shape optimization problem, both the shape calculus and the discretization by means of a modern space-time multipole method are demonstrated. For comparison reasons, also the related stationary shape optimization problem is considered.

Luise Blank, M. Hassan Farshbaf-Shaker, Harald Garcke, Christoph Rupprecht, and Vanessa Styles present, in *Multi-material phase field approach to structural topology optimization*, how to formulate and solve multi-material structural topology and shape optimization problems within a phase field approach. The first-order optimality system is determined and then numerically solved by an H^1-gradient projection method.

H. Harbrecht (✉)
Department of Mathematics and Computer Science, Universität Basel, Rheinsprung 21, 4051 Basel, Switzerland
e-mail: helmut.harbrecht@unibas.ch

G. Leugering et al. (eds.), *Trends in PDE Constrained Optimization*, International Series of Numerical Mathematics 165, DOI 10.1007/978-3-319-05083-6_12

Two-Stage Stochastic Optimization Meets Two-Scale Simulation

Sergio Conti, Benedict Geihe, Martin Rumpf, and Rüdiger Schultz

Abstract Risk averse stochastic optimization is investigated in the context of elastic shape optimization, allowing for microstructures in the admissible shapes. In particular, a two-stage model for shape optimization under stochastic loading with risk averse cost functionals is combined with a two-scale approach for the simulation of microstructured materials. The microstructure is composed of an elastic material with geometrically simple perforations located on a regular periodic lattice. Different types of microscopic geometries are investigated and compared to each other. In addition they are compared to optimal nested laminates, known to realize the optimal lower bound of compliance cost functionals. We combine this two-scale approach to elastic shapes with a two-stage stochastic programming approach to risk averse shape optimization, dealing with risk neutral and risk averse cost functionals in the presence of stochastic loadings.

Keywords Two-stage stochastic programming • Risk averse optimization • Two-scale elastic shape optimization • Microstructure optimization • Finite element method • Boundary element method

Mathematics Subject Classification (2010). 90C15, 74B05, 74P05, 74Q05, 74S05, 74S15, 49M29.

S. Conti
Institut für Angewandte Mathematik, Universität Bonn, Endenicher Allee 60, 53115 Bonn, Germany
e-mail: sergio.conti@uni-bonn.de

B. Geihe (✉) • M. Rumpf
Institut für Numerische Simulation, Universität Bonn, Endenicher Allee 60, 53115 Bonn, Germany
e-mail: benedict.geihe@ins.uni-bonn.de; martin.rumpf@ins.uni-bonn.de

R. Schultz
Fakultät für Mathematik, Universität Duisburg-Essen, Thea-Leymann-Str. 9, D-45117 Essen, Germany
e-mail: ruediger.schultz@uni-due.de

G. Leugering et al. (eds.), *Trends in PDE Constrained Optimization*, International Series of Numerical Mathematics 165, DOI 10.1007/978-3-319-05083-6_13

1 Introduction

In nature, when biological material has to resist strong mechanical loading, fine scale structures frequently characterize the material. Prominent examples are the microstructure of wood [5] or the substantia spongiosa of bones [35]. The pattern formed by these elastic structures is not uniform but varies spatially. This spatial variation seems to be adapted to the local load configuration, which supports the hypothesis of nature optimizing mechanical structures in the ontogenesis [45]. Thus, a natural question arises, what are "optimal" microstructures, which are observable in nature or can be used in the design of mechanical devices. When optimizing material structures one has to take into account that load configurations in nature and in engineering are usually not deterministic but stochastic.

This paper addresses the optimization of microstructures in elastic materials under stochastic loading. It is well known that microstructures form when minimizing compliance or tracking type cost functionals, unless a penalty on the area of material interfaces is taken into account. The optimal microstructures are well-understood and can be represented by nested laminates [1]. The laminate construction is an analytically elegant tool but can hardly be reproduced in mechanical devices, nor is it observed in optimization problems posed in nature. Thus, the question arises how close one can get to the optimal design with constructible microstructures. To this end, different types of parametrized microstructures will be investigated and compared.

2 Related Work

Shape optimization under deterministic loading has extensively been investigated in the literature. For an overview we refer to the textbooks [1, 13]. Most approaches deal with a macroscopic shape description under the assumption of sufficient shape regularity, which is usually guaranteed by an additional regularizing cost functional such as the shape perimeter. If only scale invariant cost functionals are taken into account, then in general an optimal shape will not exist. Indeed, minimizing sequences of shapes will be characterized by very fine microstructures. The theory of homogenization allows to describe the set of possible microstructures and the associated set of attainable effective material properties [20, 21, 23, 36]. The heterogeneous multi-scale method (HMM) [49, 50] depicts a very general paradigm for efficient numerical treatment of multi-scale problems using independent macroscopic and microscopic models. Homogenization theory was extended from multiphase, uniformly coercive materials to perforated structures and porous materials, see for example [24] and references therein. In [34] a two-scale adaptive finite element scheme has been proposed for elliptic problems on perforated domains.

Optimal microstructures in elasticity have first been derived by Hashin in 1962 [31] in the concentric sphere construction for hydrostatic loads. The construction was later generalized for anisotropic strains using confocal ellipsoids in [29]. The investigation of nested laminate structures dates back to the 1980s [40, 46] and was later used in a practical numerical scheme for topology optimization in [2]. In all these cases proofs of optimality rely on the Hashin-Shtrikman bounds on the attainable sets of effective elastic properties [32]. Related to the homogenization approach is the so called free material optimization method where the optimization is directly carried out on the coefficients of the elasticity tensor, as for example in [33].

Alternatively, the cost functional can be reduced by a proper design of fine scale perforations drilled into a homogeneous material. The layout of elastic structures based on this approach has been investigated already in the early 1990s [14]. The shape optimization via such mechanically feasible, periodic perforation patterns on the microscale has also been studied in [9]. Closely related to our approach is the approach by Barbarosie and Toader. In [10, 11] they optimized the geometry of fine scale perforations. The numerical method is based on a boundary tracking approach of a triangulated domain with additional remeshing steps to ensure mesh quality. In [12] this approach is extended to a two-scale setting combining a finite element scheme on the macroscale with the above treatment of locally periodic perforations on the microscale.

Shape optimization under a fixed load is rarely realistic. Multiload approaches consider a fixed (usually small) number of different loading configurations and have been developed for example in [3, 30] and references therein. In this paper, we deal with stochastic loading and risk averse optimization. Optimization under uncertainty requires an appropriate treatment of the available uncertain data information. Different approaches have been analyzed, which are appropriate for different types of risk. Robust optimization corresponds to a treatment of the worst-case [15] and is based on information about the ranges of the uncertain parameters. Applications to shape optimization can be found in [8, 22]. In stochastic optimization data uncertainty, typically quantified by probability distributions, has been largely studied in a finite-dimensional setting, both in a linear situation for mixed-integer and other nonlinear models, see for example [42]. Shape optimization with stochastic loading has been discussed previously in various contexts, for example for beam models in [38]. A number of papers addressed worst-case optimization, see for example [4, 8, 16] and the optimization scenario in aerodynamic design in [43]. A trust-region algorithm for PDE optimization under uncertainty was developed in [37]. In [25] we have proposed an efficient optimization approach for stochastic loading based on the representation of realizations of surface and volume loads as linear combinations of a few basis modes. In our previous work [25, 26, 28] we have shown how this approach can be used to effectively perform shape optimization with different treatments of the stochastic loads.

3 Elasticity of Micro Perforated Elastic Material

We consider here an elastic body composed of microstructured material and suppose that the microstructure is composed of mechanically constructible perforations. On the microscale these perforations form a regular lattice, which is not necessarily oriented parallel to the macroscopic axes. The geometry of these perforations and the orientation of the lattice vary macroscopically and are locally described by a finite set of parameters.

Before we investigate a computationally feasible two-scale formulation let us discuss the case of an elastic object with perforation on a fine scale lattice with regular lattice spacing $\delta > 0$, as illustrated in Fig. 1. We denote by $D \subset \mathbb{R}^d$ the underlying object domain, connected, with Lipschitz boundary and suppose that $\Gamma_D \subset \partial D$ is the Dirichlet boundary where the elastic object is fixated and $\Gamma_N \subset \partial D$ the Neumann boundary on which boundary forces are applied. We suppose that Dirichlet and Neumann boundaries are relatively open subsets of ∂D with Lipschitz boundary, the first one nonempty. The elastic object itself is perforated with holes of size less than δ drilled into homogeneous elastic bulk material on D and described by the perforated domain $D_\alpha^\delta = D \setminus \left(\bigcup_{x \in \delta\mathbb{Z}^d} x + \delta\mathfrak{m}_{\alpha(x)}\right)$. Here, $\mathfrak{m}_{\alpha(x)} \subset [-\gamma, \gamma]^d$ with γ fixed and $0 < \gamma < \frac{1}{2}$ describes the geometry of the perforation placed at $x \in \delta\mathbb{Z}^d$ and defined on the reference domain $[-\frac{1}{2}, \frac{1}{2}]^d$ for a parameter function $\alpha : D \to \mathbb{R}^m$ with $m \in \mathbb{N}$. We denote by $\mathcal{U}_{\text{ad}}$ a closed set of admissible parameters such that $\alpha(x) \in \mathcal{U}_{\text{ad}}$ for every $x \in D$. Let us assume that there are no perforations close to the Dirichlet and Neumann boundary, i. e. $\mathfrak{m}_{\alpha(x)} = \emptyset$ for $\text{dist}(x, \Gamma_D \cup \Gamma_N) \leq \Delta$ for some fixed $\Delta > 0$. For a displacement $u^\delta : D_\alpha^\delta \to \mathbb{R}^d$ and a boundary force density $g : \Gamma_N \to \mathbb{R}^d$ the elastic energy is given by

$$E^\delta[\alpha, u^\delta] = \frac{1}{2}\int_{D_\alpha^\delta} \mathbf{C}(x)\epsilon[u^\delta](x) : \epsilon[u^\delta](x)\,\mathrm{d}x - \int_{\Gamma_N} g(x)\cdot u^\delta(x)\,\mathrm{d}a$$

where $\mathbf{C}$ is the elasticity tensor of the homogeneous bulk material, $\epsilon[\phi] = \frac{1}{2}(\mathbf{D}\phi + \mathbf{D}\phi^T)$ denotes the strain tensor with $\mathbf{D}\phi$ being the Jacobian of the displacement ϕ and $A : B := \text{tr}(A^T B)$. If $\mathbf{C}$ is uniformly coercive and $g \in L^2(\Gamma_N, \mathbb{R}^d)$, the unique minimizer in the space $H^{1,2}_{\Gamma_D,\delta} := \{u \in H^{1,2}(D_\alpha^\delta)^d \,|\, u = 0 \text{ on } \Gamma_D\}$ is the

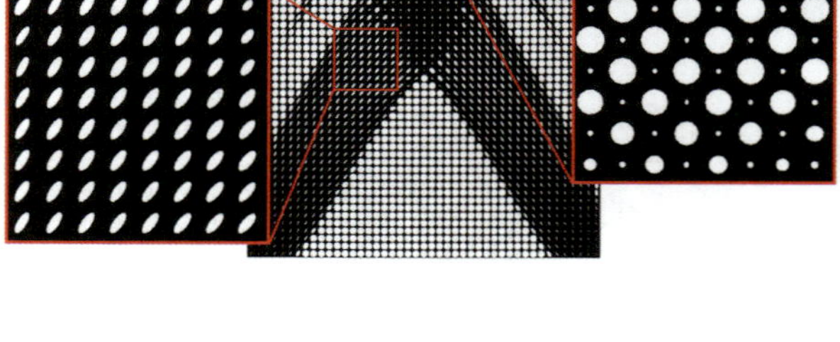

Fig. 1 Single-scale model for a carrier plate under shearing with 45 × 45 ellipsoidal holes. The two blow-ups show regions with locally (almost) periodic patterns

solution of the associated variational problem $\int_{D_\alpha^\delta} \mathbf{C}(x)\epsilon[u^\delta](x) : \epsilon[\phi^\delta](x)\,\mathrm{d}x = \int_{\Gamma_N} g(x)\cdot\phi^\delta(x)\,\mathrm{d}a$ for all $\phi^\delta \in H^{1,2}_{\Gamma_D,\delta}$. Figure 1 shows an elastic object with a perforation based on ellipsoidal holes where the parameters of the ellipses are optimized with respect to a compliance type cost functional.

A classical result from homogenization theory [21, 23] describes the elastic behavior of the material in the limit for $\delta \to 0$. Indeed a suitable extension of the elastic displacement u^δ onto the whole domain D converges to a displacement $u^\star \in H^{1,2}_{\Gamma_D} := \{u \in H^{1,2}(D)^d \mid u = 0 \text{ on } \Gamma_D\}$ which solves the variational problem

$$\int_D \mathbf{C}^\star(x)\epsilon[u^\star](x) : \epsilon[\phi](x)\,\mathrm{d}x = \int_{\Gamma_N} g(x)\cdot\phi(x)\,\mathrm{d}a \tag{3.1}$$

for all $\phi \in H^{1,2}_{\Gamma_D}$. Here, $\mathbf{C}^\star(x)$ is the effective elasticity tensor encoding the effective properties of the perforated material on the macroscale. Thereby, the underlying two-scale formulation of the limit problem is as follows. Find an effective macroscopic displacement $u^\star \in H^{1,2}_{\Gamma_D}$ and a microscopic correction $w^\star \in \mathbf{W}_\alpha$ which solve the equation

$$\int_D \int_{\mathcal{C}_\alpha(x)} \mathbf{C}(y)(\epsilon[u^\star](x) + \epsilon[w^\star](x,y)) : (\epsilon[\phi](x) + \epsilon[\psi](x,y))\,\mathrm{d}y\,\mathrm{d}x = \int_{\Gamma_N} g(x)\cdot\phi(x)\,\mathrm{d}a$$

for all $\phi \in H^{1,2}_{\Gamma_D}$ and all functions $\psi \in \mathbf{W}_\alpha$ where $\mathcal{C}_\alpha(x) := (-\frac{1}{2},\frac{1}{2})^d \setminus \mathfrak{m}_{\alpha(x)}$ and the function space of microscopic periodic displacement corrections is defined as

$$\begin{aligned}\mathbf{W}_\alpha := \{\phi : (x,y) \to \phi(x,y) \in \mathbb{R}^d \mid x \in D\,,\ y \in \mathcal{C}_\alpha(x)\,,\\ \phi(x, y+z) = \phi(x,y)\forall z \in \mathbb{Z}^d \text{ with } \|\phi\|_{\mathbf{W}_\alpha} \leq \infty\}\end{aligned}$$

where $\|\phi\|_{\mathbf{W}_\alpha} := \left(\int_D \int_{\mathcal{C}_\alpha(x)} \phi(x,y)^2 + |\mathbf{D}_y\phi(x,y)|^2\,\mathrm{d}y\,\mathrm{d}x\right)^{\frac{1}{2}}$. The effective elasticity tensor $\mathbf{C}^\star = \mathbf{C}^\star[\alpha]$ can be defined variationally

$$\mathbf{C}^\star(x)\epsilon[u](x) : \epsilon[u](x) = \int_{\mathcal{C}_\alpha(x)} \mathbf{C}(y)\epsilon[R^\star[u]](x,y) : \epsilon[R^\star[u]](x,y)\,\mathrm{d}y$$

where $R^\star[u](x,y) := u(x) + w(x,y)$ for $u \in H^{1,2}_{\Gamma_D}$ is the microscopic reconstruction with w solving the correction problem $\int_{\mathcal{C}_\alpha(x)} \mathbf{C}(y)(\epsilon[u](x) + \epsilon[w](x,y)) : \epsilon[\psi](x,y)\,\mathrm{d}y = 0$ for all $\psi \in \mathbf{W}_\alpha$. Indeed, using the symmetry assumption $\mathbf{C}^\star_{ijkl} = \mathbf{C}^\star_{jikl} = \mathbf{C}^\star_{ijlk} = \mathbf{C}^\star_{klij}$, also for the effective elasticity tensor $\mathbf{C}^\star$ one observes that

$$\mathbf{C}^\star_{ijkl} = \mathbf{C}^\star\epsilon_{ij} : \epsilon_{kl} = \mathbf{C}^\star\epsilon_{ij+kl} : \epsilon_{ij+kl} - \mathbf{C}^\star\epsilon_{ij-kl} : \epsilon_{ij-kl} \tag{3.2}$$

with $\epsilon_{ij} = \frac{1}{2}(e_i \otimes e_j + e_j \otimes e_i)$ and $\epsilon_{ij \pm kl} = \frac{1}{2}(\epsilon_{ij} \pm \epsilon_{kl})$ where e_i is the ith canonical basis vector in $\mathbb{R}^d$. Thus, for every $x \in D$ one evaluates the reconstruction $R^\star$ for a basis of affine displacements and then via the above representation the coefficients of the effective elasticity tensor $\mathbf{C}^\star_{ijkl}$ for all $1 \leq i, j, k, l \leq d$.

In shape optimization it turns out to be advantageous to allow also spatially varying orientation of the microscopic perforation pattern, for which the microscopic perforation $\mathfrak{m}_{\alpha(x)}$ is no longer contained in $[-\gamma, \gamma]^d$. This enlarges substantially the class of possible microstructures which can be achieved without increasing much the number of parameters. To this end, we allow for a rotation $Q(\alpha(x))$ of the microscopic cell $x + \delta[-\frac{1}{2}, \frac{1}{2}]^d$ depending on the local value $\alpha(x)$ of the macroscopic parameter function and use $\mathcal{C}_\alpha(x) = Q(\alpha(x))(-\frac{1}{2}, \frac{1}{2})^d \setminus \mathfrak{m}_{\alpha(x)}$ in the definition of the two-scale approach above. Furthermore, we have to adopt the definition of $\mathbf{W}_\alpha$ using the rotated periodicity assumption $\phi(x, y + z) = \phi(x, y) \forall z \in Q(\alpha(x))\mathbb{Z}^d$. Let us emphasize that in this case the fine scale problem on a scale δ need not be properly defined any longer.

In this paper we will compare the performance of different types of microscopic perforations on two dimensional domains ($d = 2$). In what follows we will describe the associated parametrization (cf. Fig. 2):

- *Cells with single ellipsoidal holes.* An ellipsoidal shaped hole is considered, parametrized by the lengths $\alpha_1, \alpha_2 \in (0, \gamma)$ of its two semiaxes and a rotation α_3 (cf. Fig. 1).
- *Cells with* 2×2 *ellipsoidal holes.* A natural extension is achieved by allowing 2×2 holes with 12 independent parameters per cell.
- *Cell structures consisting of axes-aligned trusses.* As an alternative construction we consider truss like structures along the edges and the diagonals of the cell, where the thickness α_i ($i = 1, \ldots, 6$) can be varied. Additional constraints make sure that the holes generated between the trusses maintain a triangular shape.
- *Cell structures consisting of freely rotated orthogonal trusses.* Finally two orthogonal trusses connecting midpoints of opposing edges of the cell are allowed to rotate freely. This periodic pattern is equivalently determined by the rectangular hole centered at the corners of the cell and parametrized by the half edge lengths $\alpha_1, \alpha_2 \in (0, \gamma)$ and an unconstrained rotation α_3.

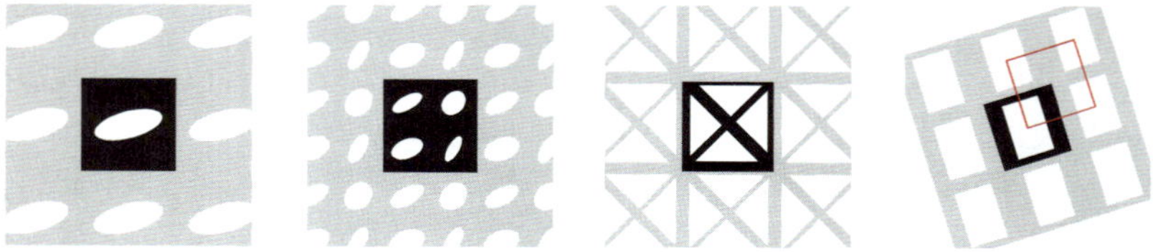

Fig. 2 Form left to right the different cells with different type of perforation are displayed: cells with single ellipsoidal holes, cells with 2×2 ellipsoidal holes, cell structures consisting of axes aligned and diagonal trusses, freely rotated cells with rectangular holes representing orthogonal trusses (see *red marking*)

All parametrizations described above lead to perforations of the fundamental cell leaving behind a certain amount of rigid constituent. The fraction $\theta(x) = \mathrm{Vol}(\mathcal{C}_\alpha(x))$ can be interpreted as the macroscopic local density of the effective material. To rule out trivial solutions to the shape optimization problem we impose a global volume constraint. The total amount of material spent $\Theta = \int_D \theta(x)\,\mathrm{d}x$ is to be kept fixed throughout the optimization procedure.

In the implementation of our two-scale simulation method we use a finite element scheme on the macroscale and a boundary element scheme on the microscale (cf. Sect. 5).

4 Two-Scale Shape Optimization Under Stochastic Loadings

For every choice of the parameter function α we can compute a corresponding macroscopic displacement $u^\star[\alpha]$. Given a cost functional $\mathbf{J}$, which may depend directly on the parameter function α and the macroscopic displacement $u^\star$, we ask for an optimal shape, described via macroscopically parametrized microscopic perforations. Precisely, we want to compute a parameter function α which minimizes $\mathcal{J}[\alpha] := \mathbf{J}[\alpha, u^\star[\alpha]]$. Before dealing with stochastic shape optimization we briefly discuss the deterministic case.

Deterministic shape optimization In this article we will focus on the compliance cost functional as a global measure of rigidity. We can write the resulting cost in the following equivalent ways (cf. (3.1)):

$$\mathbf{J}[\alpha, u] = \int_{\Gamma_N} g(x) \cdot u(x)\,\mathrm{d}a = \int_D \mathbf{C}^\star[\alpha](x)\,\varepsilon[u](x) : \varepsilon[u](x)\,\mathrm{d}x\,. \tag{4.1}$$

The derivative $\mathcal{J}'[\alpha]$, which plays in our context the role of the shape derivative, takes the form $\mathcal{J}'[\alpha] = \mathbf{J}_{,\alpha}[\alpha, u^\star[\alpha]] + \mathbf{J}_{,u}[\alpha, u^\star[\alpha]](\partial_\alpha u^\star[\alpha])$, where $\mathbf{J}_{,\alpha}[\alpha, u^\star[\alpha]] = \int_D \partial_\alpha \mathbf{C}^\star[\alpha](x)\epsilon[u^\star[\alpha]](x) : \epsilon[u^\star[\alpha]](x)\,\mathrm{d}x$. To avoid computing sensitivities $\partial_\alpha u^\star[\alpha]$ of the displacement w.r.t. the perforation parameter function α we employ the dual problem. Here, the dual solution $p^\star = p^\star[\alpha] \in H^{1,2}_{\Gamma_D}$ is defined as the weak solution of

$$\int_D \mathbf{C}^\star[\alpha](x)\epsilon[p^\star](x) : \epsilon[\phi](x)\,\mathrm{d}x = -\mathbf{J}_{,u}[\alpha, u^\star[\alpha]](\phi) \tag{4.2}$$

for all $\phi \in H^{1,2}_{\Gamma_D}$. In our case of a compliance type cost functional $p^\star[\alpha] = -2u^\star[\alpha]$. With the dual solution at hand one can rewrite the derivative of the cost

$$\begin{aligned}\mathcal{J}'[\alpha] &= \mathbf{J}_{,\alpha}[\alpha, u^\star[\alpha]] + \int_D \left(\partial_\alpha \mathbf{C}^\star[\alpha]\right)(x)\epsilon[u^\star[\alpha]](x) : \epsilon[p^\star[\alpha]](x)\,\mathrm{d}x \\ &= -\int_D \left(\partial_\alpha \mathbf{C}^\star[\alpha]\right)(x)\epsilon[u^\star[\alpha]](x) : \epsilon[u^\star[\alpha]](x)\,\mathrm{d}x\,.\end{aligned} \tag{4.3}$$

Finally, we are left to compute $\partial_\alpha \mathbf{C}^\star[\alpha](x)$ for $x \in D$. To this end, taking into account (3.2) we consider $\mathbf{C}^\star[\alpha](x)\varepsilon_{ij\pm kl} : \varepsilon_{ij\pm kl}$ for fixed i, j, k, l and for fixed $x \in D$ and define the local cost functional $j_C[\alpha] = \mathbf{j}_C[\alpha, w^\star_{ij\pm kl}[\alpha]]$ with $\mathbf{j}_C[\alpha, w] = \int_{\mathcal{C}_\alpha(x)} \mathbf{C}(y)\left(\varepsilon_{ij\pm kl} + \varepsilon[w](x,y)\right) : \left(\varepsilon_{ij\pm kl} + \varepsilon[w](x,y)\right) \mathrm{d}y$ and $w^\star_{ij\pm kl}[\alpha](x,\cdot)$ is given as the solution of the local correction problem $0 = \int_{\mathcal{C}_\alpha(x)} \mathbf{C}(y)\left(\varepsilon_{ij\pm kl} + \varepsilon[w](x,y)\right) : \varepsilon[\psi](x,y)\,\mathrm{d}y$ for all $\psi \in \mathbf{W}_\alpha$ and with x being fixed. This implies

$$\mathbf{j}_C[\alpha, w^\star_{ij\pm kl}[\alpha]] = \int_{\mathcal{C}_\alpha(x)} \mathbf{C}(y)\left(\varepsilon_{ij\pm kl} + \varepsilon[w^\star_{ij\pm kl}[\alpha]](x,y)\right) : \left(\varepsilon_{ij\pm kl} + \varepsilon[w^\star_{ij\pm kl}[\alpha]](x,y)\right) \mathrm{d}y\,.$$

From the correction problem we immediately deduce that $\partial_w \mathbf{j}_C[\alpha, w^\star_{ij\pm kl}[\alpha]] = 0$. Hence, one obtains $\partial_\alpha j_C[\alpha] = \partial_\alpha \mathbf{j}_C[\alpha, w^\star_{ij\pm kl}[\alpha]]$. Taking into account a family of perforations defined via the mapping $s \mapsto \mathfrak{m}_{\alpha(x)+s\beta}$ for $s \in \mathbb{R}$ with $|s|$ small, we then obtain for the variation of the local cost $j_C[\alpha]$ in direction $\beta \in \mathbb{R}^m$

$$\begin{aligned}\partial_\alpha j_C[\alpha](\beta) &= \frac{\mathrm{d}}{\mathrm{d}s} j_C[\alpha + s\beta]\Big|_{s=0} = \frac{\mathrm{d}}{\mathrm{d}s}\mathbf{j}_C[\alpha + s\beta, w^\star_{ij\pm kl}[\alpha]]\Big|_{s=0}\\ &= \int_{\partial\mathfrak{m}_{\alpha(x)}} (v_{\alpha,\beta}(y)\cdot n_{\partial\mathfrak{m}_{\alpha(x)}}(y))\mathbf{C}(y)\left(\varepsilon_{ij\pm kl} + \varepsilon[w^\star_{ij\pm kl}[\alpha]](x,y)\right) :\\ &\qquad \left(\varepsilon_{ij\pm kl} + \varepsilon[w^\star_{ij\pm kl}[\alpha]](x,y)\right) \mathrm{d}y\end{aligned}$$

where $n_{\partial\mathfrak{m}_{\alpha(x)}}(y)$ denotes the inner normal of the perforation $\mathfrak{m}_{\alpha(x)}$ at $y \in \partial\mathfrak{m}_{\alpha(x)}$ and $v_{\alpha,\beta}(y)$ is the velocity vector associated with the variation of $\mathfrak{m}_{\alpha(x)}$ in the direction β at position $y \in \partial\mathfrak{m}_{\alpha(x)}$. Finally, we obtain for the variation of the effective elasticity tensor in a direction $\beta \in \mathbb{R}^m$

$$\begin{aligned}&\partial_\alpha \mathbf{C}^\star_{ijkl}[\alpha](\beta) =\\ &\int_{\partial\mathfrak{m}_{\alpha(x)}} \mathbf{C}(y)\Big(\left(\varepsilon_{ij+kl} + \varepsilon[w^\star_{ij+kl}](x,y)\right) : \left(\varepsilon_{ij+kl} + \varepsilon[w^\star_{ij+kl}[\alpha]](x,y)\right) -\\ &\qquad \left(\varepsilon_{ij-kl} + \varepsilon[w^\star_{ij-kl}](x,y)\right) : \left(\varepsilon_{ij-kl} + \varepsilon[w^\star_{ij-kl}[\alpha]](x,y)\right)\Big)\\ &\qquad \cdot\, (v_{\alpha,\beta}(y)\cdot n_{\partial\mathfrak{m}_{\alpha(x)}}(y))\,\mathrm{d}y\,.\end{aligned}$$

Two-stage stochastic shape optimization In a more realistic situation the actual loading of an elastic work piece is usually not fixed but varies stochastically. Therefore we now extend the above framework and consider random surface loads $g(\omega) \in L^2(\Gamma_N;\mathbb{R}^d)$ with ω being a realization on an abstract probability space $(\Omega, \mathcal{A}, \wp)$. Here, finite-dimensional linear stochastic programs serve as blueprints for our stochastic shape optimization models. In this context a two-stage scheme of alternating decision and observation applies. The first-stage decision of a concrete shape, in our context the parameter vector α, must not anticipate future information on the random data, here the random boundary force $g(\omega)$. The second-stage decision in our context corresponds to the solution of the elastic problem and the evaluation of the cost value for a concrete realization ω and for fixed α and $g(\omega)$. The overall aim of two-stage stochastic programming is to find an α which is in a stochastic sense "optimal" under these circumstances. Different modes of ranking random variables then lead to different types of stochastic programs. In a risk neutral setting the ranking is done by taking the expectation $\mathbb{E}_\omega$. With risk aversion, see [17,41] for a recent textbook and a monograph as well as the journal publications [27,39,44], expectation is replaced by statistical parameters reflecting some perception of risk (risk measures) or stochastic dominance relations are employed. In what follows we will focus on risk measures. For a fixed realization ω and fixed parameter function α primal and dual solutions $u[\alpha](\omega)$, $p[\alpha](\omega)$ can be computed as described above in the deterministic setting. As the solutions now depend on ω so do the associated variational problems (3.1) and (4.2) as well as the cost functional (4.1) and its gradient (4.3). Altogether we obtain the random shape optimization model $\min\{\mathbf{J}[\alpha,u,\omega] : \alpha \in \mathcal{U}_{\text{ad}}\}$, which amounts to finding a "minimal" member in the family of random variables $\mathbf{J}[\alpha,u,\omega]$. Taking the expectation yields the risk neutral problem $\min\{\mathbf{Q}_{\text{EV}}[\alpha] := \mathbb{E}_\omega(\mathbf{J}[\alpha,u,\omega]) : \alpha \in \mathcal{U}_{\text{ad}}\}$. Risk averse problems are the expected excess $\min\left\{\mathbf{Q}_{\text{EE}_\eta}[\alpha] := \mathbb{E}_\omega(\max\{\mathbf{J}[\alpha,u,\omega]-\eta,0\}) : \alpha \in \mathcal{U}_{\text{ad}}\right\}$ or the excess probability $\min\left\{\mathbf{Q}_{\text{EP}_\eta}[\alpha] := \mathbb{P}_\omega(\mathbf{J}[\alpha,u,\omega] > \eta) : \alpha \in \mathcal{U}_{\text{ad}}\right\}$ over a preselected target $\eta \in \mathbb{R}$. For the numerical realization we will use smooth approximations of the max-function and the Heaviside function leading to $\mathbf{Q}^\varepsilon_{\text{EE}_\eta}[\alpha] := \mathbb{E}_\omega\left(q^\varepsilon(\mathbf{J}[\alpha,u,\omega])\right)$, where $q^\varepsilon(t) := \frac{1}{2}(\sqrt{(t-\eta)^2+\varepsilon}+(t-\eta))$ and $\mathbf{Q}^\varepsilon_{\text{EP}_\eta}[\alpha] := \mathbb{E}_\omega\left(H^\varepsilon(\mathbf{J}[\alpha,u,\omega])\right)$ with $H^\varepsilon(t) := (1+e^{-\frac{2(t-\eta)}{\varepsilon}})^{-1}$ for $\varepsilon > 0$. For actual computations $(\Omega,\mathcal{A},\wp)$ is assumed to be finite, in the sense that there are finitely many realizations ω_i and probabilities π_i, $i = 1,\ldots,N_s$. We can then rewrite $\mathbf{Q}_{\text{EV}}[\alpha] = \sum_{i=1}^{N_s}\pi_i\mathbf{J}[\alpha,u,\omega_i]$, and $\mathbf{Q}^\varepsilon_{\text{EE}_\eta}[\alpha]$ and $\mathbf{Q}^\varepsilon_{\text{EP}_\eta}[\alpha]$ accordingly. The shape derivative as derived above can directly be applied to the stochastic functionals. The chain rule yields

$$\mathbf{Q}'_{\text{EV}}[\alpha](\beta) = \sum_{i=1}^{N_s}\pi_i\,\mathbf{J}'[\alpha,u,\omega_i](\beta)\,,$$

$$(\mathbf{Q}^\varepsilon_{\text{EE}_\eta})'[\alpha](\beta) = \sum_{i=1}^{N_s}\tfrac{\pi_i}{2}\mathbf{J}'[\alpha,u,\omega_i](\beta)\left(\frac{\mathbf{J}[\alpha,u,\omega_i]-\eta}{\sqrt{(\mathbf{J}[\alpha,u,\omega_i]-\eta)^2+\varepsilon}}+1\right),$$

$$(\mathbf{Q}^\varepsilon_{\text{EP}_\eta})'[\alpha](\beta) = \sum_{i=1}^{N_s}\tfrac{2}{\varepsilon}\pi_i\mathbf{J}'[\alpha,u,\omega_i](\beta)\,\frac{e^{-\frac{2}{\varepsilon}(\mathbf{J}[\alpha,u,\omega_i]-\eta)}}{\left(1+e^{-\frac{2}{\varepsilon}(\mathbf{J}[\alpha,u,\omega_i]-\eta)}\right)^2}$$

for a direction $\beta : D \to \mathbb{R}^m$ in which α is varied. So far it seems that for every ω_i one has to compute a primal and a dual solution. The solution, however, depends linearly on the right hand side; therefore a significant amount of computational time can be spared when a large number N_s of scenarios is generated by a small set of basis surface loads $g_1, \dots, g_K$, as was discussed in [25]. The actual loads $g(\omega)$ are given as linear combinations $g(\omega) = \sum_{j=1}^{K} \lambda_j(\omega) g_j$, with random coefficients $\lambda_j(\omega) \in \mathbb{R}$, $j = 1, \dots, K$. We thus only need to solve the elasticity problem for the different basis forces. To be more precise let $u^{j,\star}[\alpha]$ be the solution of (3.1) for $g = g_j$, $j = 1, \dots, K$. Then we find $u^{\star}[\alpha](\omega) := \sum_{j=1}^{K} \lambda_j(\omega)\, u^{j,\star}[\alpha]$ to be the unique solution of (3.1) with $g = g(\omega)$. The same procedure can be taken for the dual solution if the cost functional is at most quadratic guaranteeing the linearity in the right hand side. As discussed, in our case of a compliance objective the dual problem is already trivial.

5 Implementation

The two-scale simulation is based on a finite element discretization on the macroscale and a boundary element method on the microscale. We use a regular mesh with N quadratic cells on the macroscale and piecewise biquadratic finite elements as checkerboard instabilities were reported in [18] for the related case of nested laminates when using linear ansatz functions. We use a Gaussian quadrature of consistency order 5 with 3×3 quadrature points per square cell. Within each cell the underlying microstructure is specified by a set of parameters in $\mathbb{R}^m$ as the discrete counterpart of the local parameter function α. For the cell problem a collocation type boundary element method is used to compute numerical approximations to the microscopic correction profiles. For details we refer to the corresponding discussion for the single-scale model in [6]. The design constraints described in Sect. 3 are implemented as inequality constraints in the optimization. The global volume constraint leads to an additional equality constraint. Unless otherwise noted we use a material with moderate anisotropy for the construction of microscopic geometries. It is characterized by the elasticity tensor

$$\mathbf{C}_{\text{aniso}} = \begin{pmatrix} 3 & 1 & \\ 1 & 3 & \\ & & 1.1 \end{pmatrix}$$

using Voigt's notation. For all numerical experiments we prescribe a volume fraction of 67 % as global constraint. Loads usually have magnitude 1.

Our algorithm for the two-scale shape optimization approach is written in `C++` based on the `quocmesh` library for finite element and boundary element computations and the open source software `Ipopt` [47,48] performing constrained finite-dimensional optimization.

6 Numerical Results

In this section we present numerical results for shape optimization problems both with deterministic and stochastic loadings, comparing the different microstructure models. The key scenario we consider is a carrier plate, in which the computational domain is the unit square, with homogeneous Dirichlet boundary data on the bottom and Neumann boundary conditions corresponding to a shearing on the top. To illustrate the generality of the method we also study two classical problems from the literature.

6.1 Deterministic Optimization

We start with the simplest microstructure, in which every unit cell has one ellipsoidal hole (first sketch in Fig. 2). The results for the three model problems discussed above are presented in Fig. 3 based on computations on a macroscopic grid with 64×64 square cells.

Comparing the results for the carrier plate scenario to the single-scale case illustrated in Fig. 1 a remarkable similarity is apparent. The oscillating pattern observed in the single-scale case (Fig. 1, left blow-up) for the upper middle region however seems to be gone. The apparent reason for these oscillations was to approximate criss-crossing beams. Such a construction is ruled out by the kinematics in the two-scale setting, since each unit cell only contains one hole, which then gets repeated over and over again at the microscale.

This suggests to allow for more than one hole within the fundamental cell, each with its own set of parameters. We investigated this structure using 2×2 holes on each cell, which is the microstructure sketched in the second panel of Fig. 2. In the result, see Fig. 4c, the oscillating pattern is now captured as expected while other regions keep their microstructure by just reproducing the shape of the former single hole four times.

Fig. 3 Local minima for two-scale optimization of a carrier plate under shearing, a cantilever on a square domain and a bridge scenario, all on a macroscopic regular rectangular mesh with 64×64 cells. The local configuration is drawn within each macroscopic element as a representative for the underlying microstructure. Furthermore, the same results are presented using a HSV color code: color corresponds to the rotation of the major semiaxis, saturation to the degree of anisotropy and value to the volume of the hole

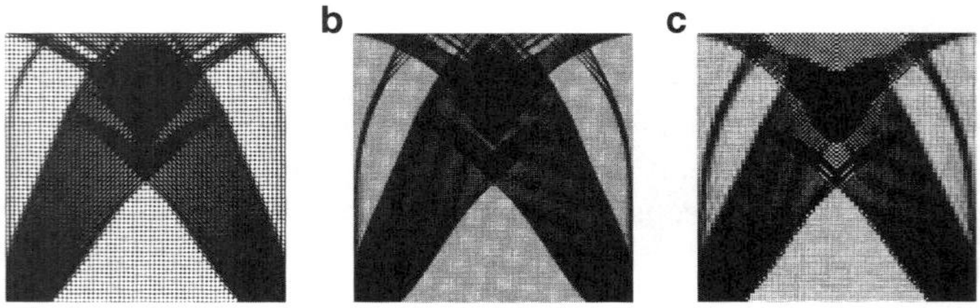

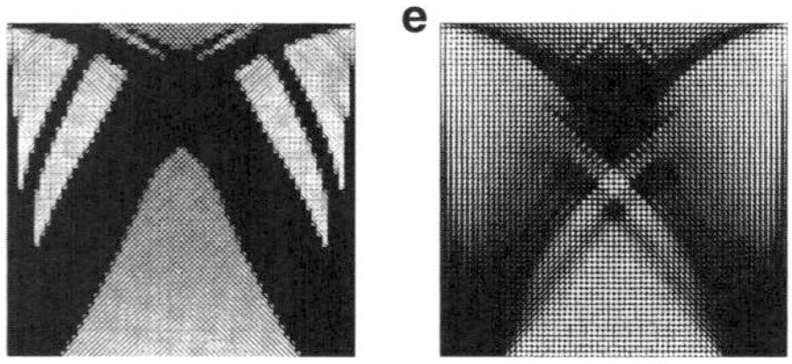

f

Structure	**J**
(a)	2.0358
(b)	2.0255
(c)	2.0276
(d)	1.8959
(e)	1.8122

Fig. 4 Local minima for two-scale optimization of a carrier plate under shearing for different configurations: (**a**) 64 × 64 macroscopic cells with 1 ellipsoidal hole each (same as in Fig. 3), (**b**) 128 × 128 macroscopic cells with 1 ellipsoidal hole each, (**c**) 64 × 64 macroscopic cells with 2 × 2 ellipsoidal holes each, (**d**) 64 × 64 macroscopic cells with 6 trusses at fixed positions each, and (**e**) 64 × 64 macroscopic cells with 2 rotated orthogonal trusses each. Final objective values are listed in (**f**)

Since we already conjectured that a framework with diagonal trusses would perform best in the upper middle region, we now explicitly consider such a microstructure. First we place six trusses at fixed positions within the fundamental cell (third panel in Fig. 2). In the resulting shape, see Fig. 4d, the expected criss-crossing pattern is found. However we now see solid trusses in the macroscopic picture. This is because the optimal shape has trusses which are not inclined by 45° and therefore cannot be reproduced by the microstructure; the optimization hence generates “macroscopic trusses” with the appropriate slope. This suggests to allow for a rotation of the whole periodic lattice (fourth panel in Fig. 2). The results are shown in Fig. 4e. The last panel of Fig. 4 gives an overview of the final objective values for the optimized designs. One can clearly see that the improvement of the shape by allowing a more complex microstructure using 2×2 holes per cell becomes manifest also in a quantitative way. The introduction of structures made from fixed trusses leads to a further significant drop in the objective functional. Finally allowing the structures to rotate freely again contributes to a substantial improvement.

So far we have successively constructed microscopic geometries that lead to a stepwise reduction in the objective value for the optimized design. It seems natural to compare the results to a microstructure that is known to be optimal a priori. For our comparison we decided to adopt the nested laminates construction as it is valid on the full range of feasible strains and explicit formulae as well as an algorithmic treatment are available in [1]. The microstructure is built up in an iterative procedure. One starts by successively layering a given rigid and a very weak material, determined by elasticity tensors B and A respectively, with proportions

m_1 and $(1 - m_1)$ in a direction e_1. The obtained material is then layered again with the rigid material B, now with proportion m_2 and in direction e_2. In our shape optimization context one considers the degenerate case in which the weak material A is replaced by voids. By a limiting process the effective elasticity tensor $\mathbf{C}^{\star}_L$ of the nested laminate can be given in closed form with the ratios m_1, m_2, the directions e_1, e_2, and the overall material density θ as degrees of freedom [1]. Moreover, these parameters depend explicitly on the local effective stress $\sigma^{\star} = \mathbf{C}^{\star}_L u^{\star}$. Indeed, the ratios m_1 and $m_2 = 1 - m_1$ are given by the ratio of the eigenvalues of $\sigma^{\star}$ and the directions e_1 and e_2 are aligned with the orthogonal system of eigenvectors. Finally using a Lagrange multiplier approach also the optimal local density θ depends, apart from elastic constants, only on the eigenvalues.

The variational structure of the problem and the values of the objective function listed in Fig. 4f give a clear ordering in the quality of the different microstructure patterns and show in particular that the rotated orthogonal trusses are superior to the other models considered, at least in the present scenario. To assess how much room for improvement is left we compare with the lower bound given by the Hashin-Shtrikman formula, focusing for simplicity on the case of an isotropic material. In the present setting this lower bound is known to be optimal and can indeed be attained by lamination [1,7]. To this end we reimplemented the alternating algorithm for the nested lamination construction proposed in [1]. The local density of the optimal structure obtained in the carrier plate scenario is compared in Fig. 5b with the result of our two-scale method for perforated isotropic material. The latter has been computed with the isotropic elasticity tensor

$$\mathbf{C}_{\text{iso}} = \begin{pmatrix} 3 & 1 & \\ 1 & 3 & \\ & & 1 \end{pmatrix}$$

in which the lower right entry was replaced by 1.0001, since we use in our boundary element scheme a fundamental solution for anisotropic elasticity. The perforation pattern in Fig. 5a is qualitatively very similar to the one obtained with anisotropic elasticity in Fig. 4e, the quantitative values of the objective function, however, differ

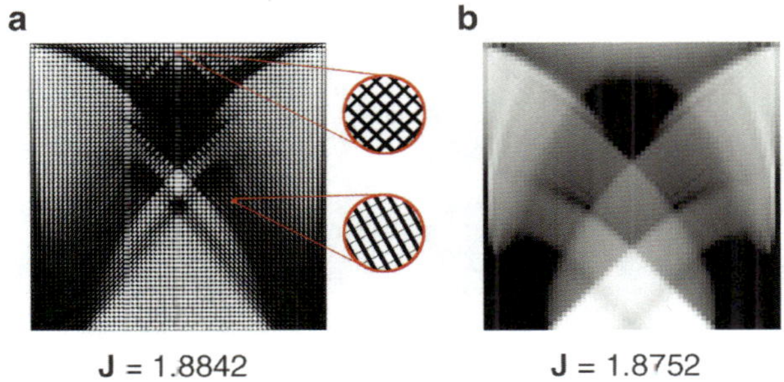

Fig. 5 A comparison of the computed optimal pattern for a carrier plate under shearing is displayed. The computations are performed on a 64 × 64 grid: (**a**) shows the microstructure composed of rotated orthogonal trusses, and (**b**) renders the density in the case of sequential laminates. The values of the objective function are listed below

significantly. The comparison of Fig. 5a, b demonstrates that the performance of the two-scale approach with rotated orthogonal trusses is indeed very close to the one of the optimal microstructure, which can be realized for example by the construction with second-order laminates. Analytically, it is known that in the low-volume-fraction limit the construction with single-scale laminates is optimal [19]. Single-scale laminates are, in the definition of [19], structures in which thin trusses with different orientations coexist without interacting; for low volume fraction and second-order laminates they correspond to our rotated trusses. The present results show that rotated trusses give almost the same objective function as laminates even for a total volume fraction of 67 %, at least in this geometry.

6.2 Risk Averse Stochastic Optimization

In this section we show that our developed two-scale algorithm can directly be applied to the more general situation of stochastic shape optimization. We consider a variant of the carrier plate scenario with sets of different loads on the upper left and upper right edge with different probabilities, as illustrated in Fig. 6 and described in the figure caption, similar to the one that was studied for single-scale stochastic shape optimization in [26]. For the optimization we consider both the risk neutral and the risk averse cost functionals introduced in Sect. 4. In this stochastic optimization we focus on the simplest choice of microstructure, the ellipsoidal holes, and on the one that performed best in the deterministic setting, the rotated trusses.

Figure 6 shows the result of the deterministic optimization using the expected value of the loads. The larger probability of the forces on the right results in a larger expected value of the force, and hence on a strong concentration of the available

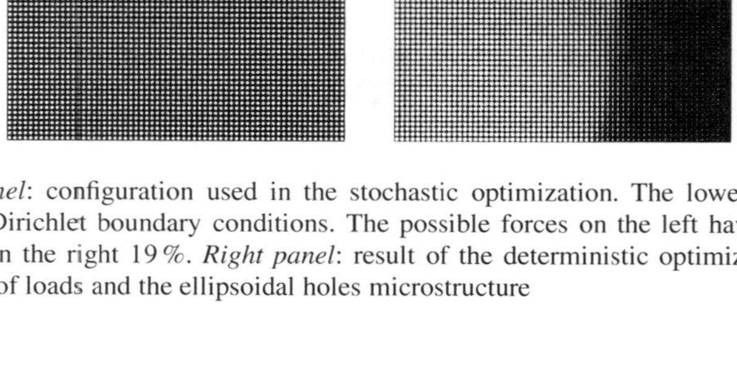

Fig. 6 *Left panel*: configuration used in the stochastic optimization. The lower boundary has homogeneous Dirichlet boundary conditions. The possible forces on the left have a probability of 1 %, those on the right 19 %. *Right panel*: result of the deterministic optimization using the expected value of loads and the ellipsoidal holes microstructure

mass on the right-hand side of the computational domain. The (on average!) minor forces on the left-hand side are dealt with by two small trusses which connect the main pillar to the other side of the domain.

In Fig. 7 we compare, using the microstructure of rotated orthogonal trusses, three different approaches to the forcing. In Fig. 7a we show the optimization of the expected value of the costs, in which both left and right forcing play a significant role. In particular, the presence of a scenario with forcing only on the left-hand side generates a substantial mass on the left-hand side of the computational domain. Figure 7b illustrates the result of optimizing w.r.t. the expected value of the loads. The two types of microstructures lead to similar shapes. In Fig. 7c we show, for a comparison, the result of the optimization in the symmetric case, with the same forces acting on both parts of the top boundary.

We now turn to the optimization of the expected excess. Figure 8 shows the results for ellipsoidal holes and rotated trusses, respectively. As in the previous case, the two types of microstructure generate similar patterns. Introducing a threshold makes the largest deviations more important, and therefore the forces on the left, which are large but have a small probability, become more important in the optimization. Indeed, for small η (and for the EV optimization) the forces on the right-hand side dominate, and correspondingly the largest structures are the vertical one on the right (which takes care of the vertical component of the forces on the right) and the diagonal from the lower left to the upper right corner (which takes care of the horizontal component of the forces on the right). With increasing η the situation becomes first symmetric, and then tilted in the other direction, with the left side and the lower right to upper left diagonal dominant at $\eta = 0.0005$.

In the case of the optimization of the excess probability, only the probability, and not the amplitude, of the large deviations plays a role. Results are shown with ellipsoidal holes for the microstructure in Fig. 9. Indeed, for small η the best result the optimization can achieve is to keep the cost functional in the scenarios corresponding to the small forces on the right-hand side below the threshold; in order to do this the small probability forces on the left-hand side are given up. The cost of these forces would, in the ideal case, be infinite (it is not due to the many

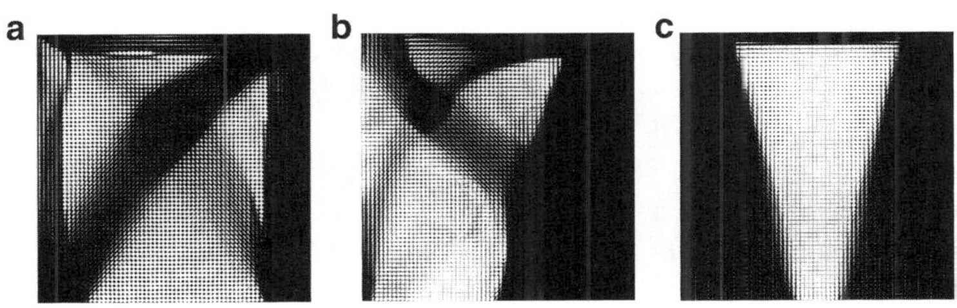

Fig. 7 Results of the two-scale shape optimization procedure using rotated trusses for the microstructure: stochastic optimization of the expected value of costs (**a**), deterministic optimization using the expected value of the loads (**b**), deterministic optimization computed for equal loads on the left and right parts (**c**)

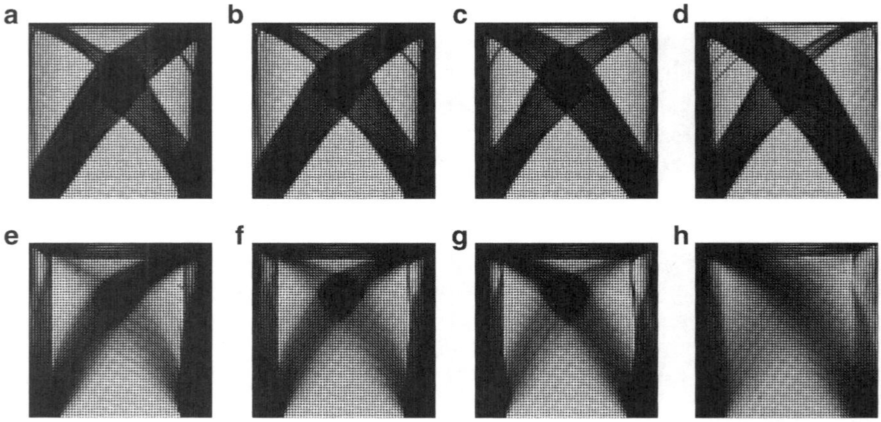

Fig. 8 Results of the two-scale stochastic optimization of the expected excess for different values of the threshold η, using ellipsoidal holes (*top*) and rotated trusses (*bottom*) for the microstructure. (**a**) EV. (**b**) $\eta = 0.0001$. (**c**) $\eta = 0.0003$. (**d**) $\eta = 0.0005$. (**e**) $\eta = 0.0001$. (**f**) $\eta = 0.0002$. (**g**) $\eta = 0.0003$. (**h**) $\eta = 0.0005$

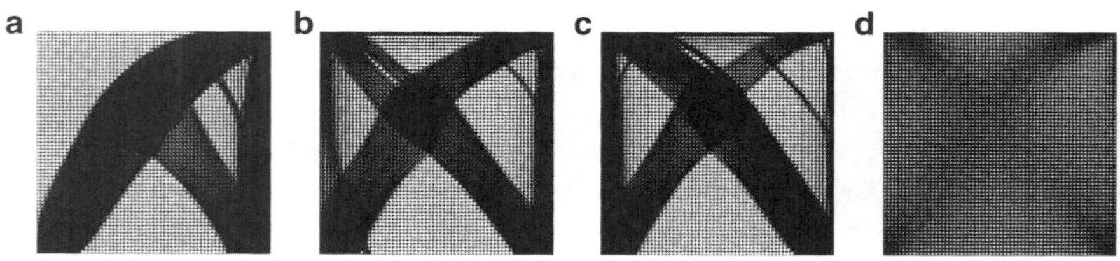

Fig. 9 Results of the two-scale stochastic optimization of the excess probability for different values of the threshold η, using ellipsoidal holes for the microstructure. (**a**) $\eta = 0.0003$. (**b**) $\eta = 0.0004$. (**c**) $\eta = 0.0005$. (**d**) $\eta = 0.0006$

numerical regularizations, including for example the fact that the volume fraction cannot be zero in any cell). This divergence does not, however, result in a divergence of the total cost functional because only the probability of these large deviations enters the optimization, not their amplitude. With increasing η, it is less important to keep the response to the small forces very small: as above, the thresholding makes the exact value of the cost functional in that case irrelevant, as long as it is below threshold. The optimization can devote material to improving the response to the forces on the left. Since they are large (but unlikely) more material is needed to bring them below threshold than for the smaller forces on the right, hence the pattern also in this case changes to a left-dominated one. When the threshold η becomes larger and larger, it is easy to keep the response to all 10 forces below it, and the problem degenerates: in a sense, there is "too much material" to achieve the aim, and the details of the shapes are not any more meaningful. We stress that the discontinuity of the excess probability has been regularized in the numerics, hence the transitions discussed are to be interpreted as gradual transitions, not as abrupt discontinuities from "above threshold" to "below threshold".

In closing we remark that, although the details of the shapes differ, the qualitative trends we discussed are very similar to the ones we had observed in the single-scale computations in [26].

Conclusions

In this paper we derived a two-scale framework for shape optimization in which the parameters of microscopic perforations on a locally periodic lattice are optimized. We compare the performance of different types of perforation geometries and demonstrate that the best performing geometry of locally rotated orthogonal trusses gets very close to the known optimal approach based on nested lamination construction on the microscale. Furthermore, we studied stochastic shape optimization in the class of two-scale materials with approximate models for expected excess and the excess probability as risk averse cost functionals.

Acknowledgements The authors would like to thank Martin Lenz for help with the boundary element method applied to microscopic cell problems. This work was supported by the Deutsche Forschungsgemeinschaft through the Schwerpunktprogramm 1253 *Optimization with Partial Differential Equations*.

References

1. G. Allaire, *Shape Optimization by the Homogenization Method*. Applied Mathematical Sciences, vol. 146 (Springer, New York, 2002)
2. G. Allaire, E. Bonnetier, G. Francfort, F. Jouve, Shape optimization by the homogenization method. Numer. Math. **76**, 27–68 (1997)
3. G. Allaire, F. Jouve, A level-set method for vibration and multiple loads structural optimization. Comput. Methods Appl. Mech. Eng. **194**(30–33), 3269–3290 (2005)
4. G. Allaire, F. Jouve, F. de Gournay, Shape and topology optimization of the robust compliance via the level set method. ESAIM Control Optim. Calc. Var. **14**, 43–70 (2008)
5. R. Astley, J. Harrington, K. Stol, Mechanical modelling of wood microstructure, an engineering approach. Ipenz Trans. **24**(1/EMCh), 21–29 (1997)
6. P. Atwal, S. Conti, B. Geihe, M. Pach, M. Rumpf, R. Schultz, On shape optimization with stochastic loadings, in *Constrained Optimization and Optimal Control for Partial Differential Equations*, ed. by G. Leugering, S. Engell, A. Griewank, M. Hinze, R. Rannacher, V. Schulz, M. Ulbrich, S. Ulbrich. International Series of Numerical Mathematics, vol. 160, ch. 2 (Springer, Basel, 2012), pp. 215–243
7. M. Avellaneda, Optimal bounds and microgeometries for elastic two-phase composites. SIAM J. Appl. Math. **47**(6), 1216–1228 (1987)

8. N.V. Banichuk, P. Neittaanmäki, On structural optimization with incomplete information. Mech. Based Des. Struct. Mach. **35**, 75–95 (2007)
9. C. Barbarosie, Shape optimization of periodic structures. Comput. Mech. **30**(3), 235–246 (2003)
10. C. Barbarosie, A.-M. Toader, Shape and topology optimization for periodic problems. I. The shape and the topological derivative. Struct. Multidiscip. Optim. **40**(1–6), 381–391 (2010)
11. C. Barbarosie, A.-M. Toader, Shape and topology optimization for periodic problems. II. Optimization algorithm and numerical examples. Struct. Multidiscip. Optim. **40**(1–6), 393–408 (2010)
12. C. Barbarosie, A.-M. Toader, Optimization of bodies with locally periodic microstructure. Mech. Adv. Mater. Struct. **19**(4), 290–301 (2012)
13. M. P. Bendsøe, *Optimization of Structural Topology, Shape, and Material* (Springer, Berlin, 1995)
14. M. P. Bendsøe, A. Díaz, N. Kikuchi, Topology and generalized layout optimization of elastic structures, in *Topology Design of Structures* (Sesimbra, 1992). NATO Advanced Science Institutes Series E: Applied Sciences, vol. 227 (Kluwer Academic, Dordrecht, 1993), pp. 159–205
15. A. Ben-Tal, L. El-Ghaoui, A. Nemirovski, *Robust Optimization* (Princeton University Press, Princeton/Oxford, 2009)
16. A. Ben-Tal, M. Kočvara, A. Nemirovski, J. Zowe, Free material design via semidefinite programming: the multiload case with contact conditions. SIAM J. Optim. **9**, 813–832 (1999)
17. J.R. Birge, F. Louveaux, *Introduction to Stochastic Programming*. Springer Series in Operations Research (Springer, New York, 1997)
18. E. Bonnetier, F. Jouve, Checkerboard instabilities in topological shape optimization algorithms, in *Proceedings of the Conference on Inverse Problems, Control and Shape Optimization (PICOF'98)*, Carthage, 1998
19. B. Bourdin, R.V. Kohn, Optimization of structural topology in the high-porosity regime. J. Mech. Phys. Solids **56**(3), 1043–1064 (2008)
20. A. Braides, A. Defranceschi, *Homogeneization of Multiple Integrals* (Claredon Press, Oxford, 1998)
21. G. Buttazzo, G. Dal Maso, Shape optimization for Dirichlet problems: relaxed formulation and optimality conditions. Appl. Math. Optim. **23**, 17–49 (1991)
22. A. Cherkaev, E. Cherkaev, Principal compliance and robust optimal design. J. Elast. **72**, 71–98 (2003)
23. D. Cioranescu, P. Donato, *An Introduction to Homogenization* (Oxford University Press, Oxford, 1999)
24. D. Cioranescu, J.S.J. Paulin, *Homogenization of Reticulated Structures* (Springer, New York, 1999)
25. S. Conti, H. Held, M. Pach, M. Rumpf, R. Schultz, Shape optimization under uncertainty – a stochastic programming perspective. SIAM J. Optim. **19**, 1610–1632 (2009)
26. S. Conti, H. Held, M. Pach, M. Rumpf, R. Schultz, Risk averse shape optimization. SIAM J. Control Optim. **49**, 927–947 (2011)
27. D. Dentcheva, A. Ruszczyński, Optimization with stochastic dominance constraints. SIAM J. Optim. **14**, 548–566 (2003)
28. B. Geihe, M. Lenz, M. Rumpf, R. Schultz, Risk averse elastic shape optimization with parametrized fine scale geometry. Math. Program. **141**(1–2), 383–403 (2013)
29. Y. Grabovsky, R.V. Kohn, Microstructures minimizing the energy of a two phase elastic composite in two space dimensions. I. The confocal ellipse construction. J. Mech. Phys. Solids **43**(6), 933–947 (1995)
30. J.M. Guedes, H.C. Rodrigues, M.P. Bendsøe, A material optimization model to approximate energy bounds for cellular materials under multiload conditions. Struct. Multidiscip. Optim. **25**, 446–452 (2003)
31. Z. Hashin, The elastic moduli of heterogeneous materials. Trans. ASME Ser. E. J. Appl. Mech. **29**, 143–150 (1962)

32. Z. Hashin, S. Shtrikman, A variational approach to the theory of the elastic behaviour of multiphase materials. J. Mech. Phys. Solids **11**, 127–140 (1963)
33. J. Haslinger, M. Kočvara, G. Leugering, M. Stingl, Multidisciplinary free material optimization. SIAM J. Appl. Math. **70**(7), 2709–2728 (2010)
34. P. Henning, M. Ohlberger, The heterogeneous multiscale finite element method for elliptic homogenization problems in perforated domains. Numer. Math. **113**(4), 601–629 (2009)
35. S.J. Hollister, N. Kikuchi, Homogenization theory and digital imaging: a basis for studying the mechanics and design principles of bone tissue. Biotechnol. Bioeng. **43**, 586–596 (1994)
36. V. Jikov, V. Zhikov, S. Kozlov, O. Oleinik, *Homogenization of Differential Operators and Integral Functionals* (Springer, Berlin/New York, 1994)
37. D.P. Kouri, M. Heinkenschloss, D. Ridzal, B.G. van Bloemen Waanders, A trust-region algorithm with adaptive stochastic collocation for PDE optimization under uncertainty. SIAM J. Sci. Comput. **35**(4), A1847–A1879 (2013)
38. R. Melchers, Optimality-criteria-based probabilistic structural design. Struct. Multidiscip. Optim. **23**(1), 34–39 (2001)
39. N. Miller, A. Ruszczyński, Two-stage stochastic linear programming: Modeling and decomposition. Oper. Res. **59**, 125–132 (2011)
40. F. Murat, L. Tartar, Calcul des variations et homogénéisation. In *Homogenization Methods: Theory and Applications in Physics* (Bréau-sans-Nappe, 1983). Collect. Dir. Études Rech. Élec. France, vol. 57 (Eyrolles, Paris, 1985), pp. 319–369
41. G.C. Pflug, W. Römisch, *Modeling, Measuring and Managing Risk* (World Scientific, Singapore, 2007)
42. A. Ruszczyński, A. Shapiro (Eds.), *Handbooks in Operations Research and Management Sciences, 10: Stochastic Programming* (Elsevier, Amsterdam, 2003)
43. V. Schulz, C. Schillings, On the nature and treatment of uncertainties in aerodynamic design. AIAA J. **47**, 646–654 (2009)
44. A. Shapiro, Minimax and risk averse multistage stochastic programming. Eur. J. Oper. Res. **219**, 719–726 (2012)
45. O. Sigmund, On the optimality of bone microstructure, in *IUTAM Symposium on Synthesis in Bio Solid Mechanics*, Copenhagen (Springer, 2002), pp. 221–234
46. L. Tartar, Estimations fines des coefficients homogénéisés, in *Ennio De Giorgi Colloquium* (Paris, 1983). Research notes in mathematics, vol. 125 (Pitman, Boston, 1985), pp. 168–187
47. A. Wächter, *An Interior Point Algorithm for Large-Scale Nonlinear Optimization with Applications in Process Engineering*. Phd thesis, Carnegie Mellon University, 2002
48. A. Wächter, L. Biegler, On the implementation of a primal-dual interior point filter line search algorithm for large-scale nonlinear programming. Math. Program. **106**(1), 25–57 (2006)
49. W. E, B. Engquist, Z. Huang, Heterogeneous multiscale method: a general methodology for multiscale modeling. Phys. Rev. B **67**(9), 092101–1–092101–4 (2003)
50. W. E, P. Ming, P. Zhang, Analysis of the heterogeneous multiscale method for elliptic homogenization problems. J. Am. Math. Soc. **18**(1), 121–156 (2005)

On Shape Optimization with Parabolic State Equation

Helmut Harbrecht and Johannes Tausch

Abstract The present paper intends to summarize the main results of Harbrecht and Tausch (Inverse Probl 27:065013, 2011; SIAM J Sci Comput 35:A104–A121, 2013) on the numerical solution of shape optimization problems for the heat equation. This is carried out by means of a specific problem, namely the reconstruction of a heat source which is located inside the computational domain under consideration from measurements of the heat flux through the boundary. We arrive at a shape optimization problem by tracking the mismatch of the heat flux at the boundary. For this shape functional, the Hadamard representation of the shape gradient is derived by use of the adjoint method. The state and its adjoint equation are expressed as parabolic boundary integral equations and solved using a Nyström discretization and a space-time fast multipole method for the rapid evaluation of thermal potentials. To demonstrate the similarities to shape optimization problems for elliptic state equations, we consider also the related stationary shape optimization problem which involves the Poisson equation. Numerical results are given to illustrate the theoretical findings.

Keywords Shape optimization • Heat equation • Boundary integral equation • Multipole method

Mathematics Subject Classification (2010). 35K05; 58J35; 65K10.

H. Harbrecht (✉)
Department of Mathematics and Computer Science, Universität Basel, Rheinsprung 21, 4051 Basel, Switzerland
e-mail: helmut.harbrecht@unibas.ch

J. Tausch
Department of Mathematics, Southern Methodist University, Dallas, TX, USA
e-mail: tausch@smu.edu

G. Leugering et al. (eds.), *Trends in PDE Constrained Optimization*, International Series of Numerical Mathematics 165, DOI 10.1007/978-3-319-05083-6_14

1 Introduction

Shape optimization is a well established mathematical and computational tool in case of an elliptic state equation, see, e.g., [2, 13, 14, 19, 25, 26, 28, 29, 32, 35] and the references therein. In contrast, the literature on shape optimization is rather limited for a parabolic state equation. Theoretical results for the latter case can be found, for instance, in [4, 20, 23, 31] and the references therein. However, the development of efficient numerical methods for shape optimization problems with a parabolic state equation is still in its beginning stages, especially for three-dimensional geometries.

With the goal to develop such efficient methods, we considered in [17, 18] shape identification problems for the heat equation. Specifically, besides the computation of the Hadamard representation of the shape gradients, we applied boundary integral equations to provide that data from the state and its adjoint which enter the shape functional and shape gradient. These boundary integral equations have been solved by multipole-based space-time boundary element methods which cluster sources in space and time have become available recently [33, 34]. That way, we were able to reconstruct unknown shapes in three dimensions on a laptop in less than half an hour computation time even though up to 1,200 design parameters and about 120,000 boundary elements have been used for the discretization of the shape optimization problem.

If one takes a closer look at [17, 18], it turns out that for a parabolic state equation both, the shape calculus and the formulation by boundary integral equations, are in principle rather similar to the case of an elliptic state equation. Besides being numerically more challenging, the main difference is that in the parabolic case singularities appear since the initial data do not generally fit the given boundary data, especially in the adjoint state equation. The similarities stem from the fact that the sought shape has not been allowed to change in time. Future research should thus go into the direction of shape optimization problems where the shape varies in time.

The present paper intends to summarize the main results of [17, 18] by focusing on a specific shape reconstruction problem with parabolic state equation. The goal is to reconstruct the shape of a heat source inside a given domain from the knowledge of the temperature and the heat flux at the boundary of the domain. Practical applications of the problem under consideration arise from the detection of any kind of heat source like e.g. fire or radioactive decay in non-accessible areas. We provide the ingredients (shape gradient, discretization of the shape, discretization of the state equation and its adjoint) for an efficient shape reconstruction algorithm and compare them with the ingredients for the related stationary problem which is obtained by letting time tend to infinity.

The paper is organized as follows. In Sect. 2, the problems under consideration are formulated. The Hadamard representation of the shape gradients is derived in Sect. 3. The following section describes the discretization of the shape. The computation of the state and the adjoint state by boundary integral equations is proposed in Sect. 5. Finally, in Sect. 6, we compare the reconstruction of shapes in case of the elliptic state equation with the reconstruction of shapes in case of the parabolic state equation.

2 Problem Formulation

The shape identification problem under consideration is as follows. Let D be a domain contained in a domain $\Omega \subset \mathbb{R}^n$, $n = 2, 3$, and consider the initial boundary value problem

$$\partial_t u - \Delta u = \chi_D \quad \text{in } \Omega \times (0, T) \tag{2.1}$$

with boundary condition

$$u = 0 \quad \text{on } \Sigma \times (0, T) \tag{2.2}$$

and initial condition

$$u = 0 \quad \text{on } \Omega \times \{0\}. \tag{2.3}$$

Here, $\Sigma = \partial\Omega$ denotes the boundary of the domain Ω, whereas we will denote the boundary of D by $\Gamma := \partial D$, see also Fig. 1. Throughout the paper, we assume that the boundaries Σ and Γ are respectively Lipschitz-continuous and C^2-smooth.

The goal is to reconstruct the discontinuous source D from measurements of the Neumann data $\partial u/\partial \mathbf{n}$ at the boundary Σ. More precisely, we will minimize the least square functional

$$J(D) = \frac{1}{2} \int_0^T \int_\Sigma \left(\frac{\partial u}{\partial \mathbf{n}} - h \right)^2 \mathrm{d}\sigma \, \mathrm{d}t \to \inf. \tag{2.4}$$

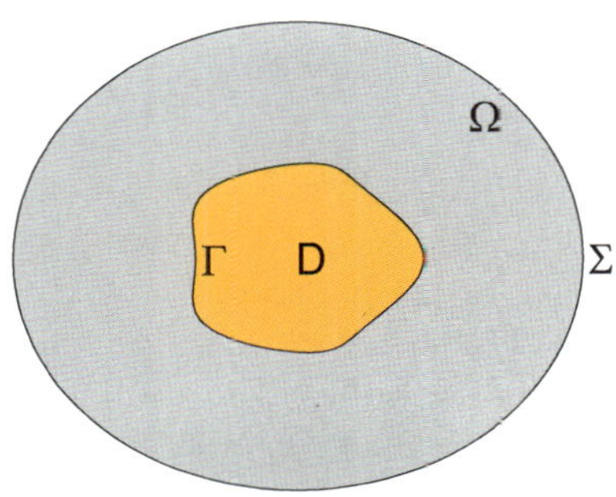

Fig. 1 The domain Ω with boundary Σ and the source D with boundary Γ

This problem has firstly been considered by Hettlich and Rundell in [22] and is known to be severely ill-posed. Since we track the Neumann data over the whole boundary Σ, uniqueness of the solution D immediately follows from [24], where the steady state case has been considered, governed by the Eq. (2.5) below. Nevertheless, uniqueness can be proven under much milder assumptions, see [22] and the references therein.

For comparison reasons, we shall also consider the steady state case which is obtained for $T \to \infty$. Then, the state equation (2.1) simplifies to the Poisson equation

$$-\Delta u = \chi_D \quad \text{in } \Omega, \tag{2.5}$$

while the initial condition (2.3) disappears and the boundary condition (2.2) becomes

$$u = 0 \quad \text{on } \Sigma. \tag{2.6}$$

The analogue of the shape functional (2.4) reads now as

$$J(D) = \frac{1}{2}\int_{\Sigma}\left(\frac{\partial u}{\partial \mathbf{n}} - h\right)^2 \mathrm{d}\sigma \to \inf. \tag{2.7}$$

To the best of our knowledge, this problem has not been considered before in the literature.

We will demonstrate the similarities between the shape calculus of the transient and the steady state case, using the shape identification problems (2.4) and (2.7). Both cost functionals (2.4) and (2.7) can be minimized by means of gradient based iterative methods. To this end, we need to compute the Hadamard representations of the shape gradients. They are obtained by applying the so-called adjoint method. The shape gradients are scalar distributions on the free boundary Γ, involving in general only information of the state and the associated adjoint state.

3 Computing the Shape Gradients

Shape calculus has to be used to derive the shape gradients of the shape optimization problems under considerations. For a general overview on shape calculus, mainly based on the perturbation of identity (Murat and Simon) or the speed method (Sokolowski and Zolesio), we refer the reader for example to [2, 27, 28, 30, 32] and the references therein.

The shape gradient of the cost functional (2.4) with parabolic state equations (2.1)–(2.3) is given in the following theorem which has been proven in [17]. Nevertheless, we present its proof here for the sake of completeness. In particular, a comparison with the proof of Theorem 3.2 reveals clearly the similarities to the derivation of the shape gradient of the associated cost functional (2.7) with elliptic state equation (2.5) and (2.6).

Theorem 3.1. *For an arbitrary boundary perturbation field* $\mathbf{V} \in C^2(\Gamma)$, *the shape gradient to the cost functional* (2.4) *with parabolic state equation* (2.1)–(2.3) *reads as*

$$\delta J(D)[\mathbf{V}] = -\int_0^T \int_\Gamma \langle \mathbf{V}, \mathbf{n}\rangle p \,\mathrm{d}\sigma\,\mathrm{d}t, \tag{3.1}$$

where p *denotes the adjoint state which satisfies the adjoint state equation*

$$\begin{aligned} -\partial_t p - \Delta p &= 0 && \text{in } \Omega \times (0,T),\\ p &= \frac{\partial u}{\partial \mathbf{n}} - h && \text{on } \Sigma \times (0,T),\\ p &= 0 && \text{on } \Omega \times \{T\}. \end{aligned} \tag{3.2}$$

Proof. Given an arbitrary boundary perturbation field $\mathbf{V} \in C^2(\Gamma)$, the directional derivative of the cost functional (2.4) is

$$\delta J(D)[\mathbf{V}] = \int_0^T \int_\Sigma \left(\frac{\partial u}{\partial \mathbf{n}} - h\right) \frac{\partial \delta u}{\partial \mathbf{n}} \,\mathrm{d}\sigma\,\mathrm{d}t$$

with $\delta u = \delta u[\mathbf{V}]$ denoting the local shape derivative. According to [22], it satisfies the following coupled initial boundary value problem

$$\begin{aligned} \partial_t \delta u_e - \Delta \delta u_e &= 0 && \text{in } (\Omega \setminus \overline{D}) \times (0,T),\\ \partial_t \delta u_i - \Delta \delta u_i &= 0 && \text{in } D \times (0,T),\\ \delta u_e &= 0 && \text{on } \Sigma \times (0,T),\\ \delta u_e = \delta u_i,\ \frac{\partial \delta u_e}{\partial \mathbf{n}} &= \frac{\partial \delta u_i}{\partial \mathbf{n}} + \langle \mathbf{V}, \mathbf{n}\rangle && \text{on } \Gamma \times (0,T),\\ \delta u_e &= 0 && \text{on } (\Omega \setminus \overline{D}) \times \{0\},\\ \delta u_i &= 0 && \text{on } D \times \{0\}. \end{aligned} \tag{3.3}$$

Observing (3.2) and (3.3), integration by parts leads to

$$\begin{aligned} 0 &= \int_0^T \int_{\Omega\setminus\overline{D}} (\partial_t \delta u_e - \Delta \delta u_e) p + (\partial_t p + \Delta p) \delta u_e \,\mathrm{d}\mathbf{x}\,\mathrm{d}t\\ &= \int_{\Omega\setminus\overline{D}} \int_0^T \{\partial_t \delta u_e p + \delta u_e \partial_t p\} \,\mathrm{d}t\,\mathrm{d}\mathbf{x} - \int_0^T \int_{\Omega\setminus\overline{D}} \{\Delta \delta u_e p - \delta u_e \Delta p\} \,\mathrm{d}\mathbf{x}\,\mathrm{d}t \end{aligned}$$

$$= \int_{\Omega\setminus\overline{D}} \underbrace{\{ \delta u_e(\cdot,T)p(\cdot,T) - \delta u_e(\cdot,0)p(\cdot,0) \}}_{=0} \,\mathrm{d}\mathbf{x}$$

$$+ \int_0^T \int_{\Sigma\cup\Gamma} \left\{ \delta u_e \frac{\partial p}{\partial \mathbf{n}} - \frac{\partial \delta u_e}{\partial \mathbf{n}} p \right\} \mathrm{d}\sigma\,\mathrm{d}t.$$

In view of $\delta u_e = 0$ on $\Sigma \times (0, T)$, this implies

$$\delta J(D)[\mathbf{V}] = \int_0^T \int_\Sigma \frac{\partial \delta u_e}{\partial \mathbf{n}} p \,\mathrm{d}\sigma\,\mathrm{d}t = \int_0^T \int_\Gamma \left\{ \frac{\partial \delta u_e}{\partial \mathbf{n}} p - \delta u_e \frac{\partial p}{\partial \mathbf{n}} \right\} \mathrm{d}\sigma\,\mathrm{d}t. \tag{3.4}$$

In complete analogy to above, we find again by integration by parts

$$0 = \int_0^T \int_D (\partial_t \delta u_i - \Delta \delta u_i) p + (\partial_t p + \Delta p)\delta u_i \,\mathrm{d}\mathbf{x}\,\mathrm{d}t$$
$$= \int_0^T \int_\Gamma \left\{ \delta u_i \frac{\partial p}{\partial \mathbf{n}} - \frac{\partial \delta u_i}{\partial \mathbf{n}} p \right\} \mathrm{d}\sigma\,\mathrm{d}t.$$

Due to the jump condition of δu at Γ, we thus conclude

$$0 = \int_0^T \int_\Gamma \left\{ \delta u_e \frac{\partial p}{\partial \mathbf{n}} + \left(\langle \mathbf{V}, \mathbf{n} \rangle - \frac{\partial \delta u_e}{\partial \mathbf{n}} \right) p \right\} \mathrm{d}\sigma\,\mathrm{d}t,$$

cf. (3.3). Inserting this equation into (3.4) yields finally (3.1). □

In case of the shape optimization problem (2.7) with elliptic state equation (2.5) and (2.6), we obtain the following shape gradient. Here, the underlying operator of the elliptic state equation is self adjoint. Thus, the adjoint state equation involves the same operator as the primal state equation.

Theorem 3.2. *For an arbitrary boundary perturbation field $\mathbf{V} \in C^2(\Gamma)$, the shape gradient to the cost functional* (2.7) *with elliptic state equation* (2.5) *and* (2.6) *reads as*

$$\delta J(D)[\mathbf{V}] = - \int_\Gamma \langle \mathbf{V}, \mathbf{n} \rangle p \,\mathrm{d}\sigma, \tag{3.5}$$

where p denotes the adjoint state which satisfies the adjoint state equation

$$\Delta p = 0 \text{ in } \Omega, \quad p = \frac{\partial u}{\partial \mathbf{n}} - h \text{ on } \Sigma. \tag{3.6}$$

Proof. The proof uses the same arguments as the proof of Theorem 3.1. For an arbitrary boundary perturbation field $\mathbf{V} \in C^2(\Gamma)$, we find the directional derivative

$$\delta J(D)[\mathbf{V}] = \int_{\Sigma} \left(\frac{\partial u}{\partial \mathbf{n}} - h \right) \frac{\partial \delta u}{\partial \mathbf{n}} \, \mathrm{d}\sigma$$

with $\delta u = \delta u[\mathbf{V}]$ satisfying the coupled boundary value problem (cf. [21])

$$\begin{aligned} \Delta \delta u_e &= 0 && \text{in } \Omega \setminus \overline{D}, \\ \Delta \delta u_i &= 0 && \text{in } D \\ \delta u_e &= 0 && \text{on } \Sigma, \\ \delta u_e = \delta u_i, \quad \frac{\partial \delta u_e}{\partial \mathbf{n}} &= \frac{\partial \delta u_i}{\partial \mathbf{n}} + \langle \mathbf{V}, \mathbf{n} \rangle && \text{on } \Gamma \end{aligned} \tag{3.7}$$

Integration by parts gives in view of (3.6) and (3.7)

$$0 = \int_{\Omega \setminus \overline{D}} \Delta p \delta u_e - \Delta \delta u_e p \, \mathrm{d}\mathbf{x} = \int_{\Sigma \cup \Gamma} \left\{ \delta u_e \frac{\partial p}{\partial \mathbf{n}} - \frac{\partial \delta u_e}{\partial \mathbf{n}} p \right\} \mathrm{d}\sigma$$

and, since $\delta u_e = 0$ on Σ, thus

$$\delta J(D)[\mathbf{V}] = \int_{\Sigma} \frac{\partial \delta u_e}{\partial \mathbf{n}} p \, \mathrm{d}\sigma = \int_{\Gamma} \left\{ \frac{\partial \delta u_e}{\partial \mathbf{n}} p - \delta u_e \frac{\partial p}{\partial \mathbf{n}} \right\} \mathrm{d}\sigma. \tag{3.8}$$

Using next integration by parts on the domain D, we likewise conclude

$$0 = \int_{D} \Delta p \delta u_i - \Delta \delta u_i p \, \mathrm{d}\mathbf{x} = \int_{\Gamma} \left\{ \delta u_i \frac{\partial p}{\partial \mathbf{n}} - \frac{\partial \delta u_i}{\partial \mathbf{n}} p \right\} \mathrm{d}\sigma.$$

The jump condition of δu at Γ (cf. (3.7)) implies

$$0 = \int_{\Gamma} \left\{ \delta u_e \frac{\partial p}{\partial \mathbf{n}} + \left(\langle \mathbf{V}, \mathbf{n} \rangle - \frac{\partial \delta u_e}{\partial \mathbf{n}} \right) p \right\} \mathrm{d}\sigma,$$

which, together with (3.8), shows finally (3.5). □

With the help of the Hadamard representations (3.1) and (3.5) of the shape gradients, we are able to develop efficient gradient based algorithms for the minimization of the cost functionals (2.4) and (2.7), respectively.

4 Discretization of the Free Boundary

In order to solve the shape optimization problems under consideration we seek a stationary point $D^{\star}$, being C^2-smooth, which satisfies

$$\delta J(D^{\star})[\mathbf{V}] = 0 \text{ for all } \mathbf{V} \in C^2(\Gamma). \tag{4.1}$$

This is called the necessary optimality condition of the shape optimization problem $J(D) \to \inf$. For related sufficient optimality conditions, we refer the reader to [9, 13] and the references therein. Nevertheless, we emphasize that, in the current context of severely ill-posed problems, sufficient optimality conditions cannot hold since the adjoint state vanishes in the optimal domain $D^\star$.

4.1 Nonlinear Ritz-Galerkin Approximation for the Shape

From now on we restrict ourselves to the practically most important case of $n = 3$ and consider the minimization of the cost functional over heat sources that are topologically equivalent to the unit sphere $\mathbb{S}^2$. Then, we can represent the heat source $D \subset \mathbb{R}^3$ by a parameterization $\boldsymbol{\gamma} = (\gamma_1, \gamma_2, \gamma_3) : \mathbb{S}^2 \to \Gamma$, which is one-to-one, preserves orientation, and the Jacobian matrix $\boldsymbol{\gamma}'(\hat{\mathbf{x}})$ is invertible for all $\hat{\mathbf{x}} \in \hat{\Gamma}$. By restricting the parameterization to a finite dimensional ansatz space V_N, we arrive at the nonlinear Ritz-Galerkin scheme for (4.1):

$$\text{Seek } \boldsymbol{\gamma}_N^\star \in V_N \text{ such that } \delta J(\boldsymbol{\gamma}_N^\star)[\mathbf{V}_N] = 0 \text{ for all } \mathbf{V}_N \in V_N. \tag{4.2}$$

For the numerical solution of the nonlinear variational problem (4.2), we apply the quasi-Newton method, updated by the inverse BFGS-rule without damping. A second order approximation is used for performing the line search update if the descent does not satisfy the Armijo rule. Since we use a gradient based iterative method, regularization is not necessary provided that we stop the iteration early enough. For all the details and a survey on available optimization algorithms, we refer to [3, 10–12] and the references therein.

Following [17, 18], we can distinguish two types of parameterizations. The first type is of the form

$$\boldsymbol{\gamma}(\hat{\mathbf{x}}) = r(\hat{\mathbf{x}}) \cdot \hat{\mathbf{x}}, \quad r \in C^2(\mathbb{S}^2) \tag{4.3}$$

and is able to represent any given star-shaped source with center in $\mathbf{0}$. The discretization of Γ is based on the ansatz

$$r_N(\hat{\mathbf{x}}) = \sum_{n=0}^{N} \sum_{m=-n}^{n} a_n^m Y_n^m(\hat{\mathbf{x}}), \quad \hat{\mathbf{x}} \in \mathbb{S}^2,$$

where $a_n^m \in \mathbb{R}$ are the design parameters and $Y_n^m \in C^\infty(\mathbb{S}^2)$ denote the spherical harmonic functions of degree n and order m. This leads to the finite dimensional parameterization

$$\boldsymbol{\gamma}_N(\hat{\mathbf{x}}) = r_N(\hat{\mathbf{x}}) \cdot \hat{\mathbf{x}}, \quad \hat{\mathbf{x}} \in \mathbb{S}^2. \tag{4.4}$$

The advantage of this approach is that the identification of the function r_N, given by the design parameters, and the heat source is one-to-one. In particular, the distance between two domains can be simply measured via the ℓ^2-norm of the difference of the associated design parameters. This approach is used in our numerical example.

The second type, also referred to as flexible shape representation, allows a more general boundary representation than the somehow restrictive approach (4.3). Namely, we choose

$$\gamma_N(\hat{\mathbf{x}}) = \sum_{n=0}^{N} \sum_{m=-n}^{n} \mathbf{a}_n^m Y_n^m(\hat{\mathbf{x}}), \quad \hat{\mathbf{x}} \in \mathbb{S}^2, \tag{4.5}$$

where $\mathbf{a}_i \in \mathbb{R}^3$ are *vector valued* design parameters. The ansatz (4.5) does not impose any restriction to the topology of the domain except for its genus. However, we lose the one-to-one correspondence between the shape of the heat source and the design parameters. Thus, a regularization of the shape function (see e.g. [13, 15, 17]) or a suitable remeshing algorithm (see e.g. [18]) needs to be applied.

4.2 Surface Mesh Generation

We shall assume that the boundary manifold $\Gamma \subset \mathbb{R}^3$ is given as a parametric surface consisting of smooth patches. More precisely, let $\square := [0, 1]^2$ denote the unit square. The manifold Γ is partitioned into a finite number of *patches*

$$\Gamma = \bigcup_{i=1}^{M} \Gamma_i, \qquad \Gamma_i = \boldsymbol{\kappa}_i(\square), \qquad i = 1, 2, \ldots, M, \tag{4.6}$$

where each $\boldsymbol{\kappa}_i : \square \to \Gamma_i$ defines a diffeomorphism of $\square$ onto Γ_i. The intersection $\Gamma_i \cap \Gamma_{i'}$, $i \neq i'$, of two patches Γ_i and $\Gamma_{i'}$ is assumed to be either $\emptyset$, or a common edge, or a common vertex. A mesh of the boundary Γ is then obtained by mapping a mesh of $\square$ to Γ via a parametrization.

The construction of the parametric representation of the moving boundary Γ should be presented in more detail. The surface of the cube $[-0.5, 0.5]^3$ consists of six patches. Each point $\mathbf{x} \in \partial([-0.5, 0.5]^3)$ can be lifted onto the boundary Γ via the operation

$$\mathbf{y}(\mathbf{x}) = \gamma\left(\frac{\mathbf{x}}{\|\mathbf{x}\|}\right) \in \Gamma. \tag{4.7}$$

In this manner, the boundary Γ is subdivided into $M = 6$ patches. The parametric representations $\boldsymbol{\kappa}_i : \square \to \Gamma_i$ can be derived easily from (4.7). Finally, we construct a mesh of Γ, required for the boundary element method, by mapping a triangular

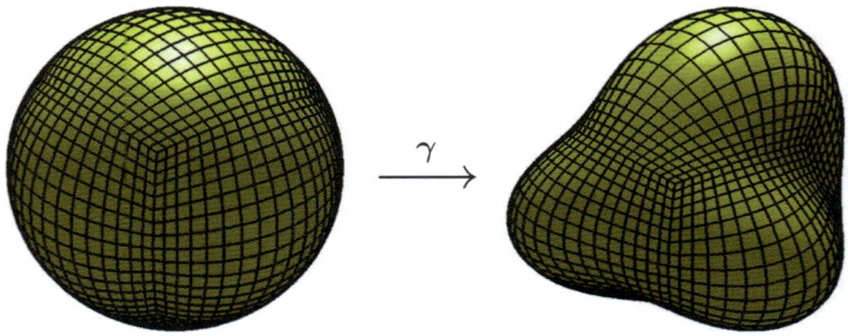

Fig. 2 Parametric representation of Γ with triangular mesh on level 4

or quadrangular mesh of the unit cube via the sphere to Γ. We refer to Fig. 2 for an illustration of the proposed parametric representation and mesh generation.

We shall finally specify how to distinguish between "nice" and "bad" parametrizations. A "nice" parametrization maps orthonormal tangents of the unit cube onto orthogonal tangents of length $\approx |\Gamma|/6$ with respect to the boundary Γ. This means that the first fundamental tensor of differential geometry, given by

$$\mathbf{S}_i(\mathbf{s}) = \left[\langle \boldsymbol{\kappa}_{i,j}(\mathbf{s}), \boldsymbol{\kappa}_{i,k}(\mathbf{s})\rangle\right]_{j,k=1,2}, \quad \mathbf{s} = [s_1, s_2]^T \in \square,$$

satisfies $\mathbf{S}_i \approx |\Gamma|/6 \cdot \mathbf{I}$. Hence, one can employ the shape functional

$$M(\gamma) = \sum_{i=1}^{6} \int_{\square} \left\| \begin{bmatrix} \langle \boldsymbol{\kappa}_{i,1}(\mathbf{s}), \boldsymbol{\kappa}_{i,1}(\mathbf{s})\rangle - \frac{|\Gamma|}{6} & \langle \boldsymbol{\kappa}_{i,1}(\mathbf{s}), \boldsymbol{\kappa}_{i,2}(\mathbf{s})\rangle \\ \langle \boldsymbol{\kappa}_{i,2}(\mathbf{s}), \boldsymbol{\kappa}_{i,1}(\mathbf{s})\rangle & \langle \boldsymbol{\kappa}_{i,2}(\mathbf{s}), \boldsymbol{\kappa}_{i,2}(\mathbf{s})\rangle - \frac{|\Gamma|}{6} \end{bmatrix} \right\|_F^2 \mathrm{d}\mathbf{s}$$

for regularizing the shape functional or as the base of a remeshing procedure.

5 Numerical Method to Compute the State and Its Adjoint

We shall discuss the numerical solution of the state equations and their adjoints by boundary element methods. With this technique only the boundaries of Ω an D need to be discretized, which avoids the complicated triangulation of the domain Ω with the varying source D. In particular, in case of the parabolic state equation, we immediately arrive at a space-time formulation. This is very advantageous since the solution's complete temporal history which enters the adjoint state, being reverse in time, is available.

5.1 Solving the Heat Equation

The thermal layer operators are given by

$$
\begin{aligned}
(\mathcal{V}g)(\mathbf{x},t) &= \int_0^t \int_\Sigma G(\|\mathbf{x}-\mathbf{y}\|, t-\tau) g(\mathbf{y},\tau)\, \mathrm{d}\sigma_{\mathbf{y}}\, \mathrm{d}\tau,\\
(\mathcal{K}g)(\mathbf{x},t) &= \int_0^t \int_\Sigma \frac{\partial G}{\partial \mathbf{n_y}}(\|\mathbf{x}-\mathbf{y}\|, t-\tau) g(\mathbf{y},\tau)\, \mathrm{d}\sigma_{\mathbf{y}}\, \mathrm{d}\tau,\\
(\mathcal{K}^\star g)(\mathbf{x},t) &= \int_0^t \int_\Sigma \frac{\partial G}{\partial \mathbf{n_x}}(\|\mathbf{x}-\mathbf{y}\|, t-\tau) g(\mathbf{y},\tau)\, \mathrm{d}\sigma_{\mathbf{y}}\, \mathrm{d}\tau,\\
(\mathcal{W}g)(\mathbf{x},t) &= -\frac{\partial}{\partial \mathbf{n_x}} \int_0^t \int_\Sigma \frac{\partial G}{\partial \mathbf{n_y}}(\|\mathbf{x}-\mathbf{y}\|, t-\tau) g(\mathbf{y},\tau)\, \mathrm{d}\sigma_{\mathbf{y}}\, \mathrm{d}\tau,
\end{aligned}
\tag{5.1}
$$

where $(\mathbf{x},t) \in \Sigma \times [0,T]$ and $G(\cdot,\cdot)$ is the heat kernel, given by

$$
G(r,t) = \frac{1}{(4\pi t)^{3/2}} \exp\left(-\frac{r^2}{4t}\right).
$$

With these boundary integral operators at hand, Green's representation formulae for the interior heat equation with homogeneous initial conditions can be written as

$$
\left(\frac{1}{2} + \mathcal{K}\right) u - \mathcal{V}\frac{\partial u}{\partial \mathbf{n}} = \mathcal{N} \quad \text{and} \quad \left(\frac{1}{2} + \mathcal{K}^\star\right)\frac{\partial u}{\partial \mathbf{n}} + \mathcal{W}u = -\frac{\partial \mathcal{N}}{\partial \mathbf{n}}, \tag{5.2}
$$

where $\mathcal{N}$ denotes the thermal Newton potential of the inhomogeneity. It is nonzero and, in accordance with [17], given by

$$
\mathcal{N}(\mathbf{x},t) := \int_0^t \int_D G(\|\mathbf{x}-\mathbf{y}\|, t-\tau)\, \mathrm{d}\mathbf{y}\, \mathrm{d}\tau = \int_\Gamma \frac{\partial H}{\partial \mathbf{n_y}}(\|\mathbf{x}-\mathbf{y}\|, t)\, \mathrm{d}\sigma_{\mathbf{y}},
$$

where the kernel $H(\cdot,\cdot)$ is defined as

$$
H(r,t) = \frac{2\sqrt{t}}{(4\pi)^{3/2}} \left[\sqrt{\pi}\, \mathrm{erfc}\left(\frac{r}{2\sqrt{t}}\right)\left(\frac{r}{2\sqrt{t}} + \frac{\sqrt{t}}{r}\right) + \exp\left(-\frac{r^2}{4t}\right)\right]. \tag{5.3}
$$

Since the temperature satisfies $u = 0$ at Σ, the unknown heat flux $\partial u/\partial \mathbf{n}$ at Σ can be derived from the boundary integral equations (5.2). Thus

$$
-\mathcal{V}\frac{\partial u}{\partial \mathbf{n}} = \mathcal{N} \quad \text{and} \quad \left(\frac{1}{2} + \mathcal{K}^\star\right)\frac{\partial u}{\partial \mathbf{n}} = -\frac{\partial \mathcal{N}}{\partial \mathbf{n}}. \tag{5.4}
$$

We will employ the second boundary integral equation for our shape reconstruction scheme. Nevertheless, the first boundary integral equation will be used to compute synthetic data in order to avoid an inverse crime.

For the computation of the solution of the state adjoint equation, we first perform the change of variables $t \mapsto T - t$ to obtain it in a more familiar form:

$$\begin{aligned} \partial_t \tilde{p} - \Delta \tilde{p} &= 0 && \text{in } \Omega \times (0, T), \\ \tilde{p} &= f && \text{on } \Sigma \times (0, T), \\ \tilde{p} &= 0 && \text{on } \Omega \times \{0\}. \end{aligned} \tag{5.5}$$

Here, $\tilde{p}(\mathbf{x}, t) = p(\mathbf{x}, T - t)$ and

$$f(\mathbf{x}, t) = \frac{\partial u}{\partial \mathbf{n}}(\mathbf{x}, T - t) - h(\mathbf{x}, T - t).$$

It is convenient to use the indirect method where the solution is written as a double layer potential

$$\tilde{p}(\mathbf{x}, t) = \int_0^t \int_\Sigma \frac{\partial G}{\partial \mathbf{n_y}}(\|\mathbf{x} - \mathbf{y}\|, t - \tau) g(\mathbf{y}, \tau) \, \mathrm{d}\sigma_\mathbf{y} \, \mathrm{d}\tau, \quad \mathbf{x} \in \Omega, \tag{5.6}$$

where g is an unknown density on Σ. By letting $\mathbf{x}$ approach the boundary surface from the inside of Ω and using the usual jump conditions, we arrive at

$$\left(-\frac{1}{2} + \mathcal{K}\right) g = f. \tag{5.7}$$

Once g has been determined, the double layer potential (5.6) must be evaluated on Γ to obtain the quantity needed in the evaluation of the shape gradient in (3.1).

The approximate solution of the boundary integral equations (5.4) and (5.7) by traditional discretization schemes poses serious difficulties since the total number of unknowns is the product of the number of spatial unknowns N_s and temporal unknowns N_t which becomes extremely large. Therefore, we proposed in [17, 18] the application of the multipole-based space-time boundary element method which has been developed in [33,34]. We then arrive at an algorithm which computes both, the state and its adjoint, in a complexity that scales essentially linearly with the total number of unknowns $N_s N_t$. We refer the reader to [17] for further details concerning the particular realization.

5.2 Solving the Poisson Equation

In case of the Laplacian, the fundamental solution is $G(r) = 1/(4\pi r)$. Hence, the standard boundary integral operators (cf. (5.1)) become

$$\left.\begin{aligned}
(\mathcal{V}g)(\mathbf{x}) &= \int_\Sigma G(\|\mathbf{x}-\mathbf{y}\|)g(\mathbf{y})\,\mathrm{d}\sigma_\mathbf{y}\\
(\mathcal{K}g)(\mathbf{x}) &= \int_\Sigma \frac{\partial G}{\partial \mathbf{n}_\mathbf{y}}(\|\mathbf{x}-\mathbf{y}\|)g(\mathbf{y})\,\mathrm{d}\sigma_\mathbf{y}\\
(\mathcal{K}^\star g)(\mathbf{x}) &= \int_\Sigma \frac{\partial G}{\partial \mathbf{n}_\mathbf{x}}(\|\mathbf{x}-\mathbf{y}\|)g(\mathbf{y})\,\mathrm{d}\sigma_\mathbf{y}\\
(\mathcal{W}g)(\mathbf{x}) &= -\frac{\partial}{\partial \mathbf{n}_\mathbf{x}}\int_\Sigma \frac{\partial G}{\partial \mathbf{n}_\mathbf{y}}(\|\mathbf{x}-\mathbf{y}\|)g(\mathbf{y})\,\mathrm{d}\sigma_\mathbf{y}
\end{aligned}\right\}\quad \mathbf{x}\in\Sigma.$$

The Dirichlet and Neumann data at Σ are again coupled by the boundary integral equations (5.2), which, in view of the homogeneous boundary conditions, results in the boundary integral equations (5.4). The Newton potential involved there can be computed as follows:

Lemma 5.1. *For $\mathbf{x} \in \mathbb{R}^3 \setminus \overline{D}$, the Newton potential admits the representation*

$$\mathcal{N}(\mathbf{x}) := \int_D G(\|\mathbf{x}-\mathbf{y}\|)\,\mathrm{d}\mathbf{y} = -\frac{1}{8\pi}\int_\Gamma \frac{\langle \mathbf{x}-\mathbf{y}, \mathbf{n}_\mathbf{y}\rangle}{\|\mathbf{x}-\mathbf{y}\|}\,\mathrm{d}\sigma_\mathbf{y}.$$

Proof. We shall write the Laplace kernel as the divergence of a radially symmetric vector field. That is, we find a scalar function $F(\cdot)$ such that

$$\operatorname{div}_\mathbf{y}\Big[F(\|\mathbf{x}-\mathbf{y}\|)(\mathbf{x}-\mathbf{y})\Big] = G(\|\mathbf{x}-\mathbf{y}|\,).$$

Simple differentiation shows that $F(\cdot)$ satisfies the differential equation in r

$$rF'(r) + 3F(r) = -G(r), \qquad r > 0.$$

A particular solution of this ordinary differential equation is $F(r) = -1/(8\pi r)$. Thus, by construction, the Gauss theorem implies the assertion:

$$\int_D G(\|\mathbf{x}-\mathbf{y}\|)\,\mathrm{d}\mathbf{x} = \int_\Gamma F(\|\mathbf{x}-\mathbf{y}\|)\langle \mathbf{x}-\mathbf{y}, \mathbf{n}_\mathbf{y}\rangle\,\mathrm{d}\sigma_\mathbf{y}.$$

□

We will employ the first boundary integral equation in (5.4) for the shape reconstruction scheme, since it is more accurate when applying a Galerkin method. Moreover, for the adjoint state, it is then more efficient to use the indirect method for the single layer potential, i.e.,

$$p(\mathbf{x}) = \int_{\Sigma} G(\|\mathbf{x} - \mathbf{y}\|)g(\mathbf{y})\, \mathrm{d}\sigma_{\mathbf{y}}, \quad \mathbf{x} \in \Omega. \tag{5.8}$$

Here, the density g is the solution of the boundary integral equation $\mathcal{V}g = f$ with the right hand $f(\mathbf{x}) := (\partial u/\partial \mathbf{n})(\mathbf{x}) - h(\mathbf{x})$.

As proposed in several earlier papers on shape optimization with elliptic state equation, see e.g. [5–8], the present boundary integral equations can be solved efficiently by the wavelet Galerkin method which has been developed in [1, 16]. Then, the computational complexity scales linearly in the number of boundary elements.

6 Numerical Results

We shall illustrate our algorithms by some numerical experiments. To that end, we choose the unit ball as computational domain Ω. The given heat source D is acorn-shaped as shown in Fig. 3. Since it is star-shaped, we employ the ansatz (4.4) with $N = 10$, that are 100 design parameters.

We apply first the reconstruction algorithm for the time interval $[0, T]$ with $T = 0.1$ and $T = 1.0$ and a noise level of 1 %. It turns out that the reconstruction for the short time interval (see Fig. 4) is quite similar but somewhat worse than for the long time interval (see Fig. 5). This has nevertheless already been observed in [17].

The reconstruction for the stationary situation is seen in Fig. 6. Its quality is clearly inferior to the time-dependent problem. Moreover, we have observed that the reconstruction is much more robust with respect to noise if the time dependent heat flux is used in the tracking functional rather than the stationary heat flux.

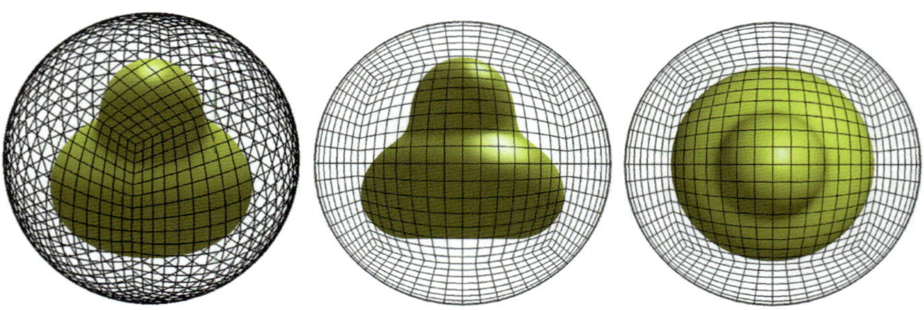

Fig. 3 The domain Ω with boundary Σ and the acorn-shaped source D with boundary Γ

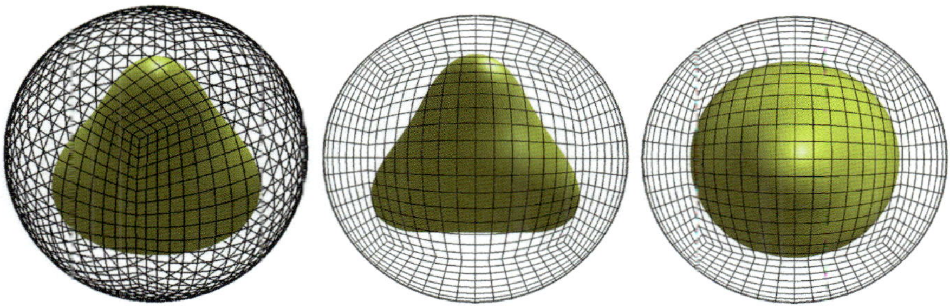

Fig. 4 The reconstruction of the heat source D in case of the heat equation and $T = 0.1$

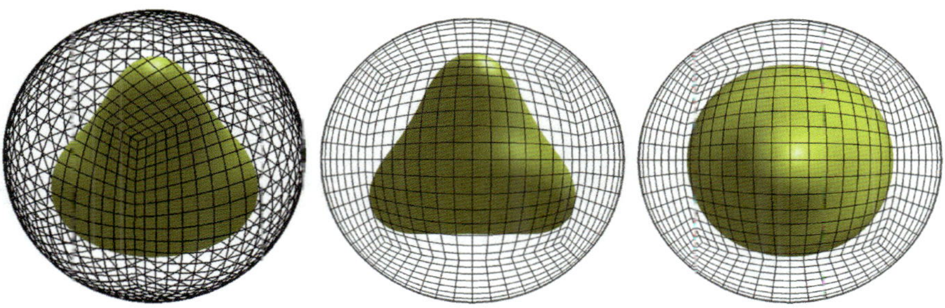

Fig. 5 The reconstruction of the heat source D in case of the heat equation and $T = 1.0$

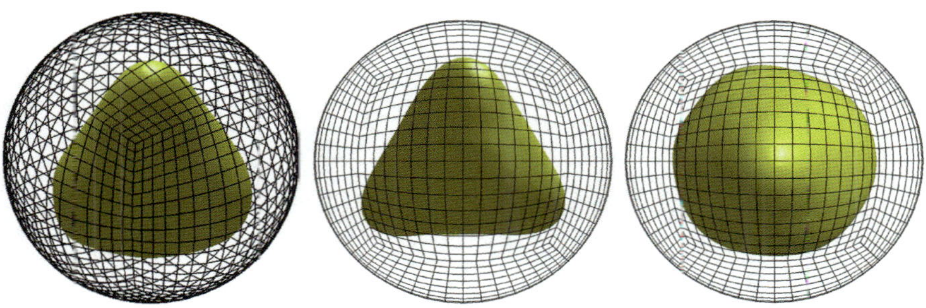

Fig. 6 The reconstruction of the heat source D in case of the stationary problem

References

1. W. Dahmen, H. Harbrecht, R. Schneider, Compression techniques for boundary integral equations. Optimal complexity estimates. SIAM J. Numer. Anal. **43**, 2251–2271 (2006)
2. M. Delfour, J.-P. Zolesio, *Shapes and Geometries* (SIAM, Philadelphia, 2001)
3. J.E. Dennis, R.B. Schnabel, *Numerical Methods for Nonlinear Equations and Unconstrained Optimization Techniques* (Prentice-Hall, Englewood Cliffs, 1983)
4. S. El Yacoubi, J. Sokolowski, Domain optimization problems for parabolic control systems. Appl. Math. Comput. Sci. **6**, 277–289 (1996)

5. K. Eppler, H. Harbrecht, Numerical solution of elliptic shape optimization problems using wavelet-based BEM. Optim. Methods Softw. **18**, 105–123 (2003)
6. K. Eppler, H. Harbrecht, A regularized Newton method in electrical impedance tomography using shape Hessian information. Control Cybern. **34**, 203–225 (2005)
7. K. Eppler, H. Harbrecht, Efficient treatment of stationary free boundary problems. Appl. Numer. Math. **56**, 1326–1339 (2006)
8. K. Eppler, H. Harbrecht, Wavelet based boundary element methods in exterior electromagnetic shaping. Eng. Anal. Bound. Elem. **32**, 645–657 (2008)
9. K. Eppler, H. Harbrecht, R. Schneider, On convergence in elliptic shape optimization. SIAM J. Control Optim. **45**, 61–83 (2007)
10. A.V. Fiacco, G.P. McCormick, *Nonlinear Programming. Sequential Unconstrained Minimization Techniques* (Wiley, New York, 1968)
11. R. Fletcher, *Practical Methods for Optimization I&II* (Wiley, New York, 1980)
12. C. Grossmann, J. Terno, *Numerik der Optimierung* (B.G. Teubner, Stuttgart, 1993)
13. H. Harbrecht, Analytical and numerical methods in shape optimization. Math. Methods Appl. Sci. **31**, 2095–2114 (2008)
14. H. Harbrecht, A Newton method for Bernoulli's free boundary problem in three dimensions. Computing **82**, 11–30 (2008)
15. H. Harbrecht, T. Hohage, Fast methods for three-dimensional inverse obstacle scattering. J. Integral Equ. Appl. **19**, 237–260 (2007)
16. H. Harbrecht, R. Schneider, Wavelet Galerkin schemes for boundary integral equations. Implementation and quadrature. SIAM J. Sci. Comput. **27**, 1347–1370 (2006)
17. H. Harbrecht, J. Tausch, An efficient numerical method for a shape identification problem arising from the heat equation. Inverse Probl. **27**, 065013 (2011)
18. H. Harbrecht, J. Tausch, On the numerical solution of a shape optimization problem for the heat equation. SIAM J. Sci. Comput. **35**, A104–A121 (2013)
19. J. Haslinger, P. Neitaanmäki, *Finite Element Approximation for Optimal Shape, Material and Topology Design*, 2nd edn. (Wiley, Chichester, 1996)
20. A. Henrot, J. Sokolowski, A shape optimization problem for the heat equation, in *Optimal Control* (Gainesville, 1997). Applied Optimization, vol. 15 (Kluwer Academic, Dordrecht, 1998), pp. 204–223
21. F. Hettlich, W. Rundell, The determination of a discontinuity in a conductivity from a single boundary measurement. Inverse Probl. **14**, 67–82 (1998)
22. F. Hettlich, W. Rundell, Identification of a discontinuous source in the heat equation. Inverse Probl. **17**, 1465–1482 (2001)
23. K.-H. Hoffmann, J. Sokolowski, Interface optimization problems for parabolic equations. Shape design and optimization. Control Cybern. **23**, 445–451 (1994)
24. V. Isakov, *Inverse Source Problems*. AMS Mathematical Surveys and Monographs, vol. 34 (American Mathematical Society, Providence, 1990)
25. K. Ito, K. Kunisch, G. Peichl, Variational approach to shape derivatives for a class of Bernoulli problems. J. Math. Anal. Appl. **314**, 126–149 (2006)
26. A.M. Khludnev, J. Sokołowski, *Modeling and Control in Solid Mechanics* (Birkhäuser, Basel, 1997)
27. F. Murat, J. Simon, Étude de problèmes d'optimal design, in *Optimization Techniques, Modeling and Optimization in the Service of Man*, ed. by J. Céa. Lecture Notes in Computer Science, vol. 41 (Springer, Berlin, 1976), pp. 54–62
28. O. Pironneau, *Optimal Shape Design for Elliptic Systems* (Springer, New York, 1983)
29. J.-R. Roche, J. Sokolowski, Numerical methods for shape identification problems. Control Cybern. **25**, 867–894 (1996)
30. J. Simon, Differentiation with respect to the domain in boundary value problems. Numer. Funct. Anal. Optim. **2**, 649–687 (1980)
31. J. Sokolowski, Shape sensitivity analysis of boundary optimal control problems for parabolic systems. SIAM J. Control Optim. **26**, 763–787 (1988)

32. J. Sokolowski, J.-P. Zolesio, *Introduction to Shape Optimization* (Springer, Berlin, 1992)
33. J. Tausch, A fast method for solving the heat equation by layer potentials. J. Comput. Phys. **224**, 956–969 (2007)
34. J. Tausch, Nyström discretization of parabolic boundary integral equations. Appl. Numer. Math. **59**, 2843–2856 (2009)
35. T. Tiihonen, Shape optimization and trial methods for free-boundary problems. RAIRO Model. Math. Anal. Numér. **31**, 805–825 (1997)

Multi-material Phase Field Approach to Structural Topology Optimization

Luise Blank, M. Hassan Farshbaf-Shaker, Harald Garcke, Christoph Rupprecht, and Vanessa Styles

Abstract Multi-material structural topology and shape optimization problems are formulated within a phase field approach. First-order conditions are stated and the relation of the necessary conditions to classical shape derivatives are discussed. An efficient numerical method based on an H^1–gradient projection method is introduced and finally several numerical results demonstrate the applicability of the approach.

Keywords Shape and topology optimization • Phase field approach • Shape sensitivity analysis • Gradient projection method

Mathematics Subject Classification (2010). Primary 49Q10; Secondary 74P05, 74P15, 90C52, 65K15.

1 Introduction

The efficient use of material and related to that the optimization of shapes and topology is of high importance for the performance of structures. Many different methods have been introduced to solve shape and topology optimization problems and we refer to Bendsoe, Sigmund [2], Sokolowski, Zolesio [14] and Allaire, Jouve, Toader [1] for details. In this paper we analyze a multi-phase field approach for

L. Blank • H. Garcke (✉) • C. Rupprecht
Fakultät für Mathematik, Universität Regensburg, 93040 Regensburg, Germany
e-mail: luise.blank@mathematik.uni-r.de; harald.garcke@mathematik.uni-r.de; christoph.rupprecht@mathematik.uni-r.de

M.H. Farshbaf-Shaker
Weierstraß-Institut, Mohrenstr. 39, 10117 Berlin, Germany
e-mail: MohammadHassanFarshbaf.Shaker@wias-berlin.de

V. Styles
Department of Mathematics, University of Sussex, Brighton BN1 9QH, UK
e-mail: v.styles@sussex.ac.uk

G. Leugering et al. (eds.), *Trends in PDE Constrained Optimization*, International Series of Numerical Mathematics 165, DOI 10.1007/978-3-319-05083-6_15

shape and topology optimization problems. This approach is related to perimeter penalizing methods. However, instead of the perimeter the Ginzburg-Landau energy

$$E^{\varepsilon}(\boldsymbol{\varphi}) := \int_{\Omega} \left(\tfrac{\varepsilon}{2}|\nabla\boldsymbol{\varphi}|^2 + \tfrac{1}{\varepsilon}\Psi(\boldsymbol{\varphi})\right), \qquad \varepsilon > 0, \tag{1.1}$$

is added to the objective functional. In (1.1) the set Ω is a given design domain, the function $\boldsymbol{\varphi}$ which takes values in $\mathbb{R}^N$ is a phase field vector, Ψ is a potential function with absolute minima which describe the different materials and the void and $\varepsilon > 0$ is a small parameter related to the interface thickness. It can be shown that (1.1) converges in the sense of Γ–limits to the perimeter functional, see Modica [12]. The phase field method has been introduced in topology optimization by Bourdin and Chambolle [7] and was subsequently used by Burger, Stainko [8], Wang, Zhou [16], Takezawa, Nishiwaki, Kitamura [15], Dedé, Borden, Hughes [9], Blank et al [3, 4] and Penzler, Rumpf, Wirth [13]. However, so far a rigorous derivation of first order conditions and an analysis of these conditions in the sharp interface limit $\varepsilon \to 0$ was missing. In this paper we not only discuss recent progress in this direction but also introduce and analyze a new efficient method to solve the constrained minimization problem.

Although in principle the phase field approach can as well be used for other shape and topology optimization problems we restrict ourselves to situations where we seek a domain Ω^M and a displacement $\boldsymbol{u}$ such that

$$\int_{\Omega^M} \boldsymbol{f}\cdot\boldsymbol{u} + \int_{\partial\Omega^M} \boldsymbol{g}\cdot\boldsymbol{u} \tag{1.2}$$

or an L^2–error to a target displacement

$$\left(\int_{\Omega^M} c|\boldsymbol{u} - \boldsymbol{u}_{\Omega}|^2\right)^{\frac{1}{2}} \tag{1.3}$$

is minimized subject to the equations of linear elasticity. Here $\boldsymbol{f}$ and $\boldsymbol{g}$ are volume and surface forces and $c \geq 0$ is a given weight function on Ω. The optimization problem (1.2) is a mean compliance minimization problem and (1.3) is an example of a compliant mechanism problem, see [1, 2] for details. In this contribution we will be brief and refer to [3] and to the forthcoming article [6] for details.

2 Setting of the Problem

In this section we introduce how structural topology optimization problems can be formulated within the phase field approach.

The goal in multi-material shape and topology optimization is to partition a given bounded Lipschitz design domain $\Omega \subset \mathbb{R}^d$ into regions occupied by either void or by $N-1$ different materials such that a given cost functional is minimized subject to given constraints. Within the phase field approach we describe the different material distributions with the help of a phase field vector $\boldsymbol{\varphi} := (\varphi^i)_{i=1}^N$, where φ^N describes the fraction of void and $\varphi^1, \ldots, \varphi^{N-1}$ describe the fractions of the $N-1$ different materials. The phase field approach allows for a certain mixing between materials and between materials and void but the mixing will be restricted to a small interfacial region. In order to ensure that the phase field vector $\boldsymbol{\varphi}$ describes fractions we require that $\boldsymbol{\varphi}$ lies pointwise in the Gibbs simplex $\boldsymbol{G} := \{\boldsymbol{v} \in \mathbb{R}^N \mid v^i \geq 0, \sum_{i=1}^N v^i = 1\}$.

In this work we prescribe the total spatial amount of the material fractions through $\fint_\Omega \boldsymbol{\varphi} = \boldsymbol{m} = (m^i)_{i=1}^N$, where it is assumed that $\sum_{i=1}^N m^i = 1$ with $m^i \in (0,1)$, $i = 1, \ldots, N$, and where $\fint_\Omega \boldsymbol{\varphi}$ denotes the mean value on Ω. We remark that in principal inequality constraints for $\fint_\Omega \boldsymbol{\varphi}$ can also be dealt with.

The potential $\Psi : \mathbb{R}^N \to \mathbb{R} \cup \{\infty\}$ is assumed to have global minima at the unit vectors $\boldsymbol{e}_i$, $i = 1, \ldots, N$, which correspond to the different materials and to the void.

In (1.1) we choose an obstacle potential $\Psi(\boldsymbol{\varphi}) = \Psi_0(\boldsymbol{\varphi}) + I_G(\boldsymbol{\varphi})$ where Ψ_0 is smooth and I_G is the indicator function of the Gibbs-simplex $\boldsymbol{G}$. Introducing $\mathcal{G} := \{\boldsymbol{v} \in H^1(\Omega, \mathbb{R}^N) \mid \boldsymbol{v}(x) \in \boldsymbol{G} \text{ a.e. in } \Omega\}$ and $\mathcal{G}^{\boldsymbol{m}} := \{\boldsymbol{v} \in \mathcal{G} \mid \fint_\Omega \boldsymbol{v} = \boldsymbol{m}\}$ we obtain

$$\hat{E}^\varepsilon(\boldsymbol{\varphi}) := \int_\Omega \left(\frac{\varepsilon}{2} |\nabla \boldsymbol{\varphi}|^2 + \frac{1}{\varepsilon} \Psi_0(\boldsymbol{\varphi}) \right) \tag{2.1}$$

and on $\mathcal{G}$ we have $E^\varepsilon(\boldsymbol{\varphi}) = \hat{E}^\varepsilon(\boldsymbol{\varphi})$.

We describe the elastic deformation with the help of the displacement vector $\boldsymbol{u} : \Omega \to \mathbb{R}^d$ and with the strain tensor $\mathcal{E} = \mathcal{E}(\boldsymbol{u}) = \frac{1}{2}(\nabla \boldsymbol{u} + (\nabla \boldsymbol{u})^T)$. The boundary $\partial\Omega$ is divided into a Dirichlet part Γ_D, a non-homogeneous Neumann part Γ_g and a homogeneous Neumann part Γ_0. Furthermore, $\mathbb{C}$ is the elasticity tensor, $\boldsymbol{f} \in L^2(\Omega, \mathbb{R}^d)$ is the volume force and $\boldsymbol{g} \in L^2(\Gamma_g, \mathbb{R}^d)$ are boundary forces.

The equations of linear elasticity which are the constraint in our optimization problem are given by

$$\begin{cases} -\nabla \cdot [\mathbb{C}(\boldsymbol{\varphi})\mathcal{E}(\boldsymbol{u})] = \left(1 - \varphi^N\right) \boldsymbol{f} & \text{in } \Omega, \\ \boldsymbol{u} = \boldsymbol{0} & \text{on } \Gamma_D, \\ [\mathbb{C}(\boldsymbol{\varphi})\mathcal{E}(\boldsymbol{u})]\, \boldsymbol{n} = \boldsymbol{g} & \text{on } \Gamma_g, \\ [\mathbb{C}(\boldsymbol{\varphi})\mathcal{E}(\boldsymbol{u})]\, \boldsymbol{n} = \boldsymbol{0} & \text{on } \Gamma_0, \end{cases} \tag{2.2}$$

where $\boldsymbol{n}$ is the outer unit normal to $\partial\Omega$. The elasticity tensor $\mathbb{C}$ is assumed to depend smoothly on $\boldsymbol{\varphi}$, $\mathbb{C}$ has to fulfill the usual symmetry condition of linear elasticity and has to be positive definite on symmetric tensors. More information and detailed literature on the theory of elasticity can be found in [3]. For the phase field approach the void is approximated by a very soft material with an elasticity tensor $\mathbb{C}^N(\varepsilon)$ depending on the interface thickness, e.g. $\mathbb{C}^N = \varepsilon^2 \tilde{\mathbb{C}}^N$ with a fixed tensor $\tilde{\mathbb{C}}^N$.

Discussions on how to interpolate the elasticity tensors $\mathbb{C}^i$, for $i = 1, \ldots, N$, given in the pure materials onto the interface can also be found in Sect. 5 and in [2, 3]. Introducing the notation $\langle \mathcal{A}, \mathcal{B} \rangle_{\mathbb{C}} := \int_\Omega \mathcal{A} : \mathbb{C}\mathcal{B}$, where for any matrices $\mathcal{A}$ and $\mathcal{B}$ the product is given as $\mathcal{A} : \mathcal{B} := \sum_{i,j=1}^d \mathcal{A}_{ij}\mathcal{B}_{ij}$, the elastic boundary value problem (2.2) can be written in the weak formulation:

Given $(\boldsymbol{f}, \boldsymbol{g}, \boldsymbol{\varphi}) \in L^2(\Omega, \mathbb{R}^d) \times L^2(\Gamma_g, \mathbb{R}^d) \times L^\infty(\Omega, \mathbb{R}^N)$ find $\boldsymbol{u} \in H^1_D(\Omega, \mathbb{R}^d)$ such that

$$\langle \mathcal{E}(\boldsymbol{u}), \mathcal{E}(\boldsymbol{\eta}) \rangle_{\mathbb{C}(\boldsymbol{\varphi})} = \int_\Omega \left(1 - \varphi^N\right) \boldsymbol{f} \cdot \boldsymbol{\eta} + \int_{\Gamma_g} \boldsymbol{g} \cdot \boldsymbol{\eta} =: F(\boldsymbol{\eta}, \boldsymbol{\varphi}), \tag{2.3}$$

which has to hold for all $\boldsymbol{\eta} \in H^1_D(\Omega, \mathbb{R}^d) := \{\boldsymbol{\eta} \in H^1(\Omega, \mathbb{R}^d) \mid \boldsymbol{\eta} = \mathbf{0} \text{ on } \Gamma_D\}$. The well-posedness of (2.3) can be shown by using the Lax-Milgram lemma and Korn's inequality, for details see [3].
Summarized, the structural optimization problem can be formulated as: Given $(\boldsymbol{f}, \boldsymbol{g}, \boldsymbol{u}_\Omega, c) \in L^2(\Omega, \mathbb{R}^d) \times L^2(\Gamma_g, \mathbb{R}^d) \times L^2(\Omega, \mathbb{R}^d) \times L^\infty(\Omega)$ and measurable sets $S_i \subseteq \Omega$, $i \in \{0, 1\}$, with $S_0 \cap S_1 = \emptyset$, we want to solve

$$(\mathcal{P}^\varepsilon) \qquad \begin{cases} \min & J^\varepsilon(\boldsymbol{u}, \boldsymbol{\varphi}) := \alpha F(\boldsymbol{u}, \boldsymbol{\varphi}) + \beta J_0(\boldsymbol{u}, \boldsymbol{\varphi}) + \gamma \hat{E}^\varepsilon(\boldsymbol{\varphi}), \\ \text{over} & (\boldsymbol{u}, \boldsymbol{\varphi}) \in H^1_D(\Omega, \mathbb{R}^d) \times H^1(\Omega, \mathbb{R}^N), \\ \text{s.t.} & \text{(2.3) is fulfilled and } \boldsymbol{\varphi} \in \boldsymbol{\mathcal{G}}^{\boldsymbol{m}} \cap \boldsymbol{U}_c, \end{cases}$$

where $\alpha, \beta \geq 0,\ \gamma, \varepsilon > 0$, $\boldsymbol{m} \in (0, 1)^N$ with $\sum_{i=1}^N m^i = 1$,

$$\boldsymbol{U}_c := \{\boldsymbol{\varphi} \in H^1(\Omega, \mathbb{R}^N) \mid \varphi^N = 0 \text{ a.e. on } S_0 \text{ and } \varphi^N = 1 \text{ a.e. on } S_1\}$$

and the functional for the compliant mechanism is given by

$$J_0(\boldsymbol{u}, \boldsymbol{\varphi}) := \left(\int_\Omega \left(1 - \varphi^N\right) c\, |\boldsymbol{u} - \boldsymbol{u}_\Omega|^2 \right)^{\frac{1}{2}}, \tag{2.4}$$

with a given non-negative weighting factor $c \in L^\infty(\Omega)$ fulfilling $|\text{supp}\, c| > 0$.

The existence of a minimizer to $(\mathcal{P}^\varepsilon)$ is shown by classical techniques of the calculus of variations in [3].

Remark 2.1. From the applicational point of view it might be desirable to fix material or void in some regions of the design domain, so the condition $\boldsymbol{\varphi} \in \boldsymbol{U}_c$ makes sense. Moreover by choosing S_0 such that $|S_0 \cap \text{supp}\, c| \neq 0$ we can ensure that it is not possible to choose only void on the support of c, i.e. in (2.4) we ensure $|\text{supp}\,(1 - \varphi^N) \cap \text{supp}\, c| > 0$.

3 Optimality System

In order to derive first-order necessary optimality conditions for the optimization problem $(\mathcal{P}^\varepsilon)$, it is essential to show the differentiability of the control-to-state operator, which is well-defined because of the well-posedness of (2.3).

Theorem 3.1. *The control-to-state operator* $S : L^\infty(\Omega, \mathbb{R}^N) \to H^1_D(\Omega, \mathbb{R}^d)$, *defined by* $S(\varphi) := \boldsymbol{u}$, *where* $\boldsymbol{u}$ *solves (2.3), is Fréchet differentiable. Its directional derivative at* $\varphi \in L^\infty(\Omega, \mathbb{R}^N)$ *in the direction* $\boldsymbol{h} \in L^\infty(\Omega, \mathbb{R}^N)$ *is given by* $S'(\varphi)\boldsymbol{h} = \boldsymbol{u}^*$, *where* $\boldsymbol{u}^*$ *denotes the unique solution of the problem*

$$\langle \mathcal{E}(\boldsymbol{u}^*), \mathcal{E}(\boldsymbol{\eta})\rangle_{\mathbb{C}(\varphi)} = -\langle \mathcal{E}(\boldsymbol{u}), \mathcal{E}(\boldsymbol{\eta})\rangle_{\mathbb{C}'(\varphi)\boldsymbol{h}} - \int_\Omega h^N \boldsymbol{f} \cdot \boldsymbol{\eta}, \qquad \forall \boldsymbol{\eta} \in H^1_D(\Omega, \mathbb{R}^d). \tag{3.1}$$

The expression (3.1) formally can be derived by differentiating the implicit state equation $\langle \mathcal{E}(S(\varphi)), \mathcal{E}(\eta)\rangle_{\mathbb{C}(\varphi)} = F(\eta, \varphi)$ with respect to $\varphi \in L^\infty(\Omega, \mathbb{R}^N)$. The proof of Theorem 3.1 can be found in [3].

With Theorem 3.1 at hand, we can now derive first order conditions. Indeed, it follows from the chain rule that the reduced cost functional $j(\varphi) := J^\varepsilon(S(\varphi), \varphi)$ is Fréchet differentiable at every $\varphi \in H^1(\Omega, \mathbb{R}^N) \cap L^\infty(\Omega, \mathbb{R}^N)$ with the Fréchet derivative $j'(\varphi)\boldsymbol{h} = J^\varepsilon_{,\boldsymbol{u}}(\boldsymbol{u}, \varphi)\boldsymbol{u}^* + J^\varepsilon_{,\varphi}(\boldsymbol{u}, \varphi)\boldsymbol{h}$. Here we have to assume that $J_0 \neq 0$ in case of $\beta \neq 0$. Owing to the convexity of $\boldsymbol{\mathcal{G}}^{\boldsymbol{m}} \cap \boldsymbol{U}_c$, we have for every minimizer $\varphi \in \boldsymbol{\mathcal{G}}^{\boldsymbol{m}} \cap \boldsymbol{U}_c$ of j in $\boldsymbol{\mathcal{G}}^{\boldsymbol{m}} \cap \boldsymbol{U}_c$ that $j'(\varphi)(\tilde{\varphi} - \varphi) \geq 0, \forall \tilde{\varphi} \in \boldsymbol{\mathcal{G}}^{\boldsymbol{m}} \cap \boldsymbol{U}_c$. We can now state the complete optimality system, see [3] for a proof.

Theorem 3.2. *Let* $\varphi \in \boldsymbol{\mathcal{G}}^{\boldsymbol{m}} \cap \boldsymbol{U}_c$ *denote a minimizer of the problem* $(\mathcal{P}^\varepsilon)$ *and* $S(\varphi) = \boldsymbol{u} \in H^1_D(\Omega, \mathbb{R}^d)$, $\boldsymbol{p} \in H^1_D(\Omega, \mathbb{R}^d)$ *are the corresponding state and adjoint variables, respectively. Then the functions* $(\boldsymbol{u}, \varphi, \boldsymbol{p}) \in H^1_D(\Omega, \mathbb{R}^d) \times (\boldsymbol{\mathcal{G}}^{\boldsymbol{m}} \cap \boldsymbol{U}_c) \times H^1_D(\Omega, \mathbb{R}^d)$ *fulfill the following optimality system consisting of the state equation*

$$\text{(SE)} \qquad \langle \mathcal{E}(\boldsymbol{u}), \mathcal{E}(\boldsymbol{\eta}_1)\rangle_{\mathbb{C}(\varphi)} = F(\boldsymbol{\eta}_1, \varphi), \ \forall \boldsymbol{\eta}_1 \in H^1_D(\Omega, \mathbb{R}^d),$$

the adjoint equation

$$\text{(AE)} \begin{cases} \langle \mathcal{E}(\boldsymbol{p}), \mathcal{E}(\boldsymbol{\eta}_2)\rangle_{\mathbb{C}(\varphi)} \\ = \alpha F(\boldsymbol{\eta}_2, \varphi) + \beta J_0^{-1}(\boldsymbol{u}, \varphi) \int_\Omega c(1 - \varphi^N)(\boldsymbol{u} - \boldsymbol{u}_\Omega) \cdot \boldsymbol{\eta}_2, \\ \forall \boldsymbol{\eta}_2 \in H^1_D(\Omega, \mathbb{R}^d), \end{cases}$$

and the gradient inequality

$$\text{(GI)} \begin{cases} \gamma\varepsilon \int_\Omega \nabla\varphi : \nabla(\tilde{\varphi} - \varphi) + \frac{\gamma}{\varepsilon} \int_\Omega \Psi_0'(\varphi) \cdot (\tilde{\varphi} - \varphi) \\ -\frac{\beta}{2} J_0^{-1}(\boldsymbol{u}, \varphi) \int_\Omega c(\tilde{\varphi}^N - \varphi^N)|\boldsymbol{u} - \boldsymbol{u}_\Omega|^2 \\ -\int_\Omega (\tilde{\varphi}^N - \varphi^N)\boldsymbol{f} \cdot (\alpha\boldsymbol{u} + \boldsymbol{p}) - \langle \mathcal{E}(\boldsymbol{p}), \mathcal{E}(\boldsymbol{u})\rangle_{\mathbb{C}'(\varphi)(\tilde{\varphi} - \varphi)} \geq 0, \\ \forall \tilde{\varphi} \in \boldsymbol{\mathcal{G}}^{\boldsymbol{m}} \cap \boldsymbol{U}_c. \end{cases}$$

4 Sharp Interface Asymptotics

In this section we present the sharp interface limit of the optimality system given in Theorem 3.2; for a detailed derivation of the sharp interface limit using the method of formally matched asymptotic expansions we refer to [3]. We now consider a more concrete form of the $\boldsymbol{\varphi}$-dependent elasticity tensor. We choose the elasticity tensor starting with constant elasticity tensors $\mathbb{C}^i$, $i \in \{1, \ldots, N-1\}$ which are defined in the pure materials, i.e. when $\boldsymbol{\varphi} = \boldsymbol{e}_i$, and model the void as a very soft material. As mentioned, a possible choice of the elasticity tensor in the void is $\mathbb{C}^N = \mathbb{C}^N(\varepsilon) = \varepsilon^2 \tilde{\mathbb{C}}^N$ where $\tilde{\mathbb{C}}^N$ is a fixed elasticity tensor. In order to model the elastic properties also in the interfacial region the elasticity tensor is assumed to be a tensor valued function $\mathbb{C}(\boldsymbol{\varphi}) := \left(\mathbb{C}_{ijkl}(\boldsymbol{\varphi})\right)_{i,j,k,l=1}^d$ which interpolate between $\mathbb{C}^1, \ldots, \mathbb{C}^{N-1}, \mathbb{C}^N(\varepsilon)$. Furthermore we assume that the weighting factor c in the compliant mechanism functional J_0 is a smooth function.

The asymptotic analysis gives that the phase field functions converge as ε tends to zero to a limit function $\boldsymbol{\varphi}$ which only takes values in $\{\boldsymbol{e}_1, \ldots, \boldsymbol{e}_N\}$. This implies that the domain Ω is partitioned into N regions Ω^i, $i \in \{1, \ldots, N\}$, which are separated by interfaces Γ_{ij}, $i < j$. We choose a unit normal at Γ_{ij} such that for $\delta > 0$ small we have $x + \delta\boldsymbol{\nu} \in \Omega^j$ and $x - \delta\boldsymbol{\nu} \in \Omega^i$. Moreover we define $[\boldsymbol{w}]_i^j := \lim_{\delta \searrow 0} (\boldsymbol{w}(x + \delta\boldsymbol{\nu}) - \boldsymbol{w}(x - \delta\boldsymbol{\nu}))$. We obtain in regions occupied by material, i.e. for $i = 1, \ldots, N-1$, that the state and the adjoint equation, respectively, have to hold

$$-\nabla \cdot \left[\mathbb{C}^i \mathcal{E}(\boldsymbol{u})\right] = \boldsymbol{f} \text{ and } -\nabla \cdot \left[\mathbb{C}^i \mathcal{E}(\boldsymbol{p})\right] = \alpha \boldsymbol{f} + \beta J_0^{-1}(\boldsymbol{u}, \boldsymbol{\varphi})(\boldsymbol{u} - \boldsymbol{u}_\Omega)c\,.$$

In case of material-material interfaces, i.e. Γ_{ij}, $i, j \in \{1, \ldots, N-1\}$ we have continuity in the variables $\boldsymbol{u}$, $\boldsymbol{p}$ and continuity for the normal stresses $\mathbb{C}\mathcal{E}(\boldsymbol{u})\boldsymbol{\nu}$ and $\mathbb{C}\mathcal{E}(\boldsymbol{p})\boldsymbol{\nu}$, i.e. for $i, j \in \{1, \ldots, N-1\}$ and $\boldsymbol{w} \in \{\boldsymbol{u}, \boldsymbol{p}\}$ we have $[\boldsymbol{w}]_i^j = \boldsymbol{0}$, $[\mathbb{C}\mathcal{E}(\boldsymbol{w})\boldsymbol{\nu}]_i^j = \boldsymbol{0}$ on Γ_{ij}. On Γ_{iN} we get $\mathbb{C}^i \mathcal{E}_i(\boldsymbol{u})\boldsymbol{\nu} = \mathbb{C}^i \mathcal{E}_i(\boldsymbol{p})\boldsymbol{\nu} = \boldsymbol{0}$. Moreover we obtain for all $i, j \neq N$

$$\begin{aligned} 0 = \gamma\sigma_{ij}\kappa - [\mathbb{C}\mathcal{E}(\boldsymbol{u}) : \mathcal{E}(\boldsymbol{p})]_i^j + [\mathbb{C}\mathcal{E}(\boldsymbol{u})\boldsymbol{\nu} \cdot (\nabla \boldsymbol{p})\boldsymbol{\nu}]_i^j \\ + [\mathbb{C}\mathcal{E}(\boldsymbol{p})\boldsymbol{\nu} \cdot (\nabla \boldsymbol{u})\boldsymbol{\nu}]_i^j + \lambda^i - \lambda^j \quad \text{on } \Gamma_{ij} \end{aligned} \tag{4.1}$$

where κ is the mean curvature of Γ_{ij} and $\boldsymbol{\lambda} \in \mathbb{R}^N$ are Lagrange multipliers. We remark that the terms involving $\boldsymbol{u}$ and $\boldsymbol{p}$ generalize the Eshelby traction known from materials science, see [3]. In addition for all $i \neq N$ it holds

$$\begin{aligned} 0 = \gamma\sigma_{iN}\kappa + \mathbb{C}^i \mathcal{E}_i(\boldsymbol{u}) : \mathcal{E}_i(\boldsymbol{p}) - \frac{\beta}{2} J_0^{-1}(\boldsymbol{u}, \boldsymbol{\varphi})c\, |\boldsymbol{u} - \boldsymbol{u}_\Omega|^2 \\ - \boldsymbol{f} \cdot (\alpha\boldsymbol{u} + \boldsymbol{p}) + \lambda^i - \lambda^N \quad \text{on } \Gamma_{iN}. \end{aligned}$$

Above the Lagrange multipliers $\lambda^1, \dots, \lambda^N$ sum up to zero and they are related to volume constraints $\fint_{\Omega^i} 1 = m^i$ which are obtained from the integral constraints $\fint_\Omega \boldsymbol{\varphi} = \boldsymbol{m}$ in the sharp interface limit.

In case that void and two or more materials appear junction points emerge, where e.g. void and two materials meet, see e.g. Fig. 7, and it might be desirable in applications to influence the angles at the junctions. By an appropriate choice of the potential Ψ the angles at the junctions can be prescribed, see [3] for details.

Remark 4.1. In the case of one material we recover the classical first order conditions for the sharp interface structural optimization problem, see e.g. Allaire, Jouve, Toader [1]. The conditions we derived above generalize the first order conditions in [1] to the multi-phase case.

5 Numerical Methods

5.1 Choice of the Potential

In the previous section we studied the Ginzburg-Landau energy with an obstacle potential which leads to an optimization problem with inequality constraints. Using instead a smooth potential would lead to equality constraints only which are usually easier to handle. However, there is a subtle problem, namely, we can not prescribe the total spatial amount of the material by $\fint_\Omega \boldsymbol{\varphi} = \boldsymbol{m}$ since the identification of pure i-th-material with $\varphi^i = 1$ does not hold any longer but the value attained in phase i depends on ε. Only in the limit for $\varepsilon \to 0$ there is a pure i-th phase at $x \in \Omega$ if $\varphi^i(x) = 1$. In Table 1 the shift of one phase is presented for a numerical experiment. The listed values are the values in areas where the values stay nearly constant, reflecting a pure phase. Therefore, the i-th material does not have approximately volume m_i by prescribing $\fint_\Omega \varphi^i = m_i$. Consequently one has to use the obstacle potential or the spatial amount has to be modelled in a different way.

5.2 Choice of the Stiffness Tensor on the Interface

The choice of the stiffness tensor on the interface also has a quite severe influence on the solution. A rough explanation in the presence of one material is the following: The stiffest structure has material everywhere. The mass

Table 1 Values for the phase identification using the double well potential

ε	0.02	0.01	0.005	0.0025	0.001
φ^1	≈1.33942	≈1.21378	≈1.13630	≈1.11450	≈1.05818

constraints prohibit this. However, since it is possible to choose $\varphi^i \in (0,1)$ on the interface, it can happen that it is best to have a large mushy region with a mixture of void and material, i.e. a broad interface, which leads to a stiffer structure. Therefore the stiffness tensor on the interface should drop down fast but smoothly from the higher stiffness to the lower stiffness. We use an quadratic interpolation of the elasticity tensors $\mathbb{C}^1, \dots, \mathbb{C}^N$ and set the directional derivative in direction from the lower to the higher stiffness at the material with the lower stiffness to zero. One possibility for N-phases is $\mathbb{C}(\boldsymbol{\varphi}) = \sum_{i,j} \mathbb{C}^{\max\{i,j\}} \varphi^i \varphi^j$ where the tensors are ordered from high to low stiffness. A similar kind of interpolation is used in the SIMP approach for one material and void [2]. The choice of the elasticity tensor on the interface influences also the speed of the numerical algorithm.

5.3 Projected H^1-Gradient Method

In this section we focus on the mean compliance problem, i.e. $\beta = 0$ and we use the reduced problem formulation

$$\min_{\boldsymbol{\varphi} \in \mathcal{G}^{\mathbf{m}}} j(\boldsymbol{\varphi}) := J^{\varepsilon}(S(\boldsymbol{\varphi}), \boldsymbol{\varphi})$$

where $\mathcal{G}^{\mathbf{m}} = \{\boldsymbol{\xi} \in H^1 \mid \fint_{\Omega} \boldsymbol{\xi} = \mathbf{m},\ \xi_i \geq 0,\ \sum \xi_i \equiv 1 \text{ a.e. in } \Omega\}$ is convex and closed and $j\colon H^1(\Omega, \mathbb{R}^N) \cap L^{\infty}(\Omega, \mathbb{R}^N) \to \mathbb{R}$ is Fréchet-differentiable, where the directional derivatives are given by:

$$j'(\boldsymbol{\varphi})\boldsymbol{\eta} = \gamma\varepsilon(\nabla\boldsymbol{\varphi}, \nabla\boldsymbol{\eta}) + \frac{\gamma}{\varepsilon}(\Psi_0'(\boldsymbol{\varphi}), \boldsymbol{\eta}) - \alpha(\mathbb{C}'(\boldsymbol{\varphi})(\boldsymbol{\eta})\mathcal{E}(\mathbf{u}), \mathcal{E}(\mathbf{u})) \,. \tag{5.1}$$

The first-order condition of a general minimization problem $\min j(\boldsymbol{\varphi})$ s.t. $\boldsymbol{\varphi} \in U$ where U is convex and closed can be rewritten as a fixed point equation: For any $\lambda > 0$ the solution is given as $\boldsymbol{\varphi} = P_H(\boldsymbol{\varphi} - \lambda \nabla_H j(\boldsymbol{\varphi}))$ where P_H is the projection onto the convex feasible set U with respect to the scalar product in H, see [10]. Based on this projected gradient methods have been developed. We propose to use the following new variant:

Algorithm 5.1. Having a current approximation $\boldsymbol{\varphi}_k$ and given a positive λ perform a line-search along the descent direction

$$\boldsymbol{v}_k := P_H(\boldsymbol{\varphi}_k - \lambda \nabla_H j(\boldsymbol{\varphi}_k)) - \boldsymbol{\varphi}_k$$

to obtain the step length β_k. Then set $\boldsymbol{\varphi}_{k+1} := \boldsymbol{\varphi}_k + \beta_k \boldsymbol{v}_k$.

Stop the iteration if $\|\boldsymbol{v}_k\|_H < \text{tol}$.

This is not the more known search along the projected gradient path $\varphi_{k+1} := P_H(\varphi_k - \beta_k \nabla_H j(\varphi_k))$ which requires in each line-search step an (expensive) projection. We can prove a global convergence result [6] which can be found for convex functions in [10].

Theorem 5.2. *Let H be a Hilbert space, $U \subset H$ be convex, closed and non-empty and $j : U \to \mathbb{R}$ be continuously Fréchet differentiable. Then, every accumulation point $\boldsymbol{\varphi}^*$ of $\{\varphi_k\}$ generated by Algorithm 5.1 is first order critical if the Armijo step length rule is used.*

The reduced cost functional j is differentiable in $H^1(\Omega, \mathbb{R}^N) \cap L^\infty(\Omega, \mathbb{R}^N)$, which is not a Hilbert-space. Nevertheless, we choose the Hilbert-space $H = \{\boldsymbol{\xi} \in H^1(\Omega, \mathbb{R}^N) \mid \fint_\Omega \boldsymbol{\xi} = \mathbf{0}\}$ with the scalar product $(\boldsymbol{\xi}, \boldsymbol{\eta})_H = (\nabla \boldsymbol{\xi}, \nabla \boldsymbol{\eta})$. The gradient does not exist in H^1. However, since

$$\tfrac{1}{2}\|(\boldsymbol{\zeta} - \boldsymbol{\varphi} + \lambda \nabla_H j(\boldsymbol{\varphi}))\|_H^2 = \tfrac{1}{2}\|\boldsymbol{\zeta} - \boldsymbol{\varphi}\|_H^2 + \lambda j'(\boldsymbol{\varphi})(\boldsymbol{\zeta} - \boldsymbol{\varphi}) + c \tag{5.2}$$

for some constant c, we do not need the H-gradient but only the directional derivatives for the projection. Hence, we define and use instead of the projection P_H the projection type operator $\mathcal{P}_H$ where $\mathcal{P}_H(\boldsymbol{\varphi}, \lambda)$ is given by the solution of

$$\begin{aligned} &\min \tfrac{1}{2}\|\boldsymbol{\zeta} - \boldsymbol{\varphi}\|_H^2 + \lambda j'(\boldsymbol{\varphi})(\boldsymbol{\zeta} - \boldsymbol{\varphi}) \\ &\text{s. t.} \quad \fint_\Omega \boldsymbol{\zeta} = \boldsymbol{m}, \quad \textstyle\sum_{i=1}^N \zeta^i \equiv 1, \quad \zeta^i \geq 0 \; \forall\, i = 1, \ldots, N\,. \end{aligned} \tag{5.3}$$

The existence and uniqueness of a solution $\mathcal{P}_H(\boldsymbol{\varphi}, \lambda)$ of (5.3) can be shown in our application, see [6]. Moreover, under some regularity conditions on j which are fulfilled for our problem, we can show the same global convergence result as in Theorem 5.2, see [6]. Numerically we solve the obstacle type problem (5.3) with a primal-dual active set approach.

5.4 Scaling

In the following we address the choice of the parameter λ in the algorithm. It turned out the scaling of the employed norm is essential for efficiency and for iteration numbers independent of the interface thickness, i.e. of ε. One can motivate this by the fact that the perimeter is approximated by the Ginzburg-Landau energy, which roughly speaking entail $\varepsilon||\nabla \boldsymbol{\varphi}_\varepsilon||_{L^2}^2 \approx const.$ for the minimizer $\boldsymbol{\varphi}_\varepsilon$. Hence we have $||\boldsymbol{\varphi}_\varepsilon||_H = O(1/\sqrt{\varepsilon})$ and $||\boldsymbol{\varphi}_\varepsilon||_{\sqrt{\varepsilon}H} = O(1)$. This is confirmed also numerically. As a consequence we choose the $\sqrt{\varepsilon}H$ metric. Since $\mathcal{P}_{\sqrt{\varepsilon}H} = \mathcal{P}_H$ this leads to the use of a scaled H-gradient since $\nabla_{\sqrt{\varepsilon}H} j = \frac{1}{\varepsilon}\nabla_H j$, respectively this emphasizes to use $\lambda = \frac{1}{\varepsilon}$. However, the iterates φ_k fulfill $||\varphi_k||_H \approx ||\varphi_\varepsilon||_H$ only when phases

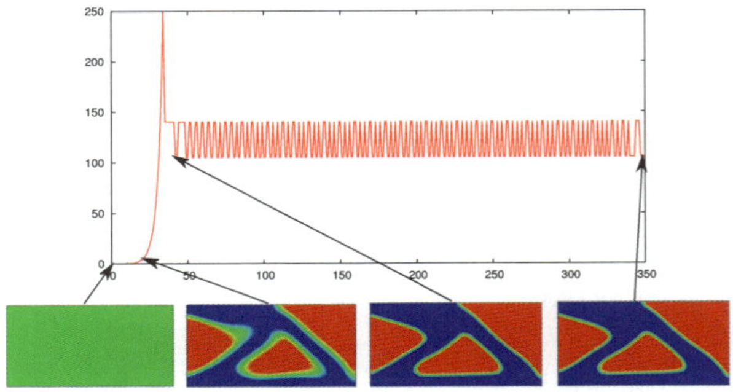

Fig. 1 Behaviour of λ and the phase distribution with respect to the iterations

are separated and interfaces are present with thickness according to ε. In the first iterations this is in general not the case. Hence, it is more appropriate to adapt λ during the iterations. As a first approach we used the following updating strategy:

Set $\lambda_0 = \frac{0.01}{\varepsilon}$ *and choose some* $0 < \bar{c} < 1$,
in the following set $\lambda_k = \lambda_{k-1}/\bar{c}$ *if* $\alpha_{k-1} = 1$ *and* $\lambda_k = \lambda_{k-1}\bar{c}$ *else.*

The changes in λ with respect to the iterations can be seen examplarily in Fig. 1, where underneath the evolution of the phases can be seen. We remark that this is no line search with respect to λ. In the following algorithm we outline one iteration step and indicate with it the cost of the method.

Algorithm 5.3. Given $\boldsymbol{\varphi}_k$ and a fixed $\sigma \in (0, 1)$

- Solve the elasticity equation (2.3) for $\mathbf{u}_k = S(\boldsymbol{\varphi}_k) : \Omega \to \mathbb{R}^d$,
- Assemble the directional derivatives $j'(\boldsymbol{\varphi}_k)\boldsymbol{\eta} \quad \forall \boldsymbol{\eta} \in H \cap L^\infty$,
- Update λ_k,
- solve the obstacle type problem (5.3) for the $\mathcal{P}_{H^1}(\boldsymbol{\varphi}_k, \lambda_k) : \Omega \to \mathbb{R}^N$,
- Set $\boldsymbol{v}_k := \mathcal{P}_{H^1}(\boldsymbol{\varphi}_k, \lambda_k) - \boldsymbol{\varphi}_k$ and stop if $\|\boldsymbol{v}_k\|_{\sqrt{\varepsilon}H} <$ tol,
- Determine the Armijo-step length $\beta_k = \sigma^{m_k}$ using back tracking
  where in each iteration we have to solve the elasticity equation
  for $\mathbf{u} = S(\boldsymbol{\varphi}_k + \beta \mathbf{v}_k) : \Omega \to \mathbb{R}^d$,
- Set $\boldsymbol{\varphi}_{k+1} := \boldsymbol{\varphi}_k + \beta_k \boldsymbol{v}_k$.

5.5 Numerical Experiments

The numerical experiments which underline the above statements are for the cantilever beam in two dimensions and with one material and void. The design domain is $\Omega = (-1, 1) \times (0, 1)$ and $\alpha = 1$. There is no volume force but a boundary force $\boldsymbol{g} \equiv (0, -250)^T$ is acting on $\Gamma_g = (0.75, 1) \times \{0\}$. The Dirichlet part is $\Gamma_D = \{-1\} \times (0, 1)$. For the stiffness tensor of the material we take $\mathbb{C}^1\mathcal{E} = 2\mu\mathcal{E} + \lambda(\mathrm{tr}\mathcal{E})I$ with Lamé constants $\mu = \lambda = 5{,}000$. Moreover we use the constant $\gamma = 0.5$ and prescribe the masses by 50 % material and 50 % void. Figure 2 displays the setting and the result for $\varepsilon = 0.03$.

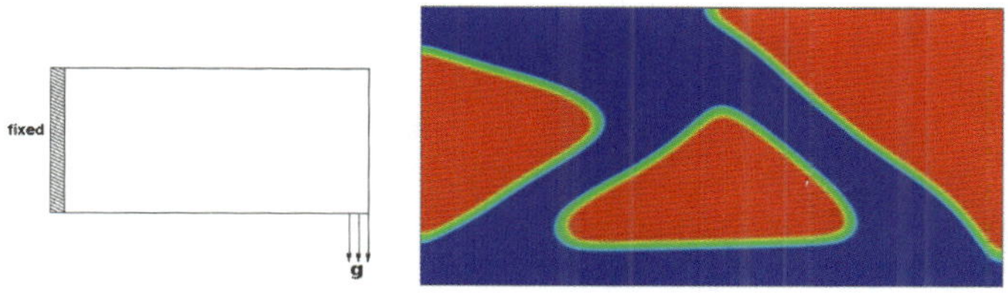

Fig. 2 Cantileaver beam, geometry (*left*) and numerical result (*right*)

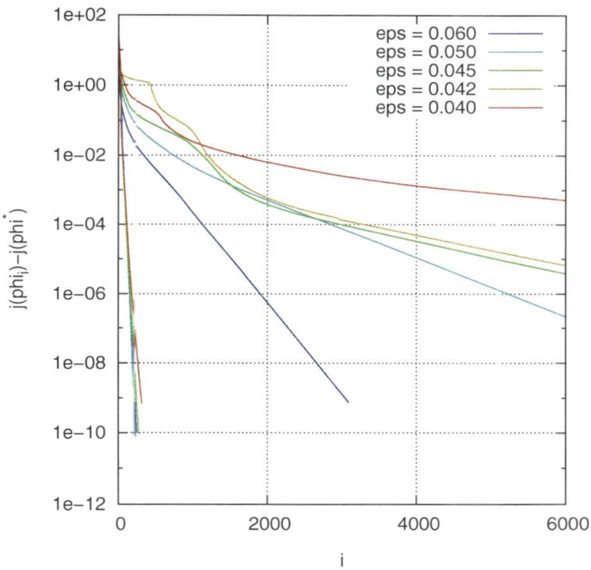

Fig. 3 With and without scaling

All computations are done using the finite element toolbox FEniCS [11]. So far we only use equidistant meshes. The elasticity equation is discretized with P1-finite elements and the arising linear systems are solved directly. In the computations with one material the problem setting is reduced to one phase field only by working with $\varphi := \varphi^2 - \varphi^1$. In Fig. 3 the upper five lines correspond to the results without scaling the gradient and shows the approximated error in the cost functional with respect to the iteration numbers. We clearly see a dependency on ε. The lower five lines correspond to the results with scaling and are nearly not distinguishable, independent of ε and lead to much better approximations for a lower number of iteration. In Fig. 4 the influence of the choice of the linear versus the quadratic interpolation of the stiffness tensor is depicted for $\varepsilon = 0.04$.

In Table 2 we study the dependency on the mesh size h and compare the approaches without scaled gradient and with linear interpolation of the elasticity tensors (called *old* in the table) to the approach using the scaled gradient and the

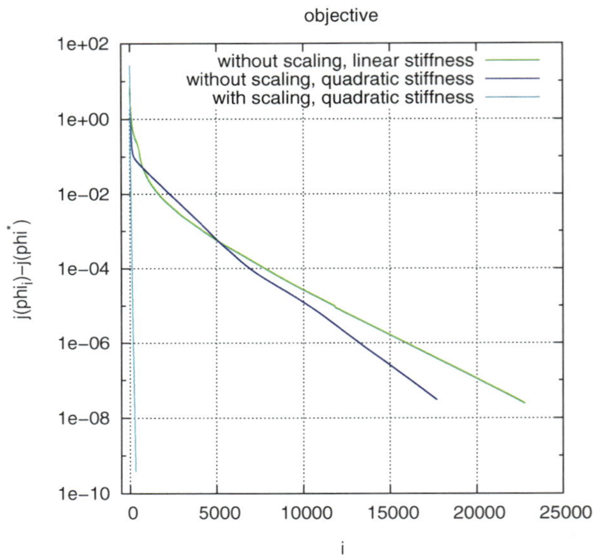

Fig. 4 Interpolated elasticity tensor

Table 2 Comparison of the previous and the new approach as well as with nested iteration for $\varepsilon = 0.04$

		Old		New		Nested	
h	DOF	CPU	Iter.	CPU	Iter.	CPU	Iter.
2^{-4}	561	12 min	9,956	5 s	112	4 s	85
2^{-5}	2,145	2 h 25 min	14,590	1 min	408	7 s	52
2^{-6}	8,385	20 h 40 min	16,936	4 min	321	14 s	24
2^{-7}	33,153	3 day 20 h 28 min	19,416	21 min	276	2 min	33
2^{-8}	131,841	**23 day 15 h 0 min**	18,891	**3 h 1 min**	270	25 min	63
						Total **28 min**	

quadratic interpolated elasticity tensor (called *new* in the table). In the last column we listed the result for the latter approach but using in addition nested iteration, i.e. using the result of the previous h as initial data for the next and solving with an decreasing tolerance *tol*. This leads to the expected speed up, here the nested approach needs roughly 15 % of the CPU-time of the *new* approach. The more severe speed up of the *old* approach is obtained by the *new* ansatz, which leads to a reduction to 0.5 % of the corresponding CPU-time of the *old* approach. Nevertheless, in any case the expected mesh independent number of iterations is confirmed. We do not list but would like to mention that in the above example the number of line search iterations stay also mesh independent and are between 1 and 3. The number of PDAS iterations are mildly mesh dependent but stay below 10 after the first few iterations.

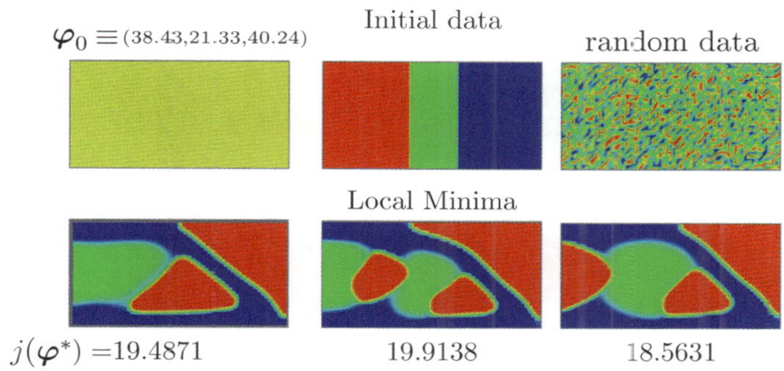

Fig. 5 Cantilever beam with two materials and void in $2d$

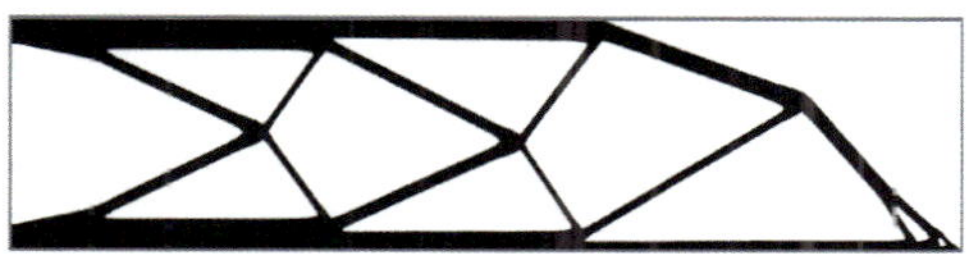

Fig. 6 A long cantilever beam with low material fraction and a low interfacial energy penalization

Fig. 7 A cantilever beam with four phases

As expected we can obtain different local minima if we start with different initial data as can be seen in Fig. 5 for a cantilever beam with two materials and void. The first column shows the result where the initial data is a constant mixture of materials and void, the second started with separated material distribution and the third with random data. The last yields the lowest value of the cost functional.

The following three Figs. 6–8 illustrate some results for a long cantilever beam with one material, for a case with three materials and void and an example for a cantilever beam in 3d with one material.

5.6 *Numerical Results for a Compliant Mechanism*

In this section we present a compliant mechanism simulation, in particular we set $\alpha = 0$ in $(\mathcal{P}^\varepsilon)$. The configuration we consider is depicted in Fig. 9, where zero

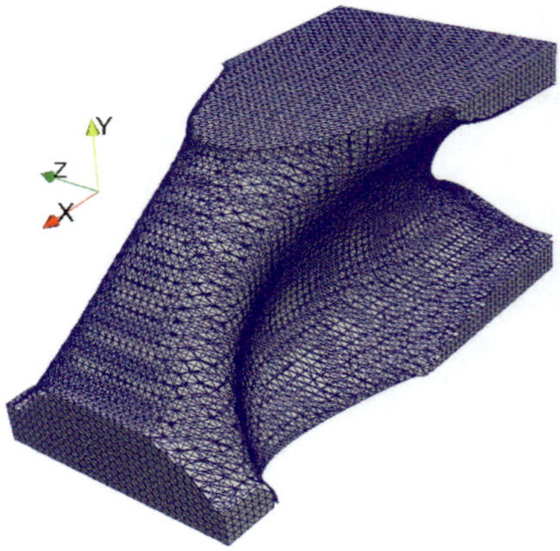

Fig. 8 A cantilever beam in three space dimensions

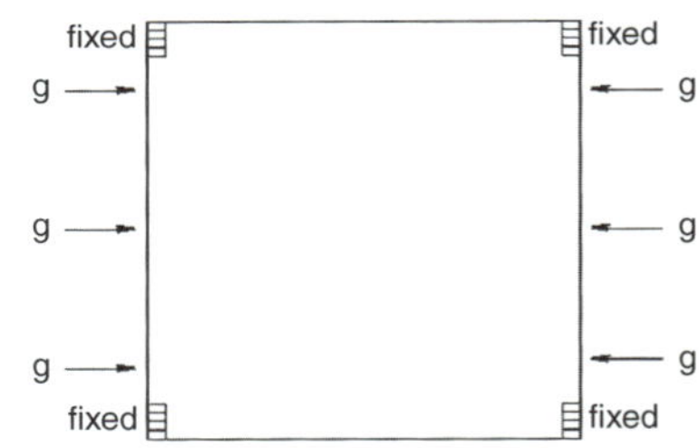

Fig. 9 Push configuration

Dirichlet boundary conditions are posed on the left and right boundaries at the top and bottom and horizontal forces are applied at sections along the left and right boundaries.

In order to solve the gradient inequality (GI) in Theorem 3.2, we use here as a first numerical approach a classical L^2-gradient flow dynamic for the reduced cost functional. The gradient flow yields the following parabolic variational inequality for all $\tilde{\boldsymbol{\varphi}} \in \boldsymbol{\mathcal{G}}^m$ and all $t > 0$:

$$\begin{aligned}
&\varepsilon \int_\Omega \frac{\partial \boldsymbol{\varphi}}{\partial t}(\tilde{\boldsymbol{\varphi}} - \boldsymbol{\varphi})dx + \gamma\varepsilon \int_\Omega \nabla\boldsymbol{\varphi} : \nabla(\tilde{\boldsymbol{\varphi}} - \boldsymbol{\varphi})dx + \frac{\gamma}{\varepsilon}\int_\Omega \Psi_0'(\boldsymbol{\varphi})\cdot(\tilde{\boldsymbol{\varphi}} - \boldsymbol{\varphi})dx \\
&-\frac{1}{2}\beta J_0(\boldsymbol{u}, \boldsymbol{\varphi})^{-1}\int_\Omega (\tilde{\varphi}^N - \varphi^N)\, c\, |\boldsymbol{u} - \boldsymbol{u}_\Omega|^2 \\
&-\int_\Omega (\tilde{\varphi}^N - \varphi^N)\boldsymbol{f}\cdot(\alpha\boldsymbol{u} + \boldsymbol{p}) - \langle \mathcal{E}(\boldsymbol{p}), \mathcal{E}(\boldsymbol{u})\rangle_{\mathbb{C}'(\boldsymbol{\varphi})(\tilde{\boldsymbol{\varphi}} - \boldsymbol{\varphi})} \geq 0. \qquad (5.4)
\end{aligned}$$

In addition, $\boldsymbol{u}$ and $\boldsymbol{p}$ have to solve the state equation (SE) and the adjoint equation (AE), see Theorem 3.2. The constraints $\varphi^N = 0$ on S_0 and $\varphi^N = 1$ on S_1 can be easily incorporated by imposing these conditions when a mesh point lies in $S_0 \cup S_1$. We replace $\frac{\partial \boldsymbol{\varphi}}{\partial t}$ in (5.4) by a time discrete approximation which corresponds to a pseudo time stepping approach. We then discretize the resulting inequality,

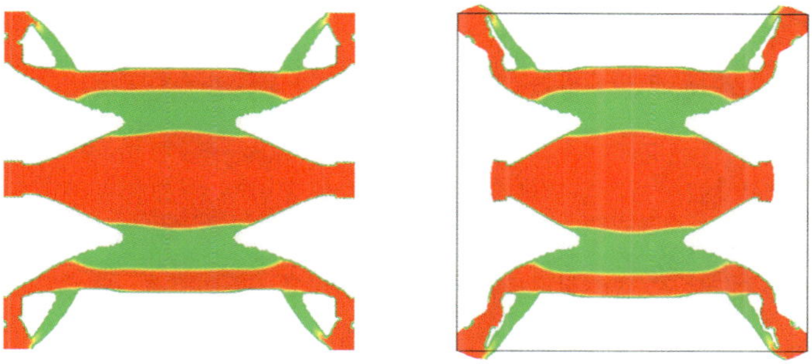

Fig. 10 Push simulation with three phases (*left*) and deformed configuration with the outline of the initial geometry (*right*)

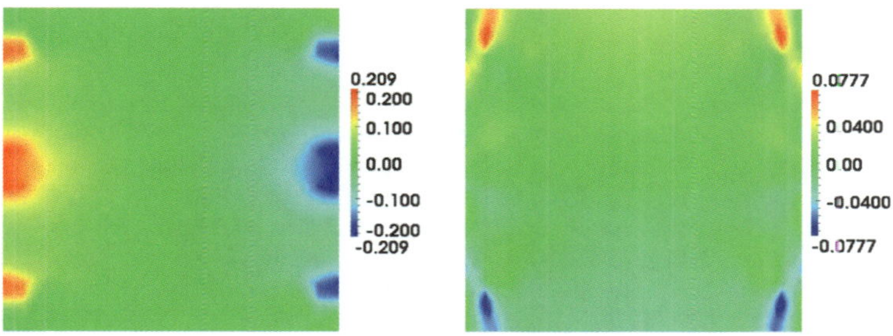

Fig. 11 Displacement vector $\mathbf{u}$, x-component (*left*), y component (*right*)

the state equation (SE) and the adjoint equation (AE) using standard finite element approximations, see [3–5].

In the computation we present we take the weighting factor $c = 2{,}000$ in $\Omega := (-1,1) \times (-1,1)$ and $\mathbf{u}_\Omega = \mathbf{0}$. We set $\Gamma_D = \{(-1,y) \cup (1,y) \in \mathbb{R}^2 : y \in [-1,-0.9] \cup [0.9,1]\}$ and $\Gamma_g = \Gamma_{g_-} \cup \Gamma_{g_+}$ with $\Gamma_{g_\pm} := \{(\pm 1, y) \in \mathbb{R}^2 : y \in [-0.8,-0.7] \cup [-0.1,0.1] \cup [0.7,0.8]\}$. We take $\boldsymbol{g} = (\pm 7, 0)^T$ on $\Gamma_{g_\pm}$ and $S_1 = \emptyset$. Since we wish to have material adjacent to the parts of the boundary that are fixed and where the forces are applied we set $S_0 = \{(x,y) \in \mathbb{R}^2 : x \in [-1,-0.9] \cup [0.9,1], y \in [-1,-0.9] \cup [-0.8,-0.7] \cup [-0.1,0.1] \cup [0.7,0.8] \cup [0.9,1]\}$. We take $N = 3$ and use an isotropic elasticity tensor $\mathbb{C}^1$ of the form $\mathbb{C}^1\mathcal{E} = 2\mu_1\mathcal{E} + \lambda_1(tr\mathcal{E})I$ with $\lambda_1 = \mu_1 = 10$ and we choose $\mathbb{C}^2 = \frac{1}{2}\mathbb{C}^1$ and $\mathbb{C}^3 = \varepsilon^2\mathbb{C}^1$ in the void. The interfacial parameters we use are $\varepsilon = \frac{1}{18\pi}$ and $\gamma = 0.2$ and we set $\beta = 10$. In addition, we choose the masses $\boldsymbol{m} = (0.35, 0.15, 0.5)^T$.

In Fig. 10 we display the optimized configuration (left hand plot) and the deformed optimal configuration together with the outline of the initial geometry (right hand plot), here hard material is shown in red and soft material in green. In Fig. 11 we display the displacement vector $\mathbf{u}$.

References

1. G. Allaire, F. Jouve, A.-M. Toader, Structural optimization using sensitivity analysis and a level set method. J. Comput. Phys. **194**, 363–393 (2004)
2. M.P. Bendsoe, O. Sigmund, *Topology Optimization* (Springer, Berlin, 2003)
3. L. Blank, M.H. Farshbaf-Shaker, H. Garcke, V. Styles, Relating phase field and sharp interface approaches to structural topology optimization, to appear in ESIAM: Control, Optimization and Calculas of Variations, DOI: 10.1051/cocv/2014006
4. L. Blank, H. Garcke, L. Sarbu, T. Srisupattarawanit, V. Styles, A. Voigt, *Phase-Field Approaches to Structural Topology Optimization*, ed. by G. Leugering. Constrained Optimization and Optimal Control for Partial Differential Equations, vol. 160 (Springer, Basel, 2012), pp. 245–255
5. L. Blank, H. Garcke, L. Sarbu, V. Styles, Non-local Allen-Cahn systems analysis and primal dual active set method. IMA J. Numer. Anal. **33**(4), 1126–1155 (2013)
6. L. Blank, C. Rupprecht, An extension of the projected gradient method to a Banach space setting with application in structural topology optimization, in progress, 2014
7. B. Bourdin, A. Chambolle, Design-dependent loads in topology optimization. ESAIM Contr. Optim. Calc. Var. **9**, 19–48 (2003)
8. M. Burger, R. Stainko, Phase-field relaxation of topology optimization with local stress constraints. SIAM J. Control Optim. **45**, 1447–1466 (2006)
9. L. Dedè, M.J. Borden, T.J.R. Hughes, Isogeometric analysis for topology optimization with a phase field model. Arch. Comput. Methods Eng. **19**(3), 427–465 (2012)
10. W.A. Gruver, E. Sachs, *Algorithmic Methods in Optimal Control* (Pitman, Boston, 1981)
11. A. Logg, K.A. Mardal, G.N. Wells, *Automated Solution of Differential Equations by the Finite Element Method* (Springer, Berlin/New York, 2012)
12. L. Modica, The gradient theory of phase transitions and minimal interface criterion. Arch. Ration. Mech. Anal. **98**, 123–142 (1987)
13. P. Penzler, M. Rumpf, B. Wirth, A phase-field model for compliance shape optimization in nonlinear elasticity. ESAIM Control Optim. Calc. Var. **18**(1), 229–258 (2012)
14. J. Sokolowski, J.P. Zolesio, *Introduction to Shape Optimization: Shape Sensitivity Analysis*. Springer Series in Computational Mathematics, vol. 10 (Springer, Berlin, 1992)
15. A. Takezawa, S. Nishiwaki, M. Kitamura, Shape and topology optimization based on the phase field method and sensitivity analysis. J. Comput. Phys. **229**(7), 2697–2718 (2010)
16. M.Y. Wang, S.W. Zhou, Multimaterial structural topology optimization with a generalized Cahn-Hilliard model of multiphase transition. Struct. Multidisc. Optim. **33**(2), 89–111 (2007)

Part III
Adaptivity and Model Reduction

Introduction to Part III Adaptivity and Model Reduction

Peter Benner and Rolf Rannacher

Despite all effort and progress in the numerical techniques to solve PDE-constrained optimization and control problems, the cost for their solution is still substantially higher than that for solving the associated forward problem for the PDE. In a practical situation, the associated computational work and memory requirement may still be too high to be acceptable, e.g., in an engineering design process. Therefore, further techniques are needed to reduce the computational cost.

In the section "*Adaptivity and Model Reduction*" of this book, two different techniques are discussed for reducing the complexity of solving PDE-constrained optimization problems numerically. One possibility is to use tailored discretizations, e.g., finite element (FE) Galerkin methods, that adapt the mesh size locally according to the optimization goal. This usually leads to meshes different from a possibly optimal, adapted finite element mesh used for solving the forward problem alone. High accuracy of the PDE solution may not be necessary in the same regions as needed for an accurate computation of the cost functional of the optimization problem and the associated (sub)optimal control. Also, a changing control input during an optimization algorithm leads to different PDE solutions that may require different locally refined meshes, necessitating the adaptation of the mesh during the optimization procedure. Therefore, mesh adaptivity should be based on error bounds taking this goal-orientation into account. A further reduction of the computational cost may be achieved by adaptive stopping criteria providing a proper balancing

P. Benner (✉)
Max Planck Institute for Dynamics of Complex Technical Systems, Sandtorstraße 1, 39106 Magdeburg, Germany
e-mail: benner@mpi-magdeburg.mpg.de

R. Rannacher
Institut für Angewandte Mathematik, Ruprecht-Karls-Universität Heidelberg, Im Neuenheimer Feld 293, 69120 Heidelberg, Germany
e-mail: rannacher@iwr.uni-heidelberg.de

G. Leugering et al. (eds.), *Trends in PDE Constrained Optimization*, International Series of Numerical Mathematics 165, DOI 10.1007/978-3-319-05083-6_16

with the discretization error for the various outer and inner algebraic iterations in solving the discretized optimization problems.

A different approach for reducing the complexity of solving PDE-constrained optimization problems consists in model order reduction, i.e., the application of mathematical methods for automatically reducing the state-space dimension of the control problem while preserving the accuracy in the map from the input function or design parameters to the optimized quantity-of-interest. Like in adaptive FE methods, reduced-order models generated to accelerate the simulation of a PDE may not be sufficient in the optimization context: in an iterative optimization algorithm, the control function changes from step to step, and this variation of input may not be covered by a snapshot-based model reduction method like proper orthogonal decomposition (POD) that is based on a pre-defined training input. On the other hand, re-computing a reduced-order model in each optimization step may become too expensive. Thus, the construction of the reduced-order model should reflect the optimization goal.

The section "*Adaptivity and Model Reduction*" consists of four papers, two dealing with local mesh adaptation for optimal control problems and two are concerned with model reduction techniques. In the survey paper "*Model reduction by adaptive discretization in optimal control*" by Rannacher, an overview is given of goal-oriented adaptive FE methods for PDE-constrained optimization problems. For problems with singularities, a quasi-uniform mesh refinement is known to result in a reduced order of convergence. In the survey "*Graded meshes in optimal control for elliptic partial differential equations*" by Apel, Pfefferer and Rösch, the strategy of local mesh grading, known to recover the full convergence order for the forward problem, is discussed for elliptic optimal control problems. Regarding model reduction, the paper "*Model order reduction for PDE-constrained optimization*" by Benner, Sachs, and Volkwein provides a survey on approaches based on reduced-order models for solving PDE-constrained optimization problems. Due to the above-mentioned shortcomings of traditional model reduction methods used in forward simulation, special model management strategies are required. These either update the reduced-order model from a previous step or determine when a new reduced-order model must be computed in an optimization loop. Two successful strategies are discussed serving these purposes, adaptive POD and trust-region POD. Moreover, the application of snapshot-free methods based on system-theoretical considerations, having a wide validity range w.r.t. input variations, to PDE-constrained optimization problems is also considered. The use of trust-region POD is also the topic of "*Adaptive trust-region POD methods in PDE-constrained optimization*" by Sachs, Schneider, and Schu. Here, this model reduction strategy is extended to solving optimization problems subject to partial integro-differential equations such as occurring in calibration problems for the pricing of financial derivatives.

Model Reduction by Adaptive Discretization in Optimal Control

Rolf Rannacher

Abstract This article gives a survey of recent developments in the economical numerical solution of PDE-based optimal control problems by adaptive methods. Systematic mesh adaptivity combined with adaptive stopping criteria for linear and nonlinear algebraic iterations is one approach to reducing the computational cost to an acceptable level. These various steps of adaptivity are driven by "goal-oriented" a posteriori error estimates derived within the general framework of the DWR (Dual Weighted Residual) method. The presented material is mainly based on results obtained within the second funding period 2011–2013 of this subproject of the DFG Priority Program 1253 "Optimization with Partial Differential Equations". In this sense it is the continuation of the article "A posteriori error estimation in PDE-constrained optimization with pointwise inequality constraints" by R. Rannacher, B. Vexler, and W. Wollner in "Constrained Optimization and Optimal Control for Partial Differential Equations" (G. Leugering et al., eds), Birkhüser, Basel, 2012.

Keywords PDE-based optimization • Model reduction • Adaptive discretization • A posteriori error estimation • Goal-oriented adaptivity • DWR method • Adaptive stopping criteria

Mathematics Subject Classification (2010). Primary 35B37, 49J20, 49M05, 49M15, 49M29; Secondary 65K10, 65M50, 65M60, 65N22, 65N30, 65N50.

1 Introduction

The use of adaptive techniques based on a posteriori error estimation is well accepted in the context of finite element discretization of partial differential equations. There are mainly two approaches in this context: error estimation with

---

R. Rannacher (✉)
Institut für angewandte Mathematik, Universität Heidelberg, Im Neuenheimer Feld 293/294, 69120 Heidelberg, Germany
e-mail: rannacher@iwr.uni-heidelberg.de

G. Leugering et al. (eds.), *Trends in PDE Constrained Optimization*, International Series of Numerical Mathematics 165, DOI 10.1007/978-3-319-05083-6_17

respect to natural norms, see Verfürth [81], Ainsworth and Oden [1], and Babuska and Strouboulis [2] for surveys, and the "goal-oriented" approach going back to Eriksson et al. [28] and Becker and Rannacher [9]; see also the surveys Becker and Rannacher [11] and Bangerth and Rannacher [4]. In the last years both approaches were extended to optimal control problems (OCPs) governed by partial differential equations. In Gaevskaya et al. [29], Hoppe et al. [49], Li et al. [58], Liu and Yan [61], and Liu [60] a posteriori error estimates are derived with respect to natural norms for elliptic OCPs with distributed or Neumann control subject to box constraints on the control variable. For an OCP with pointwise state constraints a posteriori error estimates were derived in Hoppe and Kieweg [50].

The motivation of "goal-oriented" adaptivity is the fact that in many applications the error in global norms does not provide useful bounds for the error in the quantity of physical interest. This is the idea underlying the so-called "Dual Weighted Residual (DWR) method" for a posteriori error control and mesh adaptation developed in Becker and Rannacher [9, 11] and Bangerth and Rannacher [4]. This general concept can directly be used in the special situation of OCPs, where the error control functional is chosen as the given cost functional of the OCP; Kapp [53]. In Becker and Vexler [12, 13] this approach has been extended to the estimation of the discretization error with respect to an arbitrary functional – a so called "quantity of interest" – depending on both the control and the state variable. This allows, among other things, an efficient treatment of parameter identification and model calibration problems. The extension of these results to nonstationary problems has been developed in Meidner [65] and Meidner and Vexler [68], where separate error estimators for temporal and spatial discretization errors are derived. These error contributions are balanced in the corresponding adaptive algorithm. Only recently goal-oriented error estimation has been considered for optimal control problems subject to inequality constraints. The case of pointwise control constraints has been treated in Vexler and Wollner [82] and Hintermüller and Hoppe [42]. In Becker [6] similar techniques are used explicitly to estimate the error in the control with respect to its natural norm. For problems with pointwise state constraints recent work has been done simultaneously by Guenther and Hinze [34], Wollner [84], and Benedix and Vexler [14]. For a survey see Rannacher et al. [72]. In the present article these results are completed by incorporating multiple-shooting methods for enhancing the global stability in nonstationary problems and adaptive stopping criteria for the various linear and nonlinear algebraic iterations in solving the discretized OCPs.

2 The Framework of the DWR Method in Optimal Control

We consider the minimization of a cost functional subject to the state equation (elliptic or parabolic PDE)

$$J(u,q) \quad \to \quad \text{min!} \qquad \mathcal{A}(u,q) = 0, \tag{2.1}$$

possibly accompanied by control and/or state constraints $q \in Q_{\rm ad} \subset Q$ and $g(u) \le 0$. In this general setting the state equation in (2.1) may be stationary or nonstationary. The "state space" V is usually a Sobolev Hilbert space, e.g., $V = H^1(\Omega)$ or $V = H_0^1(\Omega)$ in the stationary case, incorporating the prescribed boundary conditions, and the "control space" Q may be a function space ("distributed control"), e.g., $Q = L^\infty(\Omega)$ or $Q = L^2(\partial\Omega)$, or a discrete space $Q = \mathbb{R}^m$. In this setting the state equation is usually given in variational form

$$\langle \mathcal{A}(u,q), \delta u\rangle = 0 \quad \forall \delta u \in V, \tag{2.2}$$

where $\langle \cdot, \cdot\rangle$ denotes the duality pairing between V and its dual V^*.

We assume that the above setting allows for the application of the Euler-Lagrange approach, which yields necessary first-order conditions for optimal solutions. Within this framework, in the basic unconstrained case without control and state constraints, an optimal solution $\{u,q\} \in V \times Q$ is characterized as stationary point of the Lagrangian functional

$$\mathcal{L}(u,q,z) := J(u.q) - \langle \mathcal{A}(u,q), z\rangle$$

defined on $V\times Q\times V$, where $z \in V$ is the corresponding adjoint state ("Lagrangian multiplier"). The resulting stationarity condition for the triple $\{u,q,z\} \in V\times Q\times V$ (so-called KKT system) reads

$$\mathcal{L}'_u(u,q,z)(\delta u) = 0 \quad \forall \delta u \in V, \tag{2.3a}$$

$$\mathcal{L}'_q(u,q,z)(\delta q) = 0 \quad \forall \delta q \in Q, \tag{2.3b}$$

$$\mathcal{L}'_z(u,q,z)(\delta z) = 0 \quad \forall \delta z \in V. \tag{2.3c}$$

For the discretization of the above OCP, we consider the finite element (FE) Galerkin method. The spatial discretization is by a standard "continuous" FE method ("cG(r) method") of polynomial degree $r \ge 1$, while the time discretization uses a "discontinuous" FE method ("dG(r) method") of polynomial degree $r \ge 0$. The use of a Galerkin discretization with the corresponding Galerkin orthogonality property is essential for the special approach to adaptivity – the DWR method – considered in this article. For a mesh-size parameter $h \in \mathbb{R}_+$ let $V_h \subset V$ and $Q_h \subset Q$ be standard finite element spaces defined on meshes $\mathcal{T}_h = \{T\}$ (triangular or quadrilateral in 2D and tetrahedral or hexahedral in 3D) covering $\bar\Omega$, which satisfy the usual regularity conditions (see Bangerth and Rannacher [4], Carey and Oden [19], or Brenner and Scott [18]). Then, the discretized OCP (without control and state constraints) seeks $\{u_h, q_h\} \in V_h \times Q_h$ such that

$$J(u_h, q_h) \to \min! \qquad \langle \mathcal{A}(u_h, q_h), \delta u_h\rangle = 0 \quad \forall \delta u_h \in V_h. \tag{2.4}$$

The corresponding discrete KKT system for stationary points $\{u_h, q_h, z_h\} \in V_h \times Q_h \times V_h$ reads

$$\mathcal{L}'_u(u_h, q_h, z_h)(\delta u_h) = 0 \quad \forall \delta u_h \in V_h, \tag{2.5a}$$

$$\mathcal{L}'_q(u_h, q_h, z_h)(\delta q_h) = 0 \quad \forall \delta q_h \in Q_h, \tag{2.5b}$$

$$\mathcal{L}'_z(u_h, q_h, \lambda_h)(\delta z_h) = 0 \quad \forall \delta z_h \in V_h. \tag{2.5c}$$

In the case of a Galerkin discretization (with exact evaluation of integrals) the system (2.5) is just the Galerkin discretization of the continuous KKT system (2.3), i.e., in this case "optimization" (formation of the necessary optimality condition) and "discretization" commute. For the error resulting from the Galerkin discretization of the KKT system (2.3), without control or state constraints, there holds the following a posteriori error representation (cf. Becker et al. [7] and Becker and Rannacher [11]):

$$J(u,q) - J(u_h, q_h) = \tfrac{1}{2}\rho^u(z-\psi_h) + \tfrac{1}{2}\rho^q(q-\chi_h) + \tfrac{1}{2}\rho^z(u-\varphi_h) + \mathcal{R}_h^{(3)}, \tag{2.6}$$

with arbitrary approximations $\{\varphi_h, \chi_h, \psi_h\} \in V_h \times Q_h \times V_h$ and a remainder $\mathcal{R}_h^{(3)}$, which is cubic in the errors $e^u := u - u_h$, $e^q := q - q_h$, and $e^z := z - z_h$. The residual functionals in (2.6) are explicitly given by

$$\rho^u(\cdot) := \mathcal{L}'_u(u_h, q_h, z_h)(\cdot) = J'_u(u_h, q_h)(\cdot) - \langle \mathcal{A}'_u(u_h, q_h)\,\cdot\,, z_h\rangle,$$
$$\rho^q(\cdot) := \mathcal{L}'_q(u_h, q_h, z_h)(\cdot) = J'_q(u_h, q_h)(\cdot) - \langle \mathcal{A}'_q(u_h, q_h)\,\cdot\,, z_h\rangle,$$
$$\rho^z(\cdot) := \mathcal{L}'_z(u_h, q_h, z_h)(\cdot) = -\langle \mathcal{A}(u_h, q_h), \cdot\rangle.$$

The evaluation of the error representation (2.6) requires approximations to the unknown "interpolation errors" $z - \psi_h$, $q - \chi_h$, and $u - \varphi_h$. For this, we use postprocessing of the computed discrete solutions $\{u_h, q_h, z_h\}$ by local higher-order interpolation, e.g., on quadrilateral meshes $\pi z_h := i_{2h}^{(2)} z_h$ may be taken as the patchwise bi-*quadratic* interpolation of the computed piecewise bi-*linear* z_h as depicted in Fig. 1.
This technique has been described and analyzed in detail in Becker and Rannacher [11] and Bangerth and Rannacher [4]. Then, setting

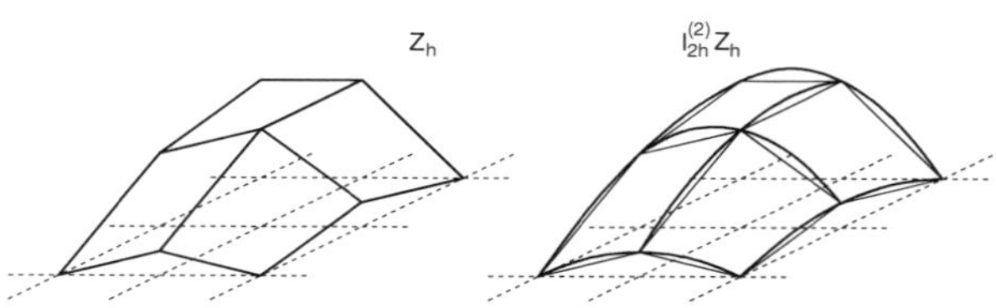

Fig. 1 Local post-processing by higher-order interpolation: "biquadratic" interpolation of computed "bilinear" nodal values

$$\eta_{\text{dis}} := \tfrac{1}{2}|\rho^u(\pi z_h - z_h) + \rho^q(\pi q_h - q_h) + \rho^z(\pi u_h - u_h)|, \tag{2.7}$$

for a prescribed error tolerance TOL the condition

$$\eta_{\text{dis}} \le \text{TOL} \quad \Rightarrow \quad \text{stop} \tag{2.8}$$

may be used as stopping criterion for the mesh adaptation process. Here, the nonlinear "cubic" remainder term $\mathcal{R}_h^{(3)}$ in (2.6) is neglected, which has turned out to be justified in most practical applications. However, there may be situations (e.g., problems close to bifurcation) in which the remainder term dominates the error representation and requires estimation.

For steering the local mesh adaptation (refinement or coarsening) the residual terms in (2.7) have to be localized to the single mesh cells $T \in \mathcal{T}_h$. A direct localization of the terms like $\rho^z(u_h, z_h)(\pi u_h - \tilde{u}_h)$ to each mesh cell leads, in general, to local contributions of wrong order (overestimation) due to oscillatory behavior of the residual terms. To overcome this problem, one may integrate in the residual terms cellwise by parts (see Becker and Rannacher [11]) or use a nodewise filtering operator (see Schmich and Vexler [77]). This results in cellwise "error indicators" $\eta_T \ge 0$, which govern the mesh adaptation by the usual rule

$$\eta_T \ge 4\frac{\text{TOL}}{N} \Rightarrow \text{refine } T, \quad \eta_T \le \frac{1}{4}\frac{\text{TOL}}{N} \Rightarrow \text{coarsen } T,$$

where $N = \dim V_h$. Obviously, this rule is rather crude and, since N is only implicitly defined, requires costly iteration. Alternatively, one may order the error indicators according to their size, $0 \le \eta_1 \le \cdots \le \eta_N$, and then refine or coarsen a fixed percentage of cells with largest or smallest indicator values, respectively. This "fixed fraction strategy" ensures that in each refinement cycle a sufficiently large number of cells is refined or the total number of cells is kept constant. For a more detailed discussion of this issue, especially of "optimal" strategies of mesh adaptation, and the derivation of local error indicators η_T, we refer to Bangerth and Rannacher [4]. It is important to use the "global" error estimator η_{dis} in the stopping criterion (2.8) rather than its usually too crude upper bound

$$|J(u,q) - J(u_h, q_h)| \approx \eta_{\text{dis}} \le \sum_{T \in \mathcal{T}_h} \{\eta_T^u + \eta_T^q + \eta_T^\lambda\}.$$

The framework of mesh adaptation described so far is the essence of the DWR method. Since its development in 1995 (Becker and Rannacher [9, 10]) this technique has been successfully applied to various problems in sciences and engineering, including problems in structural and fluid mechanics, chemically reactive flow, radiative transfer, and most recently also optimal control and parameter estimation; for overviews see Becker and Rannacher [11] and Bangerth and Rannacher [4].

Remark 2.1. In the above framework of residual-based error estimation and mesh adaptation in the approximation of OCPs the essential point is that of the concept of "admissibility" of states. Here, "admissible" is meant relatively to the target functional $J(\cdot)$ to be optimized. A computed "optimal" control q_h^{opt} may be reasonable even if the corresponding "optimal" state $u_h^{\text{opt}}(q_h^{\text{opt}})$ is admissible in an only weak sense, i.e. satisfies the state equation only approximately with possible large deviations in certain parts of the computational domain. The mesh adaptation is driven by residual- and sensitivity-based a posteriori error estimates, which is natural for this type of problems. Significant reduction of complexity and work can be achieved by adaptive discretization and parameter tuning, dynamic stabilization by multiple-shooting techniques, and balanced stopping criteria for nonlinear and linear algebraic iterations.

The development described above has resulted by 2006 in the following state-of-the-art in residual-based adaptivity in solving stationary OCPs:

- Elliptic reaction-diffusion problems in 2D (Kapp [53] and Becker et al. [7])
- Flow control problems in 2D (Becker [5])
- Parameter estimation (Becker and Vexler [12])
- Problems with control constraints (Vexler and Wollner [82])

Since then these initial results have been developed further into several directions:

1. Development of the DWR method for stationary OCPs with control and state constraints (cf. Benedix and Vexler [14] and Wollner [85–87]): In the case of control constraints of box type, $Q_{\text{ad}} := \{q \in Q \mid a \le q \le b, \text{ a.e. in } \Omega\}$, the variational *equality* (2.3b) becomes a variational *inequality* of the form

$$\mathcal{L}'_q(u_h, q_h, z_h)(\delta q_h - q_h) \ge 0 \quad \forall \delta q_h \in Q_{\text{ad},h}.$$

   A brief survey of solution methods for control-constrained OCPs is given in Herzog and Kunisch [37]. For this case an a posteriori error representation similar to (2.6) can be derived. However, due to the control constraint more care has to be taken in approximating the residual term $\rho^q(q{-}q_h)$ where in this case q_h cannot be replaced by an arbitrary element in the discrete control space $Q_{\text{ad},h}$. In this context mesh-adaptivity in the traditional norm-oriented sense has been studied in Hintermüller et al. [43] and Hintermüller and Kunisch [46]. In the case of state constraints $g(u) \le 0$ the adjoint variable is only a measure the approximation of which is troublesome (cf. Meyer et al. [70], Hintermüller and Hinze [41], Hinze and Schiela [48], and Hintermüller and Kunisch [45, 46]). In order to avoid this complication, one may use a barrier or penalty technique, i.e., the cost functional is modified like (cf. Bergounioux [15] and Rannacher et al. [72])

$$J_\gamma(u,q) := J(u,q) + \frac{\gamma}{2}\|g(u)^+\|_\Omega^2,$$

where $g(u)^+$ denotes the positive part of $g(u)$, or (see Wollner [85])

$$J_\gamma(u,q) := J(u,q) - \gamma \int_\Omega \log(g(u(x)))\,dx.$$

These techniques have also been used for incorporating higher-order state constraints of the form $|\nabla u| \le c$ with applications in elasto-plasticity (see Wollner [84–87], Schiela and Wollner [75], and also Ortner and Wollner [71]). Related work on a priori and a posteriori error estimates for state-constrained elliptic OCPs is described in Casas [21], Casas and Fernandez [22], Deckelnick [24], and Günther et al. [35].
2. Development of the DWR method for simultaneous spatial mesh and time step adaptivity in the solution of nonstationary OCPs (cf. Meidner [65], Meidner and Vexler [68, 68], and Meidner et al. [66]): In the nonstationary case (without control or state constraints), one obtains a posteriori error representations such as (2.6), in which the effects of time and space discretization are separated and can therefore also be adapted separately. For the rather technical details, we refer to Meidner and Vexler [69]. Space-time adaptivity in the framework of the DWR method has been developed in Schmich and Vexler [77] for standard parabolic problems and in Schmich [76] and Besier and Rannacher [17] for the nonstationary Navier-Stokes equations.
3. Adaptive tuning of regularization parameters for ill-posed OPCs (cf. Griesbaum et al. [31] and Kaltenbacher et al. [52]).
4. Adaptive balancing of discretization and algebraic iteration errors in solving the KKT system of OCPs (see Meidner et al. [67], Rannacher et al. [74], and Rannacher and Vihharev [73]).
5. Adaptive multiple shooting solution of the KKT system of parabolic OCPs (cf. Hesse and Kanschat [39] and Carraro et al. [20]).

The results on item (1) have been surveyed in Rannacher et al. [72], while the research on items (2) and (3) has largely be done in cooperations outside this project. In this article, we report on the recent results obtained for items (4) and (5). The numerical computations cited from the corresponding literature have been done using the following software environments:

- **GASCOIGNE** (http://www.gascoigne.de),
- **RoDoBo** (http://www.rodobo.uni-hd.de),
- **deal.II** (http://www.dealii.org),
- **DOpElib** (http://www.dopelib.net).

3 Balancing of Discretization and Algebraic Iteration Errors

The material presented in this section is based on Meidner et al. [67] and Rannacher and Vihharev [73]. The error representation (2.6) assumes that the discrete solution $\{u_h, q_h, z_h\}$ is computed exactly, which in reality is not possible. We rather obtain an approximation $\{\tilde{u}_h, \tilde{q}_h, \tilde{z}_h\}$ by an iterative solution process on the discrete level. The estimation of this "iteration error" and its balancing with the "discretization error" is the subject of the following discussion.

3.1 The "Linear" Case

We consider the linear-quadratic OCP

$$\begin{aligned} J(u,q) &:= \tfrac{1}{2}\|u-\bar{u}\|_\Omega^2 + \tfrac{1}{2}\alpha\|q\|_\Omega^2 \quad \to \quad \min! \\ -\Delta u &= f + q \quad \text{in } \Omega := (0,1)^2, \quad u = 0 \quad \text{on } \partial\Omega, \end{aligned} \tag{3.1}$$

with state u, force f, target $\bar{u}$, $\alpha = 10^{-3}$, and "distributed control" q. In this case the Euler-Lagrange method uses the Lagrangian functional (dropping the subscript Ω)

$$\mathcal{L}(u,q,z) := J(u,q) + (f+q,z) - (\nabla u, \nabla z),$$

with the adjoint variable z. For any optimal pair $\{u,q\} \in V \times Q := H_0^1(\Omega) \times L^2(\Omega)$ there exists an adjoint state $z \in V$ such that the triplet $\{u,q,z\}$ is a stationary point of the Lagrangian, i.e., it solves the *linear* KKT system (cf. Tröltzsch [79]):

$$(\nabla \delta u, \nabla z) - (u, \delta u) = -(\bar{u}, \delta u) \quad \forall \delta u \in V, \tag{3.2a}$$

$$(\delta q, z) + \alpha(\delta q, q) = 0 \quad \forall \delta q \in Q, \tag{3.2b}$$

$$(\nabla u, \nabla \delta z) - (q, \delta z) = (f, \delta z) \quad \forall \delta z \in V. \tag{3.2c}$$

Using conforming (bilinear) Q_1 functions for discretizing all three variables $\{u,q,z\}$ in FE subspaces $V_h \subset V$ and $Q_h \subset Q$ results in the discrete KKT systems

$$(\nabla \delta u_h, \nabla z_h) - (u_h, \delta u_h) = -(\bar{u}, \delta u_h) \quad \forall \delta u_h \in V_h, \tag{3.3a}$$

$$(\delta q_h, z_h) + \alpha(\delta q_h, q_h) = 0 \quad \forall \delta q_h \in Q_h, \tag{3.3b}$$

$$(\nabla u_h, \nabla \delta z_h) - (q_h, \delta z_h) = (f, \delta z_h) \quad \forall \delta z_h \in V_h. \tag{3.3c}$$

These linear saddle point problems are solved by a multigrid (MG) method using simple block iterations as smoothers. In the course of this iteration, we obtain approximate discrete solutions $\{\tilde{u}_h, \tilde{q}_h, \tilde{z}_h\}$. The following theorem provides an error estimate for the resulting combined discretization and iteration error.

Theorem 3.1 (Meidner et al. [67]). *Let $\{u, q, z\} \in V \times Q \times V$ be the solution of the KKT system and $\{\tilde{u}_h, \tilde{q}_h, \tilde{z}_h\} \in V_h \times Q_h \times V_h$ the approximative solution of the discrete KKT system on the mesh $\mathcal{T}_h$. Then,*

$$
\begin{aligned}
J(u,q) - J(\tilde{u}_h, \tilde{q}_h) &= \tfrac{1}{2}\rho^z(\tilde{u}_h, \tilde{z}_h)(u - \tilde{u}_h) + \tfrac{1}{2}\rho^q(\tilde{q}_h, \tilde{z}_h)(q - \tilde{q}_h) \\
&\quad + \tfrac{1}{2}\rho^u(\tilde{u}_h, \tilde{q}_h)(z - \tilde{z}_h) + \rho^u(\tilde{u}_h, \tilde{q}_h)(\tilde{z}_h) \qquad (3.4) \\
&=: \eta_{\text{dis}} + \eta_{\text{it}},
\end{aligned}
$$

with the residuals

$$
\begin{aligned}
\rho^z(\tilde{u}_h, \tilde{z}_h)(\cdot) &:= (\tilde{u}_h - \bar{u}, \cdot) - (\nabla \cdot, \nabla \tilde{z}_h), \\
\rho^q(\tilde{q}_h, \tilde{z}_h)(\cdot) &:= \alpha(\cdot, \tilde{q}_h) + (\cdot, \tilde{z}_h), \\
\rho^u(\tilde{u}_h, \tilde{q}_h)(\cdot) &:= (f + \tilde{q}_h, \cdot) - (\nabla \tilde{u}_h, \nabla \cdot).
\end{aligned}
$$

Remark 3.2. The proof of Theorem 3.1 is a simple variation of the corresponding argument for the unperturbed nonlinear case leading to (2.6). Using the particular projection structure of the multigrid algorithm the iteration residual term $\rho^u(\tilde{u}_h, \tilde{q}_h)(\tilde{z}_h)$ can be rewritten in a more detailed form, which provides further insight into the performance of the smoothing iteration on the different mesh levels.

3.1.1 Evaluation of the Error Representation

The evaluation of the iteration residual $\eta_{\text{it}} = \rho^u(\tilde{u}_h, \tilde{q}_h)(\tilde{z}_h)$ only involves the actually computed approximations $\tilde{u}_h, \tilde{q}_h, \tilde{z}_h$ and therefore does not require any approximation. The discretization residual η_{dis}, however, involves the unknown exact solution $\{u, q, z\}$, which needs to be approximated. For this, we employ post-processing using local higher-order interpolation as described above, which yields approximations $\{\pi\tilde{u}_h, \pi\tilde{z}_h\}$. This technique usually works satisfactorily for the primal and adjoint states. The approximation of the term $\rho^q(\tilde{q}_h, \tilde{z}_h)(q - \tilde{q}_h)$ usually requires more care. In contrast to state and adjoint state the control variable q can generally not be approximated by "local higher-order approximation" for the following reasons: Firstly, in the case of finite dimensional control space Q there is no "patch-like" structure. Secondly, if q is a distributed control, it typically does not possess sufficient smoothness (due to the inequality constraints) for the improved approximation property. Therefore in Rannacher et al. [72] an alternative approach has been proposed, which employs the projection of the control space into the set of *admissible* controls, $\pi^q : Q_h \to Q_{\text{ad}}$. This construction again results in an estimator for the discretization error similar to that in (2.7),

$$\eta_{\text{dis}} \approx \tfrac{1}{2}\rho^z(\tilde u_h,\tilde z_h)(\pi\tilde u_h-\tilde u_h) + \tfrac{1}{2}\rho^q(\tilde q_h,\tilde z_h)(\pi^q\tilde q_h-\tilde q_h) + \tfrac{1}{2}\rho^u(\tilde u_h,\tilde q_h)(\pi\tilde z_h-\tilde z_h).$$

In order to use this error estimator for guiding mesh refinement, we have to localize it to cell-wise or node-wise contributions. This may be achieved as described above.

3.1.2 Numerical Example for the Liner Case

We consider the above linear-quadratic OCP with target distribution $\bar u = \frac{2\pi^2-1}{2\pi^2}\sin(\pi x)\sin(\pi y)$ and the exact solution $u = -\frac{1}{2\pi^2}\sin(\pi x)\sin(\pi y)$, $q = \frac{1}{2\alpha\pi^2}\sin(\pi x)\sin(\pi y)$, $z = -\frac{1}{2\pi^2}\sin(\pi x)\sin(\pi y)$, and the forcing term f being accordingly adjusted. The corresponding KKT system is discretized as described above. The resulting discrete saddle point system is solved by a multigrid method with the stopping criterion

$$|\eta_{\text{it}}| \le \tfrac{1}{10}|\eta_{\text{dis}}|, \tag{3.5}$$

using the V-cycle with, firstly, $4+4$ block-ILU smoothing steps and, secondly, only one undamped block Jacobi smoothing step on each level. For measuring the quality of the above error estimates, we use the "effectivity indices"

$$I_{\text{eff}}^{\text{dis}} := \left|\frac{\eta_{\text{dis}}}{E_{\text{dis}}}\right|, \qquad I_{\text{eff}}^{\text{it}} := \left|\frac{\eta_{\text{it}}}{E_{\text{it}}}\right|,$$

where $E_{\text{dis}} := J(u,q) - J(u_h,q_h)$, $E_{\text{it}} := J(u_h,q_h) - J(\tilde u_h,\tilde q_h)$ and $E := J(u,q) - J(\tilde u_h,\tilde q_h)$ are the exact errors. Tables 1 and 2 show that the adaptive stopping criterion (3.5) leads to a good match of discretization and iteration errors and that the separate error estimators are sharp.

Table 1 Results for the linear-quadratic model problem with $\alpha = 10^{-3}$: MG method with $4+4$ block-ILU smoothing

N	E	#it	E_{dis}	η_{dis}	$I_{\text{eff}}^{\text{dis}}$	E_{it}	η_{it}	$I_{\text{eff}}^{\text{it}}$
25	9.35e−4	2	9.35e−4	1.83e−3	1.96	1.14e−7	1.97e−7	1.73
81	1.64e−4	2	1.78e−4	2.19e−4	1.22	1.42e−5	1.68e−5	1.18
289	3.75e−5	2	4.16e−5	4.39e−5	1.05	4.13e−6	4.33e−6	1.04
1,089	1.05e−5	2	1.02e−5	1.03e−5	1.01	3.48e−7	3.52e−7	1.01
3,985	2.67e−6	2	2.54e−6	2.55e−6	1.00	1.28e−7	1.28e−7	1.00
13,321	6.65e−7	2	6.48e−7	6.49e−7	1.00	1.63e−8	1.63e−8	1.00
47,201	1.76e−7	2	1.70e−7	1.69e−7	0.99	6.76e−9	6.77e−9	1.00
163,361	4.89e−8	2	4.69e−8	4.68e−8	0.99	1.97e−9	1.97e−9	1.00

Table 2 Results for the linear-quadratic model problem with $\alpha = 10^{-3}$: MG method with 4 + 4 block Jacobi smoothing

N	E	#it	E_{dis}	η_{dis}	$I_{\text{eff}}^{\text{dis}}$	E_{it}	η_{it}	$I_{\text{eff}}^{\text{it}}$
25	9.44e−4	4	1.83e−3	9.35e−4	1.96	1.55e−5	8.99e−6	1.73
81	1.84e−4	5	2.20e−4	1.78e−4	1.23	7.59e−6	6.44e−6	1.18
289	4.36e−5	5	4.40e−5	4.16e−5	1.05	2.04e−6	1.96e−6	1.04
1,089	1.10e−5	4	1.03e−5	1.02e−5	1.01	8.53e−7	8.44e−7	1.01
3,985	2.69e−6	4	2.55e−6	2.56e−6	0.99	1.31e−7	1.30e−7	1.00
13,321	6.94e−7	4	6.47e−7	6.69e−7	0.96	2.51e−8	2.51e−8	1.00
47,201	1.95e−7	4	1.69e−7	1.90e−7	0.88	4.39e−9	4.40e−9	1.00
171,969	7.24e−8	3	4.42e−8	6.93e−8	0.63	3.07e−9	3.10e−9	0.99

3.2 The "Nonlinear" Case

Next, we consider a general nonlinear optimization problem of the form

$$J(u,q) := J_1(u) + J_2(q) \quad \to \quad \min!$$

subject to the constraint

$$A(u,q)(\delta u) = 0 \quad \forall \delta u \in V,$$

with cost functional $J(\cdot,\cdot)$ and a semilinear form $A(\cdot,\cdot)(\cdot)$. We assume that there is a locally unique minimum $\{u,q\} \in V \times Q$, which corresponds to a saddle-point of the Lagrange functional

$$\mathcal{L}(u,q,z) = J(u,q) - A(u,q)(z),$$

where $z \in V$ denotes the associated adjoint state. The triplet $\{u,q,z\} \in V \times Q \times V$ is determined by the first-order optimality condition (KKT system):

$$A'_u(u,q)(\delta u, z) = J'_1(u)(\delta u) \quad \forall \delta u \in V, \tag{3.6a}$$

$$A'_q(u,q)(\delta q, z) = J'_2(q)(\delta q) \quad \forall \delta q \in Q, \tag{3.6b}$$

$$A(u,q)(\delta z) = 0 \quad \forall \delta z \in V. \tag{3.6c}$$

This nonlinear system is solved by the Newton method. Let, for $x = \{u,q,z\}$, $H(x)$ be the Hessian operator of the Lagrangian $\mathcal{L}(\cdot,\cdot,\cdot)$. Then, the Newton increment $\delta x^n = \{\delta u^n, \delta q^n, \delta\lambda^n\}$ is determined by the linear system

$$H(x^n)\delta x^n = -\mathcal{L}'(x^n) \tag{3.7}$$

and the Newton update (with damping θ_n) reads $x^{n+1} = x^n + \theta_n \delta x^n$. Usually this Newton iteration is performed in inexact form, i.e., the linear Newton equations (3.7) are solved only approximately (e.g., by an MG method). Hence, we have to balance three sources of errors:

$$\eta_{\text{dis}} \approx \eta_{\text{it}}^{\text{nonlin}} \approx \eta_{\text{it}}^{\text{lin}}. \tag{3.8}$$

This can be achieved on the basis of the following general result.

Theorem 3.2 (Rannacher and Vihharev [73]). *Let $\{\tilde{u}_h, \tilde{q}_h, \tilde{z}_h\} \in V_h \times Q_h \times V_h$ be an approximation to the solution $\{u, q, z\} \in V \times Q \times V^*$ of the KKT system obtained by any iterative process on the mesh $\mathcal{T}_h$. Then, there holds the following error representation:*

$$\begin{aligned} J(u,q) - J(\tilde{u}_h, \tilde{q}_h) &= \tfrac{1}{2}\rho^z(\tilde{u}_h, \tilde{q}_h, \tilde{z}_h)(u-\tilde{u}_h) + \tfrac{1}{2}\rho^q(\tilde{q}_h, \tilde{z}_h)(q-\tilde{q}_h) \\ &\quad + \tfrac{1}{2}\rho^u(\tilde{u}_h, \tilde{q}_h)(z-\tilde{z}_h) - \rho^u(\tilde{u}_h, \tilde{q}_h)(\tilde{z}_h) + \mathcal{R}_h^{(3)}, \end{aligned} \tag{3.9}$$

where the residual terms are given by

$$\begin{aligned} \rho^z(\tilde{u}_h, \tilde{q}_h, \tilde{z}_h)(\cdot) &:= J_1'(\tilde{u}_h)(\cdot) - A_u'(\tilde{u}_h, \tilde{q}_h)(\cdot, \tilde{z}_h), \\ \rho^q(\tilde{q}_h, \tilde{z}_h)(\cdot) &:= J_2'(\tilde{q}_h)(\cdot) - A_q'(\tilde{u}_h, \tilde{q}_h)(\cdot, \tilde{\lambda}_h), \\ \rho^u(\tilde{u}_h, \tilde{q}_h)(\cdot) &:= -A(\tilde{u}_h, \tilde{q}_h)(\cdot), \end{aligned}$$

and the remainder term $\mathcal{R}_h^{(3)}$ is cubic in the errors $e^u := u - \tilde{u}_h$, $e^q := q - \tilde{q}_h$, and $e^z := z - \tilde{z}_h$.

We use the first residual terms for estimating the discretization error of an iterated approximation $\{u_h^n, q_h^n, z_h^n\}$:

$$\eta_{\text{dis}}^n := \tfrac{1}{2}\big|\rho^z(u_h^n, q_h^n, z_h^n)(\pi u_h^n - u_h^n) + \rho^q(q_h^n, z_h^n)(\pi q_h^n - q_h^n) + \rho^u(u_h^n, q_h^n)(\pi z_h^n - z_h^n)\big|.$$

For estimating the error caused by solving the Newton equations only approximately, we consider the difference of two consecutive approximations, $x_h^n = \{u_h^n, q_h^n, z_h^n\} \in V_h \times Q_h \times V_h$, obtained be the n-th Newton iterates on the discrete level $\mathcal{T}_h$. There holds

$$\begin{aligned} J(u_h^{n+1}, q_h^{n+1}) - J(u_h^n, q_h^n) &= \langle d_h^n, \widetilde{\delta u}_h^n\rangle + \langle g_h^n, \widetilde{\delta q}_h^n\rangle - \langle r_h^n, z_h^n\rangle \\ &\quad + \rho(u_h^n, q_h^n)(z_h^n) \; + \; O\big(|\widetilde{\delta u}_h^n|^2 + |\widetilde{\delta q}_h^n|^2\big), \end{aligned} \tag{3.10}$$

with the corresponding linear iteration residuals $\langle d_h^n, \cdot\rangle$, $\langle g_h^n, \cdot\rangle$, $\langle r_h^n, \cdot\rangle$. Then, the adaptive stopping strategy uses the following error indicators:

$$\eta_{\text{it}}^{n,\text{out}} := |\rho(u_h^n, q_h^n)(z_h^n)|, \quad \eta_{\text{it}}^{n,\text{in}} := \max\big\{|\langle r_h^n, z_h^n\rangle|, |\langle d_h^n, \widetilde{\delta u}_h^n\rangle|, |\langle g^n, \widetilde{\delta q}_h^n\rangle|\big\}.$$

The "inner" linear Newton correction equations and the "outer" nonlinear Newton iteration are iterated until

$$\eta_{\rm it}^{n,{\rm in}} \le \tfrac{1}{10}\,\eta_{\rm it}^{n,{\rm out}}, \qquad \eta_{\rm it}^{n,{\rm out}} \le \tfrac{1}{10}\,\eta_{\rm dis}^{n}. \tag{3.11}$$

3.2.1 Numerical Example for the Nonlinear Case

We consider the OCP

$$J(u,q) := \tfrac{1}{2}\|u-\bar u\|_\Omega^2 + \tfrac{1}{2}\alpha\|q\|_\Omega^2 \quad \to \quad \text{min!}$$

subject to the nonlinear PDE constraint

$$-\varepsilon\Delta u + \frac{q}{(1+u)^2} = 0, \quad \text{in } \Omega := (0,1)^2, \qquad u=0, \ \text{ on } \partial\Omega, \tag{3.12}$$

and the state constraint $u_a \le u$, in $\overline{\Omega}$, with $\varepsilon = 10^{-4}$ and $u_a = -0.99$. The target distribution is given by $\bar u(x) = -u_a - 3\,u_a\,|x-(0.5,0.5)|$. For treating the state constraint a barrier approach is employed, (see Wollner [85] and Rannacher et al. [72]), i.e., the cost functional is augmented as follows:

$$J_\gamma(u,q) = J(u,q) + b_\gamma(u), \quad b_\gamma(u) := -\int_\Omega \gamma\log(u-u_a)\,dx.$$

In the tests the parameters are chosen as $\alpha = 10^{-6}$ and $\gamma = 10^{-4}$. For this modified problem the appropriate solution spaces are $V = H_0^1(\Omega)$ for the state function and $Q := L^2(\Omega)$ for the control. For any optimal solution $\{u,q\} \in V\times Q$ there exists an adjoint solution $z \in V$ such that the triplet $\{u,q,z\} \in V\times Q\times V$ solves the following KKT system (dropping again the subscript Ω):

$$(\nabla\delta u,\nabla z) - 2\,(q\,(1+u)^{-3}\,z,\delta u) - (u-\bar u,\delta u) = b_\gamma'(u)(\delta u, z), \tag{3.13a}$$

$$\alpha(\delta q, q) - (\delta q\,(1+u)^{-2}, z) = 0, \tag{3.13b}$$

$$(\nabla u, \nabla\delta z) + (q\,(1+u)^{-2},\delta z) = 0, \tag{3.13c}$$

for all $\{\delta u,\delta q,\delta z\} \in V\times Q\times V$. This KKT system is discretized again by the FEM using conforming bilinear functions for all three variables $\{u,q,z\}$. The resulting nonlinear saddle point problems are solved by an inexact Newton iteration, where in each linear substep a V-cycle MG is employed with one block-ILU pre- and post-smoothing step on each mesh level. The starting values for the Newton iteration are taken from the approximate solution on the preceding coarser mesh level. For the "algebraic" stopping criterion (iteration to convergence), we require that the initial nonlinear and linear residuals are reduced by the factor 10^{-11} (Figs. 2 and 3).

Tables 3–5 show the convergence history of the tree different approximation processes involving increasing degrees of adaptivity. The total effectivity index is defined by $I_{\text{eff}}^{\text{tot}} := (\eta_{\text{dis}} + \eta_{\text{it}})/E$. The reported computational results confirm the effectivity of the algebraic adaptive strategy using the stopping criterion (3.11).

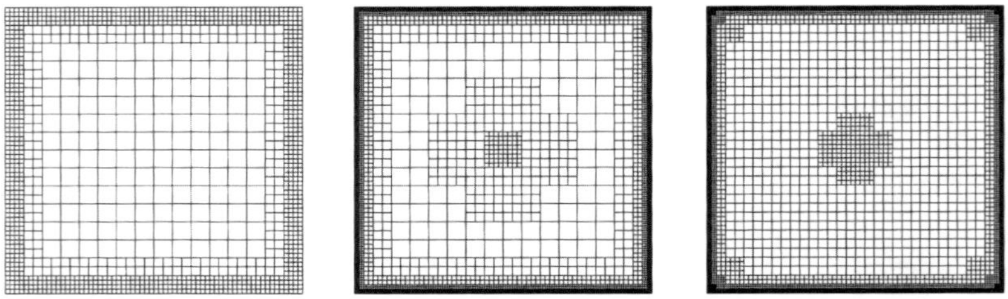

Fig. 2 Locally refined meshes

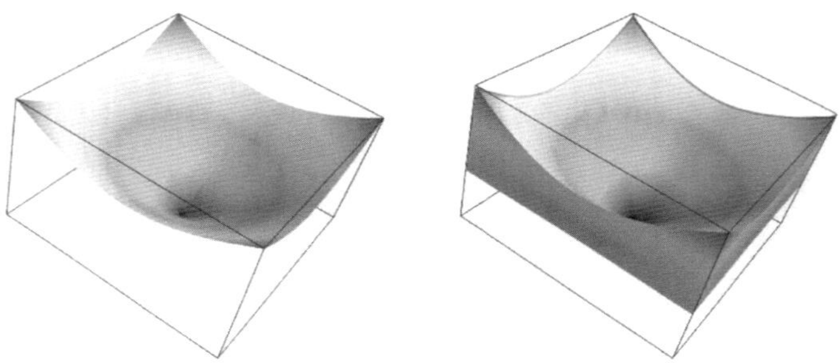

Fig. 3 Target $\bar{u}$ (*left*) and optimal state u (*right*)

Table 3 (1) Fully converged "exact" Newton iteration

N	#it	E	$\eta_{\text{dis}} + \eta_{\text{it}}$	η_{dis}	η_{it}	$I_{\text{eff}}^{\text{tot}}$
3,313	15	3.96e−03	2.54e−03	2.54e−03	1.86e−22	0.64
6,897	6	1.61e−03	1.30e−03	1.30e−03	4.15e−17	0.81
14,113	5	5.56e−04	4.50e−04	4.50e−04	2.05e−15	0.82
28,609	5	1.72e−04	9.49e−05	9.49e−05	2.37e−14	0.55
57,649	4	4.50e−05	3.45e−05	3.45e−05	2.81e−15	0.77
121,777	4	1.03e−05	9.99e−06	9.99e−06	4.09e−14	0.97

Table 4 (2) Adaptively stopped "exact" Newton iteration

N	#it	E	$\eta_{\mathrm{dis}} + \eta_{\mathrm{it}}$	η_{dis}	η_{it}	$I_{\mathrm{eff}}^{\mathrm{tot}}$
3,313	3	3.96e−03	2.68e−03	2.68e−03	2.13e−08	0.68
6,897	2	1.61e−03	1.30e−03	1.33e−03	2.78e−06	0.81
14,113	2	5.56e−04	4.53e−04	4.72e−04	6.74e−06	0.81
28,609	2	1.72e−04	9.57e−05	9.50e−05	6.68e−07	0.55
57,649	2	4.50e−05	3.45e−05	3.45e−05	8.91e−09	0.77
121,777	2	1.03e−05	9.99e−06	9.99e−06	1.08e−09	0.97

Table 5 (3) Adaptively stopped "inexact" Newton iteration

N	#it	E	$\eta_{\mathrm{dis}} + \eta_{\mathrm{it}}$	η_{dis}	η_{it}	$I_{\mathrm{eff}}^{\mathrm{tot}}$
3,313	3	3.96e−03	2.68e−03	2.68e−03	2.14e−08	0.68
6,897	2	1.61e−03	1.30e−03	1.33e−03	2.78e−06	0.81
14,113	2	5.56e−04	4.54e−04	4.73e−04	6.74e−06	0.82
28,609	2	1.72e−04	9.59e−05	9.50e−05	8.73e−07	0.55
57,649	2	4.50e−05	3.44e−05	3.45e−05	6.75e−08	0.77
121,777	2	1.03e−05	9.99e−06	9.99e−06	1.32e−09	0.97

4 Nonstationary Problems: The Multiple Shooting Method

The content of this section is based on Carraro et al. [20]. In the past decade, several approaches have been developed toward the solution of nonstationary OCPs. Standard FE methods with adaptive mesh refinement have been presented in Becker et al. [3] and Meidner and Vexler [68]. A summarizing survey of theoretical as well as practical aspects of such OCPs is given in Hinze et al. [47].

In solving nonstationary OCPs based on the corresponding KKT system, which in this case is a boundary value problem in time, one is frequently confronted with the problem of instability. The generally nonlinear state operator may be unstable if governing an initial value problem, though allowing for a stable solution of the OCP. This phenomenon may become critical in the context of a Newton iteration when the starting trajectory is still too inaccurate. To overcome this problem it is common to employ the concept of "multiple shooting" (MS). In this approach the time interval is split into finitely many subintervals such that on each local subinterval the stability properties of the state operator are sufficient for the convergence of a global Newton iteration linking the subtrajectories together and enforcing the boundary conditions. In the limit one obtains a globally admissible state, corresponding adjoint state and control. There are two different versions of the MS method applied to OCPs: (a) *direct* MS (DMS) and (b) *indirect* MS (IMS). In applying an MS method for solving a nonstationary OCP adaptivity may be used in choosing the shooting intervals for increasing stability, in solving the subproblems on the single shooting intervals by space-time FE methods, and in the outer inexact Newton iteration for installing a globally admissible solution.

For the last three decades MS methods have been extensively studied for solving ODE- or DAE-governed OCPs (see Leineweber et al. [57], Körkel et al. [55], and the literature cited therein). Early on, their capability of integrating even highly unstable systems made them an indispensable tool in the solution of complex OCPs. In the more recent field of PDE-based optimal control MS has not yet been thoroughly investigated, except in the framework of the method of lines (MOL) approach, which essentially reduces the more difficult PDE constraint to the standard, though high-dimensional, ODE case. There are some publications on special topics related to the application of MS as part of the solution process. Serban et al. [78] develop an approach toward spatial grid adaptation in the MOL framework called "structured adaptive mesh refinement" (SAMR). Heinkenschloss [36] investigates different preconditioners for time domain decomposition methods, thereby choosing MS as a representative example. All these publications are exclusively concerned with *direct* multiple shooting (DMS). To our knowledge, the only work on *indirect* multiple shooting in the PDE context is that of Hesse [38] and Hesse and Kanschat [39], which applies the DWR approach described above for dynamic spatial mesh adaptation within the IMS framework, though without going into the details of the shooting procedure itself.

In the following, we will discuss the IMS approach for the solution of PDE-based parabolic OCPs mainly without control or state constraints. The inclusion of control constraints has been discussed in Carraro et al. [20]. State constraints may be treated analogously as in the stationary case discussed above by a penalty or barrier technique. Numerical results for linear and nonlinear model problems with and without control constraints illustrate the efficient use of IMS, particularly in cases where standard methods fail.

Remark 4.3. There is another motivation of employing MS for the solution of parabolic OCPs. It enables parallel computation on the different shooting intervals. This aspect is addressed by the so-called "Parareal Method" developed in Lions et al. [59] and applied to OCPs in Maday and Turinici [62], but will not be discussed in this article. For further work in this direction, we refer to Bal [3] and Ulbrich [80]. Actually, Gander and Vandewalle [30] pointed out that the Parareal Method may be interpreted in the framework of multiple shooting.

4.1 The Parabolic OCP and Its KKT System

In the following, we will show how shooting methods may be included into the framework of OCPs with or without control constraints of box-type. The focus is on the detailed presentation of an algorithm for the IMS, thereby highlighting the particular difficulties to be overcome in the PDE context. We consider OCPs of the following form:

$$\min_{(u,q)} J(u,q), \tag{4.1}$$

subject to a parabolic PDE constraint

$$\begin{aligned} \partial_t u(x,t) + A(u)(x,t) + B(q)(x,t) &= f(x,t), \quad (x,t) \in \Omega \times I, \\ u(x,0) &= u_0(x), \quad x \in \Omega, \end{aligned} \tag{4.2}$$

supplemented by suitable spatial boundary conditions. We will also briefly discuss the treatment of additional control constraints of box type:

$$q_-(x,t) \le q(x,t) \le q_+(x,t), \quad (x,z) \in \Omega \times I. \tag{4.3}$$

Remark 4.4. Other global types of control constraints such as

$$\int_{\Omega \times I} q(x,t)\, dx\, dt \le c,$$

are not considered. It is not clear how to handle such global constraints in the IMS context since they cannot be localized to the different shooting intervals.

Following Carraro et al. [20], we explain some details of problem (4.1)–(4.3) and fix the notation used further on. The computational domain is a space-time cylinder $\Omega \times I$ with a bounded spatial domain $\Omega \subset \mathbb{R}^d$, $d \in \{1,2,3\}$, which for simplicity is assumed to be convex polygonal or polyhedral, and a finite time interval $I = (0,T]$. Additional boundary conditions of Dirichlet or Neumann type are prescribed on the boundary part $\partial\Omega \times I$. We adopt the usual Bochner space notation $W(I;Y)$ for spaces of functions that map from I to a normed function space Y. If $V \hookrightarrow H \hookrightarrow V^*$ is a Gelfand triple of Hilbert spaces of functions on Ω (V^* being the dual space of V) and R a Banach space of functions on Ω or Γ, the usual setting for the parabolic PDE (4.2) is as follows: For a given control $q \in L^2(I;R)$ and source $f \in L^2(I;V^*)$, find a function $u(x,t)$ that satisfies (4.2) and the imposed boundary conditions in a certain weak sense. The natural solution space for the state $u(x,t)$ is

$$X := \{v \mid v \in L^2(I,V), \partial_t v \in L^2(I,V^*)\},$$

which is well-known to be continuously embedded into the space $C(\bar{I};H)$ of continuous functions on the closure $\bar{I}$ with values in H (cf. Dautray and Lions [23]). As usual, we require the differential operator $A : X \to L^2(I;V^*)$ to be elliptic and coercive, linear or nonlinear, and the control operator $B : L^2(I;R) \to L^2(I;V^*)$ to be linear or simply the identity operator, where $R \hookrightarrow V^*$. These operators are understood as pointwise-in-time operators $A : V \to V^*$ and $B : R \to V^*$. Then, using the semilinear forms and scalar products

$$a_I(u)(\delta u) := \int_0^T \langle A(u), \delta u \rangle_{V^* \times V}\, dt, \quad b_I(q)(\delta u) := \int_0^T \langle B(q), \delta u \rangle_{V^* \times V}\, dt,$$

$$((u, \delta u))_I := \int_0^T (u, \delta u)_H\, dt, \quad \delta u \in X,$$

the weak formulation of (4.2) with weakly included initial condition reads

$$((\partial_t u, \delta u))_I + a_I(u)(\delta u) + b_I(q)(\delta u) + (u(0) - u_0, \delta u(0)) = ((f, \delta u))_I, \tag{4.4}$$

for all $\delta u \in X$. Further on, we will skip the interval index I in the notation $((\cdot, \cdot))_I$, $a_I(\cdot)(\cdot)$ and $b_I(\cdot)(\cdot)$ if the respective interval is given by the context.

The "cost functional" $J(u, q)$ is assumed to have the following structure:

$$J(u, q) = \kappa_1 J_1(u) + \kappa_2 J_2(u(T)) + \tfrac{\alpha}{2} \|q\|_Q^2.$$

It consists of a term J_1 distributed in time, a term J_2 evaluated at the final time T, and a usual regularization term. We will only consider one of the two terms depending on the state variable, i.e., we impose the conditions $\kappa_1, \kappa_2 \in \{0, 1\}, \kappa_1 \neq \kappa_2$, whereas $\alpha > 0$. As we will see later, it is important for the purpose of MS that the time-distributed term J_1 can be localized or split up to contributions from the different subintervals $I_j \subset I$. This is the case for functionals of tracking type:

$$J_1(u) = \int_0^T \|u(t) - \hat{u}(t)\|_V^2\, dt,$$

with some prescribed function $\hat{u} \in L^2(I; V)$. If we impose the additional control constraint (4.3), the set Q_{ad} of admissible control functions is given by

$$Q_{\mathrm{ad}} = \{q \in Q \mid q_- \le q \le q_+, \text{ a.e. in } \Omega \times I\}, \tag{4.5}$$

where $q_-, q_+ \in Q$ are given functions satisfying $q_- < q_+$. Thus, the set Q_{ad} is a convex subset of Q and may even coincide with Q in the case without control constraints. In compact form, our OCP thus reads

$$\min_{q \in Q_{\mathrm{ad}}, u \in X} J(u, q) \quad \text{subject to (4.4)}. \tag{4.6}$$

Results on the well-posedness of such OCPs can be found, e.g., in Hinze et al. [47] or Tröltzsch [79]. We will always assume the unique solvability of (4.4), which enables the definition of a solution operator $S : Q_{\mathrm{ad}} \to X$ by $S(q) = u$ and of the reduced cost functional $j(q) := J(S(q), q)$. Using this notation the OCP (4.6) can be written in short as

$$\min_{q \in Q_{\mathrm{ad}}} j(q). \tag{4.7}$$

Next, we briefly recall the first-order necessary optimality conditions for problem (4.6) with and without control constraints. The corresponding Lagrange functional $\mathcal{L} : Q \times X \times X \to \mathbb{R}$ has the form

$$\begin{aligned}\mathcal{L}(q,u,z) := J(q,u) + ((\partial_t u, z)) + a(u)(z) + b(q)(z) - ((f,z)) \\ + (u(0) - u_0, z(0)),\end{aligned} \tag{4.8}$$

where $z \in X$ denotes the adjoint variable ("Lagrange multiplier"). Then, under certain conditions, for any optimal solution $\{u,q\} \in X \times Q$ of (4.6) there exists a $z \in X$ such that the triple $\{u,q,z\}$ is stationary point of the Lagrangian $\mathcal{L}$. The corresponding PDE system comprises the "state", "adjoint" and "control equations". It can be written in the following explicit form for all variations $\delta u,\ \delta z \in X$ and $\delta q \in Q$:

$$((\partial_t u, \delta z)) + a(u)(\delta z) + b(q)(\delta z) - ((f, \delta z)) + (u(0) - u_0, \delta z(0)) = 0, \tag{4.9a}$$

$$J_u'(q,u)(\delta u) - ((\partial_t z, \delta u)) + a_u'(u)(\delta u, z) + (z(T), \delta u(T)) = 0, \tag{4.9b}$$

$$J_q'(q,u)(\delta q) + b_q'(q)(\delta q, z) = 0. \tag{4.9c}$$

The three Eqs. (4.9) form a boundary value problem in time for the state and adjoint variables. Here, the boundary values are the given initial condition $u(x,0) = u_0(x)$ for the state variable, and the condition $z(x,T) = 0$ or, if $\kappa_2 \neq 0$, $z(x,T) = J_2'(u(T))(\delta u(T))$ for the adjoint variable at end time T. The solution of the adjoint equation is running backward in time, and the state and adjoint problems are coupled by the third equation via the control variable.

Remark 4.5. In the presence of box constraints on the control of the form (4.3), we simply have to replace the control equation (4.9c) by the variational inequality

$$J_q'(q,u)(\delta q - q) + b_q'(q)(\delta q - q, z) \geq 0 \quad \forall \delta q \in Q_{\text{ad}}, \tag{4.10}$$

while the state and adjoint equations remain unchanged. This is due to the convexity of the set Q_{ad}. The direct treatment of the resulting optimality system is complicated by the control inequality, but there is a way of transforming (4.10) into several equations by using the concept of "active sets". This has already been used, e.g., by Bergounioux, Ito and Kunisch [16], Griesse and Vexler [32], and Vexler and Wollner [82], for the elliptic case. For the parabolic case, a similar procedure has been suggested by Kunisch and Rösch [56] and was used, e.g., by Griesse and Vexler [32] and Griesse and Volkwein [33]. Our numerical treatment of control constraints is also based on this concept; the technical details in the context of the IMS method are described in Carraro et al. [20].

4.2 Indirect Multiple Shooting (IMS)

There are two ways of integrating multiple shooting into the solution process of OCPs such as (4.1)–(4.3). The "direct multiple shooting" (DMS) applies multiple shooting to the side condition (4.4) in weak form, which leads to a sequence of initial value problems on subintervals of $I = [0,T]$. The "indirect multiple shooting" (IMS) method produces a sequence of intervalwise boundary value problems of type (4.13) governed by systems of optimality conditions on subintervals of I. An advantage of IMS is that one can use standard routines for the solution of the local OCPs on the subintervals. In the ODE context DMS methods have been preferred because they transform the original OCP into a finite dimensional nonlinear programming problem (NLP), which allows for the use of efficient methods for solving NLPs. In the following, we first give a brief survey of known results on convergence and stability of shooting methods and comment on the additional problems that occur in the PDE context.

4.2.1 Preliminaries: Stability and Convergence of Shooting Methods

It is well-known that even linear and well conditioned ODE boundary value problems (BVP) often become highly sensitive to perturbations in the data when reformulated as initial value problems (IVP). In the context of multiple shooting, such perturbations cannot be avoided, because one has to replace the unknown initial values by parameterized ones (in the PDE framework, even the discretization of the initial value functions entails a perturbation of the exact value). This phenomenon, examined by deHoog and Mattheij [25], is caused by a dichotomic behavior of the BVP solution $y(t)$ which determines the conditioning of the BVP. This leads to an exponential amplification over time of errors in the initial data s of the parameterized IVP (L a Lipschitz constant):

$$\|y(t;s_1) - y(t;s_2)\| \le ce^{L(t-t_0)}\|s_1 - s_2\|. \tag{4.11}$$

Local stability estimates of this type are common in IVP and can be proven, e.g., with the help of Gronwall's inequality. Reformulation of BVP as parameterized IVP is the idea behind the simple shooting method. The mentioned instability of the IVP reflects the frequently observed fact that simple shooting is in many cases an unstable algorithm. Fortunately, inequality (4.11) suggests a way how to cope with this deficiency. Therefore, we decompose the time interval $I = [0,T]$ into smaller subintervals, referred to as "shooting intervals", by choosing intermediate "shooting points" $0 = \tau_0 < \tau_1 < \cdots < \tau_{M-1} < \tau_M = T$:

$$I = \{0\} \cup \left(\cup_{j=0}^{M-1} I_j\right), \quad I_j = (\tau_j, \tau_{j+1}]. \tag{4.12}$$

By this splitting into subintervals the local exponential factors in (4.11) are reduced thus stabilizing the algorithm, which is now referred to as "multiple shooting" (MS). In the following a variant of this algorithm will be described, which is suitable for PDE-governed OCPs.

When concerned with the convergence of shooting methods, one has to distinguish between the convergence of the Newton-type iteration applied on the shooting system (see (4.13d)–(4.13g) below) and the convergence of the discrete solutions to a limit (the continuous solution). We will briefly discuss both aspects without going into detail. Conditions for the convergence of Newton's method in shooting algorithms are presented in Weiss [83] who also observes that the domain of starting values for Newton's method is enlarged with an increasing number of shooting intervals. Deuflhard [26] is concerned with globalization techniques for Newton's method. In summary, a variety of Newton-type methods are available for solving the shooting system, but there is always some trade-off between finding good starting values and exploiting the full quadratic convergence.

In the ODE context, convergence orders of multiple shooting are coupled to the orders of the IVP solvers used on the subintervals. This is standard in the linear case, whereas for nonlinear ODE boundary value problems it was examined, e.g., by Jankowski [51] and Hieu [40]. We conjecture that most results achieved for shooting methods in the ODE context carry over to the PDE framework in case of a fixed spatial mesh (this corresponds to the so-called method of lines).

Some of the most challenging aspects in the transfer of shooting methods from the ODE to the PDE context are due to the additional spatial dimensions. Shooting variables s (the local initial values) are no longer scalars or vectors in $\mathbb{R}^n$, but functions in certain Sobolev spaces. Hence the proper choice of norms in the analysis is a crucial aspect. All spatially distributed functions have to be discretized in space, which leads to large stiffness matrices at the time-points and therefore to huge linear systems that cannot be treated directly and have to be solved in a matrix-free manner by an iterative solver, e.g., a Krylov space method.

On the time domain decomposition (4.12), we consider restrictions u^j, z^j, and q^j of the global state, adjoint and control variables u, z, and q, respectively, to the single subintervals I_j, and define corresponding function spaces by $X_j = \{v^j \mid v^j \in L^2(I_j, V), \partial_t v^j \in L^2(I_j, V^*)\}$ and $Q_j \subseteq L^2(I_j, R)$. The goal is to solve a BVP of the form (4.9) on each I_j in such a way that the composition of the intervalwise solutions constitutes a solution to the original global OCP. In this way the possible global instability of the state equation can be reduced by decreasing the lengths of the shooting intervals. First, we define the extended Lagrange functional, denoted by $\bar{\mathcal{L}}$, on the single shooting intervals I_j, for $j = 0, \dots, M-1$:

$$\bar{\mathcal{L}}((q^j,u^j,z^j)_{j=0}^{M-1},(s^j,\lambda^j)_{j=0}^{M}) := \kappa_1 \sum_{j=0}^{M-1} J_1(u^j) + \kappa_2 J_2(u^{M-1}(\tau_M))$$

$$+ \frac{\alpha}{2}\sum_{j=0}^{M-1}\int_{I_j} \|q^j\|^2\,dt + \sum_{j=0}^{M-1}\big[((\partial_t u^j, z^j)) + a(u^j)(z^j) + b(q^j)(z^j) - ((f|_{I_j}, z^j))\big]$$

$$+ \sum_{j=0}^{M-1}(u^j(\tau_j) - s^j, z^j(\tau_j)) + \sum_{j=0}^{M-1}(s^{j+1} - u^j(\tau_{j+1}), \lambda^{j+1}) + (s^0 - u_0, \lambda^0),$$

involving the unknown values of the intervalwise solutions $s^j = u^j$ and $\lambda^j = z^j$ at the shooting points τ_j. The "shooting variables" $s^j, \lambda^j \in H$ are either arbitrarily chosen or guessed approximations to $u(\tau_j)$ and $z(\tau_j)$. The extended functional constitutes the Lagrangian of a new OCP that is similar to the original one (4.1)–(4.3) but is subject to additional equality constraints. The corresponding optimality system reads as follows:

$$((\partial_t u^j, \delta z)) + a(u^j)(\delta z) + b(q^j)(\delta z) - ((f|_{I_j}, \delta z)) + (u^j(\tau_j) - s^j, \delta z(\tau_j)) = 0, \tag{4.13a}$$

$$\kappa_1 J'_{1,u^j}(u^j)(\delta u) - ((\partial_t z^j, \delta u)) + a'_{u^j}(u^j)(\delta u, z^j) + (z^j(\tau_{j+1}) - \lambda^{j+1}, \delta u(\tau_{j+1})) = 0, \tag{4.13b}$$

$$\alpha((q^j, \delta q)) + b'_{q^j}(q^j)(\delta q, z^j) = 0, \tag{4.13c}$$

$$(s^0 - u_0, \delta\lambda) = 0, \tag{4.13d}$$

$$j = 1,\dots,M: \qquad (s^j - u^{j-1}(\tau_j), \delta\lambda) = 0, \tag{4.13e}$$

$$j = 0,\dots,M-1: \qquad (\lambda^j - z^j(\tau_j), \delta s) = 0, \tag{4.13f}$$

$$(\lambda^M, \delta s) = 0, \tag{4.13g}$$

where Eqs. (4.13a)–(4.13c) hold for $j = 0,\dots,M-2$. For $j = M-1$, however, in the case $\kappa_2 \neq 0$ there appears an additional term $\kappa_2 J_2'(\dots)(\dots)$ in the adjoint equation, which describes the initial value of the adjoint equation on I_{M-1}. In system (4.13), the first three equations bear the structure of the optimality conditions of the original OCP, whereas the remaining four equations represent continuity conditions for the state and adjoint variables.

Obviously, the IMS based on the extended Lagrange functional $\bar{\mathcal{L}}$ leads to a sequence of boundary value problems similar to (4.9) on the different subintervals I_j. The additional equality constraints given by (4.13d)–(4.13g) ensure global admissibility of the composed intervalwise solutions, i.e., that the prescribed initial value is matched, $s^0 - u_0 = 0$, that the jumps between the interval solutions at the interval endpoints and the initial values on the following intervals vanish,

$s^j - u^{j-1}(\tau_j) = 0$ and $\lambda^j - z^j(\tau_j) = 0$, respectively. These conditions can be deduced formally as the derivatives $\mathcal{L}'_{s^j}(\varphi)$ and $\mathcal{L}'_{z^j}(\psi)$ of $\tilde{\mathcal{L}}$ with respect to the shooting variables s^j and λ^j, respectively. The intervalwise optimality systems (4.13) show that the shooting variables s^j and λ^{j+1} are state and adjoint initial values on the subintervals I_j.

Remark 4.6. As there is no need to fit given values at the end time τ_M for the global state variable and at the end time τ_0 for the global adjoint variable, the corresponding equations in (4.13) are artificial. We therefore skip the variables s^M and λ^0 in order to decrease the size of the shooting system. Furthermore, we could also skip the variables s^0 and λ^M and replace them by the known initial values $s^0 \equiv u_0$ and $\lambda^M \equiv 0$. The main reason for keeping them in the system is the resulting simplification in the implementation of the method.

4.3 Numerical Methods and Implementation

The foregoing discussion has shown that the IMS formulation (4.13) is fully equivalent to the original OCP (4.6). Next, we will discuss the discretization of (4.13) and the practical implementation of the resulting large algebraic systems.

4.3.1 The Intervalwise Optimal Control Problems

The optimality systems (4.13a)–(4.13c) on the single subintervals bear the same structure as the optimality conditions (4.9) of the original global problem. They may be solved independently (which allows for parallelization of the MS code), and any algorithm designed for problems of type (4.9) can be employed for their solution. We follow the procedure presented in Meidner and Vexler [68], which is based on the direct minimization of the reduced cost functional $j(q)$. For that, we have to recall some well-known results concerning the gradient and Hessian of $j(q)$. A detailed presentation can also be found in Hinze et al. [47].

The first-order derivative of $j(q)$ in direction δq is given by the identity $j'(q)(\delta q) = \mathcal{L}'_q(q, u, z)(\delta q)$. The derivatives $\mathcal{L}'_z$ and $\mathcal{L}'_u$ vanish. In our context, we obtain the following representation:

$$j'(q)(\delta q) = \alpha((q, \delta q)) + b'_q(q)(\delta q, z), \quad \delta q \in Q. \tag{4.14}$$

In a similar way, the second-order derivative $j''(q)(\delta q, \tau q)$ is given by

$$\begin{aligned} j''(q)(\delta q, \tau q) = \mathcal{L}''_{qq}(q, u, z)(\delta q, \tau q) + \mathcal{L}''_{uq}(q, u, z)(\delta u, \tau q) \\ + \mathcal{L}''_{zq}(q, u, z)(\delta z, \tau q), \quad \delta q, \tau q \in Q, \end{aligned} \tag{4.15}$$

where δu and δz are the solutions of

$$\mathcal{L}''_{qz}(q,u,z)(\delta q,\varphi) + \mathcal{L}''_{uz}(q,u,z)(\delta u,\varphi) = 0, \quad (4.16a)$$

$$\mathcal{L}''_{qu}(q,u,z)(\delta q,\varphi) + \mathcal{L}''_{uu}(q,u,z)(\delta u,\varphi) + \mathcal{L}''_{zu}(q,u,z)(\delta z,\varphi) = 0, \quad (4.16b)$$

holding for all $\varphi \in X$. These equations are referred to as "tangent equation" and as additional "adjoint equation". They can be written explicitly as

$$((\partial_t \delta u, \varphi)) + a'_u(u)(\delta u, \varphi) + b'_q(q)(\delta q, \varphi) + (\delta u(0), \varphi(0)) = 0, \quad (4.17a)$$

$$\begin{aligned} J'_{uu}(q,u)(\delta u,\varphi) - ((\partial_t \delta z, \varphi)) + a''_{uu}(u)(\delta u, \varphi, z) + a''_{zu}(u)(\varphi, \delta z) \\ +(\delta z(T), \varphi(T)) = 0. \end{aligned} \quad (4.17b)$$

Then, in the context of our problem $j''(q)(\delta q, \chi)$ has the form

$$j''(q)(\delta q,\chi) = J''_{qq}(q,u)(\delta q,\chi) + b''_{qq}(q)(\delta q,\chi,z) + b''_{zq}(q)(\chi,\delta z). \quad (4.18)$$

Considering the reduced optimal control problem (4.7) is the crucial step in solving the intervalwise OCPs in an iterative process by a matrix-free Newton-CG method.

4.3.2 The System of Shooting Conditions

The application of the IMS method consist in the solution of the system of continuity conditions (4.13d)–(4.13g). This is written as $F(y) = 0$, i.e., the variable y comprises all shooting variables $\{s^0, \lambda^1, \dots, s^{M-1}, \lambda^M\}$. To find a zero of this system, we employ Newton's method, which in each iteration step requires the solution of the linear system

$$\nabla F(y_k)\delta y = -F(y_k). \quad (4.19)$$

The Jacobian ∇F is of size $4RM \times 4RM$, where M is the number of shooting intervals. The directional derivatives with respect to s and λ are obtained as solutions of certain variational equations (cf. Carraro et al. [20] for details).

The computation of the whole Jacobian ∇F with the sensitivity method (see, e.g., Hinze et al. [47]) requires for each pair of derivatives u_s^j, z_s^j and u_λ^j, z_λ^j the solution of variational equations for δs and $\delta\lambda$ running through a set of basis functions of the discrete space V_h^s. This is very costly on strongly refined spatial meshes. To avoid this, we want to solve (4.19) by a matrix-free approach. For that, we choose a Newton-CG method, which requires only the solution of two additional problems in each iteration. Similarly, for the solution of (4.19), we employ a Newton-GMRES iterative method. This approach resembles the adjoint approach for solving reduced OCP (see again Hinze et al. [47]).

Remark 4.7. With increasing number of shooting intervals, the conditioning of the Jacobian ∇F deteriorates, thus necessitating the use of a preconditioner. In Heinkenschloss [36] different preconditioners are compared in the context of the DMS method and a symmetric Gauss-Seidel block preconditioner is recommended. Numerical tests confirm that this result carries over to the described IMS method. This preconditioner is easily applied, since it only requires the solution of two additional linear boundary value problems per GMRES iteration and may easily be included in our matrix-free framework.

Remark 4.8. In the above discussion it is assumed that the OCP does not involve control or state constraints. This is mainly for simplifying the presentation. The inclusion of *local* (pointwise) control constraints is considered in Carraro et al. [20]. As in the stationary case considered above, state constraints may be treated by penalization techniques. For solving the control-constrained problem, one may use the so-called "gradient projection method, which is globally convergent but with an only linear rate (see Hinze et al. [47] and Dunn [27]). Projected Newton methods have to be used with care (see Kelley and Sachs [54]). Alternatives are the "primal-dual active set strategies" which involve both state and adjoint variables and have been extensively studied (see Bergounioux et al. [16] and Kunisch and Rösch [56]). These methods are known to be equivalent to a superlinearly convergent "semi-smooth Newton method" (cf. Hintermüller et al. [44]).

4.4 Space-Time Discretization

Finally, we briefly describe the general setting of a FE discretization in space and time for solving problem (4.1)–(4.3) together with the variational equations for the Hessian and the gradient of the reduced cost functional $j(\cdot)$.

(i) Time semi-discretization
For discretization in time, we use the so-called "discontinuous Galerkin method" of degree $r \geq 0$ (in short: "dG(r) method"). To this end, the single shooting intervals $I_j = \{\tau_j\} \cup (\tau_j, \tau_{j+1}]$ are partitioned into further subintervals $\mathcal{I}_n^j = (t_{n-1}^j, t_n^j]$ of length k_n^j with left and right end points $\tau_j = t_0^j < t_1^j < \cdots < t_{N_j}^j = \tau_{j+1}$. For the lowest-order case $r = 0$ the dG(0) method can be interpreted as the classical first-order backward Euler time-stepping method if the time integrals are evaluated by the box rule. Alternatively, one could also use the "continuous" counterpart of the dG(r) methods, the cG(r) method, of degree $r \geq 1$. For further details on the formulation of the corresponding discrete Galerkin equations, we refer to Meidner and Vexler [68], Becker et al. [8] and Carraro et al. [20].

(ii) Spatial discretization
The spatial discretization uses a standard conforming finite element method as described above. On shape regular meshes $\mathcal{T}_h$ (cf. Carey and Oden [19] or Brenner and Scott [18]) consisting of closed cells T, quadrilaterals (in 2D) or hexahedra (in 3D) the FE subspaces are given by $V_h^s := \{v_h \in V \,|\, v_{h|T} \in Q^s(T), T \in \mathcal{T}_h\}$. Here, $Q^s(T), s \in \mathbb{N}$, is the space of functions obtained by isoparametric transformations of bilinear ($s = 1$), biquadratic ($s = 2$) and in general higher order polynomials defined on a reference unit cell. For simplicity, we only consider the dG(0) method for the discretization in time and bilinear shape functions in space. Then, the full space-time discretization of the state equation seeks $u_{hk}^j \in X_{h,k}^{s,r}(I_j)$ and $s_h^j \in V_h^s$, $j = 0, \dots, M-1$, satisfying the corresponding discretized Galerkin equations. For the technical details, we again refer to Meidner and Vexler [68], Becker et al. [8] and Carraro et al. [20].

Remark 4.9. In this article, we only consider the case of a fixed mesh $\mathcal{T}_h$ for all discrete time levels. More generally, one could also allow the mesh to change in time, i.e., use meshes $\mathcal{T}_{h,n}$. In the framework of shooting methods such dynamic meshes have been used in Hesse and Kanschat [39].

4.5 Numerical Experiments

We illustrate the performance of the IMS method described above by three numerical examples. We begin with a linear-quadratic OCP that necessitates the use of a multiple shooting method. This example is then extended by introducing an additional nonlinearity into the state equation. Finally, the example is additionally supplemented by a control constraint of box type.

4.5.1 Linear Example

The following test example from Hesse and Kanschat [39] is chosen in order to demonstrate the necessity of using the *multiple* shooting method in solving general nonstationary OPCs:

$$J(u,q) := \frac{1}{2}\|u(T) - 0.5\|_\Omega^2 + \frac{\alpha}{2}\int_0^T \|q(t)\|_\Omega^2\, dt \quad \to \quad \min! \tag{4.20}$$

subject to the nonstationary Helmholtz problem

$$\begin{aligned} &\partial_t u(x,t) - \Delta u(x,t) - \omega u(x,t) = q(x,t), \quad \text{in } \Omega \times (0,T], \\ &u(x,0) = u^0(x), \quad \text{in } \Omega, \quad u(x,t) = 0, \quad \text{on } \partial\Omega \times [0,T], \end{aligned} \tag{4.21}$$

where $u^c(x) = \cos\left(\frac{1}{2}\pi x_1\right)\cos\left(\frac{1}{2}\pi x_2\right)$. We choose $\Omega = (-1,1)^2 \subset \mathbb{R}^2$, $T = 5$, and $\alpha = 10^{-2}$. The parameter ω runs through a set of integers $5 \leq \omega \leq 10$. The initial value $u_0(x)$ is the eigenfunction corresponding to the smallest frequency of the Laplacian on Ω, $\lambda_{min} = \pi^2/2 \approx 4.9348$. The goal is to fit the constant function $\hat{u}(x,5) \equiv 0.5$ at the end time $T = 5$. In Fig. 4, we see that the state variable obviously tries to match this prescribed value at the end time, but develops a boundary layer due to the incompatible homogeneous Dirichlet boundary condition. The adjoint solution resembles a regularized line Dirac function along $\partial\Omega$.

For values of ω exceeding λ_{min} there occur instabilities in the state equation and consequently the behavior of our solution method in simple shooting form may deteriorate at about $\omega = 5$. This effect is illustrated in Table 6. There, the IMS method is compared to simple shooting and to a state-of-the-art method described, e.g., in Becker et al. [8]. The latter method solves the KKT system directly, whereas simple shooting treats the problem as a BVP and uses Newton's method to solve an additional equation representing the shooting system. The comparison is made with respect to the number of Newton-GMRES steps needed for achieving about the same accuracy in the optimal values $J(q,u)$. These results were obtained on a four times globally refined spatial mesh of 256 cells and with 500 uniform time steps. For $\omega \leq 5$ all three methods yield equally good results. However, for $\omega > 5$ the simple shooting method and the state-of-the-art method are not able to solve the

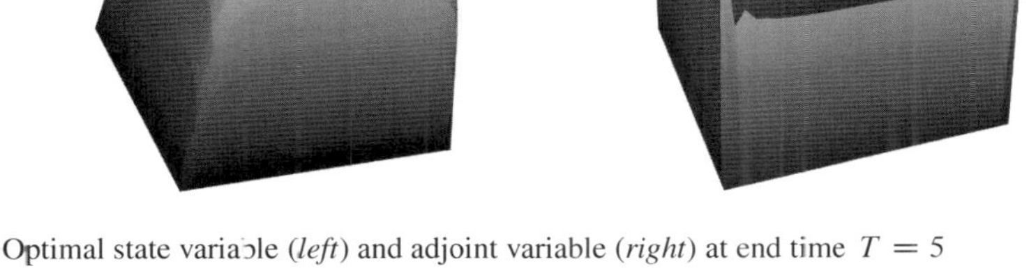

Fig. 4 Optimal state variable (*left*) and adjoint variable (*right*) at end time $T = 5$

Table 6 Comparison in terms of # of iterations of a state-of-the-art algorithm (StA), simple shooting (SimS) and multiple shooting with 5 shooting intervals (IMS_5)

ω	# it_{StA}	$J(u,q)$	# it_{SimS}	$J(u,q)$	# it_{IMS_5}	$J(u,q)$
3	36	0.0938	20	0.0938	22	0.0938
5	58	0.0794	20	0.0794	25	0.0794
6	–	–	–	–	25	0.0884
7	–	–	–	–	26	0.0971
8	–	–	–	–	52	0.1058
9	–	–	–	–	–	–

problem, whereas the IMS method still works well for increasing ω. For $\omega = 8$ two iterations of the outer Newton-type solver (with 26 GMRES iterations each) are needed to solve the problem, where normally Newton-type methods should only require one iteration for each linear subproblem. For $\omega \geq 9$, using five shooting intervals is no longer sufficient for solving the problem at all. Table 7 shows that for larger ω also the least number of required shooting intervals increases due to increasing ill-conditioning. In this case the preconditioner mentioned in Remark 4.7 turned out to be indispensable.

4.5.2 A Nonlinear Example

Next, we add the nonlinear term u^3 to the constraining Helmholtz equation. Furthermore, our goal is no longer to match a constant function at the end time T but rather to match a function $\hat{u}(x,t)$ on the whole time interval. This means that we want to solve the problem

$$J(u,q) := \frac{1}{2}\int_0^T \|u-\hat{u}\|_\Omega^2\,dt + \frac{\alpha}{2}\int_0^T \|q\|_\Omega^2\,dt \quad\rightarrow\quad \min! \tag{4.22}$$

subject to the nonstationary nonlinear Helmholtz problem

$$\begin{aligned} &\partial_t u(x,t) - \Delta u(x,t) - 7u(x,t) + u^3(x,t) = q(x,t), \quad \text{in } \Omega\times(0,T], \\ &u(x,0) = u_0(x), \text{ in } \Omega, \quad u(x,t) = 0, \text{ on } \partial\Omega\times[0,T]. \end{aligned} \tag{4.23}$$

We take $\Omega = (-1,1)^2$ and $T = 5$ as before, fix $\alpha = 0.5$ and $\omega = 7$ at a value for which simple shooting is expected to fail. Further, we choose the target function

$$\hat{u}(x,t) := \begin{cases} \frac{2}{5}t\cdot(1-x_1^{12})(1-x_2^{12}), & t \leq \frac{5}{2}, \\ (\frac{2}{5}t-2)\cdot(1-x_1^{12})(1-x_2^{12}), & t > \frac{5}{2}, \end{cases} \tag{4.24}$$

with zero boundary conditions and a maximum absolute value at the center $(0,0)$ of Ω. The initial function $u_0(x) \equiv 0$ is chosen such that it fits the value $\hat{u}(x,0)$. The computations are again carried out on a four times globally refined mesh, but this

Table 7 Minimum number of shooting intervals (SI) required for a stable time integration depending on ω

ω	# iter.	SI	$J(u,q)$	Residue
5	20	1	0.0794	$8.4e^{-10}$
6	22	2	0.0884	$7.1e^{-09}$
7	23	2	0.0971	$1.4e^{-07}$
8	43	7	0.1058	$3.4e^{-10}$
9	48	7	0.1143	$2.1e^{-09}$
10	48	7	0.1226	$7.2e^{-09}$

time, we choose 10 equally distributed shooting intervals each of which comprises 50 interior time steps. Figure 5 shows the temporal development of the state variable $u(0,0,t)$ and control $q(0,0,t)$ at different cycles of the MS procedure. In the first iteration with arbitrary initial values (solid curves), we can clearly distinguish the 10 shooting intervals. The second shooting cycle (dotted curves) is already close to convergence, but more shooting cycles are needed to reach the prescribed tolerance (dashed curves).

4.5.3 An Example Involving Control Constraints

Finally, we supplement the above OCP by the box-type control constraint $-0.5 \leq q(x,t) \leq 0.5$. Figure 6 shows the computed state and control variable at different multiple shooting iterations. We see that the control constraint is fulfilled, while there is only little difference in the optimal state variables. In this case, we need almost twice as many Newton steps for fulfilling the global continuity conditions as in the unconstrained case. This could be caused by the preconditioner proposed in Remark 4.7. For more details on the solution issue, we refer to Carraro et al. [20].

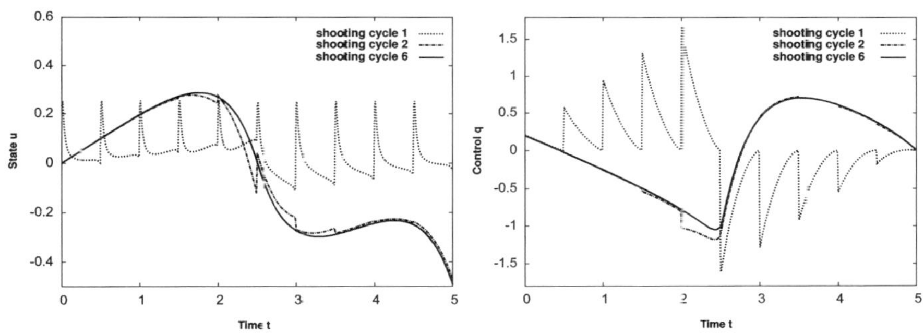

Fig. 5 State $u(0,0,t)$ and control $q(0,0,t)$ at different IMS cycles in the unconstrained case

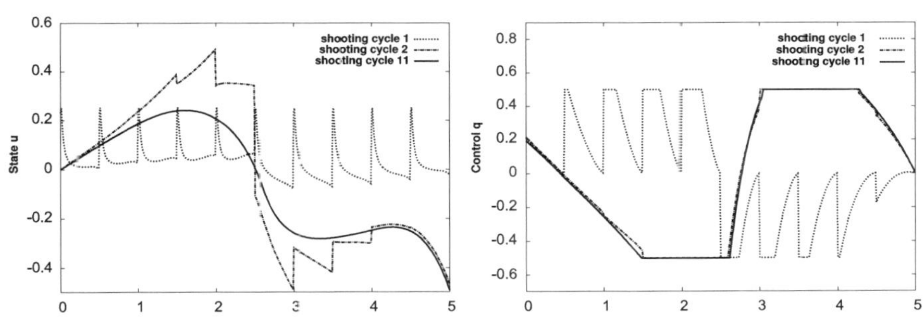

Fig. 6 State $u(0,0,t)$ and control $q(0,0,t)$ at different IMS cycles in the constrained case

4.6 Final Remarks

The concrete numerical realization of the IMS method suggested in Carraro at al. [20] is especially well-suited for problems with a high-dimensional control space, because it avoids the generation of the Hessian matrix during the solution of the intervalwise OCPs as well as building up the whole Jacobian of the system of shooting conditions. This matrix-free approach works for nonlinear problems with or without control constraints. One advantage of the IMS method is the possible use of already existing algorithms and software for the solution of PDE-based OCPs. The shooting subproblems can be solved by standard methods while an external Newton loop acts on the shooting system. A comparison to DMS methods within the framework of the DWR method for nonlinear PDE-governed OCPs is subject of current work.

The proper choice of shooting points τ_j is a critical issue in the PDE context, since with an increasing number of shooting points the dimension of the shooting system (4.19) gets ever larger leading to a significant increase in computational work. Therefore, the determination of the minimal number of shooting points and their position is crucial for the efficient solution of problems that respond very sensitively to perturbations in the data. The above numerical results have been achieved by a trial and error method. In order to avoid such rather cumbersome processes, criteria are desirable for adaptively determining the optimal total number of shooting points. However, even for ODE-based boundary value problems with solution $y(t;s)$, there are only few results concerning this question. Maier [63] develops a method that starts from a given shooting point distribution and automatically discards or inserts shooting points whenever necessary, but works only for a certain problem class. Alternatively, Mattheij and Staarink [64] suggest to impose a bound for the growth of the sensitivity matrix $G(t) := \frac{d}{ds} y(t;s)$, which is given as the solution of a matrix ODE arising from the linearization of the original ODE w.r.t. the shooting parameter s. Proceeding forward in time, whenever $\|G(t)\|$ exceeds a pre-chosen threshold value C ($\|\cdot\|$ being an arbitrary matrix norm), the current time-point t_i is taken as a new shooting point τ_j. This approach has some major deficiencies, e.g., there is no indication how to choose the bounding constant C reasonably and what to do in the nonlinear case. More importantly, its transfer to the PDE context is not clear even in the linear case. The necessity of matrix-free computation means that the sensitivity matrices are not available. We only have directional derivatives, i.e., choosing a norm of the sensitivities as bounding constant C is thus not feasible in the PDE case. This issue is currently under investigation.

Acknowledgements This work was supported during the period 2007–2013 within the DFG Priority Program 1253 "Optimization with Partial Differential Equations", grant 306/15-1, "Model reduction by adaptive discretization in optimal control".

References

1. M. Ainsworth, J.T. Oden, *A Posteriori Error Estimation in Finite Element Analysis* (Wiley-Interscience, New York, 2000)
2. I. Babuška, Th. Strouboulis, *The Finite Element Method and Its Reliability* (The Clarendon Press/Oxford University Press, New York, 2001)
3. G. Bal, On the convergence and stability of the parareal algorithm to solve partial differential equations, in *Proceedings of the 15th International Domain Decomposition Conference*. Lecture Notes in Computer Science and Engineering (Springer, Berlin, 2003), pp. 426–432
4. W. Bangerth, R. Rannacher, *Adaptive Finite Element Methods for Differential Equations*, Lectures in Mathematics ETH Zürich (Birkhäuser, Basel, 2003)
5. R. Becker, Mesh adaptation for stationary flow control. J. Math. Fluid Mech. **3**, 317–341 (2001)
6. R. Becker, Estimating the control error in discretized PDE-constraint optimization. J. Numer. Math. **14**, 163–185 (2006)
7. R. Becker, H. Kapp, R. Rannacher, Adaptive finite element methods for optimal control of partial differential equations: basic concept. SIAM J. Control Optim. **39**, 113–132 (2000)
8. R. Becker, D. Meidner, B. Vexler, Efficient numerical solution of parabolic optimization problems by finite element methods. Optim. Methods Softw. **22**, 813–833 (2007)
9. R. Becker, R. Rannacher, A feed-back approach to error control in finite element methods: basic analysis and examples. East-West J. Numer. Math. **4**, 237–264 (1996)
10. R. Becker, R. Rannacher, Weighted a-posteriori error estimates in FE methods, in *Lecture ENUMATH-95*, Paris, 1995; *Proceedings of ENUMATH-97*, Heidelberg, 1997, ed. by H.G. Bock et al. (World Science Publisher, Singapore, 1998) pp. 621–637
11. R. Becker, R. Rannacher, An optimal control approach to a posteriori error estimation, *in Acta Numerica 2001*, ed. by A. Iserles (Cambridge University Press, Cambridge, 2001), pp. 1–102
12. R. Becker, B. Vexler, A posteriori error estimation for finite element discretization of parameter identification problems. Numer. Math. **96**, 435–459 (2004)
13. R. Becker, B. Vexler, Mesh refinement and numerical sensitivity analysis for parameter calibration of partial differential equations. J. Comput. Phys. **206**, 95–110 (2005)
14. O. Benedix, B. Vexler, A posteriori error estimation and adaptivity for elliptic optimal control problems with state constraints. Comput. Optim. Appl. **44**, 3–25 (2009)
15. M. Bergounioux, A penalization method for optimal control of elliptic problems with state constraints. SIAM J. Control Optim. **30**, 305–323 (1992)
16. M. Bergounioux, K. Ito, K. Kunisch, Primal-dual strategy for constrained optimal control problems. SIAM J. Control Optim. **37**, 1176–1194 (1999)
17. M. Besier, R. Rannacher, Goal-oriented space-time adaptivity in the finite element Galerkin method for the computation of nonstationary incompressible flow. Int. J. Numer. Meth. Fluids 2011. Published online doi:10.1002/fld.2735
18. S.C. Brenner, L.R. Scott, *The Mathematical Theory of Finite Element Methods*. Texts in Applied Mathematics, vol. 15, 2nd edn. (Springer, Berlin, 2002)
19. G.F. Carey, J.T. Oden, *Finite Elements. Computational Aspects* (Prentice-Hall, Englewood Cliffs, 1984)
20. T. Carraro, M. Geiger, R. Rannacher, Indirect multiple shooting for nonlinear parabolic optimal control problems with control constraints. SIAM J. Sci. Comput. **36**, A452–A481 (2014). Online publication http://epubs.siam.org/toc/sjoce3/36/2. doi:10.1137/120895809
21. E. Casas, Control of an elliptic problem with pointwise state constraints. SIAM J. Control Optim. **24**, 1309–1318 (1986)
22. E. Casas, L.A. Fernández, Optimal control of semilinear elliptic equations with pointwise constraints on the gradient of the state. Appl. Math. Optim. **27**, 35–56 (1993)
23. R. Dautray, J.-L. Lions, *Evolution Problems I*. Mathematical Analysis and Numerical Methods for Science and Technology, vol. 5 (Springer, Berlin, 1992)
24. K. Deckelnick, A. Günther, M. Hinze, Finite element approximation of elliptic control problems with constraints on the gradient. Numer. Math. **111**, 335–350 (2008)

25. F. de Hoog, R.M.M. Mattheij, On dichotomy and well conditioning in BVP. SIAM J. Num. Anal. **24**, 89–105 (1987)
26. P. Deuflhard, A modified Newton method for the solution of ill-conditioned systems of nonlinear equations with application to multiple shooting. Numer. Math. **22**, 289–315 (1974)
27. J.C. Dunn, Global and asymptotic convergence rate estimates for a class of projected gradient processes. SIAM J. Control Optim. **19**, 368–400 (1981)
28. K. Eriksson, D. Estep, P. Hansbo, C. Johnson, *Computational Differential Equations* (Cambridge University Press, Cambridge, 1996)
29. A. Gaevskaya, R.H.W. Hoppe, Y. Iliash, M. Kieweg, Convergence analysis of an adaptive finite element method for distributed control problems with control constraints, in *Control of Coupled Partial Differential Equations*, ed. by G. Leugering et al. (Birkhäuser, Basel, 2007)
30. M. Gander, S. Vandewalle, Analysis of the parareal time-parallel time-integration method. SIAM J. Sci. Comput. **29**, 556–578 (2007)
31. A. Griesbaum, B. Kaltenbacher, B. Vexler, Efficient computation of the Tikhonov regularization parameter by goal-oriented adaptive discretization. Inverse Probl. **24**(2), 025025 (2008)
32. R. Griesse, B. Vexler, Numerical sensitivity analysis for the quantity of interest in pde-constrained optimization. SIAM J. Sci. Comput. **29**, 22–48 (2007)
33. R. Griesse, S. Volkwein, A primal-dual active set strategy for optimal boundary control of a nonlinear reaction-diffusion system. SIAM J. Control Optim. **44**, 467–494 (2005)
34. A. Günther, M. Hinze, A-posteriori error control of a state constrained elliptic control problem. J. Numer. Math. **16**, 307–322 (2008)
35. A. Günther, M. Hinze, M.H. Tber, A posteriori error representations for elliptic optimal control problems with control and state constraints, in *Constrained Optimization and Optimal Control for Partial Differential Equations*, ed. by G. Leugering et al. (Springer, Basel, 2012), pp. 303–317
36. M. Heinkenschloss, A time-domain decomposition iterative method for the solution of distributed linear quadratic optimal control problems. J. Comput. Appl. Math. **173**, 169–198 (2005)
37. R. Herzog, K. Kunisch, Algorithms for pde-constrained optimization. GAMM Rep. **33**, 163–176 (2010)
38. H.K. Hesse, Multiple shooting and mesh adaptation for PDE constrained optimization problems, doctoral dissertation, Heidelberg University, 2008
39. H.K. Hesse, G. Kanschat, Mesh adaptive multiple shooting for partial differential equations. Part 1: linear-quadratic optimal control problems. J. Numer. Math. **17**, 195–217 (2009)
40. N.T. Hieu, Remarks on the shooting method for nonlinear two-point boundary-value problems. VNU J. Sci. **3**, 18–25 (2003)
41. M. Hintermüller, M. Hinze, Moreau-Yosida regularization in state constrained elliptic control problems: error estimates and parameter adjustment. SIAM J. Numer. Anal. **47**, 1666–1683 (2008)
42. M. Hintermüller, R.H.W. Hoppe, Goal-oriented adaptivity in control constrained optimal control of partial differential equations. SIAM J. Control Optim. **47**, 1721–1743 (2008)
43. M. Hintermüller, R.H.W. Hoppe, Y. Iliash, M. Kieweg, An a posteriori error analysis of adaptive finite element methods for distributed elliptic control problems with control constraints. ESIAM Control Optim. Calc. Var. **14**, 540–560 (2008)
44. M. Hintermüller, K. Ito, K. Kunisch, The primal-dual active set strategy as a semismooth Newton method. SIAM J. Optim. **13**, 865–888 (2003)
45. M. Hintermüller, K. Kunisch, Stationary optimal control problems with pointwise state constraints. SIAM J. Optim. **20**, 1133–1156 (2009)
46. M. Hintermüller, K. Kunisch, PDE-constrained optimization subject to pointwise constraints on the control, the state, and its derivative. SIAM J. Optim. **20**, 1133–1156 (2009)
47. M. Hinze, R. Pinnau, M. Ulbrich, S. Ulbrich, *Optimization with PDE Constraints*. Mathematical Modelling: Theory and Applications, vol. 23 (Springer, New York, 2009)

48. M. Hinze, A. Schiela, Discretization of interior point methods for state constrained elliptic optimal control problems: optimal error estimates and parameter adjustment. Comput. Optim. Appl. (2009) doi:10.1007/s10589-009-9278-x
49. R.H.W. Hoppe, Y. Iliash, Ch. Iyyunni, N.H. Sweilam, A posteriori error estimates for adaptive finite element discretizations of boundary control problems. J. Numer. Math. **14**, 57–82 (2006)
50. R.H.W. Hoppe, M. Kieweg, A posteriori error estimation of finite element approximations of pointwise state constrained distributed control problems. J. Numer. Math. **17**, 219–244 (2009)
51. T. Jankowski, Approximate solutions of boundary value problems for systems of ordinary differential equations. Zh. Vychisl. Mat. Mat. Fiz. **35**, 1050–1057 (1995)
52. B. Kaltenbacher, A. Kirchner, B Vexler, Adaptive discretizations for the choice of a Tikhonov regularization parameter in nonlinear inverse problems. Inverse Probl. **27**, 125008 (2011)
53. H. Kapp, Adaptive Finite elements for optimization in partial differential equations, doctoral dissertation, Heidelberg University, 2000
54. C.T. Kelley, E.W. Sachs, Solution of optimal control problems by a pointwise projected Newton method. SIAM J. Control Optim. **33**, 1731–1757 (1995)
55. S. Körkel, E. Kostina, H.G. Bock, J.P. Schlöder, Numerical methods for optimal control problems in design of robust optimal experiments for nonlinear dynamic processes. Optim. Methods Softw. **19**, 327–338 (2004)
56. K. Kunisch, A. Rösch, Primal-dual active set strategy for a general class of constrained optimal control problems. SIAM J. Optim. **13**, 321–334 (2002)
57. D.B Leineweber, I. Bauer, H.G. Bock, and J.P. Schlöder, An efficient multiple shooting based reduced SQP strategy for large-scale dynamic process optimization. Part I: theoretical aspects. Comput. Chem. Eng. **27**, 157–166 (2003)
58. R. Li, W. Liu, H. Ma, T. Tang, Adaptive finite element approximation for distributed elliptic optimal control problems. SIAM J. Control Optim. **41**, 1321–1349 (2002)
59. J.-L Lions, Y. Maday, G. Turinici, Resolution d'edp par un schema en temps 'parareel'. C. R. Acad. Sci. Paris Ser. I **332**, 661–668 (2001)
60. W. Liu, Adaptive multi-meshes in finite element approximation of optimal control. Contemp. Math. **383**, 113–132 (2005)
61. W. Liu, N. Yan, A posteriori error estimates for distributed convex optimal control problems. Adv. Comput. Math. **15**, 285–309 (2001)
62. Y. Maday, G. Turinici, A parareal in time procedure for the control of partial differential equations. C. R. Acad. Sci. Paris Ser. I **335**, 387–392 (2002)
63. M.R. Maier, An adaptive shooting method for singularly perturbed boundary value problems. SIAM J. Sci. Stat. Comput. **7**, 418–440 (1986)
64. R.M.M. Mattheij, G.W.M. Staarink, On optimal shooting intervals. Math. Comput. **42**, 25–40 (1984)
65. D. Meidner, Adaptive space-time finite element methods for optimiza- tion problems governed by nonlinear parabolic systems, doctoral dissertation, Heidelberg University, 2008.
66. D. Meidner, R. Rannacher, B. Vexler, A priori error estimates for finite element discretizations of parabolic optimization problems with pointwise state constraints in time. SIAM J. Control Optim. **49**, 1961–1997 (2011)
67. D. Meidner, R. Rannacher, J. Vihharev, Goal-oriented error control of the iterative solution of finite element equations. J. Numer. Math. **17**, 143–172 (2009)
68. D. Meidner, B. Vexler, Adaptive space-time finite element methods for parabolic optimization problems. SIAM J. Control Optim. **46** 116–142 (2007)
69. D. Meidner, B. Vexler, A priori error estimates for space-time finite element discretization of parabolic optimal control problems. Part I: problems without control constraints. SIAM J. Control Optim. **47**, 1150–1177 (2008); Part II: problems with control constraints. SIAM J. Control Optim. **47**, 1301–1329 (2008)
70. C. Meyer, U. Prüfert, F. Tröltzsch, On two numerical methods for state-constrained elliptic control problems. Optim. Methods Softw. **22**, 871–899 (2007)
71. C. Ortner, W. Wollner, A priori error estimates for optimal control problems with pointwise constraints on the gradient of the state. Numer. Math. **118**, 587–600 (2011)

72. R. Rannacher, B. Vexler, W. Wollner, A posteriori error estimation in PDE-constrained optimization with pointwise inequality constraint, in *Constrained Optimization and Optimal Control for Partial Differential Equations*, ed. by G. Leugering et al. (Birkhüser, Basel, 2012), pp. 349–373
73. R. Rannacher, J. Vihharev, Adaptive finite element analysis of nonlinear problems: balancing of discretization and iteration errors. J. Numer. Math. **21**, 23–62 (2010)
74. R. Rannacher, A. Westenberger, W. Wollner, Adaptive finite element approximation of eigenvalue problems: balancing discretization and iteration error. J. Numer. Math. **18**, 303–327 (2010)
75. A. Schiela, W. Wollner, Barrier methods for optimal control problems with convex nonlinear gradient state constraints. SIAM J. Optim. **21**, 269–286 (2011)
76. M. Schmich, Adaptive finite element methods for computing nonstationary incompressible flow, doctoral dissertation, Heidelberg University, 2009
77. M. Schmich, B. Vexler, Adaptivity with dynamic meshes for space-time finite element discretizations of parabolic equations. SIAM J. Sci. Comput. **30**, 369–393 (2008)
78. R. Serban, S. Li, L. Petzold, Adaptive algorithms for optimal control of time-dependent partial differential-algebraic systems. Int. J. Numer. Methods Eng. **57**, 1457–1469 (2003)
79. F. Tröltzsch, *Optimal Control of Partial Differential Equations: Theory, Methods and Applications* (American Mathematical Society, Providence, 2010)
80. S. Ulbrich, generalized SQP-methods with 'parareal' time-domain decomposition for time-dependent PDE-constrained optimization, in Real-time PDE-constrained optimization (SIAM, Philadelphia, 2007), pp. 145–168
81. R. Verfürth, *A Review of A Posteriori Error Estimation and Adaptive Mesh-Refinement Techniques* (Wiley/Teubner, New York/Stuttgart, 1996)
82. B. Vexler, W. Wollner, Adaptive finite elements for elliptic optimization problems with control constraints. SIAM J. Control Optim. **47**, 509–534 (2008)
83. R. Weiss, The convergence of shooting methods. BIT **13**, 470–475 (1973)
84. W. Wollner, Adaptive FEM for PDE constrained optimization with point-wise constraints on the gradient of the state. Proc. Appl. Math. Mech. **8**, 10873 (2009)
85. W. Wollner, Adaptive methods for PDE-based optimal control with pointwise inequality constraints, doctoral dissertation, Heidelberg University, 2010
86. W. Wollner, A posteriori error estimates for a finite element discretization of interior point methods for an elliptic optimization problem with state constraints. Comput. Optim. Appl. **47**, 133–159 (2010)
87. W. Wollner, Goal-oriented adaptivity for optimization of elliptic systems subject to pointwise inequality constraints: application to free material optimization. Proc. Appl. Math. Mech. **10**, 669–672 (2010)

Graded Meshes in Optimal Control for Elliptic Partial Differential Equations: An Overview

Thomas Apel, Johannes Pfefferer, and Arnd Rösch

Abstract It is well known that singularities in the solution of boundary value problems due to corners and edges of the domain lead to a reduction of the convergence order of the standard finite element method when quasi-uniform meshes are used. It is also well known that locally graded meshes are suited to recover the optimal convergence order. Less well known are the critical angles when mesh grading becomes necessary; it is not always the same but depends on the norm in which the error is estimated. In this paper, an overview of the results is given and lacking estimates are pointed out. Since the error estimates for optimal control problems are based on those for pure boundary value problems both cases are always considered.

Keywords Elliptic partial differential equation • Finite elements • A priori error estimates • Mesh grading • Optimal control

Mathematics Subject Classification (2010). 49M25, 65N15, 65N30, 65N50.

T. Apel • J. Pfefferer
Institut für Mathematik und Bauinformatik, Universität der Bundeswehr München, 85577 Neubiberg, München, Germany
e-mail: Thomas.Apel@unibw.de; Johannes.Pfefferer@unibw.de

A. Rösch (✉)
Fakultät für Mathematik, Universität Duisburg–Essen, Thea-Leymann-Straße 9, D-45127 Essen, Germany
e-mail: arnd.roesch@uni-due.de

G. Leugering et al. (eds.), *Trends in PDE Constrained Optimization*, International Series of Numerical Mathematics 165, DOI 10.1007/978-3-319-05083-6_18

1 Introduction

Discretization and a priori error estimates of optimal control problems are well studied topics. It starts with publications from Falk, Geveci, and Malanowski [16, 18, 20]. The topic came back into the focus of the optimal control community by a paper from Arada, Casas, and Tröltzsch [14].

Until now, there is a significant number of publications concerning the a priori error analysis of discretization for optimal control problems. However, most of them deal with quasi-uniform meshes. Of course, if the domain is polygonal or polyhedral and has re-entrant corners this is not the appropriate choice. But there are also situations for convex domains where quasi-uniform meshes do not produce optimal approximation rates.

In this paper, we will give an overview in which situations locally refined meshes improve the approximation rates of finite element discretizations for elliptic optimal control problems. We introduce the idea of graded meshes in a natural way. To keep the presentation as clear as possible, we desist from defining the whole technical machinery which is needed to prove the presented results.

The paper is structured as follows. In Sect. 2, we discuss a priori error estimates for finite element discretizations of the elliptic equation

$$- \Delta y + y = f \text{ in } \Omega \tag{1}$$

with Dirichlet boundary condition

$$y = 0 \text{ on } \Gamma \tag{2}$$

or Neumann boundary condition

$$\partial_n y = g \text{ on } \Gamma. \tag{3}$$

Since we consider week formulations of the elliptic problems, we can easily discuss nonhomogeneous Neumann boundary conditions, but for simplicity we restrict ourselves to homogeneous Dirichlet boundary conditions. We assume that the domain $\Omega \subset R^2$ is polygonal. In Sect. 3, we consider different types of elliptic optimal control problems. The extension to the three-dimensional case is explained in Sect. 4.

2 Regularity and Mesh Grading

The quality of numerical results depends on different aspects. We will discuss in detail a possible singular behavior of the solution of the elliptic boundary value problem near the corners of the domain. Moreover, we will assume that the data f

and g are sufficiently smooth. Consequently, the essential analytical and numerical difficulties are caused by the corners of the domain.

A common opinion in the community is that uniform meshes are the appropriate choice for convex domains. We will see that this is only correct in special cases.

2.1 Singular Exponents

The Laplacian as main part of our elliptic partial differential equation (1) allows us to analyze the behavior in the corner in an explicit way. In order to keep the notation simple we will focus on one single corner. The polygonal domain has, of course, more than one corner; hence, a complete discussion would lead to a sum of these terms for each corner. The inner angle of the single corner is denoted by $\omega \in (0, 2\pi)$. The distance of a point $x \in \Omega$ to that corner is denoted by r, and the corresponding polar angle by ϕ. The solution y of (1), (2), or (1), (3), can be represented in a form

$$y = y_{reg} + y_{sing}$$

where y_{sing} is the singular part of the solution and y_{reg} stays for a (more) regular part. Using a cut-off function ξ, the singular part of the solution can be described by

$$y_{sing} = c_s \, \xi(r) \, r^{\lambda} \sin(\lambda\phi)$$

for Dirichlet boundary conditions and

$$y_{sing} = c_s \, \xi(r) \, r^{\lambda} \cos(\lambda\phi)$$

for Neumann boundary conditions with $\lambda = \pi/\omega$ and a stress intensity coefficient c_s. Consequently, we have $\lambda > 1/2$ for an arbitrary corner of the polygon. In any case, we have at least $y_{reg} \in H^2(\Omega)$.

The quantity λ is often called singular exponent. The singular exponent describes the behavior of the solution of the elliptic equation close to a corner. Of course in a polygonal domain we have several corners and consequently each corner has its own singular exponent depending on the size of the angle.

Remark 2.1 ([9]). Consider the more general linear elliptic equation

$$Ly = f \quad \text{in } \Omega \tag{1}$$

with

$$Ly(x) := -\sum_{i,j=1}^{2} \frac{\partial}{\partial x_i}\left(a_{ij}(x)\frac{\partial}{\partial x_j} y(x)\right) + \sum_{i=1}^{2} a_i(x)\frac{\partial}{\partial x_i} y(x) + a_0(x) y(x). \tag{2}$$

and Dirichlet or Neumann boundary conditions. The regularity of the solution y is characterized by one particular eigenvalue of an operator pencil, which is obtained by an integral transformation of the Dirichlet or Neumann boundary value problem for the equation $L_0 y = g$ in Ω, where the operator L_0 is obtained from the principal part of the operator L by freezing the coefficients in the corner point. That means, the regularity is not influenced by the lower order terms with the coefficients a_i, $i = 0, 1, 2$. Moreover, the coefficient functions $a_{ij}(x)$ are of interest only in the corner, see, for example, [26]. In that paper the eigenvalue of interest is denoted by $\lambda_- \in \mathbb{C}$. For our purposes we introduce the real quantity $\lambda = -\operatorname{Im}\lambda_-$.

In the case of the Dirichlet or Neumann problem for the Laplace operator and a two-dimensional domain with a re-entrant corner with interior angle $\omega \in (\pi, 2\pi)$, the value of λ is explicitly known, $\lambda = \pi/\omega$. That means in particular $\lambda \in (1/2, 1)$. In the more general case of an elliptic operator L we follow [24, Chap. 5] and consider the linear coordinate transformation $y_1 = x_1 + d_1 x_2$, $y_2 = d_2 x_2$, with $d_1 = -a_{12}/a_{22}$ and $d_2 = \sqrt{a_{11}a_{22} - a_{12}^2}/a_{22}$. In this way, the differential operator L_0 is transformed into a multiple of the Laplace operator and the neighborhood of the corner, a circular sector with opening ω, into another sector with opening ω'. The quantity of interest is then $\lambda = \pi/\omega'$. Since $\omega' \in (\pi, 2\pi)$ for $\omega \in (\pi, 2\pi)$ we have also in the general case $\lambda \in (1/2, 1)$.

2.2 *Quasi-uniform Meshes*

We denote the energy space by V, i.e., $V = H_0^1(\Omega)$ for Dirichlet boundary condition and $V = H^1(\Omega)$ for Neumann boundary condition. Let us assume a quasi-uniform mesh of triangles with the associated mesh size h. Moreover we will use piecewise linear finite elements. The corresponding finite element space is denoted by $V_h \subset V$. We motivate the results for the problem with Neumann boundary condition.

The finite element solution y_h is given as the solution of

$$(\nabla y_h, \nabla v_h)_\Omega + (y_h, v_h)_\Omega = (f, v_h)_\Omega + (g, v_h)_\Gamma \quad \forall v_h \in V_h \tag{3}$$

A standard finite element analysis yields

$$\|y - y_h\|_{H^1(\Omega)} \le c \inf_{v_h \in V_h} \|y - v_h\|_{H^1(\Omega)}.$$

The best approximation error is given by

$$\|y - y_h\|_{H^1(\Omega)} \le c \inf_{v_h \in V_h} \|y - v_h\|_{H^1(\Omega)} \le ch^{\min(1,\lambda)} \|y\|_{X^\lambda} \tag{4}$$

where X^λ is an appropriately chosen function space representing the characteristic singular behavior of the solution. The space X^λ may differ in every occurrence throughout this paper. We found $\lambda > 1$ for convex corners. In this case, we can set $X^\lambda = H^2(\Omega)$. Consequently, we get the approximation order h for convex domains. For non-convex corners we have $\lambda < 1$. Thus, the largest angle ω limits the approximation rate. The space X^λ can be chosen as an appropriate Besov space.

Approximation rates in the L^2-norm can be obtained by the Aubin-Nitsche trick

$$\|y - y_h\|_{L^2(\Omega)} \le ch^{2\min(1,\lambda)}\|y\|_{X^\lambda}. \tag{5}$$

The finite element error and the interpolation error have the same order in the L^2-norm for convex domains. In contrast to this, the order of the finite element error, $h^{2\lambda}$, is worse than the order of the interpolation error, $h^{1+\lambda}$. The order of the finite element error cannot be improved in the non-convex case. This can be easily seen in the case $g \equiv 0$ because of

$$\|y - y_h\|^2_{H^1(\Omega)} = a(y, y - y_h) = (f, y - y_h)_\Omega \le \|f\|_{L^2(\Omega)}\|y - y_h\|_{L^2(\Omega)}.$$

Next, we consider the approximation rate on the boundary Γ. We apply the Aubin-Nitsche trick directly: Let w be the solution of the dual problem

$$(\nabla v, \nabla w)_\Omega = (y - y_h, v)_\Gamma \quad \forall v \in V.$$

We find for an appropriately chosen interpolant $I_h w$

$$\begin{aligned}(y - y_h, y - y_h)_\Gamma &= (\nabla(y - y_h), \nabla w)_\Omega \\ &= (\nabla(y - y_h), \nabla(w - I_h w))_\Omega \\ &\le ch^{\min(1,\lambda)}\|y\|_{X^\lambda} h^{1/2}\|w\|_{H^{3/2}(\Omega)} \\ &\le ch^{1/2+\min(1,\lambda)}\|y\|_{X^\lambda}\|y - y_h\|_\Gamma\end{aligned}$$

which implies

$$\|y - y_h\|_{L^2(\Gamma)} \le ch^{1/2+\min(1,\lambda)}\|y\|_{X^\lambda}.$$

However, this simple estimate is only optimal in the non-convex case. An improvement of this estimate in the convex case was obtained in [21]. Numerical experiments indicate a better behavior of approximately

$$\|y - y_h\|_{L^2(\Gamma)} \sim h^{1/2+\min(3/2,\lambda)},$$

see [21]. Let us mention that the interpolation error has exactly this size. A corresponding error estimate with an additional logarithmic factor is contained in the upcoming PhD thesis of Johannes Pfefferer [25].

Pointwise error estimates are also of interest. The finite element error estimate

$$\|y - y_h\|_{L^\infty(\Omega)} \le c(1 + |\log h|)h^2 \tag{6}$$

can be proved for $y \in W^{2,\infty}(\Omega)$. This regularity property of the solution can only be expected if the largest angle of the polygon is at most $\pi/2$.

2.3 *Mesh Grading*

We have seen, in the last subsection, that uniform meshes can guarantee best approximation rates only up to a certain size of the largest angle. This behavior is caused by the singular part of the solution.

2.3.1 The Idea of Mesh Grading

The loss of accuracy is always connected with the factor r^λ in the singular part of the solution. A simple idea (see [23]) is to use a local transformation of coordinates via

$$r = \varrho^{1/\mu}.$$

We will use this transformation in a certain neighborhood of each corner. Let us denote by Ω_C and Ω'_C the neighborhood and the transformed one. One easily finds

$$\partial_{\varrho\varrho} y_{sing} \sim \partial_{\varrho\varrho} r^\lambda = \partial_{\varrho\varrho} \varrho^{\lambda/\mu}$$

and

$$y \in H^2(\Omega'_C) \quad \Leftrightarrow \quad \lambda/\mu > 1 \quad \Leftrightarrow \quad \mu < \lambda. \tag{7}$$

For a quasi-uniform discretization with respect to the new variable and mesh size h we can expect

$$\|y - I_h y\|_{L^2(\Omega'_C)} \le ch^2 |y|_{H^2(\Omega'_C)}$$

for an appropriate interpolant $I_h y$ and $\mu < \lambda$.

Next, we explain how this idea is realized in computations. We denote by h_T the diameter of the element T and by r_T its distance to a specific corner. We consider meshes of the form

$$\begin{aligned} c_1 h^{1/\mu} \le h_T \le c_2 h^{1/\mu} \quad & \text{for } r_T = 0, \\ c_1 h r_T^{1-\mu} \le h_T \le c_2 h r_T^{1-\mu} \quad & \text{for } 0 < r_T \le R, \\ c_1 h \le h_T \le c_2 h \quad & \text{for } r_T > R. \end{aligned}$$

The quantity R describes the radius of the refinement region. This quantity allows to have a specific mesh grading to more than one corner. Let us mention that $\mu = 1$ corresponds to quasi-uniform meshes (no mesh grading). The number of unknowns is proportional to h^{-2} like for quasi-uniform meshes.

Usually, the introduced transformation of coordinates does not appear in the papers. A very similar argumentation can be done in weighted Sobolev spaces. The corresponding estimate to (4) in the energy space is given by

$$\|y - y_h\|_{H^1(\Omega)} \le ch\|y\|_{X^\lambda} \tag{8}$$

for $\mu < \lambda$. Applying the Aubin-Nitsche trick, we get

$$\|y - y_h\|_{L^2(\Omega)} \le ch^2\|y\|_{X^\lambda}, \tag{9}$$

again for $\mu < \lambda$.

Remark 2.2. For convex corners we have $\lambda > 1$. Hence, quasi-uniform meshes represented by $\mu = 1$ can guarantee optimal error estimates in the sense of (8) or (9).

2.3.2 Pointwise Error Estimates

Here we will discuss pointwise error estimates. We start with estimate (6). As mentioned in the last section, an optimal error estimate for quasi-uniform meshes is obtained if the solution y belongs to $W^{2,\infty}(\Omega)$.

Again we use the local transformation of coordinates

$$r = \varrho^{1/\mu}.$$

From the relation

$$\partial_{\varrho\varrho} y_{sing} \sim \partial_{\varrho\varrho} r^\lambda = \partial_{\varrho\varrho} \varrho^{\lambda/\mu}$$

we find

$$y \in W^{2,\infty}(\Omega'_C) \quad \Leftrightarrow \quad \lambda/\mu > 2 \quad \Leftrightarrow \quad \mu < \lambda/2. \tag{10}$$

This derivation indicates that a mesh grading with $\mu < \lambda/2$ guarantees a pointwise approximation order of $(1 + |\log h|)h^2$.

The numerical analysis to this aspect was published in [8, 28] for Dirichlet boundary conditions where techniques from [27] were adapted to locally refined meshes. The pointwise error estimate in [28] can be written in the form

$$\|y - y_h\|_{L^\infty(\Omega)} \leq c(1 + |\log h|)h^2 \|y\|_{X^\lambda} \tag{11}$$

for $\mu < \lambda/2$.

Remark 2.3. The estimate (11) shows that mesh grading ($\mu < 1$) is necessary to obtain optimal approximation rates for all corners with interior angle $\omega > \pi/2$.

2.3.3 Error Estimates on the Boundary

Let us now discuss the accuracy of finite element solutions on the boundary. To obtain the estimate $\|y - y_h\|_{L^2(\Gamma)} \leq ch^2$ for quasi-uniform meshes we need at least H^2-regularity on each side of the polygon. This regularity can only be expected for angles $\omega < 2\pi/3$. It turns out that some weighted Sobolev space $W^{2,\infty}_{1/2}(\Omega)$ is more appropriate for the numerical analysis. The condition on the angle remains $\omega < 2\pi/3$.

Again, a different mesh grading is necessary to obtain the desired accuracy for arbitrary polygonal domains. Let us start with the local transformation of coordinates

$$r = \varrho^{1/\mu}.$$

Now the condition

$$y_{sing} \in W^{2,\infty}_{1/2}(\Omega'_C)$$

can be formulated as

$$\varrho^{1/(2\mu)} \partial_{\varrho\varrho} \varrho^{\lambda/\mu} \in L^\infty(\Omega'_C).$$

This leads to the condition

$$\frac{1}{2\mu} + \frac{\lambda}{\mu} - 2 > 0$$

which can be equivalently expressed by

$$\mu < \frac{1}{2}\left(\frac{1}{2} + \lambda\right). \tag{12}$$

Table 1 Summary of mesh grading results for different norms

Norm	Grading parameter	Approximation rate	Critical angle
$\|y - y_h\|_{H^1(\Omega)}$	$\mu < \lambda$	h	π
$\|y - y_h\|_{L^2(\Omega)}$	$\mu < \lambda$	h^2	π
$\|y - y_h\|_{L^\infty(\Omega)}$	$\mu < \lambda/2$	$(1 + \|\log h\|)h^2$	$\pi/2$
$\|y - y_h\|_{L^2(\Gamma)}$	$\mu < \frac{1}{2}\left(\frac{1}{2} + \lambda\right)$	$(1 + \|\log h\|^{3/2})h^2$	$2\pi/3$

The derivation of approximation rates can be found in [6]. The corresponding numerical experiments are published in [5]. The error estimate in [6] can be written as

$$\|y - y_h\|_{L^2(\Gamma)} \leq c(1 + |\log h|^{3/2})h^2\|y\|_{X^\lambda} \tag{13}$$

for $\mu < \frac{1}{2}\left(\frac{1}{2} + \lambda\right)$.

Remark 2.4. Mesh grading ($\mu < 1$) is necessary for all angles with $\omega \geq 2\pi/3$ to obtain optimal approximation rates in the $L^2(\Gamma)$-norm.

2.3.4 Short Overview

In Table 1 we present the principles for the choice of the grading parameter and the critical angle for quasi-uniform meshes for each specific norm. Recall that the number of elements for graded meshes has the same order as for quasi-uniform meshes.

3 Graded Meshes in Optimal Control

As we have seen, mesh grading is useful to obtain good approximation rates for the numerical solution of an elliptic partial differential equation. These results can be applied to different types of optimal control problems. We will focus mainly on results for control constrained problems. However, we will comment on problems with pointwise state constraints, too.

3.1 Distributed Control

We consider the minimization of

$$J(y, u) = \frac{1}{2}\|y - y_d\|^2_{L^2(\Omega)} + \frac{\nu}{2}\|u\|^2_{L^2(\Omega)}$$

subject to

$$(\nabla y, \nabla v)_\Omega + (y, v)_\Omega = (u, v)_\Omega \quad \forall v \in V,$$

$u \in U = L^2(\Omega)$, and

$$u_a \le u(x) \le u_b \quad \text{a.e. in } \Omega.$$

We assume $u_a < u_b$ and $\nu > 0$.

The discretized problem aims to minimize

$$J_h(y_h, u_h) = \frac{1}{2}\|y_h - y_d\|^2_{L^2(\Omega)} + \frac{\nu}{2}\|u_h\|^2_{L^2(\Omega)}$$

subject to

$$(\nabla y_h, \nabla v_h)_\Omega + (y_h, v_h)_\Omega = (u_h, v_h)_\Omega \quad \forall v_h \in V_h,$$

$u \in U_h$, and

$$u_a \le u_h(x) \le u_b \quad \text{a.e. in } \Omega.$$

The first order optimality systems contain the adjoint states p and p_h, respectively. The corresponding adjoint equations are given by

$$\begin{aligned}(\nabla v, \nabla p)_\Omega + (v, p)_\Omega &= (y - y_d, v)_\Omega \quad \forall v \in V,\\ (\nabla v_h, \nabla p_h)_\Omega + (v_h, p_h)_\Omega &= (y_h - y_d, v_h)_\Omega \quad \forall v_h \in V_h.\end{aligned}$$

In the sequel, we use a bar to indicate optimal solutions.

The choice $U_h = U$ leads to the variational approach by Hinze [19]. In this case, we can apply directly the relations (5), (6), (9), and (11). We note only the mesh grading results. We remark that the following estimates include different regularity assumptions for y_d. However, all these assumptions are satisfied for $y_d \in C^{0,\sigma}(\Omega)$ with an arbitrary $\sigma > 0$.

Lemma 3.1. *Assume $\mu < \lambda$ for the mesh grading and $U_h = U$. Then we have the estimate*

$$\|\bar u - \bar u_h\|_{L^2(\Omega)} + \|\bar y - \bar y_h\|_{L^2(\Omega)} + \|\bar p - \bar p_h\|_{L^2(\Omega)} \le ch^2.$$

A strong mesh grading $\mu < \lambda/2$ yields

$$\|\bar u - \bar u_h\|_{L^\infty(\Omega)} + \|\bar y - \bar y_h\|_{L^\infty(\Omega)} + \|\bar p - \bar p_h\|_{L^\infty(\Omega)} \le (1 + |\log h|^{3/2})ch^2.$$

Similar results can be obtained for a space U_h of piecewise constant functions and a post-processing step. Here an additional assumption is needed. The triangulation is divided into two parts. In one part of the triangulation the optimal control $\bar{u}$ is smooth, i.e., it belongs to $H^2(T_i)$ for such finite elements T_i. The control $\bar{u}$ has kinks in finite elements belonging to a second part K of the triangulation. The assumption $|K| \leq ch$ is needed to obtain the desired results. The post-processed control $\tilde{u}$ is defined by

$$\tilde{u}_h = P_{[u_a,u_b]}\left(-\frac{1}{\nu}\bar{p}_h\right).$$

The following results are contained in [8, 9].

Lemma 3.2. *Assume $\mu < \lambda$ for the mesh grading, $|K| \leq ch$, and U_h be a space of piecewise constant functions. Then the estimate*

$$\|\bar{u} - \tilde{u}_h\|_{L^2(\Omega)} + \|\bar{y} - \bar{y}_h\|_{L^2(\Omega)} + \|\bar{p} - \bar{p}_h\|_{L^2(\Omega)} \leq ch^2.$$

is valid. A strong mesh grading $\mu < \lambda/2$ yields

$$\|\bar{u} - \tilde{u}_h\|_{L^\infty(\Omega)} + \|\bar{y} - \bar{y}_h\|_{L^\infty(\Omega)} + \|\bar{p} - \bar{p}_h\|_{L^\infty(\Omega)} \leq (1 + |\log h|^{3/2})ch^2.$$

Remark 3.3. Numerical experiments show the approximation rates of Lemma 3.2 also for piecewise linear controls. However, there is no theoretical justification for that effect until now.

3.2 Neumann Boundary Control

We consider the minimization of

$$J(y,u) = \frac{1}{2}\|y - y_d\|^2_{L^2(\Omega)} + \frac{\nu}{2}\|u\|^2_{L^2(\Gamma)}$$

subject to

$$(\nabla y, \nabla v)_\Omega + (y, v)_\Omega = (f, v)_\Omega + (u, v)_\Gamma \quad \forall v \in V,$$

$u \in U = L^2(\Gamma)$, and

$$u_a \leq u(x) \leq u_b \quad \text{a.e. on } \Gamma.$$

We assume $u_a < u_b$ and $\nu > 0$.

The discretized problem aims to minimize

$$J_h(y_h, u_h) = \frac{1}{2}\|y_h - y_d\|^2_{L^2(\Omega)} + \frac{\nu}{2}\|u_h\|^2_{L^2(\Omega)}$$

subject to

$$(\nabla y_h, \nabla v_h)_\Omega + (y_h, v_h)_\Omega = (f, v_h)_\Omega + (u_h, v_h)_\Gamma \quad \forall v_h \in V_h,$$

$u_h \in U_h$, and

$$u_a \le u_h(x) \le u_b \quad \text{a.e. on } \Gamma.$$

The corresponding adjoint equations are given by

$$(\nabla v, \nabla p)_\Omega + (v, p)_\Omega = (y - y_d, v)_\Omega \quad \forall v \in V,$$
$$(\nabla v_h, \nabla p_h)_\Omega + (v_h, p_h)_\Omega = (y_h - y_d, v_h)_\Omega \quad \forall v_h \in V_h.$$

A post-processed control $\tilde{u}$ is defined by

$$\tilde{u}_h = P_{[u_a, u_b]}\left(-\frac{1}{\nu}\bar{p}_h|_\Gamma\right).$$

For the following result we refer to [6].

Lemma 3.4. *Assume $\mu < \frac{1}{2}(\frac{1}{2} + \lambda)$ for the mesh grading. Then we get for the variational approach ($U = U_h$)*

$$\|\bar{u} - \bar{u}_h\|_{L^2(\Gamma)} + \|\bar{y} - \bar{y}_h\|_{L^2(\Omega)} + \|\bar{p} - \bar{p}_h\|_{L^2(\Omega)} \le c(1 + |\log h|^{3/2})h^2.$$

If $|K| \le ch$ and U_h is a space of piecewise constant functions then we have

$$\|\bar{u} - \tilde{u}_h\|_{L^2(\Gamma)} + \|\bar{y} - \bar{y}_h\|_{L^2(\Omega)} + \|\bar{p} - \bar{p}_h\|_{L^2(\Omega)} \le c(1 + |\log h|^{3/2})h^2$$

for the post-processing approach.

3.3 Problems with Piecewise State Constraints

Here we consider the minimization of

$$J(y, u) = \frac{1}{2}\|y - y_d\|^2_{L^2(\Omega)} + \frac{\nu}{2}\|u\|^2_{L^2(\Omega)}$$

subject to

$$(\nabla y, \nabla v)_\Omega + (y, v)_\Omega = (u, v)_\Omega \quad \forall v \in V,$$

$u \in U = L^2(\Omega)$,

$$y \leq y_c \text{ a.e. in } \Omega',$$

and possibly

$$u_a \leq u(x) \leq u_b \quad \text{a.e. in } \Omega.$$

We assume $u_a < u_b$, $\nu > 0$, $y_d \in L^2(\Omega)$, $\Omega' \subset \Omega$, and $y_c \in C(\bar{\Omega}')$.

There are several publications on a priori error estimates for finite element approximations for smooth domains. Let us mention the basic papers [15, 22]. In both approaches the obtained approximation rate is the square root of the $L^\infty(\Omega')$-error of the state equation using the optimal regularity of the control $\bar{u}$. Consequently, an $L^\infty(\Omega')$-error estimate for the state equation is a key result to derive a priori error estimates for state constrained problems. In [27] we find the estimate for the boundary value problem

$$\|y - y_h\|_{L^\infty(\Omega')} \leq c((1 + |\log h|) \inf_{v_h \in V_h} \|y - v_h\|_{L^\infty(\Omega'')} + \|y - y_h\|_{L^2(\Omega)}) \tag{1}$$

implying

$$\|y - y_h\|_{L^\infty(\Omega')} \leq ch^{2-2\varepsilon}$$

with an arbitrary small $\varepsilon > 0$ which hides logarithmic terms. For smooth domains the convergence order for a piecewise linear finite element approximation is the square root of this expression, i.e.,

$$\|\bar{u} - \bar{u}_h\|_{L^2(\Omega)} + \|\bar{y} - \bar{y}_h\|_{L^2(\Omega)} \leq ch^{1-\varepsilon}. \tag{2}$$

The situation is similar for the case of polygonal domains and $\Omega' \subset\subset \Omega$. One can benefit from elliptic regularity for the first term in (1) since the state constraints are separated from the corner singularities. The second term requires a moderate mesh grading for non-convex domains, i.e., $\mu < \lambda$. For a mesh grading with $\mu < \lambda$ and $\Omega' \subset\subset \Omega$ the derivation of estimate (2) with the methods of [15, 22] seems to be possible.

The situation becomes more delicate for $\Omega' = \Omega$. A strong mesh grading with $\mu < \lambda/2$ is already needed for the optimal convergence of the state equation in the L^∞-norm. However, the derivation of an approximation result for the optimal control problem requires additional results like uniform boundedness properties of discrete solutions in appropriate spaces and additional smoothness properties. We expect that the desired approximation properties for the optimal control problem can be derived using a careful analysis in weighted Sobolev spaces but this is not done yet.

4 Results for Three-Dimensional Domains

4.1 Regularity

The solution of (1), (2), or (1), (3), can again be written as

$$y = y_{reg} + y_{sing}.$$

However, the singular part is in general not as simple as in the two-dimensional case but it consists of terms that correspond to the edges and the vertices of the domain.

In the vicinity of edges the singular part can be written in cylindrical coordinates as

$$u_{sing}(r,\phi,z) = c_s(r,z)\,\xi(r)\,r^{\lambda_e}\,\Phi(\phi) \tag{1}$$

with $\Phi(\phi) = \sin(\lambda_e\phi)$ in the case of Dirichlet boundary conditions and $\Phi(\phi) = \cos(\lambda_e\phi)$ in the case of Neumann boundary conditions. The singularity exponent is $\lambda_e = \pi/\omega$ in both cases, as in the two-dimensional case. The main difference to the two-dimensional case is that the stress intensity distribution c_s is now a function. Note that this function does in general not only depend on the edge variable z.

In the vicinity of vertices of the domain the singular part can be written in spherical coordinates as

$$u_{sing}(r,\phi,\theta) = c_s\,\xi(r)\,r^{\lambda_v}\,\Phi(\phi,\theta)$$

where (λ_v, Φ) is an eigenpair of a corresponding operator pencil. The eigenpair is explicitly known only in very special cases, in general is has to be computed numerically. The stress intensity coefficient, however, is a constant as in the two-dimensional case.

There are no singular parts if $\lambda_e > 1$ and $\lambda_v > \frac{1}{2}$, that means the global regularity can be described by

$$\lambda = \min\{\lambda_e, \lambda_v + \frac{1}{2}\}, \tag{2}$$

and we have $u \in H^s(\Omega)$ for $s < 1 + \lambda$. The minimum is taken over all (singular) edges e and all (singular) vertices v.

4.2 Discretization with Shape-Regular (Isotropic) Finite Elements

The discretization of the boundary value problems can be discussed with arguments similar to Sects. 2.2 and 2.3.

By using piecewise linear finite elements on quasi-uniform meshes we obtain

$$\|y - y_h\|_{H^1(\Omega)} \le ch^{\min(1,\lambda)-\varepsilon}\|y\|_{X^\lambda},$$

$$\|y - y_h\|_{L^2(\Omega)} \le ch^{2\min(1,\lambda)-\varepsilon}\|y\|_{X^\lambda}$$

for arbitrary $\varepsilon > 0$ and with λ from (2). Probably the estimate holds also for $\varepsilon = 0$ but an investigation would need regularity results in Besov spaces which are not known to the authors.

Graded meshes can be introduced as in the two-dimensional case by requiring that the element diameter h_T and the distance r_T to the singular edges and vertices are related by

$$\begin{aligned} c_1 h^{1/\mu} \le h_T &\le c_2 h^{1/\mu} && \text{for } r_T = 0, \\ c_1 h r_T^{1-\mu} \le h_T &\le c_2 h r_T^{1-\mu} && \text{for } 0 < r_T \le R, \\ c_1 h \le h_T &\le c_2 h && \text{for } r_T > R. \end{aligned} \tag{3}$$

As long as $\mu > \frac{1}{3}$ the complexity is the same as for quasi-uniform meshes since the number of elements is of order h^{-3}. For stronger mesh grading as it would be required in the case of mixed boundary conditions or discontinuous coefficients, the number of unknowns would increase to $O(h^{-1/\mu})$ if $\mu < \frac{1}{3}$, see [10]. For the discretization error we get

$$\|y - y_h\|_{H^1(\Omega)} \le ch\|y\|_{X^\lambda}, \tag{4}$$

$$\|y - y_h\|_{L^2(\Omega)} \le ch^2\|y\|_{X^\lambda}, \tag{5}$$

for sufficient mesh grading with parameter $\mu < \lambda$, see [10] and also [2, 17] for earlier results. The corresponding pointwise error estimates and estimates of the error on the boundary (for the Neumann problem) are work in progress.

On the basis of the estimate (5) we proved in [13] that the L^2-error estimates of Lemmas 3.1 and 3.2 (distributed control problems) hold in the three-dimensional case as well. The error estimates for boundary control problems (Lemma 3.4) are only currently proved for the three-dimensional case, see [7].

4.3 Anisotropic Discretizations in Tensor Product Domains

The structure of the solution near edges, see (1), reveals that the critical term $r^{\lambda_e}\Phi(\varphi)$ acts only in planes perpendicularly to the edge. The z-derivatives of u are much more regular than the derivatives in directions perpendicular to the z-axis. This indicates that the many elements along the edge, see Fig. 1, might be unnecessary.

The elements of the mesh on the right hand side of Fig. 1 are characterized by two size parameters. The size in edge direction is the global mesh size h, whereas only the size h_T in perpendicular direction satisfies the usual grading conditions (3). Note that these elements are not shape regular since the aspect ratio behaves like $O(h^{1-1/\mu})$ for elements with $r_T = 0$ and like $O(r_T^{\mu-1})$ when $r_T > 0$; we call them anisotropic. Note that such meshes have $O(h^{-3})$ elements independent of the value of the grading parameter $\mu \in (0, 1]$.

Since some standard error estimates for the local interpolation error do not hold for anisotropic elements, the proof of estimate (5) was proved only recently, [11], although estimate (4) was proved already about 20 years earlier, [1]. Based on this result we proved in [12] that the L^2-error estimates of Lemmas 3.1 and 3.2 (distributed control problems) hold in the three-dimensional case as well. Error estimates in the $L^\infty(\Omega)$- and the $L^2(\Gamma)$-norms are not proved for anisotropic meshes until now.

4.4 Anisotropic Discretization in General Polyhedral Domains

For treating general polyhedral domains with anisotropic grading near edges one needs to combine this with isotropic grading towards the singular corners. In [4] we devised the strategy first to split the domain into $O(1)$ macro-elements (tetrahedra) such that each macro-element touches at most one singular edge and at most one singular vertex. We described the regularity of the elliptic equation in these macro-elements and, based on this, devised a local refinement strategy for each type of macro-element. Due to limitations of the Lagrange interpolation we were, however, only able to prove the H^1-error estimates (4) for the boundary value problem. Only recently, we succeeded to construct a quasi-interpolation operator that allowed to

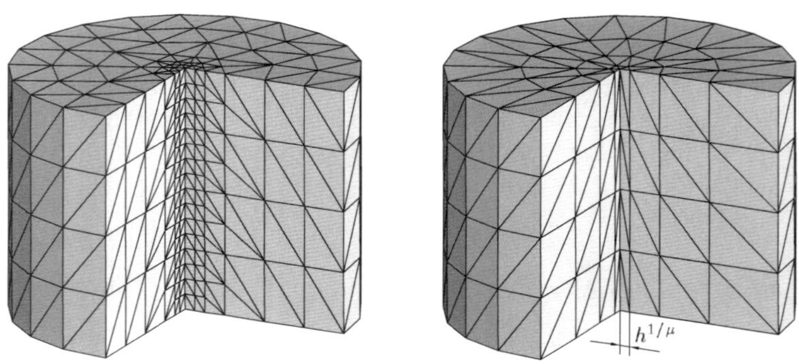

Fig. 1 Isotropic versus anisotropic mesh grading near an edge

prove the estimate (5) in the L^2-norm, see [3]. An immediate consequence for the variational discretization of distributed optimal control problems in polyhedral domains is that Lemma 3.1 holds as well for this kind of graded mesh, [3].

4.5 *Summary of Sect. 4*

In the three-dimensional case we have to consider edge and vertex singularities. Nevertheless, the solution is characterized by the quantity λ defined in (2). For the treatment of the edge singularities one has to decide between the more convenient isotropic and the more efficient anisotropic mesh grading. Both types of grading are controlled by a grading parameter μ which, as a rule of thumb, should be chosen as given in Table 1. The theory in the three-dimensional case is an ongoing research topic and not yet finished.

References

1. Th. Apel, M. Dobrowolski, Anisotropic interpolation with applications to the finite element method. Computing **47**, 277–293 (1992)
2. Th. Apel, B. Heinrich, Mesh refinement and windowing near edges for some elliptic problem. SIAM J. Numer. Anal. **31**, 695–708 (1994)
3. Th. Apel, A.L. Lombardi, M. Winkler, Anisotropic mesh refinement in polyhedral domains: error estimates with data in $L^2(\Omega)$. ESAIM: Math. Model. Numer. Anal. **48**(4), 1117–1145 (2014)
4. Th. Apel, S. Nicaise, The finite element method with anisotropic mesh grading for elliptic problems in domains with corners and edges. Math. Methods Appl. Sci. **21**, 519–549 (1998)
5. Th. Apel, J. Pfefferer, A. Rösch, Finite element error estimates for Neumann boundary control problems on graded meshes. Comput. Optim. Appl. **52**(1), 3–28 (2012)
6. Th. Apel, J. Pfefferer, A. Rösch, Finite element error estimates on the boundary with application to optimal control. Accepted for publication in Math. Comp.
7. Th. Apel, J. Pfefferer, M. Winkler, *Local Mesh Refinement for the Discretisation of Neumann Boundary Control Problems on General Polyhedra* (in preparation)
8. Th. Apel, A. Rösch, D. Sirch, L^∞-error estimates on graded meshes with application to optimal control. SIAM J. Control Optim. **48**(3), 1771–1796 (2009)
9. Th. Apel, A. Rösch, G. Winkler, Optimal control in non-convex domains: a priori discretization error estimates. Calcolo **44**(3), 137–158 (2007)
10. Th. Apel, A.-M. Sändig, J.R. Whiteman, Graded mesh refinement and error estimates for finite element solutions of elliptic boundary value problems in non-smooth domains. Math. Methods Appl. Sci. **19**(1), 63–85 (1996)
11. Th. Apel, D. Sirch, L^2-error estimates for Dirichlet and Neumann problems on anisotropic finite element meshes. Appl. Math. **56**, 177–206 (2011)
12. Th. Apel, D. Sirch, G. Winkler, *Error Estimates for Control Constrained Optimal Control Problems: Discretization with Anisotropic Finite Element Meshes*, SPP 1253. Preprint SPP1253-02-06. Accepted by Math. Program
13. Th. Apel, G. Winkler, Optimal control under reduced regularity. Appl. Numer. Math. **59**(9), 2050–2064 (2009)

14. N. Arada, E. Casas, F. Tröltzsch, Error estimates for the numerical approximation of a semilinear elliptic control problem. Comput. Optim. Appl. **23**(2), 201–229 (2002)
15. K. Deckelnick, M. Hinze, Convergence of a finite element approximation to a state-constrained elliptic control problem. SIAM J. Numer. Anal. **45**(5), 1937–1953 (2007)
16. R.S. Falk, Approximation of a class of optimal control problems with order of convergence estimates. J. Math. Anal. Appl. **44**, 28–47 (1973)
17. R. Fritzsch, P. Oswald, Zur optimalen Gitterwahl bei Finite-Elemente-Approximationen. Wissenschaftliche Zeitschrift TU Dresden **37**(3), 155–158 (1988)
18. T. Geveci, On the approximation of the solution of an optimal control problem governed by an elliptic equation. RAIRO Anal. Numér. **13**(4), 313–328 (1979)
19. M. Hinze, A variational discretization concept in control constrained optimization: the linear-quadratic case. Comput. Optim. Appl. **20**(1), 45–61 (2005)
20. K. Malanowski, Convergence of approximations vs. regularity of solutions for convex, control–constrained optimal control problems. Appl. Math. Optim. **8**, 69–95 (1981)
21. M. Mateos, A. Rösch, On saturation effects in the Neumann boundary control of elliptic optimal control problems. Comput. Optim. Appl. **49**(2), 359–378 (2011)
22. C. Meyer, Error estimates for the finite-element approximation of an elliptic control problem with pointwise state and control constraints. Control Cybern. **37**(1), 51–83 (2008)
23. L.A. Oganesjan, L. Rukhovets, Variational-difference schemes for second order linear elliptic equations in a two-dimensional region with a piecewise-smooth boundary. Ž. Vyčisl. Mat. i Mat. Fiz. **8**, 97–114 (1968). (In Russian)
24. L.A. Oganesjan, L. Rukhovets, V. Rivkind, Variational-difference methods for solving elliptic equations, part II, Vol. 8 of Differential equations and their applications, Izd. Akad. Nauk Lit. SSR, Vilnius, 1974. In Russian
25. J. Pfefferer, Numerical analysis for elliptic Neumann boundary control problems on polygonal domains, Ph.D. thesis, Universität der Bundeswehr München, 2014
26. A.-M. Sändig, Error estimates for finite element solutions of elliptic boundary value problems in non-smooth domains. Z. Anal. Anwend. **9**, 133–153 (1990)
27. A.H. Schatz, L.B. Wahlbin, Interior maximum norm estimates for finite element methods. Math. Comput. **31**, 414–442 (1977)
28. D. Sirch, Finite element error analysis for PDE-constrained optimal control problems: the control constrained case under reduced regularity, Ph.D. thesis, Technische Universität München, 2010

Model Order Reduction for PDE Constrained Optimization

Peter Benner, Ekkehard Sachs, and Stefan Volkwein

Abstract The optimization and control of systems governed by partial differential equations (PDEs) usually requires numerous evaluations of the forward problem or the optimality system. Despite the fact that many recent efforts, many of which are reported in this book, have been made to limit or reduce the number of evaluations to 5–10, this cannot be achieved in all situations and even if this is possible, these evaluations may still require a formidable computational effort. For situations where this effort is not acceptable, model order reduction can be a means to significantly reduce the required computational resources. Here, we will survey some of the most popular approaches that can be used for this purpose. In particular, we address the issues arising in the strategies discretize-then-optimize, in which the optimality system of the reduced-order model has to be solved, and optimize-then-discretize, where a reduced-order model of the optimality system has to be found. The methods discussed include versions of proper orthogonal decomposition (POD) adapted to PDE constrained optimization as well as system-theoretic methods.

This work was supported by the DFG Priority Program 1253 "Optimization with Partial Differential Equations", grants BE 2174/8-2, SA 289/20-1, and the DFG project "A-posteriori-POD Error Estimators for Nonlinear Optimal Control Problems governed by Partial Differential Equations", grant VO 1658/2-1.

P. Benner (✉)
Max Planck Institute for Dynamics of Complex Technical Systems, Sandtorstraße 1, 39106 Magdeburg, Germany
e-mail: benner@mpi-magdeburg.mpg.de

E. Sachs
FB 4 – Mathematik, University of Trier, 54286 Trier, Germany
e-mail: sachs@uni-trier.de

S. Volkwein
Department of Mathematics and Statistics, University of Konstanz, Universitätsstraße 10, 78464 Konstanz, Germany
e-mail: stefan.volkwein@uni-konstanz.de

G. Leugering et al. (eds.), *Trends in PDE Constrained Optimization*, International Series of Numerical Mathematics 165, DOI 10.1007/978-3-319-05083-6_19

Keywords Model order reduction • PDE constrained optimization • Optimal control • Proper orthogonal decomposition • Adaptive methods

Mathematics Subject Classification (2010). Primary 49M05; Secondary 93C20.

1 Introduction

Optimal control problems for partial differential equations are often hard to tackle numerically because their discretization leads to very large scale optimization problems. Therefore different techniques of model reduction were developed to approximate these problems by smaller ones that are tractable with less effort.

One popular model reduction technique for large-scale state-space systems is the *moment matching approximation* considered first in [28, 30]. This method is based on projecting the dynamical system onto Krylov subspaces computed by an Arnoldi- or Lanczos process. Krylov methods prove to be efficient for large-scale sparse systems, since only matrix-vector multiplications are required. The moment matching method shows the drawbacks that stability and passivity are not necessarily preserved in the reduced-order system and that there is no global approximation error bound; see, e.g., [7, 39]. *Balanced truncation* [92] is another well studied model reduction technique for state-space systems. This method utilizes the solutions to two Lyapunov equations, the so-called controllability and observability Gramians. The balanced truncation method is based on transforming the state-space system into a balanced form so that its controllability and observability Gramians become diagonal and equal. Moreover, the states that are difficult to reach or to observe, are truncated. The advantage of this method is that it preserves the asymptotic stability in the reduced-order system. Furthermore, a-priori error bounds are available. Recently, the theory of balanced truncation model reduction was extended to descriptor systems; see, e.g., [61] and [43]. Both the moment matching approximation and the balanced truncation approach do not rely on snapshots, which have to be taken more or less arbitrarily. For an overview we refer the reader to [3,78]. However, up to now, both strategies can be applied more or less only to linear, time-invariant dynamical systems and do not yet cover time variant or nonlinear models. There are attempts to deal with time variant equations by approximating them through piecewise constant models; see, e.g., [14].

Recently the application of *reduced-order models* to linear time varying and nonlinear systems, in particular to nonlinear control systems, has received an increasing amount of attention. The reduced-order approach is based on projecting the dynamical system onto subspaces consisting of basis elements that contain characteristics of the expected solution. This is in contrast to, e.g., finite element techniques, where the basis elements of the subspaces do not relate to the physical properties of the system that they approximate. The *reduced basis* (RB) method, as developed in [35, 66] and [51], is one such reduced-order method, where the basis elements correspond to the dynamics of expected control regimes. Let us refer to [26, 44, 53, 68] for the successful use of reduced basis method in PDE constrained optimization problems. Currently *Proper Orthogonal Decomposition* (POD) is probably the mostly used and most successful model reduction technique for nonlinear optimal control problems, where the basis functions contain information from the solutions of the dynamical system at pre-specified time-instances, so-called snapshots. Due to a possible linear dependence or almost linear dependence the snapshots themselves are not appropriate as a basis. Hence a singular value decomposition is carried out and the leading generalized eigenfunctions are chosen as a basis, referred to as the POD basis. POD is successfully used in a variety of fields including fluid dynamics, coherent structures [1, 4] and inverse problems [8]. Moreover, in [6] POD is successfully applied to compute reduced-order controllers. The relationship between POD and balancing was considered in [56, 75, 90]. An error analysis for non-linear dynamical systems in finite dimensions was carried out in [71] and a missing point estimation in models described by POD was studied in [5]. Let us also mention that POD and the reduced basis method are successfully combined by variants of the POD greedy algorithm; see [42] and [41], for instance.

Reduced-order models are used in PDE-constrained optimization in various ways; see, e.g., [37, 48, 76] for a survey. In optimal control problems it is sometimes necessary to compute a feedback control law instead of a fixed optimal control. In the implementation of these feedback laws, models of reduced order can play an important and very useful role, see [2, 6, 31, 55, 58, 72]. Another useful application is the use in optimization problems, where a PDE solver is part of the function evaluation. Obviously, thinking of a gradient evaluation or even a step-size rule in the optimization algorithm, an expensive function evaluation leads to an enormous amount of computing time. Here, the reduced-order model can replace the system given by a PDE in the objective function. It is quite common that a PDE can be replaced by a five- or ten-dimensional system of ordinary differential equations. This results computationally in a very fast method for optimization compared to the effort for the computation of a single solution of a PDE. There is a large amount of literature in engineering applications in this regard, we mention only the papers [60, 64]. Recent applications can also be found in finance using the RB model [67] and the POD model [80] in the context of calibration for models in option pricing.

To explain the reduced-order modelling we choose the following generic nonlinear optimal control problem: Minimize the cost functional

$$J(y,u) = \frac{1}{2}\int_0^T \int_\Omega |y(t,\boldsymbol{x}) - y_d(t,\boldsymbol{x})|^2 \,\mathrm{d}\boldsymbol{x}\,\mathrm{d}t + \frac{\kappa}{2}\,\|u\|_U^2 \tag{1.1a}$$

subject to the semilinear parabolic partial differential equation

$$\begin{aligned} y_t(t,\boldsymbol{x}) - \Delta y(t,\boldsymbol{x}) + f(t,\boldsymbol{x},y(t,\boldsymbol{x})) &= (\mathcal{B}u)(t,\boldsymbol{x}), && (t,\boldsymbol{x}) \in Q,\\ y(t,\boldsymbol{x}) &= 0, && (t,\boldsymbol{x}) \in \Sigma,\\ y(0,\boldsymbol{x}) &= y_0(\boldsymbol{x}), && \boldsymbol{x} \in \Omega \end{aligned} \tag{1.1b}$$

and to the control constraints

$$u \in U_{\text{ad}}. \tag{1.1c}$$

In (1.1) we suppose that $T > 0$ holds and the spatial domain $\Omega \subset \mathbb{R}^d$, $d \in \{1, 2, 3\}$, is a bounded open set with Lipschitz-continuous boundary Γ. We set $Q = (0, T) \times \Omega$ and $\Sigma = (0, T) \times \Gamma$. Furthermore, the set $U_{\text{ad}} \subseteq U$ of admissible controls is a closed and convex subset of a Hilbert space U, which is identified by its dual space U'. Let us refer to the recent paper [38], where the authors investigate a state-constrained linear-quadratic optimal control problem using proper orthogonal decomposition. Let us set $H = L^2(\Omega)$ and $V = H_0^1(\Omega)$ with dual space $V' = H^{-1}(\Omega)$. In (1.1a) we suppose that y_d belongs to $L^2(0, T; H)$ and $\kappa > 0$ holds. Let $f : Q \times \mathbb{R} \to \mathbb{R}$ contain the semilinear term and the control operator $\mathcal{B} : U \to L^2(0, T; H)$ be linear and continuous. Finally, we suppose that $y_0 \in L^2(\Omega)$. For given control $u \in U_{\text{ad}}$ a solution to (1.1b) is understood as a weak solution, i.e., y satisfies

$$\begin{aligned} &\frac{\mathrm{d}}{\mathrm{d}t} \langle y(t), \varphi \rangle_H + \int_\Omega \nabla y(t) \cdot \nabla \varphi + \big(f(t, \cdot\,, y(t)) - (\mathcal{B}u)(t)\big)\varphi \,\mathrm{d}\boldsymbol{x} = 0 \\ &\qquad\qquad\qquad\qquad\qquad\qquad\qquad\qquad \forall \varphi \in V,\ t \in (0, T], \\ &\langle y(0), \phi \rangle_H = \langle y_0, \phi \rangle_H \qquad\qquad\qquad\qquad\quad \forall \phi \in H, \end{aligned} \tag{1.1b$'$}$$

where $y(t)$ stands for $y(t, \cdot)$ as a function in the spatial variable $\boldsymbol{x}$. We suppose that (1.1b′) admits a unique weak solution $y = y(u)$ for any $u \in U_{\text{ad}}$. We refer to [19, 73] for sufficient conditions, but also for the proof that (1.1) possesses a local optimal solution $\bar{x} = (\bar{y}, \bar{u})$. Let us introduce the so called reduced cost functional[1]

$$\hat{J}(u) = J(y(u), u), \quad u \in U_{\text{ad}},$$

where $y(u)$ is the unique weak solution to (1.1b′).

The paper is organized as follows: First we consider two venues how to discretize a PDE-constrained optimization problem followed by an introduction to POD-based methods. We address various aspects about the proper choice of the POD basis in the course of the optimization. The fourth section is devoted to system-theoretic aspects such as techniques based on Krylov subspaces and system balancing.

[1] Not to be confused with the "reduced-order" terminology used in the model reduction context—here, "reduced" means that the cost functional is written in dependence of the control only, using the fact that the weak solution $y(u)$ is uniquely determined by the chosen u!

2 Optimization with Surrogate Models

The reduced-order approximation to (1.1) can be derived by a *discretize-then-optimize* or by an *optimize-then-discretize* approach. In the first case, the optimal control problem is projected onto the reduced-order subspace and the resulting low-dimensional optimal control problem is solved by appropriate optimisation methods., like, e.g., a sequential quadratic programming algorithm [65, Chapter 18]. In the second case, we start by deriving optimality conditions for (1.1) and discretize the obtained conditions by a reduced-order projection.

Next we describe both approaches for (1.1). Suppose that $\psi_1, \dots, \psi_\ell \in V$ are given linearly independent functions in V. Then, we define the subspace $V^\ell = \operatorname{span}\{\psi_1, \dots, \psi_\ell\}$. Let us mention that we avoid a discretization of the control space U; see [47].

2.1 *Discretize-Then-Optimize*

In this approach we only replace the state y by a reduced-order approximation. For that reason we introduce the affine Galerkin ansatz

$$y^\ell(t) = y_{\mathsf{p}}(t) + \sum_{i=1}^{\ell} \mathrm{y}_i^\ell(t)\psi_i, \quad t \in [0, T], \tag{2.1}$$

with a chosen particular element $y_{\mathsf{p}}(t) \in V$ and coefficient functions $\mathrm{y}_i^\ell : [0, T] \to \mathbb{R}$, $1 \le i \le \ell$. Then, the reduced-order Galerkin projection for (1.1) reads as follows:

$$\min J(y^\ell, u) = \int_0^T \int_\Omega \left| y^\ell(t) - y_d(t) \right|^2 \mathrm{d}\boldsymbol{x}\,\mathrm{d}t + \frac{\kappa}{2} \|u\|_U^2 \tag{2.2a}$$

subject to the projected nonlinear dynamical system

$$\frac{\mathrm{d}}{\mathrm{d}t} \langle y^\ell(t), \psi \rangle_H + \int_\Omega \nabla y^\ell(t) \cdot \nabla \psi + \big(f(t, \cdot, y^\ell(t)) - (\mathcal{B}u)(t)\big)\psi \,\mathrm{d}\boldsymbol{x} = 0 \tag{2.2b}$$

$$\forall \psi \in V^\ell,\ t \in (0, T],$$

$$\langle y^\ell(0), \psi \rangle_H = \langle y_0, \psi \rangle_H \qquad \forall \psi \in V^\ell$$

and

$$u \in U_{\mathsf{ad}}. \tag{2.2c}$$

Notice that (2.2b) is a finite-dimensional system of ordinary differential equations for the coefficient vector $\mathrm{y}^\ell = (\mathrm{y}_1^\ell, \dots, \mathrm{y}_\ell^\ell)^T : [0, T] \to \mathbb{R}^\ell$. Throughout this work we suppose that there exists a unique weak solution $y^\ell = y^\ell(u)$ to (2.2b) for any $u \in U_{\mathsf{ad}}$. Then, we can define the reduced cost functional associated with the reduced-order approximation as follows:

$$\hat{J}^\ell(u) = J(y^\ell(u), u), \quad u \in U_{\mathsf{ad}},$$

where $y^\ell(u)$ denotes the weak solution to (2.2b).

2.2 Optimize-Then-Discretize

Using a Lagrangian framework it is straightforward to derive first-order necessary optimality conditions for (1.1); see, e.g., [46, Chapter 1]. Assuming that $(\bar{y}, \bar{u})$ is a local optimal solution to (1.1) and that a constraint qualification condition is satisfied, there exists a *Lagrange multiplier* or *dual variable* $\bar{p}$ satisfying together with $\bar{x} = (\bar{y}, \bar{u})$ the following adjoint or dual equation [85, Chapter 6]

$$\begin{aligned}
&-\frac{\mathrm{d}}{\mathrm{d}t}\langle \bar{p}(t), \varphi\rangle_H + \int_\Omega \nabla \bar{p}(t)\cdot\nabla\varphi + f_y(t, \cdot, \bar{y}(t))\bar{p}(t)\varphi \,\mathrm{d}\boldsymbol{x} \\
&\qquad = \int_\Omega \big(y_d(t) - \bar{y}(t)\big)\varphi \,\mathrm{d}\boldsymbol{x} \qquad \forall \varphi \in V,\ t \in (0, T], \\
&\langle \bar{p}(T), \phi\rangle_H = 0 \qquad \forall \phi \in H
\end{aligned} \tag{2.3a}$$

and the *variational inequality*

$$\langle \kappa\bar{u} - \mathcal{B}^\star\bar{p}, u - \bar{u}\rangle_U \geq 0 \quad \forall u \in U_{\mathsf{ad}} \tag{2.3b}$$

where $\mathcal{B}^\star : L^2(0, T; H) \to U$ denotes the adjoint operator of $\mathcal{B}$. Now, the state equation (1.1b′) as well as the system (2.3) are discretized by a reduced-order Galerkin scheme. Here, we choose the same Galerkin ansatz functions as for the state variable which its motivated by the error analysis in [37]. Analogous to (2.1) we make the ansatz

$$p^\ell(t) = p_{\mathsf{p}}(t) + \sum_{i=1}^{\ell} \mathrm{p}_i^\ell(t)\psi_i, \quad t \in [0, T], \tag{2.4}$$

for the adjoint variable, where $p_{\mathsf{p}}(t) \in V$ is a chosen particular function and $\mathrm{p}_i^\ell : [0, T] \to \mathbb{R}$ stands for the ℓ nodal coefficient functions. Then, we arrive at the following reduced-order system for the unknown reduced-order solution $\bar{x}^\ell = (\bar{y}^\ell, \bar{u}^\ell)$ and $\bar{p}^\ell$

$$\frac{\mathrm{d}}{\mathrm{d}t}\langle \bar{y}^\ell(t), \psi\rangle_H + \int_\Omega \nabla \bar{y}^\ell(t)\cdot\nabla\psi + \big(f(t,\cdot,\bar{y}^\ell(t)) - (\mathcal{B}\bar{u}^\ell)(t)\big)\psi\,\mathrm{d}\boldsymbol{x} = 0,$$

$$\langle y^\ell(0), \psi\rangle_H = \langle y_0, \psi\rangle_H,$$

$$-\frac{\mathrm{d}}{\mathrm{d}t}\langle \bar{p}^\ell(t), \psi\rangle_H + \int_\Omega \nabla\bar{p}^\ell(t)\cdot\nabla\psi + \big(f_y(t,\cdot,\bar{y}^\ell(t))\bar{p}^\ell(t) + \bar{y}(t) - y_d(t)\big)\psi\,\mathrm{d}\boldsymbol{x} = 0,$$

$$\langle \bar{p}^\ell(T), \psi\rangle_H = 0$$

for all $\psi \in V^\ell$, $t \in [0,T]$ and

$$\langle \kappa\bar{u}^\ell - \mathcal{B}^\star\bar{p}^\ell, u - \bar{u}^\ell\rangle_U \geq 0 \quad \forall u \in U_{\mathsf{ad}}.$$

To solve the obtained reduced-order scheme for the first-order necessary optimality conditions one can apply, e.g., semismooth Newton [45] or interior point methods [77,87].

3 POD-Based Methods

Let X be either the space H or the space V. In X we denote by $\langle\cdot\,,\cdot\rangle_X$ and $\|\cdot\|_X = \langle\cdot\,,\cdot\rangle_X^{1/2}$ the inner product and the associated norm, respectively. Notice that X is separable, i.e., X has a countable dense subset. This implies that X possesses a countable orthonormal basis; see, e.g., [74, p. 47].

For fixed $n, \wp \in \mathbb{N}$ let the so-called *snapshots* $w_1^k, \ldots, w_n^k \in X$ be given for $1 \leq k \leq \wp$. To avoid a trivial case, we suppose that at least one of the w_j^k's is nonzero. Then, we introduce the finite-dimensional, linear subspace

$$\mathcal{V} = \operatorname{span}\left\{w_j^k \mid 1 \leq j \leq n \text{ and } 1 \leq k \leq \wp\right\} \subset X \tag{3.1}$$

with dimension $d \in \{1,\ldots,n\wp\}$. We call the set $\mathcal{V}$ *snapshot subspace*.The method of POD consists in choosing a complete orthonormal basis $\{\psi_i\}_{i=1}^\infty$ in X such that for every $\ell \in \{1,\ldots,d^n\}$ the mean square error between the $n\wp$ elements w_ℓ^k and their corresponding ℓ-th partial Fourier sum is minimized on average:

$$\begin{cases} \min \displaystyle\sum_{k=1}^{\wp}\sum_{j=1}^{n}\alpha_j\Big\|w_j^k - \sum_{i=1}^{\ell}\langle w_j^k, \psi_i\rangle_X\,\psi_i\Big\|_X^2 \\ \text{s.t. } \{\psi_i\}_{i=1}^\ell \subset X \text{ and } \langle\psi_i,\psi_j\rangle_X = \delta_{ij},\ 1 \leq i,j \leq \ell, \end{cases} \tag{3.2}$$

where the α_j's denote positive weighting parameters. Here, the symbol δ_{ij} denotes the Kronecker symbol satisfying $\delta_{ii} = 1$ and $\delta_{ij} = 0$ for $i \neq j$. An optimal solution $\{\bar{\psi}_i^n\}_{i=1}^\ell$ to (3.2) is called a *POD basis of rank* ℓ.

To solve (3.2) we define the linear operator $\mathcal{R} : X \to X$ as follows:

$$\mathcal{R}\psi = \sum_{k=1}^{\wp} \sum_{j=1}^{n} \alpha_j \, \langle \psi, w_j^k \rangle_X \, w_j^k \quad \text{for } \psi \in X \tag{3.3}$$

with positive weights $\alpha_1, \ldots, \alpha_n$. Then, $\mathcal{R}$ is a compact, nonnegative and selfadjoint operator. Suppose that $\{\bar{\lambda}_i\}_{i=1}^{\infty}$ and $\{\bar{\psi}_i^n\}_{i=1}^{\infty}$ denote the nonnegative eigenvalues and associated orthonormal eigenfunctions of $\mathcal{R}$ satisfying

$$\mathcal{R}\bar{\psi}_i = \bar{\lambda}_i \bar{\psi}_i, \quad \bar{\lambda}_1 \geq \ldots \geq \bar{\lambda}_d > \bar{\lambda}_{d+1} = \ldots = 0.$$

Then, for every $\ell \in \{1, \ldots, d\}$, the first ℓ eigenfunctions $\{\bar{\psi}_i\}_{i=1}^{\ell}$ solve (3.2). Moreover, the value of the cost evaluated at the optimal solution $\{\bar{\psi}_i\}_{i=1}^{\ell}$ satisfies

$$\sum_{k=1}^{\wp} \sum_{j=1}^{n} \alpha_j \left\| w_j^k - \sum_{i=1}^{\ell} \langle w_j^k, \bar{\psi}_i \rangle_X \, \bar{\psi}_i \right\|_X^2 = \sum_{i=\ell+1}^{d} \bar{\lambda}_i.$$

For more details we refer the reader to [48, 50] and [37, Chapter 2], for instance.

Remark 3.1. (a) In the context of the optimal control problem (1.1) a reasonable choice for the snapshots is $w_j^1 \approx y(t_j)$ and $w_j^2 \approx p(t_j)$ for the time grid $0 = t_1 < \ldots < t_n = T$. Utilizing new POD error estimates for evolution problems [20, 82] and optimal control problems [49, 86], convergence and rate of convergence results are derived for linear-quadratic control constrained problems in [37] for the choices $X = H$ and $X = V$.

(b) For the numerical realization, the space X has to be discretized by, e.g., finite element discretizations. In this case the Hilbert space X has to be replaced by an Euclidean space $\mathbb{R}^m$ endowed with a weighted inner product; see [37].

3.1 A-Posteriori Error Analysis

In contrast to methods of balanced truncation type, the POD method is somehow lacking a reliable a-priori error analysis. Unless its snapshots are generating a sufficiently rich state space, it is not a-priorily clear how far the optimal solution of the POD problem is from the exact one. On the other hand, the POD method is a universal tool that is applicable also to problems with time-dependent coefficients or to nonlinear equations. Moreover, by generating snapshots from the real (large) model, a space is constructed that inhibits the main and relevant physical properties of the state system. This, and its ease of use makes POD very competitive in practical use, despite of a certain heuristic flavor.

Based on a perturbation argument [27] it is derived in [53, 86] how far the suboptimal control $\bar{u}^{\ell}$, computed on the basis of the POD model, is from the (unknown) exact $\bar{u}$. Let $\mathcal{D}$ be an open, bounded subset of $\mathbb{R}^{\mathfrak{p}}$ and $U = L^2(\mathcal{D})$. By $U_{\mathrm{ad}} \subset U$ we define the closed, convex and bounded subset

$$U_{\mathrm{ad}} = \left\{u \in L^2(\mathcal{D}) \,\middle|\, u_a(s) \le u(s) \le u_b(s) \text{ for almost all } s \in \mathcal{D}\right\}$$

with $u_a, u_b \in L^2(\mathcal{D})$ satisfying $u_a \le u_b$ in $\mathcal{D}$ a.e. Suppose that $f \equiv 0$ holds, i.e., (1.1b) is a linear evolution problem. Then, the error estimate reads as follows:

$$\|\bar{u}^\ell - \bar{u}\|_U \le \frac{1}{\kappa} \|\zeta^\ell\|_U, \tag{3.4}$$

where the computable perturbation function $\zeta^\ell \in U$ is given by

$$\zeta^\ell = \begin{cases} -\min\left(0, \kappa\bar{u}^\ell - \mathcal{B}^\star \tilde{p}^\ell\right) & \text{in } \mathcal{A}_a^\ell = \left\{s \in \mathcal{D} \,\middle|\, \bar{u}^\ell = u_a \text{ a.e.}\right\}, \\ \max\left(0, \kappa\bar{u}^\ell - \mathcal{B}^\star \tilde{p}^\ell\right) & \text{in } \mathcal{A}_b^\ell = \left\{s \in \mathcal{D} \,\middle|\, \bar{u}^\ell = u_b \text{ a.e.}\right\}, \\ -\left(\kappa\bar{u}^\ell - \mathcal{B}^\star \tilde{p}^\ell\right) & \text{in } \mathcal{D} \setminus \left(\mathcal{A}_a^\ell \cup \mathcal{A}_b^\ell\right). \end{cases}$$

Furthermore, $\tilde{y}$ and $\tilde{p}$ solve

$$\begin{aligned} &\frac{\mathrm{d}}{\mathrm{d}t} \langle \tilde{y}^\ell(t), \varphi\rangle_H - \int_\Omega \nabla \tilde{y}^\ell(t) \cdot \nabla\varphi - (\mathcal{B}\bar{u}^\ell)(t)\varphi \,\mathrm{d}\boldsymbol{x} = 0, \\ &\langle \tilde{y}^\ell(0), \psi\rangle_H = \langle y_0, \psi\rangle_H, \\ -&\frac{\mathrm{d}}{\mathrm{d}t} \langle \tilde{p}^\ell(t), \varphi\rangle_H + \int_\Omega \nabla \tilde{p}^\ell(t) \cdot \nabla\varphi + \left(\tilde{y}^\ell(t) - y_d(t)\right)\varphi \,\mathrm{d}\boldsymbol{x} = 0, \\ &\langle \tilde{p}^\ell(T), \varphi\rangle_H = 0 \end{aligned}$$

for all $\varphi \in V$, $t \in [0, T]$. It is shown in [37, 86] that $\|\zeta^\ell\|_U$ tends to zero as ℓ tends to infinity. Hence, increasing the number of POD ansatz functions leads to more accurate POD suboptimal controls. This idea turns out to be numerically very efficient. For linear-quadratic problems we refer to [37, 38, 83, 84, 89]. It is able to compensate for the lack of a priori analysis for POD methods. The analysis is extended to nonlinear optimal control problems in [26, 53]. For a parameter estimation example we also refer to [57]. Unfortunately, the a-posteriori error estimates requires a lower bound for the smallest eigenvalue of the reduced Hessian, which is—unless the control space is low dimensional—usually computationally expensive. Another approach is to use the a-posteriori error estimate in an inexact sequential quadratic programming (SQP) approach, where the sufficient accuracy of each level of the SQP method is guaranteed by the error control; see [52]. Let us mention that there is related work also available in the reduced-basis literature; see, e.g., [24, 34, 63]. Here, the authors concentrate on deriving online efficient a-posteriori error estimators.

3.2 Optimality System POD

The accuracy of the reduced-order model can be controlled by the a-posteriori error analysis presented in the previous subsection. However, if the POD basis is created from a reference trajectory containing features which are quite different from those of the optimally controlled trajectory, a rather huge number of POD ansatz functions have to be included in the reduced-order model. This fact may lead to non-efficient reduced-order models and numerical instabilities. To avoid these problems, the POD basis is generated in an initialization step utilizing *optimality system POD* (OS-POD); see [54]. In OS-POD, the POD basis is updated in the direction of the minimum of the cost. Recall that the POD basis is computed from the state $y = y(u)$ with some control $u \in \mathcal{U}_{\mathsf{ad}}$. Thus, the reduced-order Galerkin projection depends on the state variable and hence on the control u at which the eigenvalue problem $\mathcal{R}\psi_i = \lambda_i \psi_i$ for $i = 1, \dots, \ell$ is solved for the basis $\{\psi_i\}_{i=1}^{\ell}$. This may deter from one of the main advantages of the POD approach for model reduction, which consists in the fact that unlike typical finite element basis functions the elements of the POD basis reflect the dynamics of the system. In optimal control this feature gets lost if the dynamics of the state corresponding to the reference control is significantly different from the trajectory corresponding to the optimal approach. Hence, we propose to consider the extended problem [54]:

$$\min J(y^\ell, u) \text{ s.t. } \begin{cases} z = (y^\ell, y, \lambda_i, \psi_i), u \in U_{\mathsf{ad}} \\ (y^\ell, u) \in H^1(0,T; V^\ell) \times U_{\mathsf{ad}} \text{ satisfy (2.2b)}, \\ (y, u) \text{ satisfy } (1.1\mathrm{b}'), \\ \mathcal{R}(y)\psi_i = \lambda_i \psi_i, \ 1 \le i \le \ell. \end{cases} \qquad (\mathbf{P}^\ell_{\mathrm{ospod}})$$

Notice that the second line of the constraints in $(\mathbf{P}^\ell_{\mathrm{ospod}})$ coincide with the constraints in (2.2), the next two are the infinite-dimensional state equation and the eigenvalue problem characterizing the POD basis. For the optimal solution the problem formulation $(\mathbf{P}^\ell_{\mathrm{ospod}})$ has the property that the associated POD reduced system is computed from the trajectory corresponding to the optimal control and thus, differently from (2.2), the problem of unmodelled dynamics is removed. Of course, $(\mathbf{P}^\ell_{\mathrm{ospod}})$ is more complicated than (2.2). For practical realization an operator splitting approach is used in [54], where also sufficient conditions are given so that $(\mathbf{P}^\ell_{\mathrm{ospod}})$ possesses a unique optimal solution $(\bar{y}^\ell, \bar{y}, \bar{\lambda}_i, \bar{\psi}_i, \bar{u}^\ell)$, which can be characterized by first-order necessary optimality conditions. Convergence results for OSPOD are studied in the Ph.D thesis [62]. The combination of OS-POD and a-posteriori error analysis is investigated in the paper [88] and the recent master thesis [36].

3.3 Trust Region POD

In PDE-constrained optimization the use of reduced order models is highly efficient, since the complex and time-consuming constraint in form of a partial differential equation is replaced by a relatively small system of ordinary differential equations. There is, however, one caveat that needs to be addressed. Suppose, one uses a reduced order model as in (2.2a)–(2.2b), where the state y_k^ℓ is based on a reduced order model at a certain control vector u_k. Then it is usually numerically very efficient to minimize this reduced order model and obtain a solution $(\bar{y}^\ell, \bar{u})$.

$$J(\bar{y}, \bar{u}) \leq J(y_k^\ell, u) \quad \text{for all} \quad (y_k^\ell, u) \quad \text{that satisfy} \quad (2.2\text{b}).$$

The problem that might occur is the fact that during this minimization one usually moves away from the original control u_k on which the reduced order model was built. In this case, it could happen that the quality of the model deteriorates and, in particular, $\bar{y}^\ell$ is no longer a good approximation of the solution $\bar{y}$ to the full model.

In [1], a direct estimate is used to monitor the accuracy of the model. In [4] a trust region approach was proposed to manage updating the reduced order model. The trust region method is a well known method for the globalization of locally convergent methods like Newton's method. Here the quadratic Taylor expansion is used as a model function for the nonlinear objective function. However, it is known that this globalization strategy also works for nonlinear models. Hence, one could use it also in this context, where a nonlinear model is replaced by another nonlinear model, e.g. POD, but of much smaller complexity than the original one.

The main key for controlling the quality of the model approximation in the trust region framework is the comparison between the predicted reduction starting from full model evaluation at (y_k, u_k)

$$v_{pred} = \hat{J}^\ell(\bar{u}) - \hat{J}(u_k)$$

and the actual reduction

$$v_{actual} = \hat{J}(\bar{u}) - \hat{J}(u_k),$$

where $\bar{y}$ is the true solution at the control $\bar{u}$. If the quotient $\rho = v_{pred}/v_{actual}$ is close to 1, then the reduced model is a good model and we accept $\bar{u}$ as u_{k+1} and likewise $\bar{y}$ as y_{k+1}. If ρ differs significantly from 1, we have to modify the model, for example, by introducing more snapshots or restricting the controls to a ball around u_k.

One drawback of this strategy is the evaluation of the full model at the control u_*. This problem can be alleviated by the introduction of a hierarchy of models, e.g. created by different fine discretizations of the PDE, and the full model in this case is replaced by the next finer model. In this way, when approaching the finest level in the hierarchy, we already are very close to the optimal point. This approach is also closely related to multifidelity optimization.

With respect to convergence in an optimization framework, the accuracy of the gradient of the model function has to be monitored. As it is shown in [80], a-priori-estimates on the accuracy of the POD approximation can be used to establish error bounds for the distance between the gradient of the model function and the original function.

Trust region POD (TRPOD) is meanwhile a well established method in applications: In the control of fluid flow problems, applications of TRPOD go back to Bergmann and Cordier [17]. More recently, Navon and co-workers [21] use this methodology in order to accomplish a 4-D VAR simulation for the shallow water equation. In [91], the TRPOD framework is combined with an efficient error bound for defining the trust region in the design optimization of vibrating structures using frequency domain formulations. An application of TRPOD to the optimization of simulated moving bed chromatography (a separation process in chemical process engineering) is discussed in [59]. Also, in financial applications, reduced order models, see [22] or [79], are gaining importance and TRPOD is a promising tool in the calibration process.

In [33] the Carter condition

$$\|\nabla \hat{J}(u_k) - \nabla \hat{J}^\ell(u_k)\|_U \leq \sigma \|\nabla \hat{J}^\ell(u_k)\|_U, \quad \sigma \in (0, 1),$$

which is essential for the convergence of the inexact trust region method, is replaced by an a posteriori error estimation in order to control the reduced-order approximation of the reduced cost functional. For that purpose error estimates for the state and the dual variable are utilized. This offers a bridge between trust region POD and the a posteriori analysis presented in Sect. 3.1.

4 System-Theoretic MOR Approaches

A disadvantage of POD-based methods is that the control function $u(t)$ has to be chosen in advance in order to compute the training set. As this reference control may differ significantly from the optimal control, the reduced order model may not be suitable for optimization, or the optimal control computed using the reduced order model may be too far away from the one obtained from the full-order model. The alternative OS-POD avoiding this problem is discussed in Sect. 3.2. As already noted there, solving the extended problem ($\mathbf{P}^\ell_{\text{ospod}}$) is computationally challenging. The advantage of using system-theoretic approaches is that these are so-called "simulation-free" methods which do not require training sets. The quality of the reduced order model thus is independent of the chosen control function $\mathrm{u}(t)$. For the purpose of deriving error bounds, it is usually assumed that $U_{\mathsf{ad}} = L^2(0, \infty; \mathbb{R}^m)$, that is, we assume m scalar input functions that are square-integrable for an infinite time horizon. It should be stressed, though, that this assumption is only required for the derivation of error bounds for the reduced order models, while the model itself is applicable of course also in the presence of control constraints or other spaces of admissible control functions.

The starting point for system-theoretic approaches to model order reduction is the description of a mathematical model as a linear or nonlinear system. As these methods were developed mainly for linear problems, and extensions to nonlinear problems exist, but are either based on heuristics or are computationally intractable, we only discuss the application to linear systems here. But significant progress in nonlinear model reduction using system-theoretic concepts can be expected in the near future and this may then lead to useful methods also in the context of PDE constrained optimization.

Thus, in the following we will consider *linear, time-invariant (LTI) systems*

$$\mathrm{M}\dot{\mathrm{y}}(t) = -\mathrm{S}\mathrm{y}(t) + \tilde{\mathrm{B}}\mathrm{u}(t) \tag{4.1a}$$

$$\mathrm{z}(t) = \mathrm{C}\mathrm{y}(t), \tag{4.1b}$$

for a given initial condition $\mathrm{y}(0) = \mathrm{y}_0 \in \mathbb{R}^N$. Here, y is the state of the system and in this context is usually the discrete vector obtained by discretization of (1.1b′), i.e., $\mathrm{y}_j(t)$ are the coefficients in the ansatz $y(t) \approx \sum_{j=1}^{N} \mathrm{y}_j(t)\psi_j$, with basis functions ψ_j of the chosen trial space, when evaluating $\langle y(t), \phi_j \rangle_H$ for the basis functions $\phi_j \in V_N$, $j = 1, \dots, N$, where $V_N \subset V$ is the chosen N-dimensional space of test functions. In this setting, M and S contain the corresponding mass and stiffness matrices and $\tilde{\mathrm{B}}$ is the discretized version of the input operator $\mathcal{B}$. The second equation in (4.1) is called the *output equation*, and $\mathrm{z}(t) \in \mathbb{R}^q$ is a vector of quantities of interest which in the setting considered here could be derived from a linear cost functional $z(t) = \mathcal{L}(y)(t)$ which might be goal of optimization. In practice, it is often a model for the possible measurements of the system, i.e., it is assumed that the full state y is not accessible for measurements. In case one assumes all state information is available, one might simply set $\mathrm{C} = I_N$, the identity matrix on $\mathbb{R}^N$. We will discuss the use of the output equation in the context of PDE constrained optimization below.

As the mass matrix M is invertible, for the clarity of presentation we work with the transformed system

$$\dot{\mathrm{y}}(t) = \mathrm{A}\mathrm{y}(t) + \mathrm{B}\mathrm{u}(t) \tag{4.2a}$$

$$\mathrm{z}(t) = \mathrm{C}\mathrm{y}(t), \tag{4.2b}$$

where $\mathrm{A} := -\mathrm{M}^{-1}\mathrm{S}$ and $\mathrm{B} := \mathrm{M}^{-1}\tilde{\mathrm{B}}$. It should be noted, though, that we do this only formally, that is, in all computations of reduced order models, one never forms A, B explicitly but works with the original $\tilde{\mathrm{B}}$ and the usually sparse matrices M, S in order to avoid fill-in as well as possible round-off errors implied by ill-conditioning of M.

The core observation of system-theoretic approaches is that the relation from inputs u to the outputs z can be represented algebraically using the Laplace transform, yielding

$$\mathrm{z}(s) = \mathrm{C}(sI_N - \mathrm{A})^{-1}\mathrm{B}\mathrm{u}(s) =: \mathrm{G}(s)\mathrm{u}(s), \tag{4.3}$$

where by abuse of notation, we denote the Laplace transforms of z, u again by z, u, and $s \in \mathbb{C}$ is the Laplace variable. The *transfer function matrix (TFM)* G is a complex $q \times m$ matrix for any s whose entries are scalar real rational functions of a complex variable and with degree less than or equal to N. The particular case of a single input, single output (SISO) system with $m = q = 1$ is often of particular interest in PDE constrained optimization as many problems can be formulated such that a single quantity is to be minimized using only a scalar control function $\mathrm{u}(t)$.

Equation (4.3) shows that the output of the system can be well approximated when we can find a reduced order model with TFM G^ℓ for which $\|\mathrm{G}(.) - \mathrm{G}^\ell(.)\|$ is small in an appropriate norm. It is convenient here to use L_2-related norms as they result in L_2- or L_∞-norm error bounds for $\mathrm{z}(.) - \mathrm{z}^\ell(.)$, where z^ℓ is the output of the reduced order model. Therefore it is the aim to find a reduced order LTI system

$$\dot{\mathrm{y}}^\ell(t) = \mathrm{A}^\ell \mathrm{y}^\ell(t) + \mathrm{B}^\ell \mathrm{u}(t), \tag{4.4a}$$

$$\mathrm{z}^\ell(t) = \mathrm{C}^\ell \mathrm{y}^\ell(t), \tag{4.4b}$$

where $\mathrm{y}^\ell(t) \in \mathbb{R}^\ell$ with $\ell \ll N$ and corresponding matrices of compatible sizes such that the TFM

$$\mathrm{G}^\ell(s) = \mathrm{C}^\ell (sI_\ell - \mathrm{A}^\ell)^{-1} \mathrm{B}^\ell \tag{4.5}$$

of (4.4) approximates G well. The strength of system-theoretic model reduction methods now shows in the fact that the reduced order model (4.4) uses the same control function $\mathrm{u}(t)$ as the full-order model (4.2). Thus, neither a discretization of the input space U nor an a priori choice of a reference control u are necessary, and the optimal control $\bar{\mathrm{u}}$ obtained from using (4.4) as surrogate in the optimization process can be directly applied in the full-order model (4.2) or even in the original PDE problem when the problem is formulated such that only the amplitude of the control signal is subject to optimization.

In the following, we will discuss the two main concepts used in model order reduction of LTI systems of the form (4.2). Both are based on Petrov-Galerkin projection, i.e., the reduced order model is computed using two full-rank matrices $V, W \in \mathbb{R}^{N \times \ell}$:

$$\mathrm{A}^\ell := W^T \mathrm{A} V, \quad \mathrm{B}^\ell := W^T \mathrm{B}, \quad \mathrm{C}^\ell := \mathrm{C} V, \tag{4.6}$$

corresponding to projecting the state-space onto range(W) and using the ansatz $\mathrm{y}(t) \approx V \mathrm{y}^\ell(t)$. The basis matrices V and W for the trial and test spaces are chosen to be bi-orthonormal, i.e., $W^T V = I_\ell$ such that VW^T becomes an oblique projector. Inserting the ansatz into (4.4) leads to a nonzero residual $V\dot{\mathrm{y}}^\ell - AV\mathrm{y}^\ell - B\mathrm{u}$ which obviously is orthogonal to range(W) as

$$W^T(V\dot{\mathrm{y}}^\ell - AV\mathrm{y}^\ell - B\mathrm{u}) = \dot{\mathrm{y}}^\ell - A^\ell \mathrm{y}^\ell - B^\ell \mathrm{u} = 0.$$

In this sense, all the considered methods in this section are Petrov-Galerkin methods and become Galerkin projection methods when $W = V$. The two model reduction techniques discussed in the following subsections only differ in the way V, W are computed, and in the resulting theoretical properties of the reduced order model. For more details on system-theoretic model order reduction techniques, consult, e.g., the recent monographs, edited volumes and survey papers [3, 10, 14, 78]. Also note that extensions to so-called *descriptor systems*, where M in (4.1) is singular, are possible, see, e.g., [14, Chapter 3].

4.1 Rational Interpolation Based Techniques

The first family of system-theoretic model order reduction methods is based on (rational) interpolation of the TFM. The interpolant is chosen as a rational matrix function of lower degree satisfying certain interpolation conditions. Hence, the original and reduced order TFMs (and some of their first derivatives) coincide:

$$\frac{d^j}{ds^j}\mathrm{G}(s_k) = \frac{d^j}{ds^j}\mathrm{G}^{\ell}(s_k), \quad k = 0, \ldots, K, \quad j = 0, \ldots, J_k, \tag{4.7}$$

where the interpolation points s_k are chosen such that $(\mathrm{A} - s_k I_N)$ is nonsingular. Practically this is usually realized by certain Krylov subspace methods.

The classical approach using $K = 0$ and a sufficiently large J_K, leading to rational Hermite interpolation at s_0, has become popular as *moment matching* or *Padé(-type) approximation* since the mid-1990s. These methods can be derived by power series expansions of the TFMs of the original and reduced order systems about s_0. The reduced order model is then determined so that the first coefficients in the series expansions match. In this context, the coefficients of the power series expansions are called moments, explaining the name "moment matching". Padé-approximation in this context means that the number of matching moments is maximized for a given degree of the approximating rational function G^{ℓ}. The observation that a reduced order model with the moment matching property is obtained by applying r steps of the (block) Arnoldi or Lanczos processes to $(\mathrm{A} - s_0 I_N)^{-1}$ or its (real) transpose with B or C^T as starting (block) vector and using bi-orthonormal bases of the resulting Krylov subspaces $\mathcal{K}_r((\mathrm{A} - s_0 I_N)^{-1}, \mathrm{B})$ and $\mathcal{K}_r((\mathrm{A}^T - s_0 I_N)^{-1}, \mathrm{C}^T)$ for V and W was the breakthrough of this approach as model reduction method—previous attempts employing explicit computations of the moments were so prone to round-off errors that they could not be used in practice, see [28, 30] for details.

For the SISO case, one obtains $r = \ell$ and $J_0 = 2r - 1$. The bi-orthonormal bases for the two Krylov subspaces are obtained automatically using the two-sided (unsymmetric) Lanczos process. One can also run the standard Arnoldi process for both Krylov subspaces independently and enforce bi-orthonormality afterwards to obtain the same results. Using only one of the Krylov subspaces and $V = W$, one has only $J_0 = r - 1$, but the computation can be performed with the stable Arnoldi process and orthogonal projection can be used to compute the reduced order model. This is beneficiary if A is, e.g., negative definite as for stiffness matrices resulting from the Galerkin finite element methods for linear diffusion-reaction problems, as in this case, negative definiteness of A^ℓ is guaranteed.

For $m, q > 1$, things become a lot more complicated, starting from the fact that V, W necessarily need to be of the same size, but the block Krylov subspaces will usually differ in dimension whenever $m \neq q$ (and often even if $m = q$ due to deflation). Computations in this case can be performed using block or band Lanczos/Arnoldi processes. For many more details on this, see the monographs and surveys already mentioned above or [29]

Also observe that the use of an expansion point $s_0 \notin \mathbb{R}$ will lead to a complex-valued reduced order system (4.4). Often, this is undesired. An easy remedy is to use two expansions point $s_0, s_1 = \overline{s_0}$ and to concatenate the resulting bases. If performed with care, the resulting V, W are then real.

Note that interpolation at $s_0 = \infty$ is also possible. In that case, one computes Krylov subspaces for (A, B) and/or (A^T, C^T). Then the moments are called *Markov parameters* and the "moment matching" problem is known as *partial realization.*

The use of a single expansion point s_0 leads to good approximation only locally. Hence, it is often useful to use more than one expansion point, yielding multi-point moment matching methods, also called *rational Krylov methods*, see, e.g., [3, 10]. By fixing $K = \ell - 1$ in (4.7), in [40] a fix-point iteration to determine locally optimal expansion points w.r.t. approximation of the TFM in the H_2-norm (basically, this is the L_2-norm of the TFM evaluated on the whole imaginary axis) is suggested. In the SISO case, if this iteration converges, the obtained rational Krylov subspaces yield basis matrices V, W such the reduced order model satisfies (4.7) for $J_k = 1$ for all $k = 0, \ldots, K$ at the mirror images (w.r.t. the imaginary axis) of the eigenvalues of A^ℓ, i.e., the poles of the reduced order TFM. This property of a reduced TFM is known to be the necessary condition for a local minimizer of $\|G - G^\ell\|_{H_2}$.

Some extensions of the presented approaches to nonlinear systems are discussed in the recent survey [10].

4.2 Methods Based on System Balancing

Balanced Truncation (BT) has been the workhorse for model reduction in systems and control theory, in particular for controller design, for the last three decades, see, e.g., [3, 10] for a thorough discussion and references to original work. Its advantages over most other methods is that for asymptotically stable models it guarantees stability of the reduced-order model—a property that none of the other methods discussed here possesses unless additional assumptions are posed or a post-processing step is included—and it has a computable a priori error bound that allows the adaptive selection of the reduced model order given a user-defined error threshold. The method is based on two main ingredients: balancing and truncation. In the following, we will briefly introduce these concepts, starting with the latter one.

The concept of *truncation* is based on finding a suitable state-space transformation $\mathcal{T}$ defined by a nonsingular matrix $\mathrm{T} \in \mathbb{R}^{n \times n}$:

$$\mathrm{T}^{-1}\dot{\mathrm{y}}(t) = (\mathrm{T}^{-1}\mathrm{A}\mathrm{T})\mathrm{T}^{-1}\mathrm{y}(t) + (\mathrm{T}^{-1}\mathrm{B})\mathrm{u}(t) \tag{4.8a}$$

$$\mathrm{z}(t) = (\mathrm{C}\mathrm{T})\mathrm{T}^{-1}\mathrm{y}(t) \tag{4.8b}$$

with partitioning

$$\mathrm{T}^{-1}\mathrm{A}\mathrm{T} = \begin{bmatrix} \mathrm{A}_{11} & \mathrm{A}_{12} \\ \mathrm{A}_{21} & \mathrm{A}_{22} \end{bmatrix}, \quad \mathrm{T}^{-1}\mathrm{B} = \begin{bmatrix} \mathrm{B}_1 \\ \mathrm{B}_2 \end{bmatrix}, \quad \mathrm{C}\mathrm{T} = [\mathrm{C}_1 \quad \mathrm{C}_2], \tag{4.9}$$

where $A_{11} \in \mathbb{R}^{\ell \times \ell}$ and the other blocks are of compatible sizes. The reduced order model is then simply obtained by truncating the states $\ell + 1, \dots, N$ of the transformed state vector $\mathrm{T}^{-1}\mathrm{y}(t)$ i.e., setting them to zero so that

$$\mathrm{A}^{\ell} = \mathrm{A}_{11}, \quad \mathrm{B}^{\ell} = \mathrm{B}_1, \quad \mathrm{C}^{\ell} = \mathrm{C}_1. \tag{4.10}$$

The challenge, of course, is to find a $\mathcal{T}$ yielding a reduced order model with good approximation properties. Here is where the second ingredient, *balancing*, comes into the game.

For the following, we assume that A is asymptotically stable, i.e., all its eigenvalues are in the left half plane. This implies that the transfer function $\mathrm{G}(s)$ of (4.2) has all its poles in the open left half plane, thus such systems are called (asymptotically) stable. Extensions to unstable systems are possible, but will not be discussed here for brevity (see [11] for some more details and references and [13] for a recent application to the control of unstable flow problems). Strictly speaking, a balancing transformation $\mathcal{T}_b$ yields a realization of an asymptotically stable LTI system (4.2) in the form (4.8) such the unique solutions $\mathrm{P}, \mathrm{Q} \in \mathbb{R}^{N \times N}$ of the *Lyapunov equations*

$$\mathrm{A}\mathrm{P} + \mathrm{P}\mathrm{A}^T + \mathrm{B}\mathrm{B}^T = 0, \quad \mathrm{A}^T\mathrm{Q} + \mathrm{Q}\mathrm{A} + \mathrm{C}^T\mathrm{C} = 0 \tag{4.11}$$

are equal and diagonal, $\mathrm{P} = \mathrm{Q} = \mathrm{diag}\{\sigma_1, \dots, \sigma_N\}$, with decaying *Hankel singular values* σ_k, i.e., $\sigma_1 \geq \sigma_2 \geq \dots \geq \sigma_N > 0$. Such a balancing transformation does not always exist, it requires some additional system-theoretic assumption (see, e.g., [3] for a full account of these), but even if it does not exist, it is possible to obtain a partial transformation that suffices to compute a reduced order model with exactly the same properties as described in the following, see [15]. If one now assumes that a balancing transformation was used to compute (4.9) and the reduced order model is then obtained via truncation as in (4.10), it holds:

1. A^{ℓ} and thus G^{ℓ} are asymptotically stable.
2. The reduced-order model satisfies the error bound

$$\|\mathrm{z} - \mathrm{z}^{\ell}\|_{L_2} \leq \sum_{k=\ell+1}^{n} \sigma_j \|\mathrm{u}\|_{L_2}.$$

As the Hankel singular values can be computed as a by-product of the balancing transformation, it is possible to use this error bound to adapt the size of the reduced order model to match a desired maximal approximation error.

It should be noted that in practical computations, neither the full transformation matrix T nor the solutions P, Q should be computed explicitly. Efficient implementations determine the matrices $\mathrm{A}^{\ell}, \mathrm{B}^{\ell}, \mathrm{C}^{\ell}$ directly from (approximate) low-rank factors of P, Q. With the advance of numerical algorithms for solving Lyapunov equations, see [16,81] for recent surveys, nowadays these techniques can be applied to any kind of system for which linear systems $\mathrm{A}x = b$ with the matrix A as above are solvable, see [11] for details.

There also exist numerous variants of BT that can be useful as model reduction techniques in PDE-constrained optimization by exchanging P, Q by other useful pairs of positive (semi-)definite matrices, see [3, 11] and [14, Chapter 1] for some of these.

4.3 Applications to PDE-Constrained Optimization

The system-theoretic model reduction methods discussed in the previous sections have so far found wide-reaching applications in simulation and control, see, e.g., [3, 10, 11, 14, 28] and many more references therein. There have also been numerous attempts to apply balanced truncation directly to PDE control problems, mostly with the target of feedback control rather than optimization. The theory for balanced truncation of distributed parameter systems, i.e., linear instationary PDE control problems, was already derived in the 1980ies, see [32]. The use of this approach for deriving robust control strategies from the reduced order model is discussed in [23]. The practical use of these approaches needs, of course, computational methods, and thus discretization at some stage of the reduction process. In [12], it is discussed how to combine BT with classical finite element discretization in the spatial variables of linear parabolic control problems while [70] goes a step further and derives an implementation of a balanced truncation algorithm where the discretization is delayed until the inner loop of the numerical algorithm is entered—the latter approach is more in the spirit of "optimize-than-discretize", while the approach in [12] can be seen as "semi-discretize-than-optimize". Balanced truncation and related methods have also been applied to flow control problems, see, e.g., [6, 13, 43, 90].

The methods based on rational interpolation have merely been used so far in simulation of dynamical systems, but also for feedback control purposes. Further reading on this includes [3, 7, 10, 14, 28, 29] and references therein. Their use for model reduction of infinite-dimensional control problems is discussed in [69] and for linear-quadratic parabolic optimal control problems in [89]. The recent paper

[18] uses rational tangential interpolation for a flow control problem. Certainly, these methods can be employed in a similar fashion as balanced truncation in PDE-constrained optimization problems, but so far this topic has received less attention.

Note that all these approaches do not discretize in time, no discretization of the controls is needed whenever the control variables are only time-dependent, and only the spatial part needs to be discretized for controls of the form $u(\boldsymbol{x},t) = v(\boldsymbol{x})\mathrm{u}(t)$. Hence, optimization based on reduced order models obtained from these methods can be performed with respect to the original U_{ad} rather than a discretized version or subset of the full space of admissible controls. The full potential of this advantageous property has yet to be explored in PDE-constrained optimization. Little work has been dedicated so far to using system-theoretic methods in an optimization context. The application of balanced truncation to a fully discrete linear-quadratic PDE optimal control problem is discussed in [9, Section 6.2]. The use of balanced truncation and rational interpolation techniques as well as POD for linear-quadratic parabolic optimal control problems is considered in [89]. Balanced truncation to accelerate a descent method for optimization of nonlinear evolution problems is analyzed in [25]. There, balanced truncation is applied to the linear adjoint system while the full state equation is solved. Speed-ups of factors 2–3 can be obtained this way compared to applying the same descent method to the full order optimality system. More research in the direction of using system-theoretic methods in a PDE-constrained optimization setting for evolution problems is certainly needed to leverage their full potential.

Conclusions

In this chapter, we have reviewed model reduction techniques for PDE-constrained optimization, where we have focused on instationary PDEs only. There is also a vast amount of literature on optimization of stationary PDEs, like linear elliptic PDEs in the simplest case. Model reduction methods as discussed here are not suitable for these problems as they rely on techniques for time-varying systems. Moreover, the computational complexity of instationary PDE-constrained optimization problems is often not so high that such a great benefit can be expected from model order reduction as for instationary problems. Nevertheless, model reduction techniques for such problems exist. These are based, e.g., on POD w.r.t. the optimization parameters rather than time, or on the Reduced Basis Method, see, e.g., [34, 52, 63, 68, 84] for some recent work on this.

We would also like to point out that there are many further possibilities for research. As already mentioned, the capabilities of system-theoretic methods in the context of PDE-constrained optimization are yet to be explored in detail. Also, the combined optimization with regard to a time-dependent control function plus one or more stationary design parameters as it occurs in many practical engineering applications is a widely open field.

Acknowledgements Our thanks go to Günter Leugering for his well-structured and focused leadership throughout the six years of existence of the DFG Priority Program 1253 "Optimization with Partial Differential Equations".

References

1. K. Afanasiev, M. Hinze, Adaptive control of a wake flow using proper orthogonal decomposition. Lect. Notes Pure Appl. Math. **216**, 317–332 (2001)
2. A. Alla, S. Volkwein, Asymptotic stability of POD based model predictive control for a semilinear parabolic PDE. Advances in Computational Mathematics, to appear (2014). http://kops.ub.uni-konstanz.de/handle/urn:nbn:de:bsz:352-253400
3. A.C. Antoulas, *Approximation of Large-Scale Dynamical Systems* (SIAM, Philadelphia, 2005)
4. E. Arian, M. Fahl, E.W. Sachs, Trust-region proper orthogonal decomposition for flow control. Technical report 2000–25, ICASE, 2000
5. P. Astrid, S. Weiland, K. Willcox, T. Backx, Missing point estimation in models described by proper orthogonal decomposition. IEEE Trans. Autom. Control **53**(10), 2237–2251 (2008)
6. J.A. Atwell, J.T. Borggaard, B.B. King, Reduced-order controllers for Burgers' equation with a nonlinear observer. Int. J. Appl. Math. Comput. Sci. **11**(6), 1311–1330 (2001)
7. Z. Bai, Krylov subspace techniques for reduced-order modeling of large-scale dynamical systems. Appl. Numer. Math. **43**, 9–44 (2002)
8. H.T. Banks, M.L. Joyner, B. Winchesky, W.P. Winfree, Nondestructive evaluation using a reduced-order computational methodology. Inverse Probl. **16**, 1–17 (2000)
9. U. Baur, Control-oriented model reduction for parabolic systems, Dissertation, TU Berlin, 2008
10. U. Baur, P. Benner, L. Feng, Model order reduction for linear and nonlinear systems: a system-theoretic perspective. Arch. Comput. Methods Eng. 2014 (electronic). DOI: 10.1007/s11831-014-9111-2
11. P. Benner, System-theoretic methods for model reduction of large-scale systems: simulation, control, and inverse problems, in *Proceedings of MathMod 2009*, Vienna, February 11–13, ed. by I. Troch, F. Breitenecker. Volume 35 of ARGESIM Report (2009), pp. 126–145
12. P. Benner, Balancing-related model reduction for parabolic control systems, in *Proceedings of the 1st IFAC Workshop on Control of Systems Governed by Partial Differential Equations* (IFAC Papers Online, Paris, 2013), pp. 257–262
13. P. Benner, J. Heiland, LQG-balanced truncation low-order controller for stabilization of laminar flows, ed. by R. King. Active Flow and Combustion Control 2014, Notes on Numerical Fluid Mechanics and Multidisciplinary Design, vol. 127 (Springer Internatinal Publishing, 2014), pp. 365–379
14. P. Benner, V. Mehrmann, D.C. Sorensen, *Dimension Reduction of Large-Scale Systems*. Lecture Notes in Computational Science and Engineering, vol. 45 (Springer, Berlin/New York, 2005)
15. P. Benner, E.S. Quintana-Ortí, G. Quintana-Ortí, Balanced truncation model reduction of large-scale dense systems on parallel computers. Math. Comput. Model. Dyn. Syst. **6**(4), 383–405 (2000)
16. P. Benner, J. Saak, Numerical solution of large and sparse continuous time algebraic matrix Riccati and Lyapunov equations: a state of the art survey. GAMM Mitt. **36**(1), 32–52 (2013)
17. M. Bergmann, L. Cordier, Optimal control of the cylinder wake in the laminar regime by trust-region methods and {POD} reduced-order models. J. Comput. Phys. **227**(16), 7813–7840 (2008)
18. J.T. Borggaard, S. Gugercin, Model reduction for DAEs with an application to flow control. Department of Mathematics, Virginia Tech, Blacksburg, 2014

19. E. Casas, Pontryagin's principle for state-constrained boundary control problems of semilinear parabolic equations. SIAM J. Control Optim. **35**, 1297–1327 (1997)
20. D. Chapelle, A. Gariah, J. Saint-Marie, Galerkin approximation with proper orthogonal decomposition: new error estimates and illustrative examples. ESAIM: Math. Model. Numer. Anal. **46**, 731–757 (2012)
21. X. Chen, S. Akella, I.M. Navon, A dual-weighted trust-region adaptive POD 4-D VAR applied to a finite-volume shallow water equations model on the sphere. Int. J. Numer. Methods Fluids **68**(3), 377–402 (2012)
22. R. Cont, N. Lantos, O. Pironneau, A reduced basis for option pricing. SIAM J. Financ. Math. **2**(1), 287–316 (2011)
23. R.F. Curtain, Model reduction for control design for distributed parameter systems, in *Research Directions in Distributed Parameter Systems*. Volume 27 of Frontiers Applied Mathematics (SIAM, Philadelphia, 2003), pp. 95–121
24. L. Dedé, Reduced basis method and a posteriori error estimation for parametrized linear-quadratic optimal control problems. SIAM J. Sci. Comput. **32**, 997–1019 (2010)
25. J.C. De Los Reyes, T. Stykel, A balanced truncation based strategy for the optimal control of evolution problems. Optim. Methods Softw. **26**(4–5), 673–694 (2011)
26. M. Dihlmann, B. Haasdonk, Certified nonlinear parameter optimization with reduced basis surrogate models. Proc. Appl. Math. Mech. **13**, 3–6 (2013)
27. A.L. Dontchev, W.W. Hager, A.B. Poore, B. Yang, Optimality, stability, and convergence in nonlinear control. Appl. Math. Optim. **31**, 297–326 (1995)
28. P. Feldmann, R.W. Freund, Efficient linear circuit analysis by Padé approximation via the Lanczos process. IEEE Trans. Computer-Aided Design **14**, 639–649 (1995)
29. R. Freund, Model reduction methods based on Krylov subspaces. Acta Numer. **12**, 267–319 (2003)
30. K. Gallivan, E. Grimme, P. Van Dooren, Asymptotic waveform evaluation via the Lanczos method. Appl. Math. Lett. **7**, 75–80 (1994)
31. J. Ghiglieri, S. Ulbrich, Optimal flow control based on POD and MPC and an application to the cancellation of Tollmien-Schlichting waves. Optim. Methods Softw. **29**(5), 1042–1074 (2014)
32. K. Glover, R.F. Curtain, J.R. Partington, Realisation and approximation of linear infinite-dimensional systems with error bounds. SIAM J. Control Optim. **26**(4), 863–898 (1988)
33. S. Rogg, Trust Region POD for Optimal Boundary Control of a Semilinear Heat Equation, Diploma thesis, Department of Mathematics and Statistics, University of Konstanz (2014)
34. M.A. Grepl, M. Kärcher, Reduced basis a posteriori error bounds for parametrized linear-quadratic elliptic optimal control problems. Comptes Rendus de l'Académie des Sciences—Series I **349**, 873–877 (2011)
35. M.A. Grepl, Y. Maday, N.C. Nguyen, A.T. Patera, Efficient reduced-basis treatment of nonaffine and nonlinear partial differential equations. ESAIM: Math. Model. Numer. Anal. **41**, 575–605 (2007)
36. E. Grimm, Optimality system POD and a-posteriori error analysis for linear-quadratic optimal control problems. Master thesis, University of Konstanz, 2013
37. M. Gubisch, S. Volkwein, Proper orthogonal decomposition for linear-quadratic optimal control (2013, submitted). http://kops.ub.uni-konstanz.de/handle/urn:nbn:de:bsz:352-250378
38. M. Gubisch, S. Volkwein, POD a-posteriori error analysis for optimal control problems with mixed control-state constraints. Comput. Optim. Appl. **58**, 619–644 (2014)
39. S. Gugercin, Projection methods for model reduction of large-scale dynamical systems. Ph.D. thesis, Rice University, Houston, 2003
40. S. Gugercin, A.C. Antoulas, C. Beattie, $\mathcal{H}_2$ Model reduction for large-scale dynamical systems. SIAM J. Matrix Anal. Appl. **30**(2), 609–638 (2008)
41. B. Haasdonk, Convergence rates of the POD-Greedy method. ESAIM: Math. Model. Numer. Anal. **47**, 859–873 (2013)
42. B. Haasdonk, M. Ohlberger, Reduced basis method for finite volume approximations of parametrized linear evolution equations. ESAIM: Math. Model. Numer. Anal. **42**, 277–302 (2008)

43. M. Heinkenschloss, D.C. Sorensen, K. Sun. Balanced truncation model reduction for a class of descriptor systems with application to the Oseen equations. SIAM J. Sci. Comput. **30**, 1038–1063 (2008)
44. C. Himpe, M. Ohlberger. Cross-gramian based combined state and parameter reduction (2013, submitted)
45. M. Hintermüller, K. Ito, K. Kunisch, The primal-dual active set strategy as a semismooth Newton method. SIAM J. Optim. **13**, 865–888 (2003)
46. M. Hinze, R. Pinnau, M. Ulbrich, M. Ulbrich, *Optimization with PDE Constraints*. Mathematical Modeling: Theory and Applications, vol. 23 (Springer, Berlin, 2009)
47. M. Hinze, F. Tröltzsch, Discrete concepts versus error analysis in PDE constrained optimization. GAMM Mitt. **33**, 148–162 (2010)
48. M. Hinze, S. Volkwein, Proper orthogonal decomposition surrogate models for nonlinear dynamical systems: error estimates and suboptimal control (Chapter 10), in *Dimension Reduction of Large-Scale Systems*, ed. by P. Benner, V. Mehrmann, D.C. Sorensen. Lecture Notes in Computational Science and Engineering, vol. 45 (Springer, Berlin/New York, 2005), pp. 261–306
49. M. Hinze, S. Volkwein, Error estimates for abstract linear-quadratic optimal control problems using proper orthogonal decomposition. Comput. Optim. Appl. **39**, 319–345 (2008)
50. P. Holmes, J.L. Lumley, G. Berkooz, C.W. Rowley, *Turbulence, Coherent Structures, Dynamical Systems and Symmetry*, 2nd edn. Cambridge Monographs on Mechanics (Cambridge University Press, Cambridge/New York, 2012)
51. K. Ito, S.S. Ravindran, A reduced basis method for control problems governed by PDEs, in *Control and Estimation of Distributed Parameter Systems*, ed. by W. Desch, F. Kappel, K. Kunisch. Proceedings of the International Conference in Vorau, vol. 126 (Birkhäuser-Verlag, Basel, 1996), pp. 153–168 (1998)
52. M. Kahlbacher, S. Volkwein, POD a-posteriori error based inexact SQP method for bilinear elliptic optimal control problems. ESAIM: Math. Model. Numer. Anal. **46**, 491–511 (2012)
53. E. Kammann, F. Tröltzsch, S. Volkwein, A method of a-posteriori error estimation with application to proper orthogonal decomposition, ESAIM: Math. Model. Numer. Anal. **47**, 555–581 (2013)
54. K. Kunisch, S. Volkwein, Proper orthogonal decomposition for optimality systems. ESAIM: Math. Model. Numer. Anal. **42**, 1–23 (2008)
55. K. Kunisch, S. Volkwein, L. Xie, HJB-POD based feedback design for the optimal control of evolution problems. SIAM J. Appl. Dyn. Syst. **3**, 701–722 (2004)
56. S. Lall, J.E. Marsden, S. Glavaski. A subspace approach to balanced truncation for model reduction of nonlinear control systems. Int. J. Robust Nonlin. Control **12**, 519–535 (2002)
57. O. Lass, S. Volkwein, Parameter identification for nonlinear elliptic-parabolic systems with application in lithium-ion battery modeling (2013, submitted) http://kops.ub.uni-konstanz.de/handle/urn:nbn:de:bsz:352-253400
58. F. Leibfritz, S. Volkwein, Reduced order output feedback control design for PDE systems using proper orthogonal decomposition and nonlinear semidefinite programming. Lin. Algebra Appl. **415**, 542–575 (2006)
59. S. Li, L. Feng, P. Benner, A. Seidel-Morgensterna, Using surrogate models for efficient optimization of simulated moving bed chromatography. Comput. Chem. Eng. (2014). doi:10.1016/j.compchemeng.2014.03.024
60. H.V. Ly, H.T. Tran, Modeling and control of physical processes using proper orthogonal decomposition. Math. Comput. Model. **33**, 223–236 (2001)
61. V. Mehrmann, T. Stykel, Balanced truncation model reduction for large-scale systems in descriptor form (Chapter 3), in *Dimension Reduction of Large-Scale Systems*, ed. by P. Benner, V. Mehrmann, D.C. Sorensen. Lecture Notes in Computational Science and Engineering, vol. 45 (Springer, Berlin/New York, 2005), pp. 83–115
62. M. Müller, Uniform convergence of the POD method and applications to optimal control. Ph.D thesis, University of Graz, 2011

63. F. Negri, G. Rozza, A. Manzoni, A. Quarteroni, Reduced basis method for parametrized elliptic optimal control problems. SIAM J. Sci. Comput. **35**(5), A2316–A2340 (2013)
64. B. Noack, K. Afanasiev, M. Morzynski, G. Tadmor, F. Thiele, A hierarchy of low-dimensional models for the transient and post-transient cylinder wake. J. Fluid. Mech. **497**, 335–363 (2003)
65. J. Nocedal, S.J. Wright, *Numerical Optimization*. Springer Series in Operations Research and Financial Engineering, 2nd edn. (Springer-Verlag, 2006)
66. A.T. Patera, G. Rozza, *Reduced Basis Approximation and A Posteriori Error Estimation for Parametrized Partial Differential Equations*. MIT Pappalardo Graduate Monographs in Mechanical Engineering (2006)
67. O. Pironneau, Calibration of options on a reduced basis. J. Comput. Appl. Math. **232**, 139–147 (2009)
68. M. Ohlberger, M. Schäfer, Error control based model reduction for parameter optimization of elliptic homogenization problems, in *Proceedings of the 1st IFAC Workshop on Control of Systems Governed by Partial Differential Equations* (IFAC Papers Online, Paris, 2013), pp. 251–256
69. M.R. Opmeer, Model order reduction by balanced proper orthogonal decomposition and by rational interpolation. IEEE Trans. Autom. Control **57**(2), 472–477 (2012)
70. M.R. Opmeer, T. Reis, W. Wollner, Finite-rank ADI iteration for operator Lyapunov equations. SIAM J. Control Optim. **51**(5), 4084–4117 (2013)
71. M. Rathinam, L. Petzold, Dynamic iteration using reduced order models: a method for simulation of large scale modular systems. SIAM J. Numer. Anal. **40**, 1446–1474 (2002)
72. S.S. Ravindran, Reduced-order adaptive controllers for fluid flows using POD. SIAM J. Sci. Comput. **15**, 457–478 (2000)
73. J.-P. Raymond, H. Zidani, Hamiltonian Pontryagin's principle for state-constrained control problems governed by semilinear parabolic equations. Appl. Math. Optim. **39**, 143–177 (1999)
74. M. Reed, B. Simon, *Methods of Modern Mathematical Physics I: Functional Analysis* (Academic, New York, 1980)
75. C.W. Rowley, Model reduction for fluids, using balanced proper orthogonal decomposition. Int. J. Bifurc. Chaos **15**, 997–1013 (2005)
76. E.W. Sachs, S. Volkwein, POD Galerkin approximations in PDE-constrained optimization. GAMM Mitt. **33**, 194–208 (2010)
77. A. Schiela, M. Weiser, Superlinear convergence of the control reduced interior point method for PDE constrained optimization. Comput. Optim. Appl. **39**, 369–393 (2008)
78. W.H.A. Schilders, H.A. van der Vorst, J. Rommes, *Model Order Reduction: Theory, Research Aspects and Applications*. Mathematics in Industry, vol. 13 (Springer, Berlin, 2008)
79. E.W. Sachs, M. Schneider, Reduced order models for the implied variance und local volatility. Technical report, Trier University, 2014
80. E.W. Sachs, M. Schu, A-priori error estimates for reduced order models in finance. ESAIM: Math. Model. Numer. Anal. **47**, 449–469 (2013)
81. V. Simoncini, Computational methods for linear matrix equations (Preprint, March 2013). http://www.dm.unibo.it/~simoncin/.
82. J.R. Singler, New POD expressions, error bounds, and asymptotic results for reduced order models of parabolic PDEs. SIAM J. Numer. Anal. **52**, 852–876 (2014)
83. A. Studinger, S. Volkwein, Numerical analysis of POD a-posteriori error estimation for optimal control. Int. Ser. Numer. Math. **164**, 137–158 (2013)
84. T. Tonn, K. Urban, S. Volkwein, Comparison of the reduced-basis and POD a-posteriori error estimators for an elliptic linear quadratic optimal control problem. Math. Comput. Model. Dyn. Syst. **17**, 355–369 (2011)
85. F. Tröltzsch, *Optimal Control of Partial Differential Equations*. American Mathematical Society, Graduate Studies Mathematics, vol. 112 (Springer, New York, 2010)
86. F. Tröltzsch, S. Volkwein, POD a-posteriori error estimates for linear-quadratic optimal control problems. Comput. Optim. Appl. **44**, 83–115 (2009)
87. M. Ulbrich, S.Ulbrich, Primal-dual interior-point methods for PDE-constrained optimization. Math. Prog. **117**, 435–485 (2008)

88. S. Volkwein, Optimality system POD and a-posteriori error analysis for linear-quadratic problems. Control Cybern. **40**, 1109–1125 (2011)
89. G. Vossen, S. Volkwein, Model reduction techniques with a-posteriori error analysis for linear-quadratic optimal control problems. Numer. Algebra Control Optim. **2**, 465–485 (2012)
90. K. Willcox, J. Peraire, Balanced model reduction via the proper orthogonal decomposition. American Institute of Aeronautics and Astronautics (AIAA) (2002), pp. 2323–2330
91. Y. Yue, K. Meerbergen, Accelerating optimization of parametric linear systems by model order reduction. SIAM J. Optim. **23**(2), 1344–1370 (2013)
92. K. Zhou, J.C. Doyle, K. Glover, *Robust and Optimal Control* (Prentice-Hall, Upper Saddle River, 1996)

Adaptive Trust-Region POD Methods in PIDE-Constrained Optimization

Ekkehard W. Sachs, Marina Schneider, and Matthias Schu

Abstract Solving optimization problems with partial integro-differential equations can be a challenging task from a theoretical but also numerical point of view. These equations arise e.g. in jump-diffusion models for the pricing of financial derivatives. We provide a formulation of calibration problems for financial market models in a proper mathematical framework. In the sequel we also address the issue of an efficient numerical solution of these problems. The main focus lies in the adequate use of reduced order models in order to achieve a computationally tractable optimization problem.

Keywords Trust region method • Proper orthogonal decomposition • Partial integro-differential equation

Mathematics Subject Classification (2010). 35Q93, 49M37, 65K10, 90C90.

1 Introduction

In the field of *partial differential equations* (PDE) or *partial integro-differential equations* (PIDE), the use of reduced order models is well-known. The idea is to replace the common finite element basis functions of a space discretization by only a few problem-dependent basis functions in a Galerkin approach. However, in PDE-constrained optimization we have in addition control or design variables which influence the solution of the PDEs. It can be observed that solutions of parameter-dependent differential equations are not arbitrary functions of the solution space. Often they can be approximated in a lower-dimensional subspace. Thus, we use more complex, problem-dependent basis functions which only span this subspace and not the whole function space. A survey of several model reduction techniques for linear dynamical systems in state space form is provided in [3, 4]. The most

E.W. Sachs (✉) • M. Schneider • M. Schu
FB IV – Department of Mathematics, University of Trier, 54286 Trier, Germany
e-mail: sachs@uni-trier.de; marina.schneider@uni-trier.de; schu@uni-trier.de

G. Leugering et al. (eds.), *Trends in PDE Constrained Optimization*, International Series of Numerical Mathematics 165, DOI 10.1007/978-3-319-05083-6_20

famous ones are *balanced truncation* and *proper orthogonal decomposition* (POD). Both methods have a close connection to *singular value decomposition* (SVD). Balanced truncation is mainly applicable to linear time-invariant systems and since POD can be even used for nonlinear systems this will be the method of our choice.

Reduced order models can be applied for the calibration of financial market models. So-called jump-diffusion models with local volatility provide many advantages concerning the pricing of financial derivatives. However, from a mathematical point of view they are quite complex since the corresponding model prices have to be calculated via a partial integro-differential equation (PIDE). The optimization problem consists of the calibration of the parameters of such a model. We compare market prices of standard European options with the model prices in a least-squares approach yielding a PIDE-constrained optimization problem. Local a priori error estimates for the reduced differential equations as well as for the corresponding reduced objective function combined with a globalizing trust-region framework yield an efficient algorithm that clearly reduces the computing time.

The application of reduced order models in financial applications, e.g. the solution of partial integro-differential equations resulting from jump-diffusion option pricing models, is a quite new issue first described in [18] where POD is used, and by [16] using a reduced basis approach with basis functions based on Black-Scholes solutions. It has been further investigated in [19] and [8]. Another application in finance can be found in [17] where reduced order models are used for the efficient computation of the implied variance in a local volatility framework as the solution of a quasilinear degenerate parabolic partial differential equation.

In this paper, we only give an overview on the results achieved in the project, for more detailed statements and proofs see the references cited throughout the paper.

2 PIDE Constrained Optimization

Partial integro-differential equations (PIDEs) arise in several fields of research. We mention for example biological applications like cell adhesion models discussed in [6, 12]. Another area of application is the area of peridynamics in continuum mechanics where these type of integral operators occur, see [24].

Here, we consider an application in finance, i.e., the calibration of option pricing models based on Lévy processes, see [9, 22]. As a financial derivative, the value of an option depends on the value of some underlying. In the past decade, several extensions of the Black-Scholes model for the pricing of options have been developed. We use a jump-diffusion model with an additional local volatility function and the following dynamics written as a stochastic differential equation

$$dS_t = \mu S_{t\_} dt + \sigma(t, S_{t\_}) S_{t\_} dW_t + S_{t\_} d\Big(\sum_{j=1}^{N_t} (e^{Z_j} - 1)\Big), \qquad (2.1)$$

where the Z_j's are independent and identically distributed. The corresponding option prices can be computed by solving a PIDE. Recently, a Dupire-like PIDE for the original equation similar to the Black-Scholes framework was derived in [1] and [15]. In view of the calibration problem that is discussed below, we focus on this Dupire-like version of the PIDE for the price of an option depending on its expiration time T and strike price K. Usually we are only interested in the call price today, so in the following – without loss of generality – we assume that the option is priced at time $t_0 = 0$ and the price of the underlying is given by S_0. For numerical reasons we apply a variable transformation $x = \ln(K/S_0)$ and obtain the following PIDE:

$$D_T - \frac{1}{2}\bar{\sigma}^2(T,x)D_{xx} + \Big(r(T) + \frac{1}{2}\bar{\sigma}^2(T,x) - \lambda\zeta\Big)D_x + \lambda(1+\zeta)D \tag{2.2}$$

$$-\lambda \int_{-\infty}^{+\infty} D(T, x-z)e^z f(z)\, dz = 0, \quad \text{in} \quad [0, T_{max}) \times (-\infty, \infty)$$

$$D(0,x) = S_0 \cdot \max\{1 - e^x, 0\} =: D_0(x), \quad x \in (-\infty, \infty).$$

This equation contains several parameters and parameter functions: The local volatility function σ, the jump intensity λ and the jump size distribution function f For the density function we use Merton's model and define $f(z) = \frac{1}{\sqrt{2\pi\sigma_J}} exp(-\frac{(z-\mu_J)^2}{2\sigma_J^2})$ for all $z \in (-\infty, \infty)$ with μ_J the expected value, σ_J the standard deviation of the normally distributed jump sizes and $\zeta = exp(\frac{\sigma_J^2}{2} + \mu_J) - 1$. Now we are able to state the corresponding calibration problem in a least-squares formulation as

$$\min_{D,\sigma,\lambda,f} J(D,\sigma,\lambda,f) := \frac{1}{2}\sum_{i=1}^{M}\Big(D(T_i, \ln(K_i/S_0)) - D_i^M\Big)^2 \tag{2.3}$$

$$\text{subject to PIDE (2.2)},$$

i.e., we adjust the parameters in such a way that the model prices fit some given market prices D_i^M at M data points (T_i, K_i). Thus, the calibration problem is a PIDE constrained optimization problem. Note that for one function evaluation of J, the PIDE constraint (2.2) has to be solved only once.

3 Numerical Solution of the PIDE

Partial integro-differential equations in contrast to PDEs do not lead to sparse systems when discretized due to the nonlocal behavior of the integral operator. Hence, their efficient numerical solution will play an important role in the solution of the calibration problem (2.3). For the discretization of the spatial variable by a finite element approach, the first step is a variational formulation of the problem. In order to formulate the problem properly, we have to set a framework for the existence and uniqueness of weak solutions of the original equation. Therefore, we define

$$W([a,b],V) := \Big\{u : u \in L^2((a,b),V),\;\; u' \in L^2((a,b),V^*)\Big\} \tag{3.1}$$

for $a, b \in \mathbb{R}$, where V is a Hilbert space with its dual V^*. Since the initial condition of (2.2) is not $L^2(\mathbb{R})$-integrable, it is necessary to introduce some appropriate function spaces.

Definition 3.1 (Weighted function spaces).

1. $L^2_{-\mu}(\mathbb{R}) := \{v \in L^1_{loc}(\mathbb{R}) : v(\cdot)e^{-\mu|\cdot|} \in L^2(\mathbb{R})\}$
   with inner product $\langle v, w\rangle_{L^2_{-\mu}} := \int_{\mathbb{R}} v(x)w(x)e^{-2\mu|x|}dx$,
2. $H^1_{-\mu}(\mathbb{R}) := \{v \in L^1_{loc}(\mathbb{R}) : v(\cdot)e^{-\mu|\cdot|}, v'(\cdot)e^{-\mu|\cdot|} \in L^2(\mathbb{R})\}$
   with inner product $\langle v, w\rangle_{H^1_{-\mu}} := \langle v, w\rangle_{L^2_{-\mu}} + \langle v', w'\rangle_{L^2_{-\mu}}$.

We state the variational formulation of the PIDE (2.2) and address its solvability:

Definition 3.2 (Weak formulation of the PIDE). Let λ, ζ be given constants and assume that $r(T)$, $\sigma(T,\cdot)$, $\sigma(T,\cdot)_x$ are continuous and bounded functions on $\mathbb{R}$. The variational formulation of the PIDE (2.2) consists of finding $D \in W([0,T_{max}], H^1_{-\mu}(\mathbb{R}))$ such that for all $T \in (0, T_{max}]$

$$\frac{d}{dT}\langle D(T,\cdot), w(\cdot)\rangle_{L^2_{-\mu}} + a^{-\mu}(T; D(T,\cdot), w(\cdot)) = 0 \quad \forall\, w \in H^1_{-\mu}(\mathbb{R}) \tag{3.2}$$

holds with initial condition

$$\langle D(0,\cdot), w(\cdot)\rangle_{L^2_{-\mu}} = \langle D_0(\cdot), w(\cdot)\rangle_{L^2_{-\mu}} \;\; \forall\, w \in H^1_{-\mu}(\mathbb{R}). \tag{3.3}$$

For each constant $\mu > 0$ and $T > 0$ the bilinear form $a^{-\mu}(T;\cdot,\cdot) : H^1_{-\mu}(\mathbb{R}) \times H^1_{-\mu}(\mathbb{R}) \to \mathbb{R}$ is defined as

$$a^{-\mu}(T;\, v, w) := \int_{\mathbb{R}} \frac{\sigma^2(T,x)}{2} v'(x)w'(x)e^{-2\mu|x|}dx \tag{3.4}$$

$$+ \int_{\mathbb{R}} \Big(r(T) - \lambda\zeta + \frac{\sigma^2(T,x)}{2}(1 + 2\mu\,\mathrm{sgn}(x)) + \frac{(\sigma^2(T,x))_x}{2}\Big)v'(x)w(x)e^{-2\mu|x|}dx$$

$$+ \int_{\mathbb{R}} \lambda(1+\zeta)v(x)w(x)e^{-2\mu|x|}dx - \lambda \int_{\mathbb{R}}\int_{\mathbb{R}} v(x-z)w(x)e^{-2\mu|x|}e^z f(z)dz\, dx.$$

This leads to the following existence and uniqueness theorem, see [19] and [20].

Theorem 3.3 (Existence and uniqueness of a weak solution). *For each $T \in [0, T_{max}]$, let $\sigma(T, \cdot)$ be continuously differentiable on $\mathbb{R}$ and positive. Let $r(\cdot)$, $\sigma(\cdot, x)$, $\sigma_x(\cdot, x)$ be uniformly Lipschitz-continuous functions in the variable T. Furthermore, assume that there exists $\mu > 0$ such that $\int_{\mathbb{R}} e^{y+\mu|y|} y f(y)\, dy < \infty$.*

Then there exists a unique solution $D \in W([0, T_{max}], H^1_{-\mu}(\mathbb{R}))$ of the problem specified in Definition 3.2.

Regarding the numerical solution of the PIDE, we use a finite element approach for the spatial variable and an implicit finite difference scheme, like Crank-Nicolson, for the time discretization. The discretization of the spatial variable via a finite element approach leads to a dense stiffness matrix due to the non-local integral term in the PIDE. Since these matrix exhibits a Toeplitz structure the matrix-vector multiplications can be computed efficiently using fast Fourier transformation (FFT). For the time discretization, an implicit method is desirable since the problem is known to be very stiff and explicit methods are restricted by strong CFL conditions. We use a Crank-Nicolson scheme with additional Rannacher smoothing of the non-smooth initial condition yielding a second-order convergence in time. The linear systems of equations for such implicit methods are dense such that we propose to use a preconditioned GMRES method for their solution. However, the repeated solution of the PIDE in the calibration problem will still be expensive. One way to deal with this issue is the use of reduced order models.

4 Model Order Reduction via POD

We briefly introduce proper orthogonal decomposition (POD), a technique to replace a large mathematical problem – in our case a discretized partial integro-differential equation – by a small one. The error between the original model and the reduced model should be small, however, the computational effort is supposed to be reduced significantly. The idea is to extract the most significant information from a set of given functions or vectors (called snapshots). To be precise, we want to find those basis functions which represent the set of snapshots better than any other basis. Mathematically, this can be written as a constrained optimization problem.

Definition 4.1 (POD basis). Given snapshots $u_1, \ldots, u_n \in H$ with $\dim(\text{span}(u_1, \ldots, u_n)) = r > 0$, find orthonormal functions $\Psi_1, \ldots, \Psi_r \in \text{span}(u_1, \ldots, u_n)$ by solving the minimization problem

$$\min_{\Psi_1,\ldots \Psi_l} \sum_{i=1}^{n} \gamma_i \left\| u_i - \sum_{j=1}^{l} \langle u_i, \Psi_j \rangle_H \Psi_j \right\|_H^2 \tag{4.1}$$

$$s.t. \quad \langle \Psi_j, \Psi_k \rangle_H = \delta_{jk} \quad \forall j, k = 1, \ldots, l$$

for all $l \in \{1, \dots, r\}$ with weights $\gamma_i > 0$, $i = 1, \dots, n$. The first l vectors $\Psi_1, \dots, \Psi_l$ are called a POD basis of rank l. The spanning subspace is denoted by $\mathcal{V}^l = \text{span}(\Psi_1, \dots, \Psi_l)$.

It is well-known in literature that the POD basis functions are given by eigenfunctions of a specific eigenvalue problem. If we only use a part of this POD basis, the average approximation error for the snapshots can be expressed by a sum over those eigenvalues λ_j whose eigenfunctions are not used. We are especially interested in error estimates of POD reduced systems in the context of parabolic differential equations and, furthermore, in optimal control problems governed by such PDEs. If we want to apply the POD technique to a parabolic differential equation – like the calibration problem with a PIDE – we choose the solution of the problem at fixed time steps $t_0, \dots, t_n$ as snapshots. We then obtain some orthonormal basis functions containing specific information about the solution of the PIDE in the above-mentioned sense. Approximating the PIDE problem via a POD approach then means that we replace the finite element basis functions by the POD basis functions calculated from the given solution. Since the original problem, the PIDE, is replaced by a smaller one, the POD approximation, we need to estimate the corresponding error. For the derivation of error estimates, we rewrite the weak form of our parabolic problem (3.2) in a more general formulation. To set up the framework we make the following assumptions (cf. [10]):

Assumption 4.2. (a) Let V and H be real, separable Hilbert spaces with the inner products $\langle \cdot, \cdot \rangle_V$ and $\langle \cdot, \cdot \rangle_H$ and the induced norms $||\cdot||_V$ and $||\cdot||_H$, respectively. With the dual spaces V^* and H^* they form a Gelfand triple $V \hookrightarrow H = H^* \hookrightarrow V^*$ with dense embeddings. Furthermore, assume an $\alpha > 0$ with $||v||_H^2 \leq \alpha ||v||_V^2$ for all $v \in V$.

(b) Let $a : [0, T] \times (V \times V) \to \mathbb{R}$ for all $t \in [0, T]$ be a uniformly continuous and coercive bilinear form. In addition let $a(\cdot; \cdot, \cdot)$ be Lipschitz continuous with respect to t.

(c) Let $L : [0, T] \times V \to \mathbb{R}$ be a linear form with $L \in L^2(V^*)$ and $c_L > 0$ such that $|L(t; v)| \leq c_L ||v||_V$ for all $t \in [0, T]$, $v \in V$.

Definition 4.3 (Continuous problem). For given initial value $y_0 \in H$ find a solution $y \in W([0, T], V)$ which satisfies

$$\frac{d}{dt}\langle y(t), v\rangle_H + a(t; y(t), v) = L(t; v) \quad \forall\, v \in V,\ t \in (0, T) \tag{4.2}$$

and initial condition $\langle y(0), v\rangle_H = \langle y_0, v\rangle_H$ for all $v \in V$.

Problem (4.2) discretized in time on a subspace $\mathcal{V}^l$ of V with equidistant time steps $t_0, \dots, t_m$ looks as follows:

Definition 4.4 (Discretized problem). For given initial value $y_0 \in H$ and some $\theta \in [0, 1]$ find $\{y_i^l\}_{i=0}^n \subset \mathcal{V}^l$ with

$$\langle \bar{\partial} y_i^l, v\rangle_H + \theta \cdot a(t_i;\, y_i^l, v) + (1-\theta)\cdot a(t_{i-1};\, y_{i-1}^l, v) = \qquad (4.3)$$
$$\theta \cdot L(t_i;\, v) + (1-\theta)\cdot L(t_{i-1}, v) \qquad \forall\, v \in \mathcal{V}^l,\ i = 1,\dots,n$$

and initial condition $\langle y_0^l, v\rangle_H = \langle y_0, v\rangle_H$ for all $v \in \mathcal{V}^l$, where $\bar{\partial}$ is an abbreviation for the finite difference quotients.

Using the stated assumptions, we can invoke an existence and uniqueness theorem in [10, pp. 512ff] to conclude that there exists a unique solution for both problems. In the following we want to address the average error between the solution $y(t)$ of problem (4.2) and the solution $y^{l,1}$ on the POD subspace, discretized in time via the θ-method in problem (4.3). The POD basis functions are calculated in the sense of (4.1) from the snapshots of the solution $y(t)$ and the corresponding difference quotients, i.e., the snapshots are $\bar{y}_i = y(t_{i-1})$ and $\bar{y}_{i+n+1} = \frac{y(t_i)-y(t_{i-1})}{t_i - t_{i-1}}$, $i = 1,\dots,n$.

Defining the H-projection Π_H^l by

$$\Pi_H^l : V \to \mathcal{V}^l \quad \Leftrightarrow \quad \langle \Pi_H^l u - u, v\rangle_H = 0 \quad \forall\, v \in \mathcal{V}^l\,, \qquad (4.4)$$

one can state the following error estimate.

Theorem 4.5. *Let $y(t)$ be the solution of (4.2), $\{y_i^{l,1}\}_{i=0}^n$ the solution of (4.3). Then with appropriate constants C_i $(i = 0,1,2)$, independent of n, we have*

$$\frac{1}{n}\sum_{i=1}^{n} ||y(t_i) - y_i^{l,1}||_H^2 \le C_0 ||y(t_0) - \Pi_H^l y(t_0)||_H^2 + C_1 \Delta t^j + C_2 ||S||_2 \sum_{j=l+1}^{r} \lambda_j$$

with $j = 2$ for the implicit Euler method assuming $y_{tt} \in L_2([0,T];H)$ and $j = 4$ for the Crank-Nicolson method, assuming $y_{ttt} \in L_2([0,T];H)$ and Δt sufficiently small.

Furthermore, for some constant C we have

$$||y(t_0) - \Pi_H^l y(t_0)||_H^2 \le n\, C \sum_{j=l+1}^{r} \lambda_j\,.$$

This result, stated in [20], is an extension of the results of [14] to the time dependent case.

Next we give estimates on the error between the POD solution compared to a discretized FE solution.

Theorem 4.6. *Let $\{y_i^{FE}\}_{i=0}^n$ be the finite element solution using the finite element space H^{n_x} in the Galerkin approximation. Let $\{y_i^{l,2}\}_{i=0}^n$ be the solution of problem (4.3) based on the FEM snapshots.*

Then with appropriate constants $\tilde{C}_0, \tilde{C}_1$ independent of n, we get for the implicit Euler method and, for sufficiently small Δt, also for the Crank-Nicolson method

$$\frac{1}{n}\sum_{i=1}^{n} ||y_i^{FE} - y_i^{l,2}||_H^2 \leq \tilde{C}_0 \, ||y_0^{FE} - \Pi_H^l y_0^{FE}||_H^2 + \tilde{C}_1 ||S||_2 \sum_{j=l+1}^{r} \lambda_j$$

where $||y_0^{FE} - \Pi_H^l y_0^{FE}||_H^2 \leq 3n \sum_{j=l+1}^{r} \lambda_j$.

5 Numerical Solution of the Calibration Problem

Given an efficient numerical method for the solution of the PIDE constraint, we turn to the numerical solution of the corresponding calibration problem (2.3). This can be rewritten in an abstract vector form, in which the PIDE is replaced by its weak formulation with state variable y and control u as:

$$\min_{y \in W, u \in \mathcal{U}} J(y,u) := \frac{1}{2}\sum_{i=1}^{D} ||Cy(\hat{t}_i) - d_i||_{\mathcal{H}}^2 \tag{5.1}$$

$$s.t. \quad \dot{y}(t) + A(u;t)y(t) - l(u;t) = 0\,, \quad t \in (0,T] \tag{5.2}$$
$$y(0) = y_0.$$

Here we made use of the fact that for a given control u and with $V := H^1_{-\mu}(\mathbb{R})$, there exist unique operators $A(T) \in L(V, V^*)$ and $l(T) \in V^*$ for all $T \in (0, T_{max}]$ such that (3.2) can be rewritten as (5.2) in the sense of $L^2(V^*)$. This form provides some advantages in terms of a simpler notation. The above problem can also be written as an unconstrained optimization problem, in the literature also known as the *reduced problem*

$$\min_{u \in \mathcal{U}} f(u) := J(y(u;\,\cdot), u). \tag{5.3}$$

Considering the discretization of the optimal control problem, there are mainly two approaches common in literature: *Optimize-then-discretize* or *discretize-then-optimize*. Regardless of whether we discretize or optimize first, we solve the optimization problem by a gradient-based method. The gradient of the problem can be calculated efficiently by means of the adjoint equation. However, second-order information is more complicated and the calculation of the exact Hessian is usually not reasonable.

Since the least-squares objective function involves pointwise observations of the PIDE solution, the corresponding adjoint equations contain delta Dirac functions.

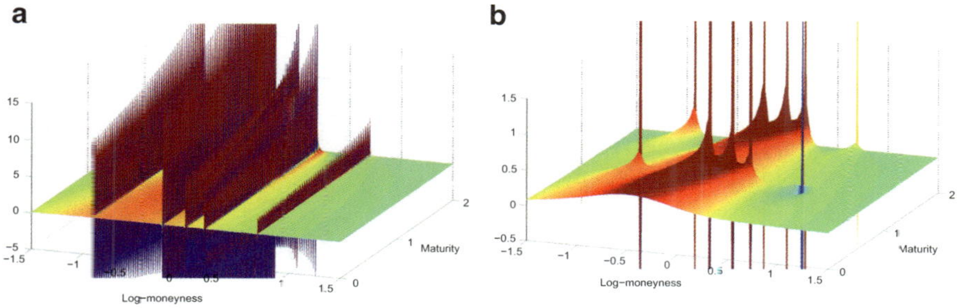

Fig. 1 First optimize: solutions of the adjoint equation ($\Delta T = 0.02$, $\Delta x = 0.0025$). (**a**) Crank-Nicolson (**b**) Rannacher time stepping

These functions will lead to high-frequency end conditions in the backward adjoint equations and thus to oscillations in the numerical solution, so it is advisable to apply a smoothed time discretization scheme, see [21]. If we first optimize, we are free in the choice of an appropriate discretization method. A standard Crank-Nicolson method applied to the adjoint equation leads to the result illustrated in Fig. 1a. The adjoint is formally divided into two parts with end conditions at the points, where market data is available. The peaks occurring in these end conditions oscillate strongly over the whole time domain. As in the case of nonsmooth initial conditions for the state equation, Rannacher smoothing can be applied at each end condition, see Fig. 1b.

Regarding the first discretize approach with a Rannacher time stepping scheme for the state equation, Fig. 2 shows the numerical solution of the corresponding discrete adjoint equation. The Rannacher method in the state equation yields a Crank-Nicolson method for the adjoint except for the last time step before $T = 0$, where four implicit Euler quarter steps are used. The peaks are not as pronounced as in the first optimize approach due to the fact that the end condition contains a kind of built-in smoothing through the elliptic operator weighted with step size. However, there are still small oscillations observable through the whole time domain, denying a quadratic convergence of the scheme.

Our main aim is the use of POD in the optimal control problem (5.1), so the error between a discretized objective function based on a finite element space, $f(u) = J(y^{FE}(u;\cdot), u)$, and an objective function based on a POD approximation with rank l, $f_l(u) = J(y^l(u;\cdot), u)$, has to be estimated. A relation between this error and the sum over the remaining eigenvalues is established in [23]. For fixed but arbitrary u and a $k_1 > 0$, we obtain

$$|f(u) - f_l(u)|^2 \leq k_1 \sum_{j=l+1}^{r} \lambda_j$$

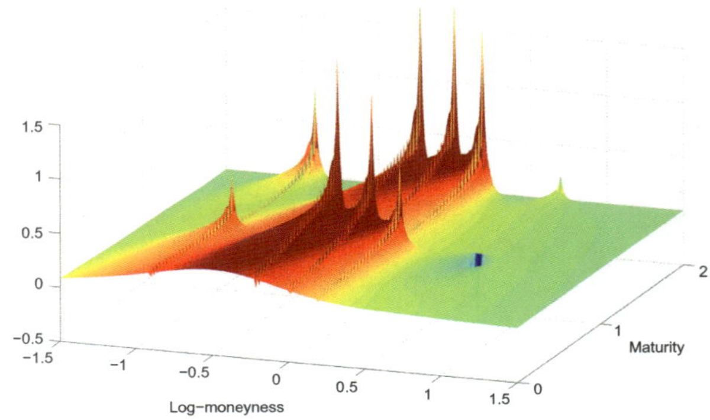

Fig. 2 First discretize: solution of the adjoint equation ($\Delta T = 0.02$, $\Delta x = 0.0025$)

if the POD basis contains the snapshots from the state solution at u. A similar result is then obtained for the gradient of f. However, we need to include snapshots from the adjoint solution as well to get for a $k_2 > 0$:

$$||\nabla f(u) - \nabla f_l(u)||^2 \leq k_2 \sum_{j=l+1}^{r} \lambda_j .$$

k_2 has several dependencies, mainly a proper weighting of adjoint and state snapshots has to be guaranteed. In general, we observe that the inclusion of adjoint snapshots in one combined basis with state snapshots leads to a far better approximation of gradients. It further has a positive effect on the otherwise strong locality of a fixed POD basis, which is illustrated in Fig. 3. First, (a) and (b) show the finite element state and adjoint solution of the PIDE for the control u. Those solutions are used as snapshots for the basis computation where we keep $l = 15$ fixed. The adjoint solution clearly shows peaks at each point where market data is available. Note that we are especially interested in the values of our state solution at these points. The pointwise difference between the full finite element state solution y for the control u and the corresponding reduced state solution based on a basis only containing state snapshots, $y^{l,S}$ is illustrated in Fig. 3c. Given the scaling of the graph this error is negligible. Figure 3d shows the same result using a basis with space and weighted adjoint snapshots to compute the POD approximation, $y^{l,SwA}$. This leads to a larger, observable error especially at the beginning for T close to zero. However, if we now make a step in steepest descent direction with step size $\Delta = 1.0\text{e-}2$ without updating the POD model, the results are quite different. Figure 3e illustrates a strongly increasing pointwise error compared to (c), unfortunately in a region of great interest. However, the basis including weighted adjoint snapshots provides further information in this region, leading to a smaller error in (f).

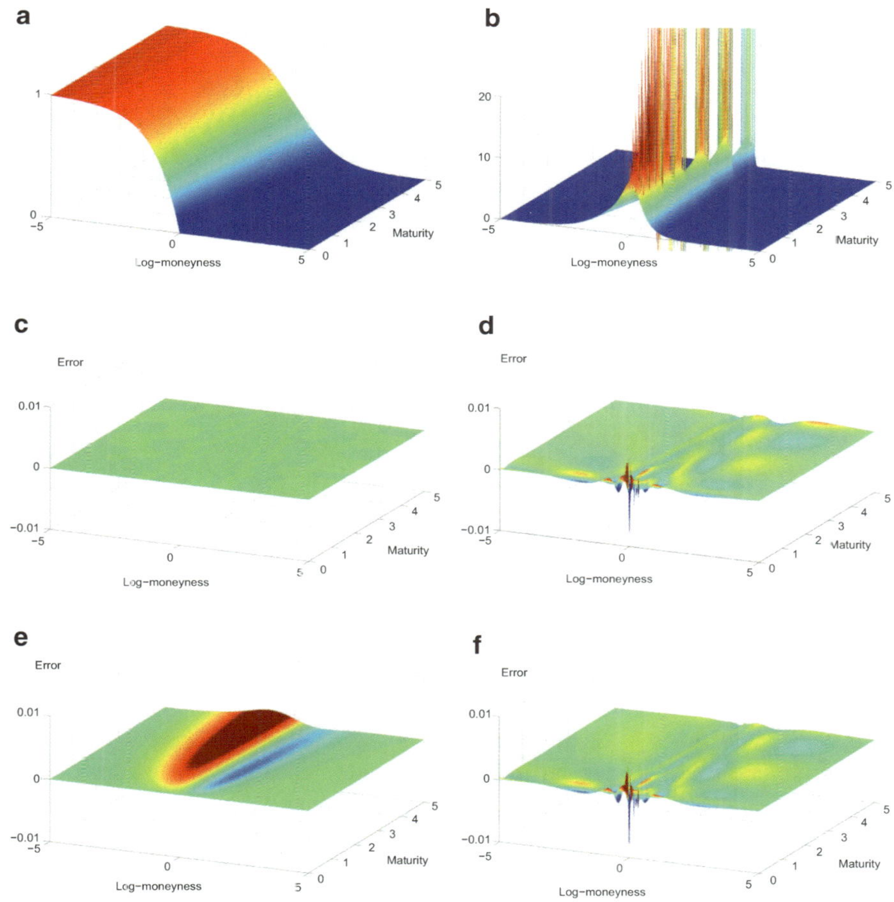

Fig. 3 Influence of adjoint snapshots on the POD state error when the control u is changed and the basis not updated. (**a**) State solution $y(u;\cdot)$. (**b**) Adjoint solution $p(u;\cdot)$. (**c**) Error $y(u;\cdot) - y^{l,S}(u;\cdot)$. (**d**) Error $y(u;\cdot) - y^{l,SwA}(u;\cdot)$. (**e**) Error $y(u_\Delta;\cdot) - y^{l,S}(u_\Delta;\cdot)$. (**f**) Error $y(u_\Delta;\cdot) - y^{l,SwA}(u_\Delta;\cdot)$

6 Trust-Region POD

We have seen in the previous section that POD is only a local model. Therefore, if we veer away from the starting parameters during the calibration process, the error estimates that hold true for unchanged parameters cannot be applied any longer. For this reason, a trust region POD algorithm for an optimal flow control problem is proposed by [5] in order to adaptively adjust the POD model. To explain the idea, we consider a discretized version of the calibration problem (5.1)

$$\min_{u \in U} f(u) = \frac{1}{2} \sum_{i=1}^{D} ||Cy^{FE}(u; t_{k_i}) - d_i||^2_{\mathcal{H}}, \tag{6.1}$$

where $\{y_k^{FE}\}_{k=0}^{n_t}$ denotes the finite element approximation to the state equation discretized with the θ-method, i.e,

$$\begin{aligned} \bar{\partial} y^{FE}(t_k) + \theta A(u;\ t_k) y^{FE}(t_k) + (1-\theta) A(u;\ t_{k-1}) y^{FE}(t_{k-1}) &= \\ \theta l(u;\ t_k) + (1-\theta) l(u;\ t_{k-1}), \qquad k = 1, \ldots, n_t& \\ y_0^{FE}(t_0) = y_0.& \end{aligned} \tag{6.2}$$

For a given control u_k, the idea of TRPOD is to use a POD model function instead of the well-known quadratic model function $m^{quad}(u_k + s) = f(u_k) + \nabla f(u_k)^T s + \frac{1}{2} s^T \nabla^2 f(u_k) s$. The POD model function is defined by

$$m_k^l(u_k + s) = \frac{1}{2} \sum_{i=1}^{D} ||Cy_{k_i}^l(u_k + s) - d_i||^2_{\mathcal{H}}, \tag{6.3}$$

where $\{y_k^l\}_{k=0}^{n_t} \subset \mathcal{V}^l$ is the POD approximation to the state equation discretized in time with the θ-method.

With this new model function one can formulate the adaptive trust-region POD method stated in Algorithm 1. Its general structure is given by a basic trust-region algorithm. In the lines 1– 3 we first calculate the state and adjoint snapshots for the current control u_k to get our model. These are then used to calculate a POD basis where the rank l is chosen such that the inequality in line 3 is satisfied. This condition is supposed to control the relative gradient error $\frac{||\nabla f(u_k) - \nabla m_k^l(u_k)||}{||\nabla m_k^l(u_k)||}$, as proposed in [7]. Lines 4– 5 describe the step calculation. The solution of the trust-region subproblem is approximated via the step determination algorithm proposed in [11, 25]. The quotient ρ_k in line 7 compares the predicted reduction of the model function with the actual reduction of the objective function. It provides a good measure for the capability of the model function, which is used in the lines 8 to 17 to decide whether the point $u_k + s_k$ is accepted as a new iterate or not. If ρ_k is greater than $\eta_1 > 0$, the step s_k is accepted. If further $\rho_k > \eta_2$, a value typically chosen to be close to one, then the model seems to be a good approximation on the trust region and the trust radius may be increased. Otherwise, the radius should be decreased. A step s_k is rejected if $\rho_k < \eta_1$. In that case the model seems to be poor on the trust region and the radius, Δ_k, has to be decreased. We want to stress that it is not necessary to compute a new POD model in case of an unsuccessful iteration.

Our main result, a global convergence proof for the algorithm above, can now be stated, see [23].

Algorithm 1 Adaptive TRPOD algorithm

Input: $\Delta_0 > 0$, $k = 0$, an initial control $u_0 \in U$ and constants $\eta_1, \eta_2, \gamma_1, \gamma_2, \gamma_3, \zeta$ satisfying
$$0 < \eta_1 \leq \eta_2 < 1, \quad 0 < \gamma_1 \leq \gamma_2 < 1 \leq \gamma_3, \quad 0 < \zeta < 1 - \eta_2.$$
1: **compute** state $y(u_k)$ and adjoint $p(u_k)$ to form the set of snapshots $\mathcal{S}$
2: **compute** for $\mathcal{S}$ the POD basis of rank l and model m_k^l such that
3: $$||\nabla f(u_k) - \nabla m_k^l(u_k)|| \leq \zeta ||\nabla m_k^l(u_k)||$$
4: **compute** an approximate solution $s_k \in U$ to
5: $$\min_{||s|| \leq \Delta_k} m_k^l(u_k + s)$$
6: **compute** $f(u_k + s_k)$ **and**
7: $$\rho_k = \frac{f(u_k) - f(u_k + s_k)}{m_k^l(u_k) - m_k^l(u_k + s_k)}$$
8: **if** $\rho_k \geq \eta_2$ **then**
9: **set** $u_{k+1} = u_k + s_k$ **and** $\Delta_{k+1} \in [\Delta_k, \gamma_3 \Delta_k]$
10: **set** $k \leftarrow k + 1$ **and** go to line 1
11: **else if** $\eta_1 \leq \rho_k < \eta_2$ **then**
12: **set** $u_{k+1} = u_k + s_k$ **and** $\Delta_{k+1} \in [\gamma_2 \Delta_k, \Delta_k]$
13: **set** $k \leftarrow k + 1$ **and** go to line 1
14: **else if** $\rho_k < \eta_1$ **then**
15: **set** $u_{k+1} = u_k$ **and** $\Delta_{k+1} \in [\gamma_1 \Delta_k, \gamma_2 \Delta_k]$
16: **set** $k \leftarrow k + 1$ **and** go to line 4
17: **end if**

Theorem 6.1 (Strong global convergence of the adaptive TRPOD). *Given problem* (6.1) *with Lipschitz continuous Fréchet derivatives A' and l' in* (6.2) *and $\zeta \in (0, 1 - \eta_2)$, let the Assumption* 4.2 *be satisfied. Further, assume that an implicit discretization scheme as the backward Euler or Crank-Nicolson method is used and the reduced order model includes adjoint information. Let $\{u_k\}$ be a sequence of iterates produced by Algorithm* 1. *Then,*

$$\lim_{k \to \infty} ||\nabla f(u_k)|| = 0.$$

In the proof of this theorem it is vital that the POD model function can fulfil the assumptions on the gradient accuracy in the sense of [7]. The size of the POD basis can be managed adaptively by comparing reduced and exact gradient until this condition is satisfied. For more details we refer to [23].

It is clear that the solution of the state and adjoint equation being discretized on a sufficiently fine finite element grid is a major part of the total computing time. To further reduce computational cost, a multi-level approach was introduced in [13]. The discretization of the PIDE is not kept fixed, but one can switch between different discretization levels. The goal is to avoid solutions on the finest grids especially when we are not near the optimal point. If we are far away from the optimal solution of the problem, we do not only use less POD basis functions to compute our model, but we also compute the snapshots, i.e. the FE solutions, on a coarser grid.

Table 1 Iterations of a TRPOD calibration run with corresponding gradient norm, function values, ratios ρ_k ,trust radius Δ_k, number of used POD basis functions l and ζ_k^l

k	$\|\|\nabla f(u_k)\|\|_2$	$f(u_k)$	$m_k(u_{k+1})$	ρ_k	Δ_k	#POD l	ζ_k^l
0	3.58e+0	1.21e+0	2.43e−1	1.00	0.10	10	0.0228
1	1.77e+0	2.44e−1	2.76e−4	1.00	0.11	16	0.0286
2	4.91e−2	3.63e−4	1.29e−5	1.00	0.11	22	0.0216
3	1.43e−3	1.38e−5	1.14e−5	0.83	0.12	28	0.1562
4	1.24e−4	1.18e−5					

Example. To illustrate the efficiency of the adaptive trust-region POD algorithm, we consider an example, where we calibrate Merton's jump-diffusion model to the market data presented in [2]. The volatility is assumed to be parameterized with 20 parameters, thus, together with the jump intensity λ, the average jump size μ_J and the mean jump size σ_J, there are 23 parameters to be calibrated. Table 1 shows the corresponding calibration run.

For each iteration it contains the norm of the 'exact' – i.e. based on the full finite element state and adjoint solution – gradient, the 'exact' function value in u_k, the value of the model function in the optimal solution of the trust-region subproblem $m_k(u_{k+1})$, the ratio between actual reduction and predicted reduction ρ_k and the corresponding trust radius Δ_k. The last two columns contain values directly connected to the POD model function: the number of POD functions l that is used in the current iteration and the ratio ζ_k^l. We observe that all iterations are successful and all ratios ρ_k are close to one. Note that in each iteration we need several evaluations of the discretized state and adjoint equations. First of all a state and adjoint solution on the full finite element grid is required to form the snapshot set for the new POD model, see lines 1– 3 of Algorithm 1. Furthermore, the approximate solution of the trust-region subproblem in lines 4– 5 involves several solutions of the differential equations discretized in the POD space for the computation of the POD model function and its gradient. The total number of evaluations of the differential equations in the TRPOD algorithm is shown in Table 2. There we compare the performance of the TRPOD algorithm, a multi-level version of the TRPOD algorithm using three different finite element grids (level 1: $\Delta x \times \Delta T = 0.01 \times 0.025$, level 2: 0.005×0.025, level 3: 0.0025×0.0125) and the quasi-Newton approach. The quasi-Newton needs 274 evaluations of the state and the adjoint equations, all computed on the full finite element grid. In contrast, the TRPOD algorithm needs 10 + 600, this means even more, evaluations. However, most of them are computed via the reduced order model and we only need five state and five adjoint solutions on the full grid. The optimal values shown in the last two columns have the same order of magnitude. Regarding the computational effort, TRPOD needs 86 s for the optimization compared to 888 s in the quasi-Newton approach corresponding to a time saving of 90 %. In the table, the total computing time is further split into the time needed for the full PIDE solutions, the time needed for the POD solutions and the time for the basis computation, i.e. the solution of

Table 2 Comparison between TRPOD, Multi-level TRPOD and a quasi-Newton algorithm

	Evaluations		Timing (sec.)				Optimal values	
Algorithm	FE	POD	Total	FE	POD	Basis	$f(u_{opt})$	$\|\|\nabla f(u_{opt})\|\|_2$
TRPOD	10	600	86	30	45	2	1.18e−5	1.24e−4
ML TRPOD	18	768	59	21	29	1	1.19e−5	2.42e−4
quasi-Newton	274	–	888	809	–	–	9.62e−6	8.29e−4

the eigenvalue problem. As expected, implementing a multi-level strategy yields a further reduction of the computing time. Although we need more iterations in this case, the main time saving is observable in the solution of the full PIDE.

Conclusion

In this work we deal with reduced order models based on proper orthogonal decomposition and their application to a PIDE constrained optimization problem arising in finance. After space and time discretization of the differential equation constraint, the dense linear systems of equations are solved by an implicit time stepping scheme using a preconditioned GMRES algorithm for each time step. However, when an optimization algorithm is applied, the repeated solution of the PIDE is still expensive. Since the PIDE is of parabolic type, POD is well suited to create a reduced order model. Theoretically, we show error estimates for the approximation quality of the reduced order models based on the work of [14]. All estimates only hold true for unchanged parameters, but a globalization has been achieved by embedding this in a trust-region framework as proposed by [5], for which we have shown convergence. The theoretical results are not only valid for the calibration problem considered here and the algorithm can be applied to a wide class of optimization problems with general parabolic constraints.

References

1. Y. Achdou, An inverse problem for a parabolic variational inequality with an integro-differential operator. SIAM J. Control Optim. **47.2**, 733–767 (2008)
2. L. Andersen, R. Brotherton-Ratcliffe, The equity option volatility smile: an implicit finite-difference approach. J. Comput. Finance **1.2**, 5–37 (1998)
3. A. Antoulas, *Approximation of Large-Scale Dynamical Systems*. Advances in Design and Control (SIAM, Philadelphia, 2005)
4. A. Antoulas, D. Sorensen, S. Gugercin, A survey of model reduction methods for large-scale systems. Contemp. Math. **280**, 193–219 (2001)
5. E. Arian, M. Fahl, E.W. Sachs, Trust-region proper orthogonal decomposition for flow control. ICASE Report 2000–25, ICASE, NASA Langley Research Center

6. N. Armstrong, K. Painter, J. Sherratt, A continuum approach to modelling cell–cell adhesion. J. Theor. Biol. **243.1**, 98–113 (2006)
7. R.G. Carter, On the global convergence of trust region algorithms using inexact gradient information. SIAM J. Numer. Anal. **28.1**, 251–265 (1991)
8. R. Cont, N. Lantos, O. Pironneau, A reduced basis for option pricing. SIAM J. Financ. Math. **2.1**, 287–316 (2011)
9. R. Cont, P. Tankov, *Financial Modelling with Jump Processes* (Chapman and Hall, London, 2004)
10. R. Dautray, J.-L. Lions, *Mathematical Analysis and Numerical Methods for Science and Technology, Vol.5: Evolution Problems I* (Springer, Berlin/New York, 1992)
11. M. Fahl, trust-region methods for flow control based on reduced order modelling, PhD thesis, University of Trier, 2000
12. A. Gerisch, On the approximation and efficient evaluation of integral terms in PDE models of cell adhesion. J. Numer. Anal. **30**, 173–194 (2010)
13. B. Kragel, Streamline diffusion POD models in optimization, PhD thesis, University of Trier, 2005
14. K. Kunisch, S. Volkwein, Galerkin proper orthogonal decomposition methods for parabolic problems. Numerische Mathematik **90.1**, 117–148 (2001)
15. O. Pironneau, Dupire-like identities for complex options. Comptes Rendus Mathematique **344.2**, 127–133 (2007)
16. O. Pironneau, Calibration of options on a reduced basis. J. Comput. Appl. Math. **232**, 139–147 (2009)
17. E.W. Sachs, M. Schneider, *Reduced Order Models for the Implied Variance under Local Volatility*, International Journal of Theoretical and Applied Finance, (in press)
18. E.W. Sachs, M. Schu, Reduced order models (POD) for calibration problems in finance, in *Numerical Mathematics and Advanced Applications*, ed. by K. Kunisch, G. Of, O. Steinbach (Springer, Berlin/Heidelberg, 2008), pp. 735–742
19. E.W. Sachs, M. Schu, Reduced order models in PIDE constrained optimization. Control Cybern. **39.3**, 661–675 (2010)
20. E.W. Sachs, M. Schu, A priori error estimates for reduced order models in finance. ESAIM: Math. Model. Numer. Anal. **47.2**, 449–469 (2011)
21. E.W. Sachs, M. Schu, Gradient computation for model calibration with pointwise observations, in *Control and Optimization with PDE Constraints*, ed. by K. Bredies, C. Clason, K. Kunisch, G. von Winckel (Springer, Basel, 2013), pp. 117–136
22. E.W. Sachs, A. Strauss, Efficient solution of a partial integro-differential equation in finance. Appl. Numer. Math. **58**, 1687–703 (2008)
23. M. Schu, Adaptive trust-region POD methods and their application in finance, PhD thesis, University of Trier, 2012
24. S.A. Silling, Reformulation of elasticity theory for discontinuities and long-range forces. J. Mech. Phys. Solids **48**, 175–209 (2000)
25. P. Toint, Global convergence of a class of trust region methods for nonconvex minimization in Hilbert spaces . IMA J. Numer. Anal. **8.2**, 231–252 (1988)

Part IV
Discretization: Concepts and Analysis

Introduction to Part IV Discretization: Concepts and Analysis

Michael Hinze

This chapter summarizes recent trends and addresses future research directions in the field of discrete concepts for PDE constrained optimization with elliptic and parabolic PDEs in the presence of pointwise constraints. It covers the range from tailored discrete concepts over adaptive a posteriori finite element approaches, to the modern algorithmical treatment of challenging optimal control applications with fluid flows.

Malte Braack, Markus Klein, Andreas Prohl and Benjamin Tews consider an optimal control problem with respect to the two-phase Navier–Stokes equations. They present different numerical schemes, in particular a level-set method, as well as an approach based an Allen-Cahn phase field model. They also consider a geometrical approach to treat the interface and address the question of convergence of numerical schemes.

Klaus Deckelnick and Michael Hinze consider the finite element approximation of an elliptic optimal control problem with pointwise bounds on the gradient of the state. They review recent results on the error analysis for various discretization approaches and prove a new bound for the problem without control constraints.

Michael Hinze, Michael Köster, and Stefan Turek present a Newton-type solver strategy for optimal flow control problems using space-time multigrid solution techniques. Based on the standard Newton approach for optimal control, they derive a space-time multigrid preconditioner, which is analyzed numerically for distributed and boundary control problems.

Kristina Kohls, Arnd Rösch and Kunibert G. Siebert summarize their findings in the analysis of adaptive finite element methods for the efficient discretization of control constrained optimal control problems. They focus on convergence of

M. Hinze (✉)
Fachbereich Mathematik Optimierung und Approximation, Universität Hamburg, Hamburg, Germany
e-mail: michael.hinze@uni-hamburg.de

G. Leugering et al. (eds.), *Trends in PDE Constrained Optimization*, International Series of Numerical Mathematics 165, DOI 10.1007/978-3-319-05083-6_21

the adaptive method, i.e., show that the sequence of adaptively generated discrete solutions converges to the true solution. At hand of a simple model problem they highlight the key components of the convergence proof and comment on generalizations of the presented result.

Thomas G. Flaig, Dominik Meidner and Boris Vexler in their paper transfer the a priori error analysis for the discretization of parabolic optimal control problems on domains allowing for H^2 regularity to a large class of nonsmooth domains. There estimation technique combines two ingredients which are used to prove the optimal convergence rates with respect to the spatial and the temporal discretization; a time discretization scheme which has the desired convergence rate in the smooth case, and a method to treat the singularities due to non-smoothness of the domain for the corresponding elliptic state equation. They demonstrate the approach with a Crank-Nicolson time discretization scheme combined with finite elements on graded meshes for the spatial discretization.

The Editor of this chapter wishes to thank all authors who contributed to this volume, and also all involved referees, whose notes and comments were very valuable in preparing this chapter.

Optimal Control for Two-Phase Flows

Malte Braack, Markus Klein, Andreas Prohl, and Benjamin Tews

Abstract We consider an optimal control problem with respect to the two-phase Navier–Stokes equations. Different numerical schemes are presented, in particular a level-set method, as well as an approach based an Allen-Cahn phase field model. We also consider a geometrical approach to treat the interface and address the question of convergence of a numerical scheme.

Keywords Two-phase • Optimal control • Stabilized finite elements • Level set • Allen-Cahn

Mathematics Subject Classification (2010). 49Q10, 76D55, 76M10, 76T05, 93C20.

1 Introduction

Multi-phase fluid dynamical problems arise in many industrial, technical and biological applications. A prototype example is the control of the interface in aluminum production, cf. [14]. The simulation of such processes becomes more important in recent years. The typical difficulties of two-phase (or multi-phase) flows are the treatment of the interface of the different fluids, because the interface should remain sufficiently sharp in order to distinguish between the two phases and to formulate surface tension forces properly. Therefore, appropriate numerical schemes are very demanding. The situation becomes even more complicated in the

M. Braack (✉) • B. Tews
Mathematical Seminar, Christian-Albrechts-University of Kiel, Ludewig-Meyn-Str. 4, 24098 Kiel, Germany
e-mail: braack@math.uni-kiel.de; tews@math.uni-kiel.de

M. Klein • A. Prohl
Mathematical Institute, Tübingen University, Auf der Morgenstelle 10, 72076 Tübingen, Germany
e-mail: klein@na.uni-tuebingen.de; prohl@na.uni-tuebingen.de

G. Leugering et al. (eds.), *Trends in PDE Constrained Optimization*, International Series of Numerical Mathematics 165, DOI 10.1007/978-3-319-05083-6_22

context of optimal control. For example, boundary control can be used in order to separate the different fluid phases in a controlled manner, i.e., to track the interface into a desired position.

In this work, we report on different numerical schemes for such optimal control processes. In particular, we present the level-set technique in the context of optimal control, an Allen-Cahn phase field model and a phase field model with a geometric interface functional. For the later one we present an analytical convergence result for spatio-temporal mesh sizes tending to zero.

1.1 Governing Equations

The domain $\Omega \subset \mathbb{R}^d$, $d \in \{2,3\}$, can be split into two time-depending subregions $\overline{\Omega} = \overline{\Omega_1(t) \dot{\cup} \Omega_2(t)}$. The interface between the two fluids is denoted by $\Gamma(t) := \overline{\Omega_1(t)} \cap \overline{\Omega_2(t)}$. The governing equations are the incompressible Navier-Stokes equations with variable density. By $\boldsymbol{v}$ we denote the velocity field, by ρ the density of the fluid. At initial time $t = 0$ the (piecewise-constant) density and the velocity are given by:

$$\rho|_{t=0} := \rho_i \text{ on } \Omega_i(0), \tag{1.1}$$

$$\boldsymbol{v}|_{t=0} := \boldsymbol{v}_0, \tag{1.2}$$

with different densities $0 < \rho_1 < \rho_2$. The time evolution of the density is described by the advection equation

$$\partial_t \rho + (\boldsymbol{v} \cdot \nabla)\rho = 0 \qquad \text{in } \Omega_T \tag{1.3}$$

on the space-time cylinder

$$\Omega_T := \{(x,t) \mid t \in I,\ x \in \Omega_1(t) \cup \Omega_2(t)\},$$

with time interval $I = [0,T]$. Due to the absence of diffusive terms in (1.3), the density remains discontinuous across the interface $\Gamma(t)$. The splitting of the domain is given implicitly by

$$\Omega_i(t) := \{\boldsymbol{x} \in \Omega \mid \rho(\boldsymbol{x},t) = \rho_i\}.$$

The momentum equation is given by

$$\rho\partial_t \boldsymbol{v} + \rho(\boldsymbol{v} \cdot \nabla)\boldsymbol{v} - \mu\Delta\, \boldsymbol{v} + \nabla p = \rho \mathbf{g} \qquad \text{in } \Omega_T, \tag{1.4}$$

where μ is the viscosity, $\mathbf{g}$ external (gravitational) forces, and p the pressure. The pressure is necessary in order to inforce the incompressibility of the fluid:

$$\operatorname{div} \boldsymbol{v} = 0 \qquad \text{in } \Omega_T . \tag{1.5}$$

The system (1.3)–(1.5) is complemented by an appropriate boundary condition on $\partial\Omega$ e.g.

$$\boldsymbol{v} = \boldsymbol{v}_D \qquad \text{on } \Gamma_D, \tag{1.6}$$

$$\mu \frac{\partial \boldsymbol{v}}{\partial \boldsymbol{n}} - p\boldsymbol{n} = \boldsymbol{0} \qquad \text{on } \Gamma_N, \tag{1.7}$$

on a Dirichlet part $\Gamma_D \subseteq \partial\Omega$ and a natural outflow part $\Gamma_N \subseteq \partial\Omega$ of the boundary. On the boundary part $\Gamma_C \subseteq \partial\Omega$ a (finite dimensional) boundary control $\boldsymbol{u}$ is applied:

$$\boldsymbol{v} = \boldsymbol{B}\boldsymbol{u} \qquad \text{on } I \times \Gamma_C, \tag{1.8}$$

with a linear and continuous operator $\boldsymbol{B} : \mathbb{R}^m \to L^2(\Gamma_C)^d$. These three boundary parts should partition the entire boundary, $\partial\Omega = \Gamma_D \dot{\cup} \Gamma_N \dot{\cup} \Gamma_C$. Furthermore, conditions on the interface $\Gamma(t)$ are needed. The velocity is assumed to be continuous and the surface of the normal stresses is balanced by surface tension forces

$$[\boldsymbol{v}] = \boldsymbol{0} \qquad \text{on } \Gamma(t), \tag{1.9}$$

$$[\mu \nabla v - p] \cdot \boldsymbol{n} = \gamma \mathcal{K} \cdot \boldsymbol{n} \qquad \text{on } \Gamma(t) . \tag{1.10}$$

Here, $[\cdot]$ denotes the jump across the interface in direction of $\boldsymbol{n}$, $\mathcal{K} = -\operatorname{div} \boldsymbol{n}$ describes the mean curvature of the interface and $\gamma \geq 0$ is a given surface tension coefficient.

In order to formulate an optimal control problem for such two-phase flows we use the notation $\boldsymbol{y} = (v, p, \rho)$ for the state variable. The continuous version of the optimal control problem is given by

$$J(\boldsymbol{y}, \boldsymbol{u}) \to \min ! \tag{1.11}$$

subject to the state equation system (1.3)–(1.5), the initial conditions (1.1)–(1.2), boundary conditions (1.6)–(1.7) and interface conditions (1.9)–(1.10). In our numerical test cases, the set $\boldsymbol{Q}$ of admissible controls will be a finite dimensional subspace of $L^2(\Omega)^d$. In practise, such control may result e.g. from an exterior magnetic field. A reasonable functional J is, e.g., of end time control with respect to the density and additional $L^2(0, T; L^2(\Omega))$ regularization of the control:

$$J(\boldsymbol{y}, \boldsymbol{u}) := \frac{1}{2} \|\rho(\cdot, T) - \rho_d\|_{\Omega}^2 + \frac{\alpha}{2} \int_0^T \|\boldsymbol{u}(\cdot, t)\|^2 dt,$$

with a positive parameter $\alpha > 0$. In the following sections we introduce different numerical methods to treat the interface and report on the corresponding numerical results. In particular, Sect. 2 addresses the level-set method in the context of optimal control and Sect. 3 the Allen-Cahn phase field approach. Finally, in Sect. 4, a geometric functional is considered which implicitly includes the interface length. These approaches exhibit different advantages and disadvantages. In particular, the issue of reinitilization in the level-set approach is delicate in the context of optimal control, and the Allen-Cahn model is not mass conservative.

2 Level Set Formulation

The level set method is one of the most established methods to capture evolution in two-phase flows. We shortly introduce the main concept of level sets and address this method in the context of optimal control. In particular, we formulate the first order necessary condition. In the numerical example we report on the effect of reinitialization which is applied to the level set function.

The main idea of the level set formulation, firstly introduced by Osher and Sethian [18] and later extended to incompressible two-phase flows by Sussman, Smereka and Osher [19], is the embedding of the interface as the zero level set of another state variable ϕ. This continuous function is a signed distance function to the interface. At initial time ϕ is set to

$$\phi(\boldsymbol{x}, 0) := \pm \text{dist}(\boldsymbol{x}, \Gamma(0)) .$$

The sign of ϕ indicates whether $\boldsymbol{x}$ belongs to $\Omega_1(0)$ or $\Omega_2(0)$. At initial time it obviously holds $|\nabla\phi(\boldsymbol{x}, 0)| = 1$. For $t \geq 0$ the interface is defined by

$$\Gamma(t) := \{\mathbf{x} \in \Omega \; : \; \phi(\mathbf{x}, t) = 0\} .$$

Since the interface moves with the fluid particles, the governing equation for the level set function is given by

$$\partial_t \phi + (\boldsymbol{v} \cdot \nabla)\phi = 0 . \tag{2.1}$$

Although the interface is only given implicitly, the normal vector of the interface can be expressed directly in terms of ϕ:

$$\boldsymbol{n} = \nabla\phi / |\nabla\phi|,$$

and therefore the curvature $\mathcal{K} = -\text{div}\, \boldsymbol{n}$ can be written in terms of derivatives of ϕ. The density can be expressed by the Heaviside function $H(\phi)$, so that the interface remains sharp. However, for optimization problems we need derivatives with respect to the density. Hence, it is numerically reasonable to use a smooth density with the help of a regularized Heaviside function H_ε:

$$H_\epsilon(\phi) := \begin{cases} 0 & \text{if } \phi < -\varepsilon, \\ 0.5(1 + \phi/\varepsilon + \pi^{-1}\sin(\pi\phi/\varepsilon)) & \text{if } -\epsilon \leq \phi \leq \varepsilon, \\ 1 & \text{if } \phi > \varepsilon, \end{cases}$$

with small parameter $\varepsilon > 0$. The smoothened density is obtained by (see [20])

$$\rho(\phi) = \rho_2 + (\rho_1 - \rho_2)H_\varepsilon(\phi)\,. \tag{2.2}$$

The interface thickness is approximately $\sim 2\varepsilon/|\nabla\phi|$. Hence, as long as $|\nabla\phi|$ remains constant, the interface thickness is almost constant, too. However, $|\nabla\phi|$ will in general not remain constant for $t > 0$. For large times this distortion will give a non-uniform thickness of the interface. In [20] a procedure, (called 'renormalization') is proposed to maintain $|\nabla\phi| = 1$. To this end, in each time step an additional nonstationary nonlinear partial differential equation is solved to steady state. However, such a renormalization is delicate in the context of gradient based algorithms to solve optimal control problems. This is due to the fact that such a procedure does not correspond to a Galerkin formulation of the state equation. Therefore, the discrete gradient of the cost functional is perturbed and may cause slow convergence, or even divergence of the optimization algorithm. We will report on this effect in the next section.

It remains to reformulate the surface tension effects, because the interface is only implicitly given. According to [13], the integral of $\mathcal{K}\cdot\boldsymbol{n}$ along Γ can be expressed in terms of a volume integral including the Dirac distribution δ. In the discrete setting, the Dirac distribution is replaced by a regularized version $\delta_\varepsilon = H'_\varepsilon$. We arrive at the following approximation of the momentum equation (1.4) including surface tension:

$$\rho\partial_t\boldsymbol{v} + \rho(\boldsymbol{v}\cdot\nabla)\boldsymbol{v} - \mu\Delta\,\boldsymbol{v} + \nabla p - \gamma\mathcal{K}(\phi)\delta_\varepsilon(\phi)\nabla\phi = \rho\mathbf{g}, \tag{2.3}$$

valid in Ω_T. In summary, the corresponding set of equations consists of (1.5)–(1.8), (2.1)–(2.3).

2.1 Variational Formulation for the Level Set Approach

In this section we express the variational formulation of the state equation [21]. By $(\cdot,\cdot)$ we denote the standard $L^2(\Omega)$ scalar product. For the weak formulation of the equations we consider homogeneous Dirichlet conditions on $\Gamma_D \subseteq \partial\Omega\setminus\Gamma_C$, and boundary control on Γ_C. We introduce the following Hilbert spaces with respect to Ω:

$$V := \{\boldsymbol{v}\in H^1(\Omega)^d : \boldsymbol{v} = \mathbf{0} \text{ on } \Gamma_D\}\,.$$

For the formulation in space-time we define:

$$
\begin{aligned}
\boldsymbol{X}^{v} &:= \{\boldsymbol{v} \in L^2(I, \boldsymbol{V}) : \partial_t \boldsymbol{v} \in L^2(I, \boldsymbol{V}^*)\}, \\
X^{\phi} &:= \{\phi \in L^2(I, H^3(\Omega)) : \partial_t \phi \in L^2(I, H^3(\Omega)^*)\}, \\
X^{p} &:= L^2(I, L_0^2(\Omega))\,.
\end{aligned}
$$

The state $\boldsymbol{y} = (\boldsymbol{v}, \phi, p)$ will be sought in the space $\boldsymbol{Y} := \boldsymbol{X}^{v} \times X^{\phi} \times X^{p}$. The higher spatial regularity for ϕ is necessary to ensure $\mathcal{K}(\phi) \in L^2(I, H^1(\Omega))$. For the test functions the corresponding space is $(\boldsymbol{w}, \tau, q) \in \boldsymbol{Y}$. The control space is a subspace of $L^2(I, \mathbb{R}^m)$: $\boldsymbol{u} \in \boldsymbol{U} \subseteq L^2(I, \mathbb{R}^m)$. Furthermore, we use the following semi-linear form and bilinear form, resp.:

$$
\begin{aligned}
A(\boldsymbol{y}; \boldsymbol{w}, \tau, \chi) &:= (\rho(\boldsymbol{v} \cdot \nabla)\boldsymbol{v} - \rho \mathbf{g}, \boldsymbol{w}) + (\mu \nabla \boldsymbol{v}, \nabla \boldsymbol{w}) - (p, \operatorname{div} \boldsymbol{w}) \\
&\quad + (\operatorname{div} \boldsymbol{v}, \chi) - (\gamma \mathcal{K}(\phi) \delta_\varepsilon(\phi) \nabla \phi, \boldsymbol{w}) + ((\boldsymbol{v} \cdot \nabla)\phi, \tau), \\
B(\boldsymbol{y}, \boldsymbol{u}; \boldsymbol{w}) &:= \int_{\Gamma_C} (\boldsymbol{v} - \boldsymbol{B}\boldsymbol{u}) \boldsymbol{w} \, ds\,.
\end{aligned}
$$

The primal problem reads in variational formulation for given $\boldsymbol{u} \in \boldsymbol{U}$: Seek $\boldsymbol{y} \in \boldsymbol{Y}$ s.t. $\boldsymbol{v}|_{t=0} = \boldsymbol{v}_0$ and

$$
\int_0^T \{\langle \rho \partial_t \boldsymbol{v}, \boldsymbol{w} \rangle + \langle \partial_t \phi, \tau \rangle + A(\boldsymbol{y}; \boldsymbol{w}, \tau, \chi) + B(\boldsymbol{y}, \boldsymbol{u}; \boldsymbol{w})\} \, dt = 0 \tag{2.4}
$$

for all $(\boldsymbol{w}, \tau, q) \in \boldsymbol{Y}$. The term $\langle \rho \partial_t \boldsymbol{v}, \boldsymbol{w} \rangle$ is well-defined in two dimensions, $d = 2$, if $\rho \in L^\infty(I, H^1(\Omega))$ due to the Sobolev embedding. Since the density is determined by (2.2) we have to ensure that $\rho(\phi(t)) \in H^1(\Omega)$ for a.e. $t \in I$.

2.2 First Order Optimality System for Level Set

Denoting the adjoint variable by $\boldsymbol{z} = (z_v, z_\phi, z_p) \in \boldsymbol{Y}$ and defining the Lagrange functional

$$
L(\boldsymbol{y}, \boldsymbol{u}, \boldsymbol{z}) := J(\boldsymbol{y}, \boldsymbol{u}) - \int_0^T \{\langle \rho \partial_t \boldsymbol{v}, z_v \rangle + \langle \partial_t \phi, z_\phi \rangle + A(\boldsymbol{y}; \boldsymbol{z}) + B(\boldsymbol{y}, \boldsymbol{u}; \boldsymbol{z})\} \, dt,
$$

we arrive at the following first order optimality system for the solution $\boldsymbol{y} \in \boldsymbol{Y}$ of the minimization problem (1.11) under the constraint (2.4):

$$
\partial_z L(\boldsymbol{y}, \boldsymbol{u}, \boldsymbol{z})(\boldsymbol{\psi}) = 0 \quad \forall \boldsymbol{\psi} \in \boldsymbol{Y}, \tag{2.5}
$$

$$
\partial_y L(\boldsymbol{y}, \boldsymbol{u}, \boldsymbol{z})(\boldsymbol{\varphi}) = 0 \quad \forall \boldsymbol{\varphi} \in \boldsymbol{Y}, \tag{2.6}
$$

$$
\partial_u L(\boldsymbol{y}, \boldsymbol{u}, \boldsymbol{z})(\lambda) = 0 \quad \forall \lambda \in \boldsymbol{U}. \tag{2.7}
$$

These optimality conditions are formally derived in [21]. These three equations consist of the *primal equation* (2.5), the *dual equation* (2.6), and the *gradient equation* (2.7). The issue of existence and uniqueness of solutions is still an open problem as it is for the simple forward Navier-Stokes system.

In order to maintain a uniform interface thickness the normalization $|\nabla\phi| \equiv 1$ is required. However, because this is not guaranteed yet, an additional reinitilization procedure was proposed in [19]. The reinitilization step consists of solving a further non-stationary, non-linear PDE after certain time steps. This procedure starts at the interface first and then moves outwards in normal direction. In our numerical examples it was not necessary to solve this equation to steady state. Usually the computation of a few time steps is sufficient in order to ensure that the level set is renormalized to a signed distance function in an ϵ-region around the interface. This numerical experience is also reported in [19]. A detailed description of this equation and its properties can be found in [21]. However, this reinitilization is not included in the optimality system (2.5)–(2.7), because it can not be formulated in variational form.

2.3 Space-Time Finite Element Formulation

For the finite-dimensional formulation, we use piecewise constant discontinuous finite elements in time (dG(0)) and piecewise biquadratic elements in space (cG(2)). The reformulation of (2.5)–(2.7) in these finite element spaces is straight-forward, except for the aspect of ensuring the inf-sup condition for the velocity-pressure coupling, and the stabilization of the convective terms. We use the local projection stabilization for both aspects, see [9]. Hence, the additional term added to the semi-linear form $A(\boldsymbol{y};\boldsymbol{z})$ reads:

$$\begin{aligned} S_h(\boldsymbol{y},\boldsymbol{z}) &= (\alpha_p \kappa_h \nabla p, \kappa_h \nabla z_p) + (\alpha_v \kappa_h (\boldsymbol{v}\cdot\nabla)\boldsymbol{v}, \kappa_h (\boldsymbol{v}\cdot\nabla) z_v) \\ &\quad + (\alpha_\phi \kappa_h (\boldsymbol{v}\cdot\nabla)\phi, \kappa_h (\boldsymbol{v}\cdot\nabla) z_\phi)\,. \end{aligned}$$

Here, $\kappa_h = id - \pi_h$ is a space fluctuation operator and π_h the local L^2-projection onto a space of patch-wise discontinuous functions. The stabilization parameters $\alpha_p, \alpha_v, \alpha_\phi$ are chosen to be mesh-dependent according to [9]. The quadratic order of the polynomials in space is necessary to capture the surface tension term $\mathcal{K}(\phi)$ in the momentum equation which contains second derivatives.

This stabilization technique is analyzed for the optimal control system with the linearized Navier-Stokes system (Oseen equation). In that case, the discrete optimality system is symmetric, so that *optimization* and *discretization* commute, see [10, 11].

As numerical solver we apply a Newton iteration to a reduced cost functional combined with an Armijo step length control. For details we refer to Becker et al. [8]. We use the finite element software packages GASCOIGNE3D [6] and RoDoBo [7].

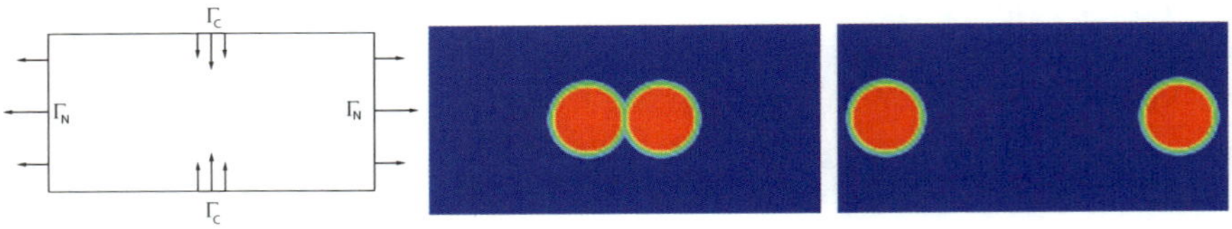

Fig. 1 Configuration of the 'kissing bubbles' (*left*), initial density $\rho|_{t=0}$ (*middle*), and desired density ρ_d

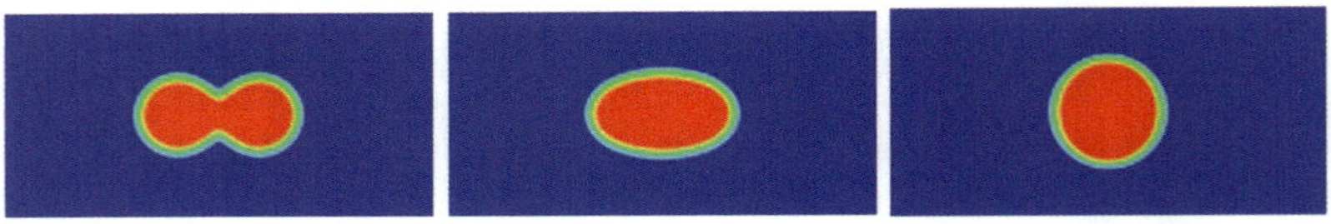

Fig. 2 Density solution without control, $u = 0$, at time steps $t = 0.1$, $t = 0.35$ and $t = 2$ (from left to right)

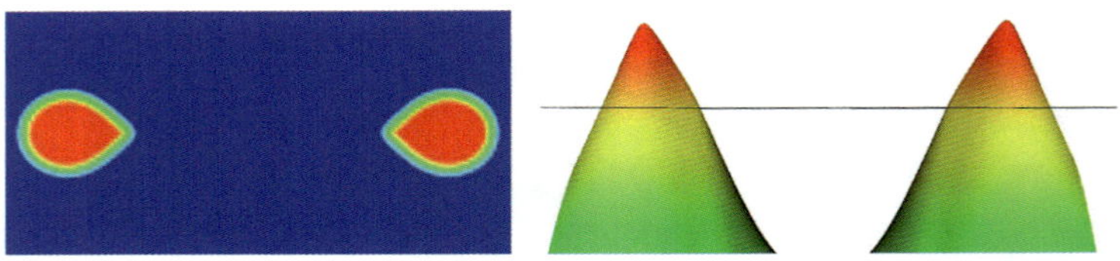

Fig. 3 Optimal density (*left*) and corresponding level set variable ϕ (*right*)

2.4 Numerical Results

In order to numerically investigate the dependency of Newton's method on the reinitialization procedure we consider a numerical example on a rectangular domain $\Omega = (0, 1) \times (0, 0.5)$; see Fig. 1. We apply a parameter control $u \in \mathbb{R}$ at the Dirichlet boundary part Γ_C for the second component of the velocities and set $v_2(x, t) = \pm u(7/16 - x)(9/16 - x)t$. At the boundary part Γ_N we impose natural outflow conditions. The densities are set to be $\rho^1 = 1$ (blue) as well as $\rho^2 = 2$ (red), and the initial distribution is given by two touching bubbles. The external force $\boldsymbol{g}$ is set to be zero, and the remaining parameter ε is set to be $\frac{1}{32}$. The optimization goal consists of separating these bubbles by tracking a desired density distribution.

The discretization parameters are given by the spatial mesh width $h = 1/64$ and the time step size is set to be 0.01. These values are sufficient to properly resolve the interface. Figure 2 shows snapshots of the density if the control is set to be zero which corresponds to homogeneous Dirichlet conditions at the upper and lower boundary part. As expected, the two bubbles merge to one big bubble due to the surface tension force. The interface width remains constant due to the reinitialization after each time step. In Fig. 3 we depicted the level set function (right) and the density distribution (left) with respect to the optimal control $u_{opt} \approx 239$. One can

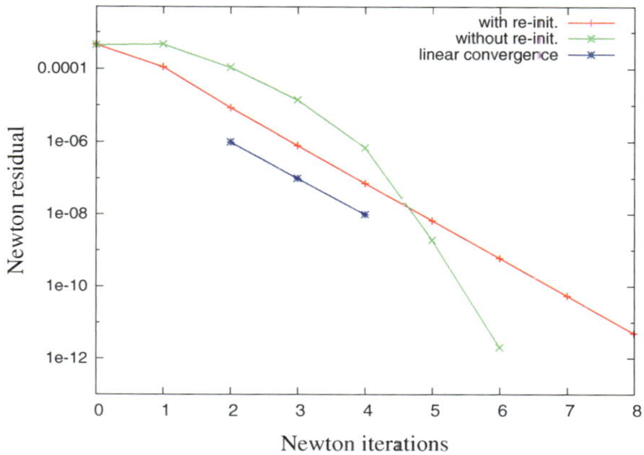

Fig. 4 Dependency of Newton's convergence on reinitialization steps

observe that the level set function equals the signed distance function nearby its zero level which causes a uniform interface thickness.

Figure 4 shows the dependency of the Newton residual corresponding to each Newton iteration on the number of reinitialization steps. Without applying these steps (green line), the Newton residual decreases with quadratic order close to the optimal point. In contrast to that, the convergence rate is reduced to only linear order if the level set function is renormalized after each time step (red line). It is also conceivably that, considering another example, the convergence order is further reduced or even that the Newton direction does not correspond to a decent direction any more. This major drawback of the level set method is due to the fact that the reinitilization step can not be expressed as a Galerkin formulation and, therefore, is not captured in the optimality system.

3 Allen-Cahn Phase Field Model

The basic idea of phase field models consists of the consideration that the two immiscible fluids do mix within a narrow interfacial region $\Gamma_\varepsilon(t) \subset \Omega$ of width $\varepsilon > 0$ around the interface $\Gamma(t)$, [12]. A phase field function $\psi : [0, T] \times \Omega \to [-1, 1]$ is introduced which can be interpreted as the volume fraction of the individual fluids. The zeros of ψ indicate the position of the interface, $\Gamma = \{\boldsymbol{x} \in \Omega \mid \psi(\boldsymbol{x}) = 0\}$. Since the interface is not given explicitly, the interface condition (1.10) is replaced by an additional volume force in the momentum equation. To be more specific, the surface tension effect is modelled by the term $\gamma_{pf} \mathrm{div}\,(\nabla\psi \otimes \nabla\psi)$. Here, the parameter γ_{pf} is proportional to the surface tension coefficient γ in (1.10), that is, $\gamma \sim \gamma_{pf}/\varepsilon$. An incompressible two-phase fluid with Allen-Cahn model, according to [16], extended to variable density can be modeled by the following system of equations:

$$\rho\partial_t \mathbf{v} + \rho(\mathbf{v}\cdot\nabla)\mathbf{v} - \mu\Delta\,\mathbf{v} + \nabla p + \gamma_{pf}\mathrm{div}\,(\nabla\psi\otimes\nabla\psi) = \rho\mathbf{g},$$
$$\mathrm{div}\,\mathbf{v} = 0,$$
$$\partial_t\psi + (\mathbf{v}\cdot\nabla)\psi - \mu_{pf}\Delta\psi + \frac{\mu_{pf}}{\varepsilon^2}F'(\psi) = 0\,.$$

The boundary condition for ψ consists of an homogeneous Neumann condition on the entire boundary:

$$\nabla\psi\cdot\boldsymbol{n} = 0 \quad \text{on } \partial\Omega\,. \tag{3.1}$$

The Allen-Cahn model uses the Ginzburg-Landau bulk energy $F(\psi) = \frac{1}{4}(\psi^2-1)^2$. The density is expressed in terms of ψ:

$$\rho(\psi) = \frac{1}{2}((1+\psi)\rho_1 + (1-\psi)\rho_2)\,. \tag{3.2}$$

The semi-linear form for the Allen-Cahn model is

$$\begin{aligned} A(\boldsymbol{y};\boldsymbol{w},\tau,\chi) := {} & (\rho(\boldsymbol{v}\cdot\nabla)\boldsymbol{v} - \rho\mathbf{g},\boldsymbol{w}) + (\mu\nabla\,\boldsymbol{v},\nabla\boldsymbol{w}) - (p,\mathrm{div}\,\boldsymbol{w}) \\ & +(\mathrm{div}\,\boldsymbol{v},\chi) + (\gamma_{pf}\nabla\psi\otimes\nabla\psi,\nabla\boldsymbol{w}) + ((\boldsymbol{v}\cdot\nabla)\psi,\tau) \\ & +(\mu_{pf}\nabla\psi,\nabla\tau) + (\mu_{pf}\varepsilon^{-2}(\psi^3-\psi),\tau)\,. \end{aligned}$$

The density is treated as a coefficient determined by (3.2). One advantage of this Allen-Cahn model is that the solution satisfies the maximum principle as mentioned in [16]. The optimality system for the optimal control problem constraint by this two-phase Allen-Cahn model is formally of the same type as (2.5)–(2.7).

However, even with boundary conditions $\boldsymbol{v} = \mathbf{0}$ on $\partial\Omega$ and (3.1) the Allen-Cahn model is not mass preserving. This can be seen by taking the test function $\tau \equiv 1$ and integration by parts. Since the convective term and the diffusive term vanish, we obtain

$$\boldsymbol{v}|_{\partial\Omega} = 0 \quad\Longrightarrow\quad \frac{\partial}{\partial t}\int_\Omega \psi\,dx = \int_\Omega \frac{\mu_{pf}}{\varepsilon^2}\psi(1-\psi^2)\,dx\,.$$

The expression on the right hand side is in general not zero. In the case of the natural outflow condition (1.7), an additional term is obtained. In order to reduce this mass error we may use local mesh refinement and choose the constant μ_{pf} to be proportional to the local mesh size h_K. Since the interface moves as the time increases, we have to allow the locally refined mesh to follow the interface propagation in time. For this purpose, we apply the goal-oriented a posteriori error estimator developed by Meidner and Vexler [17] which assesses the discretization error with respect to an arbitrary functional j. The adaptation process identifies adaptive time steps and local mesh sizes.

Since there do not occur second derivatives in the variational formulation, (bi-)linear elements can be used for the spatial discretization. We use the same (adaptive) time step and the same spatial mesh for the discretization of the forward problem and for the backward problem.

3.1 Numerical Results

In the following numerical example we are interested in the behavior of the error estimator as well as the reduction of the mass error.

The interface thickness is known to be proportional to ε. In order to resolve the interface properly, the interface thickness should be in the range of the local (in the vicinity of the interface) mesh size. Therefore, it is reasonable to choose the parameter ε mesh-size dependent, $\varepsilon|_T = \alpha h_T$, for each cell T of the triangulation. By a previous coarse grid computation we obtain with $\alpha = 0.64$ a properly resolved interface, so that we identify this value as a reasonable proportionality factor. The surface tension coefficient γ of (2.3) is equal to the ratio γ_{pf}/ε. Therefore, a fixed surface tension coefficient requires to choose γ_{pf} mesh-size dependent as well. Hence, we obtain $\gamma_{pf}|_T = \gamma \alpha h_T = 3.2 \cdot 10^{-3} h_T$. For numerical reasons (sufficient diffusion), the parameter μ_{pf} is choosen as $\mu_{pf} = 10\,\gamma_{pf}$. Since we are interested in minimizing the error with respect to the density, we choose the mean of the density, that is, $j(\psi) := \frac{1}{|\Omega_T|} \int_{\Omega_T} \rho(\psi)\, dx$, as target fuctional for mesh adaptation.

In Fig. 5 we depict the interface propagation at several time steps. The dynamically changing meshes follow the interface evolution and ensure a well-resolved

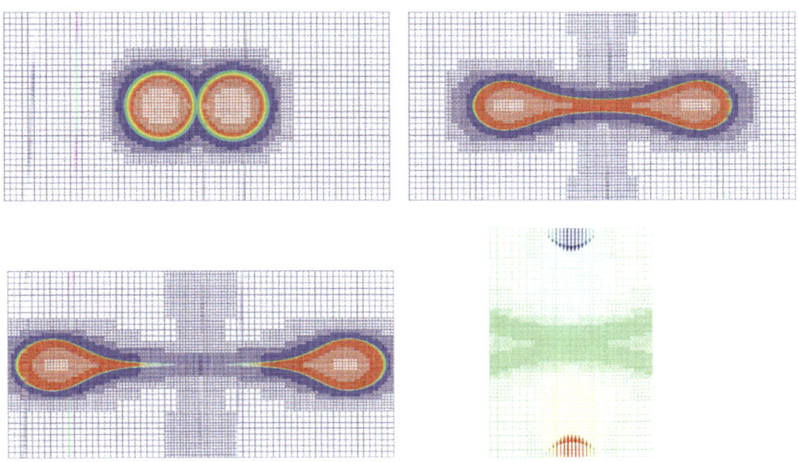

Fig. 5 Evolution of density distribution for the Allen-Cahn model with control at time $t = 0$ (*upper left*), $t = 1.48$ (*upper right*) and $t = 2$ (*lower left*). The optimal control at $t = 2$ is visualized in the *lower right* figure

Table 1 Obtained values by adaptive refinement of Allen-Cahn Navier-Stokes equations with optimal control

dofs	N_{max}	M	η_h	η_k	$\eta_h + \eta_k$	$j_{ex} - j(\rho_{kh})$	i_{eff}
109,395	2,145	50	8.63e−4	1.19e−3	2.06e−3	8.40e−3	4.08
248,073	5,673	52	1.75e−5	1.16e−3	1.18e−3	3.65e−3	3.09
1,148,253	17,121	96	−1.10e−5	5.13e−4	5.02e−4	1.42e−3	2.83
6,481,145	46,277	188	−9.58e−7	2.05e−4	2.04e−4	5.38e−4	2.64

Table 2 Obtained values of the optimal control u_{opt} and the corresponding functional value with the level set method and the Allen-Cahn model

	Level set	Allen-Cahn
u_{opt}	239.29	193.1
J_{opt}	5.12e-3	4.87e-2

interface thickness. Similar to the level set method, the control u is able to match the desired density ρ_d to a certain extent. However, the optimal solution of level set and Allen-Cahn shows differences.

Table 1 shows the behavior of the error estimator for different levels of discretization. N_{max} denotes the maximum of the spatial degrees of freedom over all time steps, M displays the number of time steps, η_k and η_h represent the estimated error in time and space, respectively. $j_{ex} = j(\rho_0)$ denotes the exact mass, that is, the domain integral of the density at initial time, and i_{eff} describes the effectivity index which is defined by the exact error divided by the estimated error. One can observe that this index remains nearly constant over all levels of mesh refinement and the mass error is reduced by the factor of approximately 16 from the coarsest to the finest discretization level. The corresponding ratio of mesh points is approximately 60.

In order to make a comparison of these results with the ones for the level-set method in Sect. 2, we list the obtained values for the optimal control u_{opt} and for the corresponding functional value J_{opt} in Table 2.

4 Phase Field Model with Geometric Interface Functional

In this section, we consider the modified geometric functional ($\epsilon > 0$),

$$\tilde{J}(\rho, \boldsymbol{u}) := J(\rho, \boldsymbol{u}) + \beta \int_{\Omega_T} \left\{ \epsilon |\nabla \rho|^2 + \frac{1}{4\epsilon}(\rho - \rho_1)^2 (\rho - \rho_2)^2 \right\} d\boldsymbol{x}dt$$

where the functional J is enriched by the β term, which is the approximation for the length of the surface interface Γ by means of a phase-field formulation (cf. [1,3] and the references therein). This geometric part of the function helps to avoid oscillatory

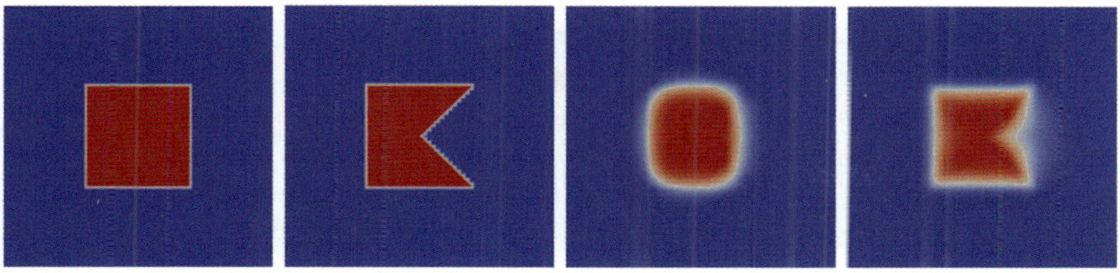

Fig. 6 Desired state ρ_d, initial condition, density at $t = T$ for $\beta \neq 0$, density at $t = T$ for $\beta = 0$ (from left to right)

effects in the interface. In Fig. 6, it is shown that $\beta = 0$ and $\beta > 0$ lead to different dynamics.

We neglect the surface tension given in (1.9)–(1.10), set $\mathbf{g} \equiv \mathbf{0}$ for simplicity and study the optimal control problem of a two-phase fluid with distributed control $\mathbf{u}$ together with homogeneous Dirichlet boundary conditions for $\mathbf{v}$. A distributed control is a mathematical idealization of a large amount of very small control devices distributed in a fine grid over the domain Ω. For instance, this is relevant in semiconductor industry, e.g. [2]. In order to have $\tilde{J}$ well-defined, and to get regular Lagrange multipliers related to the mass equation (1.3), we introduce an artificial diffusion term $-\delta\Delta\rho$. The problem of this section then reads as follows:

Minimize $\tilde{J}$ subject to

$$\frac{1}{2}(\rho\partial_t \boldsymbol{v} + \partial_t(\rho\boldsymbol{v}) - \rho(\boldsymbol{v}\cdot\nabla)\boldsymbol{v} + \operatorname{div}(\rho\boldsymbol{v}\otimes\boldsymbol{v})) - \operatorname{div}(\mu\nabla\boldsymbol{v}) + \nabla p = \rho\boldsymbol{u},$$

$$\operatorname{div}\boldsymbol{v} = 0,$$

$$\partial_t\rho + (\boldsymbol{v}\cdot\nabla)\rho - \delta\Delta\rho = 0,$$

together with boundary conditions $\boldsymbol{v} = 0$ and $\partial_{\boldsymbol{n}}\rho = 0$ on $\partial\Omega$. This multi-parameter approach accomplishes the following goals: The objective functional $\tilde{J}$ is forcing the density to approximate a given shape ρ_d, and also to minimize the perimeter of involved interfaces. In most studied situations, the fluids behave as if surface tension is present, see e.g. Fig. 6. By the parameter $\delta > 0$ in the equation, we gain regularity for solutions of the equation by making jumps in the density diffusive (this was also done in order to show existence of the equation, cf. [15]). This helps to derive optimality conditions in a rigorous way, and it helps to have the objective functional $\tilde{J}$ well-defined. We note that there is a motivation to have a proper limit function for either $\delta \to 0$ or $\epsilon \to 0$, but it is not clear what the limit would be if both parameters tend to zero simultaneously. Based on a priori estimates, a necessary condition for any convergence result is $\delta = \mathcal{O}(\epsilon)$. Experiments also indicate this relation, cf. [3].

In [3], it is shown that the above optimization problem has at least one solution, and necessary optimality conditions are derived rigorously for $\delta, \epsilon > 0$. A key ingredient for the derivation of optimality conditions are a priori estimates for the states $\boldsymbol{v}$ and ρ, both relying on the artificial diffusion.

4.1 Discrete Optimization Problem

In this section, we focus on the corresponding discrete problem to $\tilde{J}$ for $\delta, \epsilon > 0$ and the numerical analysis for positive spatial mesh size h and time step k, which will later tend to zero. Let $\mathcal{T}_h$ be a quasi-uniform triangulation of Ω with $h := \max_{T \in \mathcal{T}_h} \operatorname{diam} T$. We introduce the following spaces.

- R_h for the approximation of the density ρ, which consists of globally continuous, piecewise linear standard finite element functions.
- V_h and M_h as an inf-sup stable conforming pair (e.g. Taylor–Hood or MINI Elements) for velocity $\boldsymbol{v}$ and pressure p, involving zero Dirichlet boundary conditions for $\boldsymbol{v}$.

We use a semi-implicit Euler scheme with time step size $k > 0$, which is combined with a Galerkin-type scheme in space for the discretization of the governing equations, which is a modification of the scheme studied in [4]. The modified scheme is given in full detail in [3, Section 5]. We introduce a fully discrete version of the optimization problem with the geometric functional $\tilde{J}$, and it is easy to show that this finite-dimensional problem has at least one solution (since the fully discrete version of $\tilde{J}$ is a continuous function on some finite dimensional space and the discrete equation is solvable), as well as to derive necessary optimality conditions (by the finite dimensional Lagrange multiplier theorem).

In [3], it is shown that these discrete solutions of the fully discrete optimal control problem converge to some functions with respect to numerical parameters $h, k > 0$ (up to subsequences), and the limit functions solve the continuous optimality conditions. Moreover, it can be shown that the sequence of discrete optimal control convergences to a optimal control $\boldsymbol{u}$ of the continuous problem strongly in $L^2(0, T; L^2(\Omega))$.

In order to show this result, we proceed as follows.

1. Derive uniform bounds for the fully discrete scheme in standard parabolic norms as well as in higher norms (such that the solutions are indeed strong). By stability of the interpolation, all time-continuous interpolants inherit these bounds.
2. With these bounds, and a discrete version of Aubin–Lions' compactness theorem, it is possible to derive strong convergence for the affine and constant in time interpolants. This will lead to the convergence of the discrete scheme to the state equation (up to subsequences).
3. It remains to show that the adjoint variables are bounded in proper norms. Here, the strong coupling between the adjoint variables and the state variables, as well as the coupling between the two adjoint variables corresponding to the mass equation and the momentum equation, respectively, causes problems. We derive standard parabolic regularity properties for the discrete adjoint variables. Here, we need the regularity of the state variables and a combined argumentation: Since derivatives of both adjoint variables are present in both adjoint equations, we have to multiply the adjoint equations simultaneously with different test

functions and to consider a proper weighted sum of the resulting inequalities. At this point, it is necessary that the state variables have strong stability properties.
4. The bounds for all discrete variables are sufficient to pass to the limit in the optimality system, which proves the main convergence result.

4.2 Numerical Results

In the numerical experiments included here, we compare the behavior for $\beta = 0$ and $\beta \neq 0$, as well as the behavior of the β term for two different scenarios. A detailed explanation and many more experiments can be found in [3]. The numerical experiment was done with grid size $h = 1/64$ and time step size $k = 0.05$. In all experiments, the initial velocity is zero. The discrete optimality system is solved by a steepest descent algorithm using a Barzilai-Borwein step size for line search, cf. [5].

In Fig. 6, we can see a different evolution depending on β: If $\beta = 5 > 0$, the geometric properties of the desired state are reached. For $\beta = 0$, the L^2-distance is minimized and the shape of the desired state fits, while geometric properties (like convexity) are not the same like in the desired state.

For the next two experiments, we neglect the $\|\rho - \rho_d\|^2_{L^2(0,T;L^2(\Omega))}$-term in the functional $\tilde{J}$ and consider only the β-term. In the first experiment, the two disjoint circles are forced to join in order to reduce the β term, see Fig. 7. In the second experiment, there is a pinch-off of the connected region into two separated regions which become circular, see Figs. 8 and 9, respectively, for the optimal velocity. For comparison, we included the evolution for the non-controlled situation where the initial condition only becomes diffuse.

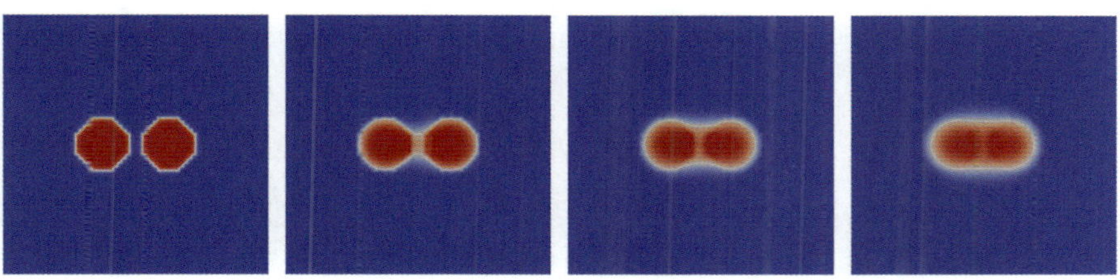

Fig. 7 Optimal solution at time $t = 0, 0.15, 0.5, 1$; only β term present

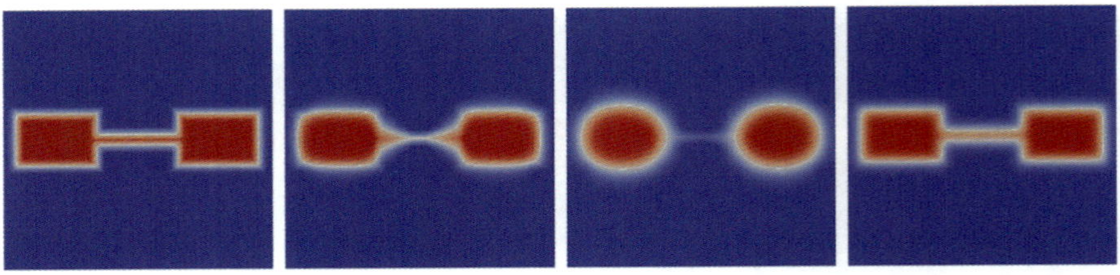

Fig. 8 Optimal solution at time $t = 0, 0.15, 0.5$, and the solution at $t = 0.5$ without control (from left to right); only β term present

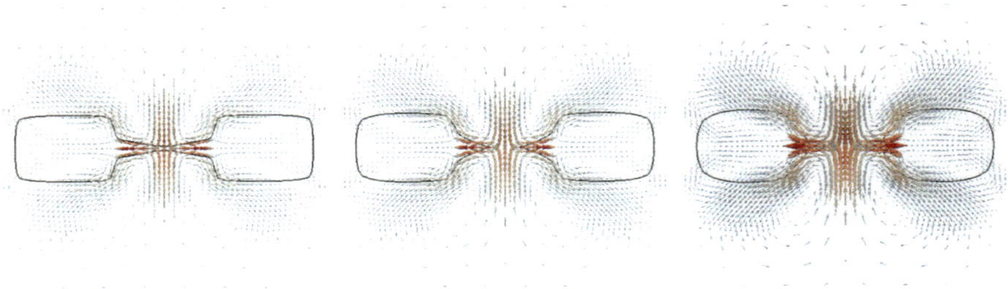

Fig. 9 Optimal velocity at time $t = 0.1, 0.15, 0.5$ (from left to right; the velocity vectors are scaled by a factor $0.1, 0.15, 0.4$, respectively); only β term present

Acknowledgements The authors acknowledge the support by the German Research Association (DFG) under grant SPP-1253, BR-3391/4-1 and PR-548/8-1.

References

1. L. Ambrosio, N. Fusco, Nicola, D. Pallara, *Functions of Bounded Variation and Free Discontinuity Problems* (Oxford University Press, Oxford, 2000)
2. J. Atecia, D.J. Beebe, Controlled microfluidic interfaces. Nature **437**, 648–655 (2004)
3. L'. Baňas, M. Klein, A. Prohl, Control of interface evolution in multi-phase fluid flows. SIAM J. Control Optim. **52**(4), 2284–2318 (2014)
4. L'. Baňas, A. Prohl, Convergent finite element discretization of the multi-fluid nonstationary incompressible magnetohydrodynamics equations. Math. Comput. **272**, 1957–1999 (2010)
5. J. Barzilai, J. Borwein, Two point step size gradient methods. IMA J. Numer. Anal. **8**, 141–148 (1988)
6. R. Becker, M. Braack et al. *Gascoigne3D, High Performance Adaptive Finite Element Toolkit*, http://www.gascoigne.de
7. R. Becker, D. Meidner, B. Vexler, RoDoBo, A C++ library for optimization with stationary and nonstationary PDEs. http://www.rodobo.org
8. R. Becker, D. Meidner, B. Vexler, Efficient numerical solution of parabolic optimization problems by finite element methods. Opt. Meth. Softw. **22**(5), 813–833 (2007)
9. M. Braack, E. Burman, Local projection stabilization for the Oseen problem and its interpretation as a variational multiscale method. SIAM J. Numer. Anal. **43**(6), 2544–2566 (2006)

10. M. Braack, B. Tews, Finite element discretizations of optimal control flow problems with boundary layers. Lect. Notes Comput. Sci. Eng. **8**, 47–55 (2011)
11. M. Braack, B. Tews, Linear-quadratic optimal control for the Oseen equations with stabilized finite elements. ESAIM: Control Optim. Calc. Var. **18**(2), 987–1004 (2012)
12. J.W. Cahn, S.M. Allen, A microscopic theory for domain wall motion and its experimental verification in Fe-Al alloy domain growth kinetics. J. Phys. Colloq. **38**, C7-51–C7-54 (1977)
13. Y.C. Chang, T.Y. Hou, T.Y. Merriman, S.J. Osher, A level set formulation of Eulerian interface capturing methods for incompressible fluid flows. J. Comput. Phys. **124**, 813–833 (1996)
14. J.-F. Gerbeau, C. Le Bris, T. Lelièvre, *Mathematical Methods for the Magnetohydrodynamics of Liquid Metals*. Numerical Mathematics and Scientific Computation (Oxford University Press, Oxford, 2006)
15. P.-L. Lions, *Mathematical Topics in Fluid Mechanics: Incompressible Models*. Oxford Lecture Series in Mathematics and Its Applications (Clarendon Press, Oxford, 1996)
16. C. Liu, J. Shen, A phase field model for the mixture of two incompressible fluids and its approximation by a Fourier-spectral method. Physica D **179**, 211–228 (2003)
17. D. Meidner, B. Vexler, Adaptive space-time finite elements methods for parabolic optimization problems. SIAM J. Control Optim. **46**, 116–142 (2007)
18. S. Osher, J.A. Sethian, Fronts propagating with curvature-dependent speed: algorithms based on Hamilton-Jacobi formulations. J. Comput. Phys. **79**, 12–49 (1988)
19. M. Sussman, P. Smereka, S.J. Osher, A level set approach for computing solutions to incompressible two-phase flow. J. Comput. Phys. **114**, 146–159 (1994)
20. M. Sussman, E. Fatemi, P. Smereka, S.J. Osher, An improved level set method for incompressible two-phase flows, J. Comput. Phys. **27**, 663–680 (1997)
21. B. Tews, Stabilized finite elements for optimal control problems in computational fluid dynamics, Dissertation, University of Kiel, 2013

A-Priori Error Bounds for Finite Element Approximation of Elliptic Optimal Control Problems with Gradient Constraints

Klaus Deckelnick and Michael Hinze

Abstract The finite element approximation of an elliptic optimal control problem with pointwise bounds on the gradient of the state is considered. We review recent results on the error analysis for various discretization approaches and prove a new bound for the problem without control constraints.

Keywords Elliptic optimal control problem • State constraints • Error estimates

Mathematics Subject Classification (2010). Primary 49J20; Secondary 65N30.

1 Introduction

The subject of this report is the finite element approximation of optimal control problems with constraints on the gradient of the state. A typical example is the optimization of a cooling process in which the temperature acts as the state variable and large temperature gradients are prohibited in order to avoid possible damage in the material. We shall restrict ourselves to a model problem which involves the optimal control of a linear elliptic partial differential equation in the presence of pointwise bounds on the gradient of the state, while the control variable can be both constrained or unconstrained. In order to discretize this problem it is common to approximate the underlying objective functional by a sequence of functionals which are obtained by discretizing the state equation with the help of a finite element method. Natural choices in this step are continuous, piecewise linear finite elements

---

K. Deckelnick
Institut für Analysis und Numerik, Otto–von–Guericke–Universität Magdeburg, Universitätsplatz 2, 39106 Magdeburg, Germany
e-mail: klaus.deckelnick@ovgu.de

M. Hinze (✉)
Schwerpunkt Optimierung und Approximation, Universität Hamburg, Bundesstraße 55, 20146 Hamburg, Germany
e-mail: michael.hinze@uni-hamburg.de

G. Leugering et al. (eds.), *Trends in PDE Constrained Optimization*, International Series of Numerical Mathematics 165, DOI 10.1007/978-3-319-05083-6_23

but also the lowest order Raviart–Thomas mixed finite element. The control variable can be handled in two ways: either by variational discretization (see [11]), which means that the first order optimality conditions give rise to an implicit discretization in terms of the discrete adjoint state; another possibility consists in discretizing the control explicitly, typically by piecewise constant functions.

This report focusses on the a-priori error analysis for the abovementioned approaches. Apart from reviewing results that have been obtained in [5, 8, 12] we prove a new bound in the case in which the control variable is unconstrained and the objective functional contains an L^r–norm ($r > 2$). In the remaining part of the paper we present a number of test calculations.

Let us close this section with a short survey of related publications. Elliptic optimal control problems with gradient constraints in nonsmooth polygonal domains are considered by Wollner in [16, 17]. While [16] is concerned with the existence of solutions, first order conditions and regularity, [17] derives a-priori error bounds for a finite element discretization. A general Moreau–Yosida framework for the treatment of elliptic optimal control problems with state and gradient constraints is presented by Hintermüller and Kunisch in [9]. Interior point approaches are investigated by Schiela and Wollner in [13]. In [15] Wollner presents an a-posteriori error analysis for an interior point approach to elliptic optimal control problems with general state constraints, including the case of pointwise bounds on the gradient of the state. A residual based adaptive approach to elliptic optimal control problems with pointwise gradient state constraints is presented by Hintermüller et al. in [10].

2 Mathematical Setting

Let $\Omega \subset \mathbb{R}^d$ $(d = 2, 3)$ be a bounded domain with a $C^{1,1}$-boundary and consider the differential operator

$$\mathcal{A}y := -\sum_{i,j=1}^{d} \partial_{x_j}\left(a_{ij} y_{x_i}\right) + a_0 y,$$

where for simplicity the coefficients a_{ij} and a_0 are assumed to be smooth functions on $\bar{\Omega}$. In what follows we assume that $a_{ij} = a_{ji}$, $a_0 \geq 0$ in Ω and that there exists $c_0 > 0$ such that

$$\sum_{i,j=1}^{d} a_{ij}(x)\xi_i \xi_j \geq c_0|\xi|^2 \quad \text{for all } \xi \in \mathbb{R}^d \text{ and all } x \in \Omega.$$

We consider the elliptic boundary value problem

$$\begin{aligned} \mathcal{A}y &= u \text{ in } \Omega \\ y &= 0 \text{ on } \partial\Omega. \end{aligned} \tag{2.1}$$

It is well–known that for every $1 < p < \infty$ (2.1) has a unique solution $y \in W^{2,p}(\Omega) \cap W_0^{1,p}(\Omega)$ with

$$\|y\|_{W^{2,p}} \le C\,\|u\|_{L^p}. \tag{2.2}$$

Here $\|\cdot\|_{L^p}$ and $\|\cdot\|_{W^{k,p}}$ denote the usual Lebesgue and Sobolev norms. If $p = 2$ we simply write $\|\cdot\| = \|\cdot\|_{L^2}$.

We consider the following optimal control problem:

$$\begin{aligned} &\min_{u\in K} J(u) = \frac{1}{2}\int_\Omega |y - y_0|^2 + \frac{\alpha}{r}\int_\Omega |u|^r \\ &\text{where } y \text{ solves (2.1) and } \nabla y \in C. \end{aligned} \tag{2.3}$$

Here, $\alpha > 0$ and $y_0 \in L^2(\Omega)$ are given, while

$$C = \{\mathbf{z} \in C^0(\bar\Omega)^d \mid |\mathbf{z}(x)| \le \delta, x \in \bar\Omega\},$$

for some given $\delta > 0$ and $|\cdot|$ denotes the Euclidian norm in $\mathbb{R}^d$. Furthermore, we consider the following two possible choices for K and r:

(I) $K = \{u \in L^\infty(\Omega) \,|\, a \le u \le b$ a.e. in $\Omega\}$, $r = 2$, where $a < b$ are given constants.

(II) $K = L^r(\Omega)$ for some $r > d$.

Note that in both cases a well–known embedding result implies that $\nabla y \in C^0(\bar\Omega)^d$ for the solution of (2.1), so that the condition $\nabla y \in C$ in (2.3) makes sense.

Existence of solutions, first order conditions as well as the structure and regularity of multipliers for control problems with pointwise constraints on the gradient of the state were investigated by Casas and Fernández in [4]. The authors allow a semilinear state equation and rather general constraints on the control and the gradient of the state. The above choices (**I**) and (**II**) fit into the framework of [4]. In order to formulate the first order optimality conditions we introduce the space of regular Borel measures $\mathcal{M}(\bar\Omega)$, which is the dual space of $C^0(\bar\Omega)$. The norm on $\mathcal{M}(\bar\Omega)$ is given by

$$\|\mu\|_{\mathcal{M}(\bar\Omega)} = \sup_{f\in C^0(\bar\Omega),|f|\le 1} \int_{\bar\Omega} f\,d\mu.$$

In case (**I**) we assume in addition that the following Slater condition holds:

$$\exists \hat u \in K \quad |\nabla \hat y(x)| < \delta,\; x \in \bar\Omega, \qquad \text{where } \hat y \text{ solves (2.1) with } u = \hat u. \tag{2.4}$$

Note that in case (**II**) one may simply choose $\hat u = 0$ to satisfy this condition.

Theorem 2.1. *An element $u \in K$ is a solution of* (2.3) *if and only if there exist $\mu \in \mathcal{M}(\bar{\Omega})^d$ and $p \in L^t(\Omega)$ $(t < \frac{d}{d-1})$ such that*

$$\int_\Omega p\mathcal{A}z - \int_\Omega (y - y_0)z - \int_{\bar{\Omega}} \nabla z \cdot d\mu = 0 \qquad \forall z \in W^{2,t'}(\Omega) \cap W_0^{1,t'}(\Omega), \tag{2.5}$$

$$\int_\Omega (p + \alpha|u|^{r-2}u)(\tilde{u} - u) \geq 0 \qquad \forall \tilde{u} \in K, \tag{2.6}$$

$$\int_{\bar{\Omega}} (\mathbf{z} - \nabla y) \cdot d\mu \leq 0 \qquad \forall \mathbf{z} \in C. \tag{2.7}$$

Here, y is the solution of (2.1) and $\frac{1}{t} + \frac{1}{t'} = 1$.

Proof. see, [4, Theorem 3] and [4, Corollary 1]. □

Remark 2.2. We may infer from (2.6) that in case

(I) $u(x) = \mathrm{Proj}_{[a,b]}(-\frac{p(x)}{\alpha})$ a.a. $x \in \Omega$,

(II) $u(x) = -\alpha^{-\frac{1}{r-1}}|p(x)|^{\frac{2-r}{r-1}}p(x)$ a.a. $x \in \Omega$.

In the latter case it is shown in [12, Corollary 5] that this relation together with (2.5) implies that $u \in W^{\frac{1-\frac{d}{r}-\epsilon}{r-1},r}(\Omega)$ for any $\epsilon > 0$. An embedding result (see [14, 4.6.1]) then yields $u \in L^{p_\epsilon}(\Omega)$, where $p_\epsilon = \frac{r-1}{1-\frac{1}{d}+\epsilon}$ for any $\epsilon > 0$.

3 Finite Element Discretization

Let $\mathcal{T}_h$ be a triangulation of Ω with maximum mesh size $h := \max_{T\in\mathcal{T}_h} \mathrm{diam}(T)$. We suppose that $\bar{\Omega}$ is the union of the elements of $\mathcal{T}_h$; boundary elements are allowed to have one curved face. In addition, we assume that the triangulation is quasi-uniform in the sense that there exists a constant $\kappa > 0$ (independent of h) such that each $T \in \mathcal{T}_h$ is contained in a ball of radius $\kappa^{-1}h$ and contains a ball of radius κh.

3.1 Piecewise Linear Approximation of the State

Let us recall the definition of the space of linear finite elements,

$$X_h := \{v_h \in C^0(\bar{\Omega}) \mid v_h \text{ is a linear polynomial on each } T \in \mathcal{T}_h\}$$

and let $X_{h0} := X_h \cap H_0^1(\Omega)$. For a given function $u \in L^2(\Omega)$ we denote by $y_h \in X_{h0}$ the solution of

$$\int_\Omega A\nabla y_h \cdot \nabla v_h + \int_\Omega a_0 y_h v_h = \int_\Omega u v_h \quad \text{for all } v_h \in X_{h0}. \tag{3.1}$$

Here, we have abbreviated $A(x) = \big(a_{ij}(x)\big)_{i,j=1}^d$. Let us define

$$C_h := \{\mathbf{c_h} : \bar{\Omega} \to \mathbb{R}^d \mid \mathbf{c_h}_{|T} \text{ is constant and } |\mathbf{c_h}_{|T}| \le \delta,\ T \in \mathcal{T}_h\}. \tag{3.2}$$

We approximate (2.3) by the following control problem depending on the mesh parameter h:

$$\begin{aligned} &\min_{u \in K} J_h(u) := \frac{1}{2}\int_\Omega |y_h - y_0|^2 + \frac{\alpha}{r}\int_\Omega |u|^r \\ &\text{subject to } \ y_h \text{ solves (3.1) and } \nabla y_h \in C_h. \end{aligned} \tag{3.3}$$

Note that the control variable is not discretized. Problem (3.3) represents a convex infinite–dimensional optimization problem of similar structure as problem (2.3), but with only finitely many constraints on the state.The following first order conditions yield an implicit discretization of the control variable in terms of the discrete adjoint state. Using (2.4) it is not difficult to see that a Slater condition holds for (3.3) provided that $0 < h \le h_0$.

Lemma 3.1. *Problem (3.3) has a unique solution $u_h \in K$. For $0 < h \le h_0$ there are $\boldsymbol{\mu}_T \in \mathbb{R}^d, T \in \mathcal{T}_h$ and $p_h \in X_{h0}$ such that*

$$\int_\Omega (A\nabla v_h \cdot \nabla p_h + a_0 v_h p_h) - \int_\Omega (y_h - y_0) v_h - \sum_{T \in \mathcal{T}_h} \nabla v_{h|T} \cdot \boldsymbol{\mu}_\mathbf{T} = 0 \quad \forall v_h \in X_{h0}, \tag{3.4}$$

$$\int_\Omega (p_h + \alpha |u_h|^{r-2} u_h)(\tilde{u} - u_h) \ge 0 \quad \forall \tilde{u} \in K, \tag{3.5}$$

$$\sum_{T \in \mathcal{T}_h} \big(\mathbf{c_h}_{|T} - \nabla y_{h|T}\big) \cdot \boldsymbol{\mu}_\mathbf{T} \le 0 \quad \forall \mathbf{c_h} \in C_h. \tag{3.6}$$

Here, $y_h \in X_{h0}$ is the solution of (3.1) with right hand side u_h.

Proof. See [4, Theorem 7] with the choices $U = L^r(\Omega)$, $K \subset U$, $C_h \subset Z := \mathbb{R}^{N_h}$, where N_h is the number of triangles in $\mathcal{T}_h$. □

Remark 3.2. Similar to Remark 2.2 we deduce from (3.5) that for

(I) $u_h(x) = \mathrm{Proj}_{[a,b]}\big(-\frac{p_h(x)}{\alpha}\big)$ a.a. $x \in \Omega$,

(II) $u_h(x) = -\alpha^{-\frac{1}{r-1}} |p_h(x)|^{\frac{2-r}{r-1}} p_h(x)$ a.a. $x \in \Omega$,

so that in both cases the discrete control is expressed implicitly in terms of the piecewise linear discrete costate p_h, the relation however being nonlinear.

For the unconstrained case (**II**), the following error bound has been proved in [8, Theorem 2.5]:

Theorem 3.3. *Let u and u_h be the solutions of (2.3) and (3.3) in case* (**II**) *respectively. Then there exists $h_0 > 0$ such that*

$$\|u - u_h\|_{L^r} \le Ch^{\frac{1}{r}(1-\frac{d}{r})}, \quad \|y - y_h\| \le Ch^{\frac{1}{2}(1-\frac{d}{r})}$$

for all $0 < h \le h_0$.

The proof relies on a careful combination of the information given by the primal and adjoint equations and we present the main ideas in the following section for a mixed finite element discretization of the state equation. The bounds in Theorem 3.3 are still satisfied if one employs a discretization of the control variable by piecewise constant functions on $\mathcal{T}_h$, see [8, Theorem 2.8]. We also remark that the above results are obtained by Ortner and Wollner in [12] without making use of adjoint information by working directly with the functionals J and J_h. Such a technique was previously used in [6] for the numerical analysis of elliptic optimal control problems with pointwise bounds on the state.

In general, both the control u and the adjoint variable p have low regularity even allowing jumps. For this reason, piecewise linear, continuous finite elements are not ideally suited for the discretization as they tend to develop oscillations near discontinuities. In the next section we present an alternative approach on the basis of a mixed finite element approach of lowest order for the state equation. This approach leads in particular to piecewise constant approximations for the state, costate and control and therefore seems to be better suited to handle discontinuities.

3.2 *Mixed Finite Element Approximation of the State*

As already mentioned we now use a mixed formulation in order to approximate the solution of (2.1). Let us introduce

$$H(\mathrm{div}, \Omega) := \{\mathbf{w} \in L^2(\Omega)^d \mid \mathrm{div}\mathbf{w} \in L^2(\Omega)\}$$

and write $(y, \mathbf{v}) = \mathcal{G}(u)$, where $\mathbf{v} = A\nabla y$ and y is the solution of (2.1).

We use a mixed finite element method based on the lowest order Raviart–Thomas element. Let

$$\mathbf{V_h} := RT_0(\Omega, \mathcal{T}_h) := \{\mathbf{w_h} \in H(\mathrm{div}, \Omega) \mid \mathbf{w_h}_{|T} \in RT_0(T) \text{ for all } T \in \mathcal{T}_h\},$$

where $RT_0(T) = \{\mathbf{w} : T \to \mathbb{R}^d \mid \mathbf{w}(x) = a + \beta x, a \in \mathbb{R}^d, \beta \in \mathbb{R}\}$. Furthermore, let

$$Y_h := \{z_h \in L^2(\Omega) \mid z_h \text{ is constant on each } T \in \mathcal{T}_h\}.$$

For a given function $u \in L^r(\Omega)$ the discrete solution $(y_h, \mathbf{v_h}) \in Y_h \times \mathbf{V_h}$ is given by

$$\int_\Omega A^{-1}\mathbf{v_h} \cdot \mathbf{w_h} + \int_\Omega y_h \,\mathrm{div}\mathbf{w_h} = 0 \qquad \forall \mathbf{w_h} \in \mathbf{V_h} \tag{3.7}$$

$$\int_\Omega z_h \,\mathrm{div}\mathbf{v_h} - \int_\Omega a_0 y_h z_h + \int_\Omega u z_h = 0 \qquad \forall z_h \in Y_h. \tag{3.8}$$

Introducing $\mathcal{G}_h(u) = (y_h, \mathbf{v_h}) \in Y_h \times \mathbf{V_h}$ as an approximation of $\mathcal{G}$ it is well–known [3] that the following error estimate holds:

$$\|y - y_h\| + \|\mathbf{v} - \mathbf{v_h}\| \le Ch\big(\|y\|_{H^1} + \|A\nabla y\|_{H^1}\big) \le Ch\|y\|_{H^2} \le Ch\|u\| \tag{3.9}$$

by (2.2). In what follows it will be crucial to control the error between $\mathbf{v}$ and $\mathbf{v_h}$ in $L^\infty(\Omega)$

Lemma 3.4. *Let $d < p < \infty$, $u \in L^p(\Omega)$ and $(y, \mathbf{v}) = \mathcal{G}(u)$, $(y_h, \mathbf{v_h}) = \mathcal{G}_h(u)$. Then there exists $h_0 > 0$ such that for $0 < h \le h_0$*

$$\|\mathbf{v} - \mathbf{v_h}\|_{L^\infty} \le Ch^{1-\frac{d}{p}} |\log h|^{1-\frac{2}{p}} \|u\|_{L^p}.$$

Proof. Let us denote by T the linear operator which assigns to u the error $\mathbf{v} - \mathbf{v_h}$. We deduce from (3.9) that

$$\|T\|_{L^2 \to L^2} \le Ch.$$

On the other hand we infer from [7, Corollary 3] that there exists $h_0 > 0$ so that for $0 < h \le h_0$

$$\|\mathbf{v} - \mathbf{v_h}\|_{L^\infty} \le Ch|\log h| \, \|u\|_{L^\infty}$$

for all $u \in L^\infty(\Omega)$, so that

$$\|T\|_{L^\infty \to L^\infty} \le Ch|\log h|.$$

The Riesz convexity theorem then implies that

$$\|T\|_{L^p \to L^p} \le \|T\|_{L^2 \to L^2}^{\frac{2}{p}} \|T\|_{L^\infty \to L^\infty}^{1-\frac{2}{p}}$$

and hence

$$\|\mathbf{v}-\mathbf{v_h}\|_{L^p} \le Ch^{\frac{2}{p}}\big(h|\log h|\big)^{1-\frac{2}{p}}\|u\|_{L^p} = h|\log h|^{1-\frac{2}{p}}\|u\|_{L^p}$$

for all $u \in L^p(\Omega)$. Denoting by I_h the usual Lagrange interpolation operator we deduce with the help of standard interpolation estimates, (2.2) and an inverse estimate that

$$\begin{aligned}\|\mathbf{v}-\mathbf{v_h}\|_{L^\infty} &\le \|\mathbf{v}-I_h\mathbf{v}\|_{L^\infty} + \|I_h\mathbf{v}-\mathbf{v_h}\|_{L^\infty}\\ &\le ch^{1-\frac{d}{p}}\|\mathbf{v}\|_{W^{1,p}} + ch^{-\frac{d}{p}}\|I_h\mathbf{v}-\mathbf{v_h}\|_{L^p}\\ &\le ch^{1-\frac{d}{p}}\|u\|_{L^p} + ch^{-\frac{d}{p}}\|\mathbf{v}-I_h\mathbf{v}\|_{L^p} + ch^{-\frac{d}{p}}\|\mathbf{v}-\mathbf{v_h}\|_{L^p}\\ &\le ch^{1-\frac{d}{p}}\|u\|_{L^p} + ch^{1-\frac{d}{p}}|\log h|^{1-\frac{2}{p}}\|u\|_{L^p}\end{aligned}$$

which yields the result. □

Similarly to (3.3) we now consider the following discrete control problem:

$$\begin{aligned}&\min_{u\in K} J_h(u) := \frac{1}{2}\int_\Omega |y_h-y_0|^2 + \frac{\alpha}{r}\int_\Omega |u|^r\\ &\text{subject to } (y_h,\mathbf{v_h}) = \mathcal{G}_h(u) \text{ and } \Big(\fint_T A^{-1}\mathbf{v_h}\Big)_{T\in\mathcal{T}_h} \in C_h,\end{aligned} \tag{3.10}$$

where C_h is as in (3.2) and $\fint_T \cdot = \frac{1}{|T|}\int_T \cdot$. We note that the control again is not discretized and that the gradient of the state variable is only constrained on average on each element. Similar to Lemma 3.1 and Remark 3.2 one has

Lemma 3.5. *Problem (3.10) has a unique solution $u_h \in K$. There exists $0 < h_1 \le h_0$ such that for $0 < h < h_1$ there are $\boldsymbol{\mu}_T \in \mathbb{R}^d, T \in \mathcal{T}_h$ and $(p_h, \boldsymbol{\chi}_\mathbf{h}) \in Y_h \times \mathbf{V_h}$ such that with $(y_h, \mathbf{v_h}) = \mathcal{G}_h(u_h)$ we have*

$$\int_\Omega A^{-1}\boldsymbol{\chi}_\mathbf{h}\cdot\mathbf{w_h} + \int_\Omega p_h\,\mathrm{div}\mathbf{w_h} + \sum_{T\in\mathcal{T}_h}\boldsymbol{\mu}_\mathbf{T}\cdot\fint_T A^{-1}\mathbf{w_h} = 0 \qquad \forall \mathbf{w_h}\in\mathbf{V_h}, \tag{3.11}$$

$$\int_\Omega z_h\,\mathrm{div}\boldsymbol{\chi}_\mathbf{h} - \int_\Omega a_0 p_h z_h + \int_\Omega (y_h-y_0)z_h = 0 \qquad \forall z_h \in Y_h, \tag{3.12}$$

$$\int_\Omega (p_h + \alpha|u_h|^{r-2}u_h)(\tilde{u}-u_h) \ge 0 \quad \forall \tilde{u}\in K, \tag{3.13}$$

$$\sum_{T\in\mathcal{T}_h}\boldsymbol{\mu}_\mathbf{T}\cdot\Big(\mathbf{c_h}_{|T} - \fint_T A^{-1}\mathbf{v_h}\Big) \le 0 \qquad \forall \mathbf{c_h}\in C_h. \tag{3.14}$$

Remark 3.6. The discrete control u_h and the discrete adjoint state p_h are related by

(**I**) $u_h(x) = \text{Proj}_{[a,b]}\big(-\dfrac{p_h(x)}{\alpha}\big)$ a.a. $x \in \Omega$,

(**II**) $u_h(x) = -\alpha^{-\frac{1}{r-1}} |p_h(x)|^{\frac{2-r}{r-1}} p_h(x)$ a.a. $x \in \Omega$.

In particular, in both cases the discrete solution u_h is piecewise constant on the triangulation $\mathcal{T}_h$.

The following a–priori estimate is crucial for the convergence analysis.

Lemma 3.7. *Let $u_h \in L^r(\Omega)$ be the optimal solution of (3.10) with corresponding state $(y_h, \mathbf{v_h}) \in Y_h \times \mathbf{V_h}$ and adjoint variables $(p_h, \boldsymbol{\chi}_\mathbf{h}) \in Y_h \times \mathbf{V_h}$, $\mu_\mathbf{T}, T \in \mathcal{T}_h$. Then*

$$\|u_h\|_{L^r} + \|y_h\| + \sum_{T \in \mathcal{T}_h} |\mu_\mathbf{T}| \le C$$

for all $0 < h \le h_1$.

Proof. The proof is carried out in [5, Lemma 3.6] for case (**I**), but the analysis can be adapted to case (**II**) in a straightforward way. □

The error analysis depends on the choice of the admissible set K and the structure of the objective functional. In case (**I**) the controls belong to $L^\infty(\Omega)$ leading to better convergence properties in the state equation. We have the following result:

Theorem 3.8. *Let u and u_h be the solutions of (2.3) and (3.10) in case (**I**) with corresponding states y and y_h respectively. Then*

$$\|u - u_h\| + \|y - y_h\| \le C h^{\frac{1}{2}} |\log h|^{\frac{1}{2}}$$

for all $0 < h \le h_1$.

Proof. See [5, Theorem 4.1]. □

Let us next turn to case (**II**), for which Theorem 3.3 gives convergence rates of $O(h^{\frac{1}{r}(1-\frac{d}{r})})$ for the control and $O(h^{\frac{1}{2}(1-\frac{d}{r})})$ for the state if a piecewise linear approximation of the state is used. Adapting the corresponding proof to the case of the Raviart–Thomas element it would be possible to derive the same convergence rates. However, as observed in [12, Remark 2], these rates are not optimal since $u \in W^{\frac{1-\frac{d}{r}-\epsilon}{r-1},r}(\Omega)$ for any $\epsilon > 0$. The following result improves these bounds and appears to be optimal as far as the control variable is concerned.

Theorem 3.9. *Let u and u_h be the solutions of (2.3) and (3.10) in case (**II**) with corresponding states y and y_h respectively. Then for every $\rho > 0$ there exists C_ρ such that*

$$\|u-u_h\|_{L^r} \le C_\rho h^{\alpha_1-\rho}, \quad \|y-y_h\| \le C_\rho h^{\alpha_2-\rho},$$

where $\alpha_1 = \frac{1-\frac{d}{r}}{r-1}, \alpha_2 = (1-\frac{d}{r})\frac{r}{2(r-1)}$.

Proof. To begin, we note that for $r \ge 2$

$$(|a|^{r-2}a - |b|^{r-2}b)(a-b) \ge 2^{2-r}|a-b|^r \qquad \forall a,b \in \mathbb{R}.$$

Hence, using (2.6) and (3.13),

$$\begin{aligned} \alpha 2^{2-r}\int_\Omega |u-u_h|^r &\le \alpha \int_\Omega (|u|^{r-2}u - |u_h|^{r-2}u_h)(u-u_h) \\ &= \int_\Omega p_h(u-u_h) + \int_\Omega p(u_h-u) \equiv I + II. \end{aligned} \tag{3.15}$$

Let us introduce $(\tilde{y}_h, \tilde{\mathbf{v}}_\mathbf{h}) = \mathcal{G}_h(u) \in Y_h \times \mathbf{V_h}$. Using (3.8) and (3.11) we infer for the first term

$$\begin{aligned} I &= -\int_\Omega p_h \operatorname{div}(\tilde{\mathbf{v}}_\mathbf{h} - \mathbf{v_h}) + \int_\Omega a_0\, p_h(\tilde{y}_h - y_h) \\ &= \int_\Omega A^{-1}\boldsymbol{\chi}_\mathbf{h}\cdot(\tilde{\mathbf{v}}_\mathbf{h} - \mathbf{v_h}) + \sum_{T\in\mathcal{T}_h} \boldsymbol{\mu}_\mathbf{T}\cdot \fint_T A^{-1}(\tilde{\mathbf{v}}_\mathbf{h} - \mathbf{v_h}) + \int_\Omega a_0\, p_h(\tilde{y}_h - y_h) \\ &= \int_\Omega A^{-1}\boldsymbol{\chi}_\mathbf{h}\cdot(\tilde{\mathbf{v}}_\mathbf{h} - \mathbf{v_h}) + \sum_{T\in\mathcal{T}_h} \boldsymbol{\mu}_\mathbf{T}\cdot\Big(P_\delta\big(\fint_T A^{-1}\tilde{\mathbf{v}}_\mathbf{h}\big) - \fint_T A^{-1}\mathbf{v_h}\Big) \\ &\quad + \int_\Omega a_0\, p_h(\tilde{y}_h - y_h) + \sum_{T\in\mathcal{T}_h} \boldsymbol{\mu}_\mathbf{T}\cdot\Big(\fint_T A^{-1}\tilde{\mathbf{v}}_\mathbf{h} - P_\delta\big(\fint_T A^{-1}\tilde{\mathbf{v}}_\mathbf{h}\big)\Big), \end{aligned}$$

where P_δ denotes the orthogonal projection onto $\bar{B}_\delta(0) = \{x \in \mathbb{R}^d \mid |x| \le \delta\}$. Note that

$$|P_\delta(x) - P_\delta(\tilde{x})| \le |x-\tilde{x}| \qquad \forall x, \tilde{x} \in \mathbb{R}^d. \tag{3.16}$$

Since by definition

$$\Big(P_\delta\big(\fint_T A^{-1}\tilde{\mathbf{v}}_\mathbf{h}\big)\Big)_{T\in\mathcal{T}_h} \in C_h$$

we deduce from (3.14) that

$$\begin{aligned} I &\le \int_\Omega A^{-1}\boldsymbol{\chi}_\mathbf{h}\cdot(\tilde{\mathbf{v}}_\mathbf{h} - \mathbf{v_h}) + \int_\Omega a_0\, p_h(\tilde{y}_h - y_h) \\ &\quad + \max_{T\in\mathcal{T}_h}\Big|\fint_T A^{-1}\tilde{\mathbf{v}}_\mathbf{h} - P_\delta\Big(\fint_T A^{-1}\tilde{\mathbf{v}}_\mathbf{h}\Big)\Big| \sum_{T\in\mathcal{T}_h}|\boldsymbol{\mu}_\mathbf{T}|. \end{aligned}$$

In order to estimate the last term we note that $\nabla y \in C$ implies that $\big(\fint_T \nabla y\big)_{T\in\mathcal{T}_h} = \big(\fint_T A^{-1}\mathbf{v}\big)_{T\in\mathcal{T}_h} \in C_h$. Using Lemma 3.4 with $(y,\mathbf{v}) = \mathcal{G}(u), (\tilde{y}_h, \tilde{\mathbf{v}}_\mathbf{h}) = \mathcal{G}_h(u)$ we infer

$$\|\tilde{\mathbf{v}}_\mathbf{h} - \mathbf{v}\|_{L^\infty} \le C h^{1-\frac{d}{p_\epsilon}} |\log h|^{1-\frac{2}{p_\epsilon}} \|u\|_{L^{p_\epsilon}} = C_\epsilon h^{1-\frac{d}{p_\epsilon}} |\log h|^{1-\frac{2}{p_\epsilon}}, \tag{3.17}$$

since $u \in L^{p_\epsilon}(\Omega)$ with $p_\epsilon = \dfrac{r-1}{1-\frac{1}{d}+\epsilon}$ $(\epsilon > 0)$ in view of Remark 2.2. As a consequence,

$$\begin{aligned}
\Big|\fint_T A^{-1}\tilde{\mathbf{v}}_\mathbf{h} - P_\delta\Big(\fint_T A^{-1}\tilde{\mathbf{v}}_\mathbf{h}\Big)\Big| &\le \Big|\fint_T A^{-1}(\tilde{\mathbf{v}}_\mathbf{h} - \mathbf{v})\Big| + \Big|P_\delta\Big(\fint_T A^{-1}\tilde{\mathbf{v}}_\mathbf{h}\Big) - P_\delta\Big(\fint_T A^{-1}\mathbf{v}\Big)\Big| \\
&\le C\|\tilde{\mathbf{v}}_\mathbf{h} - \mathbf{v}\|_{L^\infty} \le C_\epsilon h^{1-\frac{d}{p_\epsilon}} |\log h|^{1-\frac{2}{p_\epsilon}}
\end{aligned}$$

in view of (3.16) and (3.17). Combining this estimate with Lemma 3.7 we deduce

$$I \le \int_\Omega A^{-1}\boldsymbol{\chi}_\mathbf{h} \cdot (\tilde{\mathbf{v}}_\mathbf{h} - \mathbf{v}_\mathbf{h}) + \int_\Omega a_0\, p_h(\tilde{y}_h - y_h) + C_\epsilon h^{1-\frac{d}{p_\epsilon}} |\log h|^{1-\frac{2}{p_\epsilon}}.$$

The symmetry of A, (3.7) and (3.12) finally give

$$\begin{aligned}
I &\le -\int_\Omega (\tilde{y}_h - y_h)\operatorname{div}\boldsymbol{\chi}_\mathbf{h} + \int_\Omega a_0\, p_h(\tilde{y}_h - y_h) + C_\epsilon h^{1-\frac{d}{p_\epsilon}} |\log h|^{1-\frac{2}{p_\epsilon}} \\
&= \int_\Omega (y_h - y_0)(\tilde{y}_h - y_h) + C_\epsilon h^{1-\frac{d}{p_\epsilon}} |\log h|^{1-\frac{2}{p_\epsilon}}.
\end{aligned} \tag{3.18}$$

In order to analyze the second term in (3.15) we define $(y^h, \mathbf{v}^\mathbf{h}) = \mathcal{G}(u_h)$. Recalling (2.5) we have

$$\begin{aligned}
II &= \int_\Omega p(\mathcal{A}y^h - \mathcal{A}y) \\
&= \int_\Omega (y - y_0)(y^h - y) + \int_{\bar{\Omega}} (\nabla y^h - \nabla y)\cdot d\boldsymbol{\mu} \\
&= \int_\Omega (y - y_0)(y^h - y) + \int_{\bar{\Omega}} (P_\delta(\nabla y^h) - \nabla y)\cdot d\boldsymbol{\mu} + \int_{\bar{\Omega}} (\nabla y^h - P_\delta(\nabla y^h))\cdot d\boldsymbol{\mu}.
\end{aligned}$$

Since $x \mapsto P_\delta(\nabla y^h(x)) \in C$ we infer from (2.7)

$$II \le \int_\Omega (y - y_0)(y^h - y) + \max_{x\in\bar{\Omega}} |\nabla y^h(x) - P_\delta(\nabla y^h(x))|\, \|\boldsymbol{\mu}\|_{\mathcal{M}(\bar{\Omega})^d}. \tag{3.19}$$

Let $x \in \bar{\Omega}$, say $x \in T$ for some $T \in \mathcal{T}_h$. Since u_h is feasible for (3.10) we have that $\fint_T A^{-1}\mathbf{v_h} \in \bar{B}_\delta(0)$ so that (3.16) implies

$$\begin{aligned}
&\left|\nabla y^h(x) - P_\delta(\nabla y^h(x))\right| \\
&\qquad \leq \left|\nabla y^h(x) - \fint_T A^{-1}\mathbf{v_h}\right| + \left|P_\delta(\nabla y^h(x)) - P_\delta\big(\fint_T A^{-1}\mathbf{v_h}\big)\right| \\
&\qquad \leq 2\left|\nabla y^h(x) - \fint_T A^{-1}\mathbf{v_h}\right|. \qquad (3.20)
\end{aligned}$$

Using a well–known interpolation estimate (cf. [2], Corollary (4.4.7)) and (2.2) we obtain

$$\begin{aligned}
\left|\nabla y^h(x) - \fint_T A^{-1}\mathbf{v_h}\right| &= \left|A^{-1}(x)\mathbf{v^h}(x) - \fint_T A^{-1}\mathbf{v_h}\right| \\
&\leq \left|A^{-1}(x)(\mathbf{v^h} - \mathbf{v})(x) - \fint_T A^{-1}(\mathbf{v^h}(x) - \mathbf{v})\right| \\
&\quad + \left|A^{-1}(x)\mathbf{v}(x) - \fint_T A^{-1}\mathbf{v}\right| + \left|\fint_T A^{-1}(\mathbf{v^h} - \mathbf{v_h})\right| \\
&\leq Ch^{1-\frac{d}{r}}\|\mathbf{v^h} - \mathbf{v}\|_{W^{1,r}} + Ch^{1-\frac{d}{p_\epsilon}}\|\mathbf{v}\|_{W^{1,p_\epsilon}} + C\|\mathbf{v^h} - \mathbf{v_h}\|_{L^\infty} \\
&\leq Ch^{1-\frac{d}{r}}\|u_h - u\|_{L^r} + Ch^{1-\frac{d}{p_\epsilon}}\|u\|_{L^{p_\epsilon}} + C\|\mathbf{v^h} - \mathbf{v_h}\|_{L^\infty}.
\end{aligned}$$

Applying Lemma 3.4 with $u - u_h$ as well as (3.17) we infer

$$\begin{aligned}
\|\mathbf{v^h} - \mathbf{v_h}\|_{L^\infty} &\leq \|(\mathbf{v^h} - \mathbf{v}) - (\mathbf{v_h} - \tilde{\mathbf{v}}_\mathbf{h})\|_{L^\infty} + \|\mathbf{v} - \tilde{\mathbf{v}}_\mathbf{h}\|_{L^\infty} \\
&\leq Ch^{1-\frac{d}{r}}|\log h|^{1-\frac{2}{r}}\|u - u_h\|_{L^r} + C_\epsilon h^{1-\frac{d}{p_\epsilon}}|\log h|^{1-\frac{2}{p_\epsilon}},
\end{aligned}$$

which combined with (3.20) yields

$$\max_{x\in\bar{\Omega}}\left|\nabla y^h(x) - P_\delta(\nabla y^h(x))\right| \leq Ch^{1-\frac{d}{r}}|\log h|^{1-\frac{2}{r}}\|u - u_h\|_{L^r} + C_\epsilon h^{1-\frac{d}{p_\epsilon}}|\log h|^{1-\frac{2}{p_\epsilon}}.$$

Returning to (3.19) we have

$$II \leq \int_\Omega (y - y_0)(y^h - y) + Ch^{1-\frac{d}{r}}|\log h|^{1-\frac{2}{r}}\|u - u_h\|_{L^r} + C_\epsilon h^{1-\frac{d}{p_\epsilon}}|\log h|^{1-\frac{2}{p_\epsilon}}. \qquad (3.21)$$

If we insert (3.21) and (3.18) into (3.15) we finally obtain

$$\begin{aligned}
\alpha 2^{2-r}\|u - u_h\|_{L^r}^r &\leq \int_\Omega (y_h - y_0)(\tilde{y}_h - y_h) + \int_\Omega (y - y_0)(y^h - y) \\
&\quad + Ch^{1-\frac{d}{r}}|\log h|^{1-\frac{2}{r}}\|u_h - u\|_{L^r} + C_\epsilon h^{1-\frac{d}{p_\epsilon}}|\log h|^{1-\frac{2}{p_\epsilon}}
\end{aligned}$$

$$= -\int_\Omega |y - y_h|^2 + \int_\Omega \left((y_0 - y_h)(y - \tilde{y}_h) + (y - y_0)(y^h - y_h) \right)$$
$$+ Ch^{1-\frac{d}{r}} |\log h|^{1-\frac{2}{r}} \|u_h - u\|_{L^r} + C_\epsilon h^{1-\frac{d}{p_\epsilon}} |\log h|^{1-\frac{2}{p_\epsilon}}$$
$$\leq -\|y - y_h\|^2 + C\left(\|y - \tilde{y}_h\| + \|y^h - y_h\| \right)$$
$$+ \frac{\alpha 2^{2-r}}{2} \|u - u_h\|_{L^r}^r + Ch^{(1-\frac{d}{r})\frac{r}{r-1}} |\log h|^{(1-\frac{2}{r})\frac{r}{r-1}} + C_\epsilon h^{1-\frac{d}{p_\epsilon}} |\log h|^{1-\frac{2}{p_\epsilon}}$$

by Young's inequality. A simple calculation shows that

$$1 - \frac{d}{p_\epsilon} = (1 - \frac{d}{r})\frac{r}{r-1} - \frac{\epsilon d}{r-1} < (1 - \frac{d}{r})\frac{r}{r-1},$$

while $1 - \frac{2}{p_\epsilon} < (1 - \frac{2}{r})\frac{r}{r-1}$. In conclusion we obtain after another application of (3.9)

$$\|u - u_h\|_{L^r}^r + \|y - y_h\|^2 \leq C_\epsilon h^{(1-\frac{d}{r})\frac{r}{r-1} - \frac{\epsilon d}{r-1}} |\log h|^{(1-\frac{2}{r})\frac{r}{r-1}},$$

from which we deduce the result of the theorem. □

4 Numerical Examples

We consider (2.3) with the choices $\Omega = B_2(0) \subset \mathbb{R}^2$, $\alpha = 1$,

$$C = \{\mathbf{z} \in C^0(\bar{\Omega})^2 \mid |\mathbf{z}(x)| \leq \frac{1}{2}, x \in \bar{\Omega}\}$$

as well as

$$y_0(x) := \begin{cases} \frac{1}{4} + \frac{1}{2}\ln 2 - \frac{1}{4}|x|^2, & 0 \leq |x| \leq 1, \\ \frac{1}{2}\ln 2 - \frac{1}{2}\ln|x|, & 1 < |x| \leq 2. \end{cases}$$

In order to construct a test example we allow an additional right hand side f in the state equation and replace (2.1) by

$$\begin{aligned} -\Delta y &= f + u \text{ in } \Omega \\ y &= 0 \quad \text{on } \partial\Omega, \end{aligned}$$

where

$$f(x) := \begin{cases} 2, & 0 \leq |x| \leq 1, \\ 0, & 1 < |x| \leq 2. \end{cases}$$

In case (**I**) we consider $K = \{u \in L^\infty(\Omega) \mid -2 \le u \le 2 \text{ a.e. in } \Omega\}$, while in case (**II**) we choose $r = 4$. The optimization problem then has in both cases the unique solution

$$u(x) = \begin{cases} -1 \ , 0 \le |x| \le 1 \\ \ \ 0 \ , 1 < |x| \le 2 \end{cases}$$

with corresponding state $y \equiv y_0$. We note that in case (**I**) the bounds on the control are not active, so that we obtain equality in (2.6), i.e. $p = -u$. Furthermore, the action of the measure μ applied to a vectorfield $\phi \in C^0(\bar{\Omega})^2$ is given by

$$\int_{\bar{\Omega}} \phi \cdot d\mu = - \int_{\partial B_1(0)} x \cdot \phi dS.$$

In what follows we frequently use the experimental order of convergence, which is defined for an error functional $E(h)$ by

$$\text{EOC} = \frac{\ln E(h_1) - \ln E(h_2)}{\ln h_1 - \ln h_2}.$$

For the numerical solution we use the routine `fmincon` contained in the Matlab optimization toolbox. The actual calculations were carried out on a polygonal approximation of $B_2(0)$. Note that while our analysis did not take into account the approximation of the domain, the observed rates show that this error doesn't dominate.

4.1 Piecewise Linears for the State with Variational Discretization

Many existing finite element codes employ continuous, piecewise linear finite elements, so that it is natural to use this element in order to discretize the state equation in optimization problems for elliptic pdes. Numerical results for case (**II**) are reported in [8] to which we refer for details. Table 1 shows the experimental order of convergence for the error functionals

$$\|u - u_h\|_{L^4(\Omega)}, \quad \|u - u_h\|, \text{ and } \|y - y_h\|.$$

Figure 1 illustrates the optimal solution u_h and the corresponding adjoint state p_h on a mesh consisting of $nt = 512$ triangles. Note that in view of the relation $u_h(x) = -|p_h(x)|^{-\frac{2}{3}} p_h(x)$ the variational control u_h necessarily is a continuous function, while the exact control u has a jump. This inconsistency is reflected

Table 1 Errors (*top*) and EOCs for piecewise linears

nt	$\|u - u_h\|_{L^4(\Omega)}$	$\|u - u_h\|$	$\|y - y_h\|$
32	$8.34633 \cdot 10^{-1}$	1.36003	$2.20346 \cdot 10^{-1}$
128	$5.88566 \cdot 10^{-1}$	$9.04770 \cdot 10^{-1}$	$7.97200 \cdot 10^{-2}$
512	$4.84191 \cdot 10^{-1}$	$5.82014 \cdot 10^{-1}$	$3.52102 \cdot 10^{-2}$
	0.54884	0.64041	1.59745
	0.29263	0.66136	1.22499

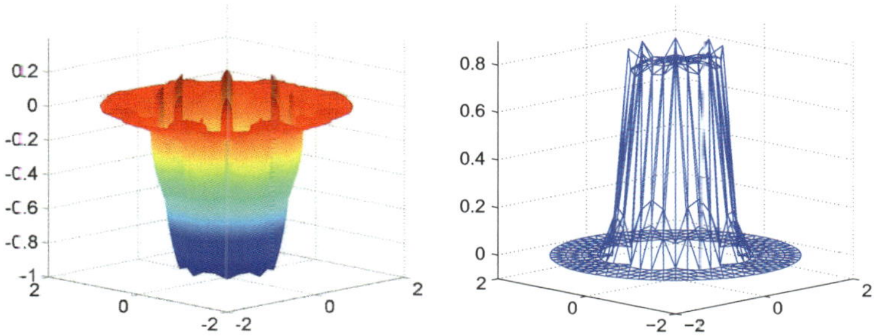

Fig. 1 Control (*left*), and adjoint state (*right*) (variational discretization)

in the appearance of oscillations near the set $\partial B_1(0)$ in Fig. 1, and also affects the performance of the optimization solvers implemented within the `fmincon` package. We conclude that variational discretization combined with continuous, piecewise linear finite elements for the state approximation is not ideally suited to control problems with gradient constraints on the state.

4.2 *Mixed Finite Element Approach with Variational Discretization*

The state equation is now approximated with the help of the lowest order Raviart–Thomas element for which we used the implementation provided by [1]. Numerical results for case (**I**) can be found in [5].

Let us report on the corresponding results for case (**II**). In Table 2 we display the experimental order of convergence for the error functionals

$$\|u - u_h\|_{L^4(\Omega)}, \quad \|u - u_h\| \text{ and } \|y - y_h\|.$$

The errors show a similar behaviour as in the case of piecewise linear finite elements and are slightly better than predicted by Theorem 3.9. Figure 2 shows the optimal state and the optimal control on a grid containing $m = 1{,}089$ gridpoints. In Table 3

Table 2 Errors and EOCs for the controls and the state with Raviart–Thomas approximation of the state

NT	$\|u - u_h\|_{L^4(\Omega)}$	$\|u - u_h\|$	$\|y - y_h\|$
32	$6.85 \cdot 10^{-1}$	1.10	$3.00 \cdot 10^{-1}$
128	$6.77 \cdot 10^{-1}$	$8.70 \cdot 10^{-1}$	$1.51 \cdot 10^{-1}$
512	$6.05 \cdot 10^{-1}$	$6.04 \cdot 10^{-1}$	$7.25 \cdot 10^{-2}$
2,048	$5.22 \cdot 10^{-1}$	$4.21 \cdot 10^{-1}$	$3.61 \cdot 10^{-2}$
8,192	$4.44 \cdot 10^{-1}$	$2.96 \cdot 10^{-1}$	$1.80 \cdot 10^{-2}$
	0.01881	0.36245	1.08340
	0.16899	0.54697	1.09552
	0.21730	0.53219	1.02287
	0.23488	0.51182	1.01139

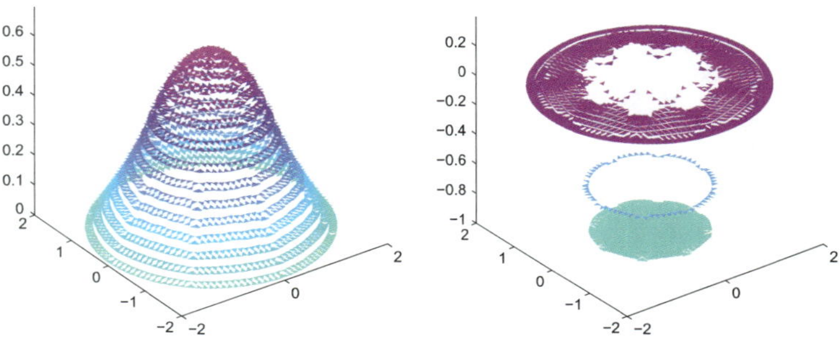

Fig. 2 Optimal state (*left*), and optimal control

Table 3 Behaviour of the discrete multipliers

NT	$\sum_{i=1}^{NT} \|\mu_T\|$
32	2.32
128	4.32
512	5.29
2,048	5.79
8,192	6.04

we display the values of $\sum_{T \in \mathcal{T}_h} |\boldsymbol{\mu}_T|$ which appear to converge to 2π, the total variation of the measure $\boldsymbol{\mu}$. The modulus of $\boldsymbol{\mu}_T, T \in \mathcal{T}_h$ as well as the set of elements T on which $\boldsymbol{\mu}_T \neq 0$ is shown in Fig. 3. It can be seen that these elements concentrate around $|x| = 1$.

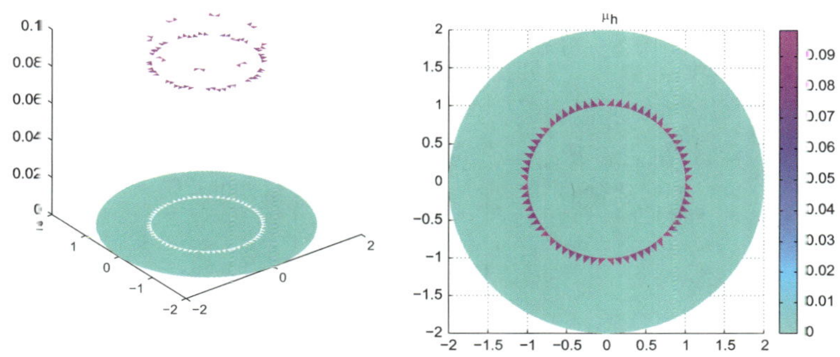

Fig. 3 $|\mu_T|$ (*left*), and support of μ_T

Acknowledgements The authors acknowledge support of the DFG priority program 1253 through grants DFG HI689/5-1 and DFG DE611/4-2.

References

1. C. Bahriawati, C. Carstensen, Three Matlab implementations of the lowest-order Raviart-Thomas MFEM with a posteriori error control. Comput. Methods Appl. Math. **5**, 333–361 (2005). Software download at www.math.hu-berlin.de/cc/download/public/software/code/Software-4.tar.gz
2. S. Brenner, R. Scott, *The Mathematical Theory of Finite Elements*, 2nd edn. (Springer, New York, 2002)
3. F. Brezzi, M. Fortin, *Mixed and Hybrid Finite Element Methods*. Springer Series in Computational Mathematics, vol. 15 (Springer, New York, 1991)
4. E. Casas, L. Fernández, Optimal control of semilinear elliptic equations with pointwise constraints on the gradient of the state. Appl. Math. Optim. **27**, 35–56 (1993)
5. K. Deckelnick, A. Günther, M. Hinze, Finite element approximation of elliptic control problems with constraints on the gradient. Numer. Math. **111**, 335–350 (2009)
6. K. Deckelnick, M. Hinze, Convergence of a finite element approximation to a state constrained elliptic control problem. SIAM J. Numer. Anal. **45**(5), 1937–1953 (2007)
7. L. Gastaldi, R.H. Nochetto, On L^∞–accuracy of mixed finite element methods for second order elliptic problems. Mat. Apl. Comput. **7**, 13–39 (1988)
8. A. Günther, M. Hinze, Elliptic control problems with gradient constraints – variational discrete versus piecewise constant controls. Comput. Optim. Appl. **49**, 549–566 (2011)
9. M. Hintermüller, K. Kunisch, PDE-constrained optimization subject to pointwise constraints on the control, the state, and its derivative. SIAM J. Optim. **20**(3), 1133–1156 (2009)
10. M. Hintermüller, M. Hinze, R.W.H. Hoppe, Weak-Duality based adaptive finite element methods for pde-constrained optimization with pointwise gradient state-constraints. J. Comput. Math. **30**, 101–123 (2012)
11. M. Hinze, A variational discretization concept in control constrained optimization: the linear-quadratic case. Comput. Optim. Appl. **30**, 45–63 (2005)
12. C. Ortner, W. Wollner, A priori error estimates for optimal control problems with pointwise constraints on the gradient of the state. Numer. Math. **118**(3), 587–600 (2011)
13. A. Schiela, W. Wollner, Barrier methods for optimal control problems with convex nonlinear gradient state constraints. SIAM J. Optim. **21**(1), 269–286 (2011)

14. H. Triebel, *Interpolation Theory, Function Spaces, Differential Operators*, 2nd edn. (Johann Ambrosius Barth, Heidelberg, 1995)
15. W. Wollner, A posteriori error estimates for a finite element discretization of interior point methods for an elliptic optimization problem with state constraints. Comput. Optim. Appl. **47**(1), 133–159 (2010)
16. W. Wollner, Optimal control of elliptic equations with pointwise constraints on the gradient of the state in nonsmooth polygonal domains. SIAM J. Control Optim. **50**(4), 2117–2129 (2012)
17. W. Wollner, A priori error estimates for optimal control problems with constraints on the gradient of the state on nonsmooth polygonal domains, in *Control and Optimization with PDE Constraints*. Springer ISNM Series, vol. 164 (Springer, Basel, 2013), pp. 193–215

Space-Time Newton-Multigrid Strategies for Nonstationary Distributed and Boundary Flow Control Problems

Michael Hinze, Michael Köster, and Stefan Turek

Abstract This paper considers a Newton-type solver strategy for optimal flow control problems using space-time multigrid solution techniques. Based on the standard Newton approach for optimal control, a space-time multigrid preconditioner is derived and numerically analysed for distributed and boundary control.

Keywords Distributed control • Boundary control • Finite elements • Time-dependent Navier–Stokes • Newton • Space-time multigrid • Optimal control

Mathematics Subject Classification (2010). 35Q30, 49K20, 49M05, 49M15, 49M29, 49M37, 65F08, 65F10, 65K05, 65M55, 65M60, 65R20, 65R32, 76D05, 76D55, 90C06, 90C30

1 Introduction

The optimal control of incompressible, nonstationary flow problems belongs to todays most challenging problems in the field of optimisation. By design, the underlying equations are of elliptic nature (in space and time), and thus, all variables in the discretised equations are fully coupled (in space and time). Such very high-dimensional discrete problems can only be solved with specialised solvers which exploit the structure of the underlying equations.

There are different approaches available to deal with nonstationary flow control, see [5,10] for an overview. In [2,3], the state of the art of multigrid methods in PDE

M. Hinze
Department of Mathematics, University of Hamburg, Bundesstrasse 55, 20146 Hamburg, Germany
e-mail: michael.hinze@uni-hamburg.de

M. Köster (✉) • S. Turek
Institute of Applied Mathematics, TU Dortmund, Vogelpothsweg 87, D-44227 Dortmund, Germany
e-mail: michael.koester@mathematik.tu-dortmund.de; stefan.turek@mathematik.tu-dortmund.de

G. Leugering et al. (eds.), *Trends in PDE Constrained Optimization*, International Series of Numerical Mathematics 165, DOI 10.1007/978-3-319-05083-6_24

constrained optimisation is summarised. Building upon [4] and the ideas proposed there, the present paper presents results for a multigrid-based solution strategy for the distributed and L^2 boundary control of nonstationary incompressible flow problems. A special Newton-type solver in the control space is developed which utilises space-time multigrid techniques for the Newton systems to enhance the efficiency.

In Sect. 2 the model problems considered in this paper are introduced. Section 3 presents a description of the standard Newton method in the control space and draws a comparison to other known methods. Section 4 introduces appropriate discretisation strategies for the space-time problems. The multigrid-based solver for the linear subproblems in the Newton approach is described in Sect. 5. Section 6 presents numerical tests regarding efficiency, and finally in Sect. 7, we draw some conclusions.

2 Model Problems

Our paper investigates nonstationary distributed as well as L^2 Dirichlet boundary control. In the following, let $\Omega \subset \mathbb{R}^d$ $(d = 2, 3)$ denote an open, bounded domain with boundary $\Gamma = \partial\Omega$ and outer unit normal vector η, $T > 0$ a final time, $Q := (0, T) \times \Omega$ the corresponding space-time domain and $\Sigma := (0, T) \times \Gamma$. Let the boundary Γ be decomposed into the three different, disjoint parts $\Gamma_D, \Gamma_N, \Gamma_C$, with $\overline{\Gamma} = \overline{\Gamma}_D \cup \overline{\Gamma}_N \cup \overline{\Gamma}_C$. Γ_D specifies the Dirichlet part of the boundary, Γ_N the Neumann part, and Γ_C the Dirichlet control part. Furthermore, let $\Sigma_N := (0, T) \times \Gamma_N$, $\Sigma_C := (0, T) \times \Gamma_C$ and $\Sigma_D := (0, T) \times \Gamma_D$.

2.1 Optimal Distributed Control of the Navier–Stokes Equations

Let $\Gamma_C = \emptyset$ in the following. Take a function $z : Q \to \mathbb{R}^d$, the so-called target function, a Dirichlet boundary condition $g : (0, T) \times \Gamma_D \to \mathbb{R}^d$ and an initial condition $y^0 : \Omega \to \mathbb{R}^d$. The aim is to find a control $u : Q \to \mathbb{R}^d$, a velocity field $y : Q \to \mathbb{R}^d$ and a pressure field $p : Q \to \mathbb{R}$ which solve the following minimisation problem,

$$J(y, u) = \frac{1}{2}||y - z||^2_{L^2(Q)} + \frac{\alpha}{2}||u||^2_{L^2(Q)} \quad \to \quad \min, \tag{2.1}$$

for a regularisation parameter $\alpha > 0$, where y, p and u are coupled through the nonstationary Navier–Stokes equations,

$$\begin{aligned}
y_t - \nu\Delta y + (y\nabla)y + \nabla p &= u && \text{in } Q,\\
-\operatorname{div} y &= 0 && \text{in } Q,\\
y &= g && \text{on } \Sigma_D,\\
\nu\partial_\eta y - p\eta &= 0 && \text{on } \Sigma_N,\\
y(0) &= y^0 && \text{in } \Omega.
\end{aligned}$$

Using the Lagrange multiplier technique, the following corresponding KKT system can be derived,

$$\begin{aligned}
y_t - \nu\Delta y + (y\nabla)y + \nabla p &= u && \text{in } Q,\\
-\operatorname{div} y &= 0 && \text{in } Q,\\
y &= g && \text{on } \Sigma_D,\\
\nu\partial_\eta y - p\eta &= 0 && \text{on } \Sigma_N,
\end{aligned}$$

$$\begin{aligned}
-\lambda_t - \nu\Delta\lambda - (y\nabla)\lambda + (\nabla y)^T\lambda + \nabla\xi &= y - z && \text{in } Q,\\
-\operatorname{div} \lambda &= 0 && \text{in } Q,\\
\lambda &= 0 && \text{on } \Sigma_D,\\
\nu\partial_\eta\lambda - \xi\eta + (y\cdot\eta)\lambda &= 0 && \text{on } \Sigma_N,
\end{aligned}$$

$$\begin{aligned}
y(0) = y^0, \qquad \lambda(T) &= 0 && \text{in } \Omega,\\
\alpha u + \lambda &= 0 && \text{in } Q. \qquad (2.2)
\end{aligned}$$

Here, $\lambda : Q \to \mathbb{R}^d$ denotes a dual velocity and $\xi : Q \to \mathbb{R}$ a dual pressure. It follows from (2.2) that in this setting, the control u lives in the same space as the dual velocity λ.

2.2 Optimal L^2 Boundary Control of the Navier–Stokes Equations

In the case of L^2 boundary control our minimisation problem reads: Find $u : \Sigma_C \to \mathbb{R}^d$, $y : Q \to \mathbb{R}^d$ and $p : Q \to \mathbb{R}$ which solve the following minimisation problem,

$$J(y,u) = \frac{1}{2}||y - z||^2_{L^2(Q)} + \frac{\alpha}{2}||u||^2_{L^2(\Sigma_C)} \quad \to \quad \min, \qquad (2.3)$$

where y, p and u are coupled through the nonstationary Navier–Stokes equations,

$$\begin{aligned} y_t - \nu\Delta y + (y\nabla)y + \nabla p &= 0 && \text{in } Q,\\ -\operatorname{div} y &= 0 && \text{in } Q,\\ y &= g && \text{on } \Sigma_D,\\ y &= u && \text{on } \Sigma_C,\\ \nu\partial_\eta y - p\eta &= 0 && \text{on } \Sigma_N,\\ y(0) &= y^0 && \text{in } \Omega. \end{aligned}$$

Using the Lagrange multiplier technique, the following corresponding KKT system can be derived,

$$\begin{aligned} y_t - \nu\Delta y + (y\nabla)y + \nabla p &= 0 && \text{in } Q,\\ -\operatorname{div} y &= 0 && \text{in } Q,\\ y &= g && \text{on } \Sigma_D,\\ y &= u && \text{on } \Sigma_C,\\ \nu\partial_\eta y - p\eta &= 0 && \text{on } \Sigma_N,\\ \\ -\lambda_t - \nu\Delta\lambda - (y\nabla)\lambda + (\nabla y)^T\lambda + \nabla\xi &= y - z && \text{in } Q,\\ -\operatorname{div}\lambda &= 0 && \text{in } Q,\\ \lambda &= 0 && \text{on } \Sigma_D \cup \Sigma_C\\ \nu\partial_\eta\lambda - \xi\eta + (y\cdot\eta)\lambda &= 0 && \text{on } \Sigma_N,\\ \\ y(0) = y^0, \qquad \lambda(T) &= 0 && \text{in } \Omega,\\ \alpha u - (\nu\partial_\eta\lambda - \xi\eta) &= 0 && \text{on } \Sigma_C. \end{aligned} \tag{2.4}$$

The control u acts only on the boundary and thus has a much smaller dimension than in the distributed case. However, for u to be computed, the fully coupled system has to be solved.

3 The Integral Equation Method for Nonlinear Problems

The integral equation approach to solve our control problems is based on the control equations of the KKT system, compare [7, 8]. At first, one defines the reduced cost functional

$$\hat{J}(u) := J(Su, u) \tag{3.1}$$

with $S : u \mapsto y$ being the solution operator that maps a control u to the solution y of the nonstationary Navier–Stokes equations. The first order optimality condition

$$(\nabla \hat{J}(u), \bar{u} - u) \geq 0 \qquad \text{for all } \bar{u} \in U_{\text{ad}}$$

leads to the Eqs. (2.2) and (2.4), respectively, with $U_{\text{ad}} := L^2(Q)$ in the distributed control and $U_{\text{ad}} := L^2(\Sigma_C)$ in the boundary control case. From these equations, one derives a Newton method which we present exemplarily for the control Eq. (2.2).

The Newton iteration The Newton iteration in the control space based on (2.2) reads

$$u_{n+1} := u_n - DF(u_n)^{-1} F(u_n), \tag{3.2}$$

i.e., expressed in two steps, with an intermediate defect d_n,

$$\text{(a) solve} \quad DF(u_n)\bar{u}_n = d_n := -F(u_n) \tag{3.3a}$$

$$\text{(b) update} \quad u_{n+1} = u_n + \bar{u}_n \tag{3.3b}$$

with

$$F(u) := \nabla \hat{J}(u) = \alpha u + \lambda \quad (\stackrel{!}{=} 0), \qquad DF(u)\bar{u} = H\hat{J}(u)\bar{u} = \alpha\bar{u} + \bar{\lambda}, \tag{3.4}$$

where H denotes the Hessian. (λ, ξ), (y, p) and $(\bar{\lambda}, \bar{\xi})$, $(\bar{y}, \bar{p})$ are the solutions of the following systems:

1. Primal/dual equation

$$\begin{aligned}
y_t - \nu\Delta y + (y\nabla)y - \nabla p &= u, && \text{in } Q,\\
-\operatorname{div} y &= 0, && \text{in } Q,\\
y &= g, && \text{on } \Sigma_D,\\
\nu\partial_\eta y - p\eta &= 0, && \text{on } \Sigma_N,
\end{aligned}$$

$$\begin{aligned}
-\lambda_t - \nu\Delta\lambda - (y\nabla)\lambda + (\nabla y)^T\lambda + \nabla\xi &= y - z && \text{in } Q,\\
-\operatorname{div}\lambda &= 0 && \text{in } Q,\\
\lambda &= 0 && \text{on } \Sigma_D,\\
\nu\partial_\eta\lambda - \xi\eta + (y\cdot\eta)\lambda &= 0 && \text{on } \Sigma_N,
\end{aligned}$$

$$y(0) = y^0, \quad \lambda(T) = 0 \qquad \text{in } \Omega,$$

2. Linearised primal equation

$$
\begin{aligned}
\bar{y}_t - \nu\Delta\bar{y} + (y\nabla)\bar{y} + (\bar{y}\nabla)y + \nabla p &= \bar{u} && \text{in } Q,\\
-\operatorname{div}\bar{y} &= 0 && \text{in } Q,\\
\bar{y} &= 0 && \text{on } \Sigma_D,\\
\nu\partial_\eta\bar{y} - \bar{p}\eta &= 0 && \text{on } \Sigma_N,\\
\bar{y}(0) &= 0 && \text{in } \Omega,
\end{aligned}
$$

3. And linearised dual equation

$$
\begin{aligned}
-\bar{\lambda}_t - \nu\Delta\bar{\lambda} - (y\nabla)\bar{\lambda} + (\nabla y)^T\bar{\lambda} + \nabla\bar{\xi} &= \bar{y} + \underbrace{(\bar{y}\nabla)\lambda - (\nabla\bar{y})^T\lambda}_{(3.5)} && \text{in } Q,\\
-\operatorname{div}\bar{\lambda} &= 0 && \text{in } Q,\\
\bar{\lambda} &= 0 && \text{on } \Sigma_D,\\
\nu\partial_\eta\bar{\lambda} - \xi\bar{\eta} + (y\cdot\eta)\bar{\lambda} &= -(\bar{y}\eta)\lambda && \text{on } \Sigma_N,\\
\bar{\lambda}(T) &= 0 && \text{in } \Omega.
\end{aligned}
$$

Remarks

(a) The calculation of $F(u_n)$ involves the simulation of a nonlinear forward and a linear backward equation in (1). The functions y_n and λ_n have to be stored.

(b) The Eq. (3.3a) is linear and can be solved with an iterative solver; in Sect. 5, we introduce a multigrid solver for this task. The iteration is based on the application of the operator $DF(\cdot)$, which involves the simulation of a linear forward and a linear backward equation (2)/(3). Both problems can be solved at roughly the same costs. Thus, each Newton iteration amounts to one nonlinear forward simulation and one linear backward iteration in (1) plus a couple of linear forward and backward iterations for the linearised equations in (2)/(3).

(c) The term (3.5) on the right-hand side in (3) was found to impose numerical difficulties in the first couple of Newton iterations. Numerical tests in this paper skip this term in the right-hand side assembly during the first one or two iterations, so the update is a kind of a mixture between a Picard and a Newton update.

4 Discretisation

The discretisation of the optimal control problem is chosen in such a way that the optimise-then-discretise approach yields the same as the discretise-then-optimise approach. We demonstrate this idea in a formal way based on (2.1). For the following illustration, we assume no-slip boundary conditions on the complete boundary, where we require $y \cdot \eta = 0$, compare the example in Sect. 6.1.

The following notations are used:

$$\begin{aligned}
A(y) &:= A(y,p) &&:= -\nu\Delta y + (y\nabla)y + \nabla p, \\
A'(y)\bar{y} &:= A'(y,p)(\bar{y},\bar{p}) &&= -\nu\Delta\bar{y} + (y\nabla)\bar{y} + (\bar{y}\nabla)y + \nabla\bar{p}, \\
A'(y)^*\lambda &:= A'(y,p)^*(\lambda,\xi) &&= -\nu\Delta\lambda - (y\nabla)\lambda + (\nabla y)^T\lambda + \nabla\xi.
\end{aligned}$$

Let $k > 0$ define a timestep size and $N \in \mathbb{N}$ the number of considered time intervals. Choosing the rectangular rule for the discretisation in time of the cost functional and the implicit Euler scheme for the discretisation of the primal equation leads to

$$J(\mathbf{y}^k, \mathbf{u}^k) = \frac{1}{2}k\sum_{i=1}^{N} ||y_i - z_i||_\Omega^2 + \frac{\alpha}{2}k\sum_{i=1}^{N} ||u_i||_\Omega^2,$$

where $\mathbf{y}^k = (y_0, \dots, y_N)$, $\mathbf{u}^k := (u_0, \dots, u_N)$ and

$$\begin{aligned}
(y_i - y_{i-1}) + kA(y_i) &= ku_i && \text{in } \Omega, \\
-\operatorname{div} y_i &= 0 && \text{in } \Omega, \\
y_0 + kA(y_0) &= y^0 + kA(y^0) && \text{in } \Omega, \\
y_i &= g(t_i) && \text{on } \Gamma.
\end{aligned}$$

Setting $\boldsymbol{\lambda}^k := (\lambda_0, \dots, \lambda_N)$ and applying formally the Lagrange multiplier technique leads to

$$\begin{aligned}
L(\mathbf{y}^k, \mathbf{u}^k, \boldsymbol{\lambda}^k) := \; & J(\mathbf{y}^k, \mathbf{u}^k) + \sum_{i=1}^{N} \big(\lambda_i, ku_i - (y_i - y_{i-1} - kA(y_i))\big) \\
& + \big(\lambda_0, (y^0 + kA(y^0)) - (y_0 - kA(y_0))\big), \qquad (4.1)
\end{aligned}$$

where boundary conditions are not shown here. From $DL(\mathbf{y}^k, \mathbf{u}^k, \boldsymbol{\lambda}^k) = 0$, one obtains the time-discretised system of equations,

$$\begin{aligned}
(y_i - y_{i-1}) + kA(y_i) &= ku_i && \text{in } \Omega, \\
-\operatorname{div} y_i &= 0 && \text{in } \Omega, \\
y_0 + kA(y_0) &= y^0 + kA(y^0) && \text{in } \Omega, \\
y_i &= g(t_i) && \text{on } \Gamma,
\end{aligned}$$

$$
\begin{aligned}
(\lambda_i - \lambda_{i+1}) + kA'(y_i)^*\lambda_i &= k(y_i - z_i) && \text{in } \Omega,\\
-\operatorname{div}\lambda_i &= 0 && \text{in } \Omega,\\
\lambda_N + kA'(y_N)^*\lambda_N &= k(y_N - z_N) && \text{in } \Omega,\\
\lambda_i &= 0 && \text{on } \Gamma,
\end{aligned}
$$

$$
\alpha u_i + \lambda_i = 0 \qquad \text{in } \Omega,
$$

The discrete counterparts of the linearised primal/dual equations are derived by taking the Fréchet derivatives of the complete system,

$$
\begin{aligned}
(\bar{y}_i - \bar{y}_{i-1}) + kA'(y_i)\bar{y}_i &= k\bar{u}_i && \text{in } \Omega,\\
-\operatorname{div} y_i &= 0 && \text{in } \Omega,\\
y_0 + kA(y_0) &= 0 && \text{in } \Omega,\\
\bar{y}_i &= 0 && \text{on } \Gamma,
\end{aligned}
$$

$$
\begin{aligned}
(\bar{\lambda}_i - \bar{\lambda}_{i+1}) + kA'(y_i)^*\bar{\lambda}_i &= k(\bar{y}_i - \bar{z}_i) - kA'(\bar{y}_i)^*\lambda_i && \text{in } \Omega,\\
-\operatorname{div}\lambda_i &= 0 && \text{in } \Omega,\\
\bar{\lambda}_N + kA'(y_N)^*\lambda_N &= k(\bar{y}_N - \bar{z}_N) && \text{in } \Omega,\\
\bar{\lambda}_i &= 0 && \text{on } \Gamma,
\end{aligned}
$$

$$
\alpha\bar{u}_i + \bar{\lambda}_i = 0 \qquad \text{in } \Omega,
$$

The fully discretised system After applying the time discretisation, a space discretisation can be used to generate the fully discretised system. In the following, the fully discretised, vector valued variables are denoted by

$$
\begin{aligned}
\mathbf{y} &:= \mathbf{y}^{k,h} := (y_0^h, \ldots, y_N^h),\\
\mathbf{u} &:= \mathbf{u}^{k,h} := (u_0^h, \ldots, u_N^h),\\
\boldsymbol{\lambda} &:= \boldsymbol{\lambda}^{k,h} := (\lambda_0^h, \ldots, \lambda_N^h),
\end{aligned}
$$

with y_i^h, u_i^h and λ_i^h vectors of degrees of freedom in the $\mathbb{R}^n$ in every timestep (for a corresponding n depending on the space). The domain Ω is approximated by a mesh Ω_h. The index h indicates the discretisation in space and the index k the discretisation in time. In this work, we employ a discretisation with the Q_2 element for the velocity and the P_1^{disc} element for the pressure, see, e.g., [14].

The discrete Newton method The discrete counterpart of the Newton iteration used in this work reads

$$1.\ \text{Solve} \quad DF_{k,h}(\mathbf{u}_n)\mathbf{g} = \mathbf{d} := -F_{k,h}(\mathbf{u}_n), \tag{4.2a}$$

$$2.\ \text{Update} \quad \mathbf{u}_{n+1} := \mathbf{u}_n + \mathbf{g}, \tag{4.2b}$$

with

$$F_{k,h}(\mathbf{u}) := \alpha\mathbf{u} + \boldsymbol{\lambda}, \qquad DF_{k,h}(\mathbf{u})\bar{\mathbf{u}} = \alpha\bar{\mathbf{u}} + \bar{\boldsymbol{\lambda}}. \tag{4.3}$$

5 Multigrid for the Control Equation

Equation (4.2a) defines a linear system for the correction $\mathbf{g}$ of the control. The linear system is defined in space and time, each component of $\mathbf{g} = \mathbf{g}^{k,h} = (g_0^h, \ldots, g_N^h)$ defines one discrete function in space at a specified point in time. In this work, a multigrid approach according to Hackbusch [6–8] is applied to (4.2a) in order to solve this equation, see also [4]. This necessitates to begin with a couple of definitions.

5.1 *The Space-Time Mesh Hierarchy*

At first, we define a hierarchy of space-time meshes as follows, which is based on a space-time fine mesh with N equidistant intervals in time and a mesh Ω_h in space.

- **Space hierarchy:** Let $\Omega_1, \Omega_2, \ldots, \Omega_M$ denote a hierarchy of $M \in \mathbb{N}$ regularly refined meshes, i.e., new vertices are generated by connecting opposite midpoints. We assume $\Omega_M = \Omega_h$ to be the finest mesh of this hierarchy.
- **Time hierarchy:** Let $T_1, \ldots, T_L$ denote a hierarchy of $L \in \mathbb{N}$ regularly refined meshes in time, defined by the following decomposition. We set $N_L := N$ and assume for convenience that there is an $N_1 \in \mathbb{N}$ with $N = 2^{L-1}N_1$. Then, for $l = 1, \ldots, L$, the mesh T_l is given by $N_l = 2^{l-1}N_1$ equidistant time intervals in $[0, T]$ with interval length $k_l := \frac{T}{N_l}$.
- **Space-time hierarchy:** A space-time hierarchy can be created by different coarsening strategies, starting from the finest combination of space and time mesh. For simplicity, we assume $L = M$ and denote by $\mathcal{Q}_1, \ldots, \mathcal{Q}_L$ a sequence of L nested space-time meshes. Typical choices for these hierarchies are, with $l = 1, \ldots, L$,
  - Coarsening in space and time: $\mathcal{Q}_l := (T_l, \Omega_l)$,
  - Semi-coarsening in time: $\mathcal{Q}_l := (T_l, \Omega_L)$,
  - Semi-coarsening in space: $\mathcal{Q}_l := (T_L, \Omega_l)$.

5.2 Discretisation and Problem Hierarchy

Secondly, we have to define a discretisation on every level. In this work, the time discretisation is carried out with the Implicit Euler scheme, while the Q_2/P_1^{disc} finite element pair is used for the space discretisation. We use the following notations:

- V^m denotes for $m = 1, \ldots, L$ the space discretisation of the control space, realised by the degrees of freedom of the underlying finite element space for the velocity.
- $W^{l,m}$ denotes for $l, m = 1, \ldots, L$ the space-time discretisation of the control space using V^m for the discretisation in space on the time mesh T_l.
- W^l defines for $l = 1, \ldots, L$ the space-time discretisation of the control space corresponding to $\mathcal{Q}_l$. As a consequence, a coarsening strategy in space in time induces $W^l = W^{l,l}$, a semi-coarsening in time $W^l = W^{l,L}$ and a semi-coasening in space $W^l = W^{L,l}$.

Problem hierarchy The space-time discretisation induces a hierarchy of problems. For $l = 1, \ldots, L$, a hierarchy of discrete equations, derived from (3.2), reads

$$F^l(\mathbf{u}^l) := \alpha \mathbf{u}^l + \boldsymbol{\lambda}^l \stackrel{!}{=} 0 \qquad \text{in } W^l, \tag{5.1}$$

with $\mathbf{u}^l \in W^l$ the discrete counterpart to $\mathbf{u}$ (to be determined), $\boldsymbol{\lambda}^l$ the discrete counterpart to $\boldsymbol{\lambda}$ and the operator F^l the discrete counterpart to F on W^l. Correspondingly, the discrete linearised system to be solved in every step of the Newton method on level l reads

$$DF^l(\mathbf{u}^l)\bar{\mathbf{u}}^l = -F^l(\mathbf{u})^l, \qquad \text{for} \quad DF^l(\mathbf{u}^l)\bar{\mathbf{u}}^l := \alpha\bar{\mathbf{u}}^l + \bar{\boldsymbol{\lambda}}^l, \tag{5.2}$$

with $\bar{\boldsymbol{\lambda}}^l = \bar{\boldsymbol{\lambda}}^l(\bar{\mathbf{y}}^l(\bar{\mathbf{u}}^l))$ the solution of the linearised discrete dual equation. The solution of the problem is sought at level L, i.e., on the finest mesh.

Primal/dual equations on lower levels Let $\mathbf{y}^L$ and $\boldsymbol{\lambda}^L$ define the solutions of the primal and dual equation on level L. For the operator DF^l to be applied on level $l < L$, corresponding primal/dual solutions $\mathbf{y}^l$ and $\boldsymbol{\lambda}^l$ are needed. They can be obtained with an L^2 projection of the finite element counterparts of $\mathbf{y}^L$ and $\boldsymbol{\lambda}^L$ in space and time to the lower level, realised approximately by a proper interpolation of the degrees of freedom.

5.3 Multigrid Components

Multigrid needs a couple of components to be properly defined in order to be effective:

Prolongation Let the prolongation operator from level l to level $l+1$ be denoted by $I_l^{l+1} : W^l \to W^{l+1}$. Depending on the choice of the coarsening strategy, the operator has a temporal and a spatial component. Each component u_i^h of a control vector $\mathbf{u}^l = (u_0^h, u_1^h, \ldots, u_{N_l}^h) \in W^l$ corresponds to a finite element function and thus, meaningful prolongation in space is the finite element prolongation. On the other hand, a prolongation in time is derived by a linear finite difference interpolation of the solutions in time, i.e.,

$$(u_0^h, u_1^h, \ldots, u_{N_l}^h) \mapsto \left(u_0^h, \frac{u_0^h + u_1^h}{2}, u_1^h, \frac{u_1^h + u_2^h}{2}, \ldots, u_{N_l}^h \right).$$

Restriction Let a restriction operator from level l to level $l-1$ be denoted by $I_l^{l-1} : W^l \to W^{l-1}$, and let $d^l := (d_0^h, \ldots, d_{N_l}^h) \in W^l$ be a defect vector. Similar to the prolongation, the restriction has a temporal and a spatial component. One possible choice for a restriction in time is a weighted mean in the sense of finite differences. However, for a discretisation with the implicit Euler, it is enough to apply a constant restriction in time, given by

$$(d_0^h, d_1^h, \ldots, d_{N_l}^h) \mapsto \left(d_0^h, \frac{d_1^h + d_2^h}{2}, d_2^h, \frac{d_3^h + d_4^h}{2}, d_4^h, \ldots, d_{N_l}^h \right).$$

This restriction is 'backward directed', thus, respects the direction of the propagation of information in time specified by the dual equation and will be shown to be effective in numerical tests.

For the restriction in space, one has to take into account that the discrete operator F^l maps $W^l \to W^l$, i.e., the operator works directly in the control space without any test functions, mass matrices or similar things involved. An appropriate choice is therefore the L^2 projection of the control space to a lower level, which is realised approximately by a simple interpolation of the degrees of freedom.

Coarse grid solver and smoother Typical choices for coarse grid solver and smoothers are iterative algorithms which only necessitate the application of the corresponding operator. For example, provided a damping parameter $0 < \omega \leq 1$, the Richardson iteration reads

$$\mathbf{g}_{\text{new}} := \mathbf{g} + \omega(\mathbf{d} - DF^l(\mathbf{u}_n)\mathbf{g}).$$

In a similar way, it is possible to apply a CG, BiCGStab or GMRES method. Applying such an algorithm on the coarse level until convergence is the usual choice for a coarse grid solver. Taking only a fixed number of iterations on any level except for the coarse level, one obtains a smoother for that level. In the following, $\mathbf{g} \mapsto S_l(\mathbf{g}, \mathbf{d}, \text{NSM})$ denotes such a smoother on level l which applies NSM smoothing steps using a right-hand side $\mathbf{d}$.

The operator to be applied in such algorithms reads DF^l and is realised by a forward-backward solving process: A forward iteration solves for $\bar{\mathbf{y}}$ and a backward iteration for $\bar{\boldsymbol{\lambda}}$. During the forward and the backward iteration, linear problems in space must be solved. In this work, we apply a multigrid solver in space for this task which provides low, level independent convergence rates. Smoothing and coarse grid solving processes in space are realised with local pressure-Schur-Complement ('Vanka'-) like techniques which process all variables in space (velocity/pressure) simultaneously.

5.4 *The Multigrid Algorithm*

With the above components, Algorithm 1 describes a basic V-cycle multigrid in the control space. For a more general implementation (also concerning other cycles, etc.), the interested reader is referred to [1, 9, 16].

Algorithm 1 Space-time multigrid

Predefined constant: NSM $\in \mathbb{N}_0$: number of (post-)smoothing steps

function SpaceTimeMultigrid($\bar{\mathbf{u}}$;$\mathbf{d}$;l)
    if $(l = 1)$ **then**
        return $DF^l(\mathbf{u}^l)^{-1}\mathbf{d}$ ▷ coarse grid solver
    end if
    while (not converged) **do**
        $\mathbf{d}^{l-1} \leftarrow I_l^{l-1}(\mathbf{d} - DF^l(\mathbf{u}^l)\bar{\mathbf{u}}) \quad \in W^{l-1}$ ▷ restriction of the defect
        $\mathbf{g}^{l-1} \leftarrow$ SpaceTimeMultigrid$(0; \mathbf{d}^{l-1}; l-1) \quad \in W^{l-1}$
        ▷ coarse grid solution
        $\bar{\mathbf{u}} \leftarrow \bar{\mathbf{u}} + I_{l-1}^l(g^{l-1})$ ▷ coarse grid correction
        $\bar{\mathbf{u}} \leftarrow S_l(\bar{\mathbf{u}}, \mathbf{d}, \text{NSM})$ ▷ postsmoothing
    end while
    return $\bar{\mathbf{u}}$ ▷ solution
end function

6 Numerical Examples

The following numerical tests focus on the optimal control of a cavity flow and a backward-facing step flow. Tests are carried out for single-grid solvers, multigrid solvers, for distributed control as well as for boundary control.

6.1 *Distributed Control for Driven–Cavity Flow*

Example We consider the optimal distributed control of the Navier–Stokes equations, see Sect. 2.1. The underlying domain is $\Omega := (0, 1)^2$ with the four boundary

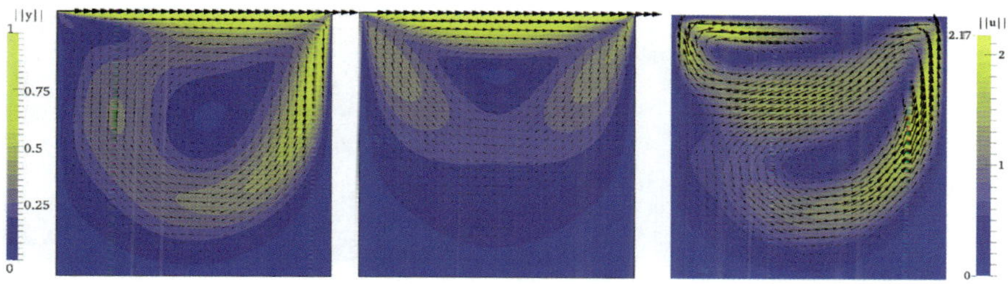

Fig. 1 'Driven–Cavity' example, velocity profile. Initial flow y^0 (*left*), target flow z (*centre*), optimal control u at $t = 0.0625$ (*right*)

Table 1 *Driven–Cavity:* Mesh statistics for distributed control, different refinement levels

Space-level	#vertices	#edges	#elements	#dof$_{pd}$	#dof$_c$
4	81	144	64	770	578
5	289	544	256	2,946	2,178
6	1,089	2,112	1,024	11,522	8,450
7	4,225	8,320	4,096	45,570	33,282

parts Γ_1, Γ_2, Γ_3 and Γ_4 on the bottom, left, top and right. The problem is set up as a pure Dirichlet problem with $y(x,t) = (0,0)$ for $x \in \Gamma_1 \cup \Gamma_2 \cup \Gamma_4$ and $y(x,t) = (1,0)$ for $x \in \Gamma_3$. The coarse grid consists of only one square element. The time interval is defined as $[0,T]$ with $T = 1$, the viscosity parameter is set to $\nu = 1/400$. The initial flow y^0 is the stationary fully developed Navier–Stokes flow at $\nu = 1/400$, while the target flow z is chosen as the fully developed, stationary Stokes flow, see Fig. 1. The regularisation parameter for the control is set to $\alpha = 0.01$.

The basic spatial coarse grid used in this test is a mesh containing one cell $[0,1]^2$ three times refined, i.e., $h = 1/8$. The basic time mesh contains 20 time intervals. Both meshes are regularly refined to generate a hierarchy of meshes. The solution is sought at the finest mesh. Table 1 presents statistical data about the space and the time mesh for different refinement levels (with '#vertices' the number of vertices, '#edges' the number of edges, '#elements' the number of elements, #dof$_{pd}$ the number of degrees of freedom in the primal and the dual space, resp., and #dof$_c$ the number of degrees of freedom in the control space). As mentioned, the spatial discretisation is carried out with Q_2/P_1^{disc}.

Solver configuration For the following tests we apply an inexact version of the described Newton algorithm above. The space-time Newton algorithm was configured to reduce the L^2 norm of the initial residual by six digits. The space-time multigrid algorithm in every Newton step reduces its residual adaptively (at least gaining two digits) such that one obtains quadratic convergence; for a description of this strategy, see, e.g., [15]. The same stopping criterion was also used for the coarse grid solver. A V-cycle is used. For smoothing, four steps of a space-time CG method are applied.

Table 2 *Driven–Cavity:* Solver statistics for distributed control. Single grid and multigrid solver applied on different coarsening strategies

S.-Lv.	#int	T_{opt}	T_{sim}	#NL	$\sum$#LIN	$\frac{T_{opt}}{T_{sim}}$	$\frac{\sum \#LIN}{\#NL}$
Single grid CG preconditioner							
5	40	0:16:35	0:00:13	4	67	79.0	16.8
6	80	2:14:23	0:01:41	4	64	79.6	16.0
7	160	15:37:20	0:11:33	4	63	81.1	15.8
Multigrid preconditioner, pure space coarsening							
5	40	0:41:32	0:00:13	4	17	197.8	4.2
6	80	3:48:29	0:01:41	4	14	135.3	3.5
7	160	25:48:41	0:11:33	4	16	134.0	4.0
Multigrid preconditioner, pure time coarsening							
5	2	0:26:47	0:00:13	4	10	127.5	2.5
6	3	3:22:01	0:01:41	4	8	119.7	2.0
7	4	21:30:05	0:11:33	4	7	111.6	1.8
Multigrid preconditioner, space-time coarsening							
5	40	0:36:46	0:00:13	4	17	175.1	4.2
6	80	3:11:02	0:01:41	4	15	113.1	3.8
7	160	21:41:53	0:11:33	4	11	112.6	2.8

Nonlinear and linear problems in space (calculated during the forward and backward loops) were solved until the l^2 norm of the residual drops below 10^{-14}; a spatial (Newton-)Multigrid solver with coarse grid solver on level four is applied in every timestep for this purpose. The local multigrid solver in space applies a local pressure Schur complement technique for smoothing and coarse grid solving, see also [11–13, 15]. This smoother is capable of processing velocity and pressure variables simultaneously, which renders it ideal for saddle-point problems.

Solver efficiency test The following test applies a single grid and a multigrid solver strategy. On different refinement levels in space and time the Newton algorithm is applied, see Table 2. 'S.-Lv.' specifies the refinement level in space, '#int' the number of intervals in time, 'T_{opt}' documents the time which was necessary for the computation of the optimisation problem and 'T_{sim}' the time which was needed for the computation of the first forward simulation, i.e., for a simulation without any control applied. '#NL' and '$\sum$#LIN' depict the number of Newton steps and the sum of all steps of the linear solver, respectively.

The numerical test is applied for four different solver configurations. The first part of the table contains the result for a single grid CG solver. In the second and third part, a multigrid solver is used where the space-time hierarchy is build up using coarsening in space only (until space level 4) or in time only (until the time mesh has

20 time intervals). The last part finally uses full space-time coarsening, i.e., coarser meshes are generated by coarsening in space and time.

One can see that for this configuration, already the single-grid solver provides linear complexity. Each refinement gives a factor of 8 in the number of unknowns and the computing time. The ratio between simulation and optimisation is a factor of about 30.

If space-time multigrid is used for preconditioning (second to fourth part of the table), the results are two-fold. The total number of multigrid steps in this test is either constant or even reducing with increasing refinement level, in particular upon increasing the number of timesteps. Counting the number of CG steps on the finest level, there are 63 CG steps for the single grid solver and 28 steps ($7 \times$ NSM, with NSM $= 4$ smoothing steps per multigrid step) for the multigrid solver with time coarsening. So multigrid successfully accellerates the convergence. A hierarchy generated from pure space-coarsening is rather ineffective.

Numerical efficiency From the viewpoint of numerical efficiency, the overhead for the space-time Newton algorithm is rather large. The ratio $\frac{T_{\text{opt}}}{T_{\text{sim}}}$ is much higher than that of a single-grid algorithm. It depends on the configuration of the coarse grid problems and the solver parameters if the approach is effective. In the above test, one can expect that the use of the multigrid approach will pay off for large problems if space-time coarsening is used, see, e.g., [4]. The convergence speeds up with higher refinement levels and the effort for solving the coarse grid problems is not too large. However it must be said, that the underlying space-time meshes for our applications in this project could not be chosen fine enough to properly work out this effect.

Possible enhancements which may help to render the approach more efficient than a single-grid solver are the choice of a different hierarchy (e.g., coarsening twice in time per space coarsening), the use alternative smoothers (e.g., GMRES) or the application of an advanced strategy to choose the stopping criteria of all the involved solver components. A detailed analysis is, however, out of the scope of this work.

Comparison to SQP In the following, a small comparison of the solver efficiency results between the Newton solver in this paper and the SQP-type solver analysed in [11, 12, 15] is drawn. The latter one applies an inexact Newton strategy in the primal/dual space where the solution vector is given as $x = (y_0, p_0, \lambda_0, \xi_0, \ldots, y_N, p_N, \lambda_N, \xi_N)$. The control is eliminated An outer space-time Newton solver reduces the L^2 norm of the nonlinear residual by the factor 10^{-6}. Linear subproblems are solved either with a one-level space-time BiCGStab(FBSimSolver) solver or a space-time multigrid solver (using V-cycle, four steps BiCGStab(FBSimSolver) for smoothing and BiCGStab(FBSimSolver) for coarse grid solving). The stopping criterion of the linear solver is configured adaptively to obtain quadratic convergence. Subproblems in space are solved with a spatial multigrid which is set up to gain two digits. The test configuration is chosen as above.

Table 3 *Driven–Cavity:* Solver statistics for distributed control. SQP-type solver in the primal/dual space (u eliminated), space-time BiCGStab and multigrid preconditioner

Single grid BiCGStab preconditioner							
S.-Lv.	#int	T_{opt}	T_{sim}	#NL	$\sum$#LIN	$\frac{T_{\text{opt}}}{T_{\text{sim}}}$	$\frac{\sum\text{\#LIN}}{\text{\#NL}}$
5	40	0:13:58	0:00:13	5	25	66.5	5.0
6	80	1:54:21	0:01:41	5	37	67.7	7.4
7	160	18:24:56	0:11:33	4	36	95.6	9.0
Multigrid preconditioner, space-time coarsening							
S.-Lv.	#int	T_{opt}	T_{sim}	#NL	$\sum$#LIN	$\frac{T_{\text{opt}}}{T_{\text{sim}}}$	$\frac{\sum\text{\#LIN}}{\text{\#NL}}$
5	40	0:15:26	0:00:13	3	6	41.5	2.0
6	80	2:04:08	0:01:41	3	7	40.3	2.3
7	160	11:24:15	0:11:33	3	6	53.9	2.0

Remark (FBSimSolver) The FBSimSolver iteration is a counterpart to the forward-backward loop applied in the computation of the space-time defect for a solution $\bar{u}$. It basically works as follows: For a given iterate $\bar{x} = (\bar{y}, \bar{p}, \bar{\lambda}, \bar{\xi})$ in the linearised primal/dual system

- Compute a new solution $(\bar{y}^{\text{new}}, \bar{p}^{\text{new}})$ using $\bar{u} = -\frac{1}{\alpha}\bar{\lambda}$ in the right-hand side,
- Compute a new solution $(\bar{\lambda}^{\text{new}}, \bar{\xi}^{\text{new}})$ using $\bar{y}^{\text{new}}$ in the right-hand side,
- Define the new iterate as $\bar{x}^{\text{new}} = (\bar{y}^{\text{new}}, \bar{p}^{\text{new}}, \bar{\lambda}^{\text{new}}, \bar{\xi}^{\text{new}})$.

For a detailed description and definition of this and the other mentioned solver components, see [12, 15].

Table 3 gives the results for the SQP solver. The solver is very stable, it basically needs only three nonlinear iterations to converge. In comparison to the Newton solver, the computing time is rather the same on low levels. For higher levels, if a space-time multigrid preconditioner is applied, the SQP solver is more efficient in this example. Space-time multigrid is indeed necessary in this case, as a single-grid solver loses efficiency on higher levels – the number of linear steps per nonlinear step Σ#LIN/#NL rises if only BiCGStab is applied. However, one should be careful with a comparison of the total time T_{opt} between both solvers, as to gain six digits in the primal/dual space does not necessarily mean to gain six digits in the control space and vice versa.

Remark: The number of nonlinear iterations differs whether the one-level preconditioner or the space-time multigrid preconditioner is applied. This is due to technical reasons. The BiCGStab solver checks the preconditioned residual while the multigrid preconditioner checks the real residual in the stopping criterion. As a consequence, BiCGStab does not solve accurately enough for the Newton to converge in three steps while multigrid does.

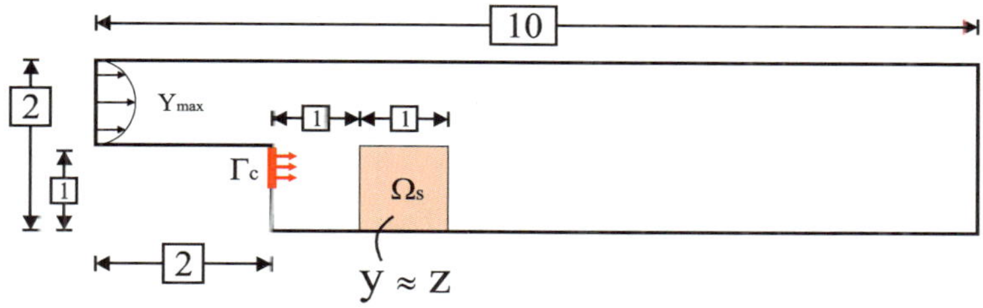

Fig. 2 Test configuration 'Backward-facing step'

Table 4 *Backward-facing step*: Mesh statistics for boundary control, different refinement levels

S.-Lv.	#vertices	#edges	#elements	$\#dof_{p\text{-}1}$	$\#dof_c$
2	97	168	72	890	1
3	337	624	288	3,362	3
4	1,249	2,400	1,152	13,058	5
5	4,801	9,408	4,608	51,458	9

6.2 L^2 Boundary Control for Backward-Facing Step

Example We consider the optimal L^2 boundary control of the Navier–Stokes equations, see Sect. 2.2. The basic domain for this test is a backward-facing step geometry, see Fig. 2, on a time interval $[0, T]$ with $T = 10$. On the left, a maximum inflow $y_{\max} = 1.5$ is prescribed, while on the right, do-nothing boundary conditions characterise the outflow; using $\nu = 1/100$, this results in a Re = 100 optimisation. The part $\Gamma_C = \{2\} \times (0.5, 1) \subset \partial\Omega$ defines a control boundary of length 0.5 on the top of the step.

The initial flow y^0 is the fully developed nonstationary Navier–Stokes flow, the target flow is the stationary Stokes flow, restricted to the observation area $\Omega_s = [3, 4] \times [0, 1]$ (which induces the right-hand side "$y - z$" of the control equation being replaced by $\chi_{\Omega_s} \cdot (y - z)$, with χ_{Ω_s} the characteristic function of Ω_s). The regularisation parameter for the control is set to $\alpha = 0.2$. Table 4 gives an overview about the problem size; the cells on the coarse mesh have a size of $h = 1.0$. Figure 3 visualises the controlled flow at $t = 1.25$ and $t = 5.0$.

Single-grid and time-multigrid test Table 5 depicts the solver statistics for a single-level CG solver and a multigrid solver, carried out on different space-time levels. Due to the small number of unknowns of the control in space, any coarsening in space would not make much sense. Therefore, for multigrid tests, pure time coarsening is applied until a space-time coarse mesh with 20 time intervals is reached.

The solver configuration is the same as in Sect. 6.1. Both types of linear solvers, the single-grid CG method as well as the time-multigrid method, converge with rather level-independent convergence rates. With four CG smoothing steps per multigrid iteration, the multigrid method needs in this example about 40 CG

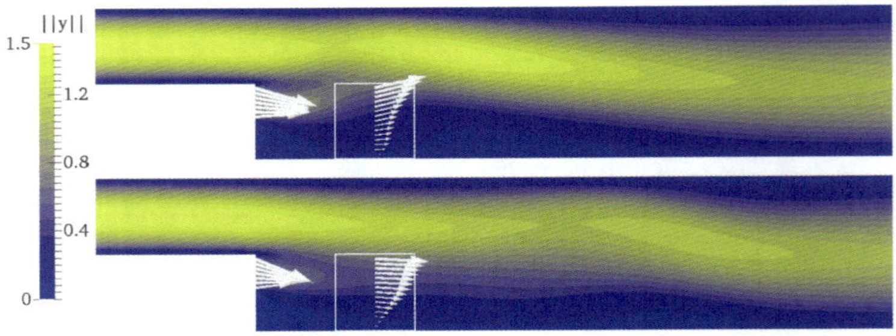

Fig. 3 Test configuration 'Backward-facing step'. Controlled flow at $t = 1.25$ and $t = 5.0$

Table 5 *Backward-facing step*: Single grid (*top*) and multigrid test (*bottom*)

Single-grid test							
S.-Lv.	#int	T_{opt}	T_{sim}	#NL	$\sum$#LIN	$\frac{T_{\text{opt}}}{T_{\text{sim}}}$	$\frac{\sum \#LIN}{\#NL}$
3	40	0:13:46	0:00:17	5	27	48.3	5.4
4	80	2:47:18	0:02:19	6	41	72.2	6.8
5	160	23:35:08	0:14:59	6	46	94.5	7.7
Multigrid test, pure time coarsening							
S.-Lv.	#int	T_{opt}	T_{sim}	#NL	$\sum$#LIN	$\frac{T_{\text{opt}}}{T_{\text{sim}}}$	$\frac{\sum \#LIN}{\#NL}$
3	40	0:32:02	0:00:17	5	8	113.7	1.6
4	80	6:18:30	0:02:19	6	10	163.9	1.7
5	160	47:50:46	0:14:59	6	10	191.6	1.7

iterations on the finest level, which is slightly less than in the one-level CG solver case. However, due the additional overhead on coarser levels and the fact that the time mesh is rather coarse, the application of the multigrid method is not really reasonable. The total time is about twice as high as a single-grid approach as the costs for solving the coarse grid problems is as large as the costs for the iteration on the finest grid – which is typical for a multigrid approach. One would need much finer time meshes until the use of multigrid will be advantageous.

7 Summary and Discussion

This paper presented the application of a space-time Newton method for optimal control of the nonstationary Navier–Stokes equations. A space-time multigrid method in the control space was used for solving linear subproblems. The basic method was described and the efficiency of the method was analysed in numerical examples using distributed and L^2 boundary control.

Concerning the numerical results, it is a fact that the Newton approach does often not need multigrid for the linear subproblems to be solved. Only on very fine time meshes in combination with distributed control, the multigrid solver seems to be advantageous as our numerical results indicate that the solver speeds up with increasing problem size.

Acknowledgements This work was financed by the program SPP1253 from the DFG, projects HI689/5-2 and TU102/24-1+2.

References

1. R.E. Bank, T.F. Dupond, An optimal order process for solving finite element equations. Math. Comput. **36**(153), 35–51 (1981)
2. A. Borzi, V. Schulz, Multigrid methods for PDE optimization. SIAM Rev. **51**(2), 361–395 (2009)
3. A. Borzi, V. Schulz, *Computational Optimization of Systems Governed by Partial Differential Equations* (SIAM, Philadelphia, 2011)
4. G. Büttner, Ein Mehrgitterverfahren zur optimalen Steuerung parabolischer Probleme. PhD thesis, Fakultät II – Mathematik und Naturwissenschaften der Technischen Universität Berlin (2004), http://edocs.tu-berlin.de/diss/2004/buettner_guido.pdf
5. M.D. Gunzburger, *Perspectives in Flow Control and Optimization* (SIAM, Philadelphia, 2003). ISBN:089871527X
6. W. Hackbusch, Fast solution of elliptic control problems. J. Opt. Theory Appl. **31**(4), 565–581 (1980)
7. W. Hackbusch, Die schnelle Auflösung der Fredholmschen Integralgleichung zweiter Art. Beiträge zur numerischen Mathematik, **9**, 47–62 (1981)
8. W. Hackbusch, Numerical solution of linear and nonlinear parabolic optimal control problems, in *Optimization and Optimal Control*, eds. by A. Auslender, W. Oettli, J. Stoer. Lecture Notes in Control and Information Science, vol. 30 (Springer, Berlin/New York, 1981), pp. 179–185
9. W. Hackbusch, *Multi-Grid Methods and Applications*. Springer Series in Computational Mathematics (Springer, Berlin, 1985). ISBN 3-540-12761-5
10. M. Hinze, Optimal and instantaneous control of the instationary Navier–Stokes equations. Habilitation thesis, Institut für Numerische Mathematik, Technische Universität Dresden, 2000
11. M. Hinze, M. Köster, S. Turek, A hierarchical space-time solver for distributed control of the Stokes equation. Preprint SPP1253-16-01, SPP1253, 2008
12. M. Hinze, M. Köster, S. Turek. A space-time multigrid solver for distributed control of the time-dependent Navier–Stokes system. Preprint SPP1253-16-02, SPP1253, 2008
13. M. Hinze, M. Köster, S. Turek, A hierarchical space-time solver for optimal distributed control of fluid flow, 2009. Proceedings of the Conference on Modeling, Simulation and Optimization of Complex Processes, Heidelberg, Accepted 21–25 July 2008
14. V. John, G. Matthies, Higher order finite element discretizations in a benchmark problem for incompressible flows. Int. J. Numer. Methods Fluids **37**, 885–903 (2001)
15. M. Köster, A hierarchical flow solver for optimisation with PDE constraints. PhD thesis, TU Dortmund, Lehrstuhl III für Angewandte Mathematik und Numerik, 2011. Slightly corrected version with an additional appendix concerning prolongation/restriction
16. H. Yserentant, Old and new convergence proofs for multigrid methods. Acta Numer. **2**, 285–326 (1993)

Convergence of Adaptive Finite Elements for Optimal Control Problems with Control Constraints

Kristina Kohls, Arnd Rösch, and Kunibert G. Siebert

Abstract We summarize our findings in the analysis of adaptive finite element methods for the efficient discretization of control constrained optimal control problems. We particularly focus on convergence of the adaptive method, i.e., we show that the sequence of adaptively generated discrete solutions converges to the true solution. We restrict the presentation to a simple model problem to highlight the key components of the convergence proof and comment on generalizations of the presented result.

Keywords Adaptive finite elements • Aposteriori error estimators • Convergence analysis • Optimal control • Control constraints

Mathematics Subject Classification (2010). 65N30, 65N12, 49J20.

1 Statement of the Main Result

In this summary we analyze adaptive finite element discretizations for control constrained optimal control problems of the form

$$\begin{aligned} &\min_{(u,y)\in\mathbb{U}^{\mathrm{ad}}\times\mathbb{Y}} \frac{1}{2}\|y - y_d\|_{\mathbb{U}}^2 + \frac{\alpha}{2}\|u\|_{\mathbb{U}}^2 \\ &\text{subject to} \qquad y \in \mathbb{Y}: \quad \mathcal{B}[y,\, v] = \langle u,\, v\rangle \qquad v \in \mathbb{Y}. \end{aligned} \tag{1.1}$$

K. Kohls • K.G. Siebert (✉)
Institut für Angewandte Analysis und Numerische Simulation, Fachbereich Mathematik, Universität Stuttgart, Pfaffenwaldring 57, D-70569 Stuttgart, Germany
e-mail: http://www.ians.uni-stuttgart.de/nmh/; kristina.kohls@ians.uni-stuttgart.de; http://www.ians.uni-stuttgart.de/nmh/; kg.siebert@ians.uni-stuttgart.de

A. Rösch
Fakultät für Mathematik, Thea-Leymann-Straße 9, D-45127 Essen, Germany
e-mail: http://www.uni-due.de/mathematik/agroesch/; arnd.roesch@uni-due.de

G. Leugering et al. (eds.), *Trends in PDE Constrained Optimization*, International Series of Numerical Mathematics 165, DOI 10.1007/978-3-319-05083-6_25

In order to highlight the basic ideas of our convergence analysis we focus on the most simple model problem in the following setting. We let $\Omega \subset \mathbb{R}^d$ be a bounded domain that is meshed exactly by some conforming initial triangulation $\mathcal{G}_0$. We consider distributed control in $\mathbb{U} = L_2(\Omega)$ with a non-empty, convex, and closed subset $\mathbb{U}^{\mathrm{ad}}$ of admissible controls. We use the $L_2(\Omega)$ scalar product $\langle \cdot, \cdot \rangle$ and write $\|\cdot\|_{\mathbb{U}} = \|\cdot\|_{2;\Omega}$ for its induced norm. The PDE constraint is given by Poisson's problem in the state space $\mathbb{Y} = \mathring{H}^1(\Omega)$ equipped with norm $\|\cdot\|_{\mathbb{Y}} = \|\nabla \cdot\|_{2;\Omega}$ and the continuous and coercive bilinear form

$$\mathcal{B}[y, v] = \langle \nabla y, \nabla v \rangle \qquad \forall y, v \in \mathbb{Y}.$$

Finally, $y_d \in L_2(\Omega)$ is a desired state and $\alpha > 0$ is some given cost parameter.

Turning to the discretization of (1.1) we denote by $\mathbb{G}$ the class of all conforming refinements of $\mathcal{G}_0$ that can be constructed using refinement by bisection [13]. For a given grid $\mathcal{G} \in \mathbb{G}$ we let $\mathbb{Y}(\mathcal{G}) \subset \mathbb{Y}$ be a conforming finite element space of piecewise polynomials of fixed degree $q \in \mathbb{N}$. We then consider the variational discretization of (1.1) by Hinze [4], i.e., we solve the discretized optimal control problem

$$\begin{aligned} &\min_{(U,Y)\in \mathbb{U}^{\mathrm{ad}}\times \mathbb{Y}(\mathcal{G})} \frac{1}{2}\|Y - y_d\|_{\mathbb{U}}^2 + \frac{\alpha}{2}\|U\|_{\mathbb{U}}^2 \\ &\text{subject to} \qquad Y \in \mathbb{Y}(\mathcal{G}): \quad \mathcal{B}[Y, V] = \langle U, V\rangle \qquad V \in \mathbb{Y}(\mathcal{G}). \end{aligned} \tag{1.2}$$

It is well-known that (1.1) as well as (1.2) admit a unique solution pair $(\hat{u}, \hat{y})$, respectively $(\hat{U}_{\mathcal{G}}, \hat{Y}_{\mathcal{G}})$; compare with [9, 15]. Below we additionally utilize the continuous and discrete adjoint states $\hat{p} \in \mathbb{Y}$, $\hat{P}_{\mathcal{G}} \in \mathbb{Y}(\mathcal{G})$, and consider the solution triplets $(\hat{u}, \hat{y}, \hat{p}) \in \mathbb{U}^{\mathrm{ad}} \times \mathbb{Y} \times \mathbb{Y}$ and $(\hat{U}_{\mathcal{G}}, \hat{Y}_{\mathcal{G}}, \hat{P}_{\mathcal{G}}) \in \mathbb{U}^{\mathrm{ad}} \times \mathbb{Y}(\mathcal{G}) \times \mathbb{Y}(\mathcal{G})$.

We use the following adaptive algorithm for approximating the true solution of (1.1). Starting with the initial conforming triangulation $\mathcal{G}_0$ of Ω we execute the standard adaptive loop

$$\mathsf{SOLVE} \longrightarrow \mathsf{ESTIMATE} \longrightarrow \mathsf{MARK} \longrightarrow \mathsf{REFINE}. \tag{1.3}$$

In practice, a stopping test is used after ESTIMATE for terminating the iteration; here we shall ignore it for notational convenience.

Assumption 1.1 (Properties of modules). For a given grid $\mathcal{G} \in \mathbb{G}$ the four used modules have the following properties.

1. The output $(\hat{U}_{\mathcal{G}}, \hat{Y}_{\mathcal{G}}, \hat{P}_{\mathcal{G}}) := \mathsf{SOLVE}(\mathcal{G}) \in \mathbb{U}^{\mathrm{ad}} \times \mathbb{Y}(\mathcal{G}) \times \mathbb{Y}(\mathcal{G})$ is the exact solution of (1.2).
2. The output $\{\mathcal{E}_{\mathcal{G}}((\hat{U}_{\mathcal{G}}, \hat{Y}_{\mathcal{G}}, \hat{P}_{\mathcal{G}}); E)\}_{E\in\mathcal{G}} := \mathsf{ESTIMATE}\big((\hat{U}_{\mathcal{G}}, \hat{Y}_{\mathcal{G}}, \hat{P}_{\mathcal{G}}); \mathcal{G}\big)$ is a *reliable* and *locally efficient* estimator for the error in the norm $\|\cdot\|_{\mathbb{U}\times\mathbb{Y}\times\mathbb{Y}}$. In Sect. 2 below we give an example of such an estimator.

3. The output $\mathcal{M} = \mathsf{MARK}(\{\mathcal{E}_\mathcal{G}((\hat{U}_\mathcal{G}, \hat{Y}_\mathcal{G}, \hat{P}_\mathcal{G}); E)\}_{E\in\mathcal{G}}, \mathcal{G})$ is a subset of elements subject to refinement. We shall allow any marking strategy such that $\mathcal{M}$ contains an element holding the maximal indicator, i.e.,

$$\max\{\mathcal{E}_\mathcal{G}((\hat{U}_\mathcal{G}, \hat{Y}_\mathcal{G}, \hat{P}_\mathcal{G}); E) \mid E \in \mathcal{G}\} \leq \max\{\mathcal{E}_\mathcal{G}((\hat{U}_\mathcal{G}, \hat{Y}_\mathcal{G}, \hat{P}_\mathcal{G}); E) \mid E \in \mathcal{M}\}.$$

   All practically relevant marking strategies do have this property.
4. The output $\mathcal{G}_+ := \mathsf{REFINE}(\mathcal{G}, \mathcal{M}) \in \mathbb{G}$ is a conforming refinement of $\mathcal{G}$ such that all elements in $\mathcal{M}$ are bisected at least once, i.e., $\mathcal{G}_+ \cap \mathcal{M} = \emptyset$.

The main contribution of this report is the following convergence result.

Theorem 1.2 (Main result). *Let* $(\hat{u}, \hat{y}, \hat{p}) \in \mathbb{U}^{\mathrm{ad}} \times \mathbb{Y} \times \mathbb{Y}$ *be the true solution of* (1.1). *Suppose that* $\{\hat{U}_k, \hat{Y}_k, \hat{P}_k\}_{k\geq 0} \subset \mathbb{U}^{\mathrm{ad}} \times \mathbb{Y} \times \mathbb{Y}$ *is any sequence of discrete solutions generated by the adaptive iteration* (1.3)*, where the modules have the properties stated in Assumption* 1.1. *Then we have*

$$\lim_{k\to\infty} \|(\hat{U}_k, \hat{Y}_k, \hat{P}_k) - (\hat{u}, \hat{y}, \hat{p})\|_{\mathbb{U}\times\mathbb{Y}\times\mathbb{Y}} = 0 \quad \textit{and} \quad \lim_{k\to\infty} \mathcal{E}_{\mathcal{G}_k}(\hat{U}_k, \hat{Y}_k, \hat{P}_k; \mathcal{G}_k) = 0.$$

The proof of this theorem uses results and ideas from the convergence proofs of Morin, Siebert, and Veeser in [12] and Siebert in [14]. It is a two step procedure presented in Sects. 3 and 4. In Sect. 3 we utilize basic *stability properties* of the algorithm to show that the sequence of discrete solutions converges to some triplet $(\hat{u}_\infty, \hat{y}_\infty, \hat{p}_\infty)$. The second step in Sect. 4 then relies on the *steering mechanisms* of (1.3), mainly encoded in properties of $\mathsf{ESTIMATE}$ and MARK, to finally prove $(\hat{u}_\infty, \hat{y}_\infty, \hat{p}_\infty) = (\hat{u}, \hat{y}, \hat{p})$.

We shortly comment on an existing convergence result for constrained optimal control problems given in [2]. It is based on some non-degeneracy assumptions on the continuous and the discrete problems and a smallness assumption on the maximal mesh-size of $\mathcal{G}_0$. Our approach does not require any of these assumptions and it is valid for a larger class of adaptive algorithms. In addition, it can easily be extended in several directions; compare with Sect. 5.

2 Aposteriori Error Estimation

In this section we shortly summarize our findings from [6, 7] providing a unifying framework for the aposteriori error analysis for control constrained optimal control problems. In what follows we shall use $a \lesssim b$ for $a \leq Cb$ with a constant C that may depend on data of (1.1) and the shape regularity of the grids in $\mathbb{G}$ but not on a and b. We shall write $a \simeq b$ whenever $a \lesssim b \lesssim a$.

2.1 First Order Optimality Systems

The analysis in [6] is based on the characterization of the solutions by the first order optimality systems. We let $S, S^*\colon \mathbb{U} \to \mathbb{Y}$ be the solution operators of the state and the adjoint equation, i.e., for any $u \in U$ we have

$$Su \in \mathbb{Y} : \qquad \mathcal{B}[Su, v] = \langle u, v\rangle \qquad \forall\, v \in \mathbb{Y} \tag{2.1}$$

and for any $g \in \mathbb{U}$ we have

$$S^*g \in \mathbb{Y} : \qquad \mathcal{B}[v, S^*g] = \langle g, v\rangle \qquad \forall\, v \in \mathbb{Y}. \tag{2.2}$$

We denote by $\Pi\colon \mathbb{U} \to \mathbb{U}^{\mathrm{ad}}$ the nonlinear projection operator such that $\Pi(p)$ is the best approximation of $-\frac{1}{\alpha}p$ in $\mathbb{U}^{\mathrm{ad}}$, i.e.,

$$\Pi(p) \in \mathbb{U}^{\mathrm{ad}} : \qquad \langle \alpha\Pi(p) + p, \Pi(p) - u\rangle \le 0 \qquad \forall u \in \mathbb{U}^{\mathrm{ad}}. \tag{2.3}$$

Utilizing these operators, the continuous solution $(\hat{u}, \hat{y}, \hat{p}) \in \mathbb{U}^{\mathrm{ad}} \times \mathbb{Y} \times \mathbb{Y}$ is the unique solution of the coupled nonlinear system

$$\hat{y} = S\hat{u}, \qquad \hat{p} = S^*(\hat{y} - y_d), \qquad \hat{u} = \Pi(\hat{p}). \tag{2.4}$$

For $\mathcal{G} \in \mathbb{G}$ we next define $S_{\mathcal{G}}, S_{\mathcal{G}}^*\colon \mathbb{U} \to \mathbb{Y}(\mathcal{G})$ to be the discrete solution operators for (2.1) and (2.2), i.e., for any $u \in \mathbb{U}$ we have

$$S_{\mathcal{G}}u \in \mathbb{Y}(\mathcal{G}) : \qquad \mathcal{B}[S_{\mathcal{G}}u, V] = \langle u, V\rangle \qquad \forall\, V \in \mathbb{Y}(\mathcal{G}), \tag{2.5}$$

and for any $g \in \mathbb{U}$ we have

$$S_{\mathcal{G}}^*g \in \mathbb{Y}(\mathcal{G}) : \qquad \mathcal{B}[V, S_{\mathcal{G}}^*g] = \langle g, V\rangle \qquad \forall\, V \in \mathbb{Y}(\mathcal{G}). \tag{2.6}$$

The discrete solution $(\hat{U}_{\mathcal{G}}, \hat{Y}_{\mathcal{G}}, \hat{P}_{\mathcal{G}}) \in \mathbb{U}^{\mathrm{ad}} \times \mathbb{Y}(\mathcal{G}) \times \mathbb{Y}(\mathcal{G})$ is then uniquely characterized by

$$\hat{Y}_{\mathcal{G}} = S_{\mathcal{G}}\hat{U}_{\mathcal{G}}, \qquad \hat{P}_{\mathcal{G}} = S_{\mathcal{G}}^*(\hat{Y}_{\mathcal{G}} - y_d), \qquad \hat{U}_{\mathcal{G}} = \Pi(\hat{P}_{\mathcal{G}}). \tag{2.7}$$

Note, that this variational discretization of Hinze requires the evaluation of the *continuous* projection operator Π for discrete functions $P \in \mathbb{Y}(\mathcal{G})$.

We have $\|S\|, \|S^*\|, \|S_{\mathcal{G}}\|, \|S_{\mathcal{G}}^*\| \le C_F$, employing coercivity of $\mathcal{B}$ with constant 1 in combination with the Friedrichs inequality $\|v\|_{2;\Omega} \le C_F\|\nabla v\|_{2;\Omega}$ for $v \in \mathring{H}^1(\Omega)$.

2.2 Basic Error Equivalence

The main obstacle in the aposteriori error analysis encountered for instance in [3,10] can be explained as follows. One would like to exploit Galerkin orthogonality in the linear state equation (2.1) and the adjoint equation (2.2). However, we observe that triplet $(\hat{U}_{\mathcal{G}}, \hat{Y}_{\mathcal{G}}, \hat{P}_{\mathcal{G}})$ is the Galerkin approximation to the triplet $(\hat{u}, \hat{y}, \hat{p})$ but $\hat{Y}_{\mathcal{G}}$ is not the Galerkin approximation to the solution $\hat{u}$ of the linear problem (2.1) since we have $\hat{y} = S\hat{u}$ but not $\hat{y} = S\hat{U}_{\mathcal{G}}$. The same argument applies to the adjoint states. This observation shows that we cannot directly employ Galerkin orthogonality for single components of (2.4) and the nonlinearity in (2.3) prevents us from making use of Galerkin orthogonality for the system (2.4). The resort to this problem is given by the following result from [6, Theorem 2.2].

Proposition 2.1 (Basic error equivalence). *If we set $\mathbb{W} = \mathbb{U} \times \mathbb{Y} \times \mathbb{Y}$ we have for $\bar{y} = S\hat{U}_{\mathcal{G}}$ and $\bar{p} = S^*(\hat{Y}_{\mathcal{G}} - y_d)$ the basic error equivalence*

$$\|(\hat{U}_{\mathcal{G}}, \hat{Y}_{\mathcal{G}}, \hat{P}_{\mathcal{G}}) - (\hat{u}, \hat{p}, \hat{y})\|_{\mathbb{W}} \simeq \|(\hat{Y}_{\mathcal{G}}, \hat{P}_{\mathcal{G}}) - (\bar{y}, \bar{p})\|_{\mathbb{Y}\times\mathbb{Y}}.$$

For the problem under consideration, the constant hidden in $\simeq$ depends on α^{-1}. For general $\mathcal{B}$ it will in addition depend on the inf-sup constant of $\mathcal{B}$. Employing this error equivalence it is sufficient to construct a reliable and efficient estimator for the right hand side $\|(\hat{Y}_{\mathcal{G}}, \hat{P}_{\mathcal{G}}) - (\bar{y}, \bar{p})\|_{\mathbb{Y}\times\mathbb{Y}}$. The functions $\bar{y}$ and $\bar{p}$ are solutions to the *linear* problems (2.1) and (2.2) with given source $\hat{U}_{\mathcal{G}}$ and $\hat{Y}_{\mathcal{G}} - y_d$, respectively. They play a similar role as the *elliptic reconstruction* used in the aposteriori error analysis of parabolic problems; compare with [11].

2.3 Aposteriori Error Estimation

We realize that $\hat{Y}_{\mathcal{G}}$ is the Galerkin approximation to $\bar{y}$ and $\hat{P}_{\mathcal{G}}$ the one to $\bar{p}$. We therefore can directly employ (existing) estimators for the linear problems (2.1) and (2.2) and their sum then constitutes an estimator for the optimal control problem; compare with [6, Theorem 3.2]. For ease of presentation we focus here on the residual estimator. If σ is an interior side we denote by $[\![\nabla y]\!]$ the flux of the normal derivative $\partial_{\vec{n}} y$ across σ. For any subset $\mathcal{G}' \subset \mathcal{G}$ we set $\Omega(\mathcal{G}') := \bigcup_{E\in\mathcal{G}'} E$ and for given $E \in \mathcal{G}$ we denote by $\mathcal{N}_{\mathcal{G}}(E) \subset \mathcal{G}$ the subset consisting of E and its direct neighbors. Finally, we indicate by $\|\cdot\|_{\mathbb{W}(\omega)}$ the natural restriction of $\|\cdot\|_{\mathbb{W}}$ to a subset $\omega \subset \Omega$. We then have the following result.

Theorem 2.2 (Aposteriori error control). *For $E \in \mathcal{G}$ we define the indicator*

$$\begin{aligned}\mathcal{E}_{\mathcal{G}}^2((\hat{U}_{\mathcal{G}}, \hat{Y}_{\mathcal{G}}, \hat{P}_{\mathcal{G}}); E) := {} & h_E{}^2\|\Delta\hat{Y}_{\mathcal{G}} + \hat{U}_{\mathcal{G}}\|_{2;E}^2 + h_E\|[\![\nabla\hat{Y}_{\mathcal{G}}]\!]\|_{2;\partial E\cap\Omega}^2 \\ & + h_E{}^2\|\Delta\hat{P}_{\mathcal{G}} + (\hat{Y}_{\mathcal{G}} - y_d)\|_{2;E}^2 - h_E\|[\![\nabla\hat{P}_{\mathcal{G}}]\!]\|_{2;\partial E\cap\Omega}^2.\end{aligned}$$

Then we have the global upper bound

$$\|(\hat{U}_{\mathcal{G}}, \hat{Y}_{\mathcal{G}}, \hat{P}_{\mathcal{G}}) - (\hat{u}, \hat{p}, \hat{y})\|_{\mathbb{W}}^2 \lesssim \mathcal{E}_{\mathcal{G}}^2((\hat{U}_{\mathcal{G}}, \hat{Y}_{\mathcal{G}}, \hat{P}_{\mathcal{G}}); \mathcal{G}) := \sum_{E \in \mathcal{G}} \mathcal{E}_{\mathcal{G}}^2((\hat{U}_{\mathcal{G}}, \hat{Y}_{\mathcal{G}}, \hat{P}_{\mathcal{G}}); E).$$

For any $E \in \mathcal{G}$ *we have the* local lower bound

$$\mathcal{E}_{\mathcal{G}}^2((\hat{U}_{\mathcal{G}}, \hat{Y}_{\mathcal{G}}, \hat{P}_{\mathcal{G}}); E) \lesssim \|(\hat{U}_{\mathcal{G}}, \hat{Y}_{\mathcal{G}}, \hat{P}_{\mathcal{G}}) - (\hat{u}, \hat{p}, \hat{y})\|_{\mathbb{W}(\Omega(\mathcal{N}_{\mathcal{G}}(E)))}^2 + \operatorname{osc}_{\mathcal{G}}^2(\hat{U}_{\mathcal{G}}, y_d; \mathcal{N}_{\mathcal{G}}(E)),$$

where

$$\operatorname{osc}_{\mathcal{G}}^2(\hat{U}_{\mathcal{G}}, y_d; E) := h_E{}^2\big(\|\hat{U}_{\mathcal{G}} - \mathbb{P}_{\mathcal{G}}\hat{U}_{\mathcal{G}}\|_{2;\Omega(\mathcal{N}_{\mathcal{G}}(E))}^2 + \|y_d - \mathbb{P}_{\mathcal{G}} y_d\|_{2;\Omega(\mathcal{N}_{\mathcal{G}}(E))}^2\big)$$

is the typical oscillation term with the L_2*-projection* $\mathbb{P}_{\mathcal{G}}$ *onto the set of discontinuous, piecewise polynomials of degree* q *over* $\mathcal{G}$.

2.4 Bounds for the Residuals

We shortly comment on the derivation of the estimators for the linear problems and thereby recording an important intermediate estimate. For given $u \in \mathbb{U}$ we set $y = Su$ and let $Y = S_{\mathcal{G}}u$ be its Galerkin-approximation in $\mathbb{Y}(\mathcal{G})$. Defining the *residual* of the state equation (2.1) by

$$\langle \mathcal{R}(S_{\mathcal{G}}u; u), v\rangle = \langle \mathcal{R}(Y; u), v\rangle := \mathcal{B}[Y, v] - \langle u, v\rangle = \mathcal{B}[Y - y, v] \qquad \forall v \in \mathbb{Y},$$

we find $\|\mathcal{R}(Y; u)\|_{\mathbb{Y}^*} \simeq \|Y - y\|_{\mathbb{Y}} = \|(S_{\mathcal{G}} - S)u\|_{\mathbb{Y}}$.

Employing Galerkin-orthogonality $\langle \mathcal{R}(Y; u), V\rangle = 0$ for all $V \in \mathbb{Y}(\mathcal{G})$ and using piecewise integration by parts we deduce for any $v \in \mathbb{Y}$ and $V \in \mathbb{V}(\mathcal{G})$ the bound

$$|\langle \mathcal{R}(Y; u), v\rangle| \le \sum_{E \in \mathcal{G}} \|\Delta Y + u\|_{2;E}\|v - V\|_{2;E} + \frac{1}{2}\|[\![\nabla Y]\!]\|_{2;\partial E \cap \Omega}\|v - V\|_{2;\partial E}.$$

Using for $v \in \mathbb{Y}$ the Scott-Zhang interpolant $V \in \mathbb{Y}(\mathcal{G})$ one obtains from interpolation estimates in H^1 by standard arguments the upper bound

$$\|Y - y\|_{\mathbb{Y}} \simeq \|\mathcal{R}(Y; u)\|_{\mathbb{Y}^*} \lesssim \Big(\sum_{E \in \mathcal{G}} h_E{}^2\|\Delta Y + u\|_{2;E}^2 + h_E\|[\![\nabla Y]\!]\|_{2;\partial E \cap \Omega}^2\Big)^{1/2}.$$

If v is smooth, i.e., $v \in H^2(\Omega) \cap \mathbb{Y}$, we may employ interpolation estimates in H^2 to obtain the *improved bound*

$$|\langle \mathcal{R}(Y; u), v\rangle| \lesssim \Big(\sum_{E\in\mathcal{G}} h_E^{\,2} \big(h_E^{\,2} \|\Delta Y + u\|_{2;E}^2 + h_E \|[\![\nabla Y]\!]\|_{2;\partial E\cap\Omega}^2\big)\Big)^{1/2} |v|_{H^2(\Omega)}. \tag{2.8}$$

Similar arguments apply to the adjoint problem. For given $g \in \mathbb{U}$ we set $p = S^* g$ and let $P = S_{\mathcal{G}}^* g$ be its Galerkin-approximation in $\mathbb{Y}(\mathcal{G})$. For the residual of (2.2), defined by

$$\langle \mathcal{R}^*(S_{\mathcal{G}}^* g; g), v\rangle = \langle \mathcal{R}^*(P; g), v\rangle := \mathcal{B}[v, P] - \langle g, v\rangle = \mathcal{B}[v, P - p] \qquad \forall v \in \mathbb{Y},$$

we have

$$|\langle \mathcal{R}^*(P; g), v\rangle| \lesssim \Big(\sum_{E\in\mathcal{G}} h_E^{\,2s} \big(h_E^{\,2} \|\Delta P + g\|_{2;E}^2 + h_E \|[\![\nabla P]\!]\|_{2;\partial E\cap\Omega}^2\big)\Big)^{1/2} |v|_{H^{s+1}(\Omega)} \tag{2.9}$$

for any $v \in H^{s+1}(\Omega) \cap \mathbb{Y}$, $s = 0, 1$. With $s = 0$ we may deduce the upper bound for $\|(S_{\mathcal{G}}^* - S^*)g\|_{\mathbb{Y}} = \|P - p\|_{\mathbb{Y}} \simeq \|\mathcal{R}^*(P; g)\|_{\mathbb{Y}^*}$ The choice $s = 1$ yields the improved estimate for the adjoint problem. Equations (2.8) and (2.9) will become important in Sect. 4 to access *local density* of adaptively generated finite element spaces; compare also with [14, Remark 3.4].

3 Convergence 1: Trusting Stability

In this section we start with the convergence analysis, where we first focus on stability properties of the algorithm that do not depend on the particular decisions taken in MARK. Hereafter, $\{\mathcal{G}_k, (\hat{U}_k, \hat{Y}_k, \hat{P}_k)\}_{k\ge 0}$ is the sequence of grids and discrete solutions generated by (1.3). For ease of notation we use for $k \ge 0$ the short hands $\mathbb{Y}_k = \mathbb{Y}(\mathcal{G}_k)$, $\hat{U}_k = \hat{U}_{\mathcal{G}_k}$, $S_k = S_{\mathcal{G}_k}$ etc.

3.1 A First Limit

Using piecewise polynomials in combination with refinement by bisection leads to nested spaces, i.e., $\mathbb{Y}_k \subset \mathbb{Y}_{k+1}$. This allows us to define the limiting space

$$\mathbb{Y}_\infty = \overline{\bigcup_{k\ge 0} \mathbb{Y}_k}^{\|\cdot\|_{\mathbb{Y}}},$$

which is exactly the space that is approximated by the adaptive iteration. It is closed in $\mathbb{Y}$ and therefore a Hilbert space. Consequently, the limiting optimal control problem

$$\begin{aligned} &\min_{(u,y)\in\mathbb{U}^{\mathrm{ad}}\times\mathbb{Y}_\infty} \frac{1}{2}\|y-y_d\|_{\mathbb{U}}^2+\frac{\alpha}{2}\|u\|_{\mathbb{U}}^2 \\ &\text{subject to} \qquad y\in\mathbb{Y}_\infty: \quad \mathcal{B}[y,v]=\langle u,v\rangle \qquad v\in\mathbb{Y}_\infty \end{aligned} \tag{3.1}$$

admits a unique solution $(\hat{u}_\infty,\hat{y}_\infty)\in\mathbb{U}^{\mathrm{ad}}\times\mathbb{Y}_\infty$. If $S_\infty, S_\infty^*\colon\mathbb{U}\to\mathbb{Y}_\infty$ denote the solution operators of the state respectively the adjoint equation in $\mathbb{Y}_\infty$ the associated first order optimality system reads

$$\hat{y}_\infty=S_\infty\hat{u}_\infty, \qquad \hat{p}_\infty=S_\infty^*(\hat{y}_\infty-y_d), \qquad \hat{u}_\infty=\Pi(\hat{p}_\infty). \tag{3.2}$$

We next show that in fact (3.1) is the limiting problem of the adaptive iteration (1.3) in that $(\hat{U}_k,\hat{Y}_k,\hat{P}_k)\to(\hat{u}_\infty,\hat{y}_\infty,\hat{p}_\infty)$. An important ingredient for this proof is the following crucial property of the adaptive algorithm shown in [1, Lemma 6.1] and [12, Lemma 4.2].

Proposition 3.1 (Convergence of solution operators). *For any $u,g\in\mathbb{U}$ we have $S_k u\to S_\infty u$ and $S_k^* g\to S_\infty^* g$ in $\mathbb{Y}$ as $k\to\infty$.*

We next show convergence $\hat{U}_k\to\hat{u}_\infty$. In this step we have to deal with the nonlinearity of the constrained optimal control problem.

Lemma 3.2 (Convergence of the controls). *The discrete controls $\{\hat{U}_k\}_{k\ge 0}$ converge strongly to $\hat{u}_\infty$, i.e.,*

$$\lim_{k\to\infty}\|\hat{U}_k-\hat{u}_\infty\|_{\mathbb{U}}=0.$$

Proof. Since both $\hat{U}_k=\Pi(\hat{P}_k)$ and $\hat{u}_\infty=\Pi(\hat{p}_\infty)$ are feasible, i.e., $\hat{U}_k,\hat{u}_\infty\in\mathbb{U}^{\mathrm{ad}}$, the definition of Π in (2.3) yields

$$\begin{aligned} \alpha\|\hat{U}_k-\hat{u}_\infty\|_{2;\Omega}^2 &= \langle\alpha\hat{u}_\infty+\hat{p}_\infty,\hat{u}_\infty-\hat{U}_k\rangle+\langle\alpha\hat{U}_k+\hat{P}_k,\hat{U}_k-\hat{u}_\infty\rangle \\ &\qquad +\langle\hat{P}_k-\hat{p}_\infty,\hat{u}_\infty-\hat{U}_k\rangle \\ &\le\langle\hat{P}_k-\hat{p}_\infty,\hat{u}_\infty-\hat{U}_k\rangle \\ &=\langle S_k^*(\hat{y}_\infty-y_d)-\hat{p}_\infty,\hat{u}_\infty-\hat{U}_k\rangle+\langle\hat{P}_k-S_k^*(\hat{y}_\infty-y_d),\hat{u}_\infty-\hat{U}_k\rangle. \end{aligned}$$

We next estimate the last two terms separately. For the first one we immediately obtain from $\hat{p}_\infty=S_\infty(\hat{y}_\infty-y_d)$ by the Cauchy-Schwarz and Young inequalities

$$\begin{aligned} \langle S_k^*(\hat{y}_\infty-y_d)-\hat{p}_\infty,\hat{u}_\infty-\hat{U}_k\rangle &= \langle(S_k^*-S_\infty^*)(\hat{y}_\infty-y_d),\hat{u}_\infty-\hat{U}_k\rangle \\ &\le\frac{\alpha}{2}\|\hat{u}_\infty-\hat{U}_k\|_{2;\Omega}^2+\frac{1}{2\alpha}\|(S_k^*-S_\infty^*)(\hat{y}_\infty-y_d)\|_{2;\Omega}^2. \end{aligned}$$

We next turn to the second term. Employing the definition of the solution operators S_k and S_k^* in (2.5) and (2.6) we use $\hat{P}_k = S_k^*(\hat{Y}_k - y_d) \in \mathbb{Y}_k$ and $\hat{y}_\infty = S_\infty \hat{u}_\infty$ to obtain

$$\begin{aligned}\langle \hat{P}_k - S_k^*(\hat{y}_\infty - y_d), \hat{u}_\infty - \hat{U}_k\rangle &= \langle \hat{u}_\infty - \hat{U}_k, S_k^*(\hat{Y}_k - \hat{y}_\infty)\rangle \\ &= \mathcal{B}[S_k(\hat{u}_\infty - \hat{U}_k), S_k^*(\hat{Y}_k - \hat{y}_\infty)] = \langle \hat{Y}_k - \hat{y}_\infty, S_k(\hat{u}_\infty - \hat{U}_k)\rangle \\ &= \langle \hat{Y}_k - \hat{y}_\infty, \hat{y}_\infty - \hat{Y}_k\rangle + \langle \hat{Y}_k - \hat{y}_\infty, (S_k - S_\infty)\hat{u}_\infty\rangle \\ &= -\|\hat{Y}_k - \hat{y}_\infty\|_{2;\Omega}^2 + \frac{1}{2}\|\hat{Y}_k - \hat{y}_\infty\|_{2;\Omega}^2 + \frac{1}{2}\|(S_k - S_\infty)\hat{u}_\infty\|_{2;\Omega}^2 \\ &\le \frac{1}{2}\|(S_k - S_\infty)\hat{u}_\infty\|_{2;\Omega}^2.\end{aligned}$$

Combining the estimates we have shown

$$\alpha\|\hat{U}_k - \hat{u}_\infty\|_{2;\Omega}^2 \le \frac{1}{\alpha}\|(S_k^* - S_\infty^*)(\hat{y}_\infty - y_d)\|_{2;\Omega}^2 + \|(S_k - S_\infty)\hat{u}_\infty\|_{2;\Omega}^2 \to 0$$

as $k \to \infty$ by Proposition 3.1. This finishes the proof. □

Convergence $(\hat{U}_k, \hat{Y}_k, \hat{P}_k) \to (\hat{u}_\infty, \hat{y}_\infty, \hat{p}_\infty)$ is now a direct consequence of the linear theory in Proposition 3.1.

Proposition 3.3 (Convergence of discrete solutions). *The Galerkin approximations* $\{(\hat{U}_k, \hat{Y}_k, \hat{P}_k)\}_{k\ge 0}$ *converge strongly to the solution* $(\hat{u}_\infty, \hat{y}_\infty, \hat{p}_\infty)$ *of* (3.1), *i.e.,*

$$\lim_{k\to\infty} \|(\hat{U}_k, \hat{Y}_k, \hat{P}_k) - (\hat{u}_\infty, \hat{y}_\infty, \hat{p}_\infty)\|_{\mathbb{U}\times\mathbb{Y}\times\mathbb{Y}} = 0.$$

Proof. We already know $\|\hat{U}_k - \hat{u}_\infty\|_{\mathbb{U}} \to 0$ from Lemma 3.2. In combination with Proposition 3.1 this yields for the discrete states

$$\begin{aligned}\|\hat{Y}_k - \hat{y}_\infty\|_{\mathbb{Y}} &= \|S_k\hat{U}_k - S_\infty\hat{u}_\infty\|_{\mathbb{Y}} \le \|S_k(\hat{U}_k - \hat{u}_\infty)\|_{\mathbb{Y}} + \|(S_k - S_\infty)\hat{u}_\infty\|_{\mathbb{Y}} \\ &\le \|S_k\|\,\|\hat{U}_k - \hat{u}_\infty\|_{\mathbb{U}} + \|(S_k - S_\infty)\hat{u}_\infty\|_{\mathbb{Y}} \to 0,\end{aligned}$$

since $\|S_k\| \le C_F$. Writing $\hat{P}_k - \hat{p}_\infty = S_k^*(\hat{Y}_k - \hat{y}_\infty) + (S_k^* - S_\infty)(\hat{y}_\infty - y_d)$ we finally deduce with the same arguments $\|\hat{P}_k - \hat{p}_\infty\|_{\mathbb{Y}} \to 0$. □

The convergence of the discrete solutions directly yields a uniform bound on the estimators. The proof follows the ideas in [14, Lemma 3.3] accounting for the situation at hand and using the following important property. Let $\mathcal{G} \in \mathbb{G}$ be given. The finite overlap of the patches $\#\mathcal{N}_{\mathcal{G}}(E) \lesssim 1$ allows us to deduce for any $g \in L_2(\Omega)$ the bound

$$\sum_{E\in\mathcal{G}} \|g\|^2_{2;\Omega(\mathcal{N}_{\mathcal{G}}(E))} = \sum_{E\in\mathcal{G}} \sum_{E'\in\mathcal{N}_{\mathcal{G}}(E)} \|g\|^2_{2;E'} \lesssim \sum_{E\in\mathcal{G}} \|g\|^2_{2;E} = \|g\|^2_{2;\Omega}. \tag{3.3}$$

The constant solely depends on shape-regularity of $\mathcal{G}$ and thus on $\mathcal{G}_0$.

Lemma 3.4 (Uniform estimator bound). *For all $k \geq 0$ we have*

$$\mathcal{E}_k((\hat{U}_{\mathcal{G}}, \hat{Y}_{\mathcal{G}}, \hat{P}_{\mathcal{G}}); \mathcal{G}_k) \lesssim 1.$$

Proof. A scaled trace inequality in combination with an inverse estimate yields for the error indicators related to the state equation

$$h_E{}^2 \|\Delta \hat{Y}_k + \hat{U}_k\|^2_{2;E} + h_E \|[\![\nabla \hat{Y}_k]\!]\|^2_{2;\partial E_k\cap\Omega} \lesssim \|\nabla \hat{Y}_k\|^2_{2;\Omega(\mathcal{N}_{\mathcal{G}}(E))} + \|\hat{U}_k\|^2_{2;E}.$$

This in turn implies by (3.3)

$$\sum_{E\in\mathcal{G}_k} h_E{}^2 \|\Delta \hat{Y}_k + \hat{U}_k\|^2_{2;E} + h_E \|[\![\nabla \hat{Y}_k]\!]\|^2_{2;\partial E\cap\Omega} \lesssim \|\nabla \hat{Y}_k\|^2_{2;\Omega} + \|\hat{U}_k\|^2_{2;\Omega} \lesssim 1,$$

since $\{\hat{U}_k, \hat{Y}_k\}_{k\geq 0}$ is bounded in $L_2(\Omega) \times \mathring{H}^1(\Omega)$. Similar arguments apply to the estimator contribution related to the adjoint problem. □

3.2 A Second Limit

We next turn to the limit of the piecewise constant mesh-size function $h_k : \Omega \to \mathbb{R}$ of $\mathcal{G}_k$ defined by $h_{k|E} = |E|^{1/d}$, $E \in \mathcal{G}$. The behavior of the mesh-size function is directly related to the decomposition

$$\mathcal{G}_k^+ := \bigcap_{\ell\geq k} \mathcal{G}_\ell = \{E \in \mathcal{G}_k \mid E \in \mathcal{G}_\ell\ \forall \ell \geq k\}, \qquad \text{and} \qquad \mathcal{G}_k^0 := \mathcal{G}_k \setminus \mathcal{G}_k^+.$$

The set $\mathcal{G}_k^+$ contains all elements that are not refined after iteration k and we observe that the sequence $\{\mathcal{G}_k^+\}_{k\geq 0}$ is nested, i.e., $\mathcal{G}_\ell^+ \subset \mathcal{G}_k^+$ for all $k \geq \ell$. The set $\mathcal{G}_k^0$ contains all elements that are refined at least once more after iteration k; in particular, $\mathcal{M}_k \subset \mathcal{G}_k^0$. Decomposing $\bar{\Omega} = \Omega_k^+ \cup \Omega_k^0 := \Omega(\mathcal{G}_k^+) \cup \Omega(\mathcal{G}_k^0)$ we have the following connection to the behavior of the mesh-size function shown in [12, Lemma 4.3 and Corollary 4.1].

Lemma 3.5 (Convergence of the mesh-size functions). *The mesh-size functions h_k converge uniformly to 0 in Ω_k^0 in the following sense*

$$\lim_{k\to\infty} \|h_k\, \chi_k^0\|_{\infty;\Omega} = \lim_{k\to\infty} \|h_k\|_{\infty;\Omega_k^0} = 0,$$

where $\chi_k^0 \in L_\infty(\Omega)$ the characteristic function of Ω_k^0.

Combining convergence of the discrete solutions with the convergence of the mesh-size functions we see that the adaptive algorithm can monitor progress in the following sense.

Lemma 3.6 (Indicators of marked elements). *All indicators of marked elements vanish in the limit, this is,*

$$\lim_{k\to\infty} \max\{\mathcal{E}_{\mathcal{G}}((\hat{U}_{\mathcal{G}}, \hat{Y}_{\mathcal{G}}, \hat{P}_{\mathcal{G}}); E) \mid E \in \mathcal{M}_k\} = 0.$$

Proof. For $k \geq 0$ pick up $E_k \in \arg\max\{\mathcal{E}_{\mathcal{G}}((\hat{U}_{\mathcal{G}}, \hat{Y}_{\mathcal{G}}, \hat{P}_{\mathcal{G}}); E) \mid E \in \mathcal{M}_k\} \neq \emptyset$. We follow [14, Lemma 3.4] and show $\mathcal{E}_k((\hat{U}_k, \hat{Y}_k, \hat{P}_k); E_k) \to 0$.

Arguing as in the proof to Lemma 3.4 we find for the indicator contribution of the state equation

$$\begin{aligned} h_E\|\Delta\hat{Y}_k + \hat{U}_k\|_{2;E_k} + {h_E}^{1/2}\|[\![\nabla\hat{Y}_k]\!]\|_{2;\partial E_k\cap\Omega} &\lesssim \|\nabla\hat{Y}_k\|_{2;\Omega(\mathcal{N}_k(E_k))} + \|\hat{U}_k\|_{2;E_k} \\ &\leq \|\nabla\hat{y}_\infty\|_{2;\Omega(\mathcal{N}_k(E_k))} + \|\hat{u}_\infty\|_{2;E_k} + \|\nabla(\hat{Y}_k - \hat{y}_\infty)\|_{2;\Omega} + \|\hat{U}_k - \hat{u}_\infty\|_{2;\Omega} \to 0 \end{aligned}$$

as $k \to \infty$ for the following reasons: By Assumption 1.1 (4) all elements in $\mathcal{M}_k$ are refined, which implies $E_k \in \mathcal{G}_k^0$. Local quasi-uniformity of $\mathcal{G}_k$ in combination with Lemma 3.5 therefore yields $|\Omega(\mathcal{N}_k(E_k))| \lesssim |E_k| \leq \|h_k\|^d_{\infty;\Omega_k^0} \to 0$. Consequently, the first two terms of the right hand side vanish by continuity of $\|\cdot\|_{2;\Omega}$ with respect to the Lebesgue measure. The last two terms converge to 0 by Proposition 3.3. The same arguments apply to the indicator contribution of the adjoint equation, which in summary yields $\mathcal{E}_{\mathcal{G}}((\hat{U}_k, \hat{Y}_k, \hat{P}_k); E_k) \to 0$ as $k \to \infty$. □

4 Convergence 2: Making the Right Decisions

In this section we verify the main result by showing $(\hat{U}_k, \hat{Y}_k, \hat{P}_k) \to (\hat{u}, \hat{y}, \hat{p})$ and $\mathcal{E}_k(\hat{U}_k, \hat{Y}_k, \hat{P}_k; \mathcal{G}_k) \to 0$. Error convergence requires appropriate decisions in the adaptive iteration, which we have summarized in Assumption 1.1. Estimator convergence is then a consequence of local efficiency as stated in Theorem 2.2.

4.1 Convergence of the Indicators

We first show that the maximal indicator of all elements vanishes in the limit.

Lemma 4.1 (Convergence of the indicators). *The maximal indicator vanishes in the limit, this is,*

$$\lim_{k\to\infty} \max\{\mathcal{E}_{\mathcal{G}}((\hat{U}_{\mathcal{G}}, \hat{Y}_{\mathcal{G}}, \hat{P}_{\mathcal{G}}); E) \mid E \in \mathcal{G}_k\} = 0.$$

Proof. Combining the assumption on marking in Assumption 1.1 (3) with the behavior of the indicators on marked elements, which we have analyzed in Lemma 3.6, we find

$$\max\{\mathcal{E}_{\mathcal{G}}((\hat{U}_k, \hat{Y}_k, \hat{P}_k); E) \mid E \in \mathcal{G}_k\}$$
$$\leq \max\{\mathcal{E}_{\mathcal{G}}((\hat{U}_k, \hat{Y}_k, \hat{P}_k); E) \mid E \in \mathcal{M}_k\} \to 0$$

as $k \to \infty$. □

4.2 Convergence of the Residuals

We next show that residuals of state and adjoint equation in the limiting first order optimality system (3.2) vanish. The proof adapts the techniques from [14, Proposition 3.1] to the situation at hand.

Proposition 4.2 (Convergence of the residual). *For the residuals $\mathcal{R}$ of* (2.1) *and $\mathcal{R}^*$ of* (2.2) *we have*

$$\mathcal{R}(\hat{y}_\infty; \hat{u}_\infty) = \mathcal{R}^*(\hat{p}_\infty; \hat{y}_\infty - y_d) = 0 \qquad \text{in } \mathbb{Y}^* = H^{-1}(\Omega).$$

Particularly, $\hat{y}_\infty = S\hat{u}_\infty$ and $\hat{p}_\infty = S^*(\hat{y}_\infty - y_d)$.

Proof. We prove the claim for $\mathcal{R}$. The assertion for $\mathcal{R}^*$ follows along the same lines. Using a density argument we only have to show $\langle \mathcal{R}(\hat{y}_\infty; \hat{u}_\infty), v\rangle = 0$ for all $v \in H^2(\Omega) \cap \mathring{H}^1(\Omega)$.

Suppose any pair $k \geq \ell$. Then we have the inclusion $\mathcal{G}_\ell^+ \subset \mathcal{G}_k^+ \subset \mathcal{G}_k$ and the sub-triangulation $\mathcal{G}_k \setminus \mathcal{G}_\ell^+$ of $\mathcal{G}_k$ covers the sub-domain $\Omega_\ell^0 = \Omega(\mathcal{G}_\ell^0)$, i.e., we can write $\Omega_\ell^0 = \Omega(\mathcal{G}_k \setminus \mathcal{G}_\ell^+)$. Moreover, $\|h_k\|_{\infty;\Omega_\ell^+} \lesssim 1$ and $\|h_k\|_{\infty;\Omega_\ell^0} \leq \|h_\ell\|_{\infty;\Omega_\ell^0}$.

Pick up any $v \in H^2(\Omega) \cap \mathring{H}^1(\Omega)$ with $|v|_{H^2(\Omega)} = 1$. We next utilize the improved bound (2.8) for $\mathcal{R}$, decompose $\mathcal{G}_k = \mathcal{G}_\ell^+ \cup (\mathcal{G}_k \setminus \mathcal{G}_\ell^+)$, and recall Lemma 3.4 to bound

$$\begin{aligned}
\langle \mathcal{R}(\hat{Y}_k; \hat{U}_k), v\rangle^2 &\lesssim \sum_{E \in \mathcal{G}_\ell^+} {h_E}^2 \big({h_E}^2 \|\Delta \hat{Y}_k + \hat{U}_k\|_{2;E}^2 + h_E \|[\![\nabla \hat{Y}_k]\!]\|_{2;\partial E \cap \Omega}^2\big) \\
&\quad + \sum_{E \in \mathcal{G}_k \setminus \mathcal{G}_\ell^+} {h_E}^2 \big({h_E}^2 \|\Delta \hat{Y}_k + \hat{U}_k\|_{2;E}^2 + h_E \|[\![\nabla \hat{Y}_k]\!]\|_{2;\partial E \cap \Omega}^2\big) \\
&\lesssim \mathcal{E}_k^2((\hat{U}_k, \hat{Y}_k, \hat{P}_k); \mathcal{G}_\ell^+) + \|h_\ell\|_{\infty;\Omega_\ell^0}^2 \mathcal{E}_k^2((\hat{U}_k, \hat{Y}_k, \hat{P}_k); \mathcal{G}_k \setminus \mathcal{G}_\ell^+) \\
&\lesssim \mathcal{E}_k^2((\hat{U}_k, \hat{Y}_k, \hat{P}_k); \mathcal{G}_\ell^+) + \|h_\ell\|_{\infty;\Omega_\ell^0}^2 \overset{!}{\leq} 2\varepsilon
\end{aligned}$$

for any $\varepsilon > 0$. This can be seen as follows: By Lemma 3.5 we may first choose ℓ large such that $\|h_\ell\|^2_{\infty;\Omega^0_\ell} \leq \varepsilon$. After fixing ℓ the "point-wise" convergence of the indicators in Lemma 4.1 allows us then to choose a suitable $k \geq \ell$ with $\mathcal{E}^2_k((\hat{U}_k, \hat{Y}_k, \hat{P}_k); \mathcal{G}^+_\ell) \leq \varepsilon$. This yields for any fixed $v \in H^2(\Omega) \cap \mathring{H}^1(\Omega)$

$$\langle \mathcal{R}(\hat{y}_\infty; \hat{u}_\infty), v\rangle = \lim_{k\to\infty} \langle \mathcal{R}(\hat{Y}_k; \hat{U}_k), v\rangle = 0,$$

observing that $\mathcal{R}$ is continuous with respect to its arguments and recalling the convergence $(\hat{U}_k, \hat{Y}_k) \to (\hat{u}_\infty, \hat{y}_\infty)$ shown in Proposition 3.3. Since v is arbitrary we have shown $\mathcal{R}(\hat{y}_\infty; \hat{u}_\infty) = 0$ in $\mathbb{Y}^*$. This in turn implies $\hat{y}_\infty = S\hat{u}_\infty$ and finishes the proof. □

4.3 Convergence of Error and Estimator

We are now in the position to prove the main result, where we again use the abbreviation $\mathbb{W} = \mathbb{U} \times \mathbb{Y} \times \mathbb{Y}$.

Proof of Theorem 1.2. Combining Propositions 2.1, 3.3, and 4.2 we obtain

$$\begin{aligned}\lim_{k\to\infty} \|(\hat{U}_k, \hat{Y}_k, \hat{P}_k) - (\hat{u}, \hat{p}, \hat{y})\|_{\mathbb{W}} &\simeq \lim_{k\to\infty} \|(\hat{Y}_k, \hat{P}_k) - (S\hat{U}_k, S^*(\hat{Y}_k - y_d))\|_{\mathbb{Y}\times\mathbb{Y}} \\ &= \|(\hat{y}_\infty, \hat{p}_\infty) - (S\hat{u}_\infty, S^*(\hat{y}_\infty - y_d)\|_{\mathbb{Y}\times\mathbb{Y}} = 0.\end{aligned}$$

This shows convergence of the error.

To show convergence of the estimator we decompose for $k \geq \ell$ as in the proof to Proposition 4.2

$$\mathcal{E}^2_k((\hat{U}_k, \hat{Y}_k, \hat{P}_k); \mathcal{G}_k) = \mathcal{E}^2_k((\hat{U}_k, \hat{Y}_k, \hat{P}_k); \mathcal{G}^+_\ell) + \mathcal{E}^2_k((\hat{U}_k, \hat{Y}_k, \hat{P}_k); \mathcal{G}_k \setminus \mathcal{G}^+_\ell).$$

We first bound the second term on the right hand side. The local lower bound of Theorem 2.2 in combination with the finite overlap of the patches $\mathcal{N}_k(E)$ allows us to bound

$$\begin{aligned}&\mathcal{E}^2_k((\hat{U}_k, \hat{Y}_k, \hat{P}_k); \mathcal{G}_k \setminus \mathcal{G}^+_\ell) \\ &\quad\lesssim \|(\hat{U}_k, \hat{Y}_k, \hat{P}_k) - (\hat{u}, \hat{p}, \hat{y})\|^2_{\mathbb{W}} + \sum_{E\in\mathcal{G}_k\setminus\mathcal{G}^+_\ell} \operatorname{osc}^2_k(\hat{U}_k, y_d; E) \\ &\quad\lesssim \|(\hat{U}_k, \hat{Y}_k, \hat{P}_k) - (\hat{u}, \hat{p}, \hat{y})\|^2_{\mathbb{W}} + \|h_\ell\|^2_{\infty;\Omega^0_\ell}\big(\|\hat{U}_k\|^2_{2;\Omega} + \|y_d\|^2_{2;\Omega}\big),\end{aligned}$$

using (3.3) and the rough estimate

$$\begin{aligned}\operatorname{osc}_k^2(\hat{U}_k, y_d; E) &= h_E^2\big(\|\hat{U}_k - \mathbb{P}_{\mathcal{G}_k}\hat{U}_k\|_{2;\Omega(\mathcal{N}_{\mathcal{G}}(E))}^2 + \|y_d - \mathbb{P}_{\mathcal{G}_k} y_d\|_{2;\Omega(\mathcal{N}_{\mathcal{G}}(E))}^2\big)\\ &\le \|h_\ell\|_{\infty;\Omega_\ell^0}^2\big(\|\hat{U}_k\|_{2;\Omega(\mathcal{N}_k(E))}^2 + \|y_d\|_{2;\Omega(\mathcal{N}_k(E))}^2\big).\end{aligned}$$

Since $\|\hat{U}_k\|_{2;\Omega}^2 + \|y_d\|_{2;\Omega}^2 \lesssim 1$ we find

$$\begin{aligned}\mathcal{E}_k^2((\hat{U}_k, \hat{Y}_k, \hat{P}_k); \mathcal{G}_k) \lesssim \mathcal{E}_k^2((\hat{U}_k, \hat{Y}_k, \hat{P}_k); \mathcal{G}_\ell^+)&\\ + \|(\hat{U}_k, \hat{Y}_k, \hat{P}_k) - (\hat{u}, \hat{p}, \hat{y})\|_{\mathbb{W}}^2 + \|h_\ell\|_{\infty;\Omega_\ell^0}^2&.\end{aligned}$$

By Lemma 3.5 the last term $\|h_\ell\|_{\infty;\Omega_\ell^0}^2$ can be made small by choosing ℓ large. After fixing ℓ we may choose as in the proof to Proposition 4.2 $k \ge \ell$ such that $\mathcal{E}_k^2((\hat{U}_k, \hat{Y}_k, \hat{P}_k); \mathcal{G}_\ell^+)$ is small. Moreover, the error convergence established above implies that the middle term $\|(\hat{U}_k, \hat{Y}_k, \hat{P}_k) - (\hat{u}, \hat{p}, \hat{y})\|_{\mathbb{W}}^2$ is small too, if we possibly increase k further. In summary, for any $\varepsilon > 0$ we find a k such that

$$\mathcal{E}_k((\hat{U}_k, \hat{Y}_k, \hat{P}_k); \mathcal{G}_k) \le \varepsilon.$$

This yields $\mathcal{E}_k((\hat{U}_k, \hat{Y}_k, \hat{P}_k); \mathcal{G}_k) \to 0$ as $k \to \infty$ and finishes the proof. □

5 Extensions and Outlook

The presented theory has been extended into several directions in the PhD thesis of the first author [5].

5.1 General Linear-Quadratic Optimal Control Problem

The abstract framework can be found in [6, §2.1] and may be summarized as follows. We can allow for continuous, non-coercive bilinear forms $\mathcal{B}\colon \mathbb{Y} \times \mathbb{Y} \to \mathbb{R}$ that satisfy an inf-sup condition. This setting includes saddle point problems like the Stokes system and other mixed formulations. More general objectives $\psi(y)$ can replace the simple tracking type functional $\|y - y_d\|_{2;\Omega}^2$. The functional ψ has to be quadratic and strictly convex. Its Fréchet-derivative ψ' has to satisfy a Lipschitz-condition. We may also consider any type of control space such that $\mathbb{Y} \hookrightarrow \mathbb{U} \hookrightarrow \mathbb{Y}^*$ is a Gelfand triple. This then covers more general cases of distributed control as well as Neumann-boundary control.

Admitting a general class of PDE constraints requires appropriate assumptions on the estimators for the linear problems (2.1) and (2.2). Quite weak assumption are summarized in [14, §2.2.3] comprising other estimators like the hierarchical estimator, an estimator based on local problems on stars, an equilibrated residual estimator, and the ZZ-estimator; compare for instance with [8] for a detailed description of the diverse estimators. We may also weaken the assumption on marking to include marking strategies that adaptively focus on specific estimator contributions, like the indicators for the error in the state or adjoint equation. Such strategies are used in a comparison of adaptive strategies for optimal control problems in [6, §6]. We refer to [14, §2.2.4 and §5] for a sufficient and necessary assumption on marking.

Most of the changes in the presented analysis are then concentrated in the proof to Lemma 3.2. This proof gets inevitably more involved due to the general structure of ψ, where one has to appropriately use convexity of ψ. All other statements can be proven using similar arguments with minor adjustments.

5.2 *Discretized Control*

Up to now we have concentrated on the variational discretization of Hinze [4]. Here, the precise structure of the set of admissible controls $\mathbb{U}^{ad}$ is not of importance. The actual computation of a discrete solution yet requires the exact computation of $\Pi(P)$ for a discrete function $P \in \mathbb{Y}(\mathcal{G})$. This typically gives restrictions on $\mathbb{U}^{ad}$, like box-constraints with piecewise constant obstacles.

Very often the control space $\mathbb{U}$ is discretized by a conforming finite element space $\mathbb{U}(\mathcal{G})$. Upon setting $\mathbb{U}^{ad}(\mathcal{G}) := \mathbb{U}^{ad} \cap \mathbb{U}(\mathcal{G})$ and assuming that $\mathbb{U}^{ad}(\mathcal{G})$ is non-empty we can define a discrete projection operator $\Pi_{\mathcal{G}}\colon \mathbb{U} \to \mathbb{U}^{ad}(\mathcal{G})$ for $p \in \mathbb{U}$ by

$$\Pi_{\mathcal{G}}(p) \in \mathbb{U}^{ad}(\mathcal{G}) : \qquad \langle \alpha \Pi_{\mathcal{G}}(p) + p, \Pi_{\mathcal{G}}(p) - U \rangle \leq 0 \qquad \forall U \in \mathbb{U}^{ad}(\mathcal{G}).$$

An efficient computation of $\Pi_{\mathcal{G}}$ benefits from a simple structure of $\mathbb{U}^{ad}$ and a suitable discrete control space $\mathbb{U}(\mathcal{G})$.

We can still consider the general setting of the previous paragraph. However, the analysis of adaptive finite elements for discretized controls gets painstakingly more laborious at several instances that we shortly list.

1. The right hand side in the basic error equivalence in Proposition 2.1 has to be extended by the term $\|\hat{U}_{\mathcal{G}} - \Pi(\hat{U}_{\mathcal{G}})\|_{\mathbb{U}}$ resulting in
$$\|(\hat{U}_{\mathcal{G}}, \hat{Y}_{\mathcal{G}}, \hat{P}_{\mathcal{G}}) - (\hat{u}, \hat{p}, \hat{y})\|_{\mathbb{W}} \simeq \|(\hat{U}_{\mathcal{G}}, \hat{Y}_{\mathcal{G}}, \hat{P}_{\mathcal{G}}) - (\Pi(\hat{P}_{\mathcal{G}}), S(\hat{U}_{\mathcal{G}}), S^* \psi'(\hat{Y}_{\mathcal{G}}))\|_{\mathbb{W}};$$
compare with [6, Theorem 2.2]
2. As a consequence, the element indicators of the estimator in Theorem 2.2 have to be enriched by a term $\|\hat{U}_{\mathcal{G}} - \Pi(\hat{P}_{\mathcal{G}})\|^2_{\mathbb{U}(E)} = \|\Pi_{\mathcal{G}}(\hat{P}_{\mathcal{G}}) - \Pi(\hat{P}_{\mathcal{G}})\|^2_{\mathbb{U}(E)}$ to grant reliability of the estimator [6, Theorem 3.2]. Frequently this term is estimated further in order to completely avoid the computation of the continuous projection

operator $\Pi(\hat{P}_{\mathcal{G}})$ [3, 10]. This typically results in a non-efficient estimator; compare with [6, Remark 6.1].

3. Nesting of spaces $\mathbb{Y}_k \subset \mathbb{Y}_{k+1}$ is essential to verify with current techniques the point-wise convergence of the discrete solution operators in Proposition 3.1, i.e., $S_k \to S_\infty$ and $S_k^* \to S_\infty^*$ as $k \to \infty$.

   Likewise, nesting $\mathbb{U}_k^{\text{ad}} \subset \mathbb{U}_{k+1}^{\text{ad}}$ of the sets of discrete admissible controls is instrumental for proving $\Pi_k(\hat{P}_k) \to \hat{u}_\infty = \Pi_\infty(\hat{p}_\infty)$ in Lemma 3.2. This nesting poses restrictions on data describing the set of admissible controls $\mathbb{U}^{\text{ad}}$. Typically, such data has to be discrete over $\mathcal{G}_0$. In the proof to Lemma 3.2 we additionally have to account for the typical situation $\hat{u}_\infty \notin \mathbb{U}_k^{\text{ad}}$. This increases substantially the complexity of the proof.

4. The finite element spaces $\{\mathbb{Y}_k\}_{k\geq 0}$ are "locally dense" in the subset "$\Omega_\infty^0 := \lim_{k\to\infty} \Omega_k^0$" of Ω in that $\min_{V\in\mathbb{Y}_k} \|v - V\|_{\mathbb{Y}(\Omega_k^0)} \to 0$ as $k \to \infty$; compare with [14, Remark 3.4]. Philosophically speaking, the improved bounds (2.8) and (2.9) for the residuals allow us to access this local density for showing that the residuals $\mathcal{R}(\hat{y}_\infty; \hat{u}_\infty)$, $\mathcal{R}^*(\hat{p}_\infty; \psi'(\hat{y}_\infty)) \in \mathbb{Y}^*$ are not supported in Ω_∞^0.

   The additional contribution for the control error requires to establish the convergence

$$\lim_{k\to\infty} \|\hat{U}_k - \Pi(\hat{P}_k)\|_{\mathbb{U}(\Omega_k^0)} = \lim_{k\to\infty} \|\Pi_k(\hat{P}_k) - \Pi(\hat{P}_k)\|_{\mathbb{U}(\Omega_k^0)} \to 0. \tag{5.1}$$

   For $\mathbb{U} = L_2$ and piece-wise constant box-constraints in combination with a discontinuous or a continuous, piecewise linear control discretization one can verify (5.1) employing local density of $\{\mathbb{U}_k\}_{k\geq 0}$ and point-wise properties of Π; compare with [5, §8.4.2 and §8.4.3]. A characterization of properties of Π and $\mathbb{U}(\mathcal{G})$ that ensure (5.1) is a challenging question and topic of future research.

5. The proof of the estimator convergence in Theorem 1.2 strongly relies on local efficiency of the indicators as stated in Theorem 2.2. For discretized control this requires $\|\hat{U}_{\mathcal{G}} - \Pi(\hat{P}_{\mathcal{G}})\|_{\mathbb{U}(E)}$ to be locally efficient, which can be shown if Π and $\Pi_{\mathcal{G}}$ are locally Lipschitz continuous with uniformly bounded Lipschitz constants. This is typically true in case of distributed control.

   In case of Neumann boundary control Lipschitz continuity of Π and $\Pi_{\mathcal{G}}$ involves the trace operator $T : H^1(\Omega) \to L_2(\partial\Omega)$. We may therefore show *global* efficiency for $\|\hat{U}_{\mathcal{G}} - \Pi(\hat{P}_{\mathcal{G}})\|_{L_2(\partial\Omega)}$ using the trace inequality on Ω. An estimate of $\|\hat{U}_{\mathcal{G}} - \Pi(\hat{P}_{\mathcal{G}})\|_{L_2(\partial E\cap\partial\Omega)}$ needs a local trace inequality on E. The typical scaling arguments yield negative powers of the local mesh-size and thereby ruling out local efficiency. As a consequence, we still can verify the error convergence of Theorem 1.2 but a proof of estimator convergence may require new techniques in that case.

References

1. I. Babuška, M. Vogelius, Feedback and adaptive finite element solution of one-dimensional boundary value problems. Numer. Math. **44**, 75–102 (1984)
2. A. Gaevskaya, R.H.W. Hoppe, Y. Iliash, M. Kieweg, Convergence analysis of an adaptive finite element method for distributed control problems with control constraints, in *Control of Coupled Partial Differential Equations*. Volume 155 of International Series of Numerical Mathematics (Birkhäuser, Basel, 2007), pp. 47–68
3. M. Hintermüller, R.H.W. Hoppe, Y. Iliash, M. Kieweg, An a posteriori error analysis of adaptive finite element methods for distributed elliptic control problems with control constraints. ESAIM Control Optim. Calc. Var. **14**, 540–560 (2008)
4. M. Hinze, A variational discretization concept in control constrained optimization: the linear-quadratics case. Comput. Optim. Appl. **30**, 45–61 (2005)
5. K. Kohls, An adaptive finite element method for control-constrained optimal control problems, PhD thesis, Fachbereich Mathematik, Unversität Stuttgart, 2013
6. K. Kohls, A. Rösch, K.G. Siebert, A posteriori error analysis of optimal control problems with control constraints. SIAM J. Control Optim. **52**(3), 1832–1861 (2014)
7. K. Kohls, A. Rösch, K.G. Siebert, A posteriori error estimators for control constrained optimal control problems, in *Constrained Optimiziation and Optimal Control for Partial Differential Equations*, ed. by Leugering et al. International Series of Numerical Mathematics, vol. 160 (Birkhäuser/Springer Basel AG, Basel, 2012), pp. 431–443
8. C. Kreuzer, K.G. Siebert, Decay rates of adaptive finite elements with Dörfler marking. Numer. Math **117**, 679–716 (2011)
9. J.L. Lions, *Optimal Control of Systems Governed by Partial Differential Equations*. Volume 170 of Die Grundlehren der Mathematischen Wissenschaften (Springer, Berlin/ New York, 1971)
10. W. Liu, N. Yan, A posteriori error estimates for distributed convex optimal control problems. Adv. Comput. Math. **15**, 285–309 (2001)
11. C. Makridakis, R.H. Nochetto, Elliptic reconstruction and a posteriori error estimates for parabolic problems. SIAM J. Numer. Anal. **41**, 1585–1594 (2003)
12. P. Morin, K.G. Siebert, A. Veeser, A basic convergence result for conforming adaptive finite elements. Math. Models Methods Appl. Sci. **18**, 707–737 (2008)
13. A. Schmidt, K.G. Siebert, *Design of Adaptive Finite Element Software*. Volume 42 of Lecture Notes in Computational Science and Engineering (Springer, Berlin, 2005). The finite element toolbox ALBERTA, With 1 CD-ROM (Unix/Linux)
14. K.G. Siebert, A convergence proof for adaptive finite elements without lower bound. IMA J. Numer. Anal. **31**, 947–970 (2011)
15. F. Tröltzsch, *Optimal Control of Partial Differential Equations*. Volume 112 of Graduate Studies in Mathematics (American Mathematical Society, Providence, 2010). Theory, methods and applications, Translated from the 2005 German original by Jürgen Sprekels

Petrov-Galerkin Crank-Nicolson Scheme for Parabolic Optimal Control Problems on Nonsmooth Domains

Thomas G. Flaig, Dominik Meidner, and Boris Vexler

Abstract In this paper we transfer the a priori error analysis for the discretization of parabolic optimal control problems on domains allowing for H^2 regularity (i.e. either with smooth boundary or polygonal and convex) to a large class of nonsmooth domains. We show that a combination of two ingredients for the optimal convergence rates with respect to the spatial and the temporal discretization is required. First we need a time discretization scheme which has the desired convergence rate in the smooth case. Secondly we need a method to treat the singularities due to non-smoothness of the domain for the corresponding elliptic state equation. In particular we demonstrate this philosophy with a Crank-Nicolson time discretization and finite elements on suitably graded meshes for the spatial discretization. A numerical example illustrates the predicted convergence rates.

Keywords Optimal control problem • Parabolic partial differential equation • Non-smooth domains • Graded meshes • Crank Nicolson scheme

Mathematics Subject Classification (2010). 49M25, 49M05, 65M15, 65M60, 49M29, 65M12.

T.G. Flaig
Institut für Mathematik und Bauinformatik, Universität der Bundeswehr München, Werner-Heisenberg-Weg 39, 85577 Neubiberg, Germany
e-mail: thomas.flaig@unibw.de

D. Meidner (✉) • B. Vexler
Fakultät für Mathematik, Lehrstuhl für Optimale Steuerung, Technische Universität München, Boltzmannstraße 3, 85748 Garching b. München, Germany
e-mail: meidner@ma.tum.de; vexler@ma.tum.de

G. Leugering et al. (eds.), *Trends in PDE Constrained Optimization*, International Series of Numerical Mathematics 165, DOI 10.1007/978-3-319-05083-6_26

1 Introduction

In this paper we extend the a priori error analysis for discretizations of a parabolic optimal control problem to the case of nonsmooth domains. The model problem under consideration is formulated as

$$\text{Minimize } J(q,u) = \frac{1}{2}\int_0^T \int_\Omega |u-\hat{u}|^2\,dx\,dt + \frac{\alpha}{2}\int_0^T \int_\Omega |q|^2\,dx\,dt, \tag{1.1a}$$

subject to the state equation

$$\begin{aligned} \partial_t u - \Delta u &= f + q && \text{in } (0,T)\times\Omega,\\ u &= 0 && \text{in } (0,T)\times\partial\Omega,\\ u(0) &= u_0 && \text{in } \Omega, \end{aligned} \tag{1.1b}$$

where $u = u(t,x)$ denotes the state variable and $q = q(t,x)$ is the control variable. A precise formulation of this problem including a functional analytic setting is given in the next section.

In the literature on a priori error estimates for this kind of problems, see, e.g., [3, 17, 18, 24–27], the domain Ω is always assumed either to have a smooth boundary $\partial\Omega$ or to be polygonal and convex. Our main contribution is an extension of the results from [26] to a more general class of domains including polygonal or polyhedral domains with (non-convex) reentrant corners.

For optimal control problems governed by elliptic equations on non-convex domains there are several contributions establishing optimal order error estimates on properly chosen graded meshes, see [5–7, 10].

Our strategy is as follows. We formulate an assumption (see Assumption 4.2) on a family of finite element meshes ensuring optimal error estimates for the elliptic Ritz projection. This assumption is satisfied for different nonsmooth domains with appropriate mesh grading, see Sect. 5 for details. Under this assumption, which is of "pure elliptic nature", we check, that all proofs from [26] can be directly extended. This means, that for getting optimal order error estimates for the parabolic optimal control problem, it is enough to check the approximation properties for the discretization of the corresponding elliptic equation. This philosophy explained here on the example of the discretization based Petrov-Galerkin Crank-Nicolson scheme in time and linear finite elements in space for the model problem mentioned above can be extended in several directions. First of all, one can include control constraints in the same fashion as in [26], also the consideration of a more general parabolic problem with variable coefficients is possible. For an extension of dG(r) (discontinuous Galerkin) discretizations, e.g., from [24, 25] an additional assumption on the L^2-projection similar to Assumption 4.2 is required. This additional assumption will be fulfilled on the same families of meshes as described in Sect. 5. Under this assumption we strongly expect that also the error estimates for a semi-linear parabolic equation, see [27], as well as for problems with state constraints, see [17, 23] can be covered.

The outline of the paper is as follows. In the next section we discuss the optimality conditions and the regularity issues for the optimal control under consideration. After the description of the discretization scheme in Sect. 3 we formulate and prove our main result on a priori error analysis in Sect. 4 under Assumption 4.2. Section 5 is devoted to the verification of this assumption for different situations. Finally, in Sect. 6 we present a numerical example illustration our results.

2 Continuous Problem

In this section, we briefly discuss the precise formulation of the optimization problem under consideration. Furthermore, we recall theoretical results on existence, uniqueness, and regularity of optimal solutions as well as optimality conditions. For this discussion, we explicitly take the possible non-smoothness of the domain Ω into account.

To set up a weak formulation of the state equation (1.1b), we introduce the following notation: For a polygonal or polyhedral Lipschitz domain $\Omega \subset \mathbb{R}^n$, $n = 2, 3$, we denote V to be $H_0^1(\Omega)$. Together with $H := L^2(\Omega)$, the Hilbert space V and its dual $V^* = H^{-1}(\Omega)$ build a Gelfand triple $V \hookrightarrow H \hookrightarrow V^*$. Here and in what follows, we employ the usual notion for Lebesgue and Sobolev spaces. Furthermore, let D_Δ be the domain of the Laplacian given by $D_\Delta := \{ v \in V \mid \Delta v \in H \}$.

Remark 2.1. If Ω is polygonal and convex or possesses a smooth boundary, then $D_\Delta = H^2(\Omega) \cap H_0^1(\Omega)$ (see e.g. [14, Theorem 4 in Section 6.3] for the case of a smooth boundary and [19, Remark 2.4.6 and Corollary 2.6.8.] for polygonal and convex domains). Since we do not assume the convexity of Ω the space D_Δ is in general not a subset of $H^2(\Omega)$.

For a time interval $I = (0, T)$ we introduce the state space

$$X := W(0,T) = \left\{ v \;\middle|\; v \in L^2(I,V) \text{ and } \partial_t v \in L^2(I,V^*) \right\}$$

and the control space $Q := L^2(I, H)$. We use the following notations for the inner products and norms on $L^2(\Omega)$ and $L^2(I, H)$:

$$(v,w) := (v,w)_{L^2(\Omega)}, \qquad (v,w)_I := (v,w)_{L^2(I,H)},$$
$$\|v\| := \|v\|_{L^2(\Omega)}, \qquad \|v\|_I := \|v\|_{L^2(I,H)}.$$

In this setting, a standard weak formulation of the state equation (1.1b) for given control $q \in Q$, $f \in L^2(I, H)$, and $u_0 \in H$ reads: Find a state $u \in X$ satisfying

$$\begin{aligned} (\partial_t u, \varphi)_I + (\nabla u, \nabla \varphi)_I &= (f + q, \varphi)_I \qquad \forall \varphi \in X, \\ u(0) &= u_0. \end{aligned} \tag{2.1}$$

Assumption 2.2. For our analysis, we will assume the following regularity properties of the data: $f, \hat{u} \in H^1(I, H)$ with $f(0), \hat{u}(T) \in V$ and $u_0 \in V$ with $\Delta u_0 \in V$.

Using this assumption, the following result on existence and regularity can be proved:

Proposition 2.3. *Under Assumption 2.2 and for fixed control $q \in Q$, there exists a unique solution $u \in X$ of problem* (2.1). *Moreover, the solution exhibits the regularity*

$$u \in L^2(I, D_\Delta) \cap H^1(I, H) \cap C(\bar{I}, V)$$

with the estimate

$$\|\Delta u\|_I + \|\partial_t u\|_I + \|\nabla u(T)\| \le C\{\|f + q\|_I + \|\nabla u_0\|\}.$$

If additionally the fixed control q is in $H^1(I, H) \subset Q$, the state u exhibits the improved regularity

$$u \in H^1(I, D_\Delta) \cap H^2(I, H)$$

and the stability estimate

$$\|\partial_t \Delta u\|_I + \|\partial_t^2 u\|_I \le C\{\|f + q\|_{H^1(I,H)} + \|\nabla (f + q)(0)\| + \|\nabla \Delta u_0\|\}$$

holds.

Proof. Existence and the regularity $u \in L^2(I, D_\Delta) \cap H^1(I, H)$ is proven in [19, Theorem 5.1.1]. Then, the assertion $u \in C(\bar{I}, V)$ and the corresponding estimates for can be obtained by choosing $-\Delta u \in L^2(I, H)$ and $\partial_t u \in L^2(I, H)$ as test function in (2.1).

The improved regularity $u \in H^1(I, D_\Delta) \cap H^2(I, H)$ can be proved as in [14] provided that the right-hand side $f + q$ exhibits the regularity $f + q \in H^1(I, H)$ with $(f + q)(0) \in V$ and the initial condition u_0 fulfills $\Delta u_0 \in V$. This is ensured by Assumption 2.2, the assumed regularity $q \in H^1(I, H)$, and the embedding $H^1(I, V) \hookrightarrow C(\bar{I}, V)$. In contrast to [14], where $H^2(\Omega)$ regularity is used, one can not expect here the state variable to lie in $H^1(I; H^2(\Omega))$. However the proof of the stated results goes through. □

The weak formulation of the optimal control problem (1.1) is given as

$$\text{Minimize } J(q, u) := \frac{1}{2}\|u - \hat{u}\|_I^2 + \frac{\alpha}{2}\|q\|_I^2 \text{ s.t. (2.1) and } (q, u) \in Q \times X, \tag{2.2}$$

where $\alpha > 0$ is the regularization parameter.

Proposition 2.4. *For $\alpha > 0$ the optimal control problem* (2.2) *admits a unique solution $(\bar{q}, \bar{u}) \in Q \times X$.*

Proof. For the standard proof we refer, e.g., to [22]. □

Utilizing the adjoint state equation for $z = z(q) \in X$ given by

$$\begin{aligned} -(\varphi, \partial_t z)_I + (\nabla\varphi, \nabla z)_I &= (\varphi, u(q) - \hat{u})_I \qquad \forall \varphi \in X, \\ z(T) &= 0, \end{aligned} \tag{2.3}$$

the optimality condition is given by

$$\bar{q} = -\alpha^{-1} z(\bar{q}). \tag{2.4}$$

Employing this optimality condition we obtain, the following regularity result:

Proposition 2.5. *Let $(\bar{q}, \bar{u}) \in Q \times X$ be the solution of the optimization problem* (2.2) *and $\bar{z} = z(\bar{q}) \in X$ be the corresponding adjoint state. Then, there holds:*

$$\bar{q}, \bar{u}, \bar{z} \in H^1(I, D_\Delta) \cap H^2(I, H).$$

Furthermore, the following stability estimates are fulfilled:

$$\|\partial_t \Delta \bar{u}\|_I + \|\partial_t^2 \bar{u}\|_I \le C\{\|f + \bar{q}\|_{H^1(I,H)} + \|\nabla(f + \bar{q})(0)\| + \|\nabla\Delta u_0\|\},$$

$$\|\partial_t \Delta \bar{q}\|_I + \|\partial_t^2 \bar{q}\|_I \le C(\alpha)\{\|\hat{u}\|_{H^1(I,H)} + \|\nabla\hat{u}(T)\| + \|f + \bar{q}\|_I + \|\nabla u_0\|\}.$$

Proof. For $\bar{q} \in Q$, Proposition 2.3 implies that $\bar{u} \in L^2(I, D_\Delta) \cap H^1(I, H) \cap C(\bar{I}, V)$. This implies that the right-hand side of the adjoint equation (2.3) fulfills $\bar{u} - \hat{u} \in H^1(I, H)$ and $\bar{u}(T) - \hat{u}(T) \in V$. Consequently, we obtain by Proposition 2.3 that $\bar{z} \in H^1(I, D_\Delta) \cap H^2(I, H)$. This implies the stated regularity of $\bar{q}$.

The stability estimates for $\bar{u}$ follows directly from Proposition 2.3. For $\bar{z}$, Proposition 2.3 applied to the adjoint equation (2.3) implies

$$\|\partial_t \Delta \bar{z}\|_I + \|\partial_t^2 \bar{z}\|_I \le C\{\|\bar{u}\|_{H^1(I,H)} + \|\hat{u}\|_{H^1(I,H)} + \|\nabla\bar{u}(T)\| + \|\nabla\hat{u}(T)\|\}$$

and the estimate for $\bar{u}$ from Proposition 2.3 together with the optimality condition (2.4) yields the assertion. □

3 Discretization

In this section, we describe the space-time finite element discretization of the optimal control problem (2.2).

3.1 Semidiscretization in Time

At first, we present the semidiscretization in time of the state equation by continuous Galerkin methods. We consider a partitioning of the time interval $\bar{I} = [0, T]$ as

$$\bar{I} = \{0\} \cup I_1 \cup I_2 \cup \cdots \cup I_M$$

with subintervals $I_m = (t_{m-1}, t_m]$ of size k_m and time points

$$0 = t_0 < t_1 < \cdots < t_{M-1} < t_M = T.$$

We define the discretization parameter k as a piecewise constant function by setting $k\big|_{I_m} = k_m$ for $m = 1, 2, \ldots, M$. Moreover, we denote by k the maximal size of the time steps, i.e., $k = \max_{m=1,2,\ldots,M} k_m$. We impose the following conditions on the time mesh:

(i) There is a constant $\kappa > 0$ (independent of k) such that for all $m = 1, 2, \ldots, M-1$

$$\kappa^{-1} \le \frac{k_m}{k_{m+1}} \le \kappa$$

holds.

(ii) There is a constant $\gamma > 0$ (independent of k) such that

$$k \le \gamma \min_{m=1,2,\ldots,M} k_m.$$

The semidiscrete trial space is given as

$$X_k = \left\{ v_k \in C(\bar{I}, V) \;\middle|\; v_k\big|_{I_m} \in \mathcal{P}_1(I_m, V),\ m = 1, 2, \ldots, M \right\},$$

while the test space consisting of discontinuous piecewise polynomials of order 0 is defined as

$$\tilde{X}_k = \left\{ v_k \in L^2(I, V) \;\middle|\; v_k\big|_{I_m} \in \mathcal{P}_0(I_m, V),\ m = 1, 2, \ldots, M,\ v_k(0) \in V \right\}.$$

Here, $\mathcal{P}_r(I_m, V)$ denotes the space of polynomials up to order r defined on I_m with values in V. We use the notations

$$(v, w)_{I_m} := (v, w)_{L^2(I_m, H)} \quad \text{and} \quad \|v\|_{I_m} := \|v\|_{L^2(I_m, H)}.$$

To define the continuous Galerkin approximation (so-called cG(1) approximation) using the spaces X_k and $\tilde{X}_k$ we use for $v_k \in X_k$ the abbreviation $v_{k,m} := v_k(t_m)$

and for $w_k \in \tilde{X}_k$ we set $w_{k,m} = \lim_{t \uparrow t_m} w_k(t)$. The bilinear form $B(\cdot, \cdot)$ for $u_k \in X_k$ and $\varphi \in \tilde{X}_k$ is then defined by

$$B(u_k, \varphi) := (\partial_t u_k, \varphi)_I + (\nabla u_k, \nabla \varphi)_I + (u_{k,0}, \varphi_0).$$

The cG(1) semidiscretization of the state equation (2.1) for a given control $q \in Q$ reads: Find a state $u_k = u_k(q) \in X_k$ such that

$$B(u_k, \varphi) = (f + q, \varphi)_I + (u_0, \varphi_0) \quad \forall \varphi \in \tilde{X}_k. \tag{3.1}$$

The existence and uniqueness of solutions to (3.1) can be directly shown by "elliptic" arguments. For the general cG(r) case we refer to [31].

The semi-discrete optimization problem for the cG(1) time discretization has the following form:

$$\text{Minimize } J(q_k, u_k) \text{ subject to (3.1) and } (q_k, u_k) \in Q \times X_k. \tag{3.2}$$

The uniquely determined optimal solution of (3.2) is denoted by $(\bar{q}_k, \bar{u}_k) \in Q \times X_k$.

Remark 3.1. Note, that the optimal control $\bar{q}_k$ is searched for in the continuous space Q, and the subscript k indicates only the usage of the semidiscretized state equation.

Similarly to the continuous case, the optimality condition can be formulated as

$$\bar{q}_k = -\alpha^{-1} z_k(\bar{q}_k),$$

where $z_k = z_k(q) \in \tilde{X}_k$ denotes the solution of the semidiscrete adjoint equation

$$B(\varphi, z_k) = (\varphi, u_k(q) - \hat{u})_I \quad \forall \varphi \in X_k.$$

This yields that $\bar{q}_k$ is piecewise constant in time, i.e., that $\bar{q}_k \in \tilde{X}_k$.

Additionally to the partition of $\bar{I}$ introduced at the beginning of this section, we consider a "dual" partition of the time interval $\bar{I}$ defined by

$$\bar{I} = \{0\} \cup I_1^* \cup I_2^* \cup \cdots \cup I_{M+1}^*$$

with $I_m^* := (t_{m-1}^*, t_m^*]$ for $m = 1, 2, \ldots, M + 1$ and

$$t_0^* := t_0, \quad t_m^* := \frac{t_{m-1} + t_m}{2} \text{ for } m = 1, 2, \ldots, M, \quad \text{and} \quad t_{M+1}^* := t_M.$$

On this partition, we define the space Q_k by

$$Q_k := \left\{ w_k \in C(\bar{I}, V) \;\middle|\; w_k\big|_{I_m^*} \in \mathcal{P}_1(I_m^*, V),\ m = 1, 2, \ldots, M + 1 \right\}$$

and the interpolation $\pi_k \colon C(\bar{I}, V) \cup \tilde{X}_k \to Q_k$ as follows:

1. For $J = I_1^* \cup I_2^*$

$$\pi_k v(t)\big|_J := v(t_1^*) + \frac{t - t_1^*}{t_2^* - t_1^*}(v(t_2^*) - v(t_1^*))$$

2. For $J = I_m^*$ with $m = 3, 4, \ldots, M - 1$:

$$\pi_k v(t)\big|_J := v(t_{m-1}^*) + \frac{t - t_{m-1}^*}{t_m^* - t_{m-1}^*}(v(t_m^*) - v(t_{m-1}^*))$$

3. For $J = I_M^* \cup I_{M+1}^*$:

$$\pi_k v(t)\big|_J := v(t_{M-1}^*) + \frac{t - t_{M-1}^*}{t_M^* - t_{M-1}^*}(v(t_M^*) - v(t_{M-1}^*)).$$

3.2 Discretization in Space

To define the finite element discretization in space, we consider a family of two or three dimensional finite element meshes $\{\mathcal{T}_h\}_{h>0}$, see, e.g., [13]. A mesh consists of triangular, quadrilateral, tetrahedral, or hexahedral cells K, which constitute a non-overlapping cover of the computational domain Ω. The corresponding mesh is denoted by $\mathcal{T}_h = \bigcup\{K\}$, where we define the discretization parameter h as a cellwise constant function by setting $h\big|_K = h_K$ with the diameter h_K of the cell K. We use the symbol h also for the maximal cell size, i.e., $h = \max h_K$.

Remark 3.2. We do not assume the family of meshes $\{\mathcal{T}_h\}_{h>0}$ to be neither shape-regular nor quasi-uniform. For dealing with corner or edge singularities, we will use graded meshes, see Sect. 5 for details.

On the mesh $\mathcal{T}_h$ we construct a conforming finite element space $V_h \subset V$ in a standard way:

$$V_h = \left\{ v \in V \;\middle|\; v\big|_K \in \mathcal{Q}_1(K) \text{ for } K \in \mathcal{T}_h \right\}.$$

Here, $\mathcal{Q}_1(K)$ consists of shape functions obtained via (bi-/tri-)linear transformations of (bi-/tri-)linear polynomials defined on the reference cell. To obtain the fully discretized versions of the time discretized state equation (3.1), we utilize the space-time finite element spaces

$$X_{k,h} = \left\{ v_{kh} \in C(\bar{I}, V_h) \;\middle|\; v_{kh}\big|_{I_m} \in \mathcal{P}_1(I_m, V_h) \right\} \subset X_k$$

and

$$\tilde{X}_{k,h} = \left\{ v_{kh} \in L^2(I, V_h) \;\middle|\; v_{kh}\big|_{I_m} \in \mathcal{P}_0(I_m, V_h) \text{ and } v_{kh}(0) \in V_h \right\} \subset \tilde{X}_k.$$

The so-called cG(1)cG(1) discretization of the state equation for given control $q \in Q$ has the form: Find a state $u_{kh} = u_{kh}(q) \in X_{k,h}$ such that

$$B(u_{kh}, \varphi) = (f + q, \varphi)_I + (u_0, \varphi_0) \quad \forall \varphi \in \tilde{X}_{k,h}. \tag{3.3}$$

Then, the corresponding optimal control problem is given as

$$\text{Minimize } J(q_{kh}, u_{kh}) \text{ subject to (3.3) and } (q_{kh}, u_{kh}) \in Q \times X_{k,h}. \tag{3.4}$$

The uniquely determined optimal solution of (3.4) is denoted by $(\bar{q}_{kh}, \bar{u}_{kh}) \in Q \times X_{k,h}$. As before, the optimality condition can be formulated as

$$\bar{q}_{kh} = -\alpha^{-1} z_{kh}(\bar{q}_{kh}), \tag{3.5}$$

where $z_{kh} = z_{kh}(q) \in \tilde{X}_{k,h}$ denotes the solution of the discrete adjoint equation

$$B(\varphi, z_{kh}) = (\varphi, u_{kh}(q) - \hat{u})_I \quad \forall \varphi \in X_{k,h}.$$

By inspection of the optimality condition (3.5), we obtain that $\bar{q}_{kh} \in \tilde{X}_{k,h}$ and so the control does not need to be discretized explicitly, cf., e.g., [20].

Finally, on the "dual" partition, we define the discrete space $Q_{k,h}$ by

$$Q_{k,h} := \left\{ w_k \in C(\bar{I}, V_h) \;\middle|\; w_k\big|_{I_m^*} \in \mathcal{P}_1(I_m^*, V_h),\ m = 1, 2, \ldots, M+1 \right\}$$

and note that $\pi_k(\tilde{X}_{k,h}) \subset Q_{k,h}$.

4 Error Analysis

In this section, we prove the main result of this article, namely an $\mathcal{O}(k^2 + h^2)$ estimate for the error $\|\bar{q} - \tilde{q}_{kh}\|_I$ between the continuous solution $\bar{q} \in Q$ of (2.2) and the postprocessed discrete solution $\tilde{q}_{kh} \in Q_{k,h}$ defined by

$$\tilde{q}_{kh} = -\alpha^{-1} \pi_k z_{kh}(\bar{q}_{kh}), \tag{4.1}$$

where $\bar{q}_{kh} \in \tilde{X}_{k,h}$ is the solution of (3.4). The asserted estimate follows directly by the triangle inequality from the Theorems 4.1 and 4.4 below.

4.1 Estimates for the Error Due to Time Discretization

Theorem 4.1. *Let Assumption 4.2 be fulfilled. For the solution $\bar{q} \in Q$ of* (2.2) *and $\tilde{q}_k$ defined by*

$$\tilde{q}_k = -\alpha^{-1}\pi_k z_k(\bar{q}_k)$$

with the solution $\bar{q}_k \in Q$ of (3.2)*, it holds*

$$\|\bar{q} - \tilde{q}_k\|_I \leq C(\alpha)k^2\{\|f + \bar{q}\|_{H^1(I,H)} + \|\hat{u}\|_{H^1(I,H)} \\ + \|\nabla(f + \bar{q})(0)\| + \|\nabla\hat{u}(T)\| + \|\nabla\Delta u_0\|\}.$$

Proof. This result can be proved following the lines of the proof of Theorem 6.6 in [26]. There, the domain is assumed to be polygonal and convex. However, the proof of Theorem 6.6 there does not exploit H^2 regularity. It requires only the regularity stated in Proposition 2.3. □

4.2 Estimates for the Error Due to Space Discretization

For the error analysis derived here, we will make use of the spatial Ritz projection $R_h : V \to V_h$ defined by

$$(\nabla R_h v, \nabla\varphi) = (\nabla v, \nabla\varphi) \quad \forall\varphi \in V_h. \tag{4.2}$$

Assumption 4.2. The family of spatial meshes $\{\mathcal{T}_h\}_h$ is constructed such that for the Ritz projection defined by (4.2) the estimate

$$h\|\nabla(v - R_h v)\| + \|v - R_h v\| \leq Ch^2\|\Delta v\|$$

holds for all $v \in D_\Delta$.

Remark 4.3. If the domain Ω is polygonal and convex, then this assumption holds on shape-regular, quasi-uniform meshes by standard finite element theory with

$$h\|\nabla(v - R_h v)\| + \|v - R_h v\| \leq Ch^2\|v\|_{H^2(\Omega)} \leq Ch^2\|\Delta v\|,$$

where the H^2 regularity is used in the last step. In the case of a non-smooth domain, however, this assumption is typically not fulfilled on quasi-uniform meshes. In Sect. 5 we discuss several situations, where this assumption holds, if the family of meshes is constructed using an appropriate mesh grading.

Theorem 4.4. *For $\tilde{q}_k$ defined by $\tilde{q}_k = -\alpha^{-1}\pi_k z_k(\bar{q}_k)$ from Theorem 4.1 with the solution $\bar{q}_k \in Q$ of (3.2) and $\tilde{q}_{kh}$ defined by (4.1) with the solution $\bar{q}_{kh} \in Q$ of (3.4), it holds under Assumption 4.2 that*

$$\|\tilde{q}_k - \tilde{q}_{kh}\|_I \leq C(\alpha)\{k^2 + h^2\}\{\|f + \bar{q}_k\|_I + \|\nabla u_0\| + \|u_0\|\} + k^2\|\partial_t \hat{u}\|_I + h^2\|\hat{u}\|_I.$$

Proof. Under Assumption 4.2 it is possible to extend the proof of Theorem 6.10 in [26] to the case of a nonsmooth domain. The main component of this proof is Lemma 5.7 in [26], which shows, that a certain discretization error can be bounded by the error with respect to the Ritz projection R_h. Using Assumption 4.2, this lemma can be directly extended to the case considered here and the above result follows. □

5 Verification of Assumption 4.2

Now we discuss cases for which the Assumption 4.2 is fulfilled. It is well known, that for convex polygonal or polyhedral domains the Assumption 4.2 holds for finite element approximations on shape-regular, quasiuniform meshes (see Remark 4.3). So we focus on examples of nonconvex domains for which the Assumption 4.2 is also fulfilled.

5.1 Nonconvex Polygonal Domains

Let $\Omega \subset \mathbb{R}^2$ be a bounded polygonal domain with one non-convex interior angel $\omega > \pi$, located at the origin. Further we introduce the distance of a finite element τ of the triangulation $\mathcal{T}_h$ to the origin as $r_\tau = \inf_{(x_1,x_2)\in\tau}\sqrt{x_1^2 - x_2^2}$. Assume that the family of shape-regular triangulation $\{\mathcal{T}_h\}_h$ fulfills the conditions

$$\left.\begin{aligned} c_1 h^{1/\nu} &\leq h_\tau \leq c_2 h^{1/\nu}, && \text{for } r_\tau = 0, \\ c_1 h r_\tau^{1-\nu} &\leq h_\tau \leq c_2 h r_\tau^{1-\nu}, && \text{for } r_\tau > 0, \end{aligned}\right\} \tag{5.1}$$

with $\nu < \frac{\pi}{\omega}$ and $h_\tau = \operatorname{diam}(\tau)$.

Remark 5.1. For meshes which fulfill the condition (5.1) the number of elements is of order h^{-2}. Therefore the number of elements is of the same order as in a quasiuniform triangulation (see e.g. [8, Remark 3.1] or [28, 29]).

On meshes fulfilling (5.1) it is known that Assumption 4.2 is fulfilled, see e.g. [11, Theorem 5.1], [28, Theorem 1] or [29, Theorem 2].

Remark 5.2. The results can be transfered to domains with a corner with interior angle $\omega > \pi$ and smooth boundary everywhere else e.g. a segment of a disk with a reentrant corner.

Remark 5.3. As the singularities show local behavior, therefore more general two dimensional domains with more than one non-convex corner can be treated in a similar fashion, as we can write the solution as the sum of a regular part and the singularity functions of each non convex corner (see e.g. [21, Section 1.4]).

5.2 Prismatic Domains

Let $\Omega = G \times Z \subset \mathbb{R}^3$ be a bounded prismatic domain, where $G \subset \mathbb{R}^2$ is a bounded polygonal domain and $Z = (0, z_0)$ is an interval. Again we assume that G has one corner with interior angel $\omega > \pi$ located at the origin.

As in [32, Section 2.3.2] we construct the triangulation of Ω in the following way: Assume that the triangulation of G is constructed such that the condition (5.1) is fulfilled. From this triangulation we get a triangulation of the domain Ω by extruding the triangles in x_3 direction quasiuniform with mesh size h. This gives a mesh of triangular prisms, to get an anisotropic thetraedal mesh each prism is divided into thetraheda. For elements of this mesh the following estimates hold

$$\left.\begin{aligned} c_1 h^{1/\nu} \le h_{\tau,i} &\le c_2 h^{1/\nu}, && \text{for } r_\tau = 0 \text{ and } i = 1,2,\\ c_1 h r_\tau^{1-\nu} \le h_{\tau,i} &\le c_2 h r_\tau^{1-\nu}, && \text{for } r_\tau > 0 \text{ and } i = 1,2,\\ c_1 h \le h_{\tau,3} &\le c_2 h, && \end{aligned}\right\}$$

where $h_{\tau,i}$ is the length of the projection of the element τ to the x_i-axis (for $=$ $1,2,3$), $r_\tau = \inf_{(x_1,x_2)\in\tau} \sqrt{x_1^2 + x_2^2}$ the distance of the element τ to the x_3-axis and $\nu < \frac{\pi}{\omega}$. On such meshes the Assumption 4.2 is fulfilled, see [9, Theorem 5.2].

Remark 5.4. In [2, Corollary 4.1] the validity of Assumption 4.2 is shown for prismatic domains with Neumann boundary conditions on $G \times \{0, z_0\}$ and Dirichlet conditions on the remaining part of $\partial\Omega$.

5.3 General Polyhedral Domains

For the solution of the Dirichlet problem for the Poisson equation on general polyhedral domains $\Omega \subset \mathbb{R}^3$ we refer to [4], where a refinement strategy is proposed that the Assumption 4.2 holds (see [4, Corollary 3.12]). As the grading strategy and construction of corresponding meshes is more complicated as in the previous cases, we omit the details here and refer for details to [4].

6 Numerical Results

For the numerical verification we consider the optimal control Problem (1.1) with $\alpha = 1$ on a L-shaped domain $\Omega = (-1,1)^2 \setminus [0,1] \times [-1,0]$ and the unit time interval $(0,T) = (0,1)$. The remaining data f and $\hat{u}$ are chosen, such that the exact solution in polar coordinates $(r,\varphi) \in \mathbb{R}_+ \times [0,2\pi)$ is given by

$$\bar{u}(r,\varphi,t) = (\mathrm{e}^{\lambda t} - 1) \cdot u_s(r,\varphi),$$

$$\bar{z}(r,\varphi,t) = (\mathrm{e}^{\lambda(1-t)} - 1) \cdot u_s(r,\varphi),$$

with $\lambda = \frac{2}{3}$ and

$$u_s(r,\varphi) = r^\lambda \sin(\lambda\varphi) \cdot (r\cos\varphi - 1)(r\cos\varphi + 1)(r\sin\varphi + 1)(r\sin\varphi - 1).$$

For the grading parameter $\nu = 0.6$ in (5.1), Fig. 1 depicts the behavior of the errors $\|\bar{q} - \tilde{q}_{kh}\|_I$ and $\|\bar{u} - \bar{u}_{kh}\|_I$ for a sequence of temporal and spacial meshes. They exhibit the proved convergence order $\mathcal{O}(k^2 + h^2)$.

Remark 6.1. For the Crank-Nicolson time stepping discretizations of [3] similar convergence results can be observed.

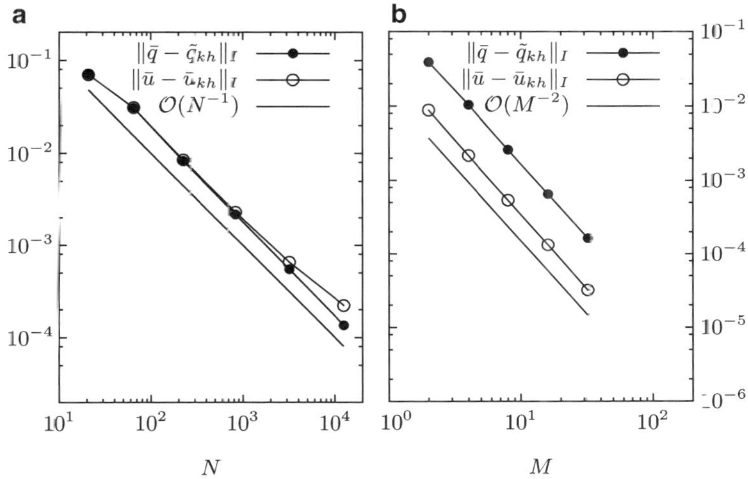

Fig. 1 Observed convergence of the numerical example. Here, N with $N^{-1} = \mathcal{O}(h^2)$ (cf. Remark 5.1) denotes the number of cells in the spatial mesh and M with $M^{-1} = k$ is the number of times steps. (**a**) Convergence with respect to spatial refinement with fixed time step size. (**b**) Convergence with respect to the time step size with fixed spatial mesh

Acknowledgements The function u_s for the numerical example in Sect. 6 was taken from a presentation by T. Apel and J. Pfefferer. The numerical experiments in Sect. 6 are carried out using both a MATLAB-FEM implementation based on [1, 12, 15] and the software packages RoDoBo [30] and GASCOIGNE [16].

References

1. J. Alberty, C. Carstensen, S.A. Funken, Remarks around 50 lines of Matlab: short finite element implementation. Numer. Algorithms **20**, 117–137 (1999)
2. T. Apel, *Anisotropic Finite Elements: Local Estimates and Applications*. Advances in Numerical Mathematics (Teubner, Stuttgart, 1999)
3. T. Apel, T.G. Flaig, Crank-Nicolson schemes for optimal control problems with evolution equations. SIAM J. Numer. Anal. **50**, 1484–1512 (2012)
4. T. Apel, A.L. Lombardi, M. Winkler, *Anisotropic mesh refinement in polyhedral domains: error estimates with data in* $L^2(\Omega)$, 2013. http://arxiv.org/abs/1303.2960
5. T. Apel, J. Pfefferer, A. Rösch, Finite element error estimates for Neumann boundary control problems on graded meshes. Comput. Optim. Appl. **52**, 3–28 (2012)
6. T. Apel, A. Rösch, D. Sirch, L^∞-error estimates on graded meshes with application to optimal control. SIAM J. Control Optim. **48**, 1771–1796 (2009)
7. T. Apel, A. Rösch, G. Winkler, Optimal control in non-convex domains: a priori discretization error estimates. Calcolo **44**, 137–158 (2007)
8. T. Apel, A.-M. Sändig, J.R. Whiteman, Graded mesh refinement and error estimates for finite element solutions of elliptic boundary value problems in non-smooth domains. Math. Methods Appl. Sci. **19**, 63–85 (1996)
9. T. Apel, D. Sirch, L^2-error estimates for Dirichlet and Neumann problems on anisotropic finite element meshes. Appl. Math. **56**, 177–206 (2011)
10. T. Apel, G. Winkler, Optimal control under reduced regularity. Appl. Numer. Math. **59**, 2050–2064 (2009)
11. I. Babuška, R. Kellogg, J. Pitkäranta, Direct and inverse error estimates for finite elements with mesh refinements. Numerische Mathematik **33**, 447–471 (1979)
12. L. Chen, C.-S. Zhang, AFEM@MATLAB: A MATLAB package of adaptive finite element methods. Technical report, University of Maryland, 2006. www.math.umd.edu/~zhangcs/paper/AFEM@matlab.pdf.gz
13. P.G. Ciarlet, *The Finite Element Method for Elliptic Problems* (North-Holland, Amsterdam, 1979)
14. L.C. Evans, *Partial Differential Equations*. Volume 19 of Graduate Studies in Mathematics (AMS, Providence, 2002)
15. S. Funken, D. Praetorius, P. Wissgott, Efficient implementation of adaptive P1-FEM in Matlab. Comput. Methods Appl. Math. **11**, 460–490 (2011)
16. *The finite element toolkit* GASCOIGNE. http://www.gascoigne.de
17. W. Gong, M. Hinze, Error estimates for parabolic optimal control problems with control and state constraints. Comput. Optim. Appl. **56**, 131–151 (2013)
18. W. Gong, M. Hinze, Z.J. Zhou, Space-time finite element approximation of parabolic optimal control problems. J. Numer. Math. **20**, 111–145 (2012)
19. P. Grisvard, *Singularities in Boundary Value Problems*. Volume 22 of Research Notes in Applied Mathematics (Springer, New York, 1992)
20. M. Hinze, A variational discretization concept in control constrained optimization: the linear-quadratic case. Comput. Optim. Appl. **30**, 45–61 (2005)
21. A. Kufner, A.-M. Sändig, *Some Applications of Weighted Sobolev Spaces* (Teubner, Leipzig, 1987)

22. J.-L. Lions, *Optimal Control of Systems Governed by Partial Differential Equations*. Volume 170 of Grundlehren der mathematischen Wissenschaften (Springer, Berlin, 1971)
23. D. Meidner, R. Rannacher, B. Vexler, A priori error estimates for finite element discretizations of parabolic optimization problems with pointwise state constraints in time. SIAM J. Control Optim. **49**, 1961–1997 (2011)
24. D. Meidner, B. Vexler, A priori error estimates for space-time finite element discretization of parabolic optimal control problems. Part I: problems without control constraints. SIAM J. Control Optim. **47**, 1150–1177 (2008)
25. D. Meidner, B. Vexler A priori error estimates for space-time finite element approximation of parabolic optimal control problems. Part II: problems with control constraints. SIAM J. Control Optim. **47**, 1301–1329 (2008)
26. D. Meidner, B. Vexler, A priori error analysis of the Petrov-Galerkin Crank-Nicolson scheme for parabolic optimal control problems. SIAM J. Control Optim. **49**, 2183–2211 (2011)
27. I. Neitzel, B. Vexler, A priori error estimates for space-time finite element discretization of semilinear parabolic optimal control problems. Numer. Math. **120**, 345–386 (2012)
28. L.A. Oganesyan, L.A. Rukhovets, Variational-difference schemes for linear second-order elliptic equations in a two-dimensional region with piecewise smooth boundary. Zh. Vychisl. Mat. Mat. Fiz. **8**, 97–114 (1968). In Russian. English translation in USSR Comput. Math. and Math. Phys. **8**, 129–152 (1968)
29. G. Raugel, Résolution numérique par une méthode d'éléments finis du problème de Dirichlet pour le laplacien dans un polygone. C. R. Acad. Sci. Paris, Sér. A **286**, 791–794 (1978)
30. RoDoBo. *A C++ library for optimization with stationary and nonstationary PDEs with interface to* GASCOIGNE *[16]*. http://www.rodobo.org.
31. F. Schieweck, A-stable discontinuous Galerkin-Petrov time discretization of higher order. J. Numer. Math. **18**, 25–57 (2010)
32. D. Sirch, finite element error analysis for PDE-constrained optimal control problems: the control constrained case under reduced regularity. Ph.D. thesis, Technische Universität München, 2010

Part V
Applications

Introduction to Part V: Applications

In this part a variety of important applications in the field of PDE constrained optimization and optimal control is presented. The results range from physical and chemical over electronic to biomedical and -technological applications. Furthermore, an introduction to a new collection of prototypical problems in PDE constrained optimization is given. This part is structured as follows:

In their article *Optimal Treatment Planning in Radiotherapy Based on Boltzmann Transport Equations* Richard C. Barnard, Martin Frank and Michael Herty study the question of finding an optimal treatment plan for radiotherapy treatment of cancer. The medical problem is that ionizing radiation shall be delivered to a targeted tissue, for example a tumor, without damaging healthy tissue around the tumor. The authors present a model for dose calculation based on the Boltzmann transport equations and derive optimal control formulations and necessary optimality conditions. Approximation methods for calculating the optimal dose and implementing the optimality conditions as well as numerical examples are presented.

In *Optimal Control of Self-Consistent Classical and Quantum Particle Systems* Martin Burger, Rene Pinnau, Marcisse Fouego and Sebastian Rau consider optimal control problems for self-consistent interacting classical and quantum particle systems both from an analytical and a numerical point of view. This topic is of great interest, for example, in designing semiconductor devices and in biomedical applications. The authors study control and design problems for the microscopic nonlinear Schrödinger-Poisson system and for the macroscopic Quantum Euler-Poisson model. Furthermore, optimization problems for Drift-Diffusion models are presented.

In *Modeling, Analysis and Optimization of Particle Growth, Nucleation and Ripening by the way of Nonlinear Hyperbolic Integro-Partial Differential Equations* Michael Gröschel, Wolfgang Peukert and Günter Leugering study the processes of particle growth, nucleation, precipitation and ripening via modeling by nonlinear 1-D hyperbolic integro-partial differential equations. The authors provide a concise

G. Leugering et al. (eds.), *Trends in PDE Constrained Optimization*, International Series of Numerical Mathematics 165, DOI 10.1007/978-3-319-05083-6_27

predictive forward modeling of the processes and establish a mathematical theory of the open-loop optimization in this context.

The stabilization of hyperbolic systems on networks is studied by Markus Dick, Martin Gugat, Michael Herty, Günter Leugering, Sonja Steffensen and Ke Wang in *Stabilization of Networked Hyperbolic Systems with Boundary Feedback*. The authors present a method to stabilize quasilinear systems on fan-shaped networks using linear feedback controls at the nodes of the networks. The evolution of the solution is measured with a Lyapunov function, for which the authors obtain the exponential decay with time. Furthermore, a numerical discretization of the Lyapunov function and a numerical analysis are presented. The results are applied to stabilize the gas flow in a fan-shaped pipe network with compressor stations.

A biomedical application in the field of optimal control is presented by Thomas Franke, Ronald H. W. Hoppe, Christopher Linsenmann, Lothar Schmid and Achim Wixforth in the article *Optimal Control of Surface Acoustic Wave Actuated Sorting of Biological Cells*. The sorting of biological cells is of great importance in many medical applications such as cancer research. The goal of the control problem studied in this article is to sort different types of biological cells in a microfluidic channel. The sorting is effected by surface acoustic waves generated by an interdigital transducer at the wall of the channel. The control variable is the time-dependent power applied to the interdigital transducer. The PDEs the control problem is based on are the incompressible Navier-Stokes equations which model the motion of the carrier fluid and the equations of motion of the boundaries of the cells.

Another example for separation processes based on PDE constrained optimal control is presented by Malte Behrens, Hans Georg Bock, Sebastian Engell, Phawitphorn Khobkhun and Andreas Potschka in *Real-Time PDE Constrained Optimal Control of a Periodic Multicomponent Separation Process*. A biotechnological separation process in which three or more different components shall be separated out of a liquid mixture is studied. The authors apply the control method of Modifier Adaptation to a PDE constrained optimization problem with periodic boundary conditions in time. As an example the separation of three amino acids in a virtual plant with real-world parameters is considered.

In *OPTPDE – A Collection of Problems in PDE-Constrained Optimization* Roland Herzog, Arnd Rösch, Stefan Ulbrich and Winnifried Wollner introduce the OPTPDE database of prototypical optimization problems with PDE constraints which can be accessed at www.optpde.net. In the article the authors describe how researchers can use the OPTPDE collection for their own work on optimization problems and how they can submit new problems for the database. Furthermore, the authors illustrate the main features of OPTPDE by an example problem.

Sebastian Engell and Günter Leugering

Optimal Treatment Planning in Radiotherapy Based on Boltzmann Transport Equations

Richard C. Barnard, Martin Frank, and Michael Herty

Abstract We look at the optimization of radiotherapy treatment planning. By using a deterministic model of dose deposition in tissue derived from the Boltzmann transport equations, we can improve on the accuracy of existing models near tissue inhomogeneities while also making use of adjoint calculus for developing necessary conditions for optimality. We describe the relevant model and consider the planning problem in an optimal control framework. Two versions of the problem are discussed, optimality conditions are derived, and numerical methods are described. Numerical examples are presented.

Keywords Kinetic equation • Optimal control • Radiotherapy

Mathematics Subject Classification (2010). 35Q20, 49J20.

1 Introduction

For the treatment of cancer, radiotherapy is, along with surgery and chemotherapy, one of the primary methods in practice today. In conjunction with those methods, radiotherapy plays a significant role in 40 % of cured patients [24]; overall, over half of cases should have radiotherapy as part of the treatment process [8, 36].

R.C. Barnard (✉)
Institute for Mathematics and Scientific Computing, University of Graz
Heinrichstr. 36, A-8010 Graz, Austria
e-mail: barnard@mathcces.rwth-aachen.de

M. Frank
MATHCCES, Department of Mathematics, RWTH Aachen University, Schinkelstr. 2, D-52062 Aachen, Germany
e-mail: frank@mathcces.rwth-aachen.de

M. Herty
IGPM, Department of Mathematics, RWTH Aachen University, Templergraben 55, D-52062 Aachen, Germany
e-mail: herty@igpm.rwth-aachen.de

G. Leugering et al. (eds.), *Trends in PDE Constrained Optimization*, International Series of Numerical Mathematics 165, DOI 10.1007/978-3-319-05083-6_28

The goal is to deliver ionizing radiation to targeted tissue, such as a tumor, while preserving, where possible, healthy tissue; the resulting damage to the affected tissue leads to cell death. Delivery of the radiative dose can be through external sources (teletherapy) as well as sources deposited in tissue (brachytherapy). In the case of teletherapy, fixed beams may be arranged around the patient and are often selected using technician/physician experience. With the advent of Intensity-Modulated Radiation Therapy (IMRT), beams may be moved during treatment and shaped by multileaf collimators. This potentially also allows for accounting for motion in the body of the patient during treatment leading to what is known as 4D Radiotherapy (4DRT) [6]. With these methods, the complexity involved requires some degree of automation via mathematical models and optimization algorithms in the treatment planning process [25].

Many dose calculation methods rely on Monte Carlo algorithms or deterministic models based on the Fermi-Eyges theory of radiation. The former in general show close correspondence to experimental results, relying on a rigorous model for the interactions of particles with human tissue. However, dose calculations can be computationally very expensive [3]. This, coupled with the need for derivative-free optimization methods, leads to difficulties in the implementation of Monte Carlo-based algorithms. The latter type of method has the benefit of being computationally efficient allowing for relatively quick dose calculations; however, near tissue inhomogeneities such as void-like regions in the lung, the physical assumptions in the model break down. One source of inaccuracy is that physical assumptions such as there only being small-angle scattering events and small angle of flight do not hold in general [22]. This leads to large errors of up to 12 % in such areas [21, 26]. In other cases, such as the irradiation of the vertebral column, dose discrepancies can be even higher [33].

Dose calculation methods based on a Boltzmann transport model have the benefit of reliance on rigorous models of particle interactions with the tissue that can be then solved exactly, in principle [17]. Furthermore, the Boltzmann based method does not make assumptions on the homogeneity of the tissue being studied. In considering the treatment planning process, optimization algorithms can exploit the deterministic nature of the dose calculation methods for derivative-based methods. This approach has been explored in recent years by various authors [1, 12, 15, 16, 18, 19, 30–32].

The remainder of the paper is structured in the following way. After describing the model for dose calculation in Sect. 2, we look at two optimal control formulations of the problem of finding an optimal treatment plan in Sect. 3 as well as the resulting necessary optimality conditions. From there, we present in Sect. 4 methods for calculating the dose and implementing the optimality conditions before presenting some brief numerical examples in Sect. 5. Finally we look at the open challenges in optimal control theory for treatment planning in Sect. 6.

2 Dose Calculation Models

We describe the transport equation modeling the distribution of electrons which leads to calculation of the radiative dose in the patient's body. We assume that $Z \subset \mathbb{R}^3$ is a convex, open bounded domain with smooth boundary which contains the relevant portion of the patient's body. The outward normal vector is denoted by n. We consider particles moving with unit velocity which do not interact with each other, only with the media (i.e. the tissue); the direction of movement for a particle is denoted by $\Omega \in S^2$ where S^2 is the unit sphere in three dimensions. We define a function ψ as the density of particles; that is, $\psi(z, \epsilon, \Omega)\cos(\theta) dA d\epsilon d\Omega$ is the number of particles passing through an area dA at the point z into an angle $d\Omega$ around Ω with θ the angle between Ω and dA. Then the linear Boltzmann equation for electron transport is

$$\Omega \cdot \nabla_z \psi(z, \epsilon, \Omega) = \left[\int_\epsilon^\infty \int_{S^2} \Sigma_s(z, \epsilon', \epsilon, \Omega' \cdot \Omega) \psi(z, \epsilon', \Omega') d\Omega' d\epsilon' \right] - \Sigma_t(z, \epsilon)\psi(z, \epsilon, \Omega) + q(z, \epsilon, \Omega). \tag{2.1}$$

Here, Σ_s and Σ_t are the scattering cross section and total cross section, respectively, and q is some source of particles. As detailed in [22, 23], through asymptotics, an approximation known as the Boltzmann Continuous Slowing Down approximation (BCSD) to (2.1) can be obtained in the following form:

$$-\partial_\epsilon \left(S_M(z, \epsilon)\psi(z, \epsilon, \Omega) \right) + \Omega \cdot \nabla_z \psi(z, \epsilon, \Omega) = \int_{S^2} \Sigma_s(z, \epsilon, \Omega' \cdot \Omega)\psi(z, \epsilon, \Omega') d\Omega' - \Sigma_t(z, \epsilon)\psi(z, \epsilon, \Omega) + q(z, \epsilon, \Omega) \tag{2.2}$$

where

$$\Sigma_s(z, \epsilon, \mu) = \int_0^\infty \Sigma_s(z, \epsilon, \epsilon', \mu) d\epsilon'.$$

In (2.1), particles travel in straight lines until collisions occur, upon which discrete changes in both direction and energy occur. In (2.2), the angular change is still discrete (and large-angle changes are still permitted) while now the energy losses from collisions are described differentially. We will assume that the stopping power is a product of the density of the tissue $\rho(z)$ and a function of energy $S(\epsilon)$. That is, we consider the BCSD in form

$$-\partial_\epsilon \left(\rho(z) S(\epsilon)\psi(z, \epsilon, \Omega) \right) + \Omega \cdot \nabla_z \psi(z, \epsilon, \Omega) = \int_{S^2} \Sigma_s(z, \epsilon, \Omega' \cdot \Omega)\psi(z, \epsilon, \Omega') d\Omega' - \Sigma_t(z, \epsilon)\psi(z, \epsilon, \Omega) + q(z, \epsilon, \Omega).$$

In general, as discussed in [14, 18] and [1], for physically reasonable cross sections, Σ_t and Σ_s are non-negative and bounded in the sup-norm, and S is nonnegative and continuous. For constant tissue density, existence of mild solutions to the previous equation has been established for physically relevant ranges of parameters [18].

Along with the BCSD, we impose boundary conditions and "terminal" conditions. After defining

$$\Gamma = \partial Z \times S^2 \text{ and } \Gamma^- = \{(x, \Omega) \in \Gamma : n(x) \cdot \Omega < 0\},$$

and some maximal allowed energy $0 < \epsilon_{max} < \infty$, we require ψ satisfy

$$\psi(x, \epsilon_{max}, \Omega) = 0 \text{ and } \psi(x, \epsilon, \Omega) = q_b(x, \epsilon, \Omega), \; (x, \Omega) \in \Gamma^-$$

where q_b denotes some boundary source, such as particles entering from external beams. We note that the energy variable acts as a "pseudo-time" variable; with this in mind, in the case of one dimensional slab-geometry, we define

$$t(\epsilon) = \int_{\epsilon}^{\epsilon_{max}} \frac{1}{S(r)} dr \text{ and } x(z) = \int_0^z \rho(\tilde{z}) d\tilde{z},$$

and – as $S > 0$ – (as detailed in [15]), this gives a change of variables. The $z - x$ transformation can in general not be extended to the 3-dimensional problem. Nevertheless, suppressing relabeling of the quantities for notational simplicity, we study the following final transport equation,

$$\begin{aligned} \partial_t \psi(t, x, \Omega) + \Omega \cdot \nabla_x \psi(t, x, \Omega) &= \int_{S^2} \Sigma_s(x, t, \Omega' \cdot \Omega) \psi(t, x, \Omega') d\Omega' \qquad (2.3) \\ &- \Sigma_t(x, t) \psi(t, x, \Omega) + q(t, x, \Omega). \\ \psi(t, x, \Omega) = q_b(t, x, \Omega), \; (x, \Omega) \in \Gamma^-, \; t \in [0, T] \\ \psi(0, x, \Omega) = 0, \; (x, \Omega) \in Z \times S^2. \end{aligned}$$

This will be the model we use for the remainder of this chapter with the reminder that it has been transformed from the BCSD model. Thus, the physics encoded by ρ and S are not lost.

With this, we now introduce the dose operator $D : L^2([0, T] \times Z \times S^2) \to L^2(Z)$, which, given a particle distribution ψ solving (2.3), gives the deposited dose

$$D(\psi)(x) := \int_0^T \int_{S^2} \psi(t, x, \Omega) d\Omega dt.$$

Despite the mesoscopic nature of both (2.2) and (2.3), we are actually interested in the macroscopic dose distribution in radiotherapy. This quantity is the primary area of interest for the optimal treatment planning problem.

3 The Optimal Treatment Problem

With the model for calculating the dose distribution established, we now turn to the problem of finding the optimal such dose. Quantifying the "best" distribution is not automatically apparent; in general, we wish to have a homogeneous level of cell death in the targeted region and low levels elsewhere. While the death of cells in the targeted tissue surrounding the tumor is clearly the primary goal (along with the preservation of healthy tissue), it is known [4] that the dose is not directly proportional to the death rate of cells. Often (e.g. [13, 14, 25]), in order to avoid difficulties in radiobiological modeling, the treatment planning problem involves tracking a desired dose distribution. The simplest version of this is as follows. We partition Z into three disjoint sections:

- Z_T: the target region, containing the tumor tissue;
- Z_R: the risk region, a critical portion of Z containing, for instance, vital organs;
- Z_N: the normal tissue, the remainder of Z.

Associated with each region, we define a positive scalar: c_T, c_R, c_N and a function $c_1 := c_T\chi_{Z_T} + c_R\chi_{Z_R} - c_N\chi_{Z_N}$ that will act as a penalty for deviation from the desired dose. This is defined simply to be $\overline{D} := \chi_{Z_T}$, the characteristic associated with Z_T This data gives, along with $c_2 > 0$, the objective functional

$$J_T(\psi, q) := \frac{1}{2}\int_Z c_1(x)\big(D(\psi)(x) - \overline{D}(x)\big)^2 dx + \frac{c_2}{2}\int_Z\int_0^T\int_{S^2} q^2(t, x, \Omega)d\Omega dx dt.$$

In light of the discussion to follow in Sect. 5, we consider problems where we require $q_b \equiv 0$. Further discussion of this constraint will be in that section. Thus J_T will only depend on q and ψ. The quadratic tracking functional is used in the following treatment planning problem

$$\mathcal{P}_T \begin{cases} \min_{\psi,q}\ J_T(\psi, q) & \text{subject to:} \\ \psi, q \in L^2([0, T] \times Z \times S^2), & \\ \psi \text{ and } q \text{ solve (2.3)}, & \\ 0 \le q(t, x, \Omega) \le f(t)g(x). & \end{cases}$$

Here $f > 0$ denotes constraints on the allowed energy levels of the source and $g > 0$ constraints on the placement of the source (see Sect. 5 for more details). In practical applications, further constraints on the source may be necessary as we do not place requirements on the structure of the source beyond these box constraints.

An alternative to J_T involves using one of the radiobiological models of cell response to radiative doses; the linear-quadratic model is well-established and studied (see, among many others, [7, 11, 20, 27–29, 35]) for describing cell death rates. We will neglect cell growth and repair rates for the following analysis. For a given dose D, the fraction of surviving cells of a single species is fitted to a linear-quadratic model

$$SF = \exp(-\alpha D - \beta D^2)$$

which gives the tumor control probability on a region of one cell type

$$TCP = \exp \int \Big(- p \exp(-\alpha_i D - \beta_i D^2) \Big) dx$$

where p is the density of tumor cells as a function of space. Assuming a single type of tumor cell assigned index 0 and assigning the other cell types in Z the indices $i = 1, \ldots, N$, and cell densities $p_i(x) \geq 0$ with $\sum_{i=0}^{N} p_i(x) = 1$, we define

$$\begin{aligned} J_{SF}(\psi, q) := a_0 & \int_Z p_0 \exp\big[- \alpha_0 D\psi - \beta_0 (D\psi)^2 \big] dx \\ & + \sum_{i=1}^{N} a_i \int_Z p_i \Big(1 - \exp\big[- \alpha_i D\psi - \beta_i (D\psi)^2 \big]\Big) dx \\ & + \frac{c_2}{2} \int_Z \int_0^T \int_{S^2} q(t, x, \Omega) d\Omega dx dt \end{aligned} \tag{3.1}$$

where we assign constant weights for each cell type $a_i > 0$ and the control $c_2 > 0$. This leads to the other optimal control problem we consider here

$$\mathcal{P}_{SF} \begin{cases} \min_{\psi, q} \; J_{SF}(\psi, q) & \text{subject to:} \\ \psi, q \in L^2([0, T] \times Z \times S^2), \\ \psi \text{ and } q \text{ solve (2.3)}, \\ 0 \leq q(t, x, \Omega) \leq f(t) g(x). \end{cases}$$

We make a few comments here about the structure of $\mathcal{P}_T$ and $\mathcal{P}_{SF}$. We denote by $\mathcal{E} : L^2([0, T] \times Z \times S^2) \to L^2([0, T] \times Z \times S^2)$ the solution operator of (2.3), that is $\mathcal{E}(q) = \psi$ if and only if ψ and q solve (2.3). Then it can be seen rather easily that $\mathcal{E}$ is linear and bounded. As the dose operator $D(\psi)$ is also linear and bounded, it is rather easy by standard arguments (such as in [34]) to show that $\mathcal{P}_T$ has a unique solution and that a necessary optimality condition will also, by the convexity of J_T, be sufficient as well [1]. However, J_{SF} is not convex and therefore, we only obtain existence of an optimal solution [1].

The optimality conditions are summed up in the following result.

Theorem 3.1. *For the radiotherapy planing problem subject to cost functionals either of tracking type or given by the surviving fractions, we refer to the first-order optimality system*

$$\partial_t \psi^* + \Omega \cdot \nabla_x \psi^* = \int_{S^2} \Sigma_s(x, \Omega' \cdot \Omega)\psi^*(t, x, \Omega')d\Omega' - \Sigma_t(x)\psi^* + q^*, \tag{3.2}$$

$$\psi^*(0, x, \Omega) = 0, \quad on\ \Gamma^-,$$

$$\psi^*(0, x, \Omega) = 0, \ (x, \Omega) \in Z \times S^2.$$

$$-\partial_t \lambda^* - \Omega \cdot \nabla_x \lambda^* = \int_{S^2} \Sigma_s(x, \Omega' \cdot \Omega)\lambda^*(t, x, \Omega')d\Omega' - \Sigma_t(x)\lambda^* + r(x). \tag{3.3}$$

$$\lambda^*(0, x, \Omega) = 0, \quad on\ \Gamma^+,$$

$$\lambda^*(T, x, \Omega) = 0.$$

$$q^*(t, x, \Omega) = \operatorname*{proj}_{[0, f(t)g(x)]} \left(-\frac{1}{c_2} \lambda^*(t, x, \Omega) \right). \tag{3.4}$$

Here, $\Gamma^+ = \{(x, \Omega) : x \in \partial Z,\ n(x) \cdot \Omega > 0\}$.

- *Given a desired dose distribution* $\overline{D} \in L^2(Z)$, *and let* ψ^*, q^* *be an optimal pair for* $\mathcal{P}_T$. *Then, there exists* $\lambda^* \in L^2([0, T] \times Z \times S^2)$ *so that the above system is satisfied with* $r(x) = c_1(D(\psi^*) - \overline{D})$.
- *Given a distribution of cells* p_i *and associated parameters* $\alpha_i, \beta_i > 0$, *if* ψ^*, q^* *is locally optimal for* $\mathcal{P}_{SF}$, *then the system is satisfied with*

$$r = \Big[(-\alpha_0 - 2\beta_0 D\psi^*)\big(a_0 p_0 \exp[-\alpha_0 D^* - \beta_0(D\psi^*)]\big)\Big]$$
$$- \sum_{i=1}^{N} \Big[(-\alpha_i - 2\beta_i D\psi^*)\big(a_i p_i \exp[-\alpha_i D^* - \beta_i(D\psi^*)]\big)\Big].$$

The stationary case has been treated in [13], whereas the full result in the case of a tracking type functional is discussed in [14] for constant cross–sections Σ_s, Σ_t and in [18] for spatially dependent cross–sections. In the same setting a rigorous result for the surviving fractions cost functionals and for $f(t) = g(x) = +\infty$ has been established in [1].

4 Approximation Methods

In order to efficiently compute solutions to equations of the form (2.3), approximation methods have been developed. The main purpose is to reduce the dimension of the state space by integration on angular velocities. This is motivated by the primary quantity of interest, namely the dose, which is independent of the particles' velocities. Hence, we are not concerned with the microscopic characterization of the particle distribution in velocity space. Introducing the notation $\langle \cdot \rangle = \int_{S^2} \cdot \, d\Omega$, we multiply (2.3) by a vector of polynomials $m(\Omega)$ and integrating, we get a system with entries of the form (denoting $\psi^{(i)} := \langle m_i \psi \rangle$):

$$\partial_t \psi^{(i)} + \nabla_x \langle \Omega m_i \psi \rangle = S(x)\psi^{(i)} - \Sigma_t \psi^{(i)} - q^{(i)}.$$

We note that this system is not closed, in that the flux depends on a quantity other than $\langle m\psi \rangle$; also, we have not specified the form that m should take. We will discuss briefly two options, the spherical harmonics method and the minimum entropy method.

If we select in one spatial dimension x the spherical harmonics in Ω as the entries and truncate after the N-th harmonic, that is we assume $\psi^{(i)}$ is zero for any $i \geq N$, we obtain a closed system with an explicit flux function which has several benefits; among them is, that an entry in the system only depends directly on the entry before and after (due to the orthonormality of the spherical harmonics functions). This approximation is known as the spherical harmonics method, denoted often by P_N. Additionally, it was shown [13] that the P_N approximation of the optimality system of $\mathcal{P}_T$ is the same as the optimality system of the control problem using P_N as the underlying approximation model. Further, the dose is simply the 0-th entry in this approximation. However, it has been known for quite a while [5] that densities ψ of particles obtained by the P_N approximation may be negative, which is clearly unphysical.

Alternatively, instead of truncating, we look for a flux that satisfies an entropy minimization principle. That is, we look for a distribution ψ_{ME} of particles that minimizes a strictly convex functional, the entropy H, for example

$$H(\psi)\langle \psi \log \psi d\omega \rangle$$

This minimization is constrained by requiring

$$\langle m_i I \rangle = \psi^{(i)}$$

for $i \leq N$. The resulting solution to this minimization problem is used to evaluate the flux function of the N-th entry in the approximate system. This is the minimum entropy, or M_N method. Clearly there is an immediate drawback to this method: the need to solve a constrained minimization problem whenever a flux needs to be computed. However, in the case of M_1, the flux can be computed explicity (see [1,9]); unfortunately, for $N > 1$, the minimizer can only be computed numerically. However, it was shown in [10] that M_1 dose calculations are accurate in tissue regions such as the vertebral column. One significant reason to use the minimum entropy approximation is that negative particle densities are avoided as well as wave propagation speeds being preserved.

5 Numerical Results

We present here a pair of examples; these test problems with different parameters have been studied in [2]. In each case, we use an optimize-then-discretize approach and then use the M_1 approximation for both forward and adjoint equations. The first is an academic example taken from [25]; a $9\,\text{cm}^2$ square of water has a C-shaped target, Z_T, and a box shaped risk region Z_R as shown in Fig. 1a. We prescribe $\overline{D} = \chi_{Z_T}$ and $c_T = 50$, $c_R = 100$, $c_N = 1$. The control is allowed to be active only on the outer edge of cells and with energy between 18 and 20 MeV; additionally we require $q^{(0)} \leq 1$. The 10 % isocurves of the resulting optimal dose $D(\psi)$ are shown in Fig. 1b. A useful tool used in evaluation of a treatment plan is the dose volume histogram (DVH); this shows the portion of the tissue receiving at least a given dose level. We plot the DVH curves for the target Z_T and organ at risk (OAR) corresponding to Z_R. The relative shapes of these curves can, naturally, be adjusted depending on the values selected for c_T and c_R.

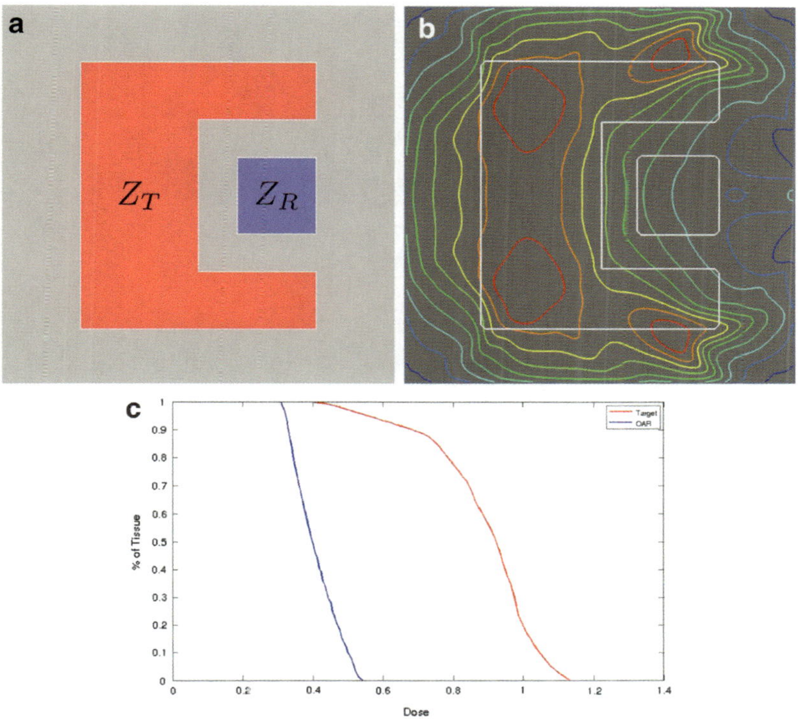

Fig. 1 C-shaped target on water phantom. (**a**) Problem layout on Water Phantom Problem. (**b**) Optimal Dose Isocurves. (**c**) Dose Volume Histogram for Target and OAR

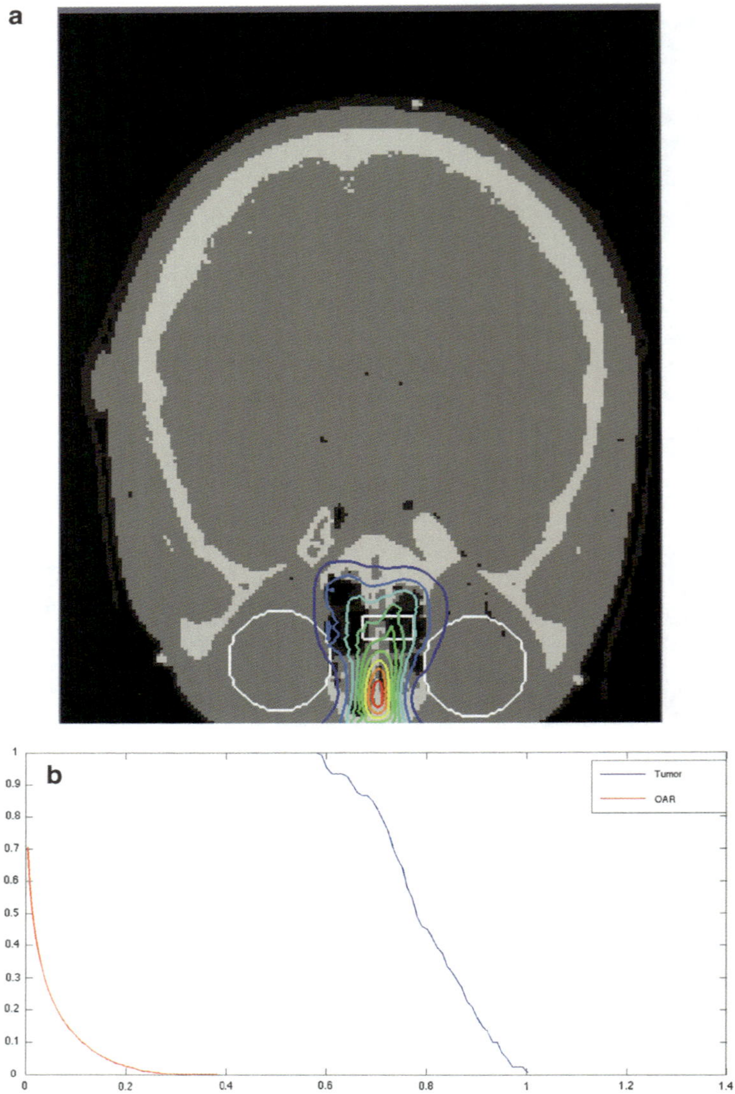

Fig. 2 Target in 2D slice of skull. (**a**) Optimal Dose Isocurves. (**b**) Dose Volume Histogram (DHV) for target and organ at risk (OAR)

We also consider a test problem involving a more realistic problem geometry. Using 2D CT data from the Visible Human Project,[1] a box target is placed in the nasal sinus cavity and as OAR, we approximate the location of the eye. Again, a beam of electrons between 18 and 22 MeV is constructed—with $q^{(0)} \leq 1$—to solve $\mathcal{P}_T$ with $c_T = 100$, $c_R = 225$, $c_N = 1$. The resulting optimal dose and associated DVH curves are displayed in Fig. 2. The OAR and target are outlined in white and the isocurves of the dose are shown in the first figure and the DVH in the second. Depending on tolerance for irradiation of the eyes, this source may be scaled in order to achieve a higher dose on the target (recalling that the dose is a linear functional of the source).

6 Outlook

Currently several challenges remain in using transport models in treatment planning. The inclusion of more highly resolved models is required to bring the accuracy of the dose calculations within levels acceptable for clinical use [10]. However, in principle, the model problems we have presented will allow for such modeling. Efficient methods using higher order moment approximations may also be needed for sufficiently improved accuracy. The current formulations of $\mathcal{P}_T$ and $\mathcal{P}_{SF}$ do not address several aspects of treatment planning that may be of practical interest. Foremost, we do not include constraints on the dose. Depending on the region of the body and organs under consideration, such constraints may require the dose to remain between a certain set of space-dependent levels. In other organs, it may be the case that we allow for high levels of dose on the OAR as long as we avoid doses high enough to lead to cell death on a given percentage of the tissue. This involves the proper formulation of constraints on the DVH curves resulting from the proposed treatment problem. Finally, we have not required that the source q be constructible. Such concerns must be addressed for transport models to be used in practical treatment planning situations.

References

1. R. Barnard, M. Frank, M. Herty, Optimal radiotherapy treatment planning using minimum entropy models. Appl. Math. Comput. **219**(5), 2668–2679 (2012)
2. R. Barnard, M. Frank, M. Herty, Treatment planning optimization for radiotherapy. Proc. Appl. Math. Mech. **13**(1), 339–340 (2013)
3. C. Börgers, Complexity of Monte Carlo and deterministic dose-calculation methods. Phys. Med. Biol. **43**, 517–528 (1998)

[1] http://www.nlm.nih.gov/research/visible/

4. C. Börgers, The radiation therapy planning problem, in *Computational Radiology and Imaging (Minneapolis, MN, 1997)*, ed. by C. Börgers, F. Natterer. Volume 110 of IMA Volumes in Mathematics and its Applications (Springer, New York, 1999), pp. 1–16
5. T. Brunner, Forms of approximate radiation transport. Sandia Report, 2002
6. M.K. Bucci, A. Bevan, M. Roach, Advances in radiation therapy: conventional to 3d, to imrt, to 4d, and beyond. CA: Cancer J. Clin. **55**(2), 117–134 (2005)
7. R.G. Dale, The application of the linear-quadratic dose-effect equation to fractionated and protracted radiotherapy. Br. J. Radiol. **58**(690), 515–528 (1985)
8. G. Delaney, S. Jacob, C. Featherstone, M. Barton, The role of radiotherapy in cancer treatment. Cancer **104**(6), 1129–1137 (2005)
9. B. Dubroca, J.-L. Feugeas, Étude théorique et numérique d'une hiérarchie de modèles aux moments pout le transfert radiatif. Analyse Numérique **329**(1), 915–920 (1999)
10. R. Duclous, B. Dubroca, M. Frank, A deterministic partial differential equation model for dose calculation in electron radiotherapy. Phys. Med. Biol. **55**(13), 3843–3857 (2010)
11. B.C. Ferreira, P. Mavroidis, M. Adamus-Górka, R. Svensson, B.K. Lind, The impact of different dose-response parameters on biologically optimized imrt in breast cancer. Phys. Med. Biol. **53**(10), 2733 (2008)
12. M. Frank, Approximate models for radiative transfer. Bull. Inst. Math. Acad. Sinica (New Series) **2**(2), 409–432 (2007)
13. M. Frank, M. Herty, M. Schäfer, Optimal treatment planning in radiotherapy based on Boltzmann transport calculations. Math. Models Methods Appl. Sci. **18**(4), 573–592 (2008)
14. M. Frank, M. Herty, A.N. Sandjo, Optimal radiotherapy treatment planning governed by kinetic equations. Math. Models Methods Appl. Sci. **20**(4), 661–678 (2010)
15. M. Frank, C. Berthon, C. Sarazin, R. Turpault, Numerical methods for balance laws with space dependent flux: application to radiotherapy dose calculation. Commun. Comput. Phys. **10**(5), 1184–1210 (2011)
16. M. Frank, M. Herty, M. Hinze, Instantaneous closed loop control of the radiative transfer equations with applications in radiotherapy. ZAMM Z. Angew. Math. Mech. **92**(1), 8–24 (2012)
17. H. Hensel, R. Iza-Teran, N. Siedow, Deterministic model for dose calculation in photon radiotherapy. Phys. Med. Biol. **51**(3), 675 (2006)
18. M. Herty, A.N. Sandjo, On optimal treatment planning in radiotherapy governed by transport equations. Math. Models Methods Appl. Sci. **21**(2):345–359 (2011)
19. M. Herty, C. Jörres, A.N. Sandjo, Optimization of a model Fokker-Planck equation. Kinet. Relat. Models 5(3), 485–503 (2012)
20. C.R. King, T.A. DiPetrillo, D.E. Wazer, Optimal radiotherapy for prostate cancer: predictions for conventional external beam, imrt, and brachytherapy from radiobiologic models. Int. J. Radiat. Oncol. Biol. Phys. **46**(1), 165–172 (2000)
21. T. Krieger, O.A Sauer, Monte carlo- versus pencil-beam-/collapsed-cone-dose calculation in a heterogeneous multi-layer phantom. Phys. Med. Biol. **50**(5), 859–868 (2005)
22. E.W. Larsen, M.M. Miften, B.A. Fraass, I.A.D. Bruinvis, Electron dose calculations using the method of moments. Med. Phys. **24**(1), 111–125 (1997)
23. G.C. Pomraning, The Fokker-Planck operator as an asymptotic limit. Math. Models Methods Appl. Sci. **2**(1):21–36 (1992)
24. U. Ringborg, D. Bergqvist, B. Brorsson, E. Cavallin-ståhl, J. Ceberg, N. Einhorn, J .erik Frödin, J. Järhult, G. Lamnevik, C. Lindholm, B. Littbrand, A. Norlund, U. Nylén, M. Rosén, H. Svensson, T.R. Möller, The swedish council on technology assessment in health care (sbu) systematic overview of radiotherapy for cancer including a prospective survey of radiotherapy practice in sweden 2001–summary and conclusions. Acta Oncologica **42**(5–6), 357–365 (2003)
25. D.M. Shepard, M.C. Ferris, G.H. Olivera, T.R. Mackie, Optimizing the delivery of radiation therapy to cancer patients. SIAM Rev. **41**, 721–744 (1999)
26. M. Sikora, J. Muzik, M. Söhn, M. Weinmann, M. Alber, Monte carlo vs. pencil beam based optimization of stereotactic lung imrt. Radiat. Oncol. **4**(64) (2009)

27. C.P. South, M. Partridge, P.M. Evans, A theoretical framework for prescribing radiotherapy dose distributions using patient-specific biological information. Med. Phys. **35**(10), 4599–4611 (2008)
28. G.G. Steel, J.D. Down, J.H. Peacock, T.C. Stephens, Dose-rate effects and the repair of radiation damage. Radiother. Oncol. **5**(4), 321–331 (1986)
29. G.G. Steel, J.M. Deacon, G.M. Duchesne, A. Horwich, L.R. Kelland, J.H. Peacock, The dose-rate effect in human tumour cells. Radiother. Oncol. **9**(4), 299–310 (1987)
30. J. Tervo, P. Kolmonen, Inverse radiotherapy treatment planning model applying boltzmann-transport equation. Math. Models. Methods. Appl. Sci. **12**, 109–141 (2002)
31. J. Tervo, P. Kolmonen, M. Vauhkonen, L.M. Heikkinen, J.P. Kaipio, A finite-element model of electron transport in radiation therapy and related inverse problem. Inv. Probl. **15**, 1345–1361 (1999)
32. J. Tervo, M. Vauhkonen, E. Boman, Optimal control model for radiation therapy inverse planning applying the Boltzmann transport equation. Lin. Alg. Appl. **428**, 1230–1249 (2008)
33. M. Treutwein, L. Bogner, Elektronenfelder in der klinischen anwendung. Strahlentherapie und Onkologie **183**, 454–458 (2007). doi:10.1007/s00066-007-1687-0
34. F. Tröltzsch, *Optimal Control of Partial Differential Equations: Theory, Methods, and Applications* Volume 112 of Graduate Studies in Mathematics (American Mathematical Society, Providence, 2010)
35. S. Webb, Optimum parameters in a model for tumour control probability including interpatient heterogeneity. Phys. Med. Biol. **39**(11), 1895–1914 (1994)
36. World Health Organization, *Radiotherapy Risk Profile* (WHO, Geneva, 2008)

Optimal Control of Self-Consistent Classical and Quantum Particle Systems

Martin Burger, René Pinnau, Marcisse Fouego, and Sebastian Rau

Abstract We study optimal control problems for self-consistent interacting classical and quantum particle systems from the analytical and numerical point of view. This involves microscopic as well as macroscopic quantum models, which have two main features in common: The control enters in a bilinear manner into the partial differential equations and in all models particle interaction takes place via a self-consistent electrostatic potential. This special model structure appears in many different types of applications, like quantum semiconductor devices, optimal quantum control or biomedical applications and it is used to construct fast optimization algorithms.

Keywords Quantum control • Schrödinger-Poisson system • Macroscopic quantum models • Drift diffusion models • Bilinear control • Optimality conditions • Adjoints

Mathematics Subject Classification (2010). 35A15, 49J20, 49J27, 49K20, 49M05, 65M50, 76Y05, 81Q93.

The authors acknowledge support from the DFG via SPP 1253/2.

M. Burger • M. Fouego
Westfälische Wilhelms-Universität Münster, Institut für Numerische und Angewandte Mathematik, Einsteinstr. 62, 48149 Münster, Germany
e-mail: martin.burger@wwu.de; fouegomarcisse@yahoo.fr

R. Pinnau (✉) • S. Rau
Technische Universität Kaiserslautern, Fachbereich Mathematik, Postfach 3049, 67653 Kaiserslautern, Germany
e-mail: pinnau@mathematik.uni-kl.de; rau@mathematik.uni-kl.de

G. Leugering et al. (eds.), *Trends in PDE Constrained Optimization*, International Series of Numerical Mathematics 165, DOI 10.1007/978-3-319-05083-6_29

1 Introduction

The optimal control of conducting quantum fluids plays a crucial role in the optimal design of quantum semiconductor devices as well as in the optimal quantum control of interacting particle systems [3, 19]. Due to the high numerical complexity of the underlying partial differential equations, the main focus of research lay during the last decades on the development of fast solvers and the investigation of approximative models, which allowed to shorten the design cycle significantly [4, 8]. The interplay of the improved performance of optimization algorithms and the increasing computing power allows now for the construction and investigation of computerbased optimization platforms [9]. Nevertheless, several challenges remained open and new ones did arise during the investigations. In this article we summarize recent results from [5, 17] showing the progress in this field.

The report is organized as follows. In Sect. 2, we present design and control problems for the microscopic nonlinear Schrödinger-Poisson (NLSP) system. The stationary design problem involves a minimization problem for the ground state, which is given by the solution of an eigenvalue problem. Further, a bilinear optimal control problem for the transient model is studied. A similar design problem is investigated in Sect. 3 for the macroscopic Quantum Euler-Poisson (QEP) model. In Sect. 4, the semiclassical limit for an optimization problem for the macroscopic Quantum Drift-Diffusion (QDD) model is performed by means of Γ-convergence results. Finally, we present in Sect. 5 constrained optimal control problems for the transient Drift-Diffusion (DD) model.

2 Optimal Design and Control Based on the NLSP Model

The most fundamental model in quantum systems is the Schrödinger equation, which needs to be self-consistently coupled with the Poisson equation for the electrical potential as well as additional interactions leading to the Nonlinear Schrödinger-Poisson system in the case of larger systems in quantum chemistry. A natural control problem in this respect is to drive a system into an appropriate stationary state for the (electron) density, by applying external lasers and potentially by internal doping as in a semiconductor device. Here we discuss a two-step structure: First of all, we focus on the design of a ground state– the ideal stationary state since minimizing the quantum energy – via optimal doping, and then control the evolution by an external field varied in time.

2.1 Designing a Ground State

We first investigate optimal control problems for the cubically nonlinear stationary nonlinear Schrödinger-Poisson equation [2]

$$\left(-\frac{\hbar^2\Delta}{2m} + V(x) + 8\pi a|\psi(x)|^2\right)\psi(x) = \mu\psi(x), \tag{2.1}$$

with

$$-\lambda^2\Delta V = W = |\psi|^2 - C, \tag{2.2}$$

supplemented with homogeneous Dirichlet boundary conditions on ψ and V. Here, (2.1) needs to be interpreted as an eigenvalue problem for a positive energy level μ and in particular we are interested in the first (smallest) eigenvalue given mass m and interaction constant a. As usual, $\hbar$ denotes the scaled Planck constant and λ a scaled Debye-length. The natural design variable is the doping profile C, but as in the case of classical semiconductor devices in [4] one easily observes that a partial decoupling can be achieved by considering the total charge density W as the design variable instead and compute $C = |\psi|^2 - W$ a-posteriori.

The ground state can also be characterized as a minimizer of the energy functional

$$E(\psi) := \int_\Omega \left(|\nabla\psi|^2 + \Omega V|\psi|^2 + 4\pi a|\psi|^4\right)\, dx \tag{2.3}$$

over the (nonconvex) admissible set

$$D = \left\{\psi \in H_0^1(\Omega) | \quad ||\psi||^2_{L^2(\Omega)} = 1\right\}. \tag{2.4}$$

With arguments following [11, 14] it can be shown that for each ground state there exists a density n such that $\psi = \sqrt{n}e^{i\theta}$ with constant phase θ. Moreover, $E(\sqrt{n})$ is strictly convex for the nonnegative real density n and the constraint becomes $\int_\Omega n\, dx = 1$, which implies that the ground state respectively the smallest eigenvalue has single multiplicity.

The wavefunction itself is of ambiguous nature in quantum mechanics, hence obtaining a specific wavefunction is not a realistic goal, but one rather looks for properties of observables such as the density $|\psi|^2$. Optimal design can reasonably be formulated via a least-squares problem of the density to some desired state, i.e. the minimization of

$$Q(|\psi|, W) = \frac{1}{2}\int_\Omega (|\psi|^2 - |\psi^*|^2)^2\, dx + \frac{\gamma}{2}\int_\Omega |W - W^*|^2\, dx, \tag{2.5}$$

over $(|\psi|, W) \in H_0^1(\Omega) \times L^2(\Omega)$, subject to

$$-\epsilon^2\Delta\psi + (8\pi a|\psi|^2 + V)\psi = \mu\psi \tag{2.6}$$

$$-\lambda^2\Delta V = W, \tag{2.7}$$

with homogeneous Dirichlet boundary conditions. As announced above we use the total charge density as a control variable, W^* is an initial density, e.g. $|\psi^*|^2 - C^*$ for an initial design of the doping and corresponding ground state. We mention that optimal control problems for the nonlinear Schrödinger equation (or Schrödinger-Poisson system) in the stationary case automatically lead to an optimal control problem constrained by an eigenvalue problem for a nonlinear operator, which is hardly treated in literature so far. In the case of optimal design of the ground state the energy formulation can be used, such that effectively the design problem is an MPEC in function spaces (cf. [7]).

The fundamental existence result is again formulated in terms of the special nonnegative real groundstate $\psi = \sqrt{n}$ and given by (cf. [5, 12] for proofs):

Theorem 2.1. *There exists a minimum*

$$(\sqrt{\overline{n}}, \overline{V}, \overline{W}) \in H_0^1(\Omega) \times H^1(\Omega) \times L^2(\Omega) \tag{2.8}$$

of (2.5).

In order to derive the first-order optimality conditions of the minimization problem (2.5) subject to (2.6)–(2.7), we use the Lagrangian

$$\mathcal{L}(\psi, V, W, \mu, p, q, r) = Q(|\psi|, W) + \mathrm{Re}\left(\int (-\epsilon^2 \Delta\psi + (h(|\psi|^2) + V)\psi - \mu\psi)\overline{p}\, dx\right)$$
$$+ \int_\Omega (-\lambda^2 \Delta V - W) q\, dx + \left(\frac{1}{2}\int_\Omega |\psi|^2\, dx - \frac{1}{2}\right) r. \tag{2.9}$$

The necessary conditions for a minimum are obtained by taking the Fréchet derivatives of $\mathcal{L}(\psi, V, W, \mu, p, q, r)$ with respect to the argument functions ψ, V, W, μ (cf. [5] for rigorous justification) and setting them to zero. The variation with respect to ψ yields the adjoint equation

$$-\epsilon^2 \Delta p + (16\pi a|\psi|^2 + V - \mu)p = -2\psi(|\psi|^2 - |\psi^*|^2) - r\psi \tag{2.10}$$

with homogeneous Dirichlet boundary condition for p. The derivative with respect to V yields an adjoint Poisson equation of the form

$$-\lambda^2 \Delta q = -\mathrm{Re}(\overline{\psi} p) \tag{2.11}$$

again with homogeneous boundary conditions for q. The variation with respect to μ yields an orthogonality relation

$$\int_\Omega \psi \overline{p}\, dx = 0 \tag{2.12}$$

and finally the optimality condition with respect to W amounts to the simple relation

$$W = q/\gamma - W^*. \tag{2.13}$$

We mention that (2.10) and (2.12) should be interpreted as a mixed system for p and the constant r rather than a single equation for p. Indeed, r can be determined by taking a scalar product of (2.10) with $\overline{\psi}$, which resembles the Fredholm alternative. With the orthogonality and (2.6) we find

$$r(\psi) = 2\int_\Omega (|\psi|^2|\psi^*|^2 - |\psi|^4)\, dx,$$

such that the adjoint equation can be rephrased as an elliptic equation for p

$$-\epsilon^2\Delta p + (16\pi a|\psi|^2 + V - \mu)p = -2\psi(|\psi|^2 - |\psi^*|^2) - r(\psi)\psi. \tag{2.14}$$

With this observation one can build an existence proof [5, 12], also uniqueness can be shown with further effort (noticing that a simple Fredholm alternative does not work due to the nonlinearity in the original eigenvalue problem):

Theorem 2.2. *Let* $(\sqrt{n}, W) \in H_0^1(\Omega) \times L^2(\Omega)$ *be given; then there exists a unique solution* $(p, q) \in H_0^1(\Omega) \times L^2(\Omega)$ *of the adjoint problem (2.14), (2.11), (2.12).*

For the numerical minimization straight-forward spatial discretization strategies can be coupled with an efficient Gummel-type iteration through the optimality system (cf. [12]) exploiting the similar structure to [4]. A representative result related to wave focusing is illustrated in Fig. 1. Here one observes the effect of the interaction strength a in the model for wave focusing, noticing that negative a corresponds to attractive interaction.

2.2 *Optimal Control of the Transient NLSP Model*

Next, we consider optimal control problems for the transient NLSP model:

$$i\varepsilon\frac{\partial\psi}{\partial t} = -\frac{\varepsilon^2}{2}\Delta\psi + (V + \alpha(t)V_e(x) + h(\psi))\,\psi, \tag{2.15a}$$

$$-\lambda^2\Delta V = |\psi|^2 \tag{2.15b}$$

with homogeneous Dirichlet conditions for ψ and V, and initial condition

$$\psi(t=0, x) = \psi_0(x) \qquad \text{for } x \in \Omega. \tag{2.15c}$$

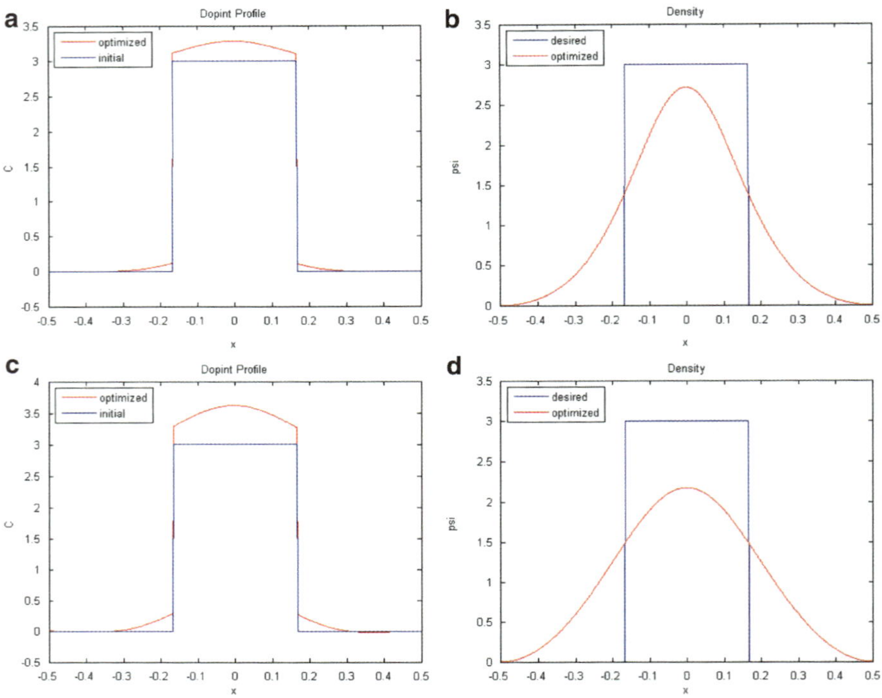

Fig. 1 Optimal design of ground states: Figures (**a**) and (**c**) show the optimized doping profile and figures (**b**) and (**d**) the electron density in the optimal state, with $\gamma = 10^{-3}$ and interaction constants $a = -0.6$ and $a = -0.1$, respectively

The positive constants ε, λ denote the scaled Planck constant and the scaled Debye length, respectively. The electrostatic potential V is induced by the electron density n defined as $n \overset{\text{def}}{=} |\psi|^2 = \psi\overline{\psi}$.

Possible potential barriers and external fields are modelled by $V_e(x)$, the function α controls the intensity of V_e and is our control parameter. The space of admissible controls is $\mathcal{U} \overset{\text{def}}{=} H_0^1(0,T)$.

For the analytical investigations we define

$$
\begin{aligned}
&Q \overset{\text{def}}{=} (0,T) \times \Omega, \quad T > 0, && \mathcal{Y} \overset{\text{def}}{=} H_0^1(\Omega),\\
&\mathcal{V} \overset{\text{def}}{=} L^2(0,T;\mathcal{Y}), && \mathcal{W} \overset{\text{def}}{=} \{v \in \mathcal{V} : v_t \in \mathcal{V}^*\},\\
&\tilde{\mathcal{Y}} \overset{\text{def}}{=} \mathcal{W} \times \mathcal{V}, && \tilde{\mathcal{Y}}_\infty \overset{\text{def}}{=} \tilde{\mathcal{Y}} \cap \left(L^\infty(0,T;H_0^1(\Omega))\right)^2,\\
&\mathcal{Z} \overset{\text{def}}{=} \mathcal{V} \times \mathcal{V} \times L^2(\Omega) && \|\cdot\|_{\tilde{\mathcal{Y}}_\infty} \overset{\text{def}}{=} \|\cdot\|_{\mathcal{V}} + \|\cdot\|_{L^\infty(0,T;H_0^1(\Omega))}.
\end{aligned}
$$

For the application of semigroup theory to show the existence of a unique solution of 2.15, we require the following assumptions to be fulfilled:

Assumption 2.3. (i) Let $\Omega \in \mathbb{R}^d, d \in \{1,2,3\}$ be a bounded domain. Furthermore, Ω is either convex or has a C^2-boundary.
(ii) Let $\psi_0 \in W_0^{1,4}(\Omega)$ and $V_e \in W_0^{1,4}(\Omega) \cap L^\infty(\Omega)$.
(iii) Let $h \in C^0(0,\infty)$ and let the mapping $\psi \mapsto h(\psi)\psi$ be locally Lipschitz continuous in $W_0^{1,4}(\Omega)$.

We note that the physically relevant function $h(u) = |u|^2$ fulfills Assumption 2.3. (iii). Using semigroup theory we obtain the following existence and uniqueness results (see also [2, 17]):

Theorem 2.4. *(i) There exists a $T > 0$ such that (2.15) has a unique solution $\psi \in C(0,T; H_0^1(\Omega))$ with*

$$\frac{d}{dt}\|\psi\|^2_{L^2(\Omega)} = 0, \tag{2.16}$$

i.e., the mass $\|\psi\|^2_{L^2(\Omega)}$ is conserved.
(ii) Let $h(u) = |u|^2$. Then, (2.15) has a unique solution $\psi \in C(\mathbb{R}; H_0^1(\Omega))$, i.e., the solution ψ exists globally in time. In addition to (2.16) the apriori estimate

$$\|\psi\|_{\mathcal{W}} \leq C \tag{2.17}$$

holds for a $C > 0$ depending on the data and $\|\alpha\|_{L^\infty(\Omega)}$.

Those results are by far sufficient to prove the existence of a minimizer for cost functionals which fulfill standard assumptions (for details see [17]).

Theorem 2.5. *There exists at least one solution $(\psi^*, V^*, \alpha^*) \in \tilde{\mathcal{Y}}_\infty \times \mathcal{U}$ of the minimization problem*

$$\mathcal{J}(\psi^*, V^*, \alpha^*) = \inf_{(\psi,V,\alpha)\in\tilde{\mathcal{Y}}_\infty\times\mathcal{U}} \mathcal{J}(\psi, V, \alpha), \tag{2.18a}$$

$$\text{s.t.} \qquad 0 = e(\psi^*, V^*, \alpha^*) \qquad \text{in } \mathcal{Z}^*, \tag{2.18b}$$

where the nonlinear operator $e \stackrel{def}{=} (e_1, e_2, e_3) : \tilde{\mathcal{Y}}_\infty \times \mathcal{U} \to \mathcal{Z}^*$ denotes the weak formulation of system (2.15).

The continuous Fréchet-differentiability of the operator e and the unique existence of adjoint variables ensure that the first-order optimality conditions are well defined (for details see [17]). This information can be used to implement a descent algorithm for the solution of the optimal control problem.

For the numerical computations we discretize the state and adjoint equations with a modified Crank-Nicholson scheme and use the BFGS-method [9].

We chose the two dimensional domain $\Omega = [0, 1]^2$ and an end time of $T = 10^{-1}$. Further, we consider the cost functional

$$\mathcal{J}^1(y, \alpha) = \frac{1}{2}\left\| |\psi(T)|^2 - |\psi_d|^2 \right\|^2_{L^2(\Omega)} + \frac{\gamma}{2}\|\alpha\|^2_{H^1(0,T)}$$

with $\gamma = 10^{-10}$. As initial distribution we choose a bell shaped distribution (see Fig. 2a), which we aim to separate into four distinct wave packages, the external potential V_e is chosen of focusing nature. The numerical results underlining the feasibility of our approach are presented in Fig. 2.

3 Optimal Control of the QEP Model

We consider the state system given by the stationary Quantum Euler-Poisson (QEP) equations stated on a bounded domain $\Omega \subset \mathbb{R}^d$ (see [10]):

$$\varepsilon^2 \Delta w = w\left(\frac{\tau_0^2}{2}|\nabla S|^2 + T_0 h(w^2) - V - V_{ext} + S\right), \tag{3.1a}$$

$$\operatorname{div}\left(w^2 \nabla S\right) = 0, \tag{3.1b}$$

$$\lambda^2 \Delta V = w^2 - C \qquad \text{in } \Omega. \tag{3.1c}$$

The unknowns are the square root of the electron density $w = \sqrt{n}$, the Fermi level S and the electrostatic potential V induced by the electron density w^2 and the background charge, the nonnegative doping profile C. Further, the potential V_{ext} allows to incorporate possible potential barriers in the model. The constants $\varepsilon, \lambda, \tau_0$ and T_0 are positive and denote the scaled Planck constant, the scaled Debye length, the scaled relaxation time and the constant temperature. The enthalpy function $h(\cdot)$ accounts for the electron-electron interaction.

Physically reasonable boundary conditions were derived in [10] and are given by

$$w = w_0, \quad S = S_0, \quad V = V_0 \qquad \text{on } \partial\Omega \tag{3.1d}$$

$$w_0 = \sqrt{C}, \quad S_0 = U, \quad V_0 = T_0 h(C) + U, \tag{3.1e}$$

where U is modelling the applied voltage.

The existence of a unique solution $y \stackrel{\text{def}}{=} (w, S, V)$ of (3.1) was shown in [10] under reasonable regularity assumptions. It lives in the state space $\mathcal{Y} \stackrel{\text{def}}{=} (w_0, S_0, V_0) + \mathcal{Y}_0$, where

$$\mathcal{Y}_0 \stackrel{\text{def}}{=} \left[\left(W^{2,p}(\Omega) \times C^{1,\gamma}(\overline{\Omega})\right) \cap \left(H_0^1(\Omega)\right)^2\right] \times H_0^1(\Omega).$$

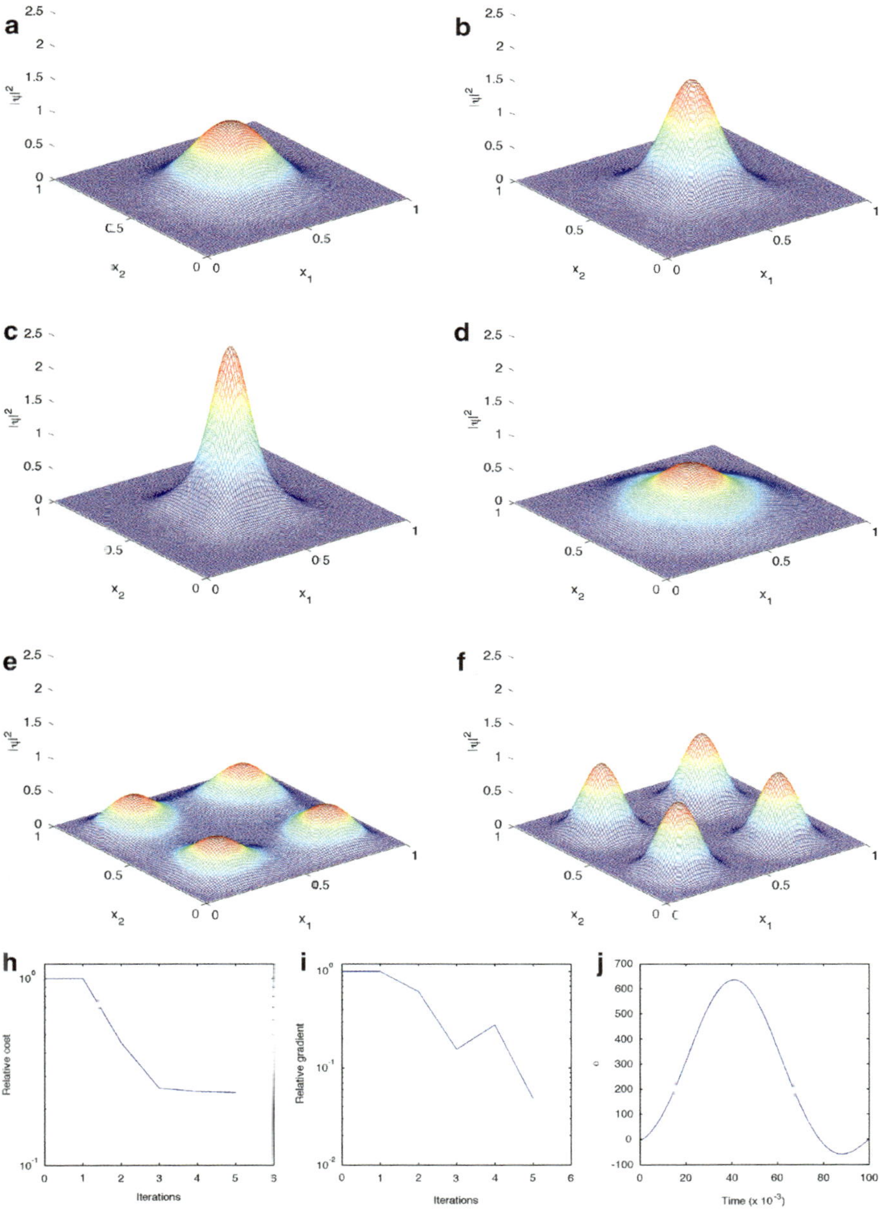

Fig. 2 Wave splitting: Figures (**a**)-(**e**) show the electron density $|\psi(t)|^2$ with optimal control α_{opt} from time $t = 0$ to the end time $t = 10^{-1}$. The desired distribution $|\psi_d|^2$ is shown in (**f**). The relative cost, relative gradient and the optimal control are shown in (**g**)-(**i**)

The set of admissible doping profiles for a unipolar device is given by

$$U_{ad} \stackrel{\text{def}}{=} \{C \in H^1(\Omega) : 0 \leq C \leq \overline{C},\ C = \hat{C} \text{ on } \partial\Omega\} \subset H^1(\Omega) \stackrel{\text{def}}{=} \mathcal{U}.$$

Here, $\hat{C} \in H^1(\Omega)$ with $0 \leq \hat{C} \leq \overline{C}$ is a given reference doping profile.

The regularity results for the state system allow to prove the existence of a minimizer [17].

Theorem 3.1. *Under standard assumptions on the cost functional there exists at least one minimizer of the minimization problem*

$$\mathcal{J}(y^*, C^*) = \inf_{(y,C)\in\mathcal{Y}\times U_{ad}} \mathcal{J}(y, C), \tag{3.2}$$

$$\text{s.t.} \qquad 0 = e(y^*, C^*) \tag{3.3}$$

where $e(y, C) : \mathcal{Y} \times U_{ad} \to \left(\left(H^1(\Omega)\right)^3\right)^*$ *denotes the weak formulation of (3.1).*

Using the Fréchet-differentiability of the operator e and the Fredholm alternative we can even show the existence of adjoint states and get the well-posedness of the first-order optimality conditions (for details see [17]). Numerical examples for the design of a MESFET device can be found in [16–18].

4 Optimal Control of the Stationary QDD Model in the Semiclassical Limit

We investigate an optimal control problem constrained by the Quantum Drift-Diffusion (QDD) model in the semiclassical limit. Stated on a bounded domain Ω the QDD model reads

$$\varepsilon^2 \frac{\Delta w}{w} = h(w) + V - S, \tag{4.1a}$$

$$\operatorname{div}(w^2 \nabla S) = 0, \tag{4.1b}$$

$$-\lambda^2 \Delta V = w^2 - C, \tag{4.1c}$$

supplemented with the boundary data

$$w = w_D, \qquad V = V_D, \qquad S = S_D \qquad \text{on } \Gamma_D, \tag{4.1d}$$

$$\frac{\partial w}{\partial \nu} = \frac{\partial V}{\partial \nu} = \frac{\partial S}{\partial \nu} = 0 \qquad \text{on } \Gamma_N, \tag{4.1e}$$

where ν denotes the outer normal along $\partial\Omega$. The boundary $\partial\Omega$ splits into two disjoint parts Γ_N and Γ_D, modelling the insulating parts and the contacts, respectively.

The variable w denotes (as in the previous sections) the square root of the electron density n, i.e., $w \stackrel{\text{def}}{=} \sqrt{n}$, S the Fermi level and V is the electrostatic potential induced by the electron density and the doping profile C, which is our control. The enthalpy function $h(w)$ accounts for electron-electron interactions and needs to fulfill some growth condition (see [1]).

For the analytical investigations we define the spaces and sets

$$\mathcal{X} \stackrel{\text{def}}{=} H^1(\Omega), \quad \mathcal{Y}_0 \stackrel{\text{def}}{=} H_0^1(\Omega \cup \Gamma_N) \cap L^\infty(\Omega), \quad \mathcal{Y} \stackrel{\text{def}}{=} w_D + \mathcal{Y}_0,$$

and the admissible set

$$U_{ad} \stackrel{\text{def}}{=} \{C \in H^1(\Omega) : -a \leq C \leq a, a > 0,\ C = \hat{C} \text{ on } \partial\Omega\}.$$

Existence of solutions of system (4.1) was discussed in [1,15], the results therein are sufficient to show the existence of a minimizer for special cost functionals (however, the special structure of the cost functionals is only needed later when we consider the semiclassical limit):

Assumption 4.1. Let $\mathcal{J} : \mathcal{X} \times U_{ad} \to \mathbb{R}$ denote a cost functional which is continuously Fréchet differentiable with Lipschitz continuous derivatives, radially unbounded with respect to C and bounded from below. Furthermore, let $\mathcal{J}$ be of separated type, i.e. we can write $\mathcal{J}(w, C) = \mathcal{J}_a(w) + \mathcal{J}_b(C)$ with $\mathcal{J}_a : \mathcal{X} \mapsto \mathbb{R}$ being weakly continuous and $\mathcal{J}_b$ being weakly lower semi-continuous in $\mathcal{X}$ and independent of ε.

Theorem 4.2. *For every $\varepsilon > 0$ there exists at least one solution $(w^*, C^*) \in \mathcal{Y} \times U_{ad}$ of the minimization problem*

$$\mathcal{J}(w^*, C^*) = \inf_{(w,C) \in \mathcal{Y} \times U_{ad}} \mathcal{J}(w, C), \tag{4.2a}$$

$$\text{s.t.} \qquad e^\varepsilon(w^*, C^*) = 0 \qquad \text{in } \mathcal{X}^*, \tag{4.2b}$$

where $e^\varepsilon(x, C) : \mathcal{Y} \times U_{ad} \to \mathcal{X}^*$ denotes the weak formulation of (4.1) for fixed ε.

Let $\mathcal{J}^\varepsilon$ denote the cost functional from (4.2) for a given $\varepsilon > 0$. For $\varepsilon = 0$, i.e., for the classical Drift-Diffusion Model, we only consider those (w, C) as possible solutions of (4.2), for which we can find a sequence $(\varepsilon_h)_{h \in \mathbb{N}}$ with $\varepsilon_h \to 0$ for $h \to \infty$ and a sequence $(w_h, C_h)_{h \in \mathbb{N}}$ with $e^{\varepsilon_h}(w_h, C_h) = 0$, which weakly converges (in the H^1-sense) to (w, C). We denote this set as $\mathcal{Z}^0$. Until now, it is not clear if all solutions of the classical Drift-Diffusion model fulfill this property. At least, this set is nonempty [1].

An interesting question arises if we consider the optimal control problem in the semiclassical limit: Can we prove (any kind of) convergence for minimizers and minima? We answer this question by using the concept of $\Gamma-$convergence and equi-coercivity and can prove the following convergence result.

Theorem 4.3 (Convergence of minima). *Let $\mathcal{J}^\varepsilon$ and $\mathcal{J}^0$ be defined as above. Then $\mathcal{J}^0(w,C)$ attains its minimum on $\mathcal{Z}^0$ and*

$$\min_{(w,C)\in\mathcal{Z}^0} \mathcal{J}^0(w,C) = \lim_{\varepsilon\to 0} \inf_{(w,C)\in\mathcal{Y}\times U_{ad}} \mathcal{J}^\varepsilon(w,C).$$

The convergence of minima allows us to also show the convergence of minimizers:

Theorem 4.4 (Convergence of minimizers). *Let $\mathcal{J}^\varepsilon$ and $\mathcal{J}^0$ be defined as above, and let $(\varepsilon_h)_{h\in\mathbb{N}}$ be a sequence with $\varepsilon_h \to 0$ as $h \to \infty$ and $(\tilde{w}_h, \tilde{C}_h)$ be a minimizer of J^{ε_h}, i.e.,*

$$\mathcal{J}^{\varepsilon_h}(\tilde{w}_h, \tilde{C}_h) = \min_{(w,C)\in\mathcal{Y}\times U_{ad}} \mathcal{J}^h(w,C).$$

Then, there exists a subsequence, again denoted by $(\varepsilon_h)_{h\in\mathbb{N}}$, such that

$$(\tilde{w}_h, \tilde{C}_h) \rightharpoonup (\tilde{w}_0, \tilde{C}_0) \qquad \text{in } \mathcal{Y}\times U_{ad}$$

and

$$\mathcal{J}^0(\tilde{w}_0, \tilde{C}_0) = \min_{(w,C)\in\mathcal{Y}\times U_{ad}} \mathcal{J}^0(w,C),$$

i.e., it is a minimizer of $\mathcal{J}^0$.

For details concerning the proofs we refer to [17].

5 Optimal Control of the Transient Drift-Diffusion Model

Finally, we provide some results on the optimal control of the transient Poisson-drift diffusion model

$$\partial_t n = \Delta n - (n\nabla V + n\nabla u) \quad \text{in } \Omega\times(0,T), \tag{5.1a}$$

$$\partial_t p = \Delta p + (p\nabla V + p\nabla u) \quad \text{in } \Omega\times(0,T), \tag{5.1b}$$

where V solves

$$\lambda^2 \Delta V = n - p - C \qquad \text{in } \Omega\times(0,T). \tag{5.1c}$$

Here $u \in C([0,T]; W^2_{2,0}) \cap L^2([0,T]; W^2_{d,0}) \cap H^1([0,T]; X)$ is the external potential and also the control variable, the doping is assumed to be fixed. State variables are the electron density n, the hole density p, and the electric potential V. The

Poisson-DD model is the most prominent and best understood transient model for semiconductor devices and a variety of interesting control problems have been analyzed and solved numerically in the last years, also in a setup of realistic devices (cf. [5, 13]).

A typical goal is the control of the current I, which can be realized by optimizing u such that

$$Q_\epsilon(n, p, W) = \int_0^T |I(t; n, p, W) - I^*(t)|^2 \, dt + R(W) \to \min_{n,p,W}, \tag{5.2}$$

where I is the computed current over a contact

$$I = \int_\Gamma (\nabla(n - p) - (n + p)\nabla(V + u)) \cdot d\nu,$$

I^* is the prescribed current realizing the optimal switching behavior, $W = V + u$ is the control variable, and

$$R(W) = \frac{\epsilon}{2} \int_0^T \int_\Omega |\Delta(W - \overline{W})|^2 \, dxdt$$

is an appropriate regularization functional on the control W. Here $\overline{W}$ is a given total charge.

There are two cases for optimization (5.2)

(a) The optimization problem without further control constraints, i.e.,

$$W = V + u \in L^\infty([0, T]; W^{1,\infty}(\Omega)) \cap L^2([0, T]; H^2(\Omega))$$

can be considered as a spatio-temporal control. In this case the Poisson equation can be disregarded for the optimization, V and u can be computed at the end from the optimal solutions for W, n, and p. Moreover, certain natural state constraints, e.g. on the total charge W or the electrical field $E = \nabla V$ (at least in one-dimension, see below) can be reformulated into control constraints. For a detailed analysis we refer to [5, 13].

(b) Additional control constraints $u \in C_{ad}$ might be considered, where C_{ad} is the restricted set of admissible control. A typical example is boundary control with voltages U_i applied on M different contacts. Those can be brought into our formulation via

$$u(x, t) = \sum_{i=1}^{M} U_i(t) V_i(x),$$

where U_i are applied voltages between contacts and the V_i are harmonic functions with special boundary values. Another example are particular penalties

R for C instead of W, e.g. a total variation penalty to enforce piecewise constant doping profiles and allow a change of edges compared to the reference profile has been considered in [6]. In the case of control constraints one cannot get rid of the Poisson equation by eliminating u in favour of W, since the reformulated constraint $W - V \in C_{ad}$ then contains the electrical potential.

As an example for constraint problems we consider the minimization problem with an L^∞-bound for the control:

$$\min_{(n, D_{ad})} Q_\epsilon(n, p, V + u) \quad \text{subject to (5.1)}, \tag{5.3}$$

where the admissible domain is given by

$$\begin{aligned} \mathcal{D}_{ad} := \{ &(n, p, V, u) \in (L^2([0,T]; H^1(\Omega)) \cap H^1([0,T]; H^{-1}(\Omega)))^2 \\ &\times L^\infty([0,T]; W^{2,\infty}(\Omega)), \times C_{ad}, \lambda^2 \Delta V = n - p - C,\ W = V + u\}. \end{aligned}$$

In this case one can easily prove the following result as in [5]:

Theorem 5.1. *Let C_{ad} be a bounded set in $L^\infty([0,T] \times \Omega)$. Then optimization problem (5.3) admits at least one solution $(\overline{n}, \overline{p}, \overline{V}, \overline{u}) \in \mathcal{D}_{ad}$.*

For a detailed statement of optimality conditions and rigorous analysis of adjoint problems, which is possible without voltage restrictions in the transient case, we refer to [13].

Finally we discuss an example of treating a natural state constraint as a control constraint by appropriate reformulation. In the setup of optimal dopant profiling consider the minimization of

$$Q_\epsilon(n, p, V, E, C) = |I(n, p, V) - I^*|^2 + \frac{\epsilon}{2} \int_\Omega |E - E^*|^2 \, dx$$

subject to the stationary Poisson-drift diffusion equations and a control constraint

$$|E(x)| \leq A \quad \text{a.e. in } \Omega \tag{5.4}$$

for electrical field $E = \nabla V$ Again in this case we eliminate the doping profile and the Poisson equation from the minimization and instead consider $\nabla V = E$ as a state equation, the doping profile is to be computed a-posteriori as $C = -\lambda^2 \nabla \cdot E - n + p$.

As an optimality condition for the optimal control problem in the one-dimensional case one can derive (cf. [5])

$$E = \text{Proj}_{[-A,A]} \left(E^* - \frac{1}{\epsilon} \Big(\int_\Omega e^V \partial_x u \partial_x \nu \, dx + \int_\Omega e^{-V} \partial_x \nu \partial_x \mu \, dx \Big) - \lambda \right),$$

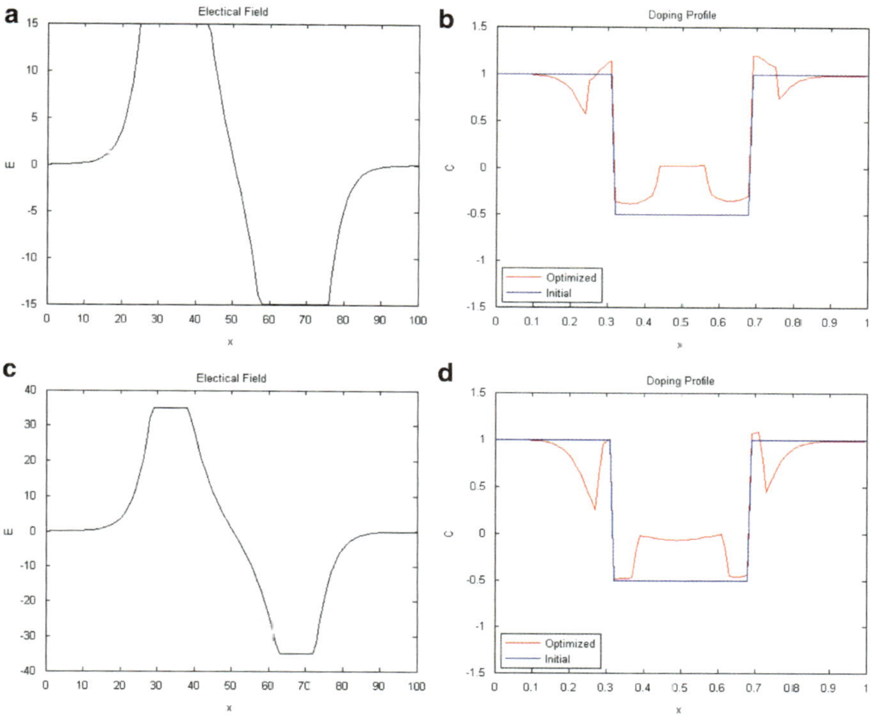

Fig. 3 Optimal control with constraint on electrical field: Figures (**a**) and (**c**) show the electrical field in the optimal state and Figures (**b**) and (**d**) the optimized doping profile, with $\epsilon = 10^{-3}$ and constraint values $A = 15$ and $A = 35$, respectively

where λ is a constant defined by the relation

$$\int_{\Omega} E \, dx = V(1) - V(0) = U$$

and the Slotboom variables u,v and adjoints ν, μ solve

$$\partial_x(e^V \partial_x u) = \partial_x(e^{-V} \partial_x v) = \partial_x(e^V \partial_x \nu) = \partial_x(e^{-V} \partial_x \mu) = 0.$$

The optimality system can be solved using a projected Gummel iteration. Results indicating the effect of changing the constraint are given in Fig. 3, one observes that the doping profile has additional jumps on the boundaries of regions where the constraint on E is active.

References

1. N.B. Abdallah, A. Unterreiter, On the stationary quantum drift diffusion model. Z. Angew. Math. Phys. **49**, 251–275 (1998)
2. A. Arnold, Mathematical properties of quantum evolution equations, in *Quantum Transport*, ed. by N. Ben Abdallah, G. Frosali. Lecture Notes in Mathematics (Springer, Berlin/Heidelberg, 2008)
3. E. Brown, H. Rabitz, Some mathematical and algorithmic challenges in the control of quantum dynamics phenomena. J. Math. Chem. **31**, 17–63 (2002)
4. M. Burger, R. Pinnau, Fast optimal design of semiconductor devices. SIAM J. Appl. Math **64**, 108–126 (2003)
5. M. Fouego, Optimal control of transient drift-diffusion and nonlinear schrödinger-poisson problems. PhD thesis, WWU Münster, 2013
6. M. Fouego, M. Burger, Optimal dopant doping profiling with tv penalty. PAMM **12**(1), 679–680 (2012)
7. M. Hintermüller, I. Kopacka, Mathematical programs with complementarity constraints in function space: C-and strong stationarity and a path-following algorithm. SIAM J. Optim. **20**(2), 868–902 (2009)
8. M. Hinze, R. Pinnau, An optimal control approach to semiconductor design. Math. Models Methods Appl. Sc. **12**(1), 89–107 (2002)
9. M. Hinze, R. Pinnau, M. Ulbrich, S. Ulbrich, *Optimization with PDE constraints* (Springer, New York, 2009)
10. A. Jüngel, A steady-state quantum euler-poisson system for potential flows. Commun. Math. Phys. **194** 463–479, (1998)
11. E.H. Lieb, R. Seiringer, J. Yngvason, Bosons in a trap: a rigorous derivation of the gross-pitaevskii energy functional, in *The Stability of Matter: From Atoms to Stars* (Springer, Berlin, 2005), pp. 759–771
12. M. Burger, M. Fouego, D. Mahrahrens, Optimal design of ground states in nonlinear schrödinger-poisson systems. Preprint, WWU Münster, 2013.
13. M. Burger, M. Fouego, R. Pinnau, Optimal control of transient drift-diffusion models. Preprint, WWU Münster, 2013
14. F. Pacard, A. Unterreiter, A variational analysis of the thermal equilibrium state of charged quantum fluids. Commun. Partial Differ. Equ. **20**(5-6), 885–900 (1995)
15. R. Pinnau, A. Unterreiter, The stationary current-voltage characteristics of the quantum drift diffusion model. SIAM J. Numer. Anal. **37**(1), 211–245 (1999)
16. R. Pinnau, S. Rau, F. Schneider, Optimal quantum semiconductor design based on the quantum Euler-Poisson model (2012, submitted)
17. S. Rau, Optimal control of interacting quantum particle systems, PhD thesis, TU Kaiserslautern, 2013
18. F. Schneider, Optimal design of quantum semiconductor devices. Master's thesis, University of Kaiserslautern, 2011
19. A. Unterreiter, S. Volkwein, Optimal control of the stationary quantum drift-diffusion model. Commun. Math. Sci. **5**, 85–111 (2007)

Modeling, Analysis and Optimization of Particle Growth, Nucleation and Ripening by the Way of Nonlinear Hyperbolic Integro-Partial Differential Equations

Michael Gröschel, Wolfgang Peukert, and Günter Leugering

Abstract We consider the processes of particle nucleation, growth, precipitation and ripening via modeling by nonlinear 1-D hyperbolic partial integro differential equations. The goal of this contribution is to provide a concise predictive forward modeling of the processes including appropriate goal functions and to establish a mathematical theory for the open-loop optimization in this context. Beyond deriving optimality conditions in the synthesis process, we present the application of a fully implicit method for Ostwald ripening of ZnO quantum dots which preserves its numeric stability even with respect to the inherent high sensitivities and wide disparity of scales. FIMOR represents an appropriate method that can be integrated to subordinate optimization studies which enables its future application in the context of continuous particle syntheses and microreaction technology (MRT).

Keywords Nonlinear integro-partial differential equation • Particle synthesis • Particle growth • Ostwald ripening • Numerical simulation • Optimization

Mathematics Subject Classification (2010). Primary 99Z99; Secondary 00A00.

M. Gröschel • G. Leugering (✉)
Institute of Applied Mathematics 2, Friedrich-Alexander University Erlangen-Nürnberg, Cauerstrasse 11, 91058 Erlangen, Germany
e-mail: guenter.leugering@fau.de

W. Peukert
Institute of Particle Technology, Friedrich-Alexander University Erlangen-Nürnberg, Cauerstrasse 4, 91058 Erlangen, Germany
e-mail: wolfgang.peukert@fau.de

G. Leugering et al. (eds.), *Trends in PDE Constrained Optimization*, International Series of Numerical Mathematics 165, DOI 10.1007/978-3-319-05083-6_30

1 Intoduction

Modeling, simulation and optimization of nucleation, growth, precipitation and ripening of nano-particles is field of growing interest. In particular, new challenges arise when we focus on the future-oriented technological field of printable electronics. Solar panels have a new competitor in applications where monocrystalline silicon wafers are too expensive or fragile. For these applications a new technology is emerging in the form of printing circuits on flexible substrates. Printable electronics will be used in creating flat panel displays for TVs, backplanes for TFT (Thin-Film Transistors), flexible circuits for OLED displays, and RFID (Radio-Frequency Identification) antennas. In this note we dwell on the recent efforts spent with in the DFG-SPP1253. Out of the vast number of potential models and application contexts we have chosen two examples leading, on the mathematical side, to 1-D hyperbolic nonlinear partial integro-differential equations. The optimization and control of such systems is a mathematical challenge and is far from being complete.

In the first part of this note we consider the synthesis of silicon nanoparticles in a reactor. We introduce the modeling and derive a nonlocal hyperbolic balance law for the particle size distribution which is subject to controls via initial data and at the boundary of the reactor, where the nucleation rate can be influenced.

After the modeling is being completed, we embark on an adjoint-based optimality analysis and derive necessary optimality conditions involving, as a novelty, a non-local adjoint equation.

The second part of this work presents the application of a Fully Implicit Method for Ostwald Ripening (FIMOR) for simulating the ripening of ZnO quantum dots (QDs). Due to its stable numerical behavior, FIMOR employs the full exponential term of the Gibbs-Thomson equation in the ripening rate which significantly outperforms the commonly used approximation of the LSW-theory in the lower nanometer regime. FIMOR preserves its numeric stability even with respect to the inherent high sensitivities and wide disparity of scales observed for the stiff PDE-ODE system at typical QD sizes below 10 nm. The implementation is consistent with experimental data for temperatures between 10 and 50 °C and yields a significant reduction by a factor of 1,000 in the computational effort compared to previous approaches. Hence, simulation time on a standard PC could be reduced from several hours to a few minutes. FIMOR represents an appropriate method that can be integrated to subordinate optimization studies which enables its future application in the context of continuous particle syntheses and microreaction technology (MRT).

2 Synthesis of Silicon Nano-particles

In order to describe the effect of growth and nucleation on the particle size distribution, we consider the partial differential equation that evolves the number distribution $q(t, x) \geq 0$ in time. One specific realization of the general population balance equation (PBE) is given in the case where the particle diameter $x \in \mathbb{R}^+$ represents the only internal variable. Assuming a constant feed rate in an ideally mixed system with regard to the cross-section of the reactor, the position of a particle transfers accordingly to its individual residence time t. This approach leads to a reduced model for the pyrolysis of monosilane

$$\frac{\partial}{\partial t}q(t,x) + \frac{\partial}{\partial x}(G(t,x)q(t,x)) = B(t,x). \tag{2.1}$$

The growth rate is denoted by G and the corresponding term to nucleation is represented by the source B on the right hand side. No initial seed particles $y(0,x) = 0,\ \forall x \in \mathbb{R}^+$ or influx boundary condition $y(t,0) = 0,\ \forall t \in [0,T]$ are assumed since all particles are formed at a critical cluster diameter through homogeneous nucleation. Due to space limitations, we dispense with a description of the full model and refer the reader to [6]. The aerosol model accounts for the formation of non-oxidized crystalline silicon nanoparticles (SiNP) via the pyrolysis of silane. Thereby, silane can undergo two kinds of possible reactions: A homogeneous gas phase reaction and a surface reaction on particles. The principally very complicated reaction network of silane decomposition and the subsequent formation of silicon hydride species is considerably simplified by taking an overall reaction equation into account. This equation is based on the studies of [13]. The equation accounts for the formation of a silylene radical due to the collision of a silane molecule with any third-body collision partner which is in the present case argon:

$$SiH_4 + Ar \rightarrow SiH_2 + H_2 + Ar \tag{2.2}$$

Therefore, the corresponding pressure dependent silane pyrolysis kinetic is given by

$$\frac{d[SiH_4]}{dt} = -k_{1a} \cdot [SiH_4] \cdot [Ar], \tag{2.3}$$

in the isothermal case. The growth rate due to chemical surface reaction G_{SiH_4} with respect to the diameter of a spherical particle

$$G_{SiH_4} = F \cdot \frac{k_p \cdot K_{SiH_4} \cdot [SiH_4]}{1 + K_{SiH_4} \cdot [SiH_4]} \cdot \frac{M_{Si}}{\rho_{Si}}.$$

The coefficients k_{1a}, k_p, K_{SiH_4} are given, F is a fitting parameter. The molar concentration of silane $[SiH_4]$ is derived from the available mass of silane m_{SiH_4} in the process. For this purpose, the temperature induced change of volume is taken into account:

$$[SiH_4] = \frac{p_{sys} \cdot m_{SiH_4}}{M_{SiH_4} \cdot c_{pVT} \cdot T}$$

assuming the thermodynamic properties of an ideal gas $c_{pVT}\, pV/\mathcal{T} = \text{const}$ which is feasible since the reactor is operated at high temperatures and low pressures. The surface growth rate therefore essentially depends on the remaining mass of silane in the system $m_{SiH_4}(t)$. By the conservation of masses, the initial mass of precursor gas $m^0_{SiH_4}$ is gradually incorporated in particles which are added to the mass of the initial particle size distribution m^0_{PSD}

$$m_{SiH_4}(t) = m^0_{SiH_4} + m^0_{PSD} - m_{PSD}(t).$$

The current total mass of all formed particles $m_{PSD}(t)$ may also be calculated by

$$m_{PSD}(t) = k_v \cdot \rho_{SiH_4} \cdot \int_{x_{min}}^{x_{max}} x^3 q(t,x) \partial x \tag{2.4}$$

using the volume shape coefficient k_v, which is equal to $\pi/6$ for spherical particles. By introducing the integral function $W_3(t,q)$ defined via

$$W_3(t,q) = \int_{x_{min}}^{x_{max}} x^3 q(t,x) \partial x,$$

the corresponding population balance equation for the growth of particles via surface reaction turns into

$$\begin{aligned}
\partial_t q(t,x) &= -G_{SiH_4}(\mathcal{T},\, W_3(t,q))\, \partial_x q(t,x) && (t,x) \in (0,T) \times (0,L) \\
q(t,0) &= u(t) && t \in (0,T) \\
q(0,x) &= q_0(x) && x \in (0,L).
\end{aligned}$$

Hereby, the assumed initial condition q_0 corresponds to present seed particles; whereas the boundary data $u(t)$ reflects for example either the nucleation rate inserting new particles or the flow rate of a feed stream. Conclusively, we made the nonlocal property of the considered system particularly explicit by including the mass balance into the population balance equation. The observed nonlocal correlation of the convective term to an integral expression evaluating the entire state variable is, however, not restricted to particle synthesis processes. Related studies in the context of highly re-entrant manufacturing systems [2] are also concerned with this kind of first-order hyperbolic initial-boundary value problems. Motivated by problems occurring in the production of semiconductors, Coron, Kawski, and Wang considered in [4] the analytic properties of this class of scalar nonlinear PDEs. Their modeling comprises in particular a nonlocal dependence of the convective term on the current solution. Since the special structure therefore covers various applications and is additionally capable of being transferred to network settings [7, 8], further investigations are indispensable for a comprehensive analysis of this important class of problems. We consider the following optimal control problem on the space time horizon $\Omega_T = (0,T) \times (0,1)$ for $T \in \mathbb{R}_{>0}$

$$\min_{w\in W, a\in A} \tfrac{1}{2}\|q(T,\cdot) - q_{\mathrm{T,d}}\|^2_{L^2(0,1)} + \tfrac{1}{2}\|\lambda(W(\cdot,q))^{\frac{1}{2}}\,(q(\cdot,1) - q_{1,\mathrm{d}})\,\|^2_{L^2(0,T)}$$

subject to

$$\begin{aligned} \dot q(t,x) &= -\lambda(W(t,q))q_x(t,x) && (t,x)\in\Omega_T \\ q(0,x) &= w(x) && x\in(0,1) \\ q(t,0) &= \tfrac{a(t)}{\lambda(W(t,q))} && t\in(0,T), \end{aligned}$$

where $W(t,q)$ is defined by the integral expression

$$W(t,q) = \int_0^1 q(t,s)\,\mathrm{d}s \qquad t\in[0,T].$$

Thereby, $\lambda \in C^1(\mathbb{R}_{\geq 0}, \mathbb{R}_{>0})$, $q_{\mathrm{T,d}} \in H^1(0,T)$ and $q_{1,\mathrm{d}} \in H^1(0,1)$ are given, such that

$$q_{\mathrm{T,d}}(T) = q_{1,\mathrm{d}}(1), \tag{2.5}$$

and the spaces W and A are to be specified. This optimal control problem applies to the setting of particle synthesis processes. Here, population balance equations represent a state-of-the-art model for the synthesis of polydisperse particulate products. Modeling the phenomena of growth coupled to a mass balance results in a partial differential equation describing the evolution of the particle size distribution in time. Particles are growing in a supersaturated solution due to homogeneous, transport dominated growth mechanisms. This results in a decrease of the residual educt concentration strongly depending on the number of currently present particles. Since the growth rate is in turn essentially determined by the current educt concentration, we have again the nonlocal correlation of the convective term to an integral expression evaluating the entire state variable. Closing this remark, we point out that in this case the assumed initial condition corresponds to present seed particles; whereas the boundary data reflects for example either the nucleation rate inserting new particles or the flow rate of a feed stream (see, e.g., [14]). For the initial boundary value problem

$$\dot q(t,x) = -\lambda(W(t,q))q_x(t,x) \qquad (t,x)\in\Omega_T \tag{2.6}$$

$$q(0,x) = w(x) \qquad x\in(0,1) \tag{2.7}$$

$$q(t,0) = \tfrac{a(t)}{\lambda(W(t,q))} \qquad t\in(0,T), \tag{2.8}$$

we obtain the following regularity result depending on the regularity of w and a:

Theorem 2.1 ($W^{1,\mathrm{p}}$ regularity for the initial boundary value problem). *For* $\mathrm{p} \in [1,\infty)$ *let* $w \in W^{1,\mathrm{p}}((0,1);\mathbb{R}_{\geq 0})$ *and* $a \in W^{1,\mathrm{p}}((0,T);\mathbb{R}_{\geq 0})$ *be given, such that* *the compatibility condition* $\frac{a(0)}{\lambda(W(0,q))} = w(0)$ *is satisfied. Then, we obtain for the regularity of* q

$$q \in C([0,T]; W^{1,\mathfrak{p}}((0,1);\mathbb{R})) \cap C([0,1]; W^{1,\mathfrak{p}}((0,T);\mathbb{R})).$$

Furthermore, q *is unique and nonnegative almost everywhere in* Ω_T*.*

Proof. The proof uses the method of characteristics and is based on the regularity study of [4]. □

Remark 2.1 ($L^{\mathfrak{p}}$ *regularity).* Note that in [4] it is shown that for $\mathfrak{p} \in [1, \infty)$,

$$w \in L^{\mathfrak{p}}((0,1);\mathbb{R}_{\geq 0}) \text{ and } a \in L^{\mathfrak{p}}((0,T);\mathbb{R}_{\geq 0})$$

we can expect a unique solution q with

$$q \in C([0,T]; L^{\mathfrak{p}}((0,1);\mathbb{R})) \cap C([0,1]; L^{\mathfrak{p}}((0,T);\mathbb{R})).$$

Let us furthermore remark that even for the weaker $L^{\mathfrak{p}}$ initial and boundary data the function $t \to W(t,q)$ for fixed q is absolutely continuous, such that also $\lambda(W(\cdot,q))$ is absolutely continuous. Since we are interested in first order optimality conditions or at least in gradient informations of the cost functional, we apply the formal Lagrange principle to deduce these conditions.

Theorem 2.2 (The first order optimality system). *The KKT-conditions are given by:*

"forward" problem

$$\dot{q}(t,x) = -\lambda(W(t,q))q_x(t,x), \qquad (t,x) \in \Omega_T \tag{2.9}$$

$$q(0,x) = w(x) \qquad x \in (0,1) \tag{2.10}$$

$$q(t,0) = \frac{a(t)}{\lambda(W(t,q))}, \qquad t \in (0,T) \tag{2.11}$$

"backward" problem

$$\dot{p}(t,x) = \lambda(W(t,q))p_x(t,x) - \lambda'(W(t,q))\int_0^1 q(t,s)p_x(t,s)\,\mathrm{d}s$$
$$- \lambda'(W(t,q))(q(t,1)^2 - q_{I,d}(t)^2), \qquad (t,x) \in \Omega_T \tag{2.12}$$

$$p(T,x) = q_{T,d}(x) - q(T,x), \qquad x \in (0,1) \tag{2.13}$$

$$p(t,1) = q_{I,d}(t) - q(t,1), \qquad t \in (0,T) \tag{2.14}$$

and the conditions

$$p(t,0) = 0, \qquad t \in (0,T) \tag{2.15}$$

$$p(0,x) = 0, \qquad x \in (0,1). \tag{2.16}$$

Proof. The proof consists of applying the formal Lagrange principle. □

In the context of the gradient information $p(\cdot,0)$ and $p(0,\cdot)$ represent the first derivative of the cost functional with respect to a and w. To guarantee the well posedness of the optimality system it is necessary to study the regularity of the adjoint equation as well.

Theorem 2.3 (Regularity of the adjoint equation with boundary and end data (2.12)–(2.14)**).** *For* $\mathrm{p} \in [1,\infty)$ *let* $p(T,\cdot) \in W^{1,\mathrm{p}}(0,1)$ *and* $p(\cdot,1) \in W^{1,\mathrm{p}}(0,T)$ *be given, such that the compatibility condition at* $(T,1)$ *is satisfied. Then there exists a unique solution* p *with regularity*

$$p \in C([0,T]; W^{1,\mathrm{p}}((0,1);\mathbb{R})) \cap C([0,1]; W^{1,\mathrm{p}}((0,T);\mathbb{R})).$$

Proof. The proof uses in particular that $\lambda'(W(t,q))\int_0^1 q(t,s)p_x(t,s)\,\mathrm{d}s$ is independent on the spatial variable x, so assuming sufficient regularity and differentiating the PDE (2.12) with respect to x leads to a linear transport equation. However, in the boundary data $\lambda'(W(t,q))\int_0^1 q(t,s)p_x(t,s)\,\mathrm{d}s$ will again emerge, such that we have to apply a fixed point theorem, to obtain the stated result. □

By the peculiar selection of the cost functional we obtain the compatibility conditions of the adjoint equation in the corner $(T,1)$, since by the assumptions in Theorem 2.1 we have $q(\cdot,1) \in W^{1,\mathrm{p}}(0,T)$ as well as $q(T,\cdot) \in W^{1,\mathrm{p}}(0,1)$ and using Eq. (2.5) we obtain

$$q_{\mathrm{T,d}}(\cdot) - q(T,\cdot) \in W^{1,\mathrm{p}}(0,1) \ \text{ and } \ q_{1,\mathrm{d}}(\cdot) - q(\cdot,1) \in W^{1,\mathrm{p}}(0,T)$$

which satisfies the compatibility condition

$$q_{\mathrm{T,d}}(1) - \rho(T,1) = q_{1,\mathrm{d}}(T) - \rho(T,1).$$

For the regularity of the adjoint PDE we can, therefore, apply Theorem 2.3 and have the well posedness of the optimality system stated in Theorem 2.2. With these results it is now possible to apply standard methods of optimization like steepest descent methods to compute a solution of the optimal control problem.

Remark 2.2 (The model in the context of supply chains). Let us remark here, that the presented model is also used to simulated supply chains and meets in this context a wide field of application (for instance see [1–4, 18]).

3 A Fully Implicit Method for Ostwald Ripening

Semiconducting zinc oxide (ZnO) nanoparticles are of great interest due to their interesting optical and electronic properties. Their use ranges from UV shielding, solar cell, information storage, and sensor applications to electronic and photonic devices [9, 22]. Nanoscaled ZnO particles also provide new, future-oriented possibilities as markers and probes in medicine, cell and molecular biology.

In this contribution we investigate the specific ripening behavior of semiconducting ZnO nanoparticles in order to simulate the effect of varying the temperature profile in the ripening process to the final particle size distribution (PSD). The numerical realization is based on a novel implicit finite volume scheme for simulating the Ostwald ripening process (FIMOR-scheme). A subsequent use of the implementation in an optimization framework will open up the possibility to identify optimal process control strategies for a precise tuning of the optical properties of the final product.

3.1 Modeling Approach

In a primary particle formation process, ZnO nanoparticles have been prepared by a controlled precipitation from zinc acetate and lithium hydroxide in alcoholic solution. In order to study the effect of the process conditions on the properties of the final product, we focus on the secondary particle formation process which is predominant in a post reaction step. Although the solution is no longer highly supersaturated after the reactants have depleted, the solubilities of the particles still vary. This effect is due to the fact that the surface tension of a particle in the lower nanometer regime depends strongly on the curvature of its surface. The solubility increases exponentially when the particle size decreases (Gibbs-Thomson effect) and small particles are dissolving [21]. After their dissolution, a local supersaturation is built up and the free molecules are incorporated at the surface of larger particles. This process is referred to as Ostwald ripening dedicated to the Nobel laureate W. Ostwald. Ripening is a thermodynamically-driven spontaneous process which occurs due to the tendency of the system to decrease the total surface area. The molar solubility c_L of a particle with diameter x is given by

$$c_L(x) = c_L^{\infty} \cdot \exp\left(\frac{4 \cdot \gamma_{\mathrm{SF}} \cdot V_{\mathrm{B}}}{\nu \cdot x \cdot k_B \cdot T}\right). \tag{3.1}$$

Thereby, c_L^{∞} is the solubility of a flat ZnO surface, γ_{SF} the surface tension, V_{B} the molecular volume, ν a stoichiometric coefficient, x the particle diameter, k_B the Boltzmann constant, and T the temperature. The interfacial energy γ_{SF} is modeled according to A. Mersmann. In [11], a theoretical approach assuming a perfect crystal is described to calculate the parameter γ_{SF} only based on material specific parameters

$$\gamma_{\mathrm{SF}} = K_\gamma \cdot \frac{k_B \cdot T}{\sqrt[3]{V_{\mathrm{B}}^2}} \cdot \ln\left(\frac{\rho_{\mathrm{ZnO}}}{c_L^\infty \cdot M}\right), \tag{3.2}$$

where N_{A} denotes the Avogadro constant, ρ_{ZnO} the density of the material, and M the molar mass of ZnO. The value of the solubility constant $K_\gamma = 0.414$ is determined in [11] from theoretical considerations and has further been validated by a comparison with experimental data [12, $K_\gamma = 0.333$]. The solubility of solids in liquids c_L^∞ is generally described by an Arrhenius law according to

$$c_L^\infty = k_L \cdot \exp\left(-\frac{\Delta H_{\mathrm{SOL}}}{k_B \cdot N_{\mathrm{A}} \cdot T}\right) \tag{3.3}$$

incorporating a positive enthalpy of dissolution ΔH_{SOL} for ZnO in ethanol and the solubility constant k_L. For the considered ZnO particles in the lower nanometer range, the relative solubility increases exponentially confirming the fact that larger particles are more energetically favored than smaller ones. The decision whether a particle dissolves or grows is determined by offsetting the present global supersaturation which promotes growth against its tendency to dissolve. The general rate equation for the diffusion controlled ripening process therefore reads [19]

$$R(t,x,c) = \frac{4 \cdot D \cdot M \cdot c_L^\infty}{\rho_{\mathrm{ZnO}} \cdot x} \cdot \left(\frac{c(t)}{c_L^\infty} - \exp\left(\frac{4 \cdot \gamma_{\mathrm{SF}} \cdot V_{\mathrm{B}}}{\nu \cdot x \cdot k_B \cdot T}\right)\right). \tag{3.4}$$

A change of sign is thus determined by the relation between the global supersaturation in the liquid and the local supersaturation at the particle surface given by the exponential expression. Moreover, D stands for the diffusion coefficient, and $c(t)$ for the current concentration of ZnO in the solution. Accordingly, depending on the temperature T and the particle diameter x, the ripening rate can be either positive (growth) or negative (dissolution). We refer to the PSD by $q(t,x)$ denoting the present density of particles with a diameter x at time t. Thus, the system describing the evolution of q in time is given by the partial differential equation (3.5) complemented by the mass balance (3.6) accounting for the required information on the concentration $c(t)$ of ZnO in the solution:

$$\frac{\partial}{\partial t} q(t,x) + \frac{\partial}{\partial x}(R(t,x,c) \cdot q(t,x)) = 0 \tag{3.5}$$

$$c(t) + k_\rho \int_{x_c}^{\infty} x^3 \cdot q(t,x)\,\mathrm{d}x = c^0 \quad \text{with} \quad k_\rho = \frac{\pi}{6}\frac{\rho_{\mathrm{ZnO}}}{V} \tag{3.6}$$

x_c thereby represents the critical diameter at which the nanoparticles are assumed to dissolve and k_ρ replaces the material specific coefficients including the volume shape factor. The total concentration c^0 of ZnO is therefore split into the amount of precipitated particles which are represented by the PSD and the concentration corresponding to ZnO molecules in the solution. the remaining boundary term

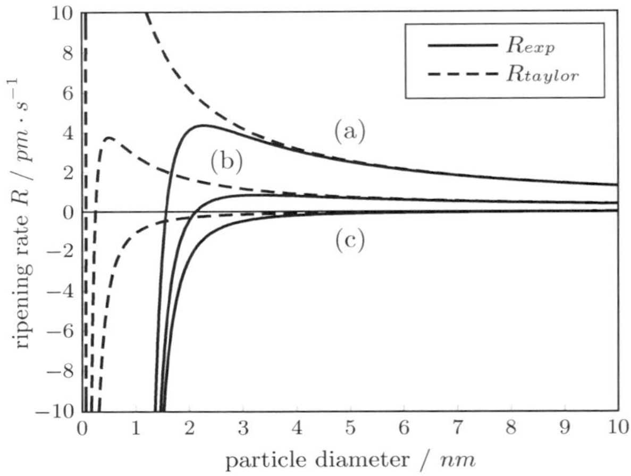

Fig. 1 Nonlinearity of ripening function at a ZnO concentration of (*a*) $10^{-8}\,\mathrm{kg\cdot m^{-3}}$, (*b*) $3\cdot 10^{-9}\,\mathrm{kg}\cdot{}^{-3}$, and (*c*) $10^{-12}\,\mathrm{kg\cdot m^{-3}}$

accounts for the increase of the concentration due to the dissolution of the smallest particles. A crucial issue related to the numerical realization of the ripening process is due to the high nonlinearities of the ripening function (see Fig. 1).

Lifshitz and Slyozov [10] and Wagner [21] have developed an asymptotic description for the evolution of a PSD during Ostwald ripening (LSW theory, see [19]). In their approach, the exponential term is replaced by a truncated Taylor series approximation (see Fig. 1). The approximated ripening functions behave numerically more friendly since the rate of dissolution, which applies for the smallest particles, is lowered significantly.

3.2 Prior Results

A detailed description on the synthesis procedure of ZnO nanoparticles from a zinc acetate precursor solution mixed with lithium hydroxide, both dissolved in ethanol, is found in [16, 17]. The produced particles include sizes between 2 and 8 nm in diameter. The particle size is obtained from Dynamic Light Scattering (DLS), Transmission Electron Microscopy analysis (TEM), UV-Vis spectroscopy (UV-Vis), and Hyper Rayleigh Scattering (HRS) measurements. The simultaneous measurement of UV-Vis and HRS measurements thereby allows to determine directly the rate of nucleation as well as the growth and ripening rates [17].

An analysis of the optical properties of the sample represents thereby a particularly favorable approach for determining the present PSD throughout the process. Due to the quantum size effect, the observed spectra which correspond to ZnO particles at a specific size differ significantly among each other in the smaller nanometer range. Since the UV-Vis spectra are furthermore easily accessible by online measurements this fact may therefore be used in order to reconstruct

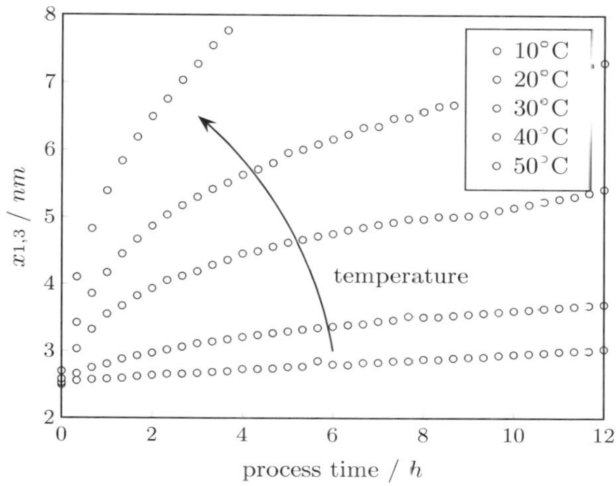

Fig. 2 Temporal evolution of the reconstructed volume weighted mean diameters $x_{1,3}$ during the ripening process at five different temperatures (Reproduced from [15, FIG. 10])

the corresponding PSD. In our group, an algorithm has been developed which decomposes the measured absorption spectra into the contributions from the different particle size fractions [16]. As a result, the evolution of the mean diameters of the PSD in the ripening process is reliably obtained. The approach is based on a precise tight-binding model which is proposed and validated in [20] describing the size dependence of the band gap. During the process, the extinction spectra between 250 and 400 nm are recorded for this purpose using a UV-Vis absorption spectrophotometer. The general effect of a variation in the temperature is depicted in Fig. 2 showing the evolution of the mean diameter in the course of five different ripening experiments. For validating the model established in Sect. 3.1 against the experimental data, an absorption spectrum is measured every 10 min. The experimentally determined starting PSD is chosen as initial condition in the simulation of the PBE (3.5). For the solution of this highly nonlinear partial differential equation, the commercially available program PARSIVAL by CiT GmbH [23] is applied. The finite element software is based on the Galerkin h-p method using Legendre polynomials. Using PARSIVAL to simulate the process, the exponential term had to be approximated by the first term of the Taylor series in order to obtain a tractable numerical problem. In Fig. 1, it becomes clear that the approximation mainly underestimates the ripening kinetics of small particles. Moreover, the root of the ripening function is shifted towards smaller particles and lower concentrations. A comparison of the simulations with the experimental data at 20 and 40 °C is presented in Fig. 3 showing the evolution of the volume weighted mean particle size $x_{1,3}$ over the process time. The variation of process temperature is qualitatively well reproduced by the simulation. As expected, the simulation is specifically inaccurate in the initial phase of the ripening procedure which is due to the use of the approximated ripening rate. However, this error has to be accepted since the finite element method used by PARSIVAL even requires extremely small time steps when solving the approximated formulation. Moreover, Fig. 4 reveals

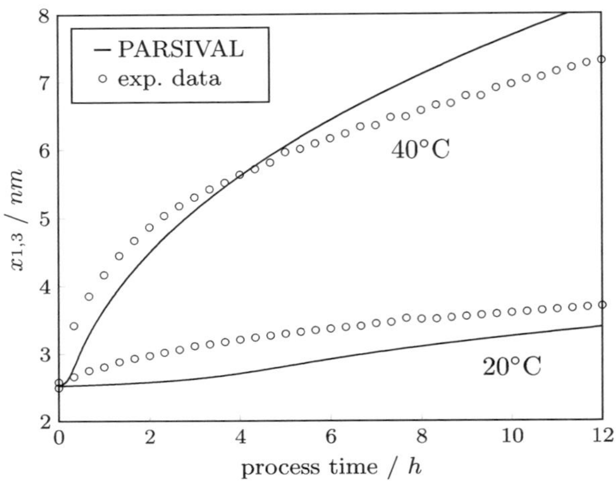

Fig. 3 Experimental data compared to PARSIVAL simulations at a ripening temperature of 20 and 40 °C, respectively (Reproduced from [15, FIG. 11])

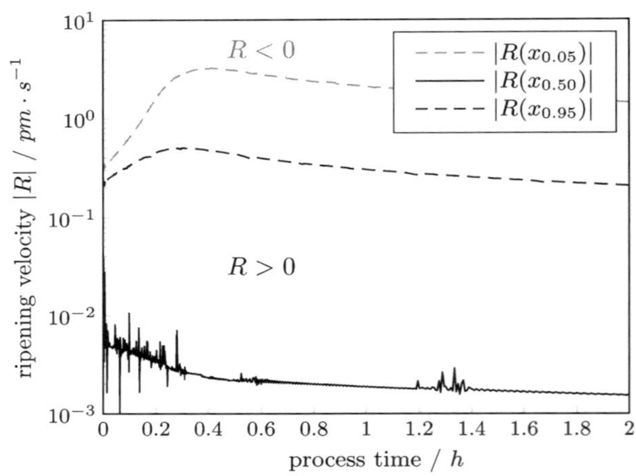

Fig. 4 Ripening velocity of particles at x_α, the α-quantile diameter, of the number density distribution $q(t, x)$ simulated in PARSIVAL (a *grey linecolor* implies a negative ripening rate; a *black linecolor* designates positive values) (Reproduced from [5, FIG. 5.48])

that the ripening rate of particles situated around the equilibrium particle size still exhibits a strong oscillatory behavior.

Practically, to now no numerical scheme is able to solve the stated model of the ripening process for ZnO particles in the lower nanometer regime within a reasonable time. In this contribution, a finite volume based fully implicit method for Ostwald ripening (FIMOR) is proposed which relies on an analytical formulation of the jacobian with respect to the discretized system. Therefore, we have to provide derivatives not only with respect to the model under consideration, but also with respect to the numerical scheme itself.

3.3 Results and Discussion

The derived FIMOR-method is based on an implicit treatment of the entire coupled PDE-ODE system. For this purpose, we extended the MUSCL-Hancock method (MHM) which is second order accurate in both space and time in its standard version by an implicit treatment of the concentration. Thereby, the integrating in time now relies on an implicit trapezoidal rule and is solved by a Newton-type iterative method. This novel approach involves the calculation of derivatives with respect to the PDE as well as concerning the numerical scheme (MHM) itself. As a result, the complete system is solved using the analytical derivation of the discrete Jacobean which drastically reduces the overall computational time by allowing for larger time steps. More precisely, the effort of a standard computation ranges from 30 s to a few minutes on a standard workstation. At the same time, the computation is more stable than the explicit FEM used in [15] and does not require any artificial diffusion term. In Fig. 5, the evolution of the ripening rates is shown, evaluated at $x_{0.05}$, $x_{0.50}$, and $x_{0.95}$ denoting the size where 5, 50, and 95 % of particles are smaller than the stated size (α-quantile diameter). Compared to the calculations based on the explicit finite element method in Fig. 4, our algorithm ensures in particular that no oscillations occur for the value $R(x_{0.50})$ which is situated near to the root of the ripening function (see Fig. 5). Furthermore, the introduced implicit framework opened up the possibility of using the original ripening rate based on the exponential term. Thus, the previously obtained parameter estimates [16] which have been presented in Sect. 3.2 are obviously not valid any more. They were based on the approximation of the ripening rate and rely therefore on the corresponding assumptions of the LSW-theory. Since the established simulation is computationally that efficient, the new implementation opens up the possibility of being usd in an optimization framework. Therefore, based on a least squares approach, a new set of parameters is derived using the evolution profiles of the mean diameter according to two different

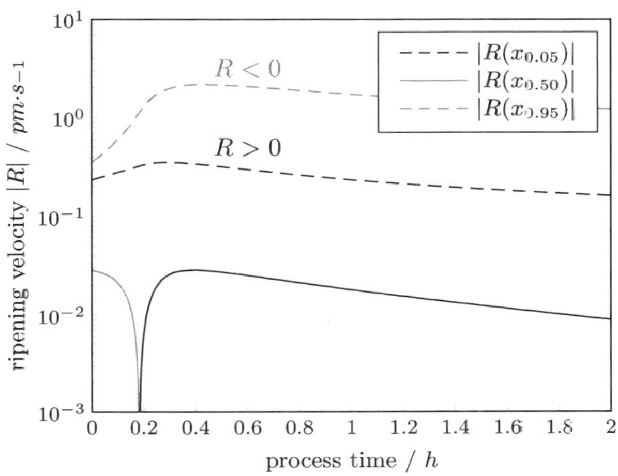

Fig. 5 Ripening velocity of particles at x_α, the α-quantile diameter, of the number density distribution $q(t, x)$ simulated by the newly introduced FIMOR-approach (a *grey linecolor* implies a negative ripening rate; a *black linecolor* designates positive values)

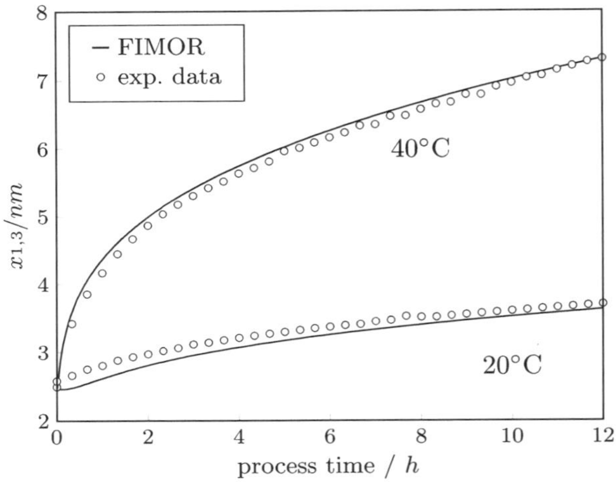

Fig. 6 Experimental data compared to the simulation at a ripening temperature of 20 and 40 °C using the proposed FIMOR-approach according to Fig. 3 (Experimental data from [15, FIG. 11])

datasets, each database thereby consisted of four different ripening experiments at varying temperatures. Figure 6 shows the evolution of the mean diameter for 20 and 40 °C in the simulations incorporating the exponential formulation of the ripening rate as well as the newly identified set of parameters.

The respective mean values depicted in Fig. 6 are calculated on the basis of the algorithm presented in [16]. A significant advantage of the introduced framework consists in the possibility of using the original formulation of the ripening rate in the simulations, i.e. without an approximation of the exponential term which describes the local solubilities. Specifically, this achievement allows for a more accurate description concerned with the dissolution behavior of the small particles. The simulation runs yield a satisfactory agreement with respect to the mean diameter of the PSDs. In particular, the simulated values provide now a better matching in the initial stage of the ripening process as well as for the final size of the particles.

Conclusion

In conclusion, we have demonstrated that mathematical modeling, simulation and optimization of nonlinear integro-partial differential equations plays an important role in the optimization of particle synthesis. In future, the combination of FIMOR and an optimization algorithm is believed to become an important tool for the prediction and optimization of particle size distributions in the context of continuous particles synthesis and microreaction technology (MRT). Thereby our methodology does not only allow for an empiric optimization of process parameters to achieve the desired electro-optical product properties but allows to account for the dispersity of the product and to rely on the physical model of Ostwald ripening in terms of the ageing mechanism.

Acknowledgements The authors acknowledge the work of the PhD-candidates A. Keimer, L. Pflug, J. Semmler and M. Walther, in particular, with respect to the numerical solutions and R. Wagner for the studies on Si formation (experiment and population balance modeling), Doris Segets and Martin Hartig for the studies with ZnO quantum dots. The authors would like to thank the German Research Council (DFG) for their financial support within the priority programs SPP 1679 (PE427/25) and SPP 1253 (LE595/23), for support within the DFG-Cluster of Excellence "Engineering of Advanced Materials" (www.eam.uni-erlangen.de) at the University of Erlangen-Nuremberg as well as for funding within the framework of the Elite Network of Bavaria: Identification, Optimization and Control with Applications in Modern Technologies.

References

1. D. Armbruster, P. Degond, C. Ringhofer, A model for the dynamics of large queuing networks and supply chains. SIAM J. Appl. Math. **66**, 896–920 (2006)
2. D. Armbruster, D.E. Marthaler, C.A. Ringhofer, K.G. Kempf, T.C. Jo, A continuum model for a re-entrant factory. Oper. Res. **54**, 933–950 (2006)
3. R.M. Colombo, M. Herty, M. Mercier, Control of the continuity equation with a non local flow. ESAIM Control Optim. Calc. Var. **17**, 353–379 (2011)
4. J.M. Coron, M. Kawski, Z. Wang, Analysis of a conservation law modeling a highly re-entrant manufacturing system. Discrete Contin. Dyn. Syst. Ser. B **14**, 1337–1359 (2010)
5. J. Gradl, Experimentelle und theoretische Untersuchungen der Bildungskinetik diffusions- sowie reaktionslimitierter Systeme am Beispiel der Nanopartikelfällung von Bariumsulfat und Zinkoxid (Cuvillier Verlag, Göttingen, 2010)
6. M. Gröschel, Optimization of particel synthesis, Ph.D. thesis, FAU Department of mathematics, 2013
7. M. Gugat, M. Herty, A. Klar, G. Leugering, Optimal control for traffic flow networks. J. Optim. Theory Appl. **126**, 589–616 (2005)
8. M. Herty, A. Klar, B. Piccoli, Existence of solutions for supply chain models based on partial differential equations. SIAM J. Math. Anal. **39**, 160 (2007)
9. C. Jagadish, S.J. Pearton (Eds.), *Zinc Oxide Bulk, Thin Films and Nanostructures: Processing, Properties, and Applications*, 1st edn. (Elsevier, Amsterdam/London, 2006)
10. I. Lifshitz, V. Slyozov, The kinetics of precipitation from supersaturated solid solutions. J. Phys. Chem. Solids **19**, 35–50 (1961)
11. A. Mersmann, Calculation of interfacial tensions. J. Cryst. Growth **102**, 841–847 (1990)
12. A. Mersmann, General prediction of statistically mean growth rates of a crystal collective. J. Cryst. Growth **147**, 181–193 (1995)
13. E.L. Petersen, M.W. Crofton, Measurements of high-temperature silane pyrolysis using SiH4 IR emission and SiH2 laser absorption. J. Phys. Chem. A **107**, 10988–10995 (2003)
14. D. Ramkrishna, *Population Balances: Theory and Applications to Particulate Systems in Engineering* (Elsevier, Burlington, 2000)
15. D. Segets, M.A.J. Hartig, J. Gradl, W. Peukert, A population balance model of quantum dot formation: oriented growth and ripening of ZnO. Chem. Eng. Sci. **70**, 4–13 (2012)
16. D.H. Segets, J. Gradl, R.K. Taylor, V. Vassilev, W. Peukert, N.S. Distribution, Analysis of optical absorbance spectra for the determination of ZnO nanoparticle size distribution, solubility, and surface energy. ACS nano **3**, 1703–1710 (2009)
17. D.H. Segets, L.M. Tomalino, J. Gradl, W. Peukert, Real-time monitoring of the nucleation and growth of ZnO nanoparticles using an optical hyper-rayleigh scattering method. J. Phys. Chem. C **113**, 11995–12001 (2009)
18. P. Shang, Z. Wang, Analysis and control of a scalar conservation law modeling a highly re-entrant manufacturing system. J. Diff. Equ. **250**, 949–982 (2011)

19. D.V. Talapin, A.L. Rogach, M. Haase, H. Weller, Evolution of an ensemble of nanoparticles in a colloidal solution: theoretical study. J. Phys. Chem. B **105**, 12278–12285 (2001)
20. R. Viswanatha, S. Sapra, B. Satpati, P.V. Satyam, B.N. Dev, D.D. Sarma, Understanding the quantum size effects in ZnO nanocrystals. J. Mater. Chem. **14**, 661 (2004)
21. C. Wagner, Theorie der Alterung von Niederschlägen durch Umlösen (Ostwald-Reifung). Zeitschrift für Elektrochemie Berichte der Bunsengesellschaft für physikalische Chemie **65**, 581–591 (1961)
22. Z.L. Wang, Zinc oxide nanostructures: growth, properties and applications. J. Phys. Condens. Matter **16**, 829–858 (2004)
23. M. Wulkow, Modeling and simulation of crystallization processes using parsival. Chem. Eng. Sci. **56**, 2575–2588 (2001)

Stabilization of Networked Hyperbolic Systems with Boundary Feedback

Markus Dick, Martin Gugat, Michael Herty, Günter Leugering, Sonja Steffensen, and Ke Wang

Abstract We summarize recent theoretical results as well as numerical results on the feedback stabilization of first order quasilinear hyperbolic systems (on networks). For the stabilization linear feedback controls are applied at the nodes of the network. This yields the existence and uniqueness of a C^1-solution of the hyperbolic system with small C^1-norm. For this solution an appropriate L^2-Lyapunov function decays exponentially in time. This implies the exponential stability of the system. A numerical discretization of the Lyapunov function is presented and a numerical analysis shows the expected exponential decay for a class of first-order discretization schemes. As an application for the theoretical results the stabilization of the gas flow in fan-shaped pipe networks with compressors is considered.

Keywords Quasilinear hyperbolic system • Feedback stabilization • Networked system • Boundary control • Lyapunov function

Mathematics Subject Classification (2010). 35L50, 93C20.

This work has been supported by DFG SPP 1253, DAAD 508846 and DAAD D/0811409 (Procope 2009/10).

M. Dick (✉) • M. Gugat • G. Leugering
Department of Mathematics, University of Erlangen-Nuremberg, Cauerstr. 11, 91058 Erlangen, Germany
e-mail: dick@math.fau.de; gugat@math.fau.de; leugering@math.fau.de

M. Herty • S. Steffensen
IGPM, Department of Mathematics, RWTH Aachen University, Templergraben 55, 52056 Aachen, Germany
e-mail: herty@igpm.rwth-aachen.de; steffensen@igpm.rwth-aachen.de

K. Wang
School of Mathematical Sciences, Fudan University, Handan Rd. 220, 200433 Shanghai, China
e-mail: kwang0815@gmail.com

G. Leugering et al. (eds.), *Trends in PDE Constrained Optimization*, International Series of Numerical Mathematics 165, DOI 10.1007/978-3-319-05083-6_31

1 Introduction

Flow processes in networks can often be modeled by coupled quasilinear hyperbolic systems. For example, the flow of natural gas in pipe networks can be modeled by the isothermal Euler equations with friction (see (5.1), (5.2)), a hyperbolic system of conservation laws (see [2, 3, 27, 29, 35, 36, 38]). The Saint-Venant equations are a model for the flow in open water canals (see [4, 5, 7, 31]). For traffic flow models on networks see [14].

In the recent literature there has been a lot of research on the stabilization and controllability of hyperbolic systems on networks (see e.g. [4–7, 9–13, 15, 17–20, 22, 23, 31, 32, 37, 38, 43, 46]). In this paper we summarize recent results about the stabilization of hyperbolic systems, numerical discretization and the isothermal Euler equations as published in [1, 9, 11, 13, 17–19]. The quasilinear system under consideration is of the form (2.2). We consider this system on a fan-shaped network, that is a tree-shaped network with exactly one node with a degree larger than two (see Fig. 1). At the nodes of the network we have coupling conditions of the form (2.12) and (2.13). Theorem 3.1 presents a result about the exponential stability of the system (2.2) on the fan-shaped network as published and proved in [9]: If we apply the feedback controls (3.6) with appropriate feedback constants $k^{(i,j)} \in (-1, 1)$ and if we have C^1-initial data (3.7) with sufficiently small C^1-norm, then the initial-boundary value problem (2.2), (2.12), (2.13), (3.6), (3.7) has a unique C^1-solution with small C^1-norm on a given finite time interval (see (3.23)). For this solution the L^2-Lyapunov function $E_\omega(t)$ from (3.5), which is the sum of weighted and squared L^2-norms of the dependent variables (see (3.3), (3.4)), decays exponentially with time (see (3.24)). In [13] the stabilization of the system (2.2) on a star-shaped network has been presented. The structure of the system (2.2) is motivated by the isothermal Euler equations with friction (5.1), (5.2) which model the gas flow in pipes. For a nonstationary solution of the Euler equations locally around a given stationary state, in terms of the characteristic variables we have a system of the form (2.2). In Corollary 5.1 we apply the stabilization method presented in Theorem 3.1 to stabilize the gas flow in a fan-shaped pipe network. This result is in detail presented in [9, 18]. Concerning the numerical results we consider the problem (2.2) and boundary conditions induced by coupling conditions of type (2.12) leading to equations of the type (3.6). Similar to the established theoretical result we present a suitable discretization of (2.2) and its corresponding Lyapunov function (3.3) and (3.4). For a single arc we show exponential decay of the discretized system for initial data having small discrete C^1-norm. The constants in Theorem 3.1 may be computed explicitly in the discrete case. The presented numerical analysis holds true for quasilinear systems for a single arc. The results have been announced in [1]. Further discussion on appropriate numerical schemes for the time integration of hyperbolic equations (including non-smooth solutions) have been investigated in [28, 30, 39]. Therein, the major focus is on high-order time integration schemes required for example in the solution of nonlinear hyperbolic equations in conservative form. If interested in optimal control, a detailed discussion on the schemes for the adjoint equation is necessary. The relation between suitable time integrators for hyperbolic systems and their adjoint equations has been investigated therein. Concerning the theoretical results on smooth solutions time delay is another important feature. The feedback laws (3.6) discussed here are without time delay. However, for the stabilization of quasilinear hyperbolic systems it is also possible to apply feedback controls with time delay. For the isothermal Euler equations this is studied in [9, 17, 19] where the delays are given by C^1-functions with bounded derivatives. In [9, 13] the time-delayed feedback stabilization of a general hyperbolic system of the form (2.2) is considered. In [8, 16, 24, 25, 34, 41, 42] the time-delayed stabilization of the wave equation is studied. Related problems of the controllability of the wave equation are considered in [44, 45]. Questions of the well-posedness and optimal control of networked hyperbolic systems are studied e.g. in [15, 21, 38].

This paper is organized as follows: In Sect. 2 we present the notation for the fan-shaped network, the quasilinear hyperbolic systems (2.2) and the coupling conditions (2.12) and (2.13). The network Lyapunov function $E_\omega(t)$ is defined in (3.5) in Sect. 3.1. In Sect. 3.2 the exponential stability of the system (2.2) is presented in Theorem 3.1. An appropriate choice of the feedback constants and the weight constants for the Lyapunov function is stated in Sect. 3.2. Section 4 contains the discretization and the result on exponential decay. In Sect. 5 we consider the isothermal Euler equations (5.1), (5.2) and present a stabilization method for the gas flow in a fan-shaped pipe network (Corollary 5.1).

2 Network Notation, Quasilinear Hyperbolic System and Coupling Conditions

2.1 *Quasilinear Hyperbolic System on a Network*

We consider a tree-shaped network with exactly one node with a degree larger than two. Such a network is called a fan-shaped network (see Fig. 1). We assume that this central node of the network has the degree N ($N \geq 3$) and call this node ω. We denote the paths of edges that meet at the node ω as *path* 1 to *path* N. Furthermore,

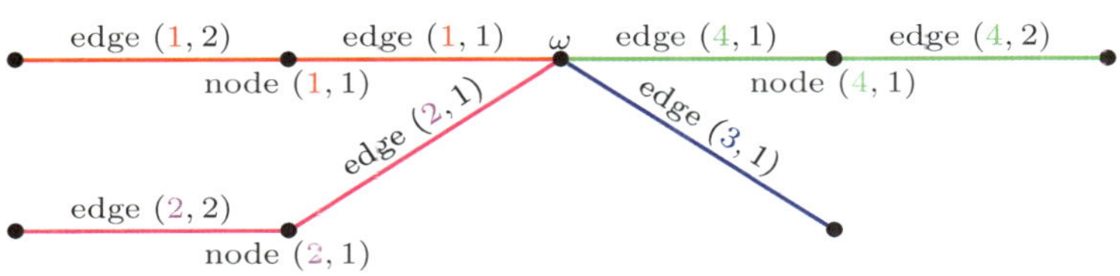

Fig. 1 Fan-shaped network. The depicted network has $M = 4$ paths which consist of $N_1 = 2$, $N_2 = 2$, $N_3 = 1$ and $N_4 = 2$ edges. The central node is denoted as ω

we assume that path i is a linear sequence of N_i ($N_i \geq 1$) edges and sequentially denote the edges in path i as *edge* $i,1$ to *edge* i, N_i where the edge $i,1$ is at the central node ω. We introduce the index sets

$$I = \{1, \ldots, N\}, \quad I_i = \{1, \ldots, N_i\}. \tag{2.1}$$

We parameterize the length of edge i, j ($i \in I$, $j \in I_i$) by the space interval $[0, L^{(i,j)}]$ with $L^{(i,j)} > 0$ such that the end $x = 0$ is closer to the node ω. The other end of edge i, j is denoted as $x = L^{(i,j)}$. For $i \in I$, $j \in I_i \backslash \{N_i\}$ we denote the node where the ends $x = L^{(i,j)}$ of edge i, j and $x = 0$ of edge $i, j+1$ meet as *node* i, j. The end $x = 0$ of edge $i, 1$ ($i \in I$) is at the node ω. We consider the system on a finite time interval $[0, T]$ with $T > 0$.

On edge i, j of the fan-shaped network we consider the following quasilinear system ($i \in I$, $j \in I_i$):

$$\begin{cases} \partial_t u_+^{(i,j)} + \Lambda_+^{(i,j)}(x, u_+^{(i,j)}, u_-^{(i,j)})\, \partial_x u_+^{(i,j)} = \Psi_+^{(i,j)}(x, u_+^{(i,j)}, u_-^{(i,j)}), \\ \partial_t u_-^{(i,j)} + \Lambda_-^{(i,j)}(x, u_+^{(i,j)}, u_-^{(i,j)})\, \partial_x u_-^{(i,j)} = \Psi_-^{(i,j)}(x, u_+^{(i,j)}, u_-^{(i,j)}) \end{cases} \tag{2.2}$$

on $[0, T] \times [0, L^{(i,j)}]$ with the dependent variables $u_+^{(i,j)}(t, x)$ and $u_-^{(i,j)}(t, x)$.

The functions $\Lambda_\pm^{(i,j)}$ in (2.2) are of the form

$$\Lambda_\pm^{(i,j)}(x, u_+^{(i,j)}, u_-^{(i,j)}) = \lambda_\pm^{(i,j)}(x) + f_\pm^{(i,j)}(x, u_+^{(i,j)}, u_-^{(i,j)}) \tag{2.3}$$

with $\lambda_\pm^{(i,j)}(x) \in C^1([0, L^{(i,j)}])$ and $f_\pm^{(i,j)}(x, u_+^{(i,j)}, u_-^{(i,j)}) \in C^1([0, L^{(i,j)}] \times \mathbb{R}^2)$. In (2.2) the right-hand sides have the form

$$\Psi_\pm^{(i,j)}(x, u_+^{(i,j)}, u_-^{(i,j)}) = -(u_+^{(i,j)} + u_-^{(i,j)})\, \psi_\pm^{(i,j)}(x) + g_\pm^{(i,j)}(x, u_+^{(i,j)}, u_-^{(i,j)}) \tag{2.4}$$

with $\psi_\pm^{(i,j)}(x) \in C^1([0, L^{(i,j)}])$ and $g_\pm^{(i,j)}(x, u_+^{(i,j)}, u_-^{(i,j)}) \in C^2([0, L^{(i,j)}] \times \mathbb{R}^2)$.

We assume that the functions $\lambda_\pm^{(i,j)}$ and $\psi_\pm^{(i,j)}$ satisfy

$$\lambda_+^{(i,j)}(x) > 0, \quad \lambda_-^{(i,j)}(x) < 0 \tag{2.5}$$

and

$$\psi_\pm^{(i,j)}(x) > 0 \tag{2.6}$$

for all $x \in [0, L^{(i,j)}]$ and that the functions $f_\pm^{(i,j)}$ and $g_\pm^{(i,j)}$ satisfy

$$f_\pm^{(i,j)}(x, 0, 0) = 0 \tag{2.7}$$

and

$$g_{\pm}^{(i,j)}(x,0,0) = \partial_{u_+^{(i,j)}} g_{\pm}^{(i,j)}(x,0,0) = \partial_{u_-^{(i,j)}} g_{\pm}^{(i,j)}(x,0,0) = 0 \tag{2.8}$$

for all $x \in [0, L^{(i,j)}]$.

If we have $|u_{\pm}^{(i,j)}| \le \varepsilon_0$ with a real number $\varepsilon_0 > 0$, from (2.7) and (2.8) together with Taylor's Theorem we obtain that there exist numbers $\kappa_f^{(i,j)} \ge 0$ and $\kappa_g^{(i,j)} \ge 0$ such that for $x \in [0, L^{(i,j)}]$ we have

$$|f_{\pm}^{(i,j)}(x,u_+,u_-)| \le \kappa_f^{(i,j)} (|u_+| + |u_-|) \quad \text{for all} \quad |u_\pm| \le \varepsilon_0 \tag{2.9}$$

and

$$|g_{\pm}^{(i,j)}(x,u_+,u_-)| \le \kappa_g^{(i,j)} (u_+^2 + u_-^2) \quad \text{for all} \quad |u_\pm| \le \varepsilon_0. \tag{2.10}$$

A detailed derivation of the estimates (2.9) and (2.10) can be found in [9]. In the stabilization method which we present in Theorem 3.1 we obtain a solution of the system (2.2) with C^1-norm smaller than $\varepsilon_1 \in (0, \varepsilon_0]$ (see (3.23)). Hence, for $\varepsilon_0 > 0$ sufficiently small, from (2.9) and (2.10) we obtain that $|f_{\pm}^{(i,j)}|$ and $|g_{\pm}^{(i,j)}|$ are small. In particular, if $\varepsilon_0 > 0$ is small enough, for $|u_{\pm}^{(i,j)}| \le \varepsilon_0$ we obtain (see (2.3), (2.5))

$$\Lambda_+^{(i,j)}(x, u_+^{(i,j)}, u_-^{(i,j)}) > 0, \quad \Lambda_-^{(i,j)}(x, u_+^{(i,j)}, u_-^{(i,j)}) < 0 \tag{2.11}$$

for all $x \in [0, L^{(i,j)}]$ and, hence, the system (2.2) is strictly hyperbolic.

2.2 Coupling Conditions

In the following we present the coupling conditions at the nodes of the fan-shaped network. The structure of the coupling conditions is motivated by the conditions for a fan-shaped network of gas pipes coupled by compressor stations which we consider in Sect. 5 (see (5.10)–(5.14)). At the central node ω, where the ends $x = 0$ of edge $i, 1$ ($i \in I$) meet, we have conditions of the form ($i \in I$, $t \in [0, T]$)

$$u_+^{(i,1)}(t,0) = \Omega^{(i)}(u_-^{(1,1)}(t,0), \ldots, u_-^{(N,1)}(t,0)) \tag{2.12}$$

with C^1-functions $\Omega^{(i)}$. At the other inner nodes of the network where the ends $x = L^{(i,j)}$ of edge i, j and $x = 0$ of edge $i, j+1$ meet ($i \in I$, $j \in I_i \backslash \{N_i\}$) we have the conditions ($t \in [0, T]$)

$$u_+^{(i,j+1)}(t,0) = \Xi^{(i,j)}(u_+^{(i,j)}(t, L^{(i,j)}), u_-^{(i,j)}(t, L^{(i,j)}), u_+^{(i,j+1)}(t,0)) \tag{2.13}$$

with C^1-functions $\Xi^{(i,j)}$. As we stabilize the system (2.2) towards the null equilibrium state, the functions $\Omega^{(i)}$ and $\Xi^{(i,j)}$ have to satisfy

$$\Omega^{(i)}(0, \dots, 0) = 0 \tag{2.14}$$

and

$$\Xi^{(i,j)}(0, \dots, 0) = 0. \tag{2.15}$$

3 Exponential Stability

In Theorem 3.1 we present a method to stabilize the system (2.2) with the coupling conditions (2.12) and (2.13) on a fan-shaped network. For the stabilization we apply the linear feedbacks (3.6) at the nodes of the network with appropriate feedback constants $k^{(i,j)} \in (-1, 1)$. For the initial-boundary value problem (2.2), (2.12), (2.13), (3.6), (3.7) we obtain the existence of a unique C^1-solution on the time interval $[0, T]$. In order to measure the system evolution, we define the network Lyapunov function $E_\omega(t)$ in (3.5) which is the sum of weighted and squared L^2-norms for $u_\pm^{(i,j)}$ (see (3.3), (3.4)). We obtain the exponential decay of the Lyapunov function with time (see (3.24)), which implies the exponential stability of the system.

3.1 Lyapunov Function and Feedback Controls

Let a finite time $T > 0$ be given. We consider a fan-shaped network and the system (2.2) with the conditions (2.3)–(2.8) and with the coupling conditions (2.12), (2.13). We define the real numbers ($i \in I$, $j \in I_i$)

$$\mu^{(i,j)} = \left(\int_0^{L^{(i,j)}} \frac{1}{\lambda_+^{(i,j)}(x)} + \frac{1}{|\lambda_-^{(i,j)}(x)|} \, dx \right)^{-1} > 0 \tag{3.1}$$

and the functions

$$h_\pm^{(i,j)}(x) = \exp\left(-\mu^{(i,j)} \int_0^x \frac{1}{\lambda_\pm^{(i,j)}(s)} \, ds \right). \tag{3.2}$$

The quotient functions $h_-^{(i,j)}/h_+^{(i,j)}$ satisfy ($x \in [0, L^{(i,j)}]$)

$$1 = \frac{h_-^{(i,j)}(0)}{h_+^{(i,j)}(0)} \leq \frac{h_-^{(i,j)}(x)}{h_+^{(i,j)}(x)} \leq \frac{h_-^{(i,j)}(L^{(i,j)})}{h_+^{(i,j)}(L^{(i,j)})} = \exp(1) = \mathrm{e}.$$

For $i \in I$, $j \in I_i$ and constants $A_{\pm}^{(i,j)} > 0$ we define the functions ($t \in [0, T]$)

$$E_{+}^{(i,j)}(t) = \int_0^{L^{(i,j)}} \frac{A_{+}^{(i,j)}}{\lambda_{+}^{(i,j)}(x)} h_{+}^{(i,j)}(x)\, (u_{+}^{(i,j)}(t,x))^2\, dx, \tag{3.3}$$

$$E_{-}^{(i,j)}(t) = \int_0^{L^{(i,j)}} \frac{A_{-}^{(i,j)}}{|\lambda_{-}^{(i,j)}(x)|} h_{-}^{(i,j)}(x)\, (u_{-}^{(i,j)}(t,x))^2\, dx \tag{3.4}$$

The network Lyapunov function is defined as ($t \in [0, T]$)

$$E_{\omega}(t) = \sum_{i \in I, j \in I_i} E_{+}^{(i,j)}(t) + E_{-}^{(i,j)}(t). \tag{3.5}$$

Weighted and squared L^2-norms of the form (3.3), (3.4) have been introduced in [6, 7] for constant $\lambda_{\pm}^{(i,j)}$ and in [20] for space dependent $\lambda_{\pm}^{(i,j)}$. The Lyapunov function $E_{\omega}(t)$ for a fan-shaped network has been presented in [9, 18].

For the stabilization we apply the following linear feedback control at the end $x = L^{(i,j)}$ of the edge i, j ($i \in I$, $j \in I_i$)

$$u_{-}^{(i,j)}(t, L^{(i,j)}) = k^{(i,j)} u_{+}^{(i,j)}(t, L^{(i,j)}) \tag{3.6}$$

for $t \in [0, T]$ with constants $k^{(i,j)} \in (-1, 1)$. Furthermore, we suppose that for the system (2.2) we have the following initial data ($i \in I$, $j \in I_i$)

$$u_{\pm}^{(i,j)}(0, x) = \phi_{\pm}^{(i,j)}(x) \tag{3.7}$$

for $x \in [0, L^{(i,j)}]$ with C^1-functions $\phi_{\pm}^{(i,j)}$ such that the C^1-compatibility conditions are satisfied at all nodes of the network. The C^1-compatibility conditions for the fan-shaped network can explicitly be found in [9].

3.2 Exponential Decay of the Network Lyapunov Function

In Theorem 3.1 we obtain the existence of a unique C^1-solution with small C^1-norm of the initial-boundary value problem (2.2), (2.12), (2.13), (3.5), (3.7) on $[0, T] \times [0, L^{(i,j)}]$. For this solution the network Lyapunov function $E_{\omega}(t)$ from (3.5) decays exponentially on $[0, T]$ (see (3.24)). For Theorem 3.1 we define the positive numbers ($i \in I$, $j \in I_i$)

$$U_{\pm}^{(i,j)} = \max_{x \in [0, L^{(i,j)}]} \left| \frac{\lambda_{\pm}^{(i,j)}(x)}{\lambda_{\mp}^{(i,j)}(x)} \right| \frac{\psi_{\mp}^{(i,j)}(x)}{\psi_{\pm}^{(i,j)}(x)} > 0 \tag{3.8}$$

and

$$V_{\pm}^{(i,j)} = \min_{x\in[0,L^{(i,j)}]} \left|\frac{\lambda_{\pm}^{(i,j)}(x)}{\lambda_{\mp}^{(i,j)}(x)}\right| \frac{\psi_{\mp}^{(i,j)}(x)}{\psi_{\pm}^{(i,j)}(x)} \left(1 + \frac{\mu^{(i,j)}}{2\psi_{\mp}^{(i,j)}(x)}\right) > 0. \tag{3.9}$$

For $i \in I$, $j \in I_i$ we assume that we have

$$\mathrm{e}\, U_{+}^{(i,j)} \le V_{+}^{(i,j)} \quad \text{or} \quad \mathrm{e}\, U_{-}^{(i,j)} \le V_{-}^{(i,j)}. \tag{3.10}$$

The condition (3.10) is in detail discussed in [9, 20]. In particular, the condition (3.10) holds if the length $L^{(i,j)}$ of edge i,j is not too long. Furthermore, for $\varepsilon_0 > 0$ and $i \in I$ with $N_i \ge 2$, $j \in I_i \setminus \{N_i\}$ we define the nonnegative numbers

$$\alpha^{(i,j)} = \max_{|\xi|\le\varepsilon_0, |\zeta|\le\varepsilon_0} \left(\partial_{u_{+}^{(i,j)}} \Xi^{(i,j)}(\xi, k^{(i,j)}\xi, \zeta) + k^{(i,j)} \partial_{u_{-}^{(i,j)}} \Xi^{(i,j)}(\xi, k^{(i,j)}\xi, \zeta)\right)^2 \tag{3.11}$$

$$\beta^{(i,j)} = \max_{|\xi|\le\varepsilon_0, |\zeta|\le\varepsilon_0} \left(\partial_{u_{-}^{(i,j+1)}} \Xi^{(i,j)}(\xi, k^{(i,j)}\xi, \zeta)\right)^2 \tag{3.12}$$

with the functions $\Xi^{(i,j)}$ from (2.13).

In Theorem 3.1 the feedback constants $k^{(i,j)} \in (-1,1)$ for the feedback laws (3.6) and the weight constants $A_{\pm}^{(i,j)} > 0$ for the Lyapunov function have to be chosen as follows (see [9, 18]): At the central node ω the constants $A_{\pm}^{(i,1)} > 0$ have to satisfy ($i \in I$)

$$A_{-}^{(i,1)} \ge 1 \ge N\, A_{+}^{(i,1)} \sum_{n\in I} \max_{|\zeta^{(\nu)}|\le\varepsilon_0\, (\nu\in I)} \left(\partial_{u_{-}^{(n,1)}} \Omega^{(i)}(\zeta^{(1)}, \dots, \zeta^{(N)})\right)^2 \tag{3.13}$$

where N is the number of edges that meet at the node ω. For $i \in I$ with $N_i \ge 2$, $j \in I_i \setminus \{N_i\}$ the inequalities

$$(k^{(i,j)})^2 h_{-}^{(i,j)}(L^{(i,j)}) A_{-}^{(i,j)} + 2\alpha^{(i,j)} A_{+}^{(i,j+1)} \le h_{+}^{(i,j)}(L^{(i,j)}) A_{+}^{(i,j)}, \tag{3.14}$$

$$2\beta^{(i,j)} A_{+}^{(i,j+1)} \le A_{-}^{(i,j+1)} \tag{3.15}$$

have to hold and for $i \in I$ at the ends $x = L^{(i,N_i)}$ the inequalities

$$\mathrm{e}\,(k^{(i,N_i)})^2\, A_{-}^{(i,N_i)} \le A_{+}^{(i,N_i)}. \tag{3.16}$$

Finally, for $i \in I$, $j \in I_i$ the following conditions have to be satisfied (see (3.10))

$$\frac{A_{+}^{(i,j)}}{A_{-}^{(i,j)}} \in [\mathrm{e}\, U_{+}^{(i,j)}, V_{+}^{(i,j)}] \quad \text{or} \quad \frac{A_{-}^{(i,j)}}{A_{+}^{(i,j)}} \in [U_{-}^{(i,j)}, \mathrm{e}^{-1} V_{-}^{(i,j)}]. \tag{3.17}$$

For a detailed discussion of the choice of $k^{(i,j)}$ and $A_{\pm}^{(i,1)}$ via an inductive scheme such that the conditions (3.13)–(3.17) are satisfied, see [9, 13, 18].

For the following theorem we define the positive numbers ($i \in I,\ j \in I_i$)

$$\gamma^{(i,j)} = \max\left\{ \max_{x\in[0,L^{(i,j)}]} \frac{\mathrm{e}A_-^{(i,j)}\lambda_+^{(i,j)}(x)}{A_+^{(i,j)}|\lambda_-^{(i,j)}(x)|},\ \max_{x\in[0,L^{(i,j)}]} \frac{A_+^{(i,j)}|\lambda_-^{(i,j)}(x)|}{A_-^{(i,j)}\lambda_+^{(i,j)}(x)} \right\}. \tag{3.18}$$

Theorem 3.1. *Consider a fan-shaped network and define the index sets I and I_i as in (2.1). Let a finite time $T > 0$ and functions $\lambda_{\pm}^{(i,j)}(x) \in C^1([0, L^{(i,j)}])$, $\psi_{\pm}^{(i,j)}(x) \in C^1([0, L^{(i,j)}])$, $f_{\pm}^{(i,j)}(x, u_+^{(i,j)}, u_-^{(i,j)}) \in C^1([0, L^{(i,j)}] \times \mathbb{R}^2)$ and $g_{\pm}^{(i,j)}(x, u_+^{(i,j)}, u_-^{(i,j)}) \in C^2([0, L^{(i,j)}] \times \mathbb{R}^2)$ be given that satisfy (2.5)–(2.8) ($i \in I,\ j \in I_i$). Consider Eqs. (2.2) on $[0, T] \times [0, L^{(i,j)}]$ with the coupling conditions (2.12), (2.13) where the C^1-functions $\Omega^{(i)}$ and $\Xi^{(i,j)}$ satisfy (2.14) and (2.15).*

Let a real number $\varepsilon_0 > 0$ be given and choose constants $\kappa_f^{(i,j)} \geq 0$ and $\kappa_g^{(i,j)} \geq 0$ such that (2.9) and (2.10) hold ($i \in I,\ j \in I_i$). Define $\mu^{(i,j)}$, $h_{\pm}^{(i,j)}(x)$, $U_{\pm}^{(i,j)}$ and $V_{\pm}^{(i,j)}$ as in (3.1), (3.2), (3.8), (3.9) and assume that (3.10) holds. Choose constants $k^{(i,j)} \in (-1, 1)$ and $A_{\pm}^{(i,j)} > 0$ such that (3.13)–(3.16) are satisfied with $\alpha^{(i,j)}$, $\beta^{(i,j)}$ as in (3.11), (3.12). Assume that (3.17) holds.

Choose $\varepsilon_1 \in (0, \varepsilon_0]$ that satisfies ($i \in I,\ j \in I_i$)

$$2\varepsilon_1 \kappa_f^{(i,j)} < \min_{x\in[0,L^{(i,j)}]} \left|\lambda_{\pm}^{(i,j)}(x)\right| \tag{3.19}$$

and

$$\varepsilon_1 \left(\kappa_f^{(i,j)} + \kappa_g^{(i,j)}\right)\left(3 + \gamma^{(i,j)}\right) < \tfrac{1}{2}\mu^{(i,j)} \tag{3.20}$$

with $\gamma^{(i,j)}$ as in (3.18). Define the real number

$$\eta = \min_{i\in I, j\in I_i} \left\{\tfrac{1}{2}\mu^{(i,j)} - \varepsilon_1\left(\kappa_f^{(i,j)} + \kappa_g^{(i,j)}\right)\left(3 + \gamma^{(i,j)}\right)\right\} > 0. \tag{3.21}$$

Then there exists $\varepsilon_2 \in (0, \varepsilon_1]$ such that the following statements hold true: Assume that we have initial conditions (3.7) with C^1-functions $\phi_{\pm}^{(i,j)}$ that satisfy ($i \in I$, $j \in I_i$)

$$||\phi_{\pm}^{(i,j)}||_{C^1([0,L^{(i,j)}])} \leq \varepsilon_2 \tag{3.22}$$

and such that the C^1-compatibility conditions are satisfied at all nodes of the network. Then the initial-boundary value problem (2.2), (2.12), (2.13), (3.6), (3.7) has a unique C^1-solution $(u_+^{(i,j)}, u_-^{(i,j)})$ on $[0, T] \times [0, L^{(i,j)}]$ that satisfies

$$||u_{\pm}^{(i,j)}||_{C^1([0,T]\times[0,L^{(i,j)}])} \leq \varepsilon_1. \tag{3.23}$$

For this solution we define the functions $E_+^{(i,j)}(t)$ and $E_-^{(i,j)}(t)$ as in (3.3), (3.4) and the network Lyapunov function $E_\omega(t)$ as in (3.5). Then we have the following inequality for $E_\omega(t)$ on $[0, T]$:

$$E_\omega(t) \le E_\omega(0)\exp(-\eta t). \tag{3.24}$$

Proof. The proof of Theorem 3.1 can be found in [9]. In particular, in [9] it is shown that for the time derivative of the network Lyapunov function $E_\omega(t)$ we have ($t \in [0, T]$)

$$\begin{aligned}\tfrac{d}{dt}E_\omega(t) \le & \sum_{i\in I, j\in I_i}\left(\tfrac{1}{2}\mu^{(i,j)} - \varepsilon_1\left(\kappa_f^{(i,j)} + \kappa_g^{(i,j)}\right)\left(3+\gamma^{(i,j)}\right)\right)E^{(i,j)}(t) \\ & + \sum_{i\in I, j\in I_i} B_0^{(i,j)}(t) + B_L^{(i,j)}(t)\end{aligned} \tag{3.25}$$

where $B_0^{(i,j)}(t)$ and $B_L^{(i,j)}(t)$ are boundary terms that appear at the end $x = 0$ and $x = L^{(i,j)}$ of edge i, j ($i \in I, j \in I_i$). The feedback laws (3.6), together with the appropriate choice of the feedback constants $k^{(i,j)}$ and the weights $A_\pm^{(i,j)}$, guarantee that the terms $B_0^{(i,j)}(t)$ and $B_L^{(i,j)}(t)$ are nonpositive for all $t \in [0, T]$. Thus, from (3.25) and the definition of $E_\omega(t)$ and η in (3.5) and (3.21) we obtain (3.24). The existence and uniqueness of a solution of (2.2) that satisfies (3.23) follows from [32, 46], where the existence of classical solutions for initial-boundary value problems with first order quasilinear hyperbolic systems is studied. □

Remark 3.2. The conditions (3.19) together with $\varepsilon_1 \in (0, \varepsilon_0]$ and (2.3), (2.5), (2.9), (3.23) guarantee that the inequalities (2.11) hold and, hence, the system (2.2) is strictly hyperbolic ($i \in I, j \in I_i$). The statements of Theorem 3.1 still hold if the constants $A_\pm^{(i,1)} > 0$ satisfy

$$A_-^{(i,1)} \ge c \ge N\, A_+^{(i,1)} \sum_{n\in I} \max_{|\zeta^{(\nu)}|\le\varepsilon_0\,(\nu\in I)} \left(\partial_{u_-^{(n,1)}}\Omega^{(i)}(\zeta^{(1)},\ldots,\zeta^{(\nu)})\right)^2$$

instead of (3.13) with an arbitrary constant $c > 0$ which is the same for all $i \in I$ (see [9, 13]).

In [6, 7, 9] the stabilization of quasilinear hyperbolic systems without a source term is considered, that is in our notation $\Psi_\pm^{(i,j)} = 0$. In this case the statements of Theorem 3.1 also hold true, where the condition (3.17) is not relevant for the choice of the weight constants $A_\pm^{(i,j)}$ and in (3.20) and (3.21) we have $\kappa_g^{(i,j)} = 0$.

4 Numerical Analysis for a Discretization of System (2.2)

Results concerning the numerical analysis do not exist to the same extend as on the continuous level. For a detailed review we refer to [1]. Here, we apply the existing results to the presented results. This requires strong simplifications of the previous presentation: We consider a single arc and a system in quasilinear form (2.2) with boundary conditions of type (3.6) at both ends $x \in \{0, L\}$. We further assume that $\Lambda_\pm$ are independent of x, no source term and assume $\lambda_\pm = \pm 1$. Dropping the superscripts (i, j) Eqs. (2.2) read with dependent variables $u_+(t, x)$ and $u_-(t, x)$ on $[0, T] \times [0, L]$, respectively,

$$\partial_t u_\pm + \Lambda_\pm(u_+, u_-)\partial_x u_\pm = 0, \; u_\pm(0, x) = \phi_\pm(x), \tag{4.1a}$$

$$u_+(t, 0) = k_\ell \, u_-(t, 0), \; u_-(t, L) = k_r \, u_+(t, L). \tag{4.1b}$$

The focus is on the discussion of numerical discretization and the corresponding L^2-stability of the discrete Lyapunov function. In order to state the stabilization result corresponding to Theorem 3.1, we introduce further notation. Let Δx denote the cell width of a uniform spatial grid and N the number of cells in the discretization of the domain $[0, L]$ such that $\Delta x N = L$ with cell centers at $x_i = (i + \frac{1}{2})\Delta x$. The left and right boundary points are $x_{-\frac{1}{2}}$ and $x_{N-\frac{1}{2}}$. The temporal grid is chosen such that the CFL condition $\lambda \frac{\Delta t}{\Delta x} \leq 1$ holds, where $\lambda = \max_{\|u\| \leq \epsilon_0} |\Lambda_\pm(u)\|$ and ϵ_0 as in Theorem 3.1. Let $t^n = n\Delta t$ and by possibly further reducing Δt assume that for some $K > 0$ we have $K\Delta t = T$. The value of $u_\pm(t, x)$ at the cell center x_i and time t^n is approximated by $u^n_{i,\pm}$. The initial condition is discretized as

$$u^0_{i,\pm} = \frac{1}{\Delta x} \int_{x_{i-\frac{1}{2}}}^{x_{i+\frac{1}{2}}} \phi^0_\pm(x) dx. \tag{4.2}$$

To compute smooth solutions we use

$$u_{\Delta,\pm} = \sum_{n=0}^{K-1} \sum_{i=0}^{N-1} u^n_{i,\pm} \chi_{[x_{i-\frac{1}{2}}, x_{i+\frac{1}{2}}) \times [t^n, t^{n+1})}(t, x),$$

$$u^n_{i,\pm} = \frac{1}{\Delta x} \int_{x_{i-\frac{1}{2}}}^{x_{i+\frac{1}{2}}} u_\pm(t^n, x) dx.$$

We discretize (4.1) by (4.2) and for $n = 0, \ldots, K-1$, and $i = 0, \ldots, N-1$

$$u^{n+1}_{i,+} = u^n_{i,+} - \frac{\Delta t}{\Delta x} \Lambda_+(u^n_{i,+}, u^n_{i,-})(u^n_{i,+} - u^n_{i-1,+}), \tag{4.3a}$$

$$u^{n+1}_{i,-} = u^n_{i,-} - \frac{\Delta t}{\Delta x} \Lambda_-(u^n_{i+1,+}, u^n_{i+1,-})(u^n_{i+1,-} - u^n_{i,-}), \tag{4.3b}$$

$$u^n_{-1,+} = k_\ell u^n_{0,-}, \; u^n_{N,-} = k_r u^n_{N-1,+}. \tag{4.3c}$$

The discrete Lyapunov function (3.5) in case of (4.1) is given by

$$E^n = \Delta x \sum_{i=0}^{N-1} A_+ \exp(-\tilde{\mu} \, x_i)(u^n_{i,+})^2 + A_- \exp(\tilde{\mu} \, x_i)(u^n_{i,-})^2 \tag{4.4}$$

which resembles the sum of (3.3) and (3.4). Note that in the case $\Lambda_\pm$ independent of x we obtain from (3.1) and (3.2)

$$-\mu \int_0^x \frac{1}{\lambda_\pm} ds = \mp x \frac{1}{2L}, \; \tilde{\mu} = \frac{1}{2L}. \tag{4.5}$$

For the numerical scheme we furthermore require that k_r and k_ℓ fulfill

$$0 < k_\ell < \sqrt{\frac{D_-^{min}}{D_+^{max}}}, \; 0 < k_r < \sqrt{\frac{D_+^{min}}{D_-^{max}}}, \tag{4.6}$$

where for ϵ_0 from Theorem 3.1 we define $D_\pm^{min} = \min\limits_{\|u_\pm\| \le \epsilon_0} \frac{\Delta t}{\Delta x} \|\Lambda_\pm(u_+, u_-)\|$ and $D_\pm^{max} = \max\limits_{\|u_\pm\| \le \epsilon_0} \frac{\Delta t}{\Delta x} \|\Lambda_\pm(u_+, u_-)\|$. Assume Δt is such that $\lambda \frac{\Delta t}{\Delta x} \le 1$ and $\lambda = \max\limits_{s \in \{+,-\}} \max\limits_{\|u_\pm\| \le \epsilon_0} \|\Lambda_s(u_+, u_-)\|$. Then, corresponding to Theorem 3.1, we obtain the discrete counterpart as Theorem 4.1.

Theorem 4.1. *Let $T > 0$ and assume* (2.5), (2.7) *and* (2.11). *Consider* (4.1) *with discretization* (4.2)–(4.4). *Let $\epsilon_0 > 0$ be given and assume that ϕ is such that $\|\phi_{i,\pm}\| \le \epsilon_0$, $\|\phi_{i,\pm} - \phi_{i-1,\pm}\| \le \Delta x \, \epsilon_0$, and that $\|u_{0,-}^n - u_{-1,-}^n\| \le \epsilon_0 \exp(t^n M)$, and $\|u_{N,+}^n - u_{N-1,+}^n\| \le \epsilon_0 \exp(t^n M)$ for $M = 2 \max\limits_{s \in \{+,-\}} \max\limits_{\|u_\pm\| \le \epsilon_0} \|\nabla_{(u_+, u_-)} \Lambda_s(u_+, u_-)\|$.*

Choose $A_\pm = 1$ and $\tilde{\mu} = 2L$ as in Eq. (4.5). *Choose k_ℓ, k_r, such that* (4.6) *and*

$$\tilde{\mu} \le \frac{1}{2} \min \left\{ \log \left(\sqrt{\frac{D_-^{\max}}{D_+^{\min}}} \|\kappa_r\| \right)^{-2}, \; \log \left(\sqrt{\frac{D_+^{\max}}{D_+^{\min}}} \|\kappa_\ell\| \right)^{-2} \right\} \tag{4.7}$$

holds true.

Then, there exists $\eta > 0$ such that

$$E^n \le \exp(-\eta \, t^n) E^0. \tag{4.8}$$

For the proof we note that the assumptions on $\Lambda_\pm$ yields two separated eigenvalues bounded away from zero for ϵ_0 sufficiently small. This is due to estimate (2.9). Further, for sufficiently small $k_\ell, k_r \in (-1, 1), k_\ell \ne 0, k_r \ne 0$ or L sufficiently large, the condition (4.7) on $\tilde{\mu}$ is always satisfied. The proof then follows by combining the results of Theorem 2.1, Remark 1 and Theorem 3.2 in [1]. Therein, the discrete counterpart to (3.25)

$$\frac{E^{n+1} - E^n}{\Delta t} \le -\eta \, E^n + R$$

for some $\eta > 0$ has been established. Due to (4.7) it can be shown that $R \le 0$ which yields the desired exponential decay of E^n.

5 Application to the Gas Flow in Networks

In this section we apply the stabilization method presented in Theorem 3.1 to the stabilization of the gas flow in a fan-shaped pipe network. The gas flow in pipes can be modeled by the isothermal Euler equations with friction, a hyperbolic system of balance laws. For pipe i, j the isothermal Euler equations have the following form ($i \in I$, $j \in I_i$) (see [2, 3, 27, 29, 35, 36, 38]):

$$\partial_t \rho^{(i,j)} + \partial_x q^{(i,j)} = 0, \tag{5.1}$$

$$\partial_t q^{(i,j)} + \partial_x \left(\frac{(q^{(i,j)})^2}{\rho^{(i,j)}} + a^2 \rho^{(i,j)} \right) = -\frac{\theta}{2} \frac{q^{(i,j)}|q^{(i,j)}|}{\rho^{(i,j)}}, \tag{5.2}$$

where $\rho^{(i,j)}(t,x) > 0$ is the density of the gas and $q^{(i,j)}(t,x) \neq 0$ the mass flux. The flow velocity is given by the quotient $q^{(i,j)}/\rho^{(i,j)}$. The constant $a > 0$ denotes the sonic speed in the gas and the constant $\theta > 0$ is the quotient of the pipe wall friction factor and the pipe diameter. In the following we study so called subsonic or subcritical states which satisfy ($i \in I$, $j \in I_i$)

$$|q^{(i,j)}|/\rho^{(i,j)} < a.$$

This is satisfied as in real gas transportation networks the absolute value of the flow velocity is much smaller than the sonic speed (see [9, 27]). Equation (5.1) states the conservation of mass and Eq. (5.2) is a momentum equation that states the loss of momentum due to the pipe wall friction. The sign of $q^{(i,j)}$ depends of the direction of the gas flow. The mass flux $q^{(i,j)}$ is positive if the gas in pipe i, j flows from the end $x = 0$ to $x = L^{(i,j)}$, it is negative if the gas flows in the other direction. For a detailed discussion of the Euler equations see [35, 36]. In [9, 11, 18, 20] the existence and behavior of stationary subcritical C^1-solutions $(\bar{\rho}^{(i,j)}(x), \bar{q}^{(i,j)}(x))$ for the system (5.1), (5.2) with given boundary data at one end of each pipe is studied. The main result presented in [9, 11, 18, 20] is that a unique stationary subcritical C^1-solution exists on a finite space interval $[0, x_0)$ with a critical length $x_0 > 0$. In typical high-pressure gas pipes the critical length is around 180 km (see [9, 11]). Furthermore, the stationary density is monotonically decreasing along the direction of the gas flow.

In the following we assume that we have a given finite time $T > 0$ and a given stationary state $(\bar{\rho}^{(i,j)}(x), \bar{q}^{(i,j)}(x)) \in C^1([0, L^{(i,j)}])$ on a fan-shaped pipe network ($i \in I$, $j \in I_i$) and consider a nonstationary solution $(\rho^{(i,j)}(t,x), q^{(i,j)}(t,x))$ of (5.1), (5.2) on $[0,T] \times [0, L^{(i,j)}]$ in a local C^1-neighborhood of the stationary state. More precisely, we assume that for the difference between the nonstationary and the stationary state we have the following estimates for the C^1-norm with a small number $\varepsilon > 0$ ($i \in I$, $j \in I_i$):

$$||\rho^{(i,j)} - \bar{\rho}^{(i,j)}||_{C^1([0,T]\times[0,L^{(i,j)}])} \le \varepsilon, \quad ||q^{(i,j)} - \bar{q}^{(i,j)}||_{C^1([0,T]\times[0,L^{(i,j)}])} \le \varepsilon. \tag{5.3}$$

By using a transformation to the characteristic variables $u_\pm^{(i,j)}(t,x)$, in [9,11,20] it is shown that the nonstationary density and mass flux can be written as ($i \in I$, $j \in I_i$)

$$\rho^{(i,j)} = \bar{\rho}^{(i,j)} \exp\left(\frac{u_-^{(i,j)} - u_+^{(i,j)}}{2a}\right), \tag{5.4}$$

$$q^{(i,j)} = \left(\bar{q}^{(i,j)} - \frac{\bar{\rho}^{(i,j)}(u_+^{(i,j)} + u_-^{(i,j)})}{2}\right) \exp\left(\frac{u_-^{(i,j)} - u_+^{(i,j)}}{2a}\right) \tag{5.5}$$

where the characteristic variables $u_\pm^{(i,j)}(t,x)$ satisfy the system (2.2) on $[0,T] \times [0, L^{(i,j)}]$. The functions $\lambda_\pm^{(i,j)}$ and $\psi_\pm^{(i,j)}$ in (2.2) can be calculated from the given stationary state as (see (2.3), (2.4))

$$\lambda_\pm^{(i,j)}(x) = \frac{\bar{q}^{(i,j)}(x)}{\bar{\rho}^{(i,j)}(x)} \pm a, \tag{5.6}$$

$$\psi_\pm^{(i,j)}(x) = \frac{\theta}{4} \frac{|\bar{q}^{(i,j)}(x)|}{\bar{\rho}^{(i,j)}(x)} \frac{2a\bar{\rho}^{(i,j)}(x) \pm \bar{q}^{(i,j)}(x)}{a\bar{\rho}^{(i,j)}(x) \pm \bar{q}^{(i,j)}(x)}. \tag{5.7}$$

The functions $f_\pm^{(i,j)}$ and $g_\pm^{(i,j)}$ are given as (see (2.3), (2.4))

$$f_\pm^{(i,j)}(x, u_+^{(i,j)}, u_-^{(i,j)}) = -\frac{1}{2}\,(u_+^{(i,j)} + u_-^{(i,j)}), \tag{5.8}$$

$$g_\pm^{(i,j)}(x, u_+^{(i,j)}, u_-^{(i,j)}) = \operatorname{sign}(\bar{q}^{(i,j)})\,\frac{\theta}{8}\,(u_+^{(i,j)} + u_-^{(i,j)})^2. \tag{5.9}$$

For a detailed derivation of Eqs. (5.6)–(5.9) we refer to [9, 11]. Note that, if $|u_\pm^{(i,j)}|$ is sufficiently small (see (3.23)), also (5.3) is satisfied. Furthermore, for $f_\pm^{(i,j)}$ and $g_\pm^{(i,j)}$ as in (5.8) and (5.9) the conditions (2.9) and (2.10) hold with $\kappa_f^{(i,j)} = \frac{1}{2}$ and $\kappa_g^{(i,j)} = \frac{\theta}{4}$.

At the central node ω of the fan-shaped pipe network we assume that we have Kirchhoff coupling conditions, that is continuity of the density and conservation of mass ($t \in [0,T]$) (see [2, 33]):

$$\rho^{(1,1)}(t,0) = \rho^{(i,1)}(t,0) \quad (i \in I\backslash\{1\}), \tag{5.10}$$

$$\sum_{i \in I} q^{(i,1)}(t,0) = 0. \tag{5.11}$$

In terms of $u_\pm^{(i,1)}(t,0)$ the conditions (5.10) and (5.11) can be written as (see [9, 11, 17])

$$\left(u_+^{(1,1)}(t,0), \ldots, u_+^{(N,1)}(t,0)\right) = \left(u_-^{(1,1)}(t,0), \ldots, u_-^{(N,1)}(t,0)\right) M_\omega \tag{5.12}$$

with an orthogonal, symmetric $(N \times N)$-matrix $M_\omega = (m_{kl})_{k,l=1}^N$. The entries of the matrix M_ω only depend on the number N of the paths of pipes in the network and are given as

$$\begin{aligned} m_{kk} &= (N-2)/N \;\; (k \in I), \\ m_{kl} &= -2/N \qquad (k,l \in I, k \neq l). \end{aligned}$$

Note that the coupling conditions (5.12) are of the form (2.12) with (2.14).

At the other nodes of the network where exactly two pipes meet, we assume that there is a compressor station (see e.g. [26,29,38,40]). The compressor increases the gas density which has decreased along the incoming pipe. The coupling conditions at the compressor connecting the end $x = L^{(i,j)}$ of pipe i, j and the end $x = 0$ of pipe $i, j+1$ ($i \in I$, $j \in I_i \backslash \{N_i\}$) are given as (see [9,11,18,20,26,29,38,40]):

$$q^{(i,j)}(t, L^{(i,j)}) = q^{(i,j+1)}(t,0), \tag{5.13}$$

$$u^{(i,j)}(t) = c^{(i,j)} \, |q^{(i,j+1)}(t,0)| \left(\left(\frac{\rho^{(i,j+1)}(t,0)}{\rho^{(i,j)}(t, L^{(i,j)})} \right)^{\mathrm{sign}(q^{(i,j+1)}(t,0)) \cdot \kappa} - 1 \right) \tag{5.14}$$

with the compressor power $u^{(i,j)}(t) \geq 0$, a compressor dependent constant $c^{(i,j)} > 0$ and the constant $\kappa \in \{\frac{2}{7}, \frac{2}{5}\}$ depending on the gas under consideration. In [9, 11, 18, 20] it is shown that the conditions (5.13), (5.14) imply that for $u_\pm^{(i,j)}(t, L^{(i,j)})$ and $u_\pm^{(i,j+1)}(t,0)$ we have a condition of the form (2.13) that satisfies (2.15) ($i \in I$, $j \in I_i \backslash \{N_i\}$).

In the following corollary, which follows from Theorem 3.1, we present a method to stabilize the isothermal Euler equations on a fan-shaped network with the coupling conditions (5.10), (5.11), (5.13) and (5.14) locally around a given stationary subcritical state.

Corollary 5.1. *Consider the Euler equations (5.1), (5.2) on a fan-shaped network with the coupling conditions (5.10), (5.11), (5.13), (5.14). Let a stationary subcritical C^1-state $(\bar{\rho}^{(i,j)}(x), \bar{q}^{(i,j)}(x))$ be given ($i \in I$, $j \in I_i$). Let a finite time $T > 0$ be given. For a nonstationary state $(\rho^{(i,j)}(t,x), q^{(i,j)}(t,x))$ as in (5.4), (5.5) with the characteristic variables $u_\pm^{(i,j)}(t,x)$ consider the system (2.2) on $[0,T] \times [0, L^{(i,j)}]$ with $\lambda_\pm^{(i,j)}$, $\psi_\pm^{(i,j)}$, $f_\pm^{(i,j)}$ and $g_\pm^{(i,j)}$ as in (5.6)–(5.9). Then there exist C^1-functions $\Omega^{(i)}$ and $\Xi^{(i,j)}$ with (2.14) and (2.15) ($i \in I$, $j \in I_i \backslash \{N_i\}$) such that in terms of $u_\pm^{(i,j)}$ the conditions (5.10), (5.11), (5.13), (5.14) can be written as (2.12), (2.13).*

Let a real number $\varepsilon_0 > 0$ be given. Define $\mu^{(i,j)}$, $h_\pm^{(i,j)}$, $U_\pm^{(i,j)}$ and $V_\pm^{(i,j)}$ as in (3.1), (3.2), (3.8), (3.9) and assume that (3.10) holds. Choose constants $k^{(i,j)} \in (-1,1)$ and $A_\pm^{(i,j)} > 0$ such that (3.13)–(3.16) are satisfied with $\alpha^{(i,j)}$, $\beta^{(i,j)}$ as in (3.11), (3.12). Assume that (3.17) holds. Choose $\varepsilon_1 \in (0, \varepsilon_0]$ that satisfies (3.19) *and (3.20) with* $\kappa_f^{(i,j)} = \frac{1}{2}$, $\kappa_g^{(i,j)} = \frac{\theta}{4}$ *and* $\gamma^{(i,j)}$ *as in (3.18). Define* $\eta > 0$ *as in (3.21).*

Then there exists $\varepsilon_2 \in (0, \varepsilon_1]$ *such that the following statements hold true: Assume that for* $u_\pm^{(i,j)}$ *we have initial conditions (3.7) with* C^1*-functions* $\phi_\pm^{(i,j)}$ *that satisfy (3.22) and the* C^1*-compatibility conditions at the nodes of the network. If we apply the feedback controls (3.6) at the nodes* i, j $(i \in I,\ j \in I_i)$*, then there exists a unique* C^1*-solution* $(u_+^{(i,j)}, u_-^{(i,j)})$ *on* $[0, L^{(i,j)}] \times [0, T]$ *that satisfies (3.23) and such that the state* $(\rho^{(i,j)}, q^{(i,j)})$ *as in (5.4), (5.5) is subcritical. For this solution the network Lyapunov function* $E_\omega(t)$ *as in (3.5) with* $E_\pm^{(i,j)}(t)$ *from (3.3), (3.4) satisfies* $(t \in [0, T])$

$$E_\omega(t) \leq E_\omega(0) \exp(-\eta t). \tag{5.15}$$

Remark 5.2. Due to (5.4) and (5.5), the nonstationary density $\rho^{(i,j)}$ and mass flux $q^{(i,j)}$ tend to the stationary state $\bar{\rho}^{(i,j)}$ and $\bar{q}^{(i,j)}$ if $u_\pm^{(i,j)}$ tend to zero. Hence, Corollary 5.1 presents a method to stabilize the gas flow locally around a given stationary state $(\bar{\rho}^{(i,j)}, \bar{q}^{(i,j)})$ on a fan-shaped pipe network.

For $i \in I$, $j \in I_i \backslash \{N_i\}$, at the end $x = L^{(i,j)}$ of pipe i, j the feedback control (3.6) can be maintained by the compressor connecting pipe i, j and pipe $i, j+1$. For $i \in I$, $j = N_i$, at the end $x = L^{(i,N_i)}$ of pipe i, N_i the feedback control can be maintained by the gas producer or consumer. In (3.6) the feedback controls are given in terms of the characteristic variables $u_\pm^{(i,j)}$. For a discussion how the feedback laws can be written in terms of the physical variables $\rho^{(i,j)}$ and $q^{(i,j)}$ see [9, 11, 20].

Acknowledgements This work has been supported by DFG GU376/7-1 and HE5386/8-1.

References

1. M.K. Banda, M. Herty, Numerical discretization of stabilization problems with boundary controls for systems of hyperbolic conservation laws. Math. Control Relat. Fields **3**, 121–142 (2013)
2. M.K. Banda, M. Herty, A. Klar, Coupling conditions for gas networks governed by the isothermal Euler equations. Netw. Heterog. Media **1**, 295–314 (2006)
3. M.K. Banda, M. Herty, A. Klar, Gas flow in pipeline networks. Netw. Heterog. Media **1**, 41–56 (2006)
4. G. Bastin, J.-M. Coron, B. d'Andréa-Novel, On Lyapunov stability of linearised Saint-Venant equations for a sloping channel. Netw. Heterog. Media **4**, 177–187 (2009)
5. R.M. Colombo, G. Guerra, M. Herty, V. Schleper, Optimal control in networks of pipes and canals. SIAM J. Control Optim. **48**, 2032–2050 (2009)
6. J.-M. Coron, *Control and Nonlinearity*. Mathematical Surveys Monographs, vol. 136 (AMS, Providence, 2007)
7. J.-M. Coron, B. d'Andréa-Novel, G. Bastin, A strict Lyapunov function for boundary control of hyperbolic systems of conservation laws. IEEE Trans. Autom. Control **52**, 2–11 (2007)

8. R. Datko, J. Lagnese, M.P. Polis, An example on the effect of time delays in boundary feedback stabilization of wave equations. SIAM J. Control Optim. **24**, 152–156 (1986)
9. M. Dick, Stabilization of the gas flow in networks: boundary feedback stabilization of quasilinear hyperbolic systems on networks. PhD thesis, University of Erlangen-Nuremberg, 2012
10. M. Dick, M. Gugat, M. Herty, S. Steffensen, On the relaxation approximation of boundary control of the isothermal Euler equations. Int. J. Control (2012), 1766–1778 85
11. M. Dick, M. Gugat, G. Leugering, Classical solutions and feedback stabilization for the gas flow in a sequence of pipes. Netw. Heterog. Media **5**, 691–709 (2010)
12. M. Dick, M. Gugat, G. Leugering, A strict H^1-Lyapunov function and feedback stabilization for the isothermal Euler equations with friction. Numer. Algebra Control Optim. **1**, 225–244 (2011)
13. M. Dick, M. Gugat, G. Leugering, Feedback stabilization of quasilinear hyperbolic systems with varying delays, in *Methods and Models in Automation and Robotics (MMAR)*, ed. by Z. Emirsajlow (IEEE Xplore, Piscataway, 2012), pp. 125–130
14. M. Garavello, B. Piccoli, *Traffic Flow on Networks: Conservation Laws Models*. AIMS Series on Applied Mathematics, vol. 1 (AIMS, Springfield, 2006)
15. M. Gugat, Optimal nodal control of networked hyperbolic systems: evaluation of derivatives. Adv. Model. Optim. **7**, 9–37 (2005)
16. M. Gugat, Boundary feedback stabilization by time delay for one-dimensional wave equations. IMA J. Math. Control Inf. **27**, 189–203 (2010)
17. M. Gugat, M. Dick, Time-delayed boundary feedback stabilization of the isothermal Euler equations with friction. Math. Control Relat. Fields **1**, 469–491 (2011)
18. M. Gugat, M. Dick, G. Leugering, Gas flow in fan-shaped networks: classical solutions and feedback stabilization. SIAM J. Control Optim. **49**, 2101–2117 (2011)
19. M. Gugat, M. Dick, G. Leugering, Stabilization of the gas flow in star-shaped networks by feedback controls with varying delay, in *System Modeling and Optimization*, ed. by D. Hömberg, F. Tröltzsch. IFIP AICT, vol. 391 (Springer, Berlin, 2013), pp. 255–265
20. M. Gugat, M. Herty, Existence of classical solutions and feedback stabilization for the flow in gas networks. ESAIM Control Optim. Calc. Var. **17**, 28–51 (2011)
21. M. Gugat, M. Herty, A. Klar, G. Leugering, V. Schleper, Well-posedness of networked hyperbolic systems of balance laws, in *Constrained Optimization and Optimal Control for Partial Differential Equations*, ed. by G. Leugering, S. Engell et al. ISNM, vol. 160 (Birkhäuser, Basel, 2012), pp. 123–146
22. M. Gugat, M. Herty, V. Schleper, Flow control in gas networks: exact controllability to a given demand. Math. Methods Appl. Sci. **34**, 745–757 (2011)
23. M. Gugat, G. Leugering, S. Tamasoiu, K. Wang, H^2-stabilization of the isothermal Euler equations: a Lyapunov function approach. Chin. Ann. Math. Ser. B. **33**, 479–500 (2012)
24. M. Gugat, M. Sigalotti, Stars of vibrating strings: switching boundary feedback stabilization. Netw. Heterog. Media **5**, 299–314 (2010)
25. M. Gugat, M. Tucsnak, An example for the switching delay feedback stabilization of an infinite dimensional system: the boundary stabilization of a string. Syst. Control Lett. **60**, 226–233 (2011)
26. M. Herty, Modeling, simulation and optimization of gas networks with compressors. Netw. Heterog. Media **2**, 81–97 (2007)
27. M. Herty, J. Mohring, V. Sachers, A new model for gas flow in pipe networks. Math. Methods Appl. Sci. **33**, 845–855 (2010)
28. M. Herty, L. Pareschi, S. Steffensen, Implicit-Explicit Runge-Kutta schemes for numerical discretization of optimal control problems. SIAM J. Numer. Anal. **51**, 1875–1899 (2013)
29. M. Herty, V. Sachers, Adjoint calculus for optimization of gas networks. Netw. Heterog. Media **2**, 733–750 (2007)
30. M. Herty, V. Schleper, Time discretizations for the numerical optimisation of hyperbolic problems. Appl. Math. Comput. **218**, 183–194 (2011)

31. G. Leugering, E.J.P.G. Schmidt, On the modelling and stabilization of flows in networks of open canals. SIAM J. Control Optim. **41**, 164–180 (2002)
32. T. Li, B. Rao, Z. Wang, Exact boundary controllability and observability for first order quasilinear hyperbolic systems with a kind of nonlocal boundary conditions. Discret. Contin. Dyn. Syst. **28**, 243–257 (2010)
33. A. Marigo, Entropic solutions for irrigation networks. SIAM J. Appl. Math. **70**, 1711–1735 (2009/2010)
34. S. Nicaise, J. Valein, E. Fridman, Stability of the heat and of the wave equations with boundary time-varying delays. Discret. Contin. Dyn. Syst. Ser. S **2**, 559–581 (2009)
35. A. Osiadacz, *Simulation and Analysis of Gas Networks* (Gulf Publishing Company, Houston, 1987)
36. A. Osiadacz, M. Chaczykowski, Comparison of isothermal and non-isothermal pipeline gas flow models. Chem. Eng. J. **81**, 41–51 (2001)
37. C. Prieur, F. Mazenc, ISS-Lyapunov functions for time-varying hyperbolic systems of balance laws. Math. Control Signals Syst. **24**, 111–134 (2012)
38. V. Schleper, Modeling, analysis and optimal control of gas pipeline networks. PhD thesis, TU Kaiserslautern, Verlag Dr. Hut, Munich, 2010
39. S. Steffensen, M. Herty, L. Pareschi, Numerical methods for the optimal control of scalar conservation laws, in *System Modeling and Optimization*, ed. by D. Hömberg, F. Tröltzsch. IFIP AICT, vol. 391 (Springer, Berlin, 2013), pp. 136–144
40. M.C. Steinbach, On PDE solution in transient optimization of gas networks. J. Comput. Appl. Math. **203**, 345–361 (2007)
41. J. Valein, E. Zuazua, Stabilization of the wave equation on 1-D networks. SIAM J. Control Optim. **48**, 2771–2797 (2009)
42. J.-M. Wang, B.-Z. Guo, M. Krstic, Wave equation stabilization by delays equal to even multiples of the wave propagation time. SIAM J. Control Optim. **49**, 517–554 (2011)
43. K. Wang, Exact boundary controllability for a kind of second-order quasilinear hyperbolic systems. Chin. Ann. Math. Ser. B **32**, 803–822 (2011)
44. K. Wang, Exact boundary controllability of nodal profile for 1-D quasilinear wave equations. Front. Math. China **6**, 545–555 (2011)
45. K. Wang, Global exact boundary controllability for 1-D quasilinear wave equations. Math. Methods Appl. Sci. **34**, 315–324 (2011)
46. Z. Wang, Exact controllability for nonautonomous first order quasilinear hyperbolic systems. Chin. Ann. Math. Ser. B **27**, 643–656 (2006)

Optimal Control of Surface Acoustic Wave Actuated Sorting of Biological Cells

Thomas Franke, Ronald H.W. Hoppe, Christopher Linsenmann, Lothar Schmid, and Achim Wixforth

Abstract The sorting of biological cells using biological micro-electro-mechanical systems (BioMEMS) is of utmost importance in various biomedical applications. Here, we consider a new type of devices featuring surface acoustic wave (SAW) actuated cell sorting in microfluidic separation channels. The SAWs are generated by an interdigital transducer (IDT) and manipulate the fluid flow such that cells of different type leave the channel through designated outflow boundaries. The operation of the device can be formulated as an optimal control problem where the objective functional is of tracking type, the state equations describe the fluid-structure interaction between the carrier fluid and the cells, and the control is the electric power applied to the IDT.

Keywords Optimal control • Biological cell sorting • Surface acoustic waves • Finite element immersed boundary method

Mathematics Subject Classification (2010). Primary 65K10; Secondary 49M05, 74F10.

The authors acknowledge support by the German National Science Foundation DFG within the DFG Priority Program SPP 1253 'Optimization with Partial Differential Equations'. The second author also acknowledges partial support by the National Science Foundation NSF (DMS-0914788 and DMS-1115658), the German Federal Ministry for Education and Research BMBF within the collaborative research projects 'FROPT' and 'MeFreSim', and the European Science Foundation ESF within the program 'OPTPDE'.

T. Franke • L. Schmid • A. Wixforth
Institute of Physics, Universität Augsburg, D-86159 Augsburg, Germany
e-mail: franketh@physik.uni-augsburg.de; lothar.schmid@physik.uni-augsburg.de; achim.wixforth@physik.uni-augsburg.de

R.H.W. Hoppe
Institute of Mathematics, Universität Augsburg, D-86159 Augsburg, Germany

Department of Mathematics, University of Houston, Houston, TX 77204-3008, USA
e-mail: hoppe@math.uni-augsburg.de; rohop@math.uh.edu

C. Linsenmann (✉)
Institute of Mathematics, Universität Augsburg, D-86159 Augsburg, Germany
e-mail: christopher.linsenmann@math.uni-augsburg.de

G. Leugering et al. (eds.), *Trends in PDE Constrained Optimization*, International Series of Numerical Mathematics 165, DOI 10.1007/978-3-319-05083-6_32

1 Introduction

We consider the optimal control of surface acoustic wave (SAW) actuated high throughput sorting of biological cells in microfluidic channels which has significant applications in basic cell biology, cancer research, clinical diagnostics, drug design in pharmacology, tissue engineering in reproductive medicine, and transplantation immunology [3, 4, 9, 13, 14].

According to [5], the experimental setup consists of a separation channel with three inlets and two outlets. The cells are injected through the middle inlet on the left and can be focused by the inflows through the other two inlets. SAWs are generated by an Interdigital Transducer (IDT) close to the lateral wall. The IDT features fingers substantially parallel to one another. A static electric field is applied to generate a strain which varies across the aperture of the IDT. The electric field is either perpendicular or parallel to the fingers and created by applying an AC voltage between two correspondingly positioned conductors. If the IDT is active, the SAWs enter the fluid filled channel and lead to a distortion of the fluid flow. Let us assume that we have cells of type A and B such that cells of type A should leave the channel through the lower outlet, whereas cells of type B are supposed to leave the channel through the upper outlet. Cells of different type can be distinguished by fluorescence. Without SAW actuation, the inflow velocities are tuned in such a way that a cell of type A leaves through the lower outlet. However, if a cell of type B is detected, the IDT is switched on and the flow is manipulated such that the cell leaves through the upper outlet (cf. Fig. 1). In an optimal control setting, the objective is

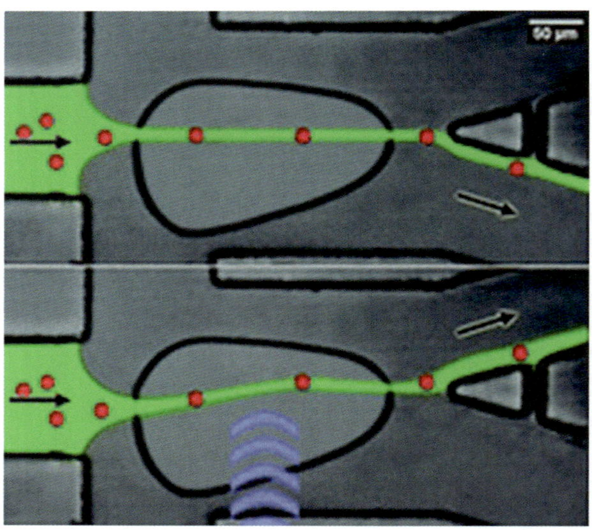

Fig. 1 Surface acoustic wave actuated cell sorting (SAWACS) in a microfluidic channel: without SAW actuation (*top*) and with SAW actuation (*bottom*) (Taken from [7])

to achieve the sorting as described above, the state equations are given by the fluid-structure interaction between the carrier fluid and the cells, and the control is the time-dependent power applied to the IDT.

For the mathematical modeling and numerical simulation of the fluid-structure interaction between the carrier fluid and the cells we will use the finite element immersed boundary (FE-IB) method [1, 2, 6] which is the finite element version of the classical immersed boundary (IB) method originally developed by Peskin (cf., e.g., [11, 12]). The FE-IB method relies on the variational formulation of a coupled system of partial differential equations consisting of the incompressible Navier-Stokes equations and the equations of motion of the boundaries of the immersed cells. As far as the spatial discretization is concerned, we use Taylor-Hood P2/P1 elements for the Navier-Stokes equations and periodic cubic splines for the equations of motion of the immersed boundaries. The discretization in time is taken care of by the backward Euler scheme for the semi-discretized Navier-Stokes equations and the forward Euler scheme for the semi-discretized equations of motion. This results in a semi-implicit scheme (Backward Euler/Forward Euler FE-IB method) which has to satisfy a CFL-type condition for stability reasons. We consider a control constrained optimal control problem for the fully discretized FE-IB method featuring an objective functional of tracking type where we prescribe desired positions of the immersed cells. Based on the necessary optimality conditions, the optimal control problem is solved by a projected gradient method with Armijo line search. Numerical results illustrate the performance of the suggested optimal control approach.

2 The Finite Element Immersed Boundary Method

The IB method comprises three groups of equations:

- The Navier-Stokes equations describing the motion of the incompressible viscous carrier fluid,
- The material elasticity equations responsible for the total elastic energy and the resulting forces exerted by the immersed cells,
- The interaction equations translating Eulerian into Lagrangian quantities and vice versa.

We denote by $\Omega \subset \mathbb{R}^2$ the Eulerian domain representing the separation channel wit boundary $\Gamma = \overline{\Gamma}_D \cup \overline{\Gamma}_N, \Gamma_D \cap \Gamma_N = \emptyset$, and by $\mathbf{v}(x,t), p(x,t)$ the velocity and the pressure of the carrier fluid in $(x,t) \in \overline{\Omega} \times [0,T], T > 0$. We further refer to $\Lambda = [0, L] \subset \mathbb{R}$ as the Lagrangian domain such that the vector valued function $\mathbf{X}(\lambda, t), \lambda \in \Lambda$, represents the closed, non self-intersecting boundary of an immersed cell at time $t \in [0, T], T > 0$.

The classical formulation of the IB equations then reads as follows: Find a triple $(\mathbf{v}, p, \mathbf{X})$ such that the incompressible Navier-Stokes equations

$$\rho\left(\frac{\partial \mathbf{v}}{\partial t}+(\mathbf{v}\cdot\nabla)\mathbf{v}\right)-2\,\eta\,\nabla\cdot\mathbf{D}(\mathbf{v})+\nabla p=\mathbf{f}_E \quad \text{in } \Omega\times(0,T] \tag{2.1a}$$

$$\nabla\cdot\mathbf{v}=0 \quad \text{in } \Omega\times(0,T] \tag{2.1b}$$

$$\mathbf{v}=\mathbf{v}_{\mathrm{D}} \quad \text{on } \Gamma_{\mathrm{D}}\times(0,T] \tag{2.1c}$$

$$(-p\,\mathbf{I}+2\,\eta\,\mathbf{D}(\mathbf{v}))\,\boldsymbol{\nu}=\mathbf{0} \quad \text{on } \Gamma_{\mathrm{N}}\times(0,T] \tag{2.1d}$$

$$\mathbf{v}(\,\cdot\,,0)=\mathbf{v}_0 \quad \text{in } \Omega \tag{2.1e}$$

are satisfied. Here, ρ and η are the density and viscosity of the carrier fluid, $\mathbf{D}(\mathbf{v})$ stands for the rate of deformation tensor $\mathbf{D}(\mathbf{v}) = (\nabla\mathbf{v} + (\nabla\mathbf{v})^T)/2$, $\mathbf{f}_E$ is a source term that will be specified in (2.3a) below, $\mathbf{v}_{\mathrm{D}}$ is a prescribed velocity, $\boldsymbol{\nu}$ denotes the exterior unit normal vector on the Neumann boundary Γ_{N}, and $\mathbf{v}_0$ refers to the initial velocity. The Navier-Stokes equations are coupled with the equations of motion of the immersed boundary

$$\frac{\partial \mathbf{X}}{\partial t}(\lambda,t)=\mathbf{v}(\mathbf{X}(\lambda,t),t)=\int_\Omega \mathbf{v}(\mathbf{x},t)\cdot\boldsymbol{\delta}(\mathbf{X}(\lambda,t)-\mathbf{x})\,\mathrm{d}\mathbf{x}, \tag{2.2a}$$

$$\mathbf{X}(\lambda,0)=\mathbf{X}_0(\lambda), \tag{2.2b}$$

where $\boldsymbol{\delta}$ stands for the Dirac delta function and $\mathbf{X}_0$ is the initial configuration of the immersed boundary. The source term $\mathbf{f}_E$ in (2.1a) is a global force density according to

$$\mathbf{f}_E(\mathbf{x},t)=\int_\Lambda \mathbf{F}_L(\lambda,t)\cdot\boldsymbol{\delta}(\mathbf{X}(\lambda,t)-\mathbf{x})\,\mathrm{d}\lambda, \tag{2.3a}$$

$$\mathbf{F}_L(\lambda,t)=-\,E'(\mathbf{X}(\,\cdot\,,t))(\lambda), \tag{2.3b}$$

where E' is the variational derivative of the elastic energy of the immersed boundary as given by

$$E(t):=E(\mathbf{X}(\,\cdot\,,t)):=\int_\Lambda \mathcal{E}^{\mathrm{e}}\Big(\frac{\partial\mathbf{X}(\lambda,t)}{\partial\lambda}\Big)\,\mathrm{d}\lambda+\int_\Lambda \mathcal{E}^{\mathrm{b}}\Big(\frac{\partial^2\mathbf{X}(\lambda,t)}{\partial\lambda^2}\Big)\,\mathrm{d}\lambda. \tag{2.4a}$$

Here, $\mathcal{E}^{\mathrm{e}}$ and $\mathcal{E}^{\mathrm{b}}$ stand for the local energy densities

$$\mathcal{E}^{\mathrm{e}}\Big(\frac{\partial\mathbf{X}(\lambda,t)}{\partial\lambda}\Big)=\frac{\kappa_{\mathrm{e}}}{2}\Big(\Big|\frac{\partial\mathbf{X}}{\partial\lambda}(\lambda,t)\Big|^2-1\Big),$$

$$\mathcal{E}^{\mathrm{b}}\Big(\frac{\partial^2\mathbf{X}(\lambda,t)}{\partial\lambda^2}\Big)=\frac{\kappa_{\mathrm{b}}}{2}\,\Big|\frac{\partial^2\mathbf{X}}{\partial\lambda^2}(\lambda,t)\Big|^2,$$

with $\kappa_{\mathrm{e}} > 0$ and $\kappa_{\mathrm{b}} > 0$ denoting the elasticity coefficients for elongation-compression and bending.

The FE-IB method relies on the variational formulation of the coupled system. We introduce the function spaces

$$
\begin{aligned}
\mathbf{V}(0,T) &:= \mathbf{H}^1((0,T),\mathbf{H}^{-1}(\Omega)) \cap \mathbf{L}^2((0,T),\mathbf{H}^1(\Omega)),\\
\mathbf{W}(0,T) &:= \{\mathbf{v} \in \mathbf{V}(0,T) \mid \mathbf{v}|_{\Gamma_D\times(0,T)} = \mathbf{v}_D\},\\
Q(0,T) &:= L^2((0,T),L^2(\Omega)),
\end{aligned}
$$

and

$$
\begin{aligned}
\mathbf{X}(0,T) &:= \mathbf{H}^1((0,T),\mathbf{L}^2(\Lambda)) \cap \mathbf{L}^2((0,T),\mathbf{H}^3_{\mathrm{per}}(\Lambda)),\\
\mathbf{H}^3_{\mathrm{per}}(\Lambda) &:= \{\mathbf{Y} \in \mathbf{H}^3(\Lambda) \mid \partial^k\mathbf{Y}(0)/\partial\lambda^k = \partial^k\mathbf{Y}(L)/\partial\lambda^k, k = 0,1,2\}.
\end{aligned}
$$

The FE-IB method amounts to the computation of a triple

$$
(\mathbf{v}, p, \mathbf{X}) \in \mathbf{W}(0,T) \times Q(0,T) \times \mathbf{X}(0,T)
$$

such that for almost all $t \in [0,T]$ and all test functions $(\mathbf{w}, q, \mathbf{Y}) \in \mathbf{H}^1_{\Gamma_D,0}(\Omega) \times L^2(\Omega) \times \mathbf{H}^3_{\mathrm{per}}(\Lambda)$ it holds

$$
\left\langle \frac{\partial \mathbf{v}}{\partial t}, \mathbf{w} \right\rangle_{\mathbf{H}^{-1},\mathbf{H}^1_{\Gamma_D,0}} + a(\mathbf{v},\mathbf{w}) - b(\mathbf{w},p) = \ell(\mathbf{w}), \tag{2.5a}
$$

$$
b(\mathbf{v},q) = 0 \tag{2.5b}
$$

$$
\mathbf{v}(\cdot,0) = \mathbf{v}_0, \tag{2.5c}
$$

$$
\left(\frac{\partial \mathbf{X}}{\partial t}, \mathbf{Y}\right)_{0,\Lambda} - \int_\Lambda \mathbf{v}(\mathbf{X}(\lambda,t),t)\cdot\mathbf{Y}(\lambda)\, \mathrm{d}\lambda = 0, \tag{2.5d}
$$

$$
\mathbf{X}(\cdot,0) = \mathbf{X}_0, \tag{2.5e}
$$

where $\langle\cdot,\cdot\rangle_{\mathbf{H}^{-1},\mathbf{H}^1_0}$ stands for the dual pairing between $\mathbf{H}^1_0(\Omega)$ and $\mathbf{H}^{-1}_{\Gamma_D,0}(\Omega)$ and $a(\cdot,\cdot), b(\cdot,\cdot)$, as well as the functional $\ell(\cdot)$ are given by

$$
a(\mathbf{v},\mathbf{w}) := (\rho(\mathbf{v}\cdot\nabla)\mathbf{v},\mathbf{w})_{0,\Omega} + (\eta\nabla\mathbf{v},\nabla\mathbf{w})_{0,\Omega} \tag{2.6a}
$$

$$
b(p,\mathbf{v}) := (p,\nabla\cdot\mathbf{v})_{0,\Omega} \quad , \quad \ell(\mathbf{w}) := \langle \mathbf{f}_E, \mathbf{w}\rangle_{\mathbf{H}^{-1},\mathbf{H}^1_0}. \tag{2.6b}
$$

For the numerical solution of (2.5) we use Taylor-Hood P2/P1 elements for the spatial discretization of (2.5a)–(2.5c) with respect to a quasi-uniform simplicial triangulation $\mathcal{T}_h(\Omega)$ of Ω that aligns with the partition of Γ and periodic cubic splines for the spatial discretization of (2.5d), (2.5e) with respect to an equidistant partition

$$
\mathcal{T}_{\Delta\lambda}(\Lambda) := \{0 = \lambda_0 < \lambda_1 < \cdots < \lambda_R = L\}
$$

of Λ into subintervals $\Lambda_r := [\lambda_{r-1}, \lambda_r]$, $1 \le r \le R$, of length $\Delta\lambda = L/R$. We note that the discrete immersed cell occupies subdomains $B_{\Delta\lambda,t} \subset \Omega$ with boundaries $\partial B_{\Delta\lambda,t}$ that are $\mathbf{C}^2$ curves described by the periodic cubic spline.

We introduce the finite element spaces

$$\begin{aligned} \mathbf{V}_h &:= \{\mathbf{v} \in \mathbf{C}^0(\Omega) \mid \mathbf{v}|_T \in \mathbf{P}_2(T)\,,\ T \in \mathcal{T}_h(\Omega)\} \\ \mathbf{V}_{\Gamma_D,h} &:= \{\mathbf{v}_h \in \mathbf{V}_h \mid \mathbf{v}_h|_{\Gamma_D} = \mathbf{v}_{h,D}\} \\ \mathbf{V}_{0,h} &:= \{\mathbf{v}_h \in \mathbf{V}_h \mid \mathbf{v}_h|_{\Gamma_D} = \mathbf{0}\} \\ Q_h &:= \{q \in L^2(\Omega) \mid q|_T \in P_1(T)\,,\ T \in \mathcal{T}_h(\Omega)\}, \end{aligned}$$

where $\mathbf{v}_{h,D}$ is a piecewise polynomial approximation of $\mathbf{v}_D$, and

$$\begin{aligned} \mathbf{S}_{\Delta\lambda} := \{\mathbf{X}_{\Delta\lambda} \in \mathbf{C}^2(\Lambda) \mid \mathbf{X}_{\Delta\lambda}|_{\Lambda_r} \in \mathbf{P}_3(\Lambda_r)\,, 1 \le r \le R, \\ \mathrm{d}^k\mathbf{X}_{\Delta\lambda}/\mathrm{d}\lambda^k(\lambda_0) = \mathrm{d}^k\mathbf{X}_{\Delta\lambda}/\mathrm{d}\lambda^k(\lambda_R)\,,\ k = 0, 1, 2\}. \end{aligned}$$

The semi-discretization of (2.5) in space requires the computation of a triple

$$(\mathbf{v}_h, p_h, \mathbf{X}_{\Delta\lambda}) \in \mathbf{C}^1((0,T), \mathbf{V}_{\Gamma_D,h}) \times L^2((0,T), Q_h) \times \mathbf{C}^1((0,T), \mathbf{S}_{\Delta\lambda})$$

such that for all $t \in [0, T]$ and all test functions $\mathbf{w}_h \in \mathbf{V}_{0,h}, q_h \in Q_h$, and $\mathbf{Y}_{\Delta\lambda} \in \mathbf{S}_{\Delta\lambda}$ it holds

$$\Big(\frac{\partial \mathbf{v}_h}{\partial t}, \mathbf{w}_h\Big)_{0,\Omega} + a(\mathbf{v}_h, \mathbf{w}_h) - b(\mathbf{w}_h, p_h) = \ell(\mathbf{w}_h), \tag{2.7a}$$

$$b(\mathbf{v}_h, q_h) = 0 \tag{2.7b}$$

$$\mathbf{v}_h(\cdot\,, 0) = \Pi_h \mathbf{v}_0, \tag{2.7c}$$

$$\Big(\frac{\partial \mathbf{X}_{\Delta\lambda}}{\partial t}, \mathbf{Y}_{\Delta\lambda}\Big)_{0,\Lambda} - \int_\Lambda \mathbf{v}_h(\mathbf{X}_{\Delta\lambda}(\lambda, t), t) \cdot \mathbf{Y}_{\Delta\lambda}(\lambda)\, \mathrm{d}\lambda = 0, \tag{2.7d}$$

$$\mathbf{X}_{\Delta\lambda}(\cdot\,, 0) = \Pi_{\Delta\lambda}\mathbf{X}_0, \tag{2.7e}$$

where Π_h and $\Pi_{\Delta\lambda}$ are the L^2-projections onto $\mathbf{V}_h$ and $\mathbf{S}_{\Delta\lambda}$, respectively.

For the algebraic formulation of (2.7) we equip $\mathbf{V}_{0,h}$, Q_h, and $\mathbf{S}_{\Delta\lambda}$ with canonical bases $\{\boldsymbol{\phi}_i\}_{i=1}^{N_1}$, $\{\psi_i\}_{i=1}^{N_2}$, and $\{\mathbf{B}_i\}_{i=1}^{N_3}$. Accordingly, we write

$$\mathbf{v}_h = \sum_{i=1}^{N_1} v_i\, \boldsymbol{\phi}_i, \quad p_h = \sum_{i=1}^{N_2} p_i\, \psi_i, \quad \mathbf{X}_{\Delta\lambda} = \sum_{i=1}^{N_3} X_i\, \mathbf{B}_i.$$

Here, the $\mathbf{B}_i$ are the B-splines with respect to the partition $\mathcal{T}_{\Delta\lambda}(\Lambda)$ and $X_1, \ldots, X_{N_3}$ are the de Boor points. As an important assumption we state that the Lagrangian force density $\mathbf{F}_L$ gets discretized by means of $\{\mathbf{B}_i\}$ as well in order to gain a useful transpose property (see (2.8a) below). Furthermore, we denote by M_L and M_E the Lagrangian and the Eulerian mass matrix, respectively, by $C(v)$ the advection matrix, by A the stiffness matrix, by B the matrix associated with the divergence operator, and by $K(X) \in \mathbb{R}^{N_1 \times N_3}$ the matrix with components $\int_\Lambda \boldsymbol{\phi}_i(\mathbf{X}_{\Delta\lambda}(\lambda)) \cdot \mathbf{B}_j(\lambda)\, \mathrm{d}\lambda$. We assume that all (Eulerian) matrices and right-hand sides are manipulated appropriately in order to enforce the Dirichlet conditions from (2.1c). Then the algebraic formulation of (2.7) reads: Find $(v, p, X) : [0, T] \to \mathbb{R}^{N_1} \times \mathbb{R}^{N_2} \times \mathbb{R}^{N_3}$, such that for almost all $t \in [0, T]$

$$M_E \frac{\mathrm{d}v}{\mathrm{d}t}(t) + C(v(t))\, v(t) + A\, v(t) + B^\top p(t) = K(X(t))^\top F_L(X(t)) \tag{2.8a}$$

$$B\, v(t) = 0\,, \tag{2.8b}$$

$$\sum_{i=1}^{N_1} v_i(0)\, \boldsymbol{\phi}_i = \Pi_h \mathbf{v}_0, \tag{2.8c}$$

$$M_L \frac{\mathrm{d}X}{\mathrm{d}t}(t) = K(X(t))\, v(t), \tag{2.8d}$$

$$\sum_{i=1}^{N_3} X_i(0)\, \mathbf{B}_i = \Pi_{\Delta\lambda} \mathbf{X}_0. \tag{2.8e}$$

3 The Semi-implicit Backward Euler/Forward Euler FE-IB Method

For the discretization in time we first consider the Backward Euler/Forward Euler FE-IB method from [6] in the sense that we discretize the Navier-Stokes equations by the backward Euler method in time and the equation of motion of the immersed boundary by the forward Euler scheme. In particular, we consider an equidistant partition

$$\mathcal{T}_{\Delta t} := \{0 =: t_0 < t_1 < \cdots < t_{\mathcal{M}} := T\},\ \mathcal{M} \in \mathbb{N},$$

of the time interval $[0, T]$ into subintervals of length $\Delta t := T/\mathcal{M}$ and set

$$\mathbf{v}_h^{(m)} := \mathbf{v}_h(\cdot, t_m), \quad p_h^{(m)} := p_h(\cdot, t_m), \quad \mathbf{X}_{\Delta\lambda}^{(m)} := \mathbf{X}_{\Delta\lambda}(\cdot, t_m).$$

We refer to

$$\mathbf{D}_{\Delta t}^{+} \mathbf{v}_h^{(m)} := (\mathbf{v}_h^{(m+1)} - \mathbf{v}_h^{(m)})/\Delta t, \quad \mathbf{D}_{\Delta t}^{-} \mathbf{v}_h^{(m)} := (\mathbf{v}_h^{(m)} - \mathbf{v}_h^{(m-1)})/\Delta t$$

as the forward and backward difference operator. We further define the total discrete energy by means of

$$E_{\Delta\lambda}(t_m) := E^{\mathrm{e}}_{\Delta\lambda}(t_m) + E^{\mathrm{b}}_{\Delta\lambda}(t_m),$$

where the discrete elastic energy $E^{\mathrm{e}}_{\Delta\lambda}(t_m)$ and the discrete bending energy $E^{\mathrm{b}}_{\Delta\lambda}(t_m)$ are given by

$$\begin{aligned} E^{\mathrm{e}}_{\Delta\lambda}(t_m) &= \frac{\kappa_{\mathrm{e}}}{2}\int_{\Lambda}\Big(\Big|\frac{\partial \mathbf{X}^{(m)}_{\Delta\lambda}}{\partial\lambda}(\lambda)\Big|^2 - 1\Big)\,\mathrm{d}\lambda \\ E^{\mathrm{b}}_{\Delta\lambda}(t_m) &= \frac{\kappa_{\mathrm{b}}}{2}\sum_{r=1}^{R}\int_{\Lambda_r}\Big|\frac{\partial^2 \mathbf{X}^{(m)}_{\Delta\lambda}}{\partial\lambda^2}(\lambda)\Big|^2\,\mathrm{d}\lambda\,. \end{aligned}$$

Observing that $\partial^3\mathbf{X}^{(m)}_{\Delta\lambda}(\lambda)/\partial\lambda^3$ is constant on Λ_r, the discrete force density takes the form

$$\begin{aligned} (\mathbf{F}^{(m)}_{L,\Delta\lambda}, \mathbf{w}_h(\mathbf{X}^{(m)}_{\Delta\lambda}))_{0,\Lambda} &= -\kappa_{\mathrm{e}}\int_{\Lambda}\frac{\partial \mathbf{X}^{(m)}_{\Delta\lambda}(\lambda)}{\partial\lambda}\cdot\nabla\mathbf{w}_h(\mathbf{X}^{(m)}_{\Delta\lambda}(\lambda))\frac{\partial \mathbf{X}^{(m)}_{\Delta\lambda}}{\partial\lambda}\,\mathrm{d}\lambda \\ &\quad + \kappa_{\mathrm{b}}\sum_{r=1}^{R}\frac{\partial^3 \mathbf{X}^{(m)}_{\Delta\lambda}}{\partial\lambda^3}\Big|_{\Lambda_r}\cdot\int_{\Lambda_r}\nabla\mathbf{w}_h(\mathbf{X}^{(m)}_{\Delta\lambda}(\lambda))\frac{\partial \mathbf{X}^{(m)}_{\Delta\lambda}}{\partial\lambda}\,\mathrm{d}\lambda\,. \end{aligned} \tag{3.1}$$

The Backward Euler/Forward Euler FE-IB reads as follows:

Given $\mathbf{v}^{(0)}_h = \Pi_h\mathbf{v}_0$ and $\mathbf{X}_{0,\Delta\lambda} = \mathbf{X}^{(0)}_{\Delta\lambda} = \Pi_{\Delta\lambda}\mathbf{X}_0$, for $m = 0,\ldots,\mathcal{M}-1$ we perform the following two steps (cf. [6]):

Algorithm 3.1.

(i) Compute $(\mathbf{v}^{(m+1)}_h, p^{(m+1)}_h) \in \mathbf{V}_{h,\Gamma_{\mathrm{D}}}\times Q_h$ such that for all $\mathbf{w}_h \in \mathbf{V}_{h,0}$

$$(\rho\mathbf{D}^{+}_{\Delta t}\mathbf{v}^{(m)}_h, \mathbf{w}_h)_{0,\Omega} + a(\mathbf{v}^{(m+1)}_h, \mathbf{w}_h) - b(p^{(m+1)}_h, \mathbf{w}_h) = \ell^{(m)}_h(\mathbf{w}_h), \tag{3.2a}$$

$$b(w_h, \mathbf{v}^{(m+1)}_h) = 0, \tag{3.2b}$$

where $\ell^{(m)}_h(\mathbf{w}_h) := (\mathbf{F}_{L,\Delta\lambda}, \mathbf{w}_h(\mathbf{X}^{(m)}_{\Delta\lambda}))_{0,\Lambda}$ is given by (3.1).

(ii) Compute $\mathbf{X}^{(m+1)}_{\Delta\lambda} \in \mathbf{S}_{\Delta\lambda}$ according to

$$\mathbf{D}^{+}_{\Delta t}\mathbf{X}^{(m)}_{\Delta\lambda} = \mathbf{v}^{(m+1)}_h(\mathbf{X}^{(m)}_{\Delta\lambda}). \tag{3.3}$$

Referring to $\partial B^{(m)}_{\Delta\lambda}$ as the boundary of the immersed cell at time t_m which consists of C^2-segments $\partial B^{(m,r)}_{\Delta\lambda}$ connecting material points $\mathbf{X}^{(m)}_{\Delta\lambda}(\lambda_{r-1})$ and $\mathbf{X}^{(m)}_{\Delta,\lambda}(\lambda_r)$, $1 \le r \le R$, one can deduce the estimate

$$\|\nabla \mathbf{v}_h^{(m+1)}\|^2_{0,\partial B_{\Delta\lambda}^{(m)}} \leq C_{\text{cell}}\, h^{-1} \|\nabla \mathbf{v}_h^{(m+1)}\|^2_{0,\Omega} \tag{3.4}$$

with a positive constant C_{cell} depending on the triangulation $\mathcal{T}_h(\Omega)$ (see (3.8) in [7]). A stability analysis reveals that the Backward Euler/Forward Euler FE-IB requires the CFL-type condition (cf. Theorem 3.1 in [7])

$$\frac{\Delta t}{h} \leq \frac{\eta}{8\, C_{\text{cell}}\, (\kappa_e \Lambda_1 + \kappa_e \Lambda_2)}, \tag{3.5}$$

where Λ_1 and Λ_2 are given by

$$\Lambda_1 := \max_{0\leq m\leq M} \max_{\lambda\in\Lambda} \left| \frac{\partial \mathbf{X}_{\Delta\lambda}^{(m)}}{\partial\lambda} \right|, \quad \Lambda_2 := \max_{0\leq m\leq M} \max_{1\leq r\leq R} \left| \frac{\partial^3 \mathbf{X}_{\Delta\lambda}^{(m)}}{\partial\lambda^3}|_{\Lambda_r} \right|.$$

The CFL-condition (3.5) for the semi-implicit scheme means a restriction of the time-step size Δt in particular depending on the amount of deformation of the immersed membrane as reflected by the quantities Λ_1 and Λ_2. For problems characterized by large values of Λ_1 and Λ_2, the time increments need to be chosen very small, leading to a high computational effort. As a remedy, a fully implicit time-stepping scheme can be used based on the application of the backward Euler scheme in time for both the Navier-Stokes equations and the equation of motion of the immersed boundary. This Backward Euler/Backward Euler FE-IB method is unconditionally stable at the expense that at each time-step a nonlinear algebraic system has to be solved. We refer to [10] for details including a predictor-corrector continuation strategy featuring an adaptive choice of the time-step size.

4 Optimal Control of the Surface Acoustic Wave Actuated Cell Sorting

In this section, following the strategy 'discretize first, then optimize', we will formulate the optimal control problem for the surface acoustic wave actuated cell sorting. The objective is to steer the immersed cells to desired positions by controlling the electric power applied to the IDT. The semi-implicit Backward Euler/Forward Euler FE-IB method from Sect. 3 serves as the state constraints.
For $z := (v, p, X)$, consider the following optimal control problem

$$\begin{cases} \min\limits_{z\in Z, u\in U} & J(z,u) \\ \text{s.t.} & S(z) = b(u) \\ & u \in U_{\text{ad}} \end{cases} \tag{4.1}$$

The objective functional J is given by

$$J(z,u) := J(X_{[1]}, X_{[2]}) := \sum_{i=1}^{2} \frac{1}{2} \left\| \mathbf{X}_{[i],\Delta\lambda}^{(\mathcal{M}(i))} - \mathbf{X}_{[i],\Delta\lambda}^{\mathrm{des}} \right\|_{0,\Lambda}^{2}, \tag{4.2}$$

where $1 \leq i \leq 2$ are the cell indices of two different biological cells and the functions $\mathbf{X}_{[i],\Delta\lambda}^{\mathrm{des}} \in \mathbf{S}_{\Delta\lambda}(\Lambda)$ mark desired final positions close to the respective outflow boundaries. The time instants $t_{\mathcal{M}(i)}$ are chosen such that the $\mathbf{x}_1$-components of barycenters of the immersed cells $\mathbf{X}_{[i],\Delta\lambda}(\Lambda, t)$ and $\mathbf{X}_{[i],\Delta\lambda}^{\mathrm{des}}(\Lambda)$ coincide. The state operator S reads

$$S(z) := \begin{pmatrix} v_0 - v^{(0)} \\ (M_E + \Delta t\, A)\, v^{(m)} + \Delta t B^{\top} p^{(m)} - \Delta t\, f_E(X^{(m-1)}) - M_E\, v^{(m-1)} \\ B\, v^{(m)} \\ X_0 - X^{(0)} \\ M_L X^{(m)} - M_L X^{(m-1)} - \Delta t\, K(X^{(m-1)})\, v^{(m)} \end{pmatrix}$$

and

$$b(u) = (0, \Delta t\, g(u^{(m-1)}), 0, 0, 0)^{\top}, \quad 1 \leq m \leq \mathcal{M}.$$

The volume force term $g(u^{(m)}) \in \mathbb{R}^{N_1}$ comprises components

$$g(u^{(m)})_i := \int_{\Omega} \mathbf{f}_{\mathrm{vol}}(u^{(m)}) \cdot \boldsymbol{\phi}_i \, \mathrm{d}\mathbf{x}, \; 1 \leq i \leq N_1,$$

where the volume force density $\mathbf{f}_{\mathrm{vol}}$ generated by the IDT is given by

$$\mathbf{f}_{\mathrm{vol}}(u^{(m)})(\mathbf{x}) := \begin{cases} (0, \beta\, u^{(m)}\, \mathrm{e}^{(-(\mathbf{x}_2 - y_0)/d)}\, k(\mathbf{x}_1, x_0, D))^{\top}, & \mathbf{x} \in \overline{\omega} \\ \mathbf{0} & , \mathbf{x} \in \Omega \setminus \omega \end{cases}$$

$$k(x, x_0, D) = \frac{\sin^2(2\pi(x - x_0)/D)}{(2\pi(x - x_0)/D)^2}.$$

Here, $\omega \subset \Omega$ denotes the subdomain where the SAW is effective, β stands for a transmission coefficient, d for the decay length, $(x_0, y_0)^{\top}$ refers to the center position of the segment at the lower lateral boundary where the SAWs enter the domain, and $D/2$ is the half width of this segment (marked green in Fig. 2 below). The function k is known as a Kirchhoff function and describes the refraction pattern of the SAW intensity.

We define the set of admissible controls by

$$U_{\mathrm{ad}} := \{u \in U := \mathbb{R}^{\mathcal{M}} \mid u^{\min} \leq u^{(m)} \leq u^{\max}, \; 1 \leq m \leq \mathcal{M}\}, \tag{4.3}$$

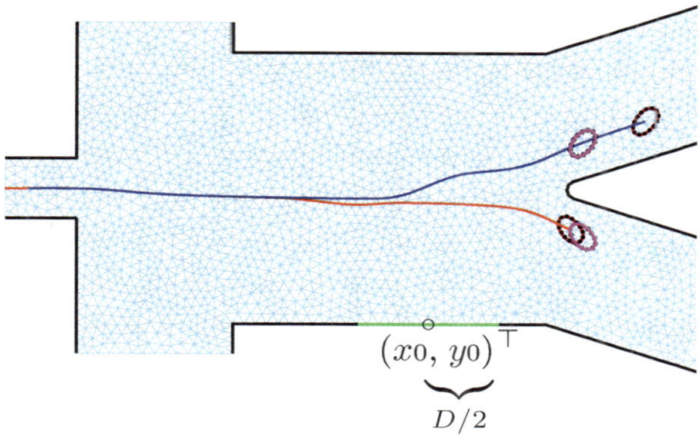

Fig. 2 Paths of two different cells under the influence of the computed optimal control. The desired positions are depicted in *magenta*

where the control $u^{(m)}$ is the power applied to the IDT at time t_m and $u^{\min}, u^{\max} \in \mathbb{R}^{\mathcal{M}}$ are given bounds. As numerical optimization scheme we use the well-known projected gradient with Armijo line search (see, e.g., [8]). To this end, we introduce the reduced objective functional

$$J_{\text{red}}(u) := J(z(u), u),$$

where $z(u) = (v(u), p(u), X(u))$ is the solution to $S(z) = b(u)$. Then problem (4.1) can be reformulated as the state-reduced optimal control problem

$$\begin{cases} \min\limits_{u\in\mathbb{R}^{\mathcal{M}}} J_{\text{red}}(u) \\ \quad \text{s.t. } u \in U_{\text{ad}} \end{cases} \tag{4.4}$$

being equivalent to (4.1). Problem (4.4) can be solved by the following scheme where $\Pi_{U_{\text{ad}}}$ denotes the projection operator onto the admissible set:

Algorithm 4.1.

(o) Let u_0 and a tolerance $\varepsilon > 0$ be given.
`for k = 0,1,2, ...`

- (i) Compute the descent direction $d_k = -\nabla J_{\text{red}}(u_k)$ via adjoint approach.
- (ii) If $\|\Pi_{U_{\text{ad}}}(u_k + d_k) - u_k\| < \varepsilon$, stop: $u^* := u_k$.
- (iii) Compute a step length α_k by Armijo line search.
- (iv) Update $u_{k+1} = u_k + \alpha_k\, d_k$, project it onto U_{ad} and go back to (i).

The computationally most challenging part is the evaluation of $\nabla J_{\text{red}}(u_k)$ which is taken care of by the adjoint approach:

For the optimization problem (4.1) we consider the Lagrangian

$$\mathcal{L}(z,u,\lambda) \;=\; J(z,u) \;+\; \lambda^\top (S(z) - b(u))\,,\quad \mathcal{L} : Z \times U \times Y \to \mathbb{R}\,.$$

The associated state equations and adjoint state equations are

$$0 = \nabla_\lambda \mathcal{L} \;=\; S(z(u)) - b(u) \tag{4.5a}$$

$$0 = \nabla_z \mathcal{L} \;=\; \nabla_z J(z(u),u) + S'(z(u))^\top \lambda(u)\,. \tag{4.5b}$$

Lemma 4.2. *Assume that* $S(z) = b(u)$ *has a unique solution* $z(u)$, $\forall u \in U$, *and that* $\lambda(u) \in Y$ *is the unique solution to* (4.5b). *Moreover assume that the mappings* $(z,u) \mapsto J(z,u)$, $z \mapsto S(z)$, $u \mapsto z(u)$, *and* $u \mapsto b(u)$ *are Fréchet-differentiable. Then there holds*

$$\nabla J_{red}(u) \;=\; \nabla_u \mathcal{L}(z(u),u,\lambda(u))\,. \tag{4.6}$$

Proof. By the chain rule we get $\nabla J_{\text{red}}(u) = \nabla z(u)\nabla_z J(z,u) + \nabla_u J(z,u)$, whence

$$\nabla J_{\text{red}}(u) - \nabla_u J(z(u),u) = -\nabla z(u)\, S'(z)^\top \lambda(u) = -\nabla b(u)\, \lambda(u).$$

□

In more detail, one has to perform the following steps to compute the reduced gradient $\nabla J_{\text{red}}(u_k)$ (for notational simplicity, only one cell is considered):

Algorithm 4.3.

(i) Compute the state $(v_k, p_k, X_k) := (v(u_k), p(u_k), X(u_k))$:
$v_k^{(0)} := v_0$, $X_k^{(0)} := X_0$ and for $1 \le m \le \mathcal{M}$

$$\begin{aligned}
(M_E + \Delta t\, A)\, v_k^{(m)} + \Delta t\, B^\top p_k^{(m)} &= \Delta t\,\big(f_E(X_k^{(m-1)}) + g(u_k^{(m-1)})\big) \\
&\quad + M_E\, v_k^{(m-1)} \\
B\, v_k^{(m)} &= 0 \\
M_L X_k^{(m)} &= M_L X_k^{(m-1)} + \Delta t\, K(X_k^{(m-1)})\, v_k^{(m)}\,.
\end{aligned}$$

(ii) Compute the adjoint state $(w_k, q_k, Y_k) := (w(u_k), q(u_k), Y(u_k))$ backward in time: $w_k^{(\mathcal{M})} := 0$, $Y_k^{(\mathcal{M})} := X^{\text{des}} - X^{(\mathcal{M})}$ and for $\mathcal{M} - 1 \ge m \ge 1$

$$\begin{aligned}
M_L Y_k^{(m)} \;=\; M_L Y_k^{(m+1)} &+ \Delta t\,\big[f_E'(X_k^{(m)})^\top w_k^{(m+1)} \\
&+ (K'(X_k^{(m)}) v_k^{(m+1)})^\top Y_k^{(m+1)}\big]\,.
\end{aligned}$$

$$\begin{aligned}(M_E + \Delta t\ A)\, w_k^{(m)} + \Delta t\ B^\top q_k^{(m)} &= \Delta t\ K(X_k^{(m-1)})^\top Y_k^{(m)} + M_E\, w_k^{(m+1)} \\ B\, w_k^{(m)} &= 0\end{aligned}$$

(iii) Set $\nabla J_{\text{red}}(u_k) := \partial J(z_k, u_k)/\partial u + \Delta t\ \sum_{m=1}^{\mathcal{M}-1} (w_k^{(m)})^\top \nabla^\top g(v_k^{(m-1)})$.

The derivatives showing up in the adjoint system represent the nontrivial terms of the adjoint operator $S'(z(u))^\top$ from (4.5b). Let us now state the optimality conditions associated with (4.1).

Theorem 4.4 (Necessary optimality conditions). *Assume the set* U_{ad} *is given by* (4.3) *and the assumptions from Lemma* 4.2 *are fulfilled. Then there exists an optimal solution* (z^*, u^*) *to* (4.1) *with associated Lagrange multiplier* λ^* *such that:* (z^*, u^*) *solves* (4.5a), λ^* *solves* (4.5b) *and*

$$\big(\nabla J_{red}(u^*)\big)_i \begin{cases} \leq 0\,, & u_i^* = u_i^{\max} \\ = 0\,, & u_i^{\min} < u_i^* < u_i^{\max} \\ \geq 0\,, & u_i^* = u_i^{\min} \end{cases} . \qquad (4.7)$$

Proof. In case of box constraints, the optimality condition for (4.4), namely $(\nabla J_{\text{red}}(u^*), u - u^*) \geq 0$, $\forall u \in U_{\text{ad}}$, can be characterized by (4.7). □

Condition (4.7) can be written in short form as $\Pi_{U_{\text{ad}}}(u^* - \nabla J_{\text{red}}(u^*)) = u^*$. This justifies the termination criterion from Algorithm 4.1, step (ii).

5 Numerical Results

As a numerical example, we consider the sorting scenario 'up – down', meaning that the first cell ($i = 1$) is supposed to take the upper outflow channel and the second cell ($i = 2$) the lower one.

The separation channel Ω is shown in Fig. 2 featuring three inflow boundaries at the left and two outflow boundaries at the right. The main part has a length of 300 μm and a width of 180 μm. The maximal inflow velocities $v_{\text{in}}^{(\text{left})}$, $v_{\text{in}}^{(\text{top})}$, and $v_{\text{in}}^{(\text{bottom})}$ have been chosen according to

$$v_{\text{in}}^{(\text{left})} = 10\ \text{mm/s}, \quad v_{\text{in}}^{(\text{top})} = 12.5\ \text{mm/s}, \quad v_{\text{in}}^{(\text{bottom})} = 10\ \text{mm/s},$$

guaranteeing that without SAW actuation a cell leaves the channel through the lower outflow boundary. As the density ρ and the dynamic viscosity η we have chosen

$$\rho = 1{,}000\ \text{kg/m}^3, \quad \eta = 7.0\ \text{mPa} \cdot \text{s}$$

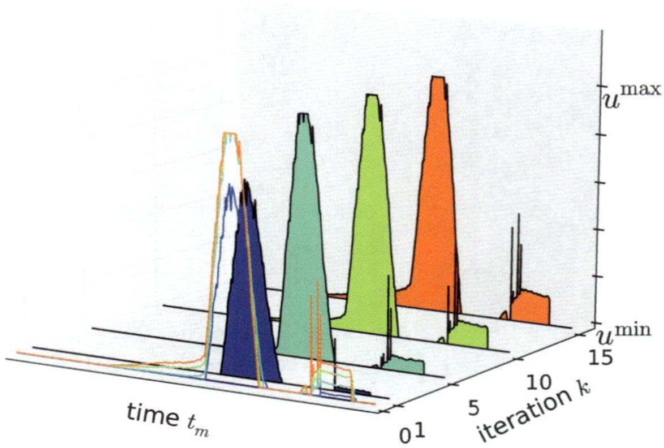

Fig. 3 Evolution of the controls u_k arising in the optimization algorithm

Table 1 Decrease of the reduced objective functional $J_{\text{red}}(u_k)$ as a function of the iteration step k of the optimization algorithm

Iteration k	0	1	5	10	15
$J_{\text{red}}(u_k)$	8.18e+01	2.53e+00	1.42e+00	1.09e+00	9.40e−01

both for the carrier fluid and the fluid enclosed by the membrane of the two cells. We note that in practice this can be achieved using density and viscosity matching by adding suitable chemicals to the carrier fluid. We have considered initially spherical cells of diameter 16 μm and moduli

$$\kappa_e = 5.0 \cdot 10^{-5}\ \text{N/m}, \quad \kappa_b = 1.0 \cdot 10^{-16}\ \text{Nm}.$$

The sorting task is complicated by setting the initial distance between the cells to 25 μm only. For the spatial discretization of the Navier-Stokes equations we have used a finite element mesh with mesh size $h = 7.5\,\mu\text{m}$, whereas for the spatial discretization of the equations of motion of the immersed boundaries we have used a partition of Λ with $\Delta\lambda = 3.6\,\mu\text{m}$. The time-step size Δt in the semi-implicit Backward Euler/Forward Euler FE-IB method has been chosen according to $\Delta t = 1/100$ ms making sure that the CFL-condition (3.5) is satisfied.

Figure 2 shows the computed paths of the cells in the separation channel along with their designated positions at final time, whereas Fig. 3 displays the computed controls of the projected gradient method.

Finally, Table 1 reflects the decrease of the reduced objective functional $J_{\text{red}}(u_k)$ as a function of the iteration step k of the optimization algorithm.

Conclusions

We have presented an optimal control approach to the sorting of different biological cells by surface acoustic wave (SAW) manipulated fluid flow in a microfluidic separation channel. The mathematical modeling and numerical simulation of the fluid-structure interaction has been taken care of by the finite element immersed boundary (FE-IB) method. The feasibility of the approach has been documented by numerical results.

References

1. D. Boffi, L. Gastaldi, A finite element approach for the immersed boundary method. Comput. Struct. **81**, 491–501 (2003)
2. D. Boffi, L. Gastaldi, L. Heltai, Numerical stability of the finite element immersed boundary method. Math. Models Methods Appl. Sci. **17**, 1479–1505 (2007)
3. J.L. Carey, J.P. McCoy, D.F. Keren, *Flow Cytometry in Clinical Diagnostics*, 4th edn. (ASCP Press, Chicago, 2007)
4. M. Eisenstein, Cell sorting: divide and conquer. Nature **441**, 1179–1185 (2006)
5. T. Franke, S. Braunmüller, L. Schmid, A. Wixforth, Surface acoustic wave actuated cell sorting (SAWACS). Lab Chip **10**, 789–794 (2010)
6. T. Franke, R.H.W. Hoppe, C. Linsenmann, L. Schmid, C. Willbold, Numerical simulation of the motion and deformation of red blood cells and vesicles in microfluidic flows. Comput. Vis. Sci. **14**, 167–180 (2011)
7. T. Franke, R.H.W. Hoppe, C. Linsenmann, K. Zeleke, Numerical simulation of surface acoustic wave actuated cell sorting. Cent. Eur. J. Math. **11**, 760–778 (2013)
8. C. Geiger, C. Kanzow, *Theorie und Numerik restringierter Optimierungsaufgaben* (Springer, Berlin, 2002)
9. T.S. Hawley, R.G. Hawley, *Flow Cytometry Protocols*, vol. 263, 2nd edn. (Humana Press, Totowa, 2004)
10. R.H.W. Hoppe, C. Linsenmann, An adaptive Newton continuation strategy for the fully implicit finite element immersed boundary method. J. Comput. Phys. **231**, 4676–4693 (2012)
11. C. Peskin, Numerical analysis of flood flow in the heart. J. Comput. Phys. **25**, 220–252 (1977)
12. C. Peskin, The immersed boundary method. Acta Numer. **11**, 479–517 (2002)
13. H.M. Shapiro, *Practical Flow Cytometry* (Wiley-Liss, New York, 2003)
14. L.A. Sklar, *Flow Cytometry for Biotechnology* (Oxford University Press, New York, 2005)

Real-Time PDE Constrained Optimal Control of a Periodic Multicomponent Separation Process

Malte Behrens, Hans Georg Bock, Sebastian Engell, Phawitphorn Khobkhun, and Andreas Potschka

Abstract We present a case study for a ternary separation process, for which structural and aleatoric model uncertainty must be taken into account. For the first time, we apply the control strategy of Modifier Adaptation to a PDE constrained optimization problem with challenging switched periodic boundary conditions in time. Numerically, real-time feasibility of the control scheme is possible by the use of a direct one-shot optimization method, whose efficiency is based on a two-grid Newton-Picard approach. We demonstrate real-time feasibility for a virtual plant with experimentally determined isotherm parameters under reasonable model-plant mismatch conditions. As a result, it is possible to drive the plant into its true optimum, i.e. to increase the productivity of the plant by 100 and 35 % in the two scenarios considered here.

Keywords Real-time control • PDE constrained optimization • Time-periodic

Mathematics Subject Classification (2010). 35Q93; 90C90; 92E20.

1 Introduction

Advanced model-based control strategies are widely used in the chemical industry for the efficient and safe production of oil-based bulk chemicals. This trend has not reached biotechnological processes nor the production of fine chemicals yet

M. Behrens • S. Engell • P. Khobkhun
Department of Biochemical and Chemical Engineering, TU Dortmund, Emil-Figge-Str. 70, 44227 Dortmund, Germany
e-mail: Malte.Behrens@bci.tu-dortmund.de; Sebastian.Engell@bci.tu-dortmund.de; Phawitphorn.Khobkhun@bci.tu-dortmund.de

H.G. Bock • A. Potschka (✉)
Interdisciplinary Center for Scientific Computing, Heidelberg University, Im Neuenheimer Feld 368, 69120 Heidelberg, Germany
e-mail: bock@iwr.uni-heidelberg.de; potschka@iwr.uni-heidelberg.de

G. Leugering et al. (eds.), *Trends in PDE Constrained Optimization*, International Series of Numerical Mathematics 165, DOI 10.1007/978-3-319-05083-6_33

due to two reasons: On the one hand, developing suitable models for small scale productions or complex biological systems may not pay off. On the other hand, the complexity of accurate models for those processes is high and their range of validity is often small. Additionally, the resulting mathematical problems require highly sophisticated algorithms to deal with the numerical challenges of large-scale nonlinear dynamic optimization problems with usually nonconvex feasible regions of small size.

In practice, the complexity of such models necessitates a trade-off between accuracy and computation time, especially when the models are applied within model based controllers. Thus, we need to address the issue of model uncertainty. One of the most successful and versatile approaches to optimization in the presence of uncertainty is feedback control (see [8]): On the basis of partial, but repeated measurements of state variables, we can estimate the system state and adapt model parameters. However, one needs to take special care that the optimization based upon this adapted model results in the optimum of the plant. The optimum of the model can be substantially suboptimal or infeasible for the true plant. A significant difference between the optimum of the updated model and the plant might occur if the mathematical structure of the model cannot describe the true behaviour.

In this article, we apply the method of Modifier Adaptation [16] for the first time to the case of PDE constrained optimization problems of a continuous process. In a nutshell, Modifier Adaptation is an iterative learning strategy to compensate structural model mismatches by fitting a Taylor expansion of an error model that augments the process model to physical measurements of the objective and constraint functions. We investigate a novel multicomponent separation process called *Multi-Column Solvent Gradient Purification (MCSGP)* [22], which is a highly complex process from the class of chromatographic separation processes used in the chemical industries. For the MCSGP process, it is possible to perform Modifier Adaptation based on desorbent consumption and purity measurements.

In order to apply the concept of Modifier Adaptation, we need to solve sequences of nonlinear, time-periodic, parabolic PDE constrained optimization subproblems. We demonstrate that a direct optimization method based on two-grid Newton-Picard inexact Sequential Quadratic Programming [19] can be employed to efficiently and reliably solve the resulting subproblems without the need to formulate adjoint equations and optimality conditions by hand. Finally, we present numerical results for several scenarios of an MCSGP process with experimental isotherm data.

2 Application Example

The MCSGP process is a continuous periodic adsorption process which is used to separate three or more components within a liquid mixture. We describe the basic principle of batch chromatography for a single column, its main drawback, and how the MCSGP overcomes this drawback by means of complex, periodically switched interconnections between several chromatographic columns.

2.1 The Chromatographic Principle

Chromatography is a tool to separate mixtures into their compounds at mild physical conditions. Chromatographic separations are based on different affinities of the single compounds of the (usually liquid) mixture to a second (mostly solid) stationary phase. The latter one is usually fixed in a column in form of a packing with porous particles. A desorbent, the so-called mobile phase, together with the stationary phase represent the thermodynamic two phase system in which the substances to be separated interact. To perform the separation the column is continuously flushed with desorbent and a certain amount of the mixture is injected for a defined period of time at the inflow port of the column. The specific interactions of the components of the mixture with the solid bed result in specific residence times and the compounds leave the column at different points in time. Compared to the initial composition of the mixture, the single components are now separated, but diluted in the mobile phase. This is the economic drawback of this method as a separation technique in chemicals production, because the final product usually should not include the mobile phase, so it has to be removed in a second step. Using as little mobile phase as possible is one economic goal of applying advanced control strategies to chromatographic separations. The operation mode described above, either collecting (preparative chromatography) or analyzing (analytic chromatography) the different fractions that are obtained at the end of the column, is called *batch* chromatography and is the most often used operation mode.

For large scale applications of chromatography and minimal product dilution and eluent consumption also continuous operating modes have been developed (see [21] for an overview). One strategy to overcome the batch characteristics is to connect different columns to a more complicated switched superstructure. For separations of two substances the so-called Simulated Moving Bed process realizes this strategy by implementing a quasi counter flow of the solid and mobile phase. Optimizing control of batch and SMB chromatography is discussed in [24]. The novel MCSGP process can be used for separations of more than two substances and is in the focus of this study.

2.2 Continuous MCSGP: The Chromatographic Column as a Building Block

In batch chromatography with three substances, the following five fractions will successively leave the column at the outlet port: Lightly adsorbing compound L,

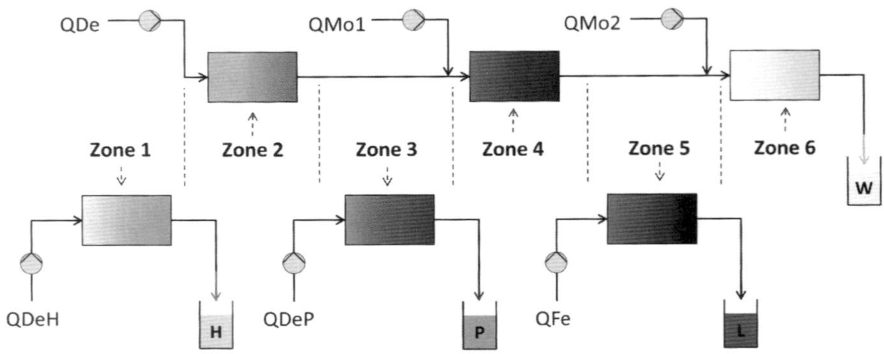

Fig. 1 Schematic of the MCSGP column interconnection. The columns in the *bottom row* are operated in batch mode (*stand alone* elution), while the columns in the *top row* are connected with two other columns (*counter current* mode)

product P contaminated with L, pure product P, product P contaminated with strongly adsorbing compound H, and finally pure H.

In the MCSGP process six columns, each corresponding to one zone, are interconnected according to Fig. 1. The columns connected in series perform a purification of the mixed fractions P/H and L/P so that the fractions L, P and H containing nearly pure substance leave the process. After each switching period the inlet and outlet ports of each column are switched in a way that simulates a movement of each column by one position to the left. Column one moves virtually to position six. In each zone – and this is one of the benefits compared to batch seperation – a different mobile phase composition can be applied. Thus, in each zone the adsorption behavior can be adapted corresponding to the separation task of that specific zone.

3 Problem Formulation

This section deals with the mathematical modeling of the MCSGP optimization problem based on the chromatographic column model and the MCSGP switching structure.

3.1 Chromatographic Column Model

For the mathematical model of the chromatographic columns, we assume that radial variations of the concentrations are negligible, thus giving rise to a system of 1D PDEs. Furthermore, we assume that the dynamics in the column are dominated by an advective term modeled assuming plug flow and a dispersive term following Fick's first law of diffusion. We assume a lumped film mass transfer between bulk and stationary phase. Because the adsorption dynamics take place on a much faster time scale than the other phenomena, we assume that the adsorption is always in equilibrium and can thus be described by algebraic isotherm equations.

Each column is of length L and represented by the computational domain $\Omega = (0, L)$. For the states we use a standard space for solutions of parabolic PDEs (see, e.g., [25])

$$W(0,T) = \{y \in L^2(0,T; H^1(\Omega)) \mid \tfrac{\partial y}{\partial t} \in L^2(0,T; (H^1(\Omega))^*)\},$$

which ensures existence of the appropriate traces in space and time. With the constants described in Table 1, the bulk phase concentrations $c_i \in W(0,T)$ and the stationary phase concentrations $q_i \in W(0,T)$ of the species $i = 1, \dots, n_{\text{species}}$ satisfy the coupled system of PDEs

$$\frac{\partial c_i}{\partial t} = -v\frac{\partial c_i}{\partial x} + \frac{vL}{\text{Pe}}\frac{\partial^2 c_i}{\partial x^2} - \frac{1-\epsilon}{\epsilon}\frac{\partial q_i}{\partial t}, \tag{1a}$$

$$\frac{\partial q_i}{\partial t} = k_{\text{eff},i}\frac{6}{d_{\text{p}}}\left(q_i^{\text{eq}}(c_1, \dots, c_{n_{\text{species}}}) - q_i\right), \tag{1b}$$

subject to the boundary conditions

$$\frac{\partial c_i}{\partial x}(t,0) = \frac{\text{Pe}}{L}\left[c_{\text{in},i}(t) - c(t,0)\right], \quad \frac{\partial c_i}{\partial x}(t,L) = 0. \tag{1c}$$

Table 1 Dynamic parameters of a chromatographic column

Symbol	Description	Value	Unit
A	Cross-sectional area	4.91	cm^2
$c_{\text{feed},i}(t)$	Feed concentration	Uncertain	g/cm^3
$c_{\text{in},i}(t)$	Inflow concentration	See Table 2	g/cm^3
ϵ	Void fraction	0.52	–
η	Dynamic viscosity	$1.2 \cdot 10^{-2}$	g/(cm s)
d_{p}	Particle diameter	0.0016	cm
L	Column length	25	cm
Pe	Péclet number	$\text{Pe} = \frac{L}{\epsilon d_{\text{p}}}\left(\frac{1}{5} + \frac{11}{1000}\text{Re}^{0.48}\right)$	–
Q_{in}	Inflow rate	see Table 2	cm^3/s
Q_{min}	Minimum flow rate	30	cm^3/s
Q_{max}	Maximum flow rate	500	cm^3/s
Re	Reynolds number	$\text{Re} = v d_{\text{p}} \rho \epsilon / \eta$	–
ρ	Fluid density	0.79	g/cm^3
v	Interstitial velocity	$v = Q_{\text{in}}/(\epsilon A)$	cm/s

Table 2 The controlled pump flow rates generate the inflow rates in the corresponding MCSGP zones with different inflow concentrations

Zone j	Pump rate	Inflow rate Q^j_{in}	Inflow concentration $c^j_{\text{in},i}$
1	Q_{DeH}	Q_{DeH}	0
2	Q_{De}	Q_{De}	0
3	Q_{DeP}	Q_{DeP}	0
4	Q_{Mo1}	$Q_{\text{De}} + Q_{\text{Mo1}}$	$c_i^2(t,L)Q^2_{\text{in}}/Q^4_{\text{in}}$
5	Q_{DeL}	Q_{DeL}	$c_{\text{feed},i}(t)$
6	Q_{Mo2}	$Q_{\text{De}} + Q_{\text{Mo1}} + Q_{\text{Mo2}}$	$c_i^4(t,L)Q^4_{\text{in}}/Q^6_{\text{in}}$

Beside the stiffness, the difficulty of solving (1) lies in the coupling of all species by the nonlinear Bi-Moreau isotherm (2a), which describes the total adsorption based on two different adsorption mechanisms. A simplification is obtained by assuming just one adsorption mechanism. (2a) then simplifies to the widely used Langmuir type isotherms (2b).

$$q_i^{\text{eq}}(c_1,\ldots,c_{n_{\text{species}}}) = \sum_{k=1}^{2} q_{ki} \frac{b_{ki}c_i + l_{ki}(b_{ki}c_i)^2}{1+\sum_{j=1}^{n_{\text{species}}}\left(2b_{kj}c_j + l_{kj}(b_{kj}c_j)^2\right)}. \tag{2a}$$

$$q_i^{\text{eq}}(c_1,\ldots,c_{n_{\text{species}}}) = H_i \frac{c_i}{1+\sum_{i=1}^{n_{\text{species}}} \beta_i c_i}. \tag{2b}$$

3.2 MCSGP Configuration and Switching

We now consider $n_{\text{zones}} = 6$ zones consisting of one chromatographic column each, and $n_{\text{species}} = 3$ components, named H, P, L for $i = 1,2,3$ (compare Sect. 2.2). We use the zone index j as a superscript to the corresponding states in (1).

The MCSGP process can be controlled by n_{zones} pumps that generate the inflow rates Q^j_{in} for each zone j according to Table 2 (see also Fig. 1). We abbreviate the controls and the free switching time T in the control vector

$$\mathbf{q} = (Q_{\text{DeH}}, Q_{\text{De}}, Q_{\text{DeP}}, Q_{\text{Mo1}}, Q_{\text{DeL}}, Q_{\text{Mo2}}, T)^{\text{T}} \in \mathbb{R}^{n_{\text{zones}}+1}.$$

The flow rate within each zone must satisfy the box constraints

$$Q_{\min} \le Q^j_{\text{in}} \le Q_{\max}, \quad \text{for } j = 1,\ldots,n_{\text{zones}}, \tag{3a}$$

and the flow rates of the pumps must satisfy

$$\mathbf{q} \ge 0. \tag{3b}$$

After one period of length T the ports of the pumps are switched. To understand the structure of the optimization problem, it suffices here to notice that there is a zone index permutation $\pi : \{1, \ldots, n_{\text{zones}}\} \to \{1, \ldots, n_{\text{zones}}\}$ such that the switched periodic boundary condition

$$c_i^{\pi(j)}(T, x) = c_i^j(0, x), \quad q_i^{\pi(j)}(T, x) = q_i^j(0, x), \tag{4}$$

is satisfied for $x \in \Omega, i = 1, \ldots, n_{\text{species}}, j = 1, \ldots, n_{\text{zones}}$. In this article, the permutation has the simple form $\pi(j) = j + 1$ for $j < n_{\text{zones}}$ and $\pi(n_{\text{zones}}) = 1$.

3.3 Optimization Problem

We collect all state variables c_i^j, q_i^j for $i = 1, \ldots, 3, , j = 1, \ldots, 6$ in a vector s. Our objective is to maximize the average productivity for product P and to penalize the consumption of the mobile phase, which can be formulated as a weighted sum minimization with weights $w_1, w_2 \geq 0$ of

$$J(\mathbf{q}, s) = -w_1 \frac{1}{T} \int_0^T \frac{Q_{\text{DeP}}}{60} c_2^3(t, L) \mathrm{d}t$$
$$+ w_2(Q_{\text{DeH}} + Q_{\text{De}} + Q_{\text{DeP}} + Q_{\text{Mo1}} + Q_{\text{DeL}} + Q_{\text{Mo2}}),$$

subject to a lower bound $\text{Pur}_{\min}$ on the final purity of P

$$G(\mathbf{q}, s) = \frac{100 y_2}{y_1 + y_2 + y_3} - \text{Pur}_{\min} \geq 0, \quad \text{where } y_i = \int_0^T c_i^3(t, L) \mathrm{d}t. \tag{5}$$

We finally end up with the PDE constrained optimization problem

$$\min_{\mathbf{q} \in \mathbb{R}^7, s \in W(0,T)^{18}} J(\mathbf{q}, s) \quad \text{s.t.} \quad G(\mathbf{q}, s) \geq 0, \ (1), \ (3), \text{ and } (4). \tag{6}$$

The formal difficulty of the free period length T entering the state space $W(0, T)$ can be easily circumvented by using a time reparametrization from $[0, T]$ to $[0, 1]$. In this case, the right-hand sides of (1a) and (1b) must be multiplied by T.

3.4 Real-Time Control with Modifier Adaptation

We consider from now on only variations in $m = 3$ manipulated variables $\mathbf{u} = (Q_{De}, Q_{DeP}, T)^{\mathrm{T}}$. This is realistic, because the feed volume flow is often determined by the upstream operations that are connected to the MCSGP process, Q_{DeH} is assumed to be on its maximum allowed flow rate due to regeneration issues for the stationary phase and $Q_{Mol,2}$ contain so-called mobile phase modifiers, also fixed in their amounts. The MCSGP process can be controlled and optimized from cycle to cylce (= 6 switches) by Optimizing Model Predictive Control (MPC) [12]. In [12] structural model mismatch was not considered. In optimizing control solutions for SMB processes structural model mismatch can be compensated by considering the difference in the absolute values of the model and plant output (bias update, see [15]). Another feedback strategy is considered here: Iterative set point optimization with Modifier Adaptation. This approach was originally proposed by [23] and was applied to chromatographic batch separations and other chemical batch processes by [9, 18]. These methods have two significant differences compared to MPC. They consider a sampling time that is larger than the settling time of the system, which is why it is termed *set point* optimization and not model predictive control. The second and most important difference is that not only a bias correction is applied to cope with plant-model mismatch but also the derivative of the plant with respect to the inputs is used as feedback information. Modifying the optimization problem using the true derivatives compensates parameteric and structural model mismatch and thus leads the process to its true optimum although the model predicts a different behaviour [18, 23]. This method is applicable to the MCSGP process because it only needs a few cycles to become stationary and the sampling rate is of the same order of magnitude. The control algorithm is visualized in Fig. 2 and mathematically described in detail in the following.

In step $k + 1$ of the optimization with previous variables $\mathbf{u}_k$, we augment the PDE constrained optimization problem by modifiers $\lambda_k, \mu_k \in \mathbb{R}^3, \varepsilon_{k1}, \varepsilon_{k2} \in \mathbb{R}$ and additional control bounds $\Delta\mathbf{u}$ according to

$$\min_{\mathbf{u}_{k+1} \in \mathbb{R}^3, s \in W(0,T)^{18}} J(q(\mathbf{u_{k+1}}), s) + \varepsilon_{k,1} + (\lambda_k)^{\mathrm{T}}(\mathbf{u_{k+1}} - \mathbf{u}_k) \tag{7a}$$

$$\text{s.t. } G(q(\mathbf{u_{k+1}}), s) + \varepsilon_{k,2} + (\mu_k)^{\mathrm{T}}(\mathbf{u_{k+1}} - \mathbf{u}_k) \geq 0, \tag{7b}$$

$$\mathbf{u}_k - \Delta\mathbf{u} \leq \mathbf{u}_{k+1} \leq \mathbf{u}_k + \Delta\mathbf{u}, \tag{7c}$$

$$(1), (3), \text{ and } (4). \tag{7d}$$

In each controller iteration, the modifiers $\varepsilon_{k1}, \varepsilon_{k2}$ are updated based on measurements of the plant mass outputs y_1, y_2, y_3 defined in (5). To obtain the modifiers λ_k and μ_k, the gradients have to be estimated by Finite Differencing with measurements from previous controller iterations. This iterative gradient correction with a sufficiently small $\Delta\mathbf{u}$ drives the plant to its optimum even if the process model is subject to considerable errors.

A critical point for the gradient estimation is the inversion of a 3-by-3 matrix U, the entries of which are the differences in $\mathbf{u}$ at the previous m set points.

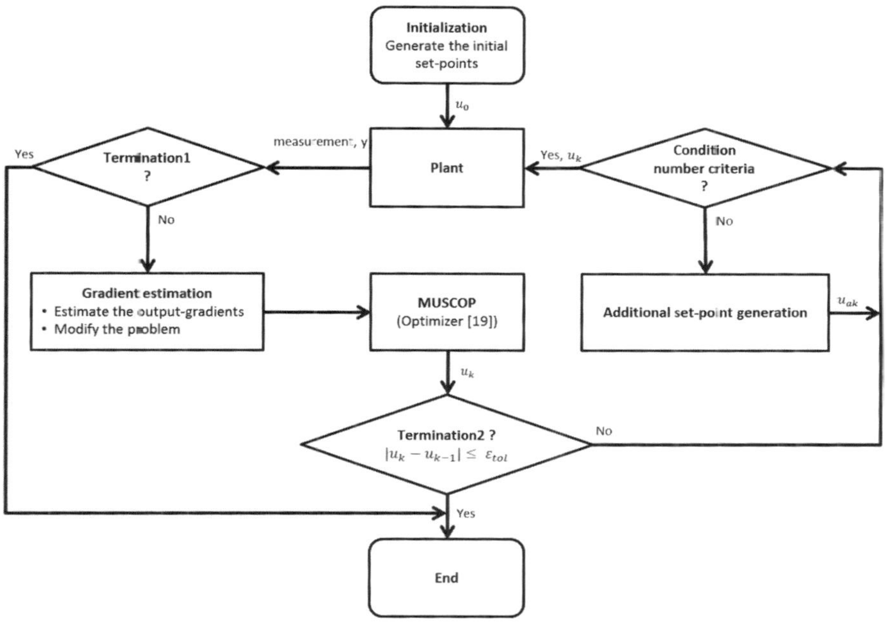

Fig. 2 Control algorithm

$$U^k = \begin{bmatrix} \mathbf{u}^k - \mathbf{u}^{k-1} \\ \vdots \\ \mathbf{u}^k - \mathbf{u}^{k-m} \end{bmatrix} \tag{8}$$

If U is ill-conditioned additional set points need to be applied to the plant in order to estimate the gradients reliably. Different strategies to obtain such a perturbed matrix U_a were compared in [2,14]. The most effective one is based on a geometrical considerartion of the estimation error combined with a rough feasibility check based on a linearized plant model. This algorithm is applied here as well. The bounds $\Delta\mathbf{u}$ are the main tuning parameters of this iterative scheme as they define the minimum number of iterations between controller and plant to get to the optimum, and also reflect the validity range of the linear error model. For other application examples in chromatography see, e.g., [2, 3, 9].

4 Numerical Solution of the Optimization Problems

One of the challenges to solve (6) and (7) lies in the fact that no fixed initial values are given for the states in PDE (1). On the basis of (1) and (4), it is in principle still possible to use a standard reduced space approach via a solution operator $S : \mathbb{R}^7 \to$

$W(0, T)^{n_{\text{species}} \times n_{\text{zones}}}$ mapping the controls to states that solve the periodic boundary value problem. The evaluation of S, however, comprises then inevitably an iterative method with at least one forward solve, e.g., in a Picard iteration, and maybe even a few derivatives in each iteration (compare, e.g., [17]). The solution becomes even more complicated when additionally derivatives of the solution of (1) and (4) need to be computed for an outer optimization loop, even though we only have seven free variables.

In this case it is beneficial to employ a *simultaneous* or *one-shot* approach (see, e.g., [5, 7, 10, 13]) in order to achieve stationarity and feasibility in one flat iteration loop instead of several nested ones. In this article, we use Direct Multiple Shooting [6] in combination with a structure-exploiting two-grid Newton-Picard inexact Sequential Quadratic Programming algorithm [19]. As an additional advantage of this approach, we do not need to formulate adjoint equations for (1), which would be rather cumbersome and error-prone to carry out by hand due to the nonlinear isotherm equation (2). Instead, adjoint derivatives can be computed on the basis of Internal Numerical Differentiation [4] and the reverse mode of Algorithmic Differentiation [11].

To this end, we use Finite Differences in space on a hierarchy of nested, equidistant meshes on Ω. For stability reasons, upwinding is employed for the convective part in (1). The resulting large-scale ODE constrained optimization problem is parameterized via local initial state values in time on a so-called shooting grid on the period horizon. The local shooting solutions must be glued together by requiring continuity in the shooting nodes as so-called *matching* constraints of the resulting large-scale finite dimensional optimization problem.

Finally, the fully discretized problem can be solved efficiently with a structure-exploiting inexact Sequential Quadratic Programming method based on two-grid Newton-Picard preconditioning [17, 20] and an extended condensing algorithm for the preprocessing of the linear-quadratic optimization subproblems [19]. The key idea of the two-grid Newton-Picard approach is to approximate the derivative matrices only on relatively coarse spatial grids, while computing the constraint residuals and the Lagrange gradient on fine grids. The width of the coarse grid determines the speed of convergence, while the quality of the fine grid determines the accuracy of the solution. This approach yields fast linear convergence, which is independent of the degrees of freedom on the fine grid. Furthermore, it is possible to estimate the convergence rate via so-called κ-estimators.

5 Case Studies

The case studies presented here were chosen to demonstrate both the effectiveness of the method described in Sect. 4 in real-time applications as well as the robustness of the proposed feedback strategy with respect to considerable model errors in a realistic application.

5.1 Scenario I

From the algorithmic point of view it is interesting how the optimization strategy performs for the most challenging situation of high purity constraints and nonlinear isotherms in the optimization problem. Therefore we consider a scenario in which high purities can be achieved and the isotherms in the model are nonlinear. The virtual plant behaves linear. Table 3 summarizes this scenario, which was solved on a grid of 201 and 438 points on the coarse and fine level.

5.2 Scenario II

In this scenario a linear adsorption model is used in a feedback controller to control and optimize a strongly nonlinear virtual plant. This case study concerns the separation of the three essential amino acids methionine, tryptophan, and phenylalanine. Amino acids behave strongly nonlinearly in many chromatographic systems, even for low concentrations. Its separation represents a challenging test case for model based control strategies if this nonlinearity is not considered in the model (structural model mismatch). Another challenge results from the fact that amino acids are partially produced by biotechnological processes in large scale and the outcomes of the fermentations may vary considerably. All substances involved may influence the adsorption behavior. The single isotherm parameters used in the simulated plant for scenario II were estimated based on experiments with a so-called C18 RP phase with a water/methanol mixture as the mobile phase. This chromatographic system is very close to the one used in [1] and the resulted isotherms are shown in Sect. 6. The thermodynamic characterization of tryptophan in [1] was assumed to be valid also for phenylalanine and methionine, all having the same strongly nonlinear behavior, based on two physically different adsorption mechanisms described by the Bi-Moreau isotherms. Furthermore we assume that the volume flows are known, as well as the temperature and the composition of the mobile phase. Besides the thermodynamics and feed composition, the uncertainty is often located in unknown deactivation effects of the stationary phase. Deactivation takes effect mathematically in the isotherms, its impact on the controlled MCSGP-process can be found in [14]. Here we present the reaction of the controller to a sudden step in the constraints and the feed concentration. The estimated isotherms

Table 3 Error scenario I: the model used in the optimization (parameters in parantheses) contains a nonlinear isotherm of the Langmuir type (2b) whereas we assume that the true plant has a linear isotherm

i	H_i	β_i	$6k_{\text{eff},i}/d_p$
1	0.4 (0.38)	0 (0)	4 (3.8)
2	3.2 (3.36)	0 (1.5)	3 (3.15)
3	6.7 (7.04)	0 (3.0)	2.5 (2.63)

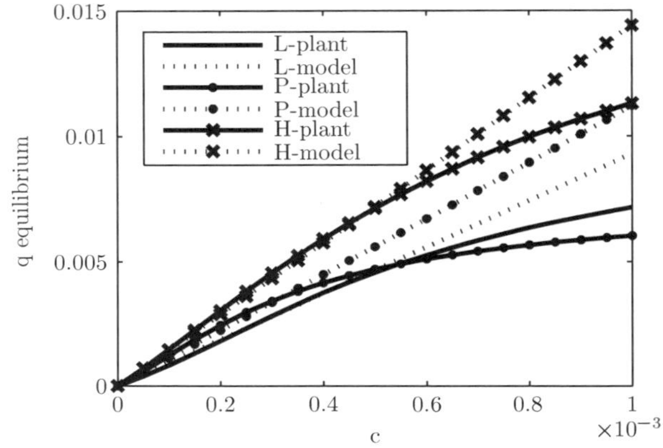

Fig. 3 The isotherms of the virtual plant and of the model in error scenario II used in the online optimization

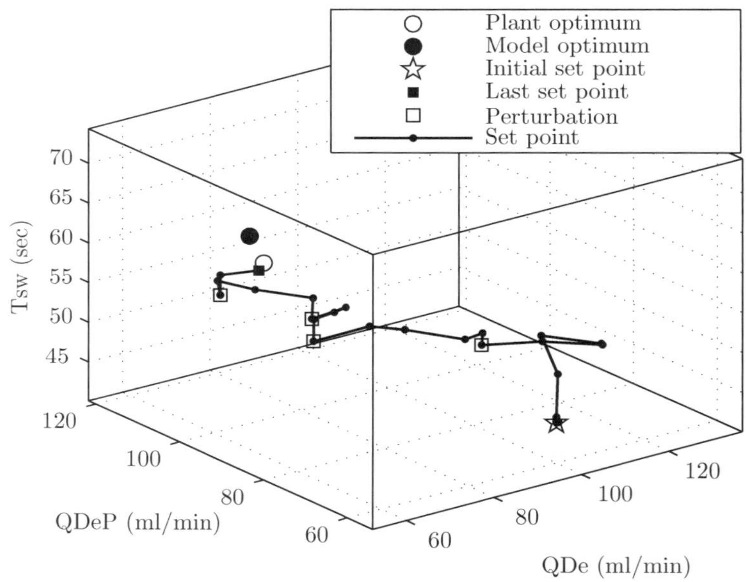

Fig. 4 Trajectory of the controlled plant and optima of plant and model in the *u* space for *Scenario I*

are assumed to represent the real plant behavior and the model is a linearization for small concentration ranges (see Fig. 3). A change in the feed concentrations is assumed to occur stepwise because of upstream batch unit operations. Such a step has a direct nonlinear influence on all purities and yields of the virtual plant and forces the controller to act immediately. Scenario II was solved on a grid of 126 and 486 points on the coarse and fine level.

6 Results

6.1 *Scenario I*

As can be seen in Figs. 4 and 5, the controller drives the plant to the true optimum despite the structural and parametric model mismatch. The productivity of the plant was improved by about 100 % without increasing the eluent consumption. The final set point is very close to the true optimum of the plant. The differences between model and plant are mainly reflected in the purity differences of up to 2 %. This is significant for high purity requirements. The optimizer terminated successfully within each iteration, i.e. it always solved the nonlinear subproblems by fulfilling the optimality conditions.

6.2 *Scenario II*

Figure 6 shows the key variables of a simulation of the controlled system. In the upper left plot in Fig. 6 the large difference between model and plant is visible. The step in the feed concentration from 0.2 to 0.25 g/l occurs between iteration 20 and 21. As expected, the controller acts immediately, forced by a sudden jump in the modifiers. The process stays feasible except for iteration 23 and the cost function is further reduced. Set point 23 was computed based on the perturbation algorithm and is not an optimization result. In the next iteration the process becomes feasible again. Prior to the concentration step, a step in the minimum purity after iteration 19 was applied The isotherm parameters of error scenario II are given in Table 4.

6.3 *Real Time Applicability*

An iterative set point optimization strategy should deliver a new set point within a computation time close to the settling time of the system. This was the case here. In scenario II and using a 2.5 GHz quad core with 15.7 GB memory the computation times were always below the settling time, which is between 50 and 90 min. If the system is close to the optimum and the model isotherms are nonlinear, this ratio gets worse, but is still in an applicable range: The final iteration for scenario I took approximately three times the settling time of the system, which is approximately 18 min. But this value strongly depends on the termination tolenrance of the optimizer, which together with other numerical parameters was not tuned for minimum computation times yet.

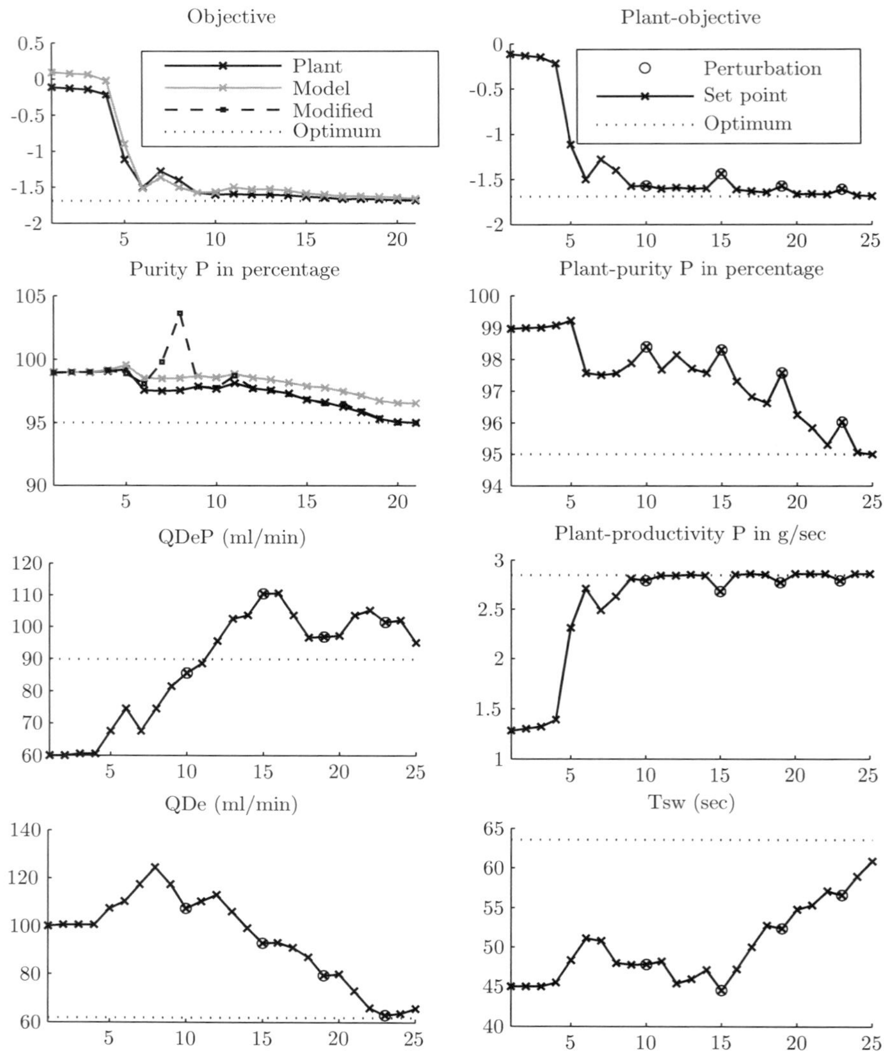

Fig. 5 Controlled and manipulated variables against iteration index for *Scenario I*. In the plots on the left in the first and second line perturbation set points are excluded

Conclusion

We applied an iterative set point optimization with gradient modification to a virtual MCSGP plant with real-world isotherm data to optimize and robustify

(continued)

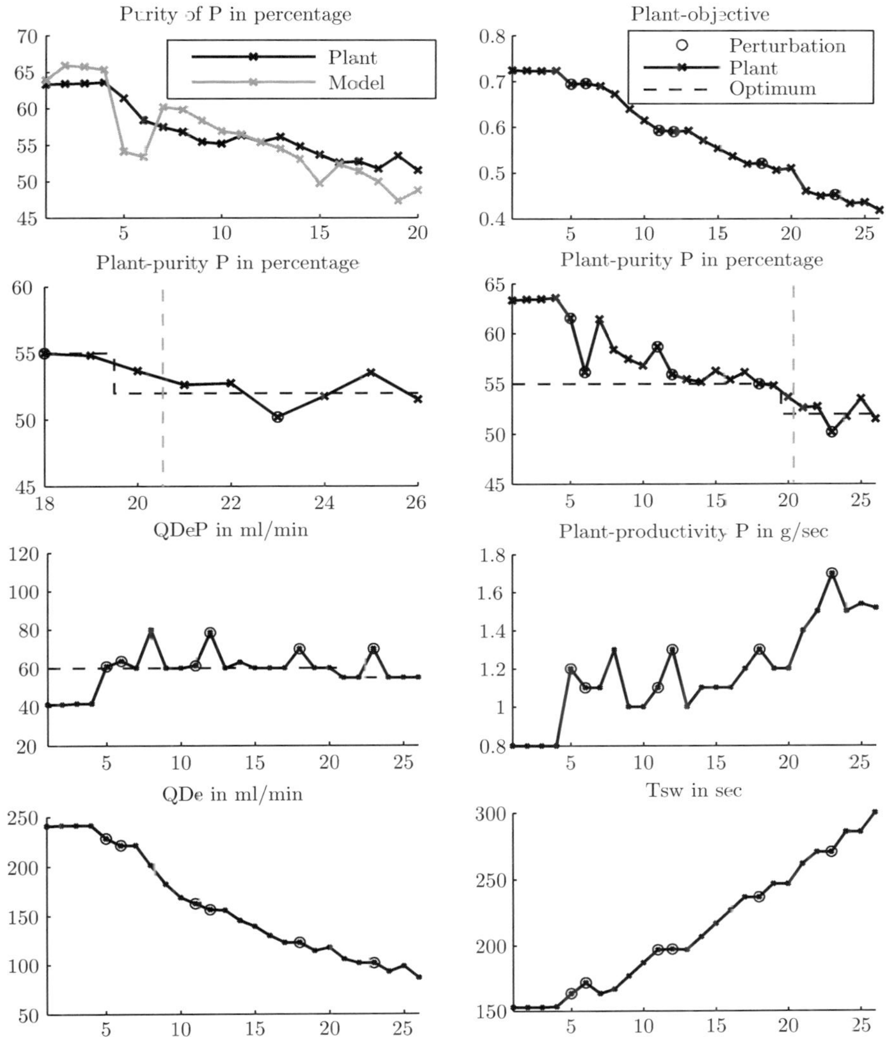

Fig. 6 Controlled and manipulated variables against iteration index for *Scenario II*. In the *upper left figure* the perturbation set points are excluded

the separation of the three essential amino acids tryptophan, methionine and phenylalanine by Modifier Adaptation feedback control. The underlying optimization problems were solved with a direct one-shot optimization approach

(continued)

Table 4 Error scenario II: the isotherm parameters of the virtual plant and the model (in parentheses) differ drastically. In particular, the model isotherm follows (2b) with $\beta_i = 0$

i	q_{1i}	q_{2i}	l_{1i}	l_{2i}	b_{1i}	b_{2i}	$6k_{\text{eff},i}/d_p$	(H_i)
1	3.55	7.21	0.46	0.70	11.6	5.5	0.26	(9.3)
2	3.74	3.74	1.38	1.38	4.5	4.5	0.35	(11.3)
3	9.55	8.9	0.73	0.71	4.14	3.54	0.69	(14.7)

on the basis of a two-grid Newton-Picard inexact Sequential Quadratic Programming method. Due to its numerical efficiency we can meet real time requirements of the process. Despite the parametric and structural model mismatch, the productivity of the virtual plant was increased by 100 and 35 % in two error scenarios while keeping the process feasible and not increasing the eluent consumption. Thus, the iterative set point optimization along with efficient optimization algorithms can be applied to a continuous chromatographic multi-column separation process with three components in real time.

References

1. T. Ahmad, G. Guichon, Effect of the mobile phase composition on the adsorption behavior of tryptophan in reversed-phase liquid chromatography. J. Chromatograph. A **1114**, 111–122 (2006)
2. M. Behrens, S. Engell, Iterative set-point optimization of continuous annular electro-chromatography, in *Proceedings of the 18th IFAC World Congress, 2011 – 28. 08 – 02.09.2011*, Milano/Italy, Paper ID 3008, 2011
3. M. Behrens, Y. Yu, S. Engell, Parameter estimation and iterative set-point optimization of continuous annular electrochromatography, in *Proceedings of the IEEE, International Conference on Industrial Technology* (Athens, Greece, *2012*), pp. 236–241, 2012
4. H.G. Bock, Numerical treatment of inverse problems in chemical reaction kinetics, in *Modelling of Chemical Reaction Systems*, ed. by K.H. Ebert, P. Deuflhard, W. Jäger. Springer Series in Chemical Physics, vol. 18 (Springer, Heidelberg, 1981), pp. 102–125
5. H.G. Bock, W. Egartner, W. Kappis, V. Schulz, Practical shape optimization for turbine and compressor blades by the use of PRSQP methods. Optim. Eng. **3**(4), 395–414 (2002)
6. H.G. Bock, K.J. Plitt, A multiple shooting algorithm for direct solution of optimal control problems, in *Proceedings of the 9th IFAC World Congress*, Budapest (Pergamon Press, 1984), pp. 242–247
7. H.G. Bock, A. Potschka, S. Sager, J.P. Schlöder, On the connection between forward and optimization problem in one-shot one-step methods, in *Constrained Optimization and Optimal Control for Partial Differential Equations*, ed. by G. Leugering, S. Engell, A. Griewank, M. Hinze, R. Rannacher, V. Schulz, M. Ulbrich, S. Ulbrich. International Series of Numerical Mathematics, vol. 160 (Springer, Basel, 2011), pp. 37–49
8. S. Engell, Feedback control for optimal process operation. J. Process Control **17**, 203–219 (2007)

9. W. Gao, S. Engell, Iterative set-point optimization of batch chromatography. Comput. Chem. Eng. **29**, 1401–1410 (2005)
10. A. Griewank, Projected Hessians for preconditioning in one-step one-shot design optimization, in *Large-Scale Nonlinear Optimization*. Nonconvex Optimization and Its Applications, vol. 83 (Springer, New York, 2006), pages 151–171
11. A. Griewank, A. Walther, *Evaluating Derivatives*, 2nd edn. (SIAM, Philadelphia, 2008)
12. C. Grossmann, G. Ströhlein, M. Morari, M. Morbidelli, Optimizing model predictive control of the chromatographic multi-column solvent gradient purification (MCSGP) process. J. Process Control **20**, 618–629 (2010)
13. S.B. Hazra, V. Schulz, J. Brezillon, N.R. Gauger, Aerodynamic shape optimization using simultaneous pseudo-timestepping. J. Comput. Phys. **204**(1), 46–64 (2005)
14. P. Khobkun, Iterative set-point optimization of the multi column solvent gradient purification process, Master's thesis, TU Dortmund University, 2013
15. A. Küpper, Optimization, state estimation, and model predictive control of simulated moving bed processes, PhD thesis, TU Dortmund University, 2010
16. D.N. Lopez, Handling uncertainties in process optimization, PhD thesis, University of Valladolid, 2013
17. K. Lust, D. Roose, A. Spence, A. R. Champneys, An adaptive Newton-Picard algorithm with subspace iteration for computing periodic solutions. SIAM J. Sci. Comput. **19**(4), 1188–1209 (1998)
18. A. Marchetti, B. Chachuat, D. Bovin, A dual modifier approach for real-time optimization. J. Process Control **20**, 1027–1037 (2010)
19. A. Potschka, A direct method for the numerical solution of optimization problems with time-periodic PDE constraints, PhD thesis, Heidelberg University, 2011
20. A. Potschka, M.S. Mommer, J.P. Schlöder, H.G. Bock, Newton-Picard-based preconditioning for linear-quadratic optimization problems with time-periodic parabolic PDE constraints. SIAM J. Sci. Comp. **34**(2), 1214–1239 (2012)
21. H. Schmidt-Traub, *Preperative Chromatography of Fine Chemicals and Pharmaceutical Agents* (Wiley-VCH, Weinheim, 2012)
22. G. Ströhlein, L. Aumann, M. Mazzotti, M. Morbidelli, A continuous, counter-current multi-column chromatographic process incorporating modifier gradients for ternary separations. J. Chromatograph. A **1126**, 338–346 (2006)
23. P. Tatjewski, Iterative optimization set-point control – the basic principle redesigned, in *Proceedings of the 15th IFAC World Congress*, Barcelona, 2002
24. A. Toumi, S. Engell, Optimization and control of chromatography. Comput. Chem. Eng. **29**, 1243–1252 (2005)
25. J. Wloka, *Partielle Differentialgleichungen: Sobolevräume u. Randwertaufgaben* (B.G. Teubner, Stuttgart, 1982)

OPTPDE: A Collection of Problems in PDE-Constrained Optimization

Roland Herzog, Arnd Rösch, Stefan Ulbrich, and Winnifried Wollner

Abstract In this article a first report on a new problem collection in PDE-constrained optimization, OPTPDE, is given. The goals of this collection are described and its current features are illustrated for a prototypical example. The entire problem collection can be accessed under www.optpde.net.

Keywords Optimization with PDE constraints • Problem collection

Mathematics Subject Classification (2010). Primary 49-00, Secondary 65-00.

1 Introduction

The present article is concerned with the OPTPDE collection [1] of problems in PDE-constrained optimization available at www.optpde.net, which was launched in February 2013. The problem collection is designed to provide scientists working on optimization problems with PDE constraints with a common reference framework of prototypical problems. The problems in this collection are hence selected to show

R. Herzog
Fakultät für Mathematik, Technische Universität Chemnitz, Reichenhainer Straße 41, 09126 Chemnitz, Germany
e-mail: roland.herzog@mathematik.tu-chemnitz.de

A. Rösch
Fachbereich Mathematik, Universität Duisburg-Essen, Forsthausweg 2, 47057 Duisburg, Germany
e-mail: arnd.roesch@uni-due.de

S. Ulbrich
Fachbereich Mathematik, Technische Universität Darmstadt, Dolivostraße 15, 64293 Darmstadt, Germany
e-mail: ulbrich@mathematik.tu-darmstadt.de

W. Wollner (✉)
Department of Mathematics, Universität Hamburg, Bundesstraße 55, 20146 Hamburg, Germany
e-mail: winnifried.wollner@uni-hamburg.de

G. Leugering et al. (eds.), *Trends in PDE Constrained Optimization*, International Series of Numerical Mathematics 165, DOI 10.1007/978-3-319-05083-6_34

various kinds of features and difficulties that have been of interest in recent years. It will be updated continuously according to current research trends.

We hope that this collection will allow researchers working on PDE-constrained optimization problems to compare their numerical results in a scientific manner with previous results without the need to search through the intense amount of literature available, and without the need of inventing new test problems that may differ only slightly from those provided by the existing literature. For complex problems with unknown solution we will also collect numerical results to provide reference values for those working on similar problems. On purpose, all problems are presented in their continuous (undiscretized) settings.

The collection consists, at present, of a database in which users can search for particular problems according to the involved differential operator, additional complications such as certain types of, e.g., state constraints, or keywords indicating the general context of the problem.

Each problem can be displayed with a short keyword like display of its main features, or with a complete description of the problem setting which includes additional data such as optimality conditions, reference values, and a short text indicating why this problem is considered in the collection. This description as well as a BIBTEX entry for the original reference are available as downloads.

Public credit is given to the person submitting the example as well as to anyone who confirms to have checked and/or contributed to the data or reference values. When you use an example from the OPTPDE collection in your own research, please give proper reference to the authors of the example, as well as to the problem collection itself. A corresponding BIBTEX entry can be found on the OPTPDE website.

2 Submissions and Reports

New problems can be included into the database, either by invitation of the editors, or by anyone submitting a new problem to editors@optpde.net. Informal inquiries whether a problem will be considered for inclusion are welcome, and a short statement describing the novelty of the proposed problem should be provided. The final submission should contain a descriptive LATEX file describing the problem including references to the original source for the example. A template file is available on the website. The submitted problem will then be subject to a review process. Successful problems need to contribute to this collection by at least one significant problem feature not yet available in the collection.

A second type of contribution to the collection is possible by providing additional material, such as optimality conditions or reference values. Of particular interest are statements regarding the verification or falsification of examples, in particular of those involving only numerical reference values.

3 An Example Problem

To illustrate the current capabilities, we demonstrate the possibility to search for problems by some of their features. The respective search mask is displayed in Fig. 1a. Here, we search for a problem that is constrained by a parabolic PDE but does not have any further constraints. When searching, we will obtain a page containing all results in the database matching our requirements, see Fig. 1b, together with a short list of the problem properties. Selecting one of these, for instance problem `rddist2`, will show an overview page with the features of the problem as stored in the database, see Fig. 1c. In the case of `rddist2`, we learn for instance that the problem is of semilinear parabolic type posed on a spatially 1D domain. The bottom of the screenshot shows who has provided the problem for the collection. Further, a link to the problem description as a PDF file is provided, and

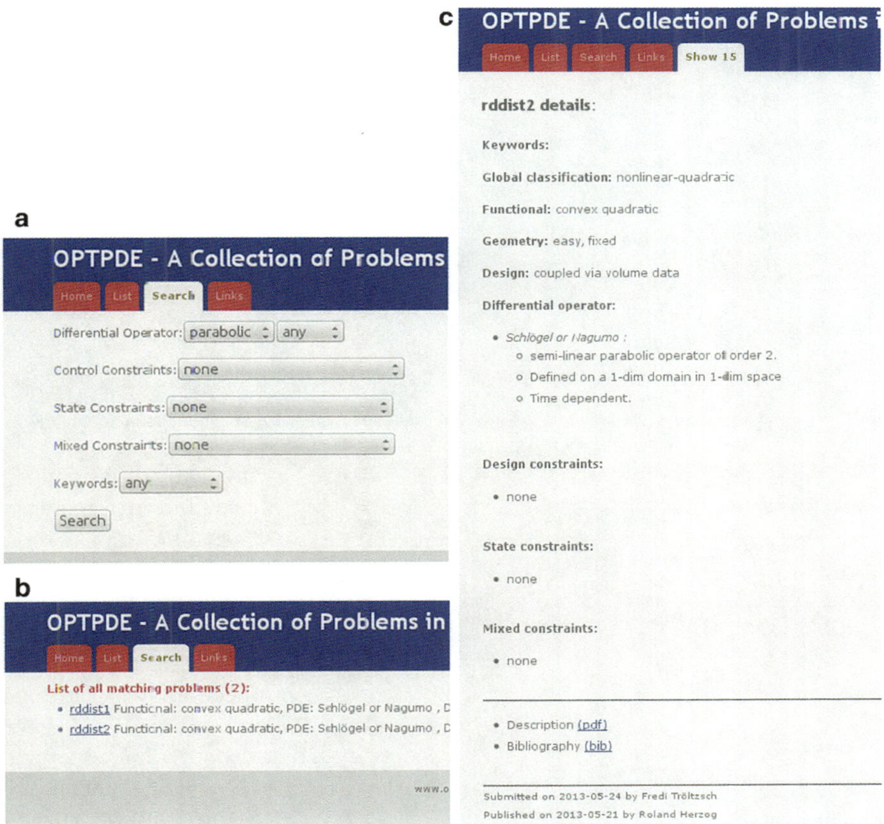

Fig. 1 Screenshots from the OPTPDE website. (**a**) Search mask. (**b**) Search results. (**c**) The displayed problem properties

RDDIST2 (2013-05-24) ID:15 OPTPDE Problem Collection

Introduction

This is a distributed optimal control problem for a semilinear 1D parabolic reaction-diffusion equation, where traveling wave fronts occur. The state equation is known as *Schlögl model* in physics and as *Nagumo equation* in neurobiology. In this context, various goals of optimization are of interest, for instance the stopping, acceleration, or extinction of a traveling wave. Here, we discuss the problem of re-directing a wave front after a certain time. The control is acting only on two subdomains near the boundary, which occupy the portion q of the entire spatial domain. This problem appears in [Buchholz et al., 2013, Section 5.2]. In the same paper, additional examples can be found which cover the other optimization goals mentioned above.

Variables & Notation

Unknowns

$f \in L^2(Q)$	control variable (forcing)
$u \in L^2(0,T;H^1(\Omega)) \cap H^1(0,T;H^1(\Omega)') \cap L^\infty(Q)$	state variable

Given Data

$\Omega = (0, L)$	spatial domain
$L = 20$	side length of domain
$Q = \Omega \times (0,T)$	computational domain
$Q_q = ((0,\delta) \cup (L-\delta, L)) \times (0,T)$	control domain
$q = 2\delta/L = 0.6$	portion of control domain
$T = 5$	terminal time
$u_0(x) = \begin{cases} 1.2\sqrt{3}, & x \in [9,11] \\ 0, & \text{elsewhere}, \end{cases}$	initial condition
$\lambda = 10^{-6}$	Tikhonov regularization parameter
$c = 450/281$	speed of the uncontrolled wave front
$u_Q(x,t) = \begin{cases} u_{nat}(x,t), & t \in [0, 2.5] \\ u_{nat}(x + ct, 2.5), & t \in (2.5, T] \end{cases}$	desired state
u_{nat}	solution of the PDE (0.1) for $f \equiv 0$.

http://www.optpde.net/rddist2 1

RDDIST2 (2013-05-24) ID:15 OPTPDE Problem Collection

The natural uncontrolled state u_{nat} is shown in Figure 0.1. In the figure, the horizontal axis shows the spatial variable x while the vertical one displays the time t. The speed c of the uncontrolled wave front was determined numerically.

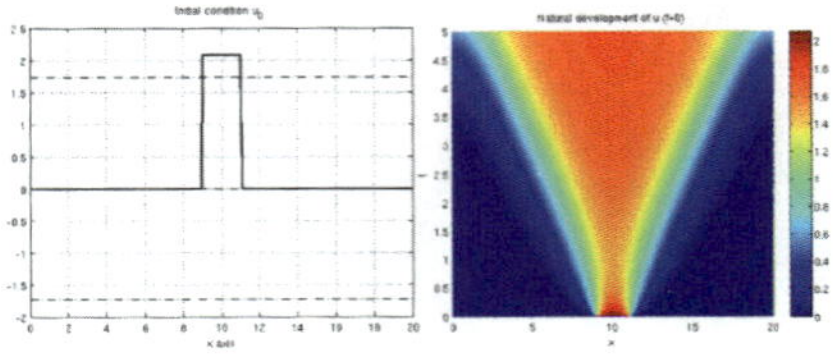

Figure 0.1: Initial state u_0 (left) and natural uncontrolled state u_{nat} (right).

Problem Description

$$\text{Minimize} \quad \frac{1}{2}\iint_Q (u(x,t) - u_Q(x,t))^2 \, dx\, dt + \frac{\lambda}{2}\iint_Q f^2(x,t)\, dx\, dt$$

$$\text{s.t.} \quad \left\{ \begin{aligned} \frac{\partial u}{\partial t}(x,t) - \frac{\partial^2 u}{\partial x^2}(x,t) + \frac{1}{3}u^3(x,t) - u(x,t) &= f(x,t) && \text{in } Q \\ u(x,0) &= u_0(x) && \text{in } \Omega \\ \frac{\partial u}{\partial x}(0,t) = \frac{\partial u}{\partial x}(L,t) &= 0 && \text{in } (0,T) \end{aligned} \right\} \quad (0.1)$$

and $f(x,t) = 0$ in $Q \setminus Q_q$.

Notice that the PDE has a non-monotone nonlinearity. The associated homogeneous elliptic (stationary) equation admits three different solutions; namely, the functions $u_1(x) \equiv -\sqrt{3}$, $u_2(x) \equiv 0$, and $u_3(x) \equiv \sqrt{3}$.

http://www.optpde.net/rddist2 2

RDDIST2 (2013-05-24) ID:15 OPTPDE Problem Collection

Optimality System

The following optimality system for the state u, the control f, and the adjoint state p, given in the strong form, represents first-order necessary optimality conditions.

$$\begin{aligned} \frac{\partial u}{\partial t}(x,t) - \frac{\partial^2 u}{\partial x^2}(x,t) + \frac{1}{3}u^3(x,t) - u(x,t) &= f(x,t) && \text{in } Q \\ u(x,0) &= u_0(x) && \text{in } \Omega \\ \frac{\partial u}{\partial x}(0,t) = \frac{\partial u}{\partial x}(L,t) &= 0 && \text{in } (0,T), \\ -\frac{\partial p}{\partial t}(x,t) - \frac{\partial^2 p}{\partial x^2}(x,t) + u^2(x,t)\,p(x,t) - p(x,t) &= u(x,t) - u_Q(x,t) && \text{in } Q \\ p(x,T) &= 0 && \text{in } \Omega \\ \frac{\partial p}{\partial x}(0,t) = \frac{\partial p}{\partial x}(L,t) &= 0 && \text{in } (0,T), \\ f(x,t) &= -\frac{1}{\lambda}p(x,t) && \text{in } Q_q \\ f(x,t) &= 0 && \text{in } Q \setminus Q_q. \end{aligned}$$

Supplementary Material

Figure 0.2 displays the desired state, as well as the optimal state and optimal control.

References

R. Buchholz, H. Engel, E. Kammann, and F. Tröltzsch. On the optimal control of the Schlögl model. *Computational Optimization and Applications*, pages 1–33, 2013. doi: 10.1007/s10589-013-9550-y.

http://www.optpde.net/rddist2 3

RDDIST2 (2013-05-24) ID:15 OPTPDE Problem Collection

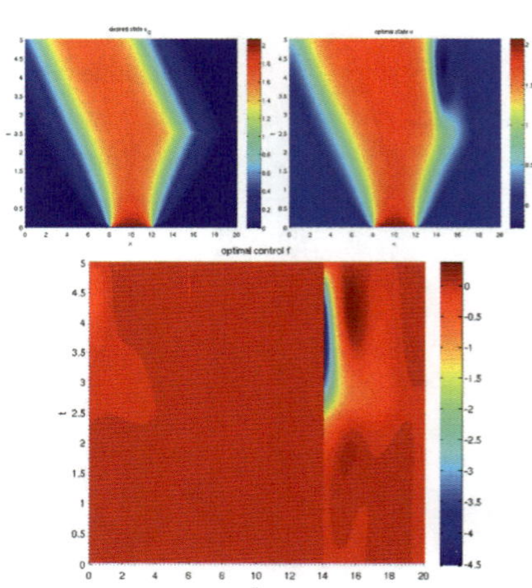

Figure 0.2: Re-directing a wave front with control support of size $q = 0.6$; desired state u_Q (top left), optimal state u (top right) and optimal control f (bottom).

http://www.optpde.net/rddist2 4

Fig. 2 The problem data for the `rddist2` example

the relevant bibliography can be downloaded as well. An HTML version (not shown here) of the problem description is also available, so that a first view on the problem is possible without following additional links.

The problem description as it appears in the PDF file for the example `rddist2`, originating from [2], is shown in Fig. 2. It is important to note that this description features a short introduction to the problem, followed by complete list of the unknowns as well as all problem data. Then the description of the optimization problem is given. As additional material, this example features both a first-order optimality system, as well as graphical representations of the optimal solution.

References

[1] OPTPDE – a collection of problems in PDE-constrained optimization. http://www.optpde.net
[2] R. Buchholz, H. Engel, E. Kammann, F. Tröltzsch, On the optimal control of the Schlögl model. Comput. Optim. Appl. **56**(1), 153–185 (2013)